SECOND EDITION

COMPREHENSIVE DICTIONARY

OF

ELECTRICAL ENGINEERING

SECOND EDITION

COMPREHENSIVE

DICTIONARY

OF

ELECTRICAL

ENGINEERING

EDITOR-IN-CHIEF
Phillip A. Laplante

Taylor & Francis
Taylor & Francis Group

Boca Raton   London   New York   Singapore

CRC PRESS, a Taylor & Francis title, part of the Taylor and Francis Group.

Published in 2005 by
CRC Press
Taylor & Francis Group
6000 Broken Sound Parkway NW, Suite 300
Boca Raton, FL 33487-2742

International Standard Book Number-10: 0-8493-3086-6 (Hardcover)
International Standard Book Number-13: 978-0-8493-3086-5 (Hardcover)
Library of Congress Card Number 2004058572

Library of Congress Cataloging-in-Publication Data

Comprehensive dictionary of electrical engineering / editor-in-chief Phillip A. Laplante.-- 2nd ed.
    p. cm.
    ISBN 0-8493-3086-6 (alk. paper)
  1. Electric engineering--Dictionaries. I. Title: Electrical engineering. II. Laplante, Phillip A.

TK9.C575 2005
621.3'03--dc22   2004058572

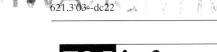

Taylor & Francis Group
is the Academic Division of T&F Informa plc.

Visit the Taylor & Francis Web site at
http://www.taylorandfrancis.com

and the CRC Press Web site at
http://www.crcpress.com

Preface to the Second Edition

Since the publication of the first edition of this dictionary more than 5 years ago, many changes in technology have occurred, particularly in the rapidly changing fields of image processing, computer electronics, fuel cells, and nanotechnology. I must say, however, that preparing the second edition of this dictionary was significantly easier than the first edition (see preface to first edition). It was easier to prepare because, fortunately, I had a set of handy resources that included terms related to these technological changes. That is, I was able to incorporate new terms from the many new CRC handbooks published within the last 2 years including, the *Fuel Cell Technology Handbook*, *Electric and Hybrid Vehicles: Design Fundamentals*, *The Computer Engineering Handbook*, *Digital Color Imaging Handbook*, *Handbook of Nanoscience Engineering and Technology*, *The RF and Microwave Handbook*, *The Power Electronics Handbook*, *Biomedical Photonics Handbook*, and *The Mechatronics Handbook*. I was also able to incorporate new terms and corrections suggested by readers of the first edition (for which I heartily thank those readers), as well as additions and corrections that are invariably needed after even the fifth read-through.

In total, more than 1500 terms were added, updated, expanded, improved, or corrected, resulting in a dictionary with over 11,000 terms and abbreviations related to electrical engineering. However, if readers discover any errors or think that any important terms have been omitted, please inform me at plaplante@psu.edu with your proposed changes. I will be happy to consider those changes for future printings and editions.

In keeping within the scope established with the first edition, most terms pertaining to computer science, information technology, and software engineering that are not directly linked to the underlying hardware were omitted. For these, please refer to *Comprehensive Dictionary of Computer Science, Engineering and Technology* (CRC Press).

Once again I want to thank Nora Konopka and the fine editorial and production staff at Taylor & Francis, especially Helena Redshaw and Amy Rodriguez, for helping to bring this second edition into being. These folks have always been a pleasure to work with.

Finally, I want to thank my family for their patience and support as I worked countless hours on this project, which most certainly would never have been completed without their blessing. Therefore, this dictionary is dedicated to Nancy, Christopher, and Charlotte.

Phillip A. Laplante, P.E., Ph.D.

Preface to the First Edition

One can only appreciate the magnitude of effort required to develop a dictionary by actually experiencing it. Although I had written nine other books, I certainly did not know what I was getting into when in January of 1996 I agreed to serve as Editor-in-Chief for this project. Now, after $2\frac{1}{2}$ years I understand.

Unlike other books that I have written, creating this dictionary was more a test of will and stamina and an exercise in project management than mere writing. And although I have managed organizations of up to 80 academics, nothing is more like "herding cats" than motivating an international collection of more than 100 distinguished engineers, scientists, and educators scattered around the globe almost entirely via email. Yet, I think there is no other way to undertake a project like this. I still marvel at how Noah Webster must have managed to construct his English Dictionary without the benefits of modern communication.

But this project, as much as it is a monument to individual will, is really the collaborative work of many brilliant and dedicated men and women. This is their dictionary and your dictionary.

Phillip A. Laplante, P.E., Ph.D.
Editor-in-Chief

Foreword

How was the dictionary constructed?

As I knew this project would require a divide-and-conquer approach with fault-tolerance, I sought to partition the dictionary by defining areas that covered all aspects of electrical engineering. I then matched these up to IEEE-defined interest areas to ensure that complete coverage was provided. This created a great deal of overlap, which was intentional. I knew that terms needed to be defined several different ways, depending on usage, and I needed to ensure that every term would be defined at least once.

The mapping of the dictionary's areas to the IEEE interest areas are as follows:

Power systems
• Power engineering
• Power electronics

Electric motors and machines
• Power engineering
• Power electronics

Digital electronics, VLSI, hardware
    description language
• Consumer electronics
• Electronic devices
• Industrial electronics
• Instruments and measurements

Microelectronics and solid state devices
• Industrial electronics
• Instruments and measurements

RF, radio, and television
• Broadcast technology

Communications and information processing
• Communications
• Information theory
• Systems, man, and cybernetics
• Reliability

Signal and image processing
• Signal processing
• Systems, man, and cybernetics

Circuits and systems
• Circuits and systems
• Instruments and measurements

Control systems
• Control systems
• Robotics and automation

Electromagnetics
• Electromagnetic compatibility
• Magnetics

Computer engineering (processors)
• Computer

Computer engineering (I/O and storage)
• Computer

Microwave systems
• Antennas and propagation
• Microwave theory and techniques

Electro-optical and lightwave systems
• Lasers and electro-optics

Illumination

Properties of materials
• Dielectrics and electrical insulation

Packaging
• Components, packaging
• Manufacturing technology

Note that software engineering was not included as an area, and most software terms have been omitted. Those that were included were done so because they relate to some aspect of assembly language programming or low-level control, or artificial intelligence and robotics. For those interested in software engineering terms, CRC's *Comprehensive Dictionary of Computer Science, Engineering, and Technology* includes those terms.

Several other IEEE interest areas were not explicitly assigned to area editors. However, after discussing this fact with the editorial board, it was decided that relevant terms of a general nature would be picked up and terms that were not tagged for the dictionary from these areas were probably too esoteric to be included.

These interest areas encompass:

Aerospace and electronic systems	Geosience and remote sensing
Education	Industry applications
Engineering in medicine and biology	Nuclear and plasma science
Engineering management	Oceanic engineering
Professional communications	Ultrasonic, ferroelectrics, and frequency control
Social implications of technology	Vehicular technology

Given the area editor structure, constructing the dictionary then consisted of the following steps:

1. Creating a terms list for each area
2. Defining terms
3. Cross-checking terms within areas
4. Cross-checking terms across areas
5. Compiling and proofing the terms and definitions
6. Reviewing compiled dictionary
7. Final proofreading

The first and most important task undertaken by the area editors was to develop a list of terms to be defined. A terms list is a list of terms (without definitions), proper names (such as important historical figures or companies), or acronyms relating to electrical engineering. What went into each terms list was left to the discretion of the area editor based on the recommendations of the contributing authors. However, lists were to include all technical terms that relate to the area (and subareas). Technical terms of a historical nature were only included if it was noted in the definition that the term is "not used" in modern engineering or that the term is "historical" only. Although the number of terms in each list varied somewhat, each area's terms list consisted of approximately 700 items.

Once the terms lists were created, they were merged and scrutinized for any obvious omissions. These missing terms were then assigned to the appropriate area editor. At this point the area editors and their contributing authors (there were 5 to 20 contributing authors per area) began the painstaking task of term definition. This process took many months. Once all of the terms and their definitions were collected, the process of converting, merging, and editing began.

The dictionary included contributions from over 100 contributors from 17 countries. Although authors were provided with a set of guidelines to write terms definitions, they were free to exercise their own judgment and to use their own style. As a result, the entries vary widely in content from short, one-sentence definitions to rather long dissertations. While I tried to provide some homogeneity in the process of editing, I neither wanted to tread on the feet of the experts and possibly corrupt the meaning of the definitions (after all, I am not an expert in any of the representative areas of the dictionary) nor did I want to interfere with the individual styles of the authors. As a result, I think the dictionary contains a diverse and rich exposition that collectively provides good insights into the areas intended to be covered by the dictionary.

Moreover, I was pleased to find the resultant collection much more lively, personal, and user-friendly than typical dictionaries.

Finally, we took advantage of the rich CRC library of handbooks, including *The Control Handbook, Electronics Handbook, Image Processing Handbook, Circuits and Filters Handbook,* and *The Electrical Engineering Handbook,* to pick up any definitions that were missing or incomplete. About 1000 terms were take from the CRC handbooks. We also borrowed, with permission from IEEE, about 40 definitions that could not be found elsewhere or could not be improved upon.

Despite the incredible support from my area editors, individual contributors, and staff at CRC Press, the final tasks of arbitrating conflicting definitions, rewording those that did not seem descriptive enough, and identifying missing ones were left to me. I hope that I have not failed you terribly in my task.

How to use the dictionary

The dictionary is organized like a standard language dictionary except that not every word used in the dictionary is defined (this would necessitate a complete embedding of an English dictionary). However, we tried to define most non-obvious technical terms used in the definition of another term.

In some cases more than one definition is given for a term. These are denoted (1), (2), (3), . . . , etc. Multiple definitions were given in cases where the term has multiple distinct meanings in differing fields, or when more than one equivalent but uniquely descriptive definition was available to help increase understanding. In a few cases, I just couldn't decide between two definitions. Pick the definition that seems to fit your situation most closely. The notation 1., 2., etc. is used to itemize certain elements of a definition and are not to be confused with multiple definitions.

Acronym terms are listed by their expanded name. Under the acronym the reader is referred to that term. For example, if you look up "RISC" you will find "See reduced instruction set computer," where the definition can be found. The only exceptions are in the cases where the expanded acronym might not make sense, or where the acronym itself has become a word (such as "laser" or "sonar").

While I chose to include some commonly used symbols (largely upon the recommendations of the contributors and area editors), this was not a principle focus of the dictionary and I am sure that many have been omitted.

Finally, we tried to avoid proprietary names and tradenames where possible. Some have crept in because of their importance, however.

Acknowledgments

A project of this scope literally requires hundreds of participants. I would like to take this moment to thank these participants both collectively and individually. I thank, in no particular order:

- The editorial board members and contributors. Although not all participated at an equal level, all contributed in some way to the production of this work.
- Ron Powers, CRC President of Book Publishing, for conceiving this dictionary, believing in me, and providing incredible support and encouragement.
- Frank MacCrory, Norma Trueblood, Nora Konopka, Carole Sweatman, and my wife Nancy for converting, typing, and/or entering many of the terms.

- Jill Welch, Nora Konopka, Ron Powers, Amy Rodriguez, Susan Fox, Karen Feinstein, Joe Ganzi, Gerry Axelrod, and others from CRC for editorial support.
- *CRC Comprehensive Dictionary of Mathematics* and *CRC Comprehensive Dictionary of Physics* editor Stan Gibilisco for sharing many ideas with me.
- My friend Peter Gordon for many of the biographical entries.
- Lisa Levine for providing excellent copy editing of the final manuscript.

Finally to my wife Nancy and children Christopher and Charlotte for their incredible patience and endurance while I literally spent hundreds of hours to enable the birth of this dictionary. This achievement is as much theirs as it is mine.

Please accept my apologies if anyone was left out—this was not intentional and will be remedied in future printings of this dictionary.

How to Report Errors/Omissions

Because of the magnitude of this undertaking and because we attempted to develop new definitions completely from scratch, we have surely omitted (though not deliberately) many terms. In addition, some definitions are possibly incomplete, weak, or even incorrect. But we wish to evolve and improve this dictionary in subsequent printings and editions. You are encouraged to participate in this collaborative, global process. Please send any suggested corrections, improvements, or new terms to be added (along with suggested definitions) to me at p.laplante@ieee.org or plaplante@psu.edu. If your submission is incorporated, you will be recognized as a contributor in future editions of the dictionary.

References

[1] Attasi, Systemes lineaires homgenes a deux indices, *IRIA Rapprot Laboria*, No. 31, Sept. 1973.
[2] Baxter, K., *Capacitive Sensors*, IEEE Press, 1997.
[3] Biey and Premoli, A., *Cauer and MCPER Functions for Low-Q Filter Design,* St. Saphorin: Georgi, 1980.
[4] Bishop, Robert, *The Mechatronics Handbook*, Boca Raton: CRC Press, 2002.
[5] Blostein, L., Some bounds on the sensitivity in RLC networks, *Proceedings of the 1st Allerton Conference on Circuits and Systems Theory,* 1963, pp. 488–501.
[6] Boutin, A.C., The misunderstood twin-T oscillator, *IEEE Circuits and Systems Magazine,* Dec. 1980, pp. 8–13.
[7] Chen, W.-K., Ed., *The Circuits and Filters Handbook,* Boca Raton, FL: CRC Press, 1995.
[8] Clarke and Hess, D.T., *Communication Circuits: Analysis and Design,* Addison-Wesley, 1971.
[9] Coultes, E. and Watson, W., Synchronous machine models by standstill frequency response tests, *IEEE Transactions on Power Apparatus and Systems,* PAS 100(4), 1480–1489, 1981.
[10] Dorf, R.C., Ed., *The Electrical Engineering Handbook,* 2nd ed., Boca Raton, FL: CRC Press, 1997.
[11] Enslow, H., Multiprocessor organization, *Computing Surveys,* 9(1), 103–129, 1977.
[12] Filanovsky, M., Piskarev, V.A., and Stromsmoe, K.A., Nonsymmetric multivibrators with an auxiliary RC-circuit, *Proc. IEEE,* 131, 141–146, 1984.
[13] Filanovsky, M. and Piskarev, V.A., Sensing and measurement of dc current using a transformer and RL-multivibrator, *IEEE Trans. Circ. Syst.,* 38, 1366–1370, 1991.

[14] Filanovsky, M., Qiu, S.-S., and Kothapalli, G., Sinusoidal oscillator with voltage controlled frequency and amplitude, *Intl. J. Electron.,* 68, 95–112, 1990.

[15] Frerking, C., *Oscillator Design and Temperature Compensation,* Van Nostrand Reinhold, 1978.

[16] Fornasini and Marchesini, G., Double-indexed dynamical systems, *Mathematical Systems Theory,* 1978, pp. 59–72.

[17] Franco, *Design with Operational Amplifiers and Analog Integrated Circuits,* McGraw-Hill, 1988.

[18] Goddard III, William A., Brenner, Donald W., Lyshevski, Sergey Edward, and Iafrate, Gerald (Eds.), *Handbook of Nanoscience Engineering and Technology,* Boca Raton: CRC Press, 2003.

[19] Golio, Mike (Ed.), *The RF and Microwave Handbook,* Boca Raton: CRC Press, 2001.

[20] Hoogers, Gregor (Ed.), *Fuel Cell Technology Handbook,* Boca Raton: CRC Press, 2003.

[21] Held, N. and Kerr, A.R., Conversion loss and noise of microwave and millimeter-wave mixers: Part 1, Theory and Part 2, Experiment, *IEEE Transactions on Microwave Theory and Techniques,* MTT-26, 49, 1978.

[22] Hennessy, L. and Patterson, D.A., *Computer Architecture: A Quantitative Approach,* 2nd ed., Kaufmann, 1996.

[23] Huelsman, P. and Allen, P.E., *Introduction to the Theory and Design of Active Filters,* McGraw-Hill, 1980.

[24] Husain, Iqbal, *Electric and Hybrid Vehicles: Design Fundamentals*, Boca Raton: CRC Press, 2003.

[25] Huising, H., Van Rossum, G.A., and Van der Lee, M., Two-wire bridge-to-frequency converter, *IEEE J. Solid-State Circuits,* SC-22, 343–349, 1987.

[26] IEEE Committee Report, Proposed excitation system definitions for synchronous machines, *IEEE Transactions on Power Apparatus and Systems,* PAS-88(8), August 1969.

[27] *IEEE Standard Dictionary of Electrical Engineering,* 6th ed., 1996.

[28] Jouppi, N.P., The nonuniform distribution of instruction-level and machine prallelism and its effect on performance, *IEEE Transactions on Computers,* 38(12), 1645–1658, Dec, 1989.

[29] Kaczorek, *Linear Control Systems,* Vol. 2, New York: John Wiley & Sons, 1993.

[30] Kaczorek, The singular general model of 2-D systems and its solution, *IEEE Transactions on Automatic Control,* AC-33(11), 1060–1061, 1988.

[31] Kaczorek, *Two-Dimensional Linear Systems,* Springer-Verlag, 1985.

[32] Kaplan, -Z., Saaroni, R., and Zuckert, B., Analytical and experimental approaches for the design of low-distortion Wien bridge oscillators, *IEEE Transactions on Instrumentation and Measurement,* IM-30,147–151, 1981.

[33] Katevenis, G.H., *Reduced Instruction Set Computer Architectures for VLSI,* MIT Press, 1985.

[34] Kurek, The general state-space model for a two-dimensional linear digital system, *IEEE Transactions on Automatic Control,* AC-30(6), 600–601, 1985.

[35] Krause, C., Wasynczuk, O., and Sudhoff, S.D., *Analysis of Electric Machinery,* IEEE Press, 1995.

[36] Levine, W.S., Ed., *The Control Handbook,* Boca Raton, FL: CRC Press, 1995.

[37] Morf, Levy, B.C., and Kung, SY., New results in 2-D systems theory, *Proc. IEEE,* 65(6), 861–872, 1977.

[38] Myers, J., *Advances in Computer Architecture,* 2nd ed., New York: John Wiley & Sons, 1982.

[39] Neubert, K.P., *Instrument Transducers,* Clarendon Press, 1975.

[40] Oklobdzija, Voijin G. (Ed.), *The Computer Engineering Handbook,* Boca Raton: CRC Press, 2002.

[41] Qiu, S.-S. and Filanovsky, I.M., Periodic solutions of the Van der Pol equation with moderate values of damping coefficient, *IEEE Transactions on Circuits and Systems,* CAS-34, 913–918, 1987.

[42] Orchard, J., Loss sensitivities in singly and doubly terminated filters, *IEEE Transactions on Circuits and Systems,* CAS-26, 293–297, 1979.

[43] Pallas-Arány and Webster, J.G., *Sensor and Signal Conditioning,* New York: John Wiley & Sons, 1991.

[44] Patterson, A. and Ditzel, D.R., The case for the RISC, *Computer Architecture News,* 8(6), 25–33, 1980.

[45] Patterson, A. and Sequin, C.H., A VLSI RISC, *IEEE Computer,* 15(9), 8–21, 1982.

[46] Pederson, O. and Mayaram, K., *Analog Integrated Circuits for Communication,* Kluwer, 1991.

[47] Radin, The 801 Minicomputer, *IBM J. Res. Devel.,* 21(3), 237–246, 1983.

[48] Ramamurthi and Gersho, A., *IEEE Transactions on Communications,* 34(1), 1105–1115, 1986.

[49] Roesser, P., A discrete state-space model for linear image processing, *IEEE Transactions on Automatic Control,* AC-20(1),1–10, 1975.

[50] Russ, J.C., Ed., *The Image Processing Handbook,* 2nd ed., Boca Raton, FL: CRC Press, 1994.

[51] Rosenbrock, H.H., *Computer-Aided Control System Design,* Academic Press, 1974.

[52] Sharma, Gaurav (Ed.), *Digital Color Imaging Handbook*, Boca Raton: CRC Press, 2003.

[53] Skvarenina, Timothy L. (Ed.), *The Power Electronics Handbook*, Boca Raton: CRC Press, 2002.

[54] Smith, *Modern Communication Circuits,* McGraw-Hill, 1986.

[55] Strauss, *Wave Generation and Shaping,* 2nd ed., McGraw-Hill, 1970.

[56] Tabak, *RISC Systems and Applications,* Research Studies Press and Wiley, 1996.

[57] Thomas and Clarke, C.A., Eds., *Handbook of Electrical Instruments and Measuring Techniques,* Prentice-Hall, 1967.

[58] Vo-Dinh, Tuan, *Biomedical Photonics Handbook*, Boca Raton: CRC Press, 2003.

[59] Whittaker, J.C., Ed., *The Electronics Handbook,* Boca Raton, FL: CRC Press, 1996.

[60] Youla and Gnavi, G., Notes of n-dimensional system theory, *IEEE Transactions on Circuits and Systems,* 26(2), 105–111, 1979.

Editor-in-Chief

Phillip A. Laplante, Ph.D. is associate professor of software engineering at Pennsylvania State's Great Valley Graduate Center. In this capacity he conducts research and teaches graduate courses in software and computer systems engineering. He also serves as the chief technology officer for the Eastern Technology Council and founded and leads its CIO community of practice, the CIO Institute.

Before joining Penn State, Dr. Laplante was president of Pennsylvania Institute of Technology, a 2-year, private college that focuses on technology training and retraining. Prior to that, he was the founding dean of the BCC/NJIT Technology and Engineering Center in southern New Jersey. He was also associate professor of computer science and chair of the Mathematics, Computer Science and Physics Department at Fairleigh Dickinson University in New Jersey.

In addition to his academic career, Dr. Laplante spent almost 8 years as a software engineer and project manager working on avionics (including the space shuttle), computer-aided design (CAD), and software test systems. He has published more than 120 papers and articles and 20 books, including *Dictionary of Computer Science, Engineering, and Technology* and *Software Engineering for Image Processing Systems*, both through CRC Press. He also edits two book series, including the Image Processing series for CRC Press, and cofounded the journal *Real-Time Imaging*, which he edited for 5 years.

Dr. Laplante earned his B.S., M.Eng., and Ph.D. in computer science, electrical engineering, and computer science, respectively, from Stevens Institute of Technology, and an M.B.A. from the University of Colorado. He is a senior member of IEEE and a member of numerous other professional societies, program committees, and boards, and is a licensed professional engineer in Pennsylvania.

Editorial Board

Eugene Veklerov
Lawrence Berkeley Labs
Editor: Signal and image processing

Janusz Zalewski
University of Central Florida
Editor: Computer engineering (processors)

Contributors

James T. Aberle
Arizona State University
Tempe, Arizona

Giovanni Adorni
Università di Parma
Parma, Italy

Ashfaq Ahmed
Purdue University
West Lafayette, Indiana

Earle M. Alexander IV
San Rafael, California

A. E. A. Almaini
Napier University
Edinburgh, Scotland

Jim Andrew
CISRA
North Ryde, Australia

James Antonakos
Broome County Community College
Binghampton, New York

Eduard Ayguade
Barcelona, Spain

Bibhuti B. Banerjee
Dexter Magnetic Materials
Fremont, California

Partha P. Banjeree
University of Alabama
Huntsville, Alabama

Ishmael ("Terry") Banks
American Electric Power Company
Athens, Ohio

Walter Banzhaf
University of Hartford
Hartford, Connecticut

Ottis L. Barron
University of Tennessee at Martin
Martin, Tennessee

Robert A. Bartkowiak
Penn State University
    at Lehigh Valley
Fogelsville, Pennsylvania

Richard M. Bass
Georgia Institute of Technology
Atlanta, Georgia

Michael R. Bastian
Brigham Young University
Provo, Utah

Jeffrey S. Beasley
New Mexico State University
Las Cruces, New Mexico

Lars Bengtsson
Halmsted University
Halmsted, Sweden

Mi Bi
Tai Seng Industrial Estate
Singapore

Edoardo Biagioni
SCS
Pittsburgh, Pennsylvania

David L. Blanchard
Purdue University Calumet
Hammond, Indiana

Wayne Bonzyk
Colman, South Dakota

R. W. Boyd
University of Rochester
Rochester, New York

M. Braae
University of Cape Town
Rondebosch, South Africa

Doug Burges
University of Wisconsin
Madison, Wisconsin

Nick Buris
Motorola
Schaumburg, Illinois

Jose Roberto Camacho
Universidade Federal de
    Uberlindia
Uberlindia, Brazil

Gerard-Andre Capolino
University of Picardie
Amiens, France

Lee W. Casperson
Portland State University
Portland, Oregon

Antonio Chella
University of Palermo
Palermo, Italy

C. H. Chen
University of Massachusetts
North Dartmouth, Massachusetts

Zheru Chi
Hong Kong Polytechnic
    University
Hung Hom, Kowloon,
    Hong Kong

Shamala Chickamenahalli
Wayne State University
Detroit, Michigan

Christos Christodoulou
University of Central Florida
Orlando, Florida

Badrul Chowdhury
University of Wyoming
Laramie, Wyoming

Dominic J. Ciardullo
Nassau Community College
Garden City, New York

Andrew Cobb
New Albany, Indiana

Christopher J. Conant
Broome County Community College
Binghamton, New York

Robin Cravey
NASA Langley Research Center
Hampton, Virginia

George W. Crawford
Penn State University
McKeesport, Pennsylvania

John K. Daher
Georgia Institute of Technology
Atlanta, Georgia

Fredrik Dahlgren
Chalmers University of Technology
Gothenburg, Sweden

E. R. Davies
University of London
Surrey, England

Ronald F. DeMara
University of Central Florida
Orlando, Florida

William E. DeWitt
Purdue University
West Lafayette, Indiana

Alex Domijan
University of Florida
Gainesville, Florida

Bob Dony
Wilfred Laurier University
Waterloo, Ontario, Canada

Tom Downs
University of Queensland
Brisbane, Australia

Marvin Drake
The MITRE Corporation
Bedford, Massachusetts

Lawrence P. Dunleavy
University of South Florida
Tampa, Florida

Scott C. Dunning
University of Maine
Orono, Maine

Andrzej Dzielinski
ISEP
Warsaw University of Technology
Warsaw, Poland

Jack East
University of Michigan
Ann Arbor, Michigan

Sandra Eitnier
San Diego, California

Samir EL-Ghazaly
Arizona State University
Tempe, Arizona

Irv Englander
Bentley College
Waltham, Massachusetts

Ivan Fair
Technical University
   of Nova Scotia
Halifax, Nova Scotia, Canada

Gang Feng
University of New South Wales
Kensington, Australia

Peter M. Fenwick
University of Auckland
Auckland, New Zealand

Paul Fieguth
University of Waterloo
Waterloo, Ontario, Canada

Igor Filanovsky
University of Alberta
Edmonton, Alberta, Canada

Wladyslau Findeisen
Warsaw University of Technology
Warsaw, Poland

Dion Fralick
NASA
Langley Research Center
Hampton, Virginia

Lawrence Fryda
Central Michigan University
Mt. Pleasant, Michigan

Mumtaz B. Gawargy
Concordia University
Montreal, Quebec, Canada

Frank Gerlitz
Washtenaw College
Ann Arbor, Michigan

Antonio Augusto Gorni
COSIPA
Cubatao, Brazil

Lee Goudelock
Laurel, Mississippi

Alex Grant
Institut für Signal- und
    Informationsverarbeitung
Zurich, Switzerland

Thomas G. Habetler
Georgia Tech
Atlanta, Georgia

Haldun Hadimioglu
Brooklyn, New York

Dave Halchin
RF MicroDevices
Greensboro, North Carolina

Thomas L. Harman
University of Houston
Houston, Texas

P. R. Hemmer
RL/EROP
Hanscom Air Force Base
Massachusetts

Vincent Heuring
University of Colorado
Boulder, Colorado

Robert J. Hofinger
Purdue University School of Technology
    at Columbus
Columbus, Indiana

Michael Honig
Northwestern University
Evanston, Illinois

Gregor Hoogers
Trier University of Applied Sciences,
    Unwelt Campus
Bierkenfeld, Germany

Yan Hui
Northern Telecom
Nepean, Ontario, Canada

Suresh Hungenahally
Griffth University
Nathan, Queensland, Australia

Iqbal Husain
University of Akron
Akron, Ohio

Eoin Hyden
Madison, New Jersey

Marija Ilic
MIT
Cambridge, Massachusetts

Mark Janos

Albert Jelalian
Jelalian Science and Engineering
Bedford, Massachusetts

Anthony Johnson
New Jersey Institute of Technology
Newark, New Jersey

C. Bruce Johnson
Phoenix, Arizona

Brendan Jones
Optus Communications
Sydney, Australia

Suganda Jutamulia
In-Harmony Technology Corporation
Petaluma, California

Richard Y. Kain
University of Minnesota
Minneapolis, Minnesota

Dikshitulu K. Kalluri
University of Massachusetts
Lowell, Massachusetts

Alex Kalu
Savannah State University
Savannah, Georgia

Gary Kamerman
FastMetrix
Huntsville, Alabama

Avishay Katz
EPRI
Palo Alto, California

Wilson E. Kazibwe
Telegyr Systems
San Jose, California

David Kelley
Penn State University
University Park, Pennsylvania

D. Kennedy
Ryerson Polytechnic Institute
Toronto, Ontario, Canada

Mohan Ketkar
University of Houston
Houston, Texas

Jerzy Klamka
Silesian Technical University
Gliwice, Poland

Krzysztof Kozlowski
Technical University of Poznan
Poznan, Poland

Ron Land
Penn State University
New Kensington, Pennsylvania

Robert D. Laramore
Cedarville College
Cedarville, Ohio

Joy Laskar
Georgia Institute of Technology
Atlanta, Georgia

Matti Latva-aho
University of Oulu
Linannmaa, Oulu, Finland

Thomas S. Laverghetta
Indiana University-Purdue University
    at Fort Wayne
Fort Wayne, Indiana

J. N. Lee
Naval Research Laboratory
Washington, D.C.

Fred Leonberger
UT Photonics
Bloomfield, Connecticut

Rodney LeRoy
Townsend and Townsend and Crew, LLP
San Francisco, California

Yilu Liu
Virginia Tech
Blacksburg, Virginia

Ging Li-Wang
Dexter Magnetic Materials
Fremont, California

Jean Jacques Loiseau
Institute Recherche en Cybernetique
Nantes, France

Harry MacDonald
San Diego, California

Chris Mack
FINLE Technologies
Austin, Texas

Krzysztov Malinowski
Warsaw University of Technology
Warsaw, Poland

S. Manoharan
University of Auckland
Auckland, New Zealand

Horacio J. Marquez
University of Alberta
Edmonton, Alberta, Canada

Francesco Masulli
University of Genoa
Genoa, Italy

Vincent P. McGinn
Northern Illinois University
DeKalb, Illinois

John A. McNeill
Worcester Polytechnic Institute
Worcester, Massachusetts

David P. Millard
Georgia Institute of Technology
Atlanta, Georgia

Monte Miller
Rockwell Semiconductor Systems
Newbury Park, California

Linn F. Mollenauer
AT&T Bell Labs
Holmdel, New Jersey

Mauro Mongiardo
University of Perugia
Perugia, Italy

Michael A. Morgan
Naval Postgraduate School
Monterey, California

Amir Mortazawi
University of Central Florida
Orlando, Florida

Michael S. Munoz
TRW Corporation

Paolo Nesi
University of Florence
Florence, Italy

M. Nieto-Vesperinas
Instituto de Ciencia de Materiales
Madrid, Spain

Kenneth V. Noren
University of Idaho
Moscow, Idaho

Behrooz Nowrouzian
University of Alberta
Edmonton, Alberta, Canada

Terrence P. O'Connor
Purdue University School of Technology
    at New Albany
New Albany, Indiana

Ben O. Oni
Tuskegee University
Tuskegee, Alabama

Thomas H. Ortmeyer
Clarkson University
Potsdam, New York

Ron P. O'Toole
Cedar Rapids, Iowa

Tony Ottosson
Chalmers University of Technology
Goteburg, Sweden

J. R. Parker
University of Calgary
Calgary, Alberta, Canada

Stefan Parkval
Royal Institute of Technology
Stockholm, Sweden

Joseph E. Pascente
Downers Grove, Illinois

Russell W. Patterson
Tennessee Valley Authority
Chattanooga, Tennessee

Steven Pekarek
University of Missouri
Rolla, Missouri

Marek Perkowski
Portland State University
Portland, Oregon

Roman Pichna
University of Oulu
Oulu, Finland

A. H. Pierson
Pierson Scientific Associates, Inc.
Andover, Massachusetts

Pragasen Pillay
Clarkson University
Potsdam, New York

Aun Neow Poo
Postgraduate School of Engineering
National University of Singapore
Singapore

Ramas Ramaswami
MultiDisciplinary Research
Ypsilanti, Michigan

Satiskuman J. Ranade
New Mexico State University
Las Cruces, New Mexico

Lars K. Rasmussen
Centre for Wireless Communications
Singapore

Walter Rawle
Ericsson, Inc.
Lynchburg, Virginia

C. J. Reddy
NASA Langley Research Center
Hampton, Virginia

Greg Reese
Dayton, Ohio

Joseph M. Reinhardt
University of Iowa
Iowa City, Iowa

Nabeel Riza
University of Central Florida
Orlando, Florida

John A. Robinson
Memorial University of Newfoundland
St. John's, Newfoundland
Canada

Eric Rogers
University of Southampton
Highfield, Southampton, England

Christian Ronse
Université Louis Pasteur
Strasbourg, France

Pieter van Rooyen
University of Pretoria
Pretoria, South Africa

Ahmed Saifuddin
Communication Research Lab
Tokyo, Japan

Robert Sarfi
ABB Power T&D Corporation
Cary, North Carolina

Simon Saunders
University of Surrey
Guildford, England

Helmut Schillinger
IOQ
Jena, Germany

Manfred Schindler
ATN Microwave
North Billerica, Massachusetts

Warren Seely
Motorola
Scottsdale, Arizona

Yun Shi
New Jersey Institute of Technology
Newark, New Jersey

Mikael Skoglund
Chalmers University of Technology
Goteborg, Sweden

Rodney Daryl Slone
University of Kentucky
Lexington, Kentucky

Keyue M. Smedley
University of California
Irvine, California

William Smith
University of Kentucky
Lexington, Kentucky

Babs Soller
University of Massachusetts Medical Center
Worcester, Massachusetts

Y. H. Song
Brunel University
Uxbridge, England

Janusz Sosnowski
Institute of Computer Science
Warsaw, Poland

Elvino Sousa
University of Toronto
Toronto, Ontario, Canada

Philip M. Spray
Amarillo, Texas

Joe Staudinger
Motorola
Tempe, Arizona

Roman Stemprok
Denton, Texas

Francis Swarts
University of the Witwatersr and
Johannesburg, South Africa

Andrzej Swierniak
Silesian Technical University
Gliwice, Poland

Daniel Tabak
George Mason University
Fairfax, Virginia

Tadashi Takagi
Mitsubishi Electric Corporation
Ofuna, Kamakura, Japan

Jaakko Talvitie
University of Oulu
Oulu, Finland

Hamid A. Toliyat
Texas A&M University
College Station, Texas

Austin Truitt
Texas Instruments
Dallas, Texas

Pieter van Rooyen
University of Pretoria
South Africa

Jonas Vasell
Chalmers University of Technology
Gateborg, Sweden

John L. Volakis
University of Michigan
Ann Arbor, Michigan

Annette von Jouanne
Oregon State University
Corvallis, Oregon

Liancheng Wang
ABB Power T&D Corporation
Cary, North Carolina

Ronald W. Waynant
FDA/CDRH
Rockville, Maryland

Larry Wear
Sacramento, California

Wilson X. Wen
AI Systems
Talstra Labs
Clayton, Australia

Barry Wilkinson
University of North Carolina
Charlotte, North Carolina

Robert E. Wilson
Western Area Power Administration
Montrose, California

Stacy S. Wilson
Western Kentucky University
Bowling Green, Kentucky

Denise M. Wolf
Lawrence Berkeley National Laboratory
Berkeley, California

E. Yaz
University of Arkansas
Fayetteville, Arkansas

Pochi Yeh
University of California
Santa Barbara, California

Jeffrey Young
University of Idaho
Moscow, Idaho

Stanislaw H. Zak
Purdue University
West Lafayette, Indiana

Qing Zhao
University of Western Ontario
London, Ontario, Canada

Jizhong Zhu
National University of Singapore
Singapore

Omar Zia
Marietta, Georgia

Special Symbols

α-**level set**   a crisp set of elements belonging to a fuzzy set A at least to a degree α

$$A_\alpha = \{x \in X \mid \mu_A(x) \geq \alpha\}$$

See also crisp set, fuzzy set.

Δf common symbol for bandwidth, in hertz.

$\epsilon_{\mathbf{rGaAs}}$ common symbol for gallium arsenide relative dielectric constant. $\epsilon_{\mathbf{rGaAs}} = 12.8$.

$\epsilon_{\mathbf{rSi}}$ common symbol for silicon relative dielectric constant. $\epsilon_{\mathbf{rSi}} = 11.8$.

ϵ_0 symbol for permitivity of free space. $\epsilon_0 = 8.849 \times 10^{-12}$ farad/meter.

ϵ_r common symbol for relative dielectric constant.

$\eta_{\mathbf{DC}}$ common symbol for DC to RF conversion efficiency. Expressed as a percentage.

η_a common symbol for power added efficiency. Expressed as a percentage.

η_t common symbol for total or true efficiency. Expressed as a percentage.

$\Gamma_{\mathbf{opt}}$ common symbol for source reflection coefficient for optimum noise performance.

μ_0 common symbol for permeability of free space constant. $\mu_0 = 1.257 \times 10^{-16}$ henrys/meter.

μ_r common symbol for relative permeability.

ω common symbol for radian frequency in radians/second. $\omega = 2 \cdot \pi \cdot$ frequency.

θ_+ common symbol for positive transition angle in degrees.

θ_- common symbol for negative transition angle in degrees.

$\theta_{\mathbf{cond}}$ common symbol for conduction angle in degrees.

$\theta_{\mathbf{sat}}$ common symbol for saturation angle in degrees.

θ_{CC} common symbol for FET channel-to-case thermal resistance in °C/watt.

θ_{JC} common symbol for bipolar junction-to-case thermal resistance in °C/watt.

A^* common symbol for Richardson's constant. $A^* = 8.7$ amperes $\cdot$ cm/°K

BV_{GD} *See* gate-to-drain breakdown voltage.

BV_{GS} *See* gate-to-source breakdown voltage.

dv/dt rate of change of voltage withstand capability without spurious turn-on of the device.

H_{ci} *See* intrinsic coercive force.

n_e common symbol for excess noise in watts.

$n_s h$ common symbol for shot noise in watts.

n_t common symbol for thermal noise in watts.

10base2   a type of coaxial cable used to connect nodes on an Ethernet network. The 10 refers to the transfer rate used on standard Ethernet, 10 megabits per second. The base means that the network uses baseband communication rather than broadband communications, and the 2 stands for

1

the maximum length of cable segment, 185 meters (almost 200). This type of cable is also called "thin" Ethernet, because it is a smaller diameter cable than the 10base5 cables.

10base5 a type of coaxial cable used to connect nodes on an Ethernet network. The 10 refers to the transfer rate used on standard Ethernet, 10 megabits per second. The base means that the network uses baseband communication rather than broadband communications, and the 5 stands for the maximum length of cable segment of approximately 500 meters. This type of cable is also called "thick" Ethernet, because it is a larger diameter cable than the 10base2 cables.

10baseT a type of coaxial cable used to connect nodes on an Ethernet network. The 10 refers to the transfer rate used on standard Ethernet, 10 megabits per second. The base means that the network uses baseband communication rather than broadband communications, and the T stands for twisted (wire) cable.

2-D Attasi model a 2-D model described by the equations

$$x_{i+1,j+1} = -A_1 A_2 x_{i,j} + A_1 x_{i+1,j}$$
$$+ A_2 x_{i,j+1} + B u_{ij}$$
$$y_{ij} = C x_{ij} + D u_{ij}$$

$i, j \in Z_+$ (the set of nonnegative integers). Here $x_{ij} \in R^n$ is the local state vector, $u_{ij} \in R^m$ is the input vector, $y_{ij} \in R^p$ is the output vector, and A_1, A_2, B, C, D are real matrices. The model was introduced by Attasi in "Systemes lineaires homogenes a deux indices," *IRIA Rapport Laboria,* No. 31, Sept. 1973.

2-D Fornasini–Marchesini model a 2-D model described by the equations

$$x_{i+1,j+1} = A_0 x_{i,j} + A_1 x_{i+1,j}$$
$$+ A_2 x_{i,j+1} + B u_{ij} \quad (1a)$$
$$y_{ij} = C x_{ij} + D u_{ij} \quad (1b)$$

$i, j \in Z_+$ (the set of nonnegative integers) here $x_{ij} \in R^n$ is the local state vector, $u_{ij} \in R^m$ is the input vector, $y_{ij} \in R^p$ is the output vector A_k ($k = 0, 1, 2$), B, C, D are real matrices. A 2-D model described by the equations

$$x_{i+1,j+1} = A_1 x_{i+1,j} + A_2 x_{i,j+1}$$
$$+ B_1 u_{i+1,j} + B_2 u_{i,j+1} \quad (2)$$

$i, j \in Z_+$ and (1b) is called the second 2-D Fornasini–Marchesini model, where x_{ij}, u_{ij}, and y_{ij} are defined in the same way as for (1), A_k, B_k ($k = 0, 1, 2$) are real matrices. The model (1) is a particular case of (2).

2-D general model a 2-D model described by the equations

$$x_{i+1,j+1} = A_0 x_{i,j} + A_1 x_{i+1,j}$$
$$+ A_2 x_{i,j+1} + B_0 u_{ij}$$
$$+ B_1 u_{i+1,j} + B_2 u_{i,j+1}$$
$$y_{ij} = C x_{ij} + D u_{ij}$$

$i, j \in Z_+$ (the set of nonnegative integers) here $x_{ij} \in R^n$ is the local state vector, $u_{ij} \in R^m$ is the input vector, $y_{ij} \in R^p$ is the output vector and A_k, B_k ($k = 0, 1, 2$), C, D are real matrices. In particular case for $B_1 = B_2 = 0$ we obtain the first 2-D Fornasini–Marchesini model and for $A_0 = 0$ and $B_0 = 0$ we obtain the second 2-D Fornasini–Marchesini model.

2-D polynomial matrix equation a 2-D equation of the form

$$AX + BY = C \quad (1)$$

where $A \in R^{k \times p}[s]$, $B \in R^{k \times q}[s]$, $C \in R^{k \times m}[s]$ are given, by a solution to (1) we mean any pair $X \in R^{p \times m}[s]$, $Y \in R^{q \times m}[s]$ satisfying the equation. The equation (1) has a solution if and only if the matrices $[A, B, C]$ and $[A, B, 0]$ are column equivalent or the greatest common left divisor of A and B is a left divisor of C. The 2-D equation

$$AX + YB = C \quad (2)$$

$A \in R^{k \times p}[s]$, $B \in R^{q \times m}[s]$, $C \in R^{k \times m}[s]$ are given, is called the bilateral 2-D polynomial matrix equation. By a solution to (2) we mean any pair $X \in R^{p \times m}[s]$, $Y \in R^{k \times q}[s]$ satisfying the equation. The equation has a solution if and only if the matrices

$$\begin{bmatrix} A & 0 \\ 0 & B \end{bmatrix} \quad \text{and} \quad \begin{bmatrix} A & C \\ 0 & B \end{bmatrix}$$

are equivalent.

2-D Roesser model   a 2-D model described by the equations

$$\begin{bmatrix} x_{i+1,j}^{h} \\ x_{i,j+1}^{v} \end{bmatrix} = \begin{bmatrix} A_1 & A_2 \\ A_3 & A_4 \end{bmatrix} \begin{bmatrix} x_{ij}^{h} \\ x_{ij}^{v} \end{bmatrix} + \begin{bmatrix} B_1 \\ B_2 \end{bmatrix} u_{ij}$$

$i, j \in Z_+$ (the set of nonnegative integers),

$$y_{ij} = C \begin{bmatrix} x_{ij}^{h} \\ x_{ij}^{v} \end{bmatrix} + D u_{ij}$$

Here $x_{ij}^{h} \in R^{n_1}$ and $x_{ij}^{v} \in R^{n_2}$ are the horizontal and vertical local state vectors, respectively, $u_{ij} \in R^{m}$ is the input vector, $y_{ij} \in R^{p}$ is the output vector and A_1, A_2, A_3, A_4, B_1, B_2, C, D are real matrices. The model was introduced by R.P. Roesser in "A discrete state-space model for linear image processing," *IEEE Trans. Autom. Contr.*, AC-20, No. 1, 1975, pp. 1-10.

2-D shuffle algorithm   an extension of the Luenberger shuffle algorithm for 1-D case. The 2-D shuffle algorithm can be used for checking the regularity condition

$$\det [E z_1 z_2 - A_0 - A_1 z_1 - A_2 z_2] \neq 0$$

for some $(z_1, z_2) \in C \times C$ of the singular general model (*See* Singular 2-D general model). The algorithm is based on the row compression of suitable matrices.

2-D Z-transform $F(z_1, z_2)$ of a discrete 2-D function f_{ij} satisfying the condition $f_{ij} = 0$ for

$i < 0$ or/and $j < 0$ is defined by

$$F(z_1, z_2) = \sum_{i=0}^{\infty} \sum_{j=0}^{\infty} f_{ij} z_1^{-i} z_2^{-j}$$

An 2-D discrete f_{ij} has the 2-D Z-transform if the sum

$$\sum_{i=0}^{\infty} \sum_{j=0}^{\infty} f_{ij} z_1^{-i} z_2^{-j}$$

exists.

2DEGFET   *See* high electron mobility transistor (HEMT).

2LG   *See* double phase to ground fault.

3-dB bandwidth   for a causal low-pass or band-pass filter with a frequency function $H(j\omega)$ the frequency at which $\mid H(j\omega) \mid_{dB}$ is less than 3 dB down from the peak value $\mid H(\omega_P) \mid$.

3-level laser   a laser in which the most important transitions involve only three energy states; usually refers to a laser in which the lower level of the laser transition is separated from the ground state by much less than the thermal energy kT. *Contrast with* 4-level laser.

3-level system   a quantum mechanical system whose interaction with one or more electromagnetic fields can be described by considering primarily three energy levels. For example, the cascade, vee, and lambda systems are 3-level systems.

4-level laser   a laser in which the most important transitions involve only four energy states; usually refers to a laser in which the lower level of the laser transition is separated from the ground state by much more than the thermal energy kT. *Contrast with* 3-level laser.

45 Mbs DPCM for NTSC color video   a codec wherein a subjectively pleasing picture is required at the receiver. This does not require transparent coding quality typical of TV signals. The output bit-rate for video matches the DS3 44.736 Megabits per second rate. The coding is done by PCM coding the NTSC composite video signal at three times the color subcarrier frequency using 8 bit per pixel. Prediction of current pixel is obtained by averaging the pixel three after current and 681 pixels before next to maintain the subcarrier phase. A leak factor is chosen before computing prediction error to main the quality of the image. For example, with a leak factor of $\frac{31}{32}$ the prediction decay is maintained

at the center of the dynamic range.

$$X_L^- = 128 + \frac{31}{32}\left(X^- - 128\right)$$

Finally, a clipper at the coder and decoder is employed to prevent quantization errors.

90% withstand voltage   a measure of the practical lightning or switching-surge impulse withstand capability of a piece of power equipment. This voltage withstand level is two standard deviations above the Basic Impulse Insulation Lever (BIL) of the equipment.

A-mode display   returned ultrasound echoes displayed as amplitude versus depth into the body.

A-site   in a ferroelectric material with the chemical formula ABO_3, the crystalline location of the A atom.

a posteriori probability   posterior statistics.

a priori probability   prior statistics.

A/D   *See* analog-to-digital converter.

AAL   *See* ATM adaptation layer.

ABC   *See* absorbing boundary condition.

ABCD   propagation of an optical ray through a system can be described by a simple 2×2 matrix. In ray optics, the characteristic of a sytem is given by the corresponding ray matrix relating the ray's position from the axis and slope at the input to those at the output.

ABCD formalism   analytic method using two-by-two ABCD matrices for propagating Gaussian beams and light rays in a wide variety of optical systems.

ABCD law   analytic formula for transforming a Gaussian beam parameter from one reference plane to another in paraxial optics, sometimes called the Kogelnik transformation. ABCD refers to the ABCD matrix.

ABCD matrix   the matrix containing ABCD parameters. *See* ABCD parameters.

ABCD parameters   a convenient mathematical form that can be used to characterize two-port networks. Sometimes referred to as chain parameters. ABCD parameters are widely used to model cascaded connections of two-port microwave networks, in which case the ABCD matrix is defined for each two-port network. ABCD parameters can also be used in analytic formalisms for propagating Gaussian beams and light rays. Ray matrices and beam matrices are similar but are often regarded as distinct.

ABC parameters have a particularly useful property in circuit analysis where the composite ABCD parameters of two cascaded networks are the matrix products of the ABCD parameters of the two individual circuits. ABCD parameters are defined as

$$\begin{bmatrix} v_1 \\ i_1 \end{bmatrix} = \begin{bmatrix} A & B \\ C & D \end{bmatrix} \begin{bmatrix} v_2 \\ i_2 \end{bmatrix}$$

where v_1 and v_2 are the voltages on ports one and two, and i_1 and i_2 are the branch currents into ports one and two.

aberration   an imperfection of an optical system that leads to a blurred or a distorted image.

abnormal event   any external or program-generated event that makes further normal program execution impossible or undesirable, resulting in a system interrupt. Examples of abnormal events include system detection of power failure; attempt to divide by 0; attempt to execute privileged instruction without privileged status; memory parity error.

abort   (1) to terminate the attempt to complete the transaction, usually because there is a deadlock or because completing the transaction would result in a system state that is not compatible with "correct" behavior, as defined by a consistency model, such as sequential consistency.

(2) in an accelerator, terminating the acceleration process prematurely, either by inhibiting the injection mechanism or by removing circulating

beam to some sort of dump. This is generally done to prevent injury to some personnel or damage to accelerator components.

ABR   *See* available bit rate.

absolute address   an address within an instruction that directly indicates a location in the program's address space. *Compare with* relative addressing.

absolute addressing   an addressing mode where the address of the instruction operand in memory is a part of the instruction so that no calculation of an effective address by the CPU is necessary.

For example, in the Motorola M68000 architecture instruction ADD 5000, D1, a 16-bit word operand, stored in memory at the word address 5000, is added to the lower word in register D1. The address "5000" is an example of using the absolute addressing mode. *See also* addressing mode.

absolute encoder   an optical device mounted to the shaft of a motor consisting of a disc with a pattern and light sources and detectors. The combination of light detectors receiving light depends on the position of the rotor and the pattern employed (typically the Gray code). Thus, absolute position information is obtained. The higher the resolution required, the larger the number of detectors needed. *See also* encoder.

absolute moment   The pth order absolute moment μ_p of a random variable $\mathbf{X}$ is the expectation of the absolute value of $\mathbf{X}$ raised to the pth power:

$$\mu_p = E[|\mathbf{X}|]^p$$

See central moment, central absolute moment, expectation.

absolute pressure   units to measure gas pressure in a vacuum chamber with zero being a perfect vacuum. Normally referred to as psia (pounds per square inch absolute).

absolute sensitivity   denoted $\mathbf{S}(y, x)$, is simply the partial derivative of y with respect to x, i.e., $\mathbf{S}(y, x) = \partial y / \partial x$, and is used to establish the relationships between absolute changes. *See* sensitivity, sensitivity measure, relative sensitivity, semi-relative sensitivity.

absolute stability   occurs when the network function $H(s)$ has only left half-plane poles.

absorber   generic term used to describe material used to absorb electromagnetic energy. Generally made of polyurethane foam and impregnated with carbon (and fire-retardant salts), it is most frequently used to line the walls, floors and ceilings of anechoic chambers to reduce or eliminate reflections from these surfaces.

absorbing boundary condition (ABC)   a fictitious boundary introduced in differential equation methods to truncate the computational space at a finite distance without, in principle, creating any reflections.

absorption   (1) process that dissipates energy and causes a decrease in the amplitude and intensity of a propagating wave between an input and output reference plane.

(2) reduction in the number of photons of a specific wavelength or energy incident upon a material. Energy transferred to the material may result in a change in the electronic structure, or in the relative movement of atoms in the material (vibration or rotation).

(3) process by which atoms or molecules stick to a surface. If a bond is formed, it is termed chemisorption, while the normal case is physisorption. The absorption process proceeds due to, and is supported by, the fact that this is a lower energy state.

absorption coefficient   (1) in a passive device, the negative ratio of the power absorbed

($p_{absorbed} = p_{in} - p_{out}$) ratioed to the power in ($p_{in} = p_{incident} - p_{reflected}$) per unit length (l), usually expressed in units of 1/wavelength or 1/meter.

(2) factor describing the fractional attenuation of light with distance traversed in a medium, generally expressed as an exponential factor, such as k in the function e^{-kx}, with units of (length)-1. Also called attenuation coefficient.

absorption cross section energy absorbed by the scattering medium, normalized to the wavenumber. It has dimensions of area.

absorption edge the optical wavelength or photon energy corresponding to the separation of valence and conduction bands in solids; at shorter wavelengths, or higher photon energies than the absorption edge, the absorption increases strongly.

absorption grating (1) a diffraction grating where alternate grating periods are opaque.

(2) an optical grating characterized by spatially periodic variation in the absorption of light. Absorption gratings are generally less efficient than phase gratings.

absorption optical fiber the amount of optical power in an optical fiber captured by defect and impurity centers in the energy bandgap of the fiber material and lost in the form of longwave infrared radiation.

AC *See* alternating current.

AC bridge one of a wide group of bridge circuits used for measurements of resistances, inductances, and capacitances, and to provide AC signal in the bridge transducers including resistors, inductors, and capacitors.

The Wheatstone bridge can be used with a sinusoidal power supply, and with an AC detector (headphones, oscilloscope), one can use essentially the same procedure for measurement of resistors as in DC applications. Only a small number of other AC bridges are used in modern electric and electronic equipment. A strong selection factor was the fact that in a standard capacitor the electrical parameter are closest to the parameters of an ideal capacitor. Hence, not only a capacitance is measured in terms of capacitance (in resistive ratio arms bridges), but the inductance as well is measured in terms of capacitance (Hay and Owen bridges).

The AC bridges with ratio arms that are tightly coupled inductances allow measurement of a very small difference between currents in these inductances, and this fact is used in very sensitive capacitance transducers.

AC circuit electrical network in which the voltage polarity and directions of current flow change continuously, and often periodically. Thus, such networks contain alternating currents as opposed to direct currents, thereby giving rise to the term.

AC coupling a method of connecting two circuits that allows displacement current to flow while preventing conductive currents. Reactive impedance devices (e.g., capacitors and inductive transformers) are used to provide continuity of alternating current flow between two circuits while simultaneously blocking the flow of direct current.

AC motor an electromechanical system that either converts alternating current electrical power into mechanical power.

AC plasma display a display that employs an internal capacitive dielectric layer to limit the gas discharge current.

AC steady-state power the average power delivered by a sinusoidal source to a network, expressed as

$$P = | V | \cdot | I | \cos(\theta)$$

where $\sqrt{2} \cdot | V |$ and $\sqrt{2} \cdot | I |$ are the peak values, respectively, of the AC steady-state voltage

7

and current at the terminals. θ represents the phase angle by which the voltage leads the current.

AC/AC converter   a power electronics device in which an AC input voltage of some magnitude, frequency, and number of phases is changed to an AC output with changes to any of the previously mentioned parameters. AC/AC converters usually rectify the input source to a DC voltage and then invert the DC voltage to the desired AC voltage.

AC/DC converter   *See* rectifier.

AC-DC integrated system   a power system containing both AC and DC transmission lines.

ACARS   aircraft communications addressing and reporting. A digital communications link using the VHF spectrum for two-way transmission of data between an aircraft and ground. It is used primarily in civil aviation applications.

ACC   *See* automatic chroma control.

accelerated testing   tests conducted at higher stress levels than normal operation but in a shorter period of time for the specific purpose to induce failure faster.

accelerating power   the excess electric power at a synchronous machine unit which cannot be transmitted to the load because of a short circuit near its terminals. This energy gives rise to increasing rotor angle.

acceleration error   the final steady difference between a parabolic setpoint and the process output in a unity feedback control system. Thus it is the asymptotic error in position that arises in a closed loop system that is commanded to move with constant acceleration. *See also* position error, velocity error.

acceleration error constant   a gain K_a from which acceleration error e_a is readily determined.

The acceleration error constant is a concept that is useful in the design of unity feedback control systems, since it transforms a constraint on the final acceleration error to a constraint on the gain of the open loop system. The relevant equations are $e_a = \frac{1}{K_a}$ and $K_a = \lim_{s \to \infty} s^2 q(s)$, where $q(s)$ is the transfer function model of the open loop system, including the controller and the process in cascade, and s is the Laplace variable. *See also* position error constant, velocity error constant.

accelerator   (1) a positive electrode in a vacuum tube to accelerate emitted electrons from its cathode by coulomb force in a desired direction.

(2) a machine used to impart large kinetic energies to charged particles such as electrons, protons, and atomic nuclei. The accelerated particles are used to probe nuclear or subnuclear phenomena in industrial and medical applications.

acceptable delay   the voice signal delay that results in inconvenience in the voice communication. A typically quoted value is 300 ms.

acceptance   in an accelerator, it defines how "large" a beam will fit without scraping into the limiting aperture of a transport line. The acceptance is the phase-space volume within which the beam must lie in order to be transmitted through an optical system without losses. From an experimenter's point of view acceptance is the phase-space volume intercepted by an experimenter's detector system.

acceptor   (1) an impurity in a semiconductor that donates a free hole to the valence band.

(2) a dopant species that traps electrons, especially with regard to semiconductors.

access channel   a channel in a communications network that is typically allocated for the purpose of setting up calls or communication sessions. Typically the users share the access channel using some multiple access algorithm such as ALOHA or CSMA.

access control   a means of allowing access to an object based on the type of access sought, the accessor's privileges, and the owner's policy.

access control list   a list of items associated with a file or other object; the list contains the identities of users that are permitted access to the associated file. There is information (usually in the form of a set of bits) about the types of access (such as read, write, or delete) permitted to the user.

access control matrix   a tabular representation of the modes of access permitted from active entities (programs or processes) to passive entities (objects, files, or devices). A typical format associates a row with an active entity or subject and a column with an object; the modes of access permitted from that active entity to the associated passive entity are listed in the table entry.

access line   a communication line that connects a user's terminal equipment to a switching node.

access mechanism   a circuit board or an integrated chip that allows a given part of a computer system to access another part. This is typically performed by using a specific access protocol.

access protocol   a set of rules that establishes communication among different parts. These can involve both hardware and software specifications.

access right   permission to perform an operation on an object, usually specified as the type of operation that is permitted, such as read, write, or delete. Access rights can be included in access control lists, capability lists, or in an overall access control matrix.

access time   the total time needed to retrieve data from memory. For a disk drive, this is the sum of the time to position the read/write head over the desired track and the time until the desired data rotates under the head. (LW)

accidental rate   the rate of false coincidences in the electronic counter experiment produced by products of the reactions of more than one beam particle within the time resolution of the apparatus.

accumulation   an increase in the majority carrier concentration of a region of semiconductor due to an externally applied electric field.

accumulator   (1) a register in the CPU (processor) that stores one of the operands prior to the execution of an operation, and into which the result of the operation is stored. An accumulator serves as an implicit source and destination of many of the processor instructions. For example, register A of the Intel 8085 is an accumulator. *See also* CPU and processor.

(2) the storage ring in which successive pulses of particles are collected in order to create a particle beam of reasonable intensity for colliding beams.

achievable rate region   for a multiple terminal communications system, a set of rate-vectors for which there exist codes such that the probability of making a decoding error can be made arbitrarily small. *See also* capacity region, multiple access channel.

achromatic   the quality of a transport line or optical system where particle momentum has no effect on its trajectory through the system. In an achromatic device or system, the output beam displacement or divergence (or both) is independent of the input beam's momentum. If a system of lenses is achromatic, all particles of the same momentum will have equal path lengths through the system.

achromatic color   perceived color devoid of hue.

ACI   *See* adjacent channel interference.

acknowledge   (1) a signal which indicates that some operation, such as a data transfer, has successfully been completed.

(2) to detect the successful completion of an operation and produce a signal indicating the success.

acoustic attenuation   the degree of amplitude suppression suffered by the acoustic wave traveling along the acousto-optic medium.

acoustic laser   a laser (or maser) in which the amplified field consists of soundwaves or phonons rather than electromagnetic waves; phonon laser or phaser.

acoustic memory   a form of circulating memory in which information is encoded in acoustic waves, typically propagated through a trough of mercury. Now obsolete.

acoustic velocity   the velocity of the acoustic signal traveling along the acousto-optic medium.

acoustic wave   a propagating periodic pressure wave with amplitude representing either longitudinal or shear particle displacement within the wave medium; shear waves are prohibited in gaseous and liquid media.

acousto-optic cell   a device consisting of a photo-elastic medium in which a propagating acoustic wave causes refractive-index changes, proportional to acoustic wave amplitude, that act as a phase grating for diffraction of light. *See also* Bragg cell.

acousto-optic channelized radiometer   *See* acousto-optic instantaneous spectrum analyzer in Bragg mode.

acousto-optic correlator   an optical system that consists of at least one acousto-optic cell, imaging optics between cells and fixed masks, and photodetectors whose outputs correspond to the correlation function of the acoustic wave signal within one cell with another signal in a second cell, or with fixed signals on a mask.

acousto-optic deflector device   device where acousto-optic interaction deflects the incident beam linearly as a function of the input frequency of the RF signal driving the device.

acousto-optic device   descriptor of acousto-optic cells of any design; generally describes a cell plus its transducer structure(s), and may encompass either bulk, guided-wave, or fiber-optic devices.

acousto-optic effect   the interaction of light with sound waves and in particular the modification of the properties of a light wave by its interactions with an electrically controllable sound wave. *See also* Brillouin scattering.

acousto-optic frequency excisor   similar to an acousto-optic spectrum analyzer where the RF temporal spectrum is spatially and selectively blocked to filter the RF signal feeding the Bragg cell.

acousto-optic instantaneous spectrum analyzer in Bragg mode   device in which the temporal spectrum of a radio frequency signal is instantaneously and spatially resolved in the optical domain using a Fourier transform lens and a RF signal-fed Bragg cell.

acousto-optic modulator   a device that modifies the amplitude or phase of a light wave by means of the acousto-optic effect.

acousto-optic processor   an optical system that incorporates acousto-optic cells configured to perform any of a number of mathematical functions such as Fourier transform, ambiguity transforms, and other time-frequency transforms.

acousto-optic scanner   a device that uses an acoustic wave in a photoelastic medium to deflect light to different angular positions based on the frequency of the acoustic wave.

acousto-optic space integrating convolver device that is the same as an acousto-optic space integrating convolver except that it implements the convolution operation.

acousto-optic space integrating correlator   an acousto-optic implementation of the correlation function where two RF signals are spatially impressed on two diffracted beams from Bragg cells, and a Fourier transform lens spatially integrates these beams onto a point sensor that generates a photo current representing the correlation function.

acousto-optic spectrum analyzer   an acousto-optic processor that produces at a photodetector output array the Fourier decomposition of the electrical drive signal of an acousto-optic device.

acousto-optic time integrating convolver   same as the acousto-optic time integrating correlator, except implements the signal convolution operation. *See* acousto-optic time integrating correlator.

acousto-optic time integrating correlator   an acousto-optic implementation of the correlation function where two RF signals are spatially impressed on two diffracted beams from Bragg cells, and a time integrating sensor generates the spatially distributed correlation results.

acousto-optic triple product processor   signal processor that implements a triple integration operation using generally both space and time dimensions.

acousto-optic tunable filter (AOTF)   an acousto-optic device that selects specific optical frequencies from a broadband optical beam, depending on the number and frequencies of acoustic waves generated in the device.

acousto-optics   the area of study of interaction of light and sound in media, and its utilization in applications such as signal processing and filtering.

ACP   *See* adjacent channel power.

acquisition   (1) in digital communications systems, the process of acquiring synchronism with the received signal. There are several levels of acquisitions, and for a given communication system several of them have to be performed in the process of setting up a communication link: frequency, phase, spreading code, symbol, frame, etc.

(2) in analog communications systems, the process of initially estimating signal parameters (for example carrier frequency offset, phase offset) required in order to begin demodulation of the received signal.

(3) in vision processing, the process by which a scene (physical phenomenon) is converted into a suitable format that allows for its storage or retrieval. *See also* synchronization.

across the line starter   a motor starter that applies full line voltage to the motor to start. This is also referred to as "hard starting" because it causes high starting currents. Larger motors require reduced voltage or "soft starting."

ACRR   *See* adjacent channel reuse ratio.

ACSR   aluminum cable, steel-reinforced. A kind of overhead electric power conductor made up of a central stranded steel cable overlaid with strands of aluminum.

ACT   *See* anticomet tail.

activation function   (1) the function that a neural element applies to its activation. The activation is usually a weighted sum of the inputs to the neural element minus a threshold value.

(2) in an artificial neural network, a function that maps the net output of a neuron to a smaller set of values. This set is usually [0, 1]. Typical

functions are the sigmoid function or singularity functions like the step or ramp.

action potential   a propagating change in the conductivity and potential across a nerve cell's membrane; a nerve impulse in common parlance.

active contour   a deformable template matching method that, by minimizing the energy function associated with a specific model (i.e. a specific characterization of the shape of an object), deforms the model in conformation to salient image features.

active device   a device that can convert energy from a DC bias source to a signal at an RF frequency. Active devices are required in oscillators and amplifiers.

active filter   a form of power electronic converter designed to effectively cancel harmonic currents by injecting currents that are equal and opposite to, or $180°$ out of phase with, the target harmonics. Active filters allow the output current to be controlled and provide stable operation against AC source impedance variations without interfering with the system impedance.

The main type of active filter is the series type in which a voltage is added in series with an existing bus voltage. The other type is the parallel type in which a current is injected into the bus and cancels the line current harmonics.

active filter   a filter that has an energy gain greater than one; that is, a filter which outputs more energy than it absorbs.

active impedance   the impedance at the input of a single antenna element of an array with all the other elements of the array excited.

active layer   *See* active region.

active learning   a form of machine learning where the learning system is able to interact with its environment so as to affect the generation of training data.

active load   a transistor connected so as to replace a function that would conventionally be performed by a passive component such as a resistor, capacitor, or inductor.

active load-pull measurement   a measurement method where transfer characteristics of a device can be measured by electrically changing the load impedance seen from the device. In an active load-pull measurement, the load impedance is defined by using an output signal from the device and an injected signal from the output of the device.

active logic   a digital logic that operates all of the time in the active, dissipative region of the electronic amplifiers from which it is constructed. The output of such a gate is determined primarily by the gate and not by the load.

active magnetic bearing   a magnetic bearing that requires input energy for stable support during operation. Generally implemented with one or more electromagnets and controllers.

active mixer   a mixer that uses three terminal devices such as FET rather than diodes as nonlinear element. One advantage of active mixers is that they can provide conversion gain.

active network   an electrical network that contains some solid state devices such as bipolar junction transistors (BJTs) or metal-oxide-silicon field effect transistors (FETs) operating in their active region of the voltage vs. current characteristic. To ensure that these devices are operating in the active region, they must be supplied with proper DC biasing.

active neuron   a neuron with a non-zero output. Most neurons have an activation threshold. The output of such a neuron has zero output until this threshold is reached.

active power *See* real power.

active power line conditioner a device which senses disturbances on a power line and injects compensating voltages or currents to restore the line's proper waveform.

active RC filter an electronic circuit made up of resistors, capacitors, and operational amplifiers that provide well-controlled linear frequency-dependent functions, e.g., low-, high-, and band-pass filters.

active redundancy a circuit redundancy technique that assures fault-tolerance by detecting the existence of faults and performing some action to remove the faulty hardware, e.g., by standby sparing.

active region semiconductor material doped such that electrons and/or holes are free to move when the material is biased. In the final fabricated device, the active regions are usually confined to very small portions of the wafer material.

active-high (1) a logic signal having its asserted state as the logic ONE state.

(2) a logic signal having the logic ONE state as the higher voltage of the two states.

active-low (1) a logic signal having its asserted state as the logic ZERO state.

(2) a logic signal having its logic ONE state as the lower voltage of the two states; inverted logic.

actuator (1) a transducer that converts electrical, hydraulic, or pneumatic energy to effective motion. For example in robots, actuators set the manipulator in motion through actuation of the joints. Industrial robots are equipped with motors that are typically electric, hydraulic, or pneumatic. *See also* industrial robot.

(2) in computers, a device, usually mechanical in nature, that is controlled by a computer, e.g., a printer paper mechanism or a disk drive head positioning mechanism.

ACTV *See* advanced compatible television.

acuity sharpness. The ability of the eye to discern between two small objects closely spaced, as on a display.

adaptability the capability of a system to change in order to suit the prevailing conditions, especially by automatic adjustment of parameters through some initialization procedure or by training.

adaptation layer control layer of a multilayer controller, situated above the direct control layer and — usually — also above the optimizing control layer, required to introduce changes into the decision mechanisms of the layer (or layers) below this adaptation layer; for example adaptation layer of the industrial controller may be responsible for adjusting the model used by the optimizing control and the decision rules used by the direct (regulation) control mechanisms.

adapter a typical term from personal computers. A circuit board containing the interface toward an additional peripheral device. For example, a graphic adapter (interface boards like EGA, VGA, CGA), a game controller, a SCSI controller, a PCMCI interface, etc.

adaptive algorithm a method for adjusting the parameters of a filter to satisfy an objective (e.g., minimize a cost function).

adaptive antenna antenna, or array of antennas, whose performance characteristics can be adapted by some means; e.g., the pattern of an array can be changed when the phasing of each of the array elements is changed.

adaptive array an array that adapts itself to maximize the reception of a desired signal and null all interfering or jamming signals. This is achieved by finding the correct weights (input excitations) to the elements comprising the array.

adaptive coding of transform coefficients coding technique that is carried out by threshold sampling and exploiting masking effects by variable quantization for different blocks. High detail blocks are coded with more quantization error than low detail blocks. This is done to take into account masking and boundary distortion effects. Transform coding becomes more attractive compared with DPCM when adaptive coding is used. The main drawback of adaptive transform coding is its sensitivity to transmission bit errors due to synchronization problems at the decoder. *See also* DPCM.

adaptive coding a coding scheme that adapts itself in some fashion to its input or output.

adaptive control a control methodology in which control parameters are continuously and automatically adjusted in response to measured/estimated process variables to achieve near-optimum system performance.

adaptive critic learning technique where the system learns to evaluate the actions of a system (usually a controller) so as to provide a reinforcement signal that is an estimate of the future value of the system's current action.

adaptive differential pulse code modulation (ADPCM) a modulation scheme in which only the difference between successive signal samples is encoded for transmission, and the quantization of the coding is adapted to the characteristics of the signal source.

adaptive filtering a filtering strategy in which filter coefficients or governing parameters evolve over time according to some updating strategy to optimize some criterion.

adaptive FIR filter a finite impulse response structure filter with adjustable coefficients. The

adjustment is controlled by an adaptation algorithm such as the least mean square (LMS) algorithm. They are used extensively in adaptive echo cancellers and equalizers in communication systems.

adaptive fuzzy system fuzzy inference system that can be trained on a data set through the same learning techniques used for neural networks. Adaptive fuzzy systems are able to incorporate domain knowledge about the target system given from human experts in the form of fuzzy rules and numerical data in the form of input–output data sets of the system to be modeled. *See also* neural network, fuzzy inference system.

adaptive intrafield predictors a technique used for picture signal prediction based on local properties of the signal or side information if portions of local properties have not been transmitted. Intrafield methods require correlation with local information for prediction purposes.

A common technique is to use a measure of the directional correlation based on local pixels that have already been transmitted. A predictor is chosen from a set to give minimum prediction error. For example, the previous line or previous pixel can be used for prediction, and the switching can then be done as follows:

$$\cap X = \text{predictor for element } X$$
$$= \begin{cases} A & \text{if } \|B - C\| < \|A - B\| \\ C & \text{otherwise} \end{cases}$$

An extension of this concept is called contour prediction where the direction of pixel A is determined by searching among E, B, C, or G.

adaptive logic network tree-structured network whose leaves are the inputs and whose root is the output. The first hidden layer consists of linear threshold units and the remaining layers are elementary logic gates, usually AND and OR gates. Each linear threshold unit is trained to fit input data in those regions of the input space where it is active (i.e., where it contributes to the overall network function).

adaptive manipulator controller   a controller that uses an adaptation process which, based on observation of the manipulator position and velocity, readjusts the parameters in the nonlinear model until the errors disappear. An adaptive manipulator controller is depicted in the figure below. Such a system would learn its own dynamic properties. The adaptive manipulator control scheme

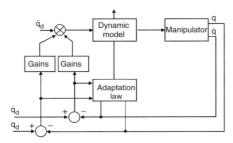

Adaptive manipulator control scheme.

presented in the figure belongs to the joint space control schemes. *See also* joint space control.

adaptive predictor   a digital filter whose coefficients can be varied, according to some error minimization algorithm, such that it can predict the value of a signal, say N, sampling time intervals into the future. The adaptive predictor is useful in many interference cancellation applications.

adaptive resonance theory (ART) network   A clustering network developed to allow the learning of new information without destroying what has already been learnt. Each cluster is represented by a prototype and learning is achieved by comparing a new input pattern with each prototype. If a prototype is found that is acceptably close to that input, the new pattern is added to that prototype's cluster and the prototype is adjusted so as to move

closer to the new input. If no prototype is acceptable, the pattern becomes a new prototype around which a new cluster may develop.

adaptive vector quantization   term that refers to methods for vector quantization that are designed to adaptively track changes in the input signal.

adaptive algorithm   an algorithm whose properties are adjusted continuously during execution with the objective of optimizing some criterion.

ADC   *See* analog-to-digital coverter.

ADCPM   *See* adaptive differential pulse code modulation.

add instruction   a machine instruction that causes two numeric operands to be added together. The operands may be from machine registers, memory, or from the instruction itself, and the result may be placed in a machine register or in memory.

adder   a logic circuit used for adding binary numbers.

additive acousto-optic processing   acousto-optic signal processing where the summation of acousto-optic modulated light waves is used to implement the signal processing operation.

additive polarity   polarity designation of a transformer in which terminals of the same polarity on the low- and high-voltage coils are physically adjacent to each other on the transformer casing. With additive polarity, a short between two adjacent terminals results in the sum of the two coil voltages appearing between the remaining terminals. Additive polarity is generally used for transformers up to 500 kVA and 34.5 kV. Larger units use subtractive polarity. See the diagram below. *See also* subtractive polarity.

15

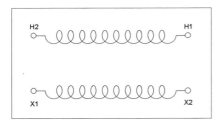

Transformer with additive polarity.

additive white Gaussian noise (AWGN) the simplest form of channel degradation in a communication system in which the source of errors in the channel can be modeled as the addition of random noise with a Gaussian distribution and a constant (white) power spectrum. *See also* thermal noise.

address a unique identifier for the place where information is stored (as opposed to the contents actually stored there). Most storage devices may be regarded by the user as a linear array, such as bytes or words in RAM or sectors on a disk. The address is then just an ordinal number of the physical or logical position. In some disks, the address may be compound, consisting of the cylinder or track and the sector within that cylinder.

In more complex systems, the address may be a "name" that is more relevant to the user but must be translated by the underlying software or hardware.

address aliasing *See* cache aliasing.

address bus the set of wires or tracks on a backplane, printed circuit board, or integrated circuit to carry binary address signals between different parts of a computer. The number of bits of address bus (the width of the bus) determines the maximum size of memory that can be addressed. Modern microchips have 32 address lines, thus 4 gigabytes of main memory can be accessed.

address decoder logic that decodes an address.

(1) A partial decoder responds to a small range of addresses and is used when recognizing particular device addresses on an I/O address bus, or when recognizing that addresses belong to a particular memory module.

(2) A full decoder takes N bits and asserts one of 2^N outputs, and is used within memories (often within RAM chips themselves).

address error an exception (error interrupt) caused by a program's attempt to access unaligned words or long words on a processor that does not accommodate such requests. The address error is detected within the CPU. This contrasts with problems that arise in accessing the memory itself, where a logic circuit external to the CPU itself must detect and signal the error to cause the CPU to process the exception. Such external problems are called bus errors. *See also* bus error.

address field the portion of a program instruction word that holds an address.

address generation interlock (AGI) a mechanism to stall the pipeline for one cycle when an address used in one machine cycle is being calculated or loaded in the previous cycle. Address generation interlocks cause the CPU to be delayed for a cycle. (AGIs on the Pentium are even more important to remove, since two execution time slots are lost).

address locking a mechanism to protect a specific memory address so that it can be accessed exclusively by a single processor.

address map a table that associates a base address in main memory with an object (or page) number.

address mapping the translation of virtual address into real (i.e., physical) addresses for memory access. *See also* virtual memory.

address register a register used primarily to hold the address of a location in memory. The location can contain an operand or an executable instruction.

16

address size prefix   a part of a machine instruction that provides information as to the length or size of the address fields in the instruction.

address space   an area of memory seen or used by a program and generally managed as a continuous range of addresses. Many computers use separate address spaces for code and data; some have other address spaces for system. An address space is usually subject to protection, with references to a space checked for valid addresses and access (such as read only).

The physical address space of a computer (2^{32} bytes, and up to 2^{64} bytes) is often larger than the installed memory. Some parts of the address range (often at extreme addresses) may be reserved for input–output device addresses. *See also* byte, memory, memory mapped I/O, and processor.

address translation   *See* address mapping.

addressing   (1) in processors: a mechanism to refer to a device or storage location by an identifying number, character, or group of characters. That may contain a piece of data or a program step.

(2) in networks, the process of identifying a network component, for instance, the unique address of a node on a local area network.

addressing fault   an error that halts the mapper when it cannot locate a referenced object in main memory.

addressing mode   a form of specifying the address (location) of an operand in an instruction. Some of the addressing modes found in most processors are direct or register direct, where the operand is in a CPU register; register indirect (or simply indirect), where a CPU register contains the address of the operand in memory; immediate, where the operand is a part of the instruction. *See also* central processing unit.

addressing range   numbers that define the number of memory locations addressable by the CPU. For a processor with one address space, the range is determined by the number of signal lines on the address bus of the CPU.

adequate service   in terms of the blocking probability, term associated with a fixed blocking. A typically quoted value may be 2. *See also* blocking.

adiabatic   a system that has no heat transfer with the environment.

adiabatic cooling   a process where the temperature of a system is reduced without any heat being exchanged between the system and its surroundings. In particle beam acceleration this term is used to describe the process in the particle source storage ring where beam emittances are reduced without affecting beam energy.

adiabatic following   an approximation made when some states in a quantum mechanical system respond to perturbations more quickly than the other states. In this approximation the rapidly responding states are assumed to depend only on the instantaneous values of the other states and are said to "follow" those states.

adiabatic passage   a technique for the creation of a long-lived coherence in a quantum mechanical system by manipulating electromagnetic field intensities so that the system always remains in an eigenstate. In practice, this involves changing field strengths on a time scale slower than the inverse of the energy spacing between relevant eigenstates of the system. For example, consider a lambda system in which only one field is present initially and all population starts out in the uncoupled ground state. If a field is gradually turned on to couple this initial state to the excited state, the system can remain transparent by evolving in such a way that it is always mathematically equivalent to the dark state that would be produced by coherent population trapping. Adiabatic passage is often used for selective transfer of population between two long-lived states of a multistate system, especially in cases where the two-step process of absorption followed by spontaneous decay (optical pumping) would tend to populate many other states.

adjacency graph   a graph in which each node represents an object, component or feature in an image. An edge between two nodes indicates two components that are touching or connected in the image.

adjacent channel interference (ACI)   the interference caused by an adjacent frequency band, e.g., in a system with frequency division duplex (FDD). Classified as either in-band or out-of-band adjacent channel interference (ACI). The in-band ACI occurs when the center frequency of interfering signal falls within the band of the desired signal. The out-of-band ACI occurs when the center frequency of interfering signal falls outside the bandwidth of the desired signal.

adjacent channel leakage power   *See* adjacent channel power.

adjacent channel power (ACP)   a power of distortion components generated in adjacent channel, which is caused by a nonlinearity of high-power amplifier amplifying a digitally modulated signal such as QPSK, QAM, etc. Adjacent channel power is defined as a ratio of signal power in channel and leakage power in adjacent channel.

adjacent channel reuse ratio (ACRR)   the reuse ratio between radio communication cells using adjacent radio channels. *See also* reuse ratio.

adjacent channels   radio channels occupying radio frequency allocations n and $n \pm 1$.

adjoint network   a network with an identical structure to the original one, but with possibly different elements. As an example, for a network described by the nodal admittance matrix, its adjoint network is represented by the transposed admittance matrix of the original network. The adjoint network is a basic tool in the computer-aided sensitivity analysis of electronic and microwave circuits.

adjustable-speed drive   *See* variable speed DC drive or variable speed AC drive.

admissible matrix   a matrix M^- that can be obtained by fixing the free parameters of the matrix M at some particular values. M^- is said to be admissible with respect to M.

admittance   the reciprocal of the impedance of an electric circuit.

admittance inverter   an idealized device or set of matrix parameters that functions electrically like a quarter-wave lossless transmission line of characteristic impedance J at each frequency, thus transforming the load admittance (Y_{LOAD}) by +90 degrees and modifying the magnitude, resulting in an input admittance (Y_{in}).

$$Y_{in} = \frac{J^2}{Y_{load}}$$

admittance matrix   the inverse of the impedance matrix in the method of moments.

ADP   *See* ammonium dihydrogen phosphate.

ADPCM   *See* adaptive differential pulse-code modulation.

ADSL   *See* asymmetric digital subscriber line.

adsorbent   the material of an adsorber, for example silica gel, alumina and charcoal. Adsorbent materials are characterized by high surface to volume ratio.

adsorber   (1) condensation of a gas on the solid material.

(2) material that attracts and holds (by Van der Waal forces) molecular layers of dense gases (i.e., very near condensation temperatures) on porous high surface/volume ratio materials.

ADTV   *See* advanced digital television.

advanced compatible television (ACTV)   an extended definition television system that can operate with existing bandwidths on existing receivers and is compatible with the NTSC broadcasting system. The ACTV system was proposed by the Advanced Television Research Consortium and was the first high definition television (HDTV) system. The HDTV system was tested by the FCC July 17, 1992. The additional picture information needed to increase the picture width and to increase the resolution to the HDTV format is transmitted in an augmented channel as an alternative to simulcast transmission. *See* Advanced Television Research Consortium.

advanced digital television (ADTV)   a high definition television (HDTV) digital transmission television system was proposed to the Federal Communications Commission by the Advanced Television Research Consortium. The ADTV system introduced a layered system to separately describe the digital transmission system, the video compression system, and the data packet transport system. The video compression method uses a MPEG++ standard that provides for compatibility with multimedia computing. *See* Advanced Television Research Consortium.

advanced mobile phone system (AMPS)   a standard for a cellular radio communications network originally developed in the 1970s by AT&T and later adopted as an industry standard by the U.S.-based Telecommunications Industries Association (TIA). It is the first cellular standard widely deployed in North America. It is also referred to as the analog cellular system. Frequency modulation with 30 kHz channels is used.

Advanced Television Research Consortium   an organization consisting of David Sarnoff Research Center, Thompson Consumer Electronics, North American Philips Corporation, NBC, and Compression Laboratories.

aeolian vibration   a high-frequency mechanical vibration of electric power lines caused by wind.

aerial cable   any fully-insulated electric power cable which is carried overhead upon poles, as opposed to the use of the more usual overhead bare conductors.

aerodynamic head   *See* disk head.

AFC   *See* automatic frequency control, alkaline fuel cell.

affine transform   a geometric image transformation including one or more translations, rotations, scales and shears that is represented by a 4×4 matrix allowing multiple geometric transformations in one transform step. Affine transformations are purely linear and do not include perspective or warping transformations.

AFM   *See* atomic force microscope.

AFT   *See* automatic fine tuning.

AFV   *See* audio follow-video switcher.

AGC   *See* automatic gain control or automatic generation control.

agent   a computational entity that acts on behalf of other entities in an autonomous fashion.

agent-based system   an application whose component are agents. *See also* agent.

aggregation   an operation performed on system variables whose purpose is to collect them in a way enabling order and/or uncertainty reduction. For linear systems both continuous-time and discrete-time state aggregation is obtained by linear transformation of the original state represented by an aggregation matrix G endowed with the following properties:

$$GA = A^*G; GB = B^*; CG' = C^*;$$

where A, B, C are original system matrices (respectively state, input, and output ones) and A^*, B^*, C^* are aggregated system matrices. The aggregation is an eigenvalues-preservation approach and it provides order reduction by neglecting some of the system modes.

For uncertainties, the aggregation defines some deterministic measures for a set of uncertain variables. For stochastic model of uncertainty the aggregation may be given by mean value, higher stochastic models or other statistical characteristics, while set membership uncertainties could be aggregated by their maximal or minimal values, mass center of the set or higher inertial moments.

AGI   *See* address generation interlock.

Aiken, Howard Hathaway   (1900–1973)
Born: Hoboken, New Jersey

Best known as the inventor of the Mark I and Mark II computers. While not commercially successful, these machines were significant in the development of the modern computer. The Mark I was essentially a mechanical computer. The Mark II was an electronic computer. Unlike UNIVAC (*See* Eckert, John Presper) these machines had a stored memory. Aiken was a professor of mathematics at Harvard. He was given the assignment to develop these computers by the Navy department. Among his colleagues in this project were three IBM scientists and Grace Hopper. It was while working on the Mark I that Grace Hopper pulled the first "bug" from a computer.

air bridge   a bridge made of metal strip suspended in air that can connect components on an integrated circuit in such a way as to cross over another strip. Air bridges are also used to suspend

metalization in spiral inductors off of the semiconducting substrate in a way that can lead to improved performance in some cases.

air capacitor   a fixed or variable capacitor in which air is the dielectric material between the capacitor's plates.

air circuit breaker   a power circuit breaker where the power contacts operate in air. Some versions employ an air blast to extend and clear the arc on contact opening, while others employ arc chutes with magnetic or thermal assists.

air core transformer   two or more coils placed so that they are linked by the same flux with an air core. With an air core the flux is not confined.

air gap   *See* magnetic recording air gap.

air ionization chamber   a device used to monitor neutron flux.

air line   a coaxial transmission line in which the volume between the inner and outer conductors are air-filled.

air terminal   a lightning rod; any device which extends upward into the air from a structure for purposes of lightning protection.

air-blast circuit breaker   a circuit breaker in which the arc which forms between the contacts on opening is extinguished with a blast of high-pressure air.

air-gap line   the line that is obtained by continuing the linear portion of the saturation curve of a synchronous machine or a DC machine. The figure shows a plot of generated voltage vs. field current at constant machine speed. Initially, an increase in field current yields a linear increase in the generated voltage, but as the iron becomes saturated, the

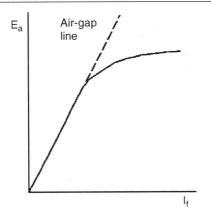

Plot of generated voltage vs. field current at constant machine speed.

voltage rolls off. The air-gap line gives the voltage that would be obtained without saturation.

air-gap voltage   the internal voltage of a synchronous machine that is generated by the air gap flux. Also referred to as the voltage behind leakage reactance.

airline   a precision coaxial transmission line with air dielectric used in a variety of calibration techniques and measurements as an impedance standard and to establish a reference plane.

airy disk   the central portion of the far-field optical diffraction pattern.

AlAs   aluminum arsenide.

albedo   the ratio between the total scattered intensity and the whole extracted from the incident light by scattering and absorption.

ALC   *See* automatic level control.

AlGaAs   symbol for aluminum gallium arsenide.

algebraic reconstruction   the process of reconstructing an image **x** from a noise-corrupted and blurred image **y**. An arbitrary image is selected as the initial condition of an iterative algorithm for solving a set of linear equations. A set of linear constraints is specified. In each iteration one constraint is applied to a linear equation. The constraints are repeated in a cyclic fashion until convergence is reached. The linear constraints are vectors in a vector space with specified basis images for the type of problem to be solved.

algorithm   (1) A systematic and precise, step-by-step procedure (such as a recipe, a program, or set of programs) for solving a certain kind of problem or accomplishing a task, for instance converting a particular kind of input data to a particular kind of output data, or controlling a machine tool. An algorithm can be executed by a machine.

(2) in image processing, algorithms can be either sequential, parallel or ordered. In sequential algorithms, pixels are scanned and processed in a particular raster-scan order. As a given pixel is processed, all previously scanned pixels have updated (processed) values, while all pixels not yet scanned have old (unprocessed) values. The algorithm's result will in general depend on the order of scanning. In a parallel algorithm, each pixel is processed independently of any changes in the others, and its new value is written in a new image, such that the algorithm's result does not depend on the order of pixel processing. In an ordered algorithm, pixels are put in an ordered queue, where priority depends on some value attached to each pixel. At each time step, the first pixel in the queue is taken out of it and processed, leading to a possible modification of priority of pixels in the queue. By default, an algorithm is usually considered as parallel, unless stated otherwise.

algorithmic state machine (ASM)   a sequential logic circuit whose design is directly specified by the algorithm for the task the machine is to accomplish.

aliasing   (1) in signal processing, distortion introduced in a digital signal when it is undersampled.

In all digital systems the signals should be filtered before they are sampled to eliminate signal components with frequencies above the Nyquist frequency,

$$\omega_N = \omega_s/2 = \pi/T,$$

where T is a sampling time, are eliminated. If this filtering is not done, signal components with frequencies

$$\omega > \omega_N$$

will appear as low-frequency components with the frequency

$$\omega_a = |((\omega + \omega_N) \mod \omega_s) - \omega_N|$$

The prefilters introduced before a sampler are called anti-aliasing filters (common choices are second- or fourth-order Butterworth, integral time absolute error (ITAE), or Bessel filters).

(2) in computer graphics, distortion due to the discrete nature of digital images that causes straight lines to appear jagged.

(3) in computer software, a single object having two different identities, such as names in memory space. Aliasing can make it difficult to determine whether two names (or access paths to reach an object) that appear to be different really access the identical object; a system designed to find parallelism when two accesses really reach different objects will have trouble achieving correct (functional) operation if aliasing is present.

alignment   (1) the requirement that a datum (or block of data) be mapped at an address with certain characteristics, usually that the address modulo the size of the datum or block be zero. For example, the address of a naturally aligned long word is a multiple of four.

(2) the act of positioning the image of a specific point on a photomask to a specific point on the wafer to be printed.

(3) the process of determining the time or phase shift of a certain signal so that part of it may be matched with another signal. image registration.

alkaline fuel cell (AFC)   A fuel cell that uses hydrogen fuel and can generate less than 5 kW of power. Because of its relatively low power output, the AFC is used in niche military and space applications. *See* proton exchange membrane fuel cell, phosphoric acid fuel cell, molten carbonate fuel cell, solid oxide fuel cell.

all pass system   a system with unit magnitude and poles and zeroes that are complex conjugate reciprocals of each other. An all-pass system with a pole at $z = a$ and a zero at $z = 1a^*$ is

$$H_a p(z) = z^{-1} - a^* 1 - a z^{-1}$$

all-digital synchronization   synchronization algorithm, where the analog-to-digital conversion takes place as early as possible to assist digital implementation of the synchronizer. In most cases, an all-digital synchronization approach leads to optimal maximum likelihood algorithms.

alley arm   a crossarm meant for use in an alleyway or other confined area in which poles must be placed close to buildings. *See* crossarm.

allocate   to create a block of storage of a given size in some memory, which is not to be used for any other purpose until expressly freed.

allocation   the act of allocating. *See also* allocate.

allocation of authority   process by which the authority (scope of competence) is allocated to various decision units; this allocation may result form the natural reasons or be a product of system partitioning.

all-optical network   an optical communications network where the role of electronics is reduced to basic supervisory and control functions. All-optical devices are used exclusively between the nodes to re-configure the network which enables the greatest use of fiber bandwidth.

all-optical switch   an optically addressed device whose optical transmission can be switched between two possible states by changes in the incident optical power.

almost sure convergence   for a stochastic process, the property of the sample values converging to a random variable with probability one (for almost all sample paths).

alnico   a permanent magnet material consisting mainly of aluminum, nickel, cobalt, and iron, which has a relatively low-energy product and high residual flux density. An alnico is most suitable for high-temperature applications.

ALOHA   a random access, multiple access protocol, originally developed by Norman Abramson at the University of Hawaii in 1970. A given user transmits a message when the message is generated without regard for coordination with the other users sharing the channel. Messages involved in collisions are retransmitted according to some retransmission algorithm. Literally, "aloha" is a greeting in the Hawaiian native language.

alpha channel   a grayscale image associated with the color channels of an image that dictates the opacity/transparency of the corresponding color channel pixels. If the color channels are multiplied by the alpha channel when stored, the image is referred to as premultiplied, otherwise it is known as unpremultiplied.

alpha particle   a subatomic particle emitted by ceramic packaging materials that causes soft errors in memory integrated circuits.

alpha particle noise   this type of noise occurs exclusively in small semiconductor capacitors, when an energetic alpha particle, either from cosmic rays or from the packaging or substrate itself, traverses the capacitor, discharging it, thereby creating an error in the stored charge. Such an accumulation of errors in a digital system has the effect of creating a noise signal.

alpha-cut   the set of all crisp, or nonfuzzy, elements whose membership function in A is greater than or equal to a given value, α.

alphanumeric mode   relates to alphabetic characters, digits, and other characters such as punctuation marks. Alphanumeric is a mode of operation of a graphic terminal or other input/output device. The graphics terminal should toggle between graphic and alphanumeric data.

alternate channel power   a measure of the linearity of a digitally modulated system. The amount of energy from a digitally transmitted RF signal that is transferred from the intended channel to one which is two channels away. It is the ratio (in decibels) of the power measured in the alternate channel to the total transmitted power.

alternating current (AC)   a periodic current the average value of which over a period is zero.

alternating current machine   an electromechanical system that either converts alternating current electrical power into mechanical power (AC motor), or converts mechanical power into alternating current electrical power (AC generator, or alternator). Some AC machines are designed to perform either of these functions, depending on the energy source to the dynamo.

alternator-rectifier exciter   a source of field current of a synchronous machine derived from the rectified output voltage of an alternator. The components of the exciter consist of the alternator and the power rectifier (including possible gate circuitry), exclusive of all input control elements. The rectifier circuits may be stationary, or rotate with the alternator, which may be driven by a motor, prime mover, or by the shaft of the synchronous machine.

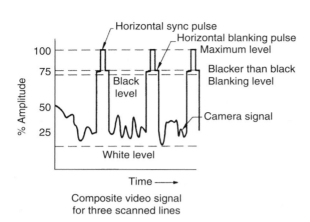

Composite video signal
for three scanned lines

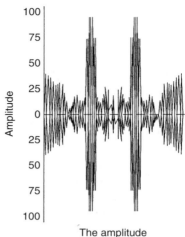

The amplitude
modulated carrier wave

ALU   *See* arithmetic and logic unit.

AM   *See* amplitude modulation.

AM to PM conversion   phase variations of an output signal, due to passing through an active device, where the phase of the output signal varies in response with the amplitude of the input signal.

AM video   the amplitude modulated video carrier wave is produced by an amplitude modulated video transmitter where the amplitude of the wave form varies in step with the video signal similar to that shown in the figure.

amateur radio   The practice and study of electronic communications as an avocation; most often referring to those persons possessing a license earned by examination (in the U.S., the Federal Communications Commission grants such licenses).

ambient field   the background magnetic field level existing in the environment, without contribution from specific magnetic field sources.

ambient temperature   the temperature of the air or liquid surrounding any electrical part or device. Usually refers to the effect of such temperature in aiding or retarding removal of heat by radiation and convection from the part or device in question.

ambiguity   in artificial intelligence, the presence of more than one meaning or possibility.

Amdahl's law   states that the speedup factor of a multiprocessor system is given by

$$S(n) = \frac{n}{1 + (n-1)f}$$

where there are n processors and f is the fraction of computational that must be performed sequentially (by one processor alone). The remaining part of the computation is assumed to be divided into n equal parts each executed by a separate processor but simultaneously. The speedup factor tends to $1/f$ as $n \to \infty$, which demonstrates that under the assumptions given, the maximum speedup is constrained by the serial fraction.

American National Standards Institute (ANSI)
The U.S. organization that recommends standards

for metrology, drawing symbology and numerous other facets for products and industries.

American standard code for information interchange (ASCII) a binary code comprised of seven digits, originally used to transmit telegraph signal information.

ammeter an instrument for measuring electric current in amperes.

ammonia maser first maser, invented by Charles H. Townes. Such a maser operates at microwave frequencies.

ammonium dihydrogen phosphate (ADP) a strong linear electro-optic material. Its chemical formula is $NH_4H_2PO_4$. *See also* potassium dihydrogen phosphate (KDP).

amorphous alloy a ferromagnetic material with very low coercive force (i.e., a narrow hysteresis loop). The material is formed as a very thin ribbon, by freezing the molting alloy before it can crystallize, thus providing a random molecular orientation.

amortisseur winding *See* damper winding.

ampacity the maximum current which can be safely carried by a conductor under specified conditions.

ampere interrupting rating the interrupting rating of a device expressed in amps (often rms symmetrical amps). *See also* MVA interrupting rating.

Ampere's Law a fundamental relationship in electromagnetic theory. In a fairly general form it is expressed by one of Maxwell's equations,

$$\nabla \times \mathbf{H}(\mathbf{r}, t) = \frac{\partial \mathbf{D}(\mathbf{r}, t)}{\partial t} + \mathbf{J}(\mathbf{r}, t)$$

where t is the time, $\mathbf{r}$ is the coordinate vector, and the other vectors are defined as $\mathbf{D}(\mathbf{r}, t)$ electric

displacement; $\mathbf{H}(\mathbf{r}, t)$, magnetic field strength; $\mathbf{J}(\mathbf{r}, t)$, electric current density.

Ampere, Andre Marie (1775–1836) Born: Lyon, France

Best known for his pioneering work in the field of Electrodynamics. During his emotionally troubled life, he held several professorships: at Bourg, Lyon, and at the Ecole Polytechnic in Paris. While Ampere worked in several sciences, the work of the Danish physicist Hans Christian Oerstad on the electric deflection of a compass needle, as demonstrated to him by Dominique Arago, caused Ampere's great interest in electromagnetism. His seminal work, *Notes on the Theory of Electrodynamic Phenomena Deduced Solely from Experiment,* established the mathematical formulations for electromagnetics including what is now known as Ampere's Law. It can be said that Ampere founded the field of electromagnetics. He is honored for this by the naming of the unit of electric current as the ampere.

amperometric sensor an electrochemical sensor that determines the amount of a substance by means of an oxidation–reduction reaction involving that substance. Electrons are transferred as a part of the reaction, so that the electrical current through the sensor is related to the amount of the substance seen by the sensor.

amplidyne a special generator that acts like a DC power amplifier by using compensation coils and a short circuit across its brushes to precisely and quickly control high powers with low level control signals.

amplified spontaneous emission spontaneous emission that has been enhanced in amplitude and perhaps modified in spectrum by propagation through an amplifying medium, usually the medium in which it was first generated.

amplifier a circuit element that has a linear input-output signal relationship, with gain in

ASCII Code Chart

Hex	Char	Hex	Char	Hex	Char	Hex	Char
00	nul	20	sp	40	@	60	`
01	soh	21	!	41	A	61	a
02	stx	22	"	42	B	62	a
03	etx	23	#	43	C	63	c
04	eot	24	$	44	D	64	d
05	enq	25	%	45	E	65	e
06	ack	26	&	46	F	66	f
07	bel	27	'	47	G	67	g
08	bs	28	(	48	H	68	h
09	ht	29	)	49	I	69	i
0A	lf	2A	*	4A	J	6A	j
0B	vt	2B	+	4B	K	6B	k
0C	ff	2C	,	4C	L	6C	l
0D	cr	2D	-	4D	M	6D	m
0E	so	2E	.	4E	N	6E	n
0F	si	2F	/	4F	O	6F	o
10	dle	30	0	50	P	70	p
11	dc1	31	1	51	Q	71	q
12	dc2	32	2	52	R	72	r
13	dc3	33	3	53	S	73	s
14	dc4	34	4	54	T	74	t
15	nak	35	5	55	U	75	u
16	syn	36	6	56	V	76	v
17	etb	37	7	57	W	77	w
18	can	38	8	58	X	78	x
19	em	39	9	59	Y	79	y
1A	sub	3A	:	5A	Z	7A	z
1B	esc	3B	;	5B	[	7B	{
1C	fs	3C	<	5C	\	7C		
1D	gs	3D	=	5D	]	7D	}
1E	rs	3E	>	5E	^	7E	~
1F	us	3F	?	5F	_	7F	

voltage, current, and/or power. *See also* balance amplifier, feedback amplifier, feedforward amplifier, laser amplifier, maser amplifier, optical amplifier, single-ended amplifier.

amplitron a classic crossed-field amplifier in which output current is obtained primarily by secondary emission from the negative electrode that serves as a cathode throughout all or most of the interaction space.

amplitude descriptor of the strength of a wave disturbance such as an electromagnetic or acoustic wave.

amplitude equations a form of the Schrödinger equation that describes the evolution of a quantum mechanical system in terms of only the coefficients of the preferred basis states. These coefficients are known as quantum mechanical amplitudes and contain both magnitude and phase information. Amplitude equations are often used to gain physical insight into interactions of quantum systems with electromagnetic fields. *See also* Schrödinger wave equation (SWE).

amplitude linearity qualitative measure of the extent to which the output amplitude of a device is a faithful reproduction of its input, with no new frequency harmonics added. A perfectly linear device would output a scaled version of its input, where the shape of the input waveform has been unaltered (i.e., there is no distortion of the input waveform). Viewed in the frequency domain, the output signal would contain only those spectral components found in the input signal, and each frequency line would be scaled by the same amount (i.e., by the gain of the device).

amplitude modulation (AM) the process of modulating a signal $x(t)$ by a carrier wave $c(t)$ for transmission:

$$y(t) = c(t)x(t),$$

where $y(t)$ is the signal to be transmitted. $c(t)$ is either a complex exponential of the form

$$c(t) = e^{j(\omega_c t + \theta_c)}$$

or a sinusoidal signal of the form

$$c(t) = \cos(\omega_c t + \theta_c).$$

ω_c is referred to as the *carrier frequency*. AM has the effect of shifting the frequency spectrum of $x(t)$ by ω_c. The signal is recovered by shifting the spectrum of $x(t)$ back to its original form. *See* frequency modulation.

amplitude response the magnitude of the steady-state response of a fixed, linear system to a unit-amplitude input sinusoid.

amplitude spectrum the magnitude of the Fourier transform $|F(\omega)|$, $-\infty < \omega < \infty$ of a signal $f(t)$. For example, the amplitude spectrum of a rectangular pulse of unit width is given in the following figure: *See also* Fourier transform.

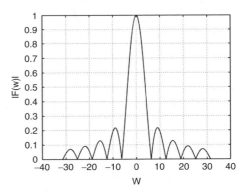

Amplitude spectrum.

amplitude stabilization circuit a circuit used to obtain a precise oscillation amplitude of oscillators. These circuits are used in instrumentation when it is required to increase the purity of output signal and reduce the frequency depression (especially in Meachem-bridge oscillator with crystal) of the main harmonic by higher harmonics (van der Pol effect). Three types of circuits are used:

1. An element of large inertia (tungsten lamp, thermistor) is included in the circuit at a point where it can change the magnitude of feedback, but not affect the frequency.

2. A controlled resistor (usually an FET operating in a triode regime) that is also part of the feedback circuit (the DC control signal is obtained with a rectifier and a filter of large time constant).

3. An automatic gain control circuit where the DC control signal obtained from a rectifier and filter is used to change the bias of oscillator active element.

amplitude-modulated link a transmitter–receiver system that utilizes amplitude-modulation for the transmission of signal frequencies.

amplitude-shift keying (ASK) a modulation technique in which each group of source bits determines the amplitude of the modulated carrier.

AMPS *See* advanced mobile phone system.

AMR *See* automated meter reading.

analog *See* analog signal, analog data.

analog data data represented in a continuous form with respect to continuous time, as contrasted with digital data represented in a discrete (discontinuous) form in a sequence of time instant.

analog multiplier a device or a circuit that generates an analog output signal that is proportional to the product or multiplication of two analog input signals.

analog optical computing optical computing that involves two-dimensional analog operations such as correlation and complex spatial frequency filtering primarily based on the property of the lens

to perform two-dimensional Fourier transform. In analog optical computing, operations to be performed are matched with and based on already known optical phenomena.

analog signal a signal represented in a continuous form with respect to continuous time, as contrasted with digital signal represented in a discrete (discontinuous) form in a sequence of time instant. *See also* analog data.

analog signal conditioning an interface between the sensor or transducer output, which represents an analog or physical world, and the analog-to-digital converter.

analog-to-digital A/D conversion a method by which a continuously varying signal (voltage) is sampled at regularly occurring intervals. Each sample is quantized to a discrete value by comparisons to preestablished reference levels. These quantized samples are then formatted to the required digital output (e.g., binary pulse code words). The A/D converter is "clocked" to provide updated outputs at regular intervals. In order not to lose any baseband information, sampling must occur at a rate higher than twice the highest incoming signal frequency component. *See also* Nyquist rate.

analog-to-digital A/D converter a device that changes an analog signal to a digital signal of corresponding magnitude. This device is also called an encoder, ADC, or A/C converter.

analysis filter a filter in the analysis section of a sub-band analysis and synthesis system.

analysis-by-synthesis coding refers to the class of source coding algorithms where the coding is based on parametric synthetization of the source signal at the encoder. The synthesized signal is analyzed, and the parameters that give the "best" result are chosen and then transmitted (in coded form). Based on the received parameters the speech is resynthesized at the receiver.

analyte   the substance being measured by a chemical or bioanalytical sensor and instrumentation system.

analytic signal   refers to a signal that has a Fourier transform that is zero valued for negative frequencies; i.e., the signal has a one-sided spectrum.

analytical Jacobian   a mathematical representation computed via differentiation of the direct kinematic equation with respect to the joint variables q. Formally one can write $\dot{x} = [\begin{smallmatrix}\dot{\phi}\\\dot{p}\end{smallmatrix}] = [\begin{smallmatrix}J_{\phi}(q)\\J_{p}(q)\end{smallmatrix}]\dot{q} = J_A(q)\dot{q}$ where the analytical Jacobian is $J_A(q) = \frac{\partial k(q)}{\partial q}$. *See* external space for notation used in these equations. The analytical Jacobian is different from the geometric Jacobian, since the end-effector angular velocity with respect to the base frame is not given by $\dot{\phi}$. Both Jacobians are related as $J = T_A(\phi)J_A$ where $T_A(\phi)$ is a matrix that depends on the particular representation of the orientation representation. In particular $T_A(\phi)$ is an identity matrix when equivalent axis of rotation in the task space is the same as the equivalent axis of rotation of the end-effector. *See also* geometric Jacobian.

anamorphic lenses   a lens system having a difference in optical magnification along the two mutually perpendicular axes (vertical plane or tilt vs. horizontal plane or panorama).

and   *See* AND.

AND   the Boolean operator that implements the conjunction of two predicates. The truth table for

$\wedge \equiv X$ and Y is

X	Y	$X \wedge Y$
F	F	F
F	T	F
T	F	F
T	T	T

n-ary ands can be obtained as conjunction of binary ands.

AND gate   a device that implements the Boolean AND operation. *See* AND.

angle diversity   a diversity technique used in radio communications based on receiving a signal over multiple arrival angles. The signal components are typically affected by uncorrelated fading processes and are combined in the receiver to improve performance. The main combining methods are selection diversity, equal gain combining, and maximal ratio combining.

angle modulation   a type of modulation where either the frequency (FM) or the phase (PM) of a carrier are varied.

angle of arrival (AOA)   the direction to a source emitting a signal impinging on a sensor array. Also called direction of arrival (DOA).

angstrom   popular unit not officially recognized as part of the SI unit system. Equal to 10^{-10} meters. Abbreviated Å. Named after Anders Ångström (1814–1874).

angular alignment loss   the optical power loss in an optical connection between two optical fibers, between an optical source and a fiber, or between an optical fiber and a detector caused by the angular misalignment of the axes of the source and fiber, the two fibers, or the fiber and detector.

angular frequency   the rate of change of the phase of a wave in radians per second.

anisotropic   direction-dependent.

anisotropic diffraction   diffraction when the refractive indices for the incident and diffracted optical waves are different.

anisotropic diffusion   a process of progressive image smoothing as a function of a time variable t, such that the degree and orientation of smoothing at a point varies according to certain parameters measured at that point (e.g., gray-level gradient, curvature, etc.) in order to smooth image noise while preserving crisp edges. The progressively smoothed image $I(x, y, t)$ (where x, y are spatial coordinates and t is time) satisfies the differential equation $\partial I / \partial t = div(c \nabla I)$, *where the diffusion factor c is a decreasing function of* ∇I. When c is constant, this reduces to the heat diffusion equation $\partial I / \partial t = c \Delta I$. Other mathematical formulations have been given where edge preserving smoothing is realized by a selective diffusion in the direction perpendicular to the gradient. *See* multiresolution analysis, mathematical morphology.

anisotropic etch   an etch with an etch rate that is direction-dependent. In wet etching, the direction dependence has to do with crystallographic axis – some planes etch at different rates than others.

anisotropic medium   (1) a medium in which the index of refraction varies with the light propagation direction within the medium. In such a medium, the constitutive relation involves a tensor.

(2) a medium that exhibits anisotropy. Examples are anisotropic crystals, ferrites in the presence of a static magnetic field, and plasma in the presence of a static magnetic field.

anisotropic scatterer   inhomogeneous medium, usually consisting of suspension of anisotropic molecules, capable of producing effects like birefringence or dichroism. As such, its dielectric permittivity is a tensor acting differently upon each component of the electromagnetic field.

anisotropy   (1) the degree of variation in a property such as index of refraction with light propagation direction.

(2) dependence of the response of a medium on the direction of the fields, for example, the x component of the electric displacement might depend in part on the y component of the fields.

annealing   a process often used in semiconductor processing to cause a change in materials or device properties to improve the circuit performance and/or reliability. *See also* simulated annealing.

annealing schedule   specifies the sequence of temperature values that are to be used in an application of simulated annealing and also specifies the number of parameter changes that are to be attempted at each temperature.

annihilation   a process in which a particle and its anti-particle meet and convert spontaneously into photons.

annul bit   a bit that is used to reduce the effect of pipeline breaks by executing the instruction after a branch instruction. The annul bit in a branch allows one to ignore the delay-slot instruction if the branch goes the wrong way. With the annul bit not set, the delayed instruction is executed. If it is set, the delayed instruction is annulled.

annular cathode   a cathode of a vacuum tube with the shape of the emitting surface of the cathode is annular. The annular cathode can produce a hollow electron beam.

annular illumination   a type of off-axis illumination where a doughnut-shaped (annular) ring of light is used as the source.

anode   the positive electrode of a device. *Contrast with* cathode.

anomalous dispersion   decrease of the index of refraction with increasing frequency; tends to occur near the center of absorbing transitions or in the wings of amplifying transitions.

ANSI   American National Standards Institute, a body which administers numerous industrial standards in the the USA including several which pertain to electric utility construction practices. *See* American National Standards Institute.

antenna   a device used to couple energy from a guiding structure (transmission line, waveguide, etc.) into a propagation medium, such as free space, and vice versa. It provides directivity and gain for the transmission and reception of electromagnetic waves.

antenna beamwidth   the effective angular extent of the antenna radiation pattern usually specified between points of fixed amplitude relative to the main lobe gain (e.g., −3 dB points).

antenna diversity   a diversity technique based on the use of multiple antennas either at the receiver (receiver antenna diversity) or at the transmitter (transmitter antenna diversity) in a radio communication link. If the separation of antennas is sufficient, the signal components are affected by different fading processes and are combined in the receiver to improve performance. *See also* RAKE receiver. *Contrast with* angle diversity.

antenna gain   the maximum ratio of an antenna's ability to focus or receive power in a given direction relative to a standard; the standard is usually an isotropic radiator or a dipole. The gain includes the efficiency of the antenna.

antenna noise temperature   the effective noise temperature of the antenna radiation resistance appearing at the antenna terminals. At a given frequency, the antenna noise temperature, $T_a(K)$, can be calculated as $\frac{P_n}{kB}$ where P_n is the noise power available at the antenna terminals (W), k is Boltzmann's constant (1.38×10^{-23} J/^{0}K), and B is the bandwidth (Hz). The antenna noise is the result of thermal noise generated in ohmic losses in the antenna structure and noise received by the antenna from external radiating sources.

antenna pattern   graph or chart representing the absolute or normalized antenna gain as a function of angle (typically azimuth or elevation) and used to describe the directional properties of an antenna. In the near field, the antenna pattern is a function of the distance from the antenna whereas in the far field, the pattern is independent of distance from the antenna.

antenna Q   ratio of the energy stored to the energy dissipated (ohmically or via radiation) per cycle.

antenna synthesis   the process of determining or designing an antenna to yield a given radiation pattern. Several synthesis methods exist. Some are closed form solutions and some use numerical techniques.

anthropomorphic manipulator   a manipulator that consists of two shoulder joints, one for rotation about a vertical axis and one for elevation out of the horizontal plane, an elbow joint with axis parallel to the shoulder elevation joint, and two or three wrist joints at the end of the manipulator (see figure). An anthropomorphic manipulator is sometimes called a jointed, elbow, or articulated manipulator.

antialiasing filter   typically, a filter which provides a prefiltering operation to ensure that the frequency components of a signal above the Nyquist frequency are sufficiently attenuated so that, when aliased, they will cause a negligible distortion to the sampled signal. *See* aliasing, Nyquist frequency.

anticollision radar   a type of radar, generally operating in the millimeter wave frequency range, used to prevent collision between moving vehicles.

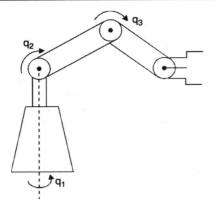

An anthropomorphic manipulator.

anticomet tail (ACT)   a special type of electron gun designed to handle highlights by increasing beam current with a defocused beam during line retrace.

antidependency   a potential conflict between two instructions when the second instruction alters an operand which is read by the first instruction. For correct results, the first instruction must read the operand before the second alters it. Also called a write-after-read hazard.

antidots   regions of repulsive potential, but which are configured so that particles (usually electrons) can pass around the potential and proceed past it. In the limiting case, a repulsive Coulomb potential is the simplest antidot structure.

antiferromagnetic   materials in which the internal magnetic moments line up antiparallel, resulting in permeabilities slightly greater than unity; unlike paramagnetic substances, these materials exhibit hysteresis and have a Curie temperature. Examples include manganese oxide, nickel oxide, and ferrous sulfide.

antifuse   a fuse-like device that when activated becomes low-impedance.

antiparticle   a particle having the same mass as a given fundamental particle, but whose other properties, while having the same magnitude, may be of opposite sign. Each particle has a partner called an antiparticle. For example, electrical charge in the case of the electron and positron, magnetic moment in the case of the neutron and antineutron. On collision a particle and its antiparticle may mutually annihilate with the emission of radiation. Some properties of the antiparticle will be identical in magnitude but opposite in sign to the particle it is paired with.

anti-plugging   a feature to prevent a motor from reversing direction directly across the line. The purpose of the anti-plugging coil and contact is to prevent the motor from starting in the opposite direction until the speed has slowed enough where the current and torque surges are within acceptable levels when changing direction.

antipodal   symmetry created by simultaneously mirroring an object in both the X and Y axes.

antiproton   antiparticle to the proton. It is a strongly interacting baryon carrying unit negative charge. It has mass of 938 MeV and carries spin 1/2.

antireflection coating   *See* antireflective coating.

antireflective coating (ARC)   a coating placed on top or below the layer of photoresist to reduce the reflection of light, and hence reduce the detrimental effects of standing waves or thin film interference.

anti-Stokes scattering   the scattering of light accompanied by a shift to higher frequencies. *Contrast with* Stokes Law of light scattering.

AOA   *See* angle of arrival.

AOTF   *See* acousto-optic tunable filter.

APART/PADE   a computer code for analysis of stray light in optical systems developed by the University of Arizona and BRO, Inc.

APC-7 connector   common term for amphenol precision connector - 7mm. A "sexless" coaxial connector with butt contact between both the inner and outer conductors capable of low standing wave ratios to frequencies up to 18 GHz.

APD   *See* avalanche photodiode.

aperiodic convolution   the convolution of two sequences. *See* convolution.

aperiodic signal   a signal that is not periodic, i.e., one for which $x(t) \neq x(t + T)$. This means that the signal $x(t)$ has a property that is changed by a time shift T. *See also* periodic signal.

aperiodic waveform   this phrase is used to describe a waveform that does not repeat itself in a uniform, periodic manner. *Compare with* periodic waveform.

aperture   (1) an opening to a cavity, or waveguide, from which radiation is either received or transmitted. Typically used as antenna or a coupling element.

(2) a physical space available for beam to occupy in a device. Aperture limitations are the physical size of the vacuum chamber, a magnetic field anomaly may deflect the beam so that the full available aperture cannot be used.

aperture antenna   an antenna with a physical opening, hole, or slit. *Contrast with* a wire antenna.

aperture correction   signal compensation used to correct the distortion caused by the non-zero aperture of a scanning electron beam. A standardized measure of the selectivity of a circuit or system. The -3 dB (or half-power) band width

is taken to be the difference between the upper (f_2) and lower (f_1) frequencies where the gain vs. frequency response curve has decreased -3 dB from the passband reference gain. Note that f_1 and f_2 define the response passband by marking the points at which the output power has decreased to one-half the value of the input power. For band widths extending down to DC, the upper -3 dB frequency is cited as the 3 dB bandwidth.

aperture coupling   a method of coupling a transmission line to an antenna in which fields leak through an aperture in a metallic ground plane separating the line from the antenna.

aperture efficiency   a figure of merit that determines how much of the incident energy is captured by an aperture. It depends on the physical dimensions of the aperture.

aperture problem   given a sequence of images over time we would like to infer the motion (optical flow) field. Based on local image information (i.e., based on the values of those pixels falling within some aperture) only the component of motion along the gray-level gradient can be inferred; the fact that the component of motion perpendicular to the gray-level gradient can only be known by resorting to global methods is known as the aperture problem. *See* optical flow, optical flux.

APL   *See* average picture level.

APLC   *See* active power line conditioner.

apodization   (1) a deliberate variation in the transmission of an optical aperture as a function of distance from the center or edges, in order to control optical transfer functions.

(2) a deliberate variation in the strength of a signal with time.

apparent concurrency   within an interval of time more than one process executes on a computer, although at the instruction level, instructions from only one process run at any single point in time. *See also* concurrency.

apparent mean thermal conductivity the effective thermal conductivity of an assemblage of material (Pearlite, super insulation) between specified temperatures.

apparent power (1) in an AC system, the product of voltage, E and current, I. Apparent power (or total power) is composed of two mutually independent components — an active component (real power), and a reactive component (imaginary power). Apparent power is denoted by S, and has the unit of voltamperes.

(2) the scalar product of the voltage and current delivered to the load. It can also be expressed as the vector $S = P + jQ$, where P = real power and Q = reactive power.

application-specific integrated circuit (ASIC) an integrated circuit designed for one particular application.

appropriate technology the technology that will accomplish a task adequately given the resources available. Adequacy can be verified by determining that increasing the technological content of the solution results in diminishing gains or increasing costs.

approximate coding a process, defined with respect to exact coding, that deals with irreversible and information-lossy processing of two-level pictures to improve compression ratio with significant degradation of picture quality. Exact coding schemes depend on the ability to predict the color of a pixel or the progression of a contour from line to line. Irreversible processing techniques try to reduce prediction errors by maintaining the continuity of the contours from line to line. With predictive coding the number of pixels can be changed to reduce those having nonzero prediction error. With block coding the compression efficiency can be improved by increasing the probability of

occurrence of the all zero block. The third approximate block coding scheme is pattern matching. In this scheme the identification codes of the repeated patterns are transmitted to the receiver. A library of patterns is maintained for continuous checking. *See also* exact coding.

approximate reasoning an inference procedure used to derive conclusions from a set of fuzzy if-then rules and some conditions (facts). The most used approximate reasoning methods are based on the generalized modus ponens. *See also* fuzzy if-then rule, generalized modus ponens, linguistic variable.

approximately controllable system an infinite-dimensional stationary linear dynamical system where the attainable set K_∞ is dense in the infinite-dimensional state space X. The set is said to be approximately controllable in $[0, T]$ if the attainable set $K(0, T)$ is dense in the infinite-dimensional state space X.

Approximate controllability in $[0, T]$ always implies approximate controllability. The converse statement is not always true.

Ar + laser laser in which the active medium consists of singly ionized argon atoms. Ar+ lasers have several wavelengths in the visible portion of the spectrum.

Arago, Dominique Francois (1786–1853)
Born: Estagel, France

Best known for the breadth and the volume of his contribution to the study of light and for his work with Ampere on the development of electrodynamics. Arago discovered that iron could be magnetized by the passage of current through a wire and, the phenomenon of magnetic rotation. It was left to Michael Faraday to properly explain this phenomenon. Arago spent a significant amount of time involved in politics and succeeded Jean Fourier as the permanent secretary to The Academy of Sciences in 1830. It has been suggested that Arago's enthusiasm and work ethic were an inspiration to many contemporary scientists.

arbiter   a unit that decides when multiple requestors may have access to a shared resource.

arbitrary reference frame   a two-dimensional space that rotates at an unspecified angular velocity ω. In electric machines/power system analysis, an orthogonal coordinate axis is established in this space upon which fictitious windings are placed. A linear transformation is established in which the physical variables of the system (voltage, current, flux linkage) are referred to variables of the fictitious windings. *See also* transformation equations, rotor reference frame, stationary reference frame, synchronous reference frame.

arbitration   *See* bus arbitration.

ARC   *See* antireflective coating.

arc detector   a device placed within a microwave power tube or within one or more of the external cavities of a microwave power tube whose purpose is to sense the presence of an overvoltage arc.

arc lamp   lamp made by driving a high current across a gap between two electrodes. Some types operate in air consuming the electrode, for example, a carbon arc in which the electrode material is made as a rod and fed into the discharge to replace what is consumed. Others operate in a vacuum envelope that reduces the electrode consumption.

arc resistance   period of time that the surface of an insulating material can be submitted to the action of an electrical arc without becoming conductive.

architecture   *See* computer architecture.

arcing fault   *See* arcing ground.

arcing ground   a ground fault on a power line which alternately clears and restrikes, causing high, repetitive voltage surges.

ARCP   *See* auxiliary resonant commutated pole converter.

area   *See* control area.

areal density   a measure for the improvement in the capacity of a disk. It is the product of the number of tracks per inch and the number of bits per inch, i.e., it is the number of bits per square inch.

argon ion laser   *See* Ar+ laser.

argument   (1) an address or value that is passed to a procedure or function call, as a way of communicating cleanly across procedure/function boundaries.
(2) a piece of data given to a hardware operator block.

arithmetic and logic unit (ALU)   a combinational logic circuit that can perform basic arithmetic and logical operations on n-bit binary operands.

arithmetic coding   a method (due to Elias, Pasco, Rissanen and others) for lossless data compression. This incremental coding algorithm works efficiently for long block lengths and achieves an average length within one bit of the entropy for the block. The name comes from the fact that the method utilizes the structures of binary expansions of the real numbers in the unit interval.

arithmetic instruction   a machine instruction that performs computation, such as addition or multiplication.

arithmetic operation   any of the following operations and combination thereof: addition, subtraction, multiplication, division.

arithmetic radian center frequency the linear radian center frequency, it is the midpoint between the higher (ω_H) and lower (ω_L) band edges, expressed in units of radians/second. The band edges are usually defined as the highest and lowest frequencies within a contiguous band of interest at which the loss equals L_{Amax}, the maximum attenuation loss across the band.

$$\omega_{oa} = \frac{\omega_H + \omega_L}{2}$$

arithmetic shift a shift in which it is assumed that the data being shifted is integer arithmetic in nature; as a result, the sign bit is not shifted, thereby maintaining the arithmetic sign of the shifted result. *See also* logical shift.

arithmetic–logic unit *See* arithmetic and logic unit (ALU).

arm a part of a robot. A robot is composed of an arm (or mainframe) and a wrist plus a tool. For many industrial robots the arm subassembly can move with three degrees of freedom. Hence, the arm subassembly is the positioning mechanism. *See also* industrial robot.

arm pin a pin insulator.

ARMA *See* auto-regressive moving-average model.

armature the magnetic circuit of a rotating electrical machine, including the main current carrying winding, in which an alternating voltage is induced by the magnetic field.

armature circuit components of the machine that carry armature current. For example, in a DC machine the armature circuit could consist of the armature windings, brushes, series field winding, compensating windings, interpoles, starting resistor(s), main-line contacts, and overload sensor.

armature current limiting a condition wherein the stator currents are clamped at the maximum allowable limit due to excessive heating of the stator.

armature reaction (1) in DC machines, a distortion of the field flux caused by the flux created by the armature current. Armature reaction in a DC machine causes lower flux at one pole-tip and higher flux at the other, which may lead to magnetic saturation. It also shifts the neutral axis, causing sparking on the commutator.

(2) in AC synchronous machines, a voltage "drop" caused by the armature current. In the steady state model of the synchronous machine, the armature reaction is accounted for by a component of the synchronous reactance.

armature voltage control a method of controlling the speed of a DC motor by varying the voltage applied to the armature while keeping the voltage applied to the field circuit constant.

armature winding an arrangement of coils carrying the main current, typically wound on the stator of a synchronous machine or the rotor of a DC machine, in which an alternating voltage is induced by the magnetic field.

armless construction a method of distribution line construction, often used for aesthetic purposes, in which pin insulators are mounted on steel brackets bolted directly to a utility pole without the use of a crossarm.

Armstrong oscillator Hartley oscillators are usually not used at VHF of higher frequencies. Similarly, the circuit is avoided at very low audio frequencies. It is important to distinguish the Hartley oscillator from the Armstrong topology. In the Armstrong oscillator, no ohmic connection exists between the two inductors. Instead, coupling is entirely magnetic.

Armstrong, Edwin Howard (1890–1954) Born: New York, New York.

Best known as the developer of frequency modulation (FM) radio and inventor of the superheterodyne receiver. Armstrong spent most of his career at Columbia University. During his life, his inventions made him quite wealthy. The superheterodyne receiver was purchased as a way for the military to detect the spark plug ignitions of approaching aircraft. Patent fights with Lee De-Forest and the difficulty in promoting FM radio led to bitterness and frustration which many felt led to his suicide.

ARQ *See* automatic repeat request.

array several antennas arranged together in space and interconnected to produce a desired radiation pattern.

array factor in antenna theory, the resulting radiation pattern of an array when each antenna in the array is replaced by an isotropic radiator.

array processor an array of processor elements operating in lockstep in response to a single instruction and performing computations on data that are distributed across the processor elements.

array signal processing signal processing techniques used for extracting information based on signals from several (identical) sensors, for example an antenna array consisting of several antenna elements.

arrester discharge current the current in an arrester during a surge.

arrester discharge voltage the voltage in an arrester during a surge.

ART network *See* adaptive resonance theory network.

artifact an error or aberration in a signal that is the result of aliasing, a quantization error, some form of noise, or the distorting effects of some type of processing. *See* outlier.

artificial constraint an additional constraint in accordance with the natural constraints to specify desired motion or force application. An artificial constraint occurs along the tangents and normals of the constraint surface. An artificial force constraint is specified along surface normals, and an artificial position constraint along tangents and hence consistency with the natural constraints is preserved. *See also* natural constraint.

artificial dielectric a dielectric material that has been modified to alter its properties. Common modifications include micromachining to remove material from the substrate under planar patch antenna to improve radiation properties and the fabrication of periodic arrays of holes to realize guiding or photonic bandgap structures.

artificial intelligence the study of computer techniques that emulate aspects of human intelligence, such as speech recognition, logical inference, and ability to reason from partial information.

artificial neural network a set of nodes called neurons and a set of connections between the neurons that is intended to perform intellectual operations in a manner not unlike that of the neurons in the human brain. In particular, artificial neural networks have been designed and used for performing pattern recognition operations. *See also* pattern recognition, perceptron.

artificial neuron an elementary analog of a biological neuron with weighted inputs, an internal threshold, and a single output. When the activation of the neuron equals or exceeds the threshold, the output takes the value $+1$, which is an analog of the firing of a biological neuron. When the activation is less than the threshold, the output takes on the value 0 (in the binary case) or -1 (in the bipolar case) representing the quiescent state of a biological neuron.

artificial skin artificial skin is a device which, when pressed against the surface by an object, causes local deformations that are measured as continuous resistance variations. The latter are transformed into electrical signals whose amplitude is proportional to the force being applied to any given point on the surface of the material of the device.

ASAP/RABET   acronym for a computer code for optical systems by BRO, Inc., for standard optical analysis and stray-light analysis such as light-scattering.

ASCII   *See* American standard code for information interchange.

ASCR   *See* asymmetrical silicon controlled rectifier.

ASDL   *See* asymmetric digital subscriber line.

ASIC   *See* application-specific integrated circuit.

ASK   *See* amplitude-shift keying.

askarel   a trade name for an insulating oil.

ASM   *See* algorithmic state machine.

aspect ratio   (1) the size invariant ratio of length to width for a rectangular box enclosing a shape, the orientation of the box being chosen to maximize the ratio. This measure is used to characterize object shapes as a preliminary to, or as a quick procedure for, object recognition.
(2) the ratio of width to height for an image or display.

aspheric   description of optical elements whose curved surfaces are not spherical, often used to reduce aberrations in optical systems.

assembler   (1) a computer program that translates an assembly-code text file to an object file suitable for linking.

(2) a program for converting assembly language into machine code.

assembly language   a programming language that represents machine code in a symbolic, easier-to-read form. *See also* assembler.

assert   (1) raising the voltage on a wire to the "high" state, usually as a signal to some other unit.
(2) to make an assertion.

assertion   (1) a Boolean expression for stating the right behavior of the program or, if hardware implemented, of a circuit.
(2) a logical expression specifying a program state that must exist or a set of conditions that program variables must satisfy at a particular point during program execution.

associate mode   an operating mode of content addressable memories, in which a stored data item is retrieved that contains a field that matches a given key.

associated reference directions   a method assigning the current and voltage directions to an electrical element so that a positive current-voltage product always means that the element is absorbing power from the network and a negative product always means that the element is delivering power to the network. This method of assigning directions is used in most circuit simulation programs.

associative memory   a memory in which each storage location is selected by its contents and then an associated data location can be accessed. Requires a comparative with each storage location and hence is more complex than random access memory. Used in fully associative cache memory and in some translation look-aside buffers or page translation tables of the hardware to support virtual memory. Given the user-space address of a page it returns the physical address of that page in main memory. Also called content addressable memory (CAM).

associative processor a parallel processor consisting of a number of processing elements, memory modules, and input–output devices under a single control unit. The capability of the processing elements is usually limited to the bit-serial operations.

associativity In a cache, the number of lines in a set. An n-way set associative cache has n lines in each set. (Note: the term "block" is also used for "line.")

astable multivibrator the circuit that is obtained from a closed-loop regenerative system that includes two similar amplifiers of high gain connected with each other via coupling circuits with reactance elements. More frequently are used RC-coupling circuits (free-running RC-multivibrators, emitter-coupled multivibrators), yet RL-circuits, usually as transformer coils, may be used as well (magnetic multivibrators).

astigmatism a defect associated with optical and electrostatic lenses where the magnification is not the same in two orthogonal planes; common where beam propagation is not along the axis of rotation of the system.

asymmetric digital subscriber lines (ADSL) a digital subscriber line (DSL) in which the rate from central switching office (CO) to the customer is much faster than the rate from customer to the CO.

asymmetric multiprocessor (1) a machine with multiple processors, in which the time to access a specific memory address is different depending on which processor performs the request.

(2) in contrast with a symmetric multiprocessor, asymmetric multiprocessor is a multiprocessor in which the processors are not assigned equal tasks. The controller (master) processor(s) are assigning tasks to (slave) processors and controlling I/O for them.

asymmetric multivibrator a multivibrator where the output voltage represents a train of narrow pulses. Most asymmetric multivibrators use a slow charge of a large timing capacitor by a small current (or via a large resistor) and a fast discharge of this capacitor via a switch. The charge process determines the duration of space; the mark duration, which coincides with the time allowed for discharge of the timing capacitor, is usually determined by a small time constant of the circuit controlling the switch. Asymmetric multivibrators find applications in voltage-to-frequency converters. Also called multivibrators with a small mark/space ratio.

asymmetric resonator standing-wave resonator in which either the reflectivities or the curvatures of the primary mirrors are unequal.

asymmetrical silicon controlled rectifier (ASCR) (1) an inverter grade SCR fabricated to have limited reverse voltage capability. Fabrication with asymmetrical voltage blocking capability in the forward and reverse direction permits reduction of turn-on time, turn-off time, and conduction drop.

(2) a thyristor that has limited conduction in the reverse direction to gain increased switching speed and low forward voltage drop. *See also* silicon controlled rectifier (SCR).

asymptotic 2-D observer a system described by the equations

$$z_{i+1,j+1} = F_1 z_{i+1,j} + F_2 z_{i,j+1}$$
$$+ G_1 u_{i+1,j} + G_2 u_{i,j+1}$$
$$+ H_1 y_{i+1,j} + H_2 y_{i,j+1}$$
$$\hat{x}_{i,j} = L z_{i,j} + K y_{i,j}$$

$i, j \in Z_+$ (the set of nonnegative integers) is called a full-order asymptotic observer of the second generalized Fornasini–Marchesini 2-D model

$$E x_{i+1,j+1} = A_1 x_{i+1,j} + A_2 x_{i,j+1}$$
$$+ B_1 u_{i+1,j} + B_2 u_{i,j+1}$$
$$y_{i,j} = C x_{i,j} + D u_{i,j}$$

$i, j \in Z_+$ if

$$\lim_{i,j \to \infty} [x_{i,j} - \hat{x}_{i,j}] = 0$$

for any $u_{i,j}$, $y_{i,j}$ and boundary conditions x_{i0} for $i \in Z_+$ and x_{0j} for $j \in Z_+$ where $z_{i,j} \in R^n$ is the local state vector of the observer at the point (i, j), $u_{ij} \in R^m$ is the input, $y_{i,j} \in R^p$ is the output, and $x_{i,j} \in R^n$ is the local semistate vector of the model, $F_1, F_2, G_1, G_2, H_1, H_2, L, K, E, A_1, A_2, B_1, B_2, C, D$ are real matrices of appropriate dimensions with E possibly singular or rectangular. In a similar way a full-order asymptotic observer can be defined for other types of the 2-D generalized models.

asymptotic stability   (1) an equilibrium state of a system of ordinary differential equations or of a system of difference equations is asymptotically stable (in the sense of Lyapunov) if it is stable and the system trajectories converge to the equilibrium state as time goes to infinity, that is, the equilibrium $\mathbf{x}_{eq}$ is asymptotically stable if it is stable and

$$\mathbf{x}(t) \to \mathbf{x}_{eq} \quad \text{as} \quad t \to \infty.$$

(2) a measure of system damping with regard to a power system's ability to reach its original steady state after a disturbance.

asymptotic tracking   refers to the ability of a unity feedback control to follow its setpoint exactly with zero error once all transients have decayed away. Clearly this is only achieved by stable systems.

asymptotically stable equilibrium   a stable equilibrium point such that all solutions that start "sufficiently close," approach this point in time. *See also* stable equilibrium.

asymptotically stable in the large   the equilibrium state of a stable dynamic system described by a first-order vector differential equation is said to be asymptotically stable in the large if its region

of attraction is the entire space $\Re^n$. *See also* region of attraction.

asymptotically stable state   the equilibrium state of a dynamic system described by a first-order vector differential equation is said to be asymptotically stable if it is both convergent and stable. *See also* stable state and convergent state.

asynchronous   not synchronous.

asynchronous AC systems   AC systems either with different operating frequencies or that are not in synchronism.

asynchronous bus   a bus in which the timing of bus transactions is achieved with two basic "handshaking" signals, a request signal from the source to the destination and an acknowledge signal from the destination to the source. The transaction begins with the request to the destination. The acknowledge signal is generated when the destination is ready to accept the transaction. Avoids the necessity to know system delays in advance and allows different timing for different transactions. *See also* synchronous bus.

asynchronous circuit   (1) a sequential logic circuit without a system clock.
(2) a circuit implementing an asynchronous system.

asynchronous demodulation   a technique for extracting the information-carrying waveform from a modulated signal without requiring a phase-synchronized carrier for demodulation. *See* synchronous demodulation.

asynchronous machine   *See* induction machine.

asynchronous operation   a term to indicate that a circuit can operate or a communication system can transmit information when ready without having to wait for a synchronizing clock pulse.

asynchronous system   a (computer, circuit, device) system in which events are not executed in a regular time relationship, that is, they are timing-independent. Each event or operation is performed upon receipt of a signal generated by the completion of a previous event or operation, or upon availability of the system resources required by the event or operation.

asynchronous transfer mode (ATM)   method of multiplexing messages onto a channel in which channel time is divided into small, fixed-length slots or cells. In ATM systems the binding of messages to slots is done dynamically, allowing dynamic bandwidth allocation. ATM is asynchronous in the sense that the recurrence of cells containing information from an individual user is not necessarily periodic.

asynchronous updating   one unit at a time is selected from within a neural network to have its output updated. Updating an output at any time is achieved by determining the value of the unit's activation function at that time.

AT bus   bus typically used in personal computer IBM AT for connecting adapters and additional memory boards. It is called also 16 bit ISA bus since it presents a data bus at 16 bit. It presents an additional connector with respect to the classical ISA bus (at 8 bit) of IBM PCs based on Intel 8088. *See also* EISA.

Atanasoff, John Vincent   (1903–1995) Born: Hamilton, New York.

Best known for his invention, along with Clifford Berry, of the first digital computer, known as the ABC (Atanasoff–Berry Computer). Unlike the many World War II computer pioneers, Atanasoff's interest in the topic dated to his Ph.D. thesis research at the University of Wisconsin. After graduation Atanasoff taught physics and mathematics at Iowa State College and continued to work on the problem of solving lengthy calculation by electronic means. Legend has it that

Atanasoff worked out the basic structure for his new machine while having a drink at an Illinois road house. Clifford Berry, an electrical engineer joined Atanasoff to help with the construction of the device based on Atanasoff's ideas. John Mauchly, another computer pioneer often visited and consulted with Atanasoff. These discussions resulted in a later lawsuit that established Atanasoff as the first person to build an electronic digital computer.

ATM   *See* asynchronous transfer mode.

ATM adaptation layer (AAL)   a layer in the ATM protocol hierarchy that adapts the (small) cell-sized payloads to a form more suitable for use by higher layer protocols. For example, AAL5 performs segmentation and reassembly to map between 48-byte payloads and variable length data segments.

atmosphere   a convenient measure of pressure. 1 standard atmosphere = 14.696 psia (pounds per square inch absolute).

atmospheric attenuation   decrease in the amplitude of a signal propagating through the atmosphere, due primarily to absorption and scatter.

atmospheric duct   a thin layer of atmosphere near the earth that acts as a waveguide, the electromagnetic field, trapped within the duct, can travel over long distances with very little attenuation.

atom   a particle of matter indivisible by chemical means, which is chemically neutral. It is the fundamental building block of the chemical elements.

atomic beam   a source of atoms traveling primarily in one direction. In practice, atomic beams are usually realized by the expansion of an atomic vapor into a vacuum through a small aperture. The resulting expanding cloud of atoms is usually made nearly unidirectional by a collimator that blocks or otherwise removes all atoms not propagating within a narrow range of angles.

atomic force microscope (AFM) a microscope in which a sharp probe tip is scanned across a surface, with piezoelectric ceramics being used to control position in three dimensions. The lateral (in-plane, or x-y) positions are raster scanned, while the vertical dimension is controlled by a feedback circuit that maintains constant force. The image produced is a topograph showing surface height as a function of position in the plane.

atomic instruction an instruction that consists of discrete operations that are all executed as a single and indivisible unit, without interruption by other system events. *See also* test-and-set instruction, atomic transaction.

atomic transaction the same as an atomic instruction, except that the notion of being atomic applies to a transaction, which may be a sequence of operations, no intermediate states of which may be seen or operated upon by another transaction. *See also* atomic instruction.

atomic transition coupling of energy levels in an atom by means of absorption or emission processes.

atomic vapor a material composed of atoms that preferentially exist as monomers in the vapor phase.

ATRC *See* Advanced Television Research Consortium.

attachment one of the events which precede a lightning stroke to the earth. Attachment occurs when the stepped leader from the thundercloud makes contact with one of several streamers which emanate from the ground or structures on the earth. The return stroke follows immediately. *See* streamer, stepped leader, return stroke.

attachment process a process that occurs in lightning when one or more stepped leader branches approach within a hundred meters or so of the ground and the electric field at the ground increases above the critical breakdown field of the surrounding air. At that time one or more upward-going discharges is initiated. After traveling a few tens of meters, one of the upward discharges, which is essentially at ground potential, contracts the tip of one branch of the stepped leader, which is at a high potential, completing the leader path to ground.

attainable set for discrete system the set of all the possible ends of system trajectories at time t_1 starting from zero initial conditions at time t_0. Denoted $K(t_0, t_1)$.

$K(t_0, t_1)$ is defined for zero initial state as follows

$$K(k_0, k_1) = \left\{ x \in R^n : \right.$$

$$x = \sum_{j=k_0}^{j=k_1-1} F(k_1, j+1) B(j) u(j) :$$

$$\left. u(j) \in R^m \right\}$$

Therefore, controllability in $[k_0, k_1]$ for discrete dynamical system is equivalent to the condition

$$K(t_0, t_1) = R^n$$

Using the concept of the attainable set it is possible to express the remaining types of controllability for discrete system.

attenuated total reflection the phenomenon associated with the appearance of a reflection minimum identified with the generation of surface waves at the metal — air interface in a prism, air, metal arrangement.

attenuation the exponential decrease with distance, in the amplitude of an electric signal traveling along a very long transmission line due to losses in the supporting medium. In electromagnetic systems attenuation is due to conductor and dielectric losses. In fiber optic systems attenuation arises from intrinsic material properties (absorption and Rayleigh scattering) and from waveguide properties such as bending, microbending, splices, and connectors.

attenuation coefficient   *See* absorption coefficient.

attenuation constant   the real part of the complex propagation constant for an electromagnetic wave.

attenuator   a device or network that absorbs part of a signal while passing the remainder with minimal distortion.

attractor   an asymptotic state of a dynamical system of which there are three basic types. Either (i) the system comes to rest and the attractor is a fixed point in state space, (ii) the system settles into a periodic motion known as a limit cycle, or (iii) the system enters a chaotic motion, in which case the attractor is called strange.

attribute   a special function in Pawlak's information system. Pawlak's information system S is a pair (U, A) where the set U is called the universe and has n members denoted x_i, while the set A consists of m functions on the universe U. These functions are called the attributes and denoted $\mathbf{a}_j$. The attributes are vector-valued functions that may be interpreted, for example, as issues under negotiation by the members of the universe U. An example of an attribute is a function of the form

$$\mathbf{a}_j : U \to \{-1, 0, 1\}^n.$$

attribute set   a set of vectors (signals) lying in metric space that possess prescribed properties.

audio   science of processing signals that are within the frequency range of hearing, that is,

roughly between 20 hertz and 20 kilohertz. Also, name for this kind of signal.

audio channels   the portion of the circuit containing frequencies that correspond to the audible sound waves. Audio frequencies range from approximately 15 hertz to 20,000 hertz.

audio coding   the process of compressing an audio signal for storage on a digital computer or transmission over a digital communication channel.

audio follow-video switcher (AFV)   a switcher that simultaneously switches the video and audio information. The term is associated with the action of the audio signal and corresponding video signal switching together.

augmented code   a code constructed from another code by adding one or more codewords to the original code.

aural subcarrier   in a composite television signal, the frequency division multiplexed carrier placed outside the visual passband that carries the audio modulation. In the NTSC (United States) system, it is placed 4.5 Mhz higher than the visual carrier.

autoassociative backpropagation   network a multilayer perceptron network which is trained by presenting the same data at both the input and output to effect a self-mapping. Such networks may be used for dimensional reduction by constraining a middle, hidden layer to have fewer neurons than the input and output layers.

autobank   an array of autotransformers.

auto-correlation   a measure of the statistical dependence between two samples of the same random process. For a random process $X(t)$, the auto-correlation is the expectation

$$R_{xx}(t_1, t_2) = E\left[X(t_1)X(t_2)\right].$$

See also cross-correlation.

auto-regressive moving-average model (ARMA)
the discrete-time input–output model in which
the current output depends both on its past values
(auto-regressive part) and the present and/or past
values of the input (moving-average part).

autoconfiguration   a process that determines
what hardware actually exists during the current
instance of the running kernel at static configuration time. It is done by the autoconfiguration
software that asks the devices to identify themselves and accomplishes other tasks associated
with events occurring during the autoconfiguration of devices. For instance, PCI devices have
autoconfiguration capabilities and do not have to
be configured by users.

autocorrelation function   the expected value
of the product of two random variables generated
from a random process for two time instants; it represents their interdependence. The Fourier transform of the autocorreclation function is the power
spectrum (power spectral density) for the random
process.

autocorrelator   a circuit that computes the autocorrelation function.

autocovariance   (1) for a random process $f(t)$,
a measure of the variability of the mean-removed
process:

$$C_f(t_1, t_2) = E\left[f(t_1)f(t_2)^T\right]$$
$$-E[f(t_1)]E\left[f(t_2)^T\right].$$

(2) for a random vector x, a measure of the
mean-square variability of a random vector x
about its mean:

$$\Lambda_x = E\left[(x - E[x])(x - E[x])^T\right].$$

See also autocorrelation, covariance.

autodecrementing   (1) an addressing mode in
which the value in a register is decremented by
one word when used as an address.

(2) in high-level languages, operation

$$i - - \Rightarrow i = i - 1$$

where i is arbitrary variable, register or memory
location.

(3) in machine code, more generally, the processor decrements the contents of the register by
the size of the operand data type; then the register contains the address of the operand. The register may be decremented by 1, 2, 4, 8, or 16
for byte, word, longword, quadword, or octaword
operands, respectively.

autoincrementing   (1) an addressing mode in
which the value in a register is incremented by one
word when used as an address.

(2) in high-level languages: operation

$$i + + \Rightarrow i = i + 1$$

where i is arbitrary variable, register or memory
location.

(3) in machine code, after evaluating the
operand address contained in the register, the processor increments the contents of the register by
1, 2, 4, 8, or 16 for a byte, word, longword, quadword, or octaword, respectively.

automated meter reading (AMR)   the use of
meters which have the capability of transmitting
at the least consumption information to the utility through some means of electronic communication.

automatic   (1) property pertaining to a process
or a device that functions without intervention by
a human operator under specified conditions.

(2) a spring-loaded tension sleeve into which a
conductor or other wire is inserted for tensioning
and attachment to a pole or other fixture.

automatic allocation   allocation of memory
space to hold one or more objects whose lifetimes
match the lifetime of the activation of a module,

such as a subroutine. Automatic allocations are usually made upon entry to a subroutine.

automatic black-level control   electronic circuitry used to maintain the black levels of the video signal at a predetermined level. The black level reference is either derived from the image or from the back porch of the horizontal blanking interval.

automatic chroma control (ACC)   ACC is used to correct the level of the input chroma signal. Typically, the ACC circuitry makes corrections to the chroma, based on the relative degeneration of the color burst reference signal, since this signal will have been subjected to the same degradation.

automatic circuit recloser   *See* recloser.

automatic fine tuning (AFT)   one of the input circuits of a color television receiver specifically designed to maintain the correct oscillator frequency of the tuner for best color reproduction of the picture. The circuit is sometimes called the automatic frequency control. *See also* automatic frequency control (AFC).

automatic focusing   on an optical disk, the process in which the distance from the objective focal plane of the disk is continuously monitored and fed back to the disk control system in order to keep the disk constantly in focus.

automatic frequency control (AFC)   electronic circuitry used to keep the received signal properly placed within the desired IF frequency range. In televisions, the AFC circuitry is also called the AFT or "automatic fine tuning" section. The AFC circuit will generate an error signal if the input frequency to the IF drifts above or below the IF frequency. The error signal is fed back to vary the local oscillator frequency in the tuner section. *See also* automatic fine tuning (AFT).

automatic frequency control (AFC)   an automatic feedback control system that is used to maintain active power balance by means of the speed governor system. In an interconnected system, scheduled power interchanges are maintained by means of controlling area generations.

automatic gain control (AGC)   a method to control the power of the received signal in order to be able to use the full dynamic range of the receiver and to prevent receiver saturation.

automatic generation control (AGC)   phrase describing the computer-based process by which electric utilities control individual generating stations to maintain system frequency and net interchange of power on a highly interconnected transmission grid. Automatic generation control (AGC) systems monitor grid frequency, actual and scheduled power flows, and individual plant output to maintain balance between actual and scheduled power production, both within transmission control areas and at individual generating stations. Control is generally accomplished by adjusting the speed control (or droop) characteristics of individual generating units. Control actions are determined by planned production schedules and power exchange agreements among participating utilities.

automatic level control (ALC)   a feedback system where an RF signal from a source is sampled, detected, and sent to a voltage controlled attenuator to maintain a constant amplitude output over a specified band of frequencies.

automatic repeat request (ARQ)   an error control scheme for channels with feedback. The transmitted data is encoded for error detection and a detected error results in a retransmission request.

automatic tracking   on an optical disk, the process in which the position of the disk head relative to the disk surface is constantly monitored and fed back to the disk control system in order to keep the read/write beam constantly on track.

45

automatic transfer switch   a self-acting switch which transfers one or more load conductor connections from one power source to another.

automatic voltage regulator (AVR)   an automatic feedback control system that is responsible for maintaining a scheduled voltage either at the terminals of a synchronous generator or at the high-side bus of the generator step-up transformer. The control is brought about by changing the level of excitation.

automation   refers to the bringing together of machine tools, materials handling process, and controls with little worker intervention, including

(1) a continuous flow production process that integrates various mechanisms to produce an item with relatively few or no worker operations, usually through electronic control;

(2) self-regulating machines (feedback) that can perform highly precise operations in sequence; and

(3) electronic computing machines.

In common use, however, the term is often used in reference to any type of advanced mechanization or as a synonym for technological progress; more specifically, it is usually associated with cybernetics.

automaton   (1) a fundamental concept in mathematics, computer engineering, and robotics.

(2) a machine that follows sequence of instructions.

(3) any automated device (robots, mechanical and electromechanical chess automata). Automata (plural of automaton) theory studies various types of automata, their properties and limitations. *See also* cellular automaton, finite state machine (FSM).

autonomic   that part of the nervous system which controls the internal organs.

autonomous operation   operation of a sequential circuit in which no external signals, other than clock signals, are applied. The necessary logic inputs are derived internally using feedback circuits.

autonomous system   a dynamic system described by a first-order vector differential equation that is unforced and stationary. In other words, such a system is governed by an equation of the form

$$\dot{x}(t) = f(x(t))$$

See also unforced system and stationary system.

autoregressive (AR)   a pth order autoregressive process is a discrete random process that is generated by passing white noise through an all-pole digital filter having p poles. Alternatively, $x[n]$ is a pth order AR process if

$$x[n] = \sum_{i=n-p}^{n-1} \alpha[i]x[i] + q[n].$$

autoregressive   processes are often used to model signals since they exhibit several useful properties. *See* moving average.

autotransformer   a power transformer that has a single continuous winding per phase, part of this winding being common to both the primary and the secondary sides. As a result, these voltages are not isolated but the transformer is reduced in weight and size. Autotransformers are most suited for relatively small changes in voltage. Three phase autotransformers are by necessity connected in a wye configuration.

autotransformer starter   a single three-phase autotransformer or three single phase transformer used to start induction motors at a reduced voltage.

auxiliary memory   *See* secondary memory.

auxiliary relay   a relay employed in power system protection schemes that does not directly sense fault presence and location. Typical auxiliary relays include lockout relays, reclosing relays, and circuit breaker anti-pump relays.

auxiliary winding   a winding designed to be energized occasionally for a specific purpose, such as starting a single-phase motor. The power to the winding may be controlled by various means including a timer, centrifugal switch, current sensing relay, or voltage (counter EMF) sensing relay.

availability   the probability that a system is operating correctly and is available to perform list functions at the instant of time t. Also defined as the value

$$1 - \text{outage}$$

See also outage.

available bit rate (ABR)   ATM congestion control algorithm that enables a source to discover the bit rate available between it and a destination in a network. The source transmits a resource manager cell containing the desired bit rate; each switch this cell passes through adjusts the bit rate down to what it can support. Upon reaching the destination, the cell contains the available bit rate and is returned to the source.

available power gain   ratio of power available from a network to the power available from the source.

avalanche breakdown   process that occurs in a semiconductor space charge region under a sufficiently high voltage such that the net electron/hole generation rate due to impact ionization exceeds certain critical value, causing the current to rise indefinitely due to a positive feedback mechanism. The I-R heating caused during this process can permanently degrade or destroy the material.

avalanche injection   the physics whereby electrons highly energized in avalanche current at a semiconductor junction can penetrate into a dielectric.

avalanche photodiode (APD)   a photodiode (detector) that provides internal current gain. Used in optical communication systems when there is limited optical power at the receiver.

average optical power   time average of the optical power carried by a non-CW optical beam.

average picture level (APL)   describes the average (mean) changes in a video signal due to a changing brightness of the visual image. The APL is typically expressed in terms of a percentage (10-15% for dark pictures and 75–90% for bright pictures). Changes in the APL can effect linearity unless DC restoration or clamping circuits are included in the video circuitry.

average power   the average value, taken over an interval in time, of the instantaneous power. The time interval is usually one period of the signal.

average-value model   a mathematical representation in which the average value of variables are used to model a system. In electric machines and drives, system variables are typically averaged over various switching intervals. This eliminates the high-frequency dynamics, but preserves the slower dynamics of the system.

averaging   the sum of N samples, images or functions, followed by division of the result by N. Has the effect of reducing noise levels. *See* blurring, image smoothing, mean filter, noise smoothing, noise suppression, smoothing.

AVR   *See* automatic voltage regulator.

AWG   American Wire Gauge, a system of wire sizing used in the USA especially in smaller conductors used in residential and commercial wiring.

AWGN   *See* additive white Gaussian noise.

axon   the conducting portion of a nerve fiber — a roughly tubular structure whose wall is composed of the cellular membrane and is filled with an ionic medium.

Ayrton, William Edward (1847–1908) Born: London, England

Best known as the inventor of a number of electrical measurement devices and as an engineering educator. Ayrton's early work was with the Indian Telegraph Service, after which he studied with William Thomson (Lord Kelvin) in Glasgow. After several more telegraph assignments Ayrton traveled to Tokyo, where he established the first electrical engineering teaching laboratory at the Imperial Engineering College. Among his many inventions he is credited with the ammeter and an improved voltmeter. His wife Bertha was also an active researcher and became the first woman to be admitted to the Institute of Electrical Engineers.

azimuth recording a recording scheme whereby the data is recorded at an acute angle from the direction of movement of the recording medium. Used in the recording scheme of video information, FM radio, and audio in VCRs.

B

B coefficient *See* loss coefficient.

B-ISDN *See* broadband integrated services digital network.

B-mode display returned ultrasound echoes displayed as brightness or gray-scale levels corresponding to the amplitude versus depth into the body.

B-site in a ferroelectric material with the chemical formula ABO_3, the crystalline location of the B atom.

B-spline the shortest cubic spline consisting of different three-degree polynomial on four intervals; It can be obtained by convolving four box functions.

Babbage, Charles (1792–1871) Born: Totnes, England

Best known for his ideas on mechanical computation. Babbage is said to have been disgusted with the very inaccurate logarithm tables of his day, as well as appalled by the amount of time and people it took to compute them. Babbage attempted to solve the problem by building mechanical computing engines. The government-funded Difference Engine was beyond the technology of the craftsman who attempted to build it. Undeterred, Babbage followed this failure with the larger and more complex Analytical Engine (also unfinished). The ideas behind the Analytical Engine formed the basis for Howard Aiken's 1944 Mark I computer. Babbage's assistant, Ada Augusta, the Countess of Lovelace and the poet Lord Byron's daughter, is honored as the first programmer for her work and because her meticulous notes preserved the descriptions of Babbage's machines.

Babinet principle principle in optics that states that the diffraction patterns produced by complementary screens are the same except for the central spot. It can be rigorously proved both for acoustic and electromagnetic waves. The Babinet's principle for scalar fields is the following: let p be the resultant field in $z > 0$ due to the incident field p_i from $z < 0$ and let p_t be the total field when the same incident wave falls on the complementary screen. Then, in $z > 0$,

$$p + p_t = p_i$$

back in a motor, the end that supports the major coupling or driving pulley.

back EMF *See* counter-EMF.

back end that portion of the nuclear fuel cycle which commences with the removal of spent fuel from the reactor.

back porch a 4.7 microsecond region in the horizontal blanking interval of the NTSC composite video signal that contains a burst of eight to ten cycles of the 3.579545 MHz (3.58 MHz) color subcarrier. The back porch occupies 7% of the total horizontal line time; starting at the end of the horizontal line sync signal and ending with the start of the video.

backbone wiring that runs within and between floors of a building and connects local-area network segments together.

backfeed in power distribution work, power which flows from the secondary lines into the primary lines through the distribution transformer, *e.g.*,from an emergency generator connected to customer load.

backflash an arc which forms along a tower during a lightning strike due to high tower or footing impedance.

background (1) refers to the received vector power level of an electromagnetic measurement (usually radar cross section) with no target present. The background includes the collective unwanted power received from sources other than the desired target under test such as positioners, foam columns, fixtures necessary to support a target, and the room or ground environment. The background level is vectorially subtracted from the received level with the target present to obtain the raw data set for a particular target.

(2) any unwanted signal. The background is a lower limit on the detection of small signals when devices are used to make a measurement in an experimental set up. The measurement is a superposition of events from the experiment itself and events from all other sources including the background.

background noise the noise that typically affects a system but is produced independent of the system. This noise is typically due to thermal effects in materials, interpreted as the random motion of electrons, and the intensity depends on the temperature of the material. In radio channels, background noise is typically due to radiation that is inherent to the universe and due mainly to radiation from astronomical bodies. There is a fundamental lower bound on the intensity of such noise which is solely dependent on the universe and independent of antenna and receiver design. *See also* thermal noise, noise temperature, noise figure.

background subtraction for images, the removal of stationary parts of a scene by subtracting two images taken at different times. For 1–D functions, the subtraction of a constant or slowly-varying component of the function to better reveal rapid changes.

backing memory the largest and slowest level of a hierarchical or virtual memory, usually a disk. It is used to store bulky programs or data (or parts thereof) not needed immediately, and need not be placed in the faster but more expensive main memory or RAM. Migration of data between RAM and backing memory is under combined hardware and software control, loading data to RAM when it is needed and returning it to the backing storage when it has been unused for a while.

backing storage *See* backing memory.

backoff a technique used in amplifiers when operated near saturation that reduces intermodulation products for multiple carriers. In its implementation, the drive signal is reduced or backed off. Input backoff is the difference in decibels between the input power required for saturation and that employed. Output backoff refers to the reduction in output power relative to saturation.

backplane *See* backplane bus.

backplane bus a special data bus especially designed for easy access by users and allowing the connection of user devices to the computer. It is usually a row of sockets, each presenting all the signals of the bus, and each with appropriate guides so that printed circuit cards can be inserted. A backplane differs from a motherboard in that a backplane normally contains no significant logic circuitry and a motherboard contains a significant amount of circuitry, for example, the processor and the main memory.

backprojection an operator associated with the Radon transform

$$g(s, \theta) = \int_{-\infty}^{+\infty} \int_{-\infty}^{+\infty} f(x, y) \\ \times \delta(x \cos \theta + y \sin \theta - s) dx dy.$$

The backprojection operator is defined as

$$b(x, y) = \int_{0}^{\pi} g(x \cos \theta + y \sin \theta, \theta) d\theta.$$

$b(x, y)$ is called the backprojection of $g(s, \theta)$. $b(x, y)$ is the sum of all rays that pass through the point (x, y).

backplane optical interconnect   *See* board-to-board optical interconnect.

backpropagation   the way in which error terms are propagated in a multilayer neural network. In a single layer feed-forward network, the weights are changed if there are differences between the computed outputs and the training patterns. For multiple layer networks, there are no training patterns for the outputs of intermediate ('hidden') layer neurons. Hence the errors between the outputs and the training patterns are propagated to the nodes of the intermediate neurons. The amount of error that is propagated is proportional to the strength of the connection.

backpropagation algorithm   a supervised learning algorithm that uses a form of steepest descent to assign changes to the weights in a feed-forward network so as to reduce the network error for a particular input or set of inputs. Calculation of the modifications to be made to the weights in the output layer allows calculation of the required modifications in the preceding layer, and modifications to any further preceding layers are made a layer at a time proceeding backwards toward the input layer; hence the name of the algorithm.

backscatter   energy from a reflected electromagnetic wave. In optics, the optical energy that is scattered in the reverse direction from the transmitted optical energy in an optical fiber transmission link or network. The backscattered energy comes from impurities in the fiber; mechanical or environmental effects that cause changes in the attenuation in the fiber; connectors, splices, couplers, and other components inserted into the optical fiber network; and faults or breaks in the optical fiber.

backscattering   the reflection of a portion of an electromagnetic wave back in the direction of the wave source. *See also* backscatter.

backside bus   a term for a separate bus from the processor to the second level cache (as opposed to the frontside bus connecting to the main memory).

backward error recovery   a technique of error recovery (also called rollback) in which the system operation is resumed from a point, prior to error occurrence, for which the processing was backed up.

backward wave interaction   interaction between backward propagating microwave electric fields against an electron stream and the electron in the electron beam. The direction of propagating microwaves and the direction of motion of electrons in the beam are opposite each other.

backward wave oscillator (BWO)   a microwave oscillator tube that is based on a backward wave interaction.

balanced   *See* balanced line.

balanced amplifier   an amplifier in which two single-ended amplifiers are operated in parallel with 90 degree hybrid. Balanced amplifiers feature a low voltage standing wave ratio because of an absorption of reflected power at the terminating resistor of the hybrids.

balanced code   a binary line code that ensures an equal number of logic ones and logic zeros in the encoded bit sequence. Also called a DC-free code because the continuous component of the power spectral density of a balanced encoded sequence falls to zero at zero frequency.

balanced line   symmetric multiconductor transmission line in which the voltage on each conductor along the transmission line has the same magnitude, but the phases are such that the voltage would sum to zero. In a two conductor transmission line, the voltages would be equal and 180 degrees out of phase. This is the equivalent of a virtual ground plane or zero E-field plane at the geometric center plane of the transmission line

51

cross section, or balanced with respect to virtual ground. Balanced wiring configurations are often used to prevent noise problems such as ground loops. *Contrast with* unbalanced line.

balanced load   a load on a multi-phase power line in which each line conductor sees the same impedance.

balanced mixer   a nonlinear 3-port device (two inputs, one output) used to translate an input signal's frequency component either up or down the frequency spectrum by generating the sum and difference of two or more frequencies present at its inputs. The three ports are termed RF (radio frequency), LO (local oscillator), and IF (intermediate frequency). A balanced mixer translates the frequency components found in the RF input signal to the IF output in such a manner as to minimize the amount of LO noise arriving at the IF. This reduces the mixer's overall noise figure and increases its sensitivity. Other advantages of these mixers include improved local oscillator isolation and linearity and higher power handling ability.

balanced modulator   a modulator in which the carrier and modulating signal are introduced so that the output contains the two sidebands without the carrier.

balanced operation   in n-phase circuits ($n > 1$), an operating condition in which the voltages (currents) of the phases are equal-amplitude sinusoids with phase-angles displaced by a specific angle ϕ. The angle (ϕ) is a function of the number of phases (n). For $n = 2$, $\phi = 90$ degrees, for $n = 3$, $\phi = 120$ degrees. In machine analysis the term "balanced" is also used to describe a machine that has symmetrical phase windings.

balanced slope detector   an arrangement of two detectors designed to convert an FM signal to AM for detection. This is accomplished by setting the IF center frequency so that it falls on the most linear portion of the response curve. Frequency

changes (FM) will result in corresponding amplitude changes that are then sent to an AM detector. The balanced version is two slope detectors connected in parallel and 180 degrees out of phase.

ball grid array (BGA)   a modern high I/O count packaging method. It reduces the package size and its pin-to-pin trace gap in order to integrate more functions and reliability in a single space. It can have as many as 324 pins. BGA sockets are high speed, high reliability, surface-mountable, and can be installed without soldering. The related terms are PBGA—plastic ball grid array, CBGA—ceramic ball grid array, TBGA—tape automated bonded ball grid array. The disadvantage of BGA packaging is that new tools and skills are required to mount or replace the chipset manually for repair purposes.

ballast   a starting and control mechanism for fluorescent and other types of gas-discharge lamps. Initially a ballast supplies the necessary starting (or striking) voltage in order to ionize the gas to establish an arc between the two filaments in the lamp. Once the gas is ionized, the ballast controls the input power and thus the light output to maximize the efficiency and life of the lamp.

balun   a network for the transformation from an unbalanced transmission line, system or device to a balanced line, system or device. Baluns are also used for impedance transformation. Derived from "balanced to unbalanced."

In antenna systems, baluns are used to connect dipole-type antennas to coaxial cable, to balance the current on dipole armatures, and to prevent currents from exciting the external surface of the coaxial shield.

See also balanced, unbalanced.

band   reference name for a range of frequencies. Current defined bands include the following.

Band Name	Reference Range
L-band	1.12–1.7 GHz
X	8.2–12.4 GHz
Ku	12.4–18 GHz
Ka	26.5–40 GHz
V	50–75 GHz
W	75–110 GHz

band gap   the energetic gap between the conduction and valence band edges of a material (usually referred to semiconductors).

band limited signal   a signal $x(t)$ is said to be band limited if its Fourier transform $X(\omega)$ is zero for all frequencies $\omega > \omega_c$, where ω_c is called the cutoff frequency.

band stop filter   filter that exhibits frequency selective characteristic such that frequency components of input signals pass through unattenuated from input to output except for those frequency components coincident with the filter stop-band region, which are attenuated. The stop-band region of the filter is defined as a frequency interval over which frequency components of the input signal are attenuated.

band structure   the energy versus momentum relationship for an electron in a periodic crystal.

band-pass filter   (1) a circuit whose transfer function, or frequency response, $H(\omega)$ is zero or is very small for frequencies not in a specified frequency band. In a strict sense $H(\omega) = 0$ for $|\omega| \not\in [\omega_1, \omega_2]$ for some $0 < \omega_1 < \omega_2$. *Compare with* low-pass filter, high-pass filter, notch filter, band reject filter.

(2) an electronic or electrical circuit which has the response shown in the figure below. There are two cut-off frequencies, ω_L and ω_H. In the pass-band $\omega_L < \omega < \omega_H$, $|N(j\omega)|$ is constant. In the stopbands, $\omega < \omega_L$ and $\omega > \omega_H$, $|N(j\omega)|$ is very small and there is practically no transmission of the signal.

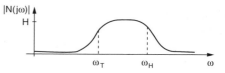

Band-pass filter response.

band-pass network   a configuration of solely passive components or combination of active and passive components that will attenuate all signals outside of the desired range of frequency.

band-pass signal   a signal whose Fourier transform or spectrum approaches zero outside a given frequency band. Ideally, the spectrum should equal zero outside the band, but this is difficult to achieve in practice. This may be described mathematically as follows: let $X(\omega)$ be the Fourier transform of the signal. Then, for a band-pass signal we have $X(\omega) = 0$ for $|\omega| \not\in [\omega_1, \omega_2]$, for some $0 < \omega_1 < \omega_2$.

band-reject filter   *see* band-stop filter.

band-stop filter   a filter which attenuates only within a finite frequency band and leaves the signal unaffected outside this band. *See* band-pass filter, high-pass filter, low-pass filter.

bandgap energy   in materials with band energy levels, the minimum energy needed to excite a charge carrier from a lower to an upper band. *See also* absorption edge.

bandgap engineering   in materials such as compound semiconductors and superlattice structures, the fabrication of materials with specific bandgap energies by varying the fractional proportions of the constituents and by varying superlattice layer thicknesses.

bandgap narrowing   reduction of the forbidden energy gap of a semiconducting material due to the narrowing influence of impurities.

bandgap reference   a voltage reference based on the 1.205 V bandgap voltage of silicon.

bandgap wavelength   the optical wavelength corresponding to a photon energy equal to the bandgap energy.

bandlimited   a waveform is described as band-limited if the frequency content of the signal is constrained to lie within a finite band of frequencies. This band is often described by an upper limit, the Nyquist frequency, assuming frequencies from DC up to his upper limit may be present. This concept can be extended to frequency bands that do not include DC.

bandwidth   (1) the frequency range of a message or information processing system measured in hertz.

(2) width of the spectral region over which an amplifier (or absorber) has substantial gain (or loss); sometimes represented more specifically as, for example, full width at half maximum.

(3) the property of a control system or component describing the limits of sinusoidal input frequencies to which the system/component will respond. It is usually measured at the half-power points, which are the upper and lower frequencies at which the output power is reduced by one half. Bandwidth is one measure of the frequency response of a system, i.e., the manner in which it performs when sine waves are applied to the input.

(4) the lowest frequency at which the ratio of the output power to the input power of an optical fiber transmission system decreases by one half (3 dB) compared to the ratio measured at approximately zero modulation frequency of the input optical power source. Since signal distortion in an optical fiber increases with distance in an optical fiber, the bandwidth is also a function of length and is usually given as the bandwidth-distance product for the optical fiber in megahertz per kilometer. *See also* signal distortion and bandwidth-distance product.

bandwidth efficiency   the ratio of the information rate in bits per second to the required bandwidth in hertz for any digital modulation technique.

bandwidth improvement (BI)   a dB reading that is a comparison of the RF bandwidth of a receiver to the IF bandwidth. Designated as BI, it is $10 \log B_{rf}/B_{if}$.

bandwidth-distance product   a measure of the information carrying capacity of an optical fiber which emphasizes that the bandwidth is a function of distance. For example, an optical fiber with a specification of 500 MHz-km bandwidth-distance product would have a 500 MHz bandwidth over 1 km, a 50 MHz bandwidth over 10 km or a 1 GHz bandwidth over 0.5 km. *See also* bandwidth, optical fiber.

bang-bang control   control action achieved by a command to the actuator that tells it to operate in either one direction or the other at any time with maximum energy.

Bang-bang control is an optimal or suboptimal piecewise constant control whose values are defined by bounds imposed on the amplitude of control components. The control changes its values according to the switching function which may be found using Pontryagin maximum principle. The discontinuity of the bang-bang control leads to discontinuity of a value function for the considered optimal control problem. Typical problems with bang-bang optimal control include time-optimal control for linear and bilinear control systems.

Bardeen, John   (1908–1991) Born: Madison, Wisconsin, U.S.A.

Bardeen is best known as one of the few persons to receive two Nobel Prizes. The first prize he received was in 1956 for his development at Bell Labs, along with Walter Brattain and William Schockley, of the first transistor. When the three applied for a patent for the device in 1948 they called it a germanium transfer resistance unit;

hence the name transistor. This device was a significant step in the development of integrated circuits. Bardeen's second Nobel, which he shared with Leon Cooper and John R. Schrieffer, was for his work at the University of Illinois in describing the theory of superconductivity.

bare-hand   refers to a method of servicing energized overhead conductors in which the line worker's body is maintained at the same potential as the conductor on which he is working, thus enabling the conductor to be contacted without danger of shock.

BARITT   barrier injection transit time, a microwave transit time device that uses injection over a forward biased barrier and transit time delay through a reverse biased junction to produce negative resistance at microwave frequencies, useful in low power and self-oscillating mixer applications.

Barkhausen criterion   two conditions placed on a feedback oscillator necessary for sustained oscillation. The Barkhausen criterion states

(1) The total loop amplitude transmission factor must be at least unity.

(2) The frequency of oscillation will be that frequency characterized by a total loop phase transmission factor of $N2\pi$ radians. N is either zero or an integer. Simply, for sustained oscillation, a disturbance that makes a complete trip around the feedback loop of the oscillator must be returned at least as strong as the original disturbance and in phase with that disturbance.

Barkhausen effect   the series of irregular changes in magnetization that occur when a magnetic material is subjected to a change in magnetizing force.

Barkhausen noise   noise arising in magnetic read heads because the interlocking magnetic domains cannot rotate freely in response to an applied field. The response to an external magnetic field is randomly discontinuous as domains

"stick," and then release. Barkhausen noise is particularly important in very small heads and thin-film heads where very few domains are involved; in larger heads the effects of many domains tend to average out and Barkhausen noise is relatively less important.

barrel distortion   a geometric distortion of a raster display in which vertical lines appear to bow outward away from the display center line. The bowing of the vertical lines increases as the distance from the vertical center increases. The appearance of these vertical lines is similar to the staffs of a barrel. Barrel distortion is a result of the overcorrection for pincushion distortion.

barrel shifter   an implementation of a shifter, which contains $\log_2$(max number of bits shifted) stages, where each stage shifts the input by a different power of two number of positions. It can be implemented as a combinational array with compact layout that can shift the data by more than one bit using only one gate. For instance, for a 4-bit word, it can execute instructions shl, shl2, shl3, and shl4. This shifter lends itself well to being pipelined.

barrier layer   layer of deposited glass adjacent to the inner tube surface to create a barrier against OH diffusion.

barrier voltage   a voltage that develops across the junction due to uncovered immobile ions on both sides of the junction. Ions are uncovered due to the diffusion of carriers across the junction.

Bartlett window   a triangular window $w[n]$ of width $2M$ defined as follows:

$$w[n] = \begin{cases} 1/2\,[1 + \cos(\pi\,n/M)], & -M \le n \le M \\ 0, & \text{otherwise} \end{cases}$$

multiplying a signal, $x[n]$ by the finite duration window signal $w[n]$ leads to the triangularly scaled, finite duration signal $z[n] = x[n]w[n]$, which is then processed. Windowing is used in

55

the spectral analysis of measured signals and in the design of finite impulse response, linear time invariant systems.

baryon   a collective term for all strongly interacting particles with masses greater than or equal to the mass of the proton. These include the proton, neutron, and hyperons.

base   (1) the number of digits in a number system (10 for decimal, 2 for binary).

(2) one of the three terminals of a bipolar transistor.

(3) a register's value that is added to an immediate value or to the value in an index register in order to form the effective address for an instruction such as LOAD or STORE.

base address   (1) an address to which an index or displacement is added to locate the desired information. The base address may be the start of an array or data structure, the start of a data buffer, the start of page in memory, etc.

(2) as a simpler alternative to a full virtual memory, the code space or data space of a program can be assumed to start at a convenient starting address (usually 0) and relocated in its entirety into a continuous range of physical memory addresses. Translation of the addresses is performed by adding the contents of an appropriate base address register to the user address.

baseband   in communication systems, the information-carrying signal which is modulated onto a carrier for transmission.

baseband signal   in digital communications, a signal that appears in the transmitter prior to passband modulation. For example, in the case of pulse amplitude modulation, $s(t) = \sum_i b_i p(t - iT)$ is a baseband signal, where b_i is the transmitted symbol at time i, and $p(t)$ is the baseband pulse shape (e.g., raised-cosine). *See also* low-pass signal.

base dynamic parameters   a set of dynamic parameters that appear in the canonical equations of motion. Canonical equations of motion of robot dynamics do not include linearly dependent equations. These are eliminated by making use of various procedures. As a result, dynamic equations of motion contain only independent equations which are used for the purpose of control. Each base dynamic parameter is a linear combination of the inertial parameters of the individual links. Base dynamic parameters are subject to the identification in adaptive control schemes applied to robot control. *See also* inertial parameters.

base frame   a frame attached to the non-moving base of the manipulator. Sometimes the base frame is called the reference frame.

base quantity   *See* per unit system.

base register   the register that contains the component of a calculated address that exists in a register before the calculation is performed (the register value in "register + immediate" addressing mode, for example).

base register addressing   addressing using the base register. Base register is the same as base address register, i.e., a general-purpose register that the programmer chooses to contain a base address.

base speed   corresponds to speed at rated torque, rated current, and rated voltage conditions at the temperature rise specified in the rating. It is the maximum speed at which a motor can operate under constant torque characteristics or the minimum speed to operate at rated power.

base station   the fixed transceiver in a mobile communication system. *See also* fixed station (FS).

base vector   a unit vector in a coordinate direction.

basic impulse insulation level (BIL)   a measurement of the impulse withstand capability of a piece of electric power equipment based on its ability to withstand 50% of impulses applied at the BIL voltage.

basic input–output system (BIOS)   part of a low-level operating system that directly controls input and output devices.

basic lightning impulse level (BIL)   the strength of insulation in terms of the withstand voltage crest value using a standard voltage level impulse.

basin of attraction   the region in state space from which a dynamical system moves asymptotically toward a particular attractor.

basis function   (1) one of a set of functions used to represent a function $f(x)$ as a series of the form

$$\sum_{n=1}^{N} a_n g_n(x)$$

where the $g_n(x)$ are the basis functions and the a_n are constant coefficients.

(2) one of a set of functions used in the transformation or representation of some function of interest. A linear transformation T of continuous functions is of the form

$$y(s) = T\{x(t)\} = \int_{-\infty}^{+\infty} x(t)b(s,t)dt.$$

where $b(s,t)$ is a basis function. For discrete sequences T would be of the form

$$y[k] = T\{x[n]\} = \sum_{n=-\infty}^{+\infty} x[n]b[k,n].$$

The function to be transformed is projected onto the basis function corresponding to the specified value of the index variable s or k. $y(s)$ is the inner product of $x(t)$ and the basis function $b(s,t)$. For the Laplace transform $b(s,t) = e^{-st}$,

and for the Fourier transform $b(\omega,t) = e^{-j\omega t}$. For the discrete-time Fourier transform $b[k,n] = e^{j2\pi knN}$, and for the Z- transform $b[z,n] = z^{-n}$.

BaTiO$_3$ (barium titanate)   a ferroelectric crystalline material that is particularly useful for photorefractive and optical multibeam coupling.

battery   one or more cells connected so as to produce energy.

baud   the signaling rate, or rate of state transitions, on a communications medium. One baud corresponds to one transition per second. It is often confused with the data transmission rate, measured in bits per second.

Numerically, it is the reciprocal of the length (in seconds) of the shortest element in a signaling code. For very low-speed modems (up to 1200 bit/s) the baud rate and bit rate are usually identical. For example, at 9600 baud, each bit has a duration of 1/9600 seconds, or about 0.104 milliseconds.

Modems operating over analog telephone circuits are bandwidth limited to about 2500 baud; for higher user data speeds each transition must establish one or more decodable states according to amplitude or phase changes. Thus, if there are 16 possible states, each can encode 4 bits of user data and the bit rate is 4 times the baud rate.

At high speeds, the reverse is true, with run-length controlled codes needed to ensure reliable reception and clock recovery. For example FDDI uses a 4B/5B coding in which a "nibble" of 4 data bits is encoded into 5 bits for transmission. A user data rate of 100 Mbit/s corresponds to transmission at 125 Mbaud.

baud rate   *See* baud.

Baum–Welch algorithm   the algorithm used to learn from examples the parameters of hidden Markov models. It is a special form of the EM algorithm.

Bayes envelope function   given the a priori distribution of a parameter Θ and a decision function ϕ, the Bayes envelope function $\rho(F_\Theta)$ is defined as

$$\rho(F_\Theta) = \min \phi r(F_\Theta, \phi),$$

where $r(F_\Theta, \phi)$ is the Bayes risk function evaluated with the a priori distribution of the parameter Θ and decision rule ϕ.

Bayes risk function   with respect to a prior distribution of a parameter Θ and a decision rule ϕ, the expected value of the loss function with respect to the prior distribution of the parameter and the observation X.

$$r(F_\Theta, \phi) = \int_\Theta \int_X L[\theta, \phi(x)] \\ \times f_{X|\Theta}(x|\theta) f_{|\text{Theta}}(\theta) dx d\theta.$$

the loss function is the penalty incurred for estimating the parameter Θ incorrectly. The decision rule $\phi(x)$ is the estimated value of the parameter based on the measured observation x.

Bayes' rule   relates the conditional probability of an event A given B and the conditional probability of the event B given A:

$$P(A \mid B) = \frac{P(B \mid A)P(A)}{P(B)}.$$

Bayesian classifier   (1) a classifier based on the Bayesian theory.
(2) a Bayesian classifier is a function of a realization of an observed random vector $\mathbf{X}$ and returns a classification w. The set of possible classes is finite. A Bayesian classifier requires the conditional distribution function of $\mathbf{X}$ given w and the prior probabilities of each class. A Bayesian classifier returns the w_i such that $P(w_i|\mathbf{X})$ is maximized. By Bayes' rule

$$P(w_i|\mathbf{X}) = \frac{P(\mathbf{X}|w_i)P(w_i)}{P(\mathbf{X})}.$$

Since $P(\mathbf{X})$ is the same for all classes, it can be ignored and the w_i that maximizes $P(\mathbf{X}|w_i)P(w_i)$ is returned as the classification.

Bayesian detector   a detector that minimizes the average of the false-alarm and miss probabilities, weighted with respect to prior probabilities of signal-absent and signal-present conditions.

Bayesian estimation   an estimation scheme in which the parameter to be estimated is modeled as a random variable with known probability density function.

Bayesian estimator   an estimator of a given parameter Θ, where it is assumed that Θ has a known distribution function and a related random variable X that is called the observation. X and Θ are related by a conditional distribution function of X given Θ. With $P(X|\Theta)$ and $P(\Theta)$ known, an estimate of Θ is made based on an observation of X. $P(\Theta)$ is known as the *a priori* distribution of Θ.

Bayesian mean square estimator   for a random variable X and an observation Y, the random variable

$$\hat{X} = E[X \mid Y],$$

where the joint density funtion $f_{XY}(x, y)$ is known. *See* mean square estimation, linear least squares estimator.

Bayesian reconstruction   an algorithm in which an image u is to be reconstructed from a noise-corrupted and blurred version v.

$$v = f(Hu) + \eta.$$

A prior distribution $p(u \mid v)$ of the original image is assumed to be known. The equation

$$\hat{u} = \mu_u + R_u H^T D R_\eta^{-1}[v - f(H\hat{u})],$$

where R_u is the covariance of the image u, R_η is the covariance of the noise η, and D is the diagonal matrix of partial derivatives of f evaluated at $\hat{u}$. An initial point is chosen and a gradient descent algorithm is used to find the closest $\hat{u}$ that minimizes the error. Simulated annealing is often used to avoid local minima.

Bayesian theory theory based on Bayes' rule, which allows one to relate the *a priori* and *a posteriori* probabilities. If $P(c_i)$ is the *a priori* probability that a pattern belongs to class c_i, $P(\mathbf{x_k})$ is the probability of pattern $\mathbf{x_k}$, $P(\mathbf{x_k} \mid c_i)$ is the class conditional probability that the pattern is $\mathbf{x_k}$ provided that it belongs to class c_i, $P(c_i \mid \mathbf{x_k})$ is the *a posteriori* conditional probability that the given pattern class membership is c_i, given pattern $\mathbf{x_k}$, then

$$P(c_i \mid \mathbf{x_k}) = \frac{P(\mathbf{x_k} \mid c_i)\, P(c_i)}{P(\mathbf{x_k})}.$$

The membership of the given pattern is determined by

$$\max_{c_i} P(c_i \mid \mathbf{x_k}) = \max_{c_i} P(\mathbf{x_k} \mid c_i)\, P(c_i).$$

Hence, the *a posteriori* probability can be determined as a function of the *a priori* probability.

BCD *See* binary coded decimal.

BCH code cyclic block forward error control codes developed by Bose and Chaudhuri, and independently by Hocquenghem. These codes are a superset of the Hamming codes, and allow for correction of multiple errors.

BCLA *See* block carry lookahead adder.

beam (1) transverse spatial localization of the power in a wave field.
(2) a slender unidirectional stream of particles or radiation.

beam cooling the process by which a particle beam's phase space volume is reduced, while conserving Liouville's theorem (empty spaces between particles exist). Beam cooling is manifest by a reduction in the transverse beam size (betatron cooling) or by a smaller momentum spread (momentum cooling).

beam divergence the geometric spreading of a radiated electromagnetic beam as it travels through space.

beam hardening a phenomenon that occurs when a polychromatic X-ray beam passes through a material. Lower energy photons are absorbed more readily than higher energy photons, increasing the effective energy of the beam as it propagates through the material.

beam intensity the average number of particles in a beam passing a given point during a certain time interval. For example the number of protons (electrons) per pulse or protons per second.

beam loading the beam being accelerated by an RF cavity and it changes the gradient and phase of the RF in the cavity.

beam mode confined electromagnetic field distributions of a propagating wave that match the boundary conditions imposed by a laser or aperture. For example, Hermite–Gaussian or Laguerre–Gaussian.

beam parameter one of several complex numbers employed to characterize the propagation of a beam; most common parameter combines in its real and imaginary parts the phase front curvature and spot size of a Gaussian beam.

beam pulsing a method used to control the power output of a klystron in order to improve the operating efficiency of the device.

beam roll a periodic change in horizontal and/or vertical positions during spill. This does not include changes caused by humans.

beam solid angle a parameter that qualitatively describes the angular distribution of radiated power from an antenna. The values range from very small numbers for very focused antennas to 4π steradians for an isotropic radiator.

beam stop   a thick metal shield that moves into the beam line in order to prevent beam from entering a specific area.

beam toroid   a device used for measuring beam intensities by measuring the magnetic field fluctuations produced by the passing beam. The magnetic field fluctuations produce a current in a coil, which is wound around a closed circular ring (torus) through which the beam passes.

beam waist   position at which a beam is most highly confined; for Gaussian beams in real media the position at which the phase fronts are flat.

beamformers   system commonly used for detecting and isolating signals that are propagating in a particular direction.

beamforming   a form of filtering in spatial rather than time domain to obtain a desired spatial impulse response in order to suppress or to reject signal components coming from certain directions. The technique involves directing one or more beams in certain directions by adjusting, for example, the element excitation of an array antenna. Used in communications applications to suppress other signals than the desired source signal. Also termed spatial filtering.

beamline   a series of magnets placed around a vacuum pipe which carry the proton beam from one portion of the accelerator to another. Also known as transport line.

beamsplitter   any of a number of passive optical devices that divide an optical wavefront into two parts. Wavefront division may be according to intensity, polarization, wavelength, spatial position, or other optical properties.

beamwidth   the angular width of the major lobe of a radiation pattern. It is usually at the half-power level, i.e., 3 dB below the peak of the major lobe. It can also be specified as the width between the nulls on either side of the major lobe (BWFN).

bearing currents   current flow in the bearings of electrical machines, because of electromagnetic unbalance in the machine or from using high $\frac{dv}{dt}$ inverters. The latter is able to charge up the stray capacitance present between the stator and rotor and between the rotor and shaft and thus allows motor bearing currents to flow, with resulting bearing damage.

beat frequencies   the two frequencies, sum and difference frequencies, generated during the heterodyning process or during the amplitude-modulating process. For example, if a 500 kHz carrier signal is amplitude-modulated with a 1 kHz frequency, the beat frequencies are 499 kHz and 501 kHz.

beat frequency oscillator   an adjustable oscillator used in superheterodyne receivers generating a frequency when combined with the final IF produces a difference or beat frequency in audio range.

becky   a knot used to secure a handline.

bed of nails   a test fixture for automated circuit qualification in which a printed wiring board is placed in contact with a fixture that contacts the board at certain nodes required for exercising the assembly.

bel   *See* decibel.

Bell, Alexander Graham   (1847–1922) Born: Edinburgh, Scotland

Best known as the first patent holder for a device to electronically transmit human speech. Bell's early interest in the mechanisms of speech come from living with his grandfather, a London speech tutor. Work with the deaf was to be a life-long vocation for Bell. Bell's inventions were not limited to the telephone. He was the first person to transmit speech without wires, he invented the gramophone, an early tape recorder, an air-cooling system, an iron lung, and he had several patents in telegraphy.

bell insulator   a type of strain insulator, shaped like saucer with ribs on its lower side and frequently used in insulator strings.

Bello functions   a group of alternative methods of characterizing a wideband communication channel, named after their proposer, P. Bello. The four functions characterizing deterministic channels are the Input Delay-spread Function, the Output Doppler-spread function, the Time-variant Transfer Function and the Delay Doppler-spread function.

BEM   *See* boundary-element method.

benchmark   standard tests that are used to compare the performance of computers, processors, circuits, or algorithms.

bending loss   in a fiber depends exponentially on the bend radius R. It is proportional to $\exp(-R/R_c)$ where the critical radius

$$R_c = \frac{a}{2n\,(n_{co} - n_{cl})},$$

a is the fiber radius, n_{co} is the refractive index of the core, and n_{cl} is the refractive index of the cladding.

BER   *See* bit error rate.

Bernoulli distribution   a random variable X with alphabet $\{0, 1\}$ and parameter α such that its probability mass function is

$$p(x) = (1 - \alpha)^x \alpha^{1-x}.$$

bias the systematic (as opposed to random) error of an estimator.

See bilateral Laplace transform a Laplace transform of the form

$$L\{f\} = \int_{-\infty}^{\infty} f(t)e^{-st}dt,$$

where s is a complex number. *See* Laplace transform.

Bernoulli process   a binary valued, discrete-time random process defined on an index set corresponding to fixed increments in time. A typical example is a sequence of coin tosses where the values of the process are denoted as "Heads" or "Tails" depending on the outcome of the tosses. The output values of the process is a sequence of statistically independent random variables with the same probability distribution. The two outcomes may or may not have equal probabilities.

Berry, Clifford Edward   (1918–1963) Born: Gladbrook, Iowa

Best known as the co-developer, along with John Vincent Atanasoff, of the first functioning electronic digital computer. Berry was recommended to Atanasoff by the Dean of Engineering at Iowa State College as a most promising student who understood the electronics well enough to help Atanasoff implement his ideas for a computing machine. Unfortunately, Berry's contributions as a computing pioneer were not honored until after his death.

beryllium oxide   a compound commonly used in the production of ceramics for electrical applications and whose dust or fumes are toxic.

Bessel beam   transverse wave amplitude distribution in which the radial variation is approximately describable in terms of truncated Bessel functions; collimation for Bessel beams is sometimes considered better than for more usual polynomial-Gaussian beams.

Bessel functions   a collection of functions, denoted as $J_v(x)$ and $Y_v(x)$, that satisfy Bessel's equation

$$x^2 \frac{d^2 f}{dx^2} + x \frac{df}{dx} + \left(x^2 - \nu^2\right) f = 0,$$

where f is equal to either J_ν or Y_ν; ν is the order of the function and x is its argument. Typically, Bessel functions arise in boundary value problems that are based upon a cylindrical coordinate system.

best-fit memory allocation   a memory allocator for variable-size segments must search a table of available free spaces to find memory space for a segment. In "best-fit" allocation, the free spaces are linked in increasing size and the search stops at the smallest space of sufficient size. *Compare with* buddy memory allocation.

beta function   a measure of beam width. The beta function details how the beam changes around the accelerator. There are separate beta functions for the x and y planes. The square root of bx is proportional to the beam's x-axis extent in phase space.

beta particle   an electron or positron emitted from a radioactive source.

betatron oscillation   stable oscillations about the equilibrium orbit in the horizontal and vertical planes. First studied in betatron oscillators, betatron oscillation is the transverse oscillation of particles in a circular accelerator about the equilibrium orbit. The restoring force for the oscillation is provided by focusing components in the magnetic field which act to bend a particle that is off the equilibrium orbit back toward it.

Beverage antenna   simple traveling wave antenna consisting of an electrically long horizontal wire above ground with a termination resistance between the end of the wire and ground equal to the characteristic impedance of the wire/ground transmission line.

Bezout identity of 2-D polynomial matrices   a systems identity defined as follows: let $N_R(z_1, z_2)$, $D_R(z_1, z_2)$ ($N_L(z_1, z_2)$, $D_L(z_1, z_2)$) be two right (left) coprime polynomial matrices, then there exists a polynomial matrix in z_2, say $E_R(z_2)$, ($E_L(z_2)$) and two polynomial matrices $X_R(z_1, z_2)$, $Y_R(z_1, z_2)$ ($X_L(z_1, z_2)$, $Y_L(z_1, z_2)$) such that

$$X_R(z_1, z_2) D_R(z_1, z_2)$$
$$+ Y_R(z_1, z_2) N_R(z_1, z_2) = E_R(z_2)$$

$$(N_L(z_1, z_2) Y_L(z_1, z_2)$$
$$+ D_L(z_1, z_2) X_L(z_1, z_2) = E_L(z_2))$$

BGA   *See* ball grid array.

BI   *See* bandwidth improvement.

bi-anisotropic media   (1) a class of material in which the electric and magnetic flux densities, **D** and **B**, are each linearly related to both the electric and magnetic field intensities, **E** and **H**, via dyadic constitutive parameters. The permittivity, permeability, and magnetoelectric coupling parameters are tensor quantities.
(2) media for which the electric and magnetic fields displacements, **D** and **B** respectively, are related to the electric and magnetic field strength **E** and **H** by general dyadics.

bi-isotropic media   media for which the electric and magnetic fields displacements, **D** and **B**, respectively, are scalarly dependent by both the electric and magnetic field strength **E** and **H**. For these media the constitutive relations are

$$\mathbf{D} = \epsilon \mathbf{E} + (\chi - j\kappa) \sqrt{\mu_0 \epsilon_0}\, \mathbf{H}$$
$$\mathbf{B} = \epsilon \mathbf{H} + (\chi + j\kappa) \sqrt{\mu_0 \epsilon_0}\, \mathbf{E}$$

where ϵ is the permittivity, μ the permeability, and the subscript 0 refers to free-space. Bi-isotropic media can be reciprocal ($\chi = 0$) or nonreciprocal ($\chi \neq 0$); nonchiral ($\kappa = 0$) or chiral ($\kappa \neq 0$).

bi-stable   pertaining to a device with two stable states, e.g., bi-stable multivibrator; circuit that has two possible output states and that will remain in its current state without requiring external inputs; a flip-flop.

bi-stable device   *See* flip-flop.

bias current   the arithmetic average of the currents that flow in the input leads of an op-amp.

bias lighting   technique used in video tubes to correct for undesirable artifacts such as lag. Applying a uniform light source to the surface of the tube (the photoconductive layer) will create a bias current in the tube, thereby minimizing the undesirable characteristics.

bias network   a key aspect of microwave circuit design is to apply the proper DC bias to the appropriate terminals of transistors (e.g., FETs) without disturbing the AC microwave operation of the circuit. In some cases, "on-chip" DC circuitry needs to be designed so as to provide stable bias voltage/current conditions for the device even when the chip DC supply voltages vary (due to weakening batteries, etc.). The other aspect of bias network design is to isolate the DC network from interfering with the AC or RF/microwave operation of the circuit, and vice-versa. In a lumped element design, this is generally accomplished by a combination of spiral inductors and MIM capacitors.

bias voltage or current   the DC power applied to a transistor allowing it to operate as an active amplifying or signal generating device. Typical voltage levels in GaAs FETs used in receivers are 1 to 7 volts between the drain and source terminals, and 0 to -5 volts on, or between, the gate and source terminals. For microwave systems, DC voltages and currents, provided by batteries or AC/DC converters required to "bias" transistors to a region of operation where they will either amplify, mix or frequency translate, or generate (oscillators) microwave energy. Since energy can be neither created nor destroyed, microwave energy amplification or creation is accomplished at the expense of DC energy.

biasing   the technique of applying a direct-current voltage to a transistor or an active network to establish the desired operating point.

bible   nickname for the National Electrical Code.

BIBO stability   *See* bounded-input bounded-output stability.

BIBO stability of 2-D linear system   a system described by the equation

$$y_{i,j} = \sum_{k=0}^{i} \sum_{l=0}^{j} g_{i-k,j-l} u_{k,l}$$

$i, j \in Z_+$ (the set of nonnegative integers) is said to be bounded-input bounded-output (BIBO) stable if for every constant $M > 0$ there exist a constant $N > 0$ such that if $\|u_{k,l}\| \leq M$ for all $k, l \in Z_+$, then $\|y_{i,j}\| \leq N$ for all $i, j \in Z_+$ where $u_{k,l} \in R^m$ is the input, $y_{i,j} \in R^p$ is the output, $g_{i,j} \in R^{p \times m}$ is the matrix impulse response of the system and $\|v\|$ denotes a norm of the vector v. The system is BIBO stable if and only if

$$\sum_{i=0}^{\infty} \sum_{j=0}^{\infty} \|g_{i,j}\| < \infty$$

BIBS   *See* bounded-input bounded-state stability.

BiCMOS   integrated circuit technology/process that incorporates bipolar and complementary metal oxide semiconductor devices on the same die.

63

bidirectional bus   a bus that may carry information in either direction but not in both simultaneously.

bi-directional laser   a ring laser with both clockwise and counter-clockwise circulating waves. Useful as a rotation rate sensor.

bidirectional pattern   a microphone pickup pattern resembling a figure eight, in which the device is most sensitive to sounds on either side of the pickup element.

bi-directional resonator   a standing-wave resonator or a ring-resonator in which the electromagnetic waves circulate in both the clockwise and counteclockwise directions.

bidirectional transducer   a surface acoustic wave (SAW) transducer which launches energy from both acoustic ports which are located at either end of the transducer structure.

bidirectional transmission distribution function (BTDF)   the optical scattering function for transmissive optics. The scattering function vs. angle is normalized to signal at zero degrees and with respect to solid angle of detector, including obliquity factor.

bifilar winding   a two-wire winding. It is often utilized in stepper motors to permit a unipolar power supply to produce alternating magnetic poles by energizing only half of the bifilar winding at any one time.

bifurcation   a term from Chaos Theory referring to a sudden change in the qualitative behavior of the solutions.

bifurcation diagram   a diagram where the sampled variable is plotted versus a parameter. The sampling period is equal to the source period. Similar to a Poincaré map.

big endian   a storage scheme in which the most significant unit of data or an address is stored at the lowest memory address. For example, in a 32-bit, or four-byte word in memory, the most significant byte would be assigned address i, and the subsequent bytes would be assigned the addresses: $i + 1$, $i + 2$, and $i + 3$. Thus, the least significant byte would have the highest address of $i + 3$ in a computer implementing the big endian address assignment. "Big endian" computers include IBM 360, MIPS R2000, Motorola M68000, SPARC, and their successors.

The little endian approach stores the least significant unit at the lowest address. (The terms big endian and little endian are taken from Jonathan Swift's satirical story, *Gulliver's Travels*.)

See also little endian.

BIL   *See* basic lightning impulse level and basic impulse insulation level.

bilateral Z-transform   a Z-transform f the form

$$Z\{x\} = \sum_{n=-\infty}^{+\infty} x[n]z^{-n}.$$

bilinear control systems   a class of nonlinear control system models that are linear in state, output and control variables treated separately but they contain the products of those variables. Such models arose naturally in modeling the number of chemical processes where the controls are flow rates that appear in the system equations as products with state variables. The bilinear control systems may also be used to model population dynamics perturbed by control actions which enter growth equations as multipliers of state variables. Bilinear control systems can arise also in connection with adaptive control nominally linear systems where uncertain parameters regarded as additional state variables leads to bilinear terms in model equations. Bilinear time-continuous control systems may be represented by the state equations having the form

$$\dot{x} = Ax + Bu + \sum_{i=1}^{m} D_i u_i x$$

where x is the state vector, u the control vector with components $u_i, i = 1, 2, \ldots, m$, A, B, D_i are matrices of the appropriate dimensions. *See also* population dynamics.

bilinear interpolation   interpolation of a value in 2–D space from four surrounding values by fitting a hyperbolic paraboloid. The value at (x, y), denoted $f(x, y)$ is interpolated using $f(x, y) = ax + by + cxy + d$, where a, b, c and d are obtained by substituting the four surrounding locations and values into the same formula and solving the system of four simultaneous equations so formed.

bilinear transformation   (1) conformal mapping of the complex plane of the form $f(z) = az + bcz + d$, where the real values a, b, c, d satisfy $ad - bc \neq 0$. Also called linear fractional transformation or Möbius transformation.

(2) a special case of (1) is a mapping from the $j\omega$ axis in the s-plane to the unit circle $|z| = 1$ in the z-plane, given by $x = 2T 1 - z^{-1} 1 + z^{-1}$, where T is the time interval between samples. Such bilinear transformations are used in the design of recursive digital filters from equivalent analogue filters in the following procedure:

- define characteristic digital frequencies Ω_i.
- prewarp these to analog frequencies ω_i using $\omega_i = \frac{2}{T} \tan\left(\frac{\Omega_i T}{2}\right); 1 \leq i \leq k$.
- design a suitable analog filter with frequencies ω_i.
- use the bilinear transformation to replace s in the analog filter with $s = \frac{2}{T} \frac{1-z^{-1}}{1+z^{-1}}$.

bimetal overload device   an overload device that employs a bimetal strip as the actuating element. The bimetal strip consists of two metals bonded together. When heated, the bimetal strip will bend due to the different coefficients of linear expansion of the two metals. The bending operates a set of contacts that automatically removes the affected load from the source of electrical power. *See also* overload heater, overload relay.

bimodal histogram   a histogram with two main groupings of values, such as the sum of two displaced Gaussians. *See* histogram.

binary   (1) a signal or other information item that has two possible states.

(2) representation of quantities in base 2.

binary code   a code, usually for error control, in which the fundamental information symbols which the codewords consist of are two-valued or binary and these symbols are usually denoted by either "1" or "0," Mathematical operations for such codes are defined over the finite or Galois field consisting of two elements denoted by GF(2). The mathematical operations for such a Galois field are addition and multiplication. For addition over GF(2) one finds that

$$1 + 1 = 0, \; 1 + 0 = 1 \quad \text{and} \quad 0 + 0 = 0.$$

For multiplication over GF(2) one finds that

$$1 \cdot 1 = 1, \; 1 \cdot 0 = 0 \quad \text{and} \quad 0 \cdot 0 = 0.$$

See also block coding, convolutional coding, error control coding.

binary-coded decimal (BCD)   a weighted code using patterns of four bits to represent each decimal position of a number. Decimal digits 0 to 9, encoded by their four-bit binary representation. Thus: $0 = 0000, 1 = 0001, 2 = 0010, 3 = 0011, 4 = 0100, 5 = 0101, 6 = 0110, 7 = 0111, 8 = 1000, 9 = 1001$.

binary erase channel   a channel where an error detecting circuit is used and the erroneous data is rejected as erasure asking for retransmission. The inputs are binary and the outputs are ternary, i.e.,

0, 1 and erasure. Used for ARQ (automatic request for retransmission) type data communication.

binary hypothesis testing a special two-hypothesis case of the M-ary hypothesis testing problem. The problem is to assess the relative likelihoods of two hypotheses H_1, H_2, normally given prior statistics $P(H_1)$, $P(H_2)$, and given observations $\mathbf{y}$ whose dependence $p(\mathbf{y} \mid H_1)$, $p(\mathbf{y} \mid H_2)$ on the hypotheses is known. The receiver operating characteristic is an effective means to visualize the possible decision rules. m-ary hypothesis testing. *See* conditional statistic, a priori statistics, a posteriori statistics, Neyman-Pearson detector, receiver operating characteristic.

binary image an image whose pixels can have only two values, 0 or 1 (i.e., "off" or "on"). The set of pixels having value 1 ("on") is called the figure or foreground, while the set of pixels having value 0 ("off") is called the background.

binary image coding compression of two-level (black/white) images, typically documents. Bilevel coding is usually lossless and exploits spatial homogeneity by runlength, relative address, quadtree, or chain coding. *Also called* bilevel image coding.

binary notation *See* binary.

binary operator any mathematical operator that requires two data elements with which to perform the operation. Addition and Logical-AND are examples of binary operators; in contrast, negative signs and Logical-NOT are examples of unary operators.

binary optics optical filters constructed with only two amplitude or two phase values to perform the functions of bulk optical components such as lenses.

binary phase frequency modulation converting signals from a binary-digit pattern [pulse form] to a continuous wave form. FM is superseded by

MFM (modified frequency modulation) is an encoding method used in floppy disk drives and older hard drives. A competing scheme, known as RLL (run length limited), produces faster data access speeds and can increase a disk's storage capacity by up to 50 percent. MFM is superseded by RLL, which is used on most newer hard drives.

binary phase grating a diffraction grating where alternating grating lines that alter the optical phase by $180°$ more than neighboring lines.

binary signal a signal that can only have two values: off and on, low and high, or zero and one.

binary symmetric channel the binary-input, binary-output symmetric channel, where the channel noise and other disturbances cause statistically independent errors in the transmitted binary sequence with average probability. The channel is memoryless.

binary tree recursively defined as a set of nodes $(n_1, \ldots n_k)$ one of which is designated the root and the remaining $k - 1$ nodes form at most two sub-trees.

binary tree predictive coding predictive image coding scheme in which pixels are ordered in a pyramid of increasingly dense meshes. The sparsest mesh consists of subsamples of the original image on a widely-spaced square lattice; succeeding meshes consist of the pixels at the centers of the squares (or diamonds) formed by all preceding meshes. Each mesh has twice the number of pixels as its predecessor. Pixel values are predicted by non-linear adaptive interpolation from surrounding points in preceding meshes. The prediction errors, or differences, are quantized, ordered into a binary tree to provide efficient coding of zeros, and are then entropy coded.

binaural attribute   psychoacoustic effects (e.g., cocktail-party effect) that depend on the fact that we have two ears.

binocular imaging   the formation of two images of a scene from two different positions so that binocular vision can be performed, in a similar manner to the way humans deploy two eyes.

binocular vision   the use of two images of a scene, taken (often simultaneously) from two different positions, to estimate depth of various point features, once correspondences between pairs of image features have been established.

binomial coefficients   the coefficients of the polynomial resulting from the expansion of $(a + b)^n$. These coefficients are equal to

$$\binom{n}{k} \frac{n!}{k!\,(n-k)!}$$

where n is the order of the polynomial and k is the index of the coefficient. The kth coefficient is multiplied by the term $a^k b^{n-k}$.

binomial distribution   the binomial distribution is the distribution of a random variable Y that is the sum of n random variables that are Bernoulli distributed.

$$Y = X_1 + X_2 + \cdots + X_n.$$

The probability mass function of such a Y is

$$p_y(k) = \binom{n}{k} p^k (1 - p^{n-k})$$

where p is the parameter of the Bernoulli distribution of any X_i.

bioanalytical sensor   a special case of a chemical sensor for determining the amount of a biochemical substance. This type of sensor usually makes use of one of the following types of biochemical reactions: enzyme-substrate, antigen-antibody, or ligand-receptor.

bioluminescence   *See* luminescence.

biomass   General term used for wood, wood wastes, sewage, cultivated herbaceous and other energy crops, and animal wastes.

biomedical sensor   a device for interfacing an instrumentation system with a biological system such as biological specimen or an entire organism. The device serves the function of detecting and measuring in a quantitative fashion a physiological property of the biological system.

biometric verifier   device that helps authenticate by measuring human characteristics.

biorthogonal filter bank   a filter bank that satisfies the perfect reconstruction condition, i.e., the product of the polyphase transfer function of the analysis and synthesis filters is a pure delay. In general, the analysis and synthesis filters are different, as opposed to the situation for an orthogonal filter bank.

biorthogonal wavelet   a generalization of orthogonal wavelet bases, where two dual basis functions span two sets of scaling spaces, V_j and $\hat{V}_j$, and two sets of wavelet spaces, W_j and $\hat{W}_j$, with each scaling space orthogonal to the dual wavelet space, *i.e.* $V_j \perp \hat{W}_j$ and $\hat{V}_j \perp W_j$ where "$\perp$" represents "orthogonal to." *See* biorthogonal filter bank.

BIOS   *See* basic input–output system.

bipolar   (1) a type of transistor that uses both polarities of carriers (electrons and holes) in its operation as a junction transistor.
(2) a type of data encoding that uses both positive and negative voltage excursions.

bipolar device   *See* bipolar.

bipolar junction transistor (BJT)   a three-terminal nonlinear device composed of two

67

bipolar junctions (collector-base, base-emitter) in close proximity. In normal operation, the voltage between base and emitter terminals is used to control the emitter current. The collector current either equals this (with BC junction in reverse bias), or goes into saturation (the BC junction goes into forward bias). Used for medium power (700 A) and medium speed (10 kHz) applications.

In power electronics applications, BJTs are typically operated as switches, in either their fully on or off states, to minimize losses. The base current flowing into the middle of the device controls the on–off state, where continuous base current is required to be in the on state. A disadvantage is the low current gain.

The base current is generally much smaller than collector and emitter currents, but not negligible as in MOSFETs.

bipolar memory memory in which a storage cell is constructed from bipolar junction transistors. *See also* static random access memory (SRAM).

bipolar neuron a neuron with a signal between -1 and $+1$.

bipolar transistor *See* bipolar junction transistor.

bipole DC system with two conductors, one positive and the other negative polarity. The rated voltage of a bipole is expressed as ±100 kV, for example.

biquad an active filter whose transfer function comprises a ratio of second-order numerator and denominator polynomials in the frequency variable.

biquadratic transfer function a rational function that comprises a ratio of second-order numerator and denominator polynomials in the frequency variable.

bird's beak feature seen in cross-sectional photomicrographs of silicon gate transistors caused by encroachment of oxide under the gate.

birefringence The property of certain materials to display different values of the refractive index for different polarizations of a light beam.

birefringent fiber optical fiber that has different speeds of propagation for light launched along its (two distinct) polarization axes.

birefringent material material that can be described by two or more refractive indices along the directions for principle axes.

birthmark a stamp on a wooden utility pole which denotes its manufacturer, date of manufacture, size, and method of preservation.

bispectra computation of the frequency distribution of the EEG exhibiting nonlinear behavior.

bispectrum the Fourier transform of the triple correlation function. It preserves phase information and uniquely represents a given process in the frequency domain. It can be used to identify different types of nonlinear system response.

BIST *See* built-in self-test.

bistable pertaining to a device with two stable states. Examples: bistable multivibrator, flip-flop. *See also* bistable system.

bistable device *See* bistable.

bistable optical device a device whose optical transmission can take on two possible values.

bistable system an optical system where the transmission can take on two possible values. *See also* bistable.

bistatic scattering a measure of the reradiated power (back-scattered) from an illuminated target in the direction other than that of the illuminating source.

bit   (1) the fundamental unit of information representation in a computer, short for "binary digit" and with two values usually represented by "0" and "1." Bits are usually aggregated into "bytes" (7 or 8 bits) or "words" (12–60 bits).

A single bit within a word may represent the coefficient of a power of 2 (in numbers), a logical TRUE/FALSE quantity (masks and Boolean quantities), or part of a character or other compound quantity. In practice, these uses are often confused and interchanged.

(2) in Information Theory, the unit of information. If an event E occurs with a probability $P(E)$, it conveys information of $\log_2(1/P(E))$ binary units or bits. When a bit (binary digit) has equiprobable 0 and 1 values, it conveys exactly 1.0 bit (binary unit) of information; the average information is usually less than this.

bit allocation   the allocation of bits to symbols with the aim of achieving some compression of the data. Not all symbols occur with the same frequency. Bit allocation attempts to represent frequently occurring symbols with fewer bits and assign more bits to symbols that rarely appear, subject to a constraint on the total number of bits available. In this way, the average string requires fewer bits. The chosen assignment of bits is usually the one that minimizes the corresponding average coding distortion of the source over all possible bit assignments that satisfy the given constraint. Typically sub-sources with larger variances or energy are allocated more bits, corresponding to their greater importance. *See* transform coding.

bit energy   the energy contained in an information-bearing signal received at a communications receiver per information bit. The power of an information bearing signal at a communications receiver divided by the information bit rate of the signal. Usually denoted by E_b as in the signal to noise ratio E_b/N_0.

bit error rate (BER)   the probability of a single transmitted bit being incorrectly determined upon reception.

bit line   used in, for example, RAM memory devices (dynamic and static) to connect all memory cell outputs of one column together using a shared signal line. In static RAM, the "bit" line together with its complemented signal "-bit" feeds a "sense amplifier" (differential in this case) at the bottom of the column serving as a driver to the output stage. The actual cell driving the bit line (and -bit) is controlled via an access transistor in each cell. This transistor is turned on/off by a "word" line, a signal run across the cells in each row.

bit parallel   a method to transmit or process information in which several bits are transmitted in parallel. Examples: a bit parallel adder with 4-bit data has 8 input ports for them (plus an initial carry bit); an 8-bit parallel port includes true 8-bit bi-directional data lines. (MP)

bit per second (bps)   measure of transfer rate of a modem or a bus or any digital communication support. (*See also* baud and baud rate. bps and baud are not equivalent since bps is a low-level measure and media; thus, it includes the number of bits sent for the low-level protocol, while baud is typically referred to a higher level of transmission).

bit period   the time between successive bits in data transmission or data recording. At the transmitter (or recorder) the timing is established by a clock. At the receiver (or reader) an equivalent clock must be recovered from the bit stream.

bit plane   the binary $N \times N$ image formed by selecting the same bit position of the pixels when the pixels of an $N \times N$ image are represented using k bits.

bit plane encoding   lossless binary encoding of the bit planes is termed bit plane encoding.

69

The image is decomposed into a set of k, $N \times N$ bit planes from the least significant bit to $k - 1$ most significant bits and then encoded for image compression.

bit rate   a measure of signaling speed; the number of bits transmitted per second. Bit rate and baud are related but not identical. Bit rate is equal to baud times the number of bits used to represent a line state. For example, if there are sixteen line states, each line state encodes four bits, and the bit rate is thus four times the baud. *See also* baud.

bit serial   processing of one bit per clock cycle. If word length is W, then one sample or word is processed in W clock cycles. In contrast, all W bits of a word are processed in the same clock cycle in a bit-parallel system.

For example: a bit serial adder with 4-bit data has one input signal for each of data them, one bit for carry-in, and two 4-bit shift registers for data.

bit-line capacitance   the equivalent capacitance experienced in each "bit line" in a RAM or ROM device. *See also* bit line.

bitmapped image   a digital image composed of pixels. Bitmapped images are resolution-dependent, i.e., if the image is stretched, the resolution changes. Also called a raster image. image, pixel, vector image.

bit-oriented block transfer (bitBLT)   a type of processing used mainly for video information characterized by minimal operations performed on large data blocks; a processor designed for such operations. bitBLT operations include transfers, masking, exclusive-OR, and similar logical functions.

bitBLT   *See* bit-oriented block transfer.

bits per pixel   the number of bits used to represent each pixel in a digital image. Typical grayscale images have 8 bits per pixel, giving 256 different gray levels. True color images have 24 bits per pixel, or 8 bits for each of the red, green and blue pixels. Compressed image sizes are often represented in bits per pixel, i.e. the total number of bits used to represent the compressed image divided by the total number of pixels.

bit-serial system   a system that uses bit serial data transfer.

bit-slice processor   a processor organization that performs separate computations (via multiple processing units) separately upon subsections of an incoming channel.

BJT   *See* bipolar junction transistor.

black burst   a TV black video signal containing horizontal and vertical sync, color burst, and setup (i.e., a composite video black signal). Black burst is also called "color black." A black burst signal is often used in the video studio to provide synchronizing pulses.

black level   the portion of the video signal pertaining to the lower luminance (brightness) levels.

black start   the task of re-starting an isolated power system which is completely de-energized. Most generating plants require substantial external electric power to start. Thus a black start may be initiated by hand-starting gas turbine generators or by opening the gates of a hydroelectric generator somewhere in the system.

blackbody   theoretically contrived object that gives rise to the so-called "black body radiation." One might imagine a closed surface object (say of metal) possessing one opening that connects the interior surface with the outside world. When the object is heated, the opening becomes a perfect "black" radiator. Such radiation depends on temperature only.

blackout   total loss of power to the entire power system.

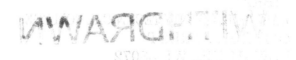

blanket   an insulating rubber mat which is fitted temporarily over energized conductors to protect nearby workers.

blanking   the electronic control circuitry that blanks the television raster during horizontal and vertical retrace.

blanking time   the short time interval when both switches in a leg of an inverter bridge must be off in order to prevent short circuiting the DC input. This is necessary because non-ideal switches cannot turn on and off instantaneously. Thus, after one switch is turned off in an inverter leg, the complimentary switch is not turned on until the designated blanking time has elapsed.

blind deconvolution   the recovery of a signal $x[n]$ from $y[n]$ — the convolution of the signal with an unkown system $h[n]$:

$$y[n] = h[n] * x[n].$$

Occasionally some knowledge of $h[n]$ is available (e.g., that it is a high-pass or low-pass filter). Frequently, detailed knowledge is available about the structure of x. *See* convolution.

blind via   a via connected to either the preliminary side or secondary side and one or more internal layers of a multilayer packaging and interconnecting structure.

blink   in computer display systems, a technique in which a pixel is alternatively turned on and off.

Bloch vector   a set of linear combinations of density matrix elements, written in vector form, that can often be related to specific observables in a quantum mechanical system. For example, in two-level systems the Bloch vector components are $2Re(\rho_{12})$, $2Im(\rho_{12})$, and $\rho_{11} - \rho_{22}$, which are related to nonlinear refractive index, absorption, and population differences, respectively. The time evolution of two-level systems can be described in terms of rotations of the Bloch vector.

block   a group of sequential locations held as one unit in a cache and selected as whole. Also called a line. *See also* memory block.

block cipher   an encryption system in which a successive number of fundamental plaintext information symbols, usually termed a block of plaintext information, are encrypted according to the encryption key. All information blocks are encrypted in the same manner according to the transformation determined by the encryption key. This implies that two identical blocks of plaintext information will always result in the same ciphertext when a particular block cipher is employed for encryption. *See also* encryption, stream cipher.

block code   a mapping of k input binary symbols into n output symbols.

block coding   (1) an error control coding technique in which a number of information symbols, and blocks, are protected against transmission errors by adding additional redundant symbols. The additional symbols are usually calculated according to a mathematical transformation based on the so-called generator polynomial of the code. A block code is typically characterized by the parameters (n, k), where k is the number of information symbols per data word, and n the final number of symbols in the code word after the addition of parity symbols or redundant symbols. The rate of a block code is given by k/n.

Typically, the lower the rate of a code the greater the number of errors detectable and correctable by the code. Block codes in which the block of information symbols and parity symbols are readily discernable are known as systematic block codes. The receiver uses the parity symbols to determine whether any of the symbols were received in error and either attempts to correct errors or requests a retransmission of the information. *See also* automatic repeat request, binary code, convolutional coding, error control coding.

(2) refers to (channel) coding schemes in which the input stream of information symbols

is split into nonoverlapping blocks which then are mapped into blocks of encoded symbols (codewords). The mapping only depends on the current message block. *Compare with* trellis coding.

block diagram a diagrammatic representation of system components and their interconnections. In elementary linear systems, the blocks are often defined by transfer functions or state space equations while the interconnecting signals are given as Laplace transformations. Although the system blocks and signals have the same mathematical form, the blocks represent operators that act on the incoming signals while the signals represent functions of time.

block matching the process of finding the closest match between a block of samples in a signal and a block of equal size in another signal (or a different part of the same signal) over a certain search range. Closeness is measured by correlation or an error metric such as mean square error. Used in data compression, motion estimation, vector quantization, and template matching schemes.

block multiplexer channel an I/O channel can be assigned to more than one data transfer at a time. It always transfers information in blocks, with the channel released for competing transfers at the end of a block. *See also* byte multiplexer channel, selector channel.

block transfer the transmission of a significantly larger quantity of data than the minimum size an interconnect is capable of transmitting, without sending the data as a number of small independent transmissions (the goal being to reduce arbitration and address overhead).

block transform a transform that divides the image into several blocks and treats each block as an independent image. The transform is then applied to each block independently. This occurs in the JPEG standard image compression algorithm, where an image is divided into 8×8 blocks and the DCT is applied independently to each block. Usually the blocks do not overlap each other: that is, they have no signal samples in common. *See* transform matrix, transform coding, lapped orthogonal transform.

block truncation coding (BTC) technique whereby an image is segmented into $n \times n$ nonoverlaping blocks of pixels, and a two-level quantizer is designed for each block. Encoding is essentially a local binarization process consisting of a $n \times n$ bit map indicating the reconstruction level associated with each pixel. Decoding is a simple process of associating the reconstructed value at each pixel as per the bit map.

block carry lookahead adder (BCLA) an adder that uses two levels of carry lookahead logic.

block-diagram simulator a simulator that allows the user to simulate systems as a combination of block diagrams, each of which performs a specific function. Each function is described using a mathematical equation or a transfer function.

blocked state *See* blocking.

blocked-rotor current *See* locked-rotor current.

blocked-rotor test an induction motor test conducted with the shaft held so it cannot rotate. Typically about 25% of rated voltage is applied, often at reduced frequency and the current is measured. The results are used to determine the winding impedances referred to the stator.

blocking state entered if a new user finds all channels or access mechanisms busy and hence is denied service. Generally accompanied by a busy signal. The call blocking probability may be given by the Erlang B or Erlang C formula. *See also* adequate service, multiple access interference.

blocking artifact the visibility in an image of rectangular subimages or blocks after certain types of image processing. *Also called* blocking effect distortion.

blocks world   a visual domain, typical of early studies on machine vision, in which objects are light, plane-faced solids over a dark background.

blooming   an area of the target that is unstable due to insufficient beam current. The area normally appears as a white puddle without definition. Insufficient beam currently may be the result of low beam control setting.

blow up   a relatively sudden and usually catastrophic increase in beam size generally caused by some magnetic field error driving the beam to resonance.

Blumlein   a water-filled transmission line that serves as a pulse generator using a wave propagation principle. The line is folded over on itself and is capable of voltage doubling across its load due to having initially both sides of the load on high potential.

Blumlein bridge   an AC bridge, two arms of which are two serially connected tightly coupled inductive coils. The point of connection of these coils is usually grounded, and the coupling is arranged in such a way that for the currents simultaneously entering or leaving the other ends of the coils the voltage drop between the ends is close to zero. If one of the currents is entering and another is leaving, then the voltage drop is essential. This creates a sensitive current-comparing bridge having application in capacitance transducers.

blurring   (1) the defocusing effect produced by the attenuation of high-frequency components, e.g., obtained by local averaging operators, possibly applied directionally (motion blurring).

(2) the broadening of image features, relative to those which would be seen in an ideal image, so that features partly merge into one another, thereby reducing resolution. The effect also applies to 1–D and other types of signal.

BNC connector   "Baby" N connector. Commonly used coaxial connector with both male and female versions used below microwave frequencies.

board   the physical structure that houses multiple chips, and connects them with traces (busses).

board-to-board optical interconnect   optical interconnection in which the source and the detector are connected to electronic elements in two separate boards.

BOB   *See* break-out box.

Bode diagram   *See* Bode plot.

Bode plot   a graphical characterization of the system frequency response: the magnitude of the frequency response $|H(j\omega)|$, $-\infty < \omega < \infty$ in decibels, and the phase angle $\angle H(j\omega)$, $-\infty < \omega < \infty$, are plotted. For example, a system described by the transfer function

$$H(s) = \frac{Y(s)}{F(s)} = \frac{s+1}{(s+2)(s+3)}$$

has the Bode plot shown in the following figure. *See also* frequency response.

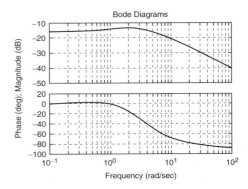

Bode plot.

Bode–Fano criteria a set of rules for determining an upper limit on the bandwidth of an arbitrary matching network.

boiler a steam generator that converts the chemical energy stored in the fuel (coal, gas, etc.) to thermal energy by burning. The heat evaporates the feedwater and generates high-pressure steam.

boiling water reactor a nuclear reactor from which heat is transferred in the form of high-pressure steam.

bolted fault a short circuit fault with no fault resistance. Bolted faults deliver the highest possible fault current for a given location and system configuration, and are used in selecting equipment withstand and interrupting ratings and in the setting of protective relays.

Boltzmann machine in its simplest form, a discrete time Hopfield network that employs stochastic neurons and simulated annealing in its procedure for updating output values. More generally it can have hidden units and be subjected to supervised training so as to learn probabilities of different outputs for each class of inputs.

Boltzmann relation relates the density of particles in one region to that in an adjacent region, with the potential energy between both regions.

bond that which binds two atoms together.

bond pad areas of metallization on the IC die that permit the connection of fine wires or circuit elements to the die.

bonded magnet a type of magnet consisting of powdered permanent magnet material, usually isotropic ceramic ferrite or neodymium-iron-boron, and a polymer binder, typically rubber or epoxy, this magnet material can be molded into complex shapes.

bonding the practice of ensuring a low-resistance path between metallic structures such as water lines, building frames, and cable armor for the purpose of preventing lightning arcs between them.

Boolean an operator or an expression of George Boole's algebra (1847). A Boolean variable or signal can assume only two values: TRUE or FALSE. This concept has been ported in the field of electronic circuits by Claude Shannon (1938). He had the idea to use the Boole's algebra for coding the status of circuit: TRUE/FALSE as HIGH/LOW as CLOSE/OPEN, etc.

Boolean algebra the fundamental algebra at the basis of all computer operations. *See also* the other definitions with Boolean as the first word.

Boolean expression an expression of the Boole's algebra, in which can appear Boolean variables/signals and Boolean operators. Boolean expressions are used for describing the behavior of digital equipments or stating properties/conditions in programs.

Boolean function common designation for a binary function of binary variables.

Boolean logic the set of rules for logical operations on binary numbers.

Boolean operator the classical Boolean operators are AND, OR, NOT. Other operators such as XOR, NAND, NOR, etc., can be easily obtained based on the fundamental ones. In hardware these are implemented with gates, see for example AND gate.

boost converter a circuit configuration in which a transistor is switched by PWM trigger pulses and a diode provides an inductor-current continuation path when the transistor is off. During the transistor on-time, the current builds up in the inductor. During the transistor off-time, the

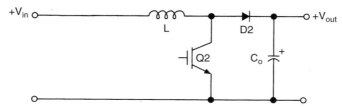

Boost converter.

voltage across the inductor reverses and adds to the input voltage, as a result, the output voltage is greater than the input voltage.

A boost converter can be viewed as a reversed buck converter. The output voltage v_o is related to the input voltage v_i by $v_o = v_i/(1-d)$ and it can be controlled by varying the duty ratio d. Its main application is in regulated DC power supplies and the regenerative braking of DC motors. Also called a step-up converter.

boot   *See* bootstrap.

boot record   structure at the beginning of a hard disk that specifies information needed for the start up and initialization of a computer and its operating system. This record is kept and displayed by the booting program.

bootstrap   (1) a technique using positive feedback to change the effective impedance at a node, for example, to reduce capacitance.

(2) to initialize a computer system into a known beginning state by loading the operating system from a disc or other storage to computer's working memory. This is done by a firmware boot program. Also called boot for short.

boson   an integral spin particle to which Bose-Einstien statistics apply. Such particles do not follow the Pauli exclusion principle. Photons, pions, alpha particles, and nuclei of even mass numbers are examples of bosons.

bottle   slang for glass insulator.

bottom antireflective coating   an antireflective coating placed just below the photoresist to reduce reflections from the substrate.

bottom-up development   an application development methodology that begins creating basic building blocks and uses them to build more complex blocks for higher levels of the system.

bound mode   a type of mode of limited spatial extension. Open waveguides can support, apart from a continuous spectrum, also a few modes, which do not extend up to infinity since they decay exponentially outside of a certain region. In an optical waveguide this is a mode whose field decays monotonically in the direction transverse to propagation and which does not lose power to radiation. Bound modes can also be interpreted in terms of guided rays and total internal reflection. Note: Except in a monomode fiber, the power in bound modes is predominantly contained in the core of the fiber. *See also* continuous spectrum.

boundary   a curve that separates two sets of points.

boundary bus   one of a set of buses which define the boundary between the portion of a power system to be analyzed and the rest of the system. Boundary buses are connected to both the internal and external systems.

boundary condition   (1) the conditions satisfied by a function at the boundary of its interval of definition. They are generally distinguished in hard or soft also called Neumann (the normal derivative of the function is equal to zero) or Dirichlet (the function itself is equal to zero).

(2) the conditions satisfied from the electromagnetic field at the boundary between two different media.

(3) rules that govern the behavior of electromagnetic fields as they move from one medium into another medium.

boundary layer a method of smoothing out a discontinuous controller or a sliding mode state estimator. For example, a boundary layer version of the discontinuous controller

$$u = -Us(e)/|s(e)| = -U\,\mathrm{sign}(s(e)),$$

where e is the control error and s is a function of e, may have the form

$$u = \begin{cases} -U\,\mathrm{sign}(s(e)) & \text{if } |s(e)| > \nu \\ -Us(e)/\nu & \text{if } |s(e)| \leq \nu, \end{cases}$$

where $\nu > 0$ is called the boundary layer width.

boundary layer controller *See* boundary layer.

boundary layer observer *See* boundary layer state estimator.

boundary layer state estimator a continuous version of a sliding mode type state estimator, that is, a sliding mode type state estimator in which the right-hand side of the differential equation describing the estimator is continuous due to the introduction of a boundary layer to smooth out the discontinuous part of the estimator's dynamics.

boundary scan a technique for applying scan design concepts to control/observe values of the signal pins of IC components by providing a dedicated boundary-scan register cell for each signal I/O pin.

boundary scan interface a serial clocked interface used to shift in test pattern or test instruction and to shift out test responses in the test mode. Boundary scan interface comprises shift-in, shift-out, clock, reset, and test select mode signals.

boundary scan path a technique that uses a standard serial test interface to assure easy access to chip or board test facilities such as test registers (in an external or internal scan paths) or local BIST. In particular it assures complete controllability and observability of all chip pins via shift in and shift out operations.

boundary scan test a technique for applying scan design concepts to control/observe values of signal pins of IC components by providing a dedicated boundary-scan register cell for each signal I/O pin.

boundary value problem a mathematical problem in which the unknown is a solution to a partial differential equation and is subject to a set of boundary conditions on the problem domain.

boundary values of 2-D general model let $x_{i,j}$ be a solution (semistate vector) to the 2-D generalized model

$$\begin{aligned} x_{i+1,j+1} = A_0 x_{i,j} &+ A_1 x_{i+1,j} \\ &+ A_2 x_{i,j+1} + B_0 u_{i,j} \\ &+ B_1 u_{i+1,j} + B_2 u_{i,j+1} \end{aligned}$$

$i, j \in Z_+$ (the set of nonnegative integers) where $u_{i,j} \in R^m$ is the input and A_k, B_k ($k = 0, 1, 2$) are real matrices of the model. The vectors $x_{i,j} \in R^n$ whose indices lie on the boundary of the rectangle $[0, N_1] \times [0, N_2]$, i.e., $x_{i,0}, x_{i,N_2}$ for $1 \leq i \leq N_1$ and $x_{0,j}, x_{N_1,j}$ for $0 \leq j \leq N_2$, are called boundary values of the solution $x_{i,j}$ to the 2-D general model. The boundary values may be also given in other ways.

boundary-element method (BEM) a numerical method (integral equation technique) well suited to problems involving structures in which the dielectric constant does not vary with space.

bounded control   *See* saturating control.

bounded distance decoding   decoding of an imperfect t-error correcting forward error correction block code in which the corrected error patterns are limited to those with t or fewer errors, even though it would be possible to correct some patterns with more than t errors.

bounded function   a function $x \in \mathcal{X}_e$ is said to be bounded if it belongs also to the original (unextended) space $\mathcal{X}$, where $\mathcal{X}$ is a space of functions with its corresponding extension $\mathcal{X}_e$. *See also* extended space and truncation.

bounded state   an equilibrium state x_e of a dynamic system is said to be bounded if there exists a real number $B = B(x_0, t_0)$, where x_0 and t_0 represent the initial values of the state and time, respectively, such that

$$\| x(t) \| < B \qquad \forall t \geq t_0$$

See also stable state.

bounded-input bounded-output (BIBO)   a signal that has a certain value at a certain instant in time, and this value does not equal infinity at any given instant of time. A bounded output is the signal resulting from applying the bounded-input signal to a stable system. See diagram below.

Bounded-input bounded-output system.

bounded-input bounded-output stability   a linear dynamic system where a bounded input yields a bounded zero-state response. More precisely, let be a bounded-input with as the least upper bound (i.e., there is a fixed finite constant such that for every t or k), if there exists a scalar such that for every t (or k), the output satisfies, then the system is said to be bounded-input bounded-output stable.

bounded-input bounded-state (BIBS) stability   if for every bounded input (*See* BIBO stability), and for arbitrary initial conditions, there exists a scalar such that the resultant state satisfies, then the system is said to be bounded-input bounded-state stable.

bounds fault   a memory management error that occurs when an offset requested in a memory object exceeds the object's size.

Boyle macromodel   A SPICE computer model for an op amp. Developed by G. R. Boyle in 1974.

Boys camera   a rotating camera used to photograph lightning and establish the multiplicity of individual flashes in a lightning stroke.

BPI   bits per inch.

bps   *See* bit per second.

Bragg angle   the required angle of incidence for light into a Bragg cell to produce a single diffraction order of maximum intensity. The sine of the Bragg angle is approximately the light wavelength divided by the grating.

Bragg cell   an acousto-optic cell designed where only a single diffraction order is produced, generally by making the acoustic column thick along the light propagation direction.

Bragg cell radiometer   similar to an acousto-optic spectrum analyzer in the Bragg mode, but with generally much longer photo-integration times such as via a long integration time photo detector array.

Bragg diffraction   the interaction of light with a thick grating or acoustic wave, producing a single diffraction order with maximum intensity.

Bragg diffraction regime   regime where the acoustic beam width is sufficiently wide to produce only two diffracted beams, i.e., the undiffracted main beam (also called the zero order or DC beam), and the principal diffracted beam.

Bragg scattering   the scattering of light from a periodically varying refractive index variation in a thick medium, so-called by analogy to the Bragg scattering of X-rays from the atomic arrays in a crystal. For instance, an acousto-optic modulator can be said to operate in the Bragg regime or alternatively in the Raman–Nath regime. *See also* Raman–Nath diffraction regime.

braking   operating condition in an electric motor in which the torque developed between the stator and rotor coils opposes the direction of rotation of the rotor. Typical braking methods in DC machines include "plugging" in which the polarity of either the field or the armature coil, but not both, is reversed while the rotor is turning, "dynamic braking" in which generator action in the armature is used to dissipate rotor energy through a braking resistor, and "regenerative braking" in which generator action in the rotor is used to dissipate rotor energy by returning electric power to the power source as the rotor slows. Typical braking methods in AC machines include switching of the phase sequence of the supply voltage, dynamic braking through the armature coils, and varying the frequency of the AC supply voltage. *See also* phase sequence.

braking resistor   resistive elements which can be switched into the electrical system to create additional load in the event of a transient disturbance, thus limiting the generator rotor acceleration such that the system can more readily return to synchronism.

branch address   the address of the instruction to be executed after a branch instruction if the conditions of the branch are satisfied. Also called a branch target address.

branch circuit   the three components of an electrical circuit are source, load, and interconnecting circuit conductors. A branch circuit is an electrical circuit designed to deliver power to the lowest-order load(s) served on a facility. It includes the overcurrent device, circuit conductors, and the load itself.

branch current   the current in a branch of a circuit.

branch history table   a hardware component that holds the branch addresses of previously executed branch instructions. Used to predict the outcome of branch instructions when these instructions are next encountered. Also more accurately called a branch target buffer.

branch instruction   an instruction is used to modify the instruction execution sequence of the CPU. The transfer of control to another sequence of instructions may be unconditional or conditional based on the result of a previous instruction. In the latter case, if the condition is not satisfied, the transfer of control will be to the next instruction in sequence. It is equivalent to a jump instruction, although the range of the transfer may be limited in a branch instruction compared to the jump. *See also* jump instruction.

branch line coupler   coupler comprised of four transmission lines, each of 90° electrical length, arranged in a cascaded configuration with the end of the last transmission line section connected to the beginning of the first transmission line to form a closed path. The input, coupled, direct, and isolated ports are located at the connection point of one transmission line with the next one.

branch penalty   the delay in a pipeline after a branch instruction when instructions in the pipeline must be cleared from the pipeline and other instructions fetched. Occurs because instructions are fetched into the pipeline one after the other and before the outcome of branch instructions are known.

branch prediction   a mechanism used to predict the outcome of branch instructions prior to their execution.

branch relation   the relationship between voltage and current for electrical components. Common branch relations are Ohm's Law and the lumped equations for capacitors and inductors. More complex branch relationships would be transistor models.

branch target buffer (BTB)   a buffer that is used to hold the history of previous branch paths taken during the execution of individual branch instructions. The BTB is used to improve prediction of the correct branch path whenever a branch instruction is encountered.

The *branch target buffer* or *branch target cache* contains the address of each recent branch instruction (or the instructions themselves), the address of the branch "target" and a record of recent branch directions. The Pentium BTB is organized as an associative cache memory, with the address of the branch instruction as a tag; it stores the most recent destination address plus a two-bit history field representing the recent history of the instruction.

branch target cache   *See* branch target buffer.

branch voltage   the voltage across a branch of a circuit.

Branly, Edouard Eugene   (1844–1940) Born: Amiens, France

Branly is best known for his work in wireless telegraphy. Branly invented the coherer, a detection device for radio waves. Branly did much theoretical work in electrostatics, electrodynamics, and magnetism. He did not, however, develop the practical side of his work, hence Marconi and Braun received the Nobel Prize for work Branly had pioneered.

Brattain, Walter   (1902–1987) Born: Amoy, China

Best known as one of the developers of the transistor. In 1956 Brattain, along with John Bardeen and William Shockley, received the Nobel Prize for their development of the point-contact transistor. It was Brattain who, along with Bardeen, observed the significant increase in power output from a metal contact resulting from a small increase in current applied through a second contact attached to the same germanium surface. This research led to the development of integrated circuits.

Braun, Karl Ferdinand   (1850–1918) Born: Fulda, Germany

Best known for his invention of the oscilloscope and for improvements to Marconi's telegraph. Braun was to share the Nobel Prize in Physics with Marconi in 1909. Braun held a number of teaching posts throughout Germany. His research resulted in the principle of magnetic coupling, which allowed significant improvements in radio transmission. He discovered crystal rectifiers, which were a significant component in early radio sets.

breadboard   a preliminary, experimental circuit, board, device or group of them. It is built only to investigate, test, analyze, evaluate, validate, determine feasibility, develop technical data, and to demonstrate the technical principles related to a concept, device, circuit, equipment, or system. It is designed in a rough experimental form, only for laboratory use, and without regard to final physical appearance of a product.

breadth-first search   a search strategy for tree or trellis search where processing is performed breadth first, i.e., the processing for the entire breadth of the tree/trellis is completed before starting the processing for the next step forward.

break frequency   the critical frequency in a frequency - dependent response: especially that frequency which may separate two modes of the response, e.g. the frequency that defines where the low frequency region ends and the midband response begins.

break point   *See* breakpoint.

break-out box (BOB)   a testing device that allows the designer to switch, cross, and tie interface leads. It often has LEDs to permit monitoring of the leads. Typical use is for RS-232 interfaces.

breakaway points of the root loci   breakaway points on the root loci correspond to multiple-order roots of the equation.

breakaway torque   minimum torque needed to begin rotating a stationary load. Breakaway torque represents the absolute minimum starting torque specification for a motor used to drive the load.

breakdown   as applied to insulation (including air), the failure of an insulator or insulating region to prevent conduction, typically because of high voltage.

breakdown strength   voltage gradient at which the molecules of medium break down to allow passage of damaging levels of electric current.

breakdown torque   maximum torque that can be developed by a motor operating at rated voltage and frequency without experiencing a significant and abrupt change in speed. Sometimes also called the stall torque or pull-out torque.

breakdown voltage   the reverse biased voltage across a device at which the current begins to dramatically deviate and increase relative to the current previously observed at lower voltages close to the breakdown voltage. This effect is attributed to avalanche or zener breakdown. It is usually specified at a predetermined value of current.

In a diode, applying a voltage greater than the breakdown voltage causes the diode to operate in the reverse breakdown region.

breakpoint   (1) an instruction address at which a debugger is instructed to suspend the execution of a program.

(2) a critical point in a program, at which execution can be conditionally stopped to allow examination if the program variables contain the correct values and/or other manipulation of data. Breakpoint techniques are often used in modern debuggers, which provide nice user interfaces to deal with them. *See also* breakpoint instruction.

breakpoint instruction   a debugging instruction provided through hardware support in most microprocessors. When a program hits a break point, specified actions occur that save the state of the program, and then switch to another program that allows the user to examine the stored state. The user can suspend the execution of a program, examine the registers, stack, and memory, and then resume the program's execution, which is very helpful in a program's debugging.

breath noise   the noise that is commonly produced when talking at the microphone. It is due to breathing.

breeder reactor   a nuclear reactor in which a non-fissile isotopes are converted to fissile isotopes by irradiation. Ideally, such a reactor produces more fissile products than it consumes.

Bremsstrahlung   electromagnetic radiation, usually in the X-ray region of the spectrum produced by electrons in a collision with the nucleus of an atom. Bremsstrahlung radiation is produced in regions of high electric potential such as areas surrounding electrostatic septa and RF cavities. Bremsstrahlung is German for breaking.

Brewster angle   the angle from normal at which there is no reflection at a planar interface between two media. The Brewster angles for perpendicular and parallel polarizations are different. For non-magnetic media, in which the relative permeability is unity, the Brewster angle for perpendicular polarization does not exist.

Brewster mode a bound radiative surface mode when one of the media is a plasma medium and has a positive dielectric function.

Brewster window transmission window oriented at Brewster's angle with respect to an incident light beam; light polarized in the plane of incident experiences no reflection.

bridge a simple device that connects two or more physical local-area networks (LANs). It forwards packets of data from one LAN segment to another without changing it, and the transfer is based on physical addresses only. The separate LAN segments bridged this way must use the same protocol.

bridge balance condition represents the relationship between bridge circuit components when the current in the balance indicator is absent. Most of the technically useful bridges include a regular connection (series, parallel, series-parallel, or parallel-series) of two two-ports. The condition of balance can be reformulated in terms of two-port parameters, so that depending on structure, the sum of two forward transfer parameters or the sum of one forward and another backward transfer parameter is equal to zero.

bridge calibration used in bridge transducer applications. It is achieved connecting two auxiliary circuits to the bridge. One circuit including two resistors and a potentiometer is connected in parallel to the bridge power supply diagonal, and the potentiometer tap and one end of the detector are connected to the same bridge node. Sliding the tap, one can eliminate the bridge offset. Another circuit, usually including a constant and a variable resistor, is connected in series with power supply. This circuit allows one to change the voltage applied to the bridge, and to establish the correspondence between the maximal deflection of the detector and maximum of the physical variable applied to the bridge resistors playing the role of active gauges.

bridge circuit the circuit that includes four lateral impedances, Z_1, Z_2, Z_3, Z_4, a diagonal impedance Z_o, and a voltage source E_g of the output impedance Z_g is an example of the so-called bridge circuit. This and other similar circuits are characterized by the bridge balance condition, which represents a relationship between the bridge elements when the current in the diagonal impedance is absent (in the case shown this condition is $Z_1 Z_3 = Z_2 Z_4$). The bridge circuits find application in instrumentation and transducers.

bridge linearization necessary design concern in transducer application of the bridge circuits. It is achieved by reduction of the bridge sensitivity in the bridges where only one arm is a transducer. Linearization can also be achieved with two transducers providing the signals of opposite signs and connected in the opposite arms of the bridge or using a current source instead of the voltage source as a bridge power supply.

bridge rectifier a full-wave rectifier to convert ac to dc, that contains four rectifying elements for single phase, and six elements for three phase, connected as the arm of a bridge circuit.

bridge sensitivity the ratio of the variation of the voltage or the current through the detector to the variation of the component that causes the disbalance of the bridge circuit.

bridge-controlled multivibrators using switches in a two-operational amplifier or in an amplifier-comparator multivibrator so that the bridge is "rotated" each half of the period, one can obtain control of the oscillation frequency by detuning a resistive bridge. The circuit can be applied in sensors with limited number of access wires.

bridging using bridges for local-area networks.

brightness the perceived luminance or apparent intensity of light. This is often different from the actual (physical) luminance, as demonstrated by brightness constancy, Mach band and simultaneous contrast.

brightness adaptation   the ability of the human visual system (HVS) to shift the narrow range in which it can distinguish different light intensities over a large span of luminances. This permits the overall sensivity of the HVS to gray levels to be very large even though the number of gray levels that it can simulateously differentiate is fairly small. grey level, human visual system (HVS), luminance.

brightness constancy   the perception that an object has the same brightness despite large changes in its illumination. Thus a piece of paper appears to be approximately as white in moonlight as in sunlight, even though the illumination from the sun may be one million times greater than that from the moon. *See* brightness, human visual system (HVS), illumination, simultaneous contrast.

Brillouin flow   a stream of electron beam emitted from an electron gun that is not exposed to a focusing magnetic field.

Brillouin frequency shift   the frequency shift that a wave experiences in undergoing Brillouin scattering. The shift can be to either lower or higher frequency, and typically has a value in the range 0.1 to 10 GHz. *See also* Stokes scattering, anti-Stokes scattering.

Brillouin laser   acoustic maser in which the amplification mechanism is considered to be Brillouin scattering.

Brillouin scattering   the scattering of light from sound waves. Typically in Brillouin scattering the sound waves have frequencies in the range 0.1 to 10 GHz, whereas in acousto-optics the sound waves have frequencies <0.1 GHz. Brillouin scattering can be either spontaneous or stimulated. *See also* acousto-optic effect, spontaneous light scattering, stimulated light scattering.

broadband   a service or system requiring transmission channels capable of supporting bit rates greater than 2 Mbit/s.

broadband antenna   an antenna whose characteristics (such as input impedance, gain, and pattern) remain almost constant over a wide frequency band. Two such types of antennas are the log periodic and the biconical.

broadband emission   an emission having a spectral distribution sufficiently broad in comparison to the response of a measuring receiver.

broadband integrated services digital network (B-ISDN)   a generic term that generally refers to the future network infrastructure that will provide ubiquitous availability of integrated voice, data, imagery, and video services.

broadband system   a broadband communication system is one that employs a high data transmission rate. In radio terminology it implies that the system occupies a wide radio bandwidth.

broadcast   (1) the transfer of data to multiple receiver units simultaneously rather than to just one other subsystem.
(2) a bus-write operation intended to be recognized by more than one attached device.

broadcast channel   a single transmitter, multiple receiver system in which identical information is transmitted to each receiver, possibly over different channels. *See also* interference channel, multiple access channel.

broadcast channel allocations   a frequency of a width prescribed by a nation's communications governing agency that are standardized throughout the country for use in one-way electronic communication.

broadcast picture quality   the acceptable picture performance for NTSC terrestrial telecast signals. A panel of untrained observers subjectively evaluates the NTSC received picture and sound quality as signal impairments are inserted into the broadcast signal. The evaluation scores are used to determine the values for objectionable signal impairment levels. The signal impairments tested are the video and audio signal-to-noise ratios, the interference due to adjacent channel signals, the interference due to co-channel signals, and the echoes (ghosts) caused by multipath signals effects.

broadcasting   sending a message to multiple receivers.

broadside   when the pattern factor is maximum in the H plane (for a dipole antenna along the z axis this is the plane where theta = 90 degrees).

broadside array   an array where the main beam of the array is directed perpendicular to the array axis. In many applications it is desirable to have the maximum radiation of an array directed normal to the axis of the array.

broadside coupled microstrip lines   microstrip lines that share the same ground plane but separated from each other in normal direction to the ground plane. Both the microstrip lines are aligned at their centers along the normal direction to the ground plane.

Brown book   *See* IEEE Color Books.

Brownian motion   a stochastic process with independent and stationary increments. The derivative of such a process is a white noise process. A Brownian motion process X_t is the solution to a stochastic differential equation of the form

$$\frac{dX}{dt} = b(t, X_t) + \sigma(t, X_t) \cdot W_t,$$

where W_t is a white noise process.

brownout   an intentional lowering of utility voltage to reduce loading on the system.

brush   a conductor, usually carbon or a carbon–copper mixture, that makes sliding electrical contact to the rotor of an electrical machine. Brushes are used with sliprings on a synchronous machine to supply the DC field and are used with a commutator on a DC machine.

brush rigging   the components used to hold the brushes of a rotating machine in place, and to insure proper brush tension is applied.

brush tension   the force required on the brushes of a rotating machine to insure proper contact between the brush and the commutator or slipring. Proper brush tension is usually provided by springs, and is specified in the manufacturer's technical manual of the machine.

brushless DC motor   *See* electronically commutated machine.

brushless exciter   *See* rotating-rectifier exciter.

brushless rotary flux compressor   a rotating machine designed to deliver pulsed output (1 MJ in 100 μs). The stator coils are excited by an external capacitor bank. The rotor is a salient structure that compresses the flux resulting in amplification of the electric pulse, by converting the rotating kinetic energy of the rotor to electrical energy.

BSO   abbreviation for bismuth silicon oxide, Bi_4SiO_{20}. A photoconductive insulating crystal that exhibits photorefractive effects. Useful in applications such as multibeam coupling and phase conjugation.

BTB   *See* branch target buffer.

BTC   *See* block truncation coding.

BTDF   *See* bidirectional transmission distribution function.

BTMA   *See* busy tone multiple access. *See also* ISMA.

bubble chamber   an instrument for rendering visible the tracks of ionizing particles. It is characterized by a vessel filled with a superheated transparent liquid, commonly hydrogen or deuterium. The passage of an ionizing particle through this liquid is marked by the appearance of a series of bubbles along the particle trajectory. If the liquid is subjected to a magnetic field, as is usually the case, the charged particle trajectories will be curved, the curvature providing information about the particles' charge and momentum.

buck converter   a transistor is switched by PWM trigger pulses and a diode provides a current continuation path when the transistor is off, thus the input voltage is chopped. A lowpass LC filter is used to attenuate the switching ripple at the output. The input current to a basic buck converter is discontinuous; therefore, in many applications an LC prefilter is applied to reduce EMI. The output voltage v_o is related to the input voltage v_i by $v_o = v_i d$ and it can be controlled by varying the duty ratio d. Isolated version of a buck converter include forward, pushpull, halfbridge, and bridge converters. Also called chopper or step-down converter.

buck-boost converter   *See* buck-boost transformer.

buck-boost transformer   a special purpose 2- or 4-coil transformer used to produce modest increases or decreases in the utilization voltage at a load site. The low-voltage coil(s), which typically have rated voltages of 5% to 15% of the high-voltage coils, and in use, the high- and low-voltage coils are connected in series to produce an autotransformer arrangement. If primary voltage is applied to the high voltage coil and load voltage is taken from the series coil combination, the low-voltage coil adds to, or boosts, the load utilization voltage. Conversely, reductions in load utilization voltage occur when these primary and secondary connections are reversed causing the low-voltage coil to buck the supply voltage. A typical 4-coil buck-boost transformer would have two 120 V primary coils and two 12 V secondary coils, which could be used to produce voltage ratios of (120/132), (120/144), (240/252), and (240/264).

In a basic buck-boost converter, the inductor accumulates energy from the input voltage source when the transistor is on and releases energy to the output when the transistor is off. It can be viewed as a buck converter followed by a boost converter with topologic simplification. In a buck-boost converter, the output voltage v_o is related to the input voltage v_i by $v_o = v_i d/(1 - d)$ and it can be controlled by varying the duty ratio d. Note that the output voltage is opposite polarity to the input. Also called a buck-boost converter, up-down transformer or up-down converter. *See also* flyback converter.

bucket   a stable phase space area where the particle beam may be captured and accelerated. An RF bucket is the stable region in longitudinal phase space. The bucket width gives the maximum phase error or timing error at the RF cavity, which a particle may have and still complete the whole acceleration cycle. The bucket height is the corresponding limit on momentum error.

bucket truck   a motor truck equipped with a shell or bucket at the end of a hydraulically-operated insulated arm. A line worker stands in the bucket and is thus raised to gain access to overhead conductors.

bucking fields   *See* differentially compounded.

buddy memory allocation   a memory allocation system based on variable sized segments will usually allocate space for a new segment from a free area somewhat larger than necessary, leaving an unallocated fragment of the original space. In "buddy" allocation, this fragment cannot be

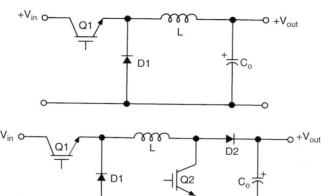

Buck converter.

Buck-boost converter.

used until its adjacent allocated space is released. Buddy allocation reduces memory fragmentation by ensuring that available areas cannot be repeatedly subdivided.

Buff book   *See* IEEE color books.

buffer   a temporary data storage area in memory that compensates for the different speeds at which different elements are transferred within a system. Buffers are used when data transfer rates and/or data processing rates between sender and receiver vary, for instance, a printer buffer, which is necessary because the computer sends data to the printer faster than the data can be physically printed. *See also* buffered input/output.

buffered input/output   input/output that transfers data through a "buffer," or temporary storage area. The main purpose of the buffer is to reduce time dependencies of the data and to decouple input/output from the program execution. Data may be prepared or consumed at an irregular rate, whereas the transfer to or from disk is at a much higher rate, or in a burst.

A buffer is used in "blocked files," where the record size as seen by the user does not match the physical record size of the device.

buffering   (1) the process of moving data into or out of buffers or to use buffers to deal with input/output from devices. *See also* buffer, buffered input/output.

(2) in optics, material surrounding the optical fiber that provides the first layer of protection from physical and environmental damage. The buffering is usually surrounded by one or more layers of jacketing material for additional physical protection of the fiber.

bug   (1) an error in a programmed implementation (may be either hardware or software). Bugs may refer to errors in correctness or performance.

(2) a syntactical or logical error in a computer program. A name attributed to early computers and electronic testing.

built-in logic block observer   technique that combines the basic features of scan designs, pseudo-random test pattern generation, and test result signature analysis.

built-in self-test (BIST)   special hardware embedded into a device (VLSI chip or a board) used to perform self testing. On-line BIST assures testing concurrently with normal operation (e.g., accomplished with coding or duplication techniques).

Off-line BIST suspends normal operation and is carried out using built-in test pattern generator and test response analyzer (e.g., signature analyzer).

bulb generator   a free-standing generator contained in a streamlined, waterproof bulb-shaped enclosure and driven by a water-wheel resembling a ship's propeller on a shaft which extends from one end of the enclosure. They are used in tidal power installations. *See* tidal power.

bulk power   a term inclusive of the generation and transmission portions of the power system.

bulk scattering   scattering at the volume of an inhomogeneous medium, generally also possessing rough boundaries. It is due to inhomogeneities in the refractive index.

bulk substation   a substation located on a high-voltage transmission line which supplies bulk power to a non-generating utility.

bulldog   an attachment for a wire or hoist.

bump   a localized orbit displacement created by vertical or horizontal correction element dipoles used to steer a beam through an available aperture or around obstacles.

bunch   a group of particles captured in a phase space bucket.

bundle   the practice of paralleling several conductors per phase in an overhead transmission line for the purpose of increasing ampacity and decreasing inductive reactance.

bundle spacer   a rigid structure which is used to maintain the spacing of wires in a bundled conductor on an overhead electric power transmission line *See* bundle.

bundled services   utility services which are sold together, like power transmission and distribution services in non-deregulated electric utilities.

buried via   a via connected to neither the primary side nor the secondary side of a multilayer packaging and interconnecting structure, i.e., it connects only internal layers.

burndown   breakage of an overhead electric power line due to heating from excess current.

burnup   a measure (e.g., megawattt-days/ton) of the amount of energy extracted from each unit of fissile material invested in a nuclear reactor.

burn-in   component testing where infant mortality failures (defective or weak parts) are screened out by testing at elevated voltages and temperatures for a specified length of time.

burst refresh   in DRAM, carrying out all required refresh actions in one continuous sequence—a *burst. See also* distributed refresh.

burst transfer   the sending of multiple related transmissions across an interconnect, with only one initialization sequence that takes place at the beginning of the burst.

burstiness factor   used in traffic description, the ratio of the peak bit rate to the average bit rate.

bus   (1) a data path connecting the different subsystems or modules within a computer system. A computer system will usually have more than one bus; each bus will be customized to fit the data transfer needs between the modules that it connects.

(2) a conducting system or supply point, usually of large capacity. May be composed of one or more conductors, which may be wires, cables, or metal bars (busbars).

(3) a node in a power system problem

(4) a heavy conductor, typically used with generating and substation equipment.

bus acquisition   the point at which a bus arbiter grants bus access to a specific requestor.

bus admittance matrix   *See* Y-bus.

bus arbiter   (1) the unit responsible for choosing which subsystem will be given control of the bus when two or more requests for control of the bus happen simultaneously. Some bus architectures, such as Ethernet, do not require a bus arbiter.

(2) the device that performs bus arbitration. *See also* bus arbitration.

bus arbitration   the process of determining which competing bus master should be granted control of the bus. The act of choosing which subsystem will be given control of the bus when two or more requests for control of the bus happen simultaneously. The element that make the decision is usually called the bus arbiter. *See also* bus priority.

bus architecture   a computer system architecture in which one or more buses are used as the communication pathway between I/O device controllers, the CPU, and memory. *See also* channel architecture.

bus bandwidth   (1) the data transfer rate in bits per second or bytes per second. In some instances the bandwidth average rate is given and in others the maximum rate is given. It is approximately equal to the width of the data bus, multiplied by the transfer rate in bus data words per second. Thus a 32 bit data bus, transferring 25 million words per second (40 ns clock) has a bandwidth of 800 Mb/s.

The useful bandwidth may be lowered by the time to first acquire the bus and possibly transfer addresses and control information.

(2) the transfer rate that is guaranteed that no user will exceed.

bus bar   a heavy conductor, typically without insulation and in the form of a bar of rectangular cross-section.

bus broadcast   *See* broadcast.

bus controller   the logic that coordinates the operation of a bus.

A device connected to the bus will issue a bus request when it wishes to use the bus. The controller will arbitrate among the current requests and grant one requester access. The bus controller also monitors possible errors, such as use of an improper address, a device not releasing the bus, and control errors.

Bus control logic may reside in multiple subsystems, distributed control, or may be centralized in a subsystem. *See also* bus cycle, bus master.

bus cycle   the sequence of steps involved in a single bus operation. A complete bus cycle may require that several commands and acknowledgments are sent between the subsystems in addition to the actual data that is sent.

For example,

1. the would-be bus master requests access to the bus
2. the bus controller grants the requester access to the bus as bus master
3. the bus master issues a read command with the read address
4. the *bus slave* responds with data
5. the master acknowledges receipt of the data
6. the bus master releases the bus.

The first two steps may be overlapped with the preceding data transfer.

See also bus controller, bus master.

bus differential relay   a differential relay specifically designed to protect high power buses with multiple inputs.

bus driver   the circuits that transmit a signal across a bus.

bus grant   an output signal from a processor indicating that the processor has relinquished control of the bus to a DMA device.

bus hierarchy   a network of buses linked together (usually multiple smaller buses connected to one or more levels of larger buses), used to increase the number of elements that may be connected to a high-performance bus structure.

bus idle   the condition that exists when the bus is not in use.

bus impedance matrix   *See* Z-bus.

bus interface unit   in modern CPU implementations, the module within the CPU directly responsible for interactions between the CPU and the memory bus.

bus line   one of the wires or conductors that constitute a bus. A bus line may be used for data, address, control, or timing.

bus locking   the action of retaining control of a bus after an operation which would normally release the bus at completion. In the manipulation of memory locks, a memory read must be followed by a write to the same location with a guarantee of no intervening operation. The bus must be locked from the initial read until after the update write to give an indivisible read/write to memory.

bus master   a bus device whose request is granted by the bus controller and thereby gains control of the bus for one or more cycles or transfers. The bus master may always reside with one subsystem, or may be transferred between subsystems, depending on the architecture of the bus control logic. *See also* bus controller, bus cycle.

bus owner   the entity that has exclusive access to a bus at a given time.

bus phase   a term applying especially to synchronous buses, controlled by a central clock, with alternating "address" and "data" transfers. A single transfer operation requires the two phases to transfer first the address and then the associated data. Bus arbitration may be overlapped with preceding operations.

bus priority   rules for deciding the precedence of devices in having bus requests honored.

Devices issue requests on one of several bus request lines, each with a different bus priority. A high priority request then "wins" over a simultaneous request at a lower priority.

The request grant signals then "daisy chains" through successive devices along the bus or is sent directly to devices in appropriate order. The requesting device closest to the bus controller then accepts the grant and blocks its propagation along the bus.

Buses may have handle interrupts and direct memory accesses with separate priority systems.

bus protocol   (1) a set of rules that two parties use to communicate.

(2) the set of rules that define precisely the bus signals that have to be asserted by the master and slave devices in each phase of a bus operation.

bus request   an input signal to a processor that requests access to the bus; a hold signal. Competing bus requests are resolved by the bus controller. *See also* bus controller.

bus slave   a device that responds to a request issued by the bus master. *See also* bus master.

bus snooping   the action of monitoring all traffic on a bus, irrespective of the address. Bus snooping is required where there are several caches with the same or overlapping address ranges. Each cache must then "snoop" on the bus to check for writes to addresses it holds; conflicting addresses may be updated or may be purged from the cache.

Bus snooping is also useful as a diagnostic tool.

bus state triggering   a data acquisition mode initiated when a specific digital code is selected.

bus tenure   the time for which a device has control of the bus, so locking out other requesters. In most buses, the bus priority applies only when a device completes its tenure; even a low priority device should keep its tenure as short as possible to avoid interference with higher priority devices. *See also* bus priority.

bus transaction   the complete sequence of actions in gaining control of a bus, performing some action, and finally releasing the bus. *See also* bus cycle.

bus watching   *See* bus snooping.

bus width   the number of data lines in a given bus interconnect.

bus-connected reactor   *See* shunt reactor.

bus yard   an area of a generating station or substation in which bus bars *cf* and switches are located.

Bush, Vannevar   (1890–1974) Born: Everett, Massachusetts

Best known as the developer of early electromechanical analog computers. His "differential analyzer," as it was called, arose from his position as a professor of power engineering at the Massachusetts Institute of Technology. Transmission problems involved the solution of first- and second-order differential equations. These equations required long and laborious calculations. His interest in mechanical computation arose from this problem. Bush's machines were used by the military during World War II to calculate trajectory tables for artillery. Vannevar Bush was also responsible for inventing the antecedent of our modern electric meter. He was also scientific advisor to President Roosevelt on the Manhattan Project.

bushing   a rigid, hollow cylindrical insulator which surrounds a conductor and which extends through a metal plate such as a the wall of a transformer tank so as to insulate the conductor from the wall.

bushing transformer   a potential transformer which is installed in a transformer bushing so as to take advantage of the insulating qualities of that bushing.

busway   a specialized raceway which holds uninsulated bus bars in a building.

busy tone multiple access (BTMA)   synonym for idle tone multiple access. *See also* idle tone multiple access.

busy waiting   a processor state in which it is reading a lock and finding it busy, so it repeats the read until the lock is available, without attempting to divert to another task. The name derives from the fact that the program is kept busy with this waiting and is not accomplishing anything else while it waits. The entire "busy loop" may be only 2 or 3 instructions.

"Busy waiting" is generally deplored because of the waste of processing facilities.

Butler matrix   a feed system (also called beamforming system), that can excite an antenna array so that it produces several beams, all offset from each other by a finite angle. The system makes use of a number of input ports connected through a combination of hybrid junctions and fixed phase shifters.

Butterworth alignment   a common filter alignment characterized by a maximally flat, monotonic frequency response.

Butterworth filter   an IIR (infinite impulse response) lowpass filter with a squared magnitude of the form:

$$|H(\omega)|^2 = 11 + (j\omega \, j\omega_c{}^{2N}).$$

buzz stick   a tester for insulators, especially strain insulators in a string. It consists of a pair of probes connected to each side of a small sphere gap. When the probes are touched to each terminal of a good insulator, the gap will break down and emit a buzzing sound.

BV$_{GD}$ common notation for FET gate-to-drain reverse breakdown voltage.

BV$_{GS}$ common notation for FET gate-to-source reverse breakdown voltage.

bw common notation for radian bandwith in radians per second.

bw$_a$ common notation for fractional arithmetic mean radian bandwidth in radians per second.

bw$_g$ common notation for fractional geometric mean radian bandwidth in radians per second.

BWO *See* backward wave oscillators.

BX cable a flexible, steel-armored cable used in residential and industrial wiring.

bypass *See* forwarding.

bypass switch a manually-operated switch used to connect load conductors when an automatic transfer switch is disconnected.

byte in most computers, the unit of memory addressing and the smallest quantity directly manipulated by instructions. The term "byte" is of doubtful origin, but was used in some early computers to denote any field within a word (e.g., DEC PDP-10). Since its use on the IBM "Stretch" computer (IBM 7030) and especially the IBM System/360 in the early 1960s, a byte is now generally understood to be 8 bits, although 7 bits is also a possibility.

byte multiplexer channel an I/O channel that can be assigned to more than one data transfer at a time and can be released for another device following each byte transfer. (In this regard, it resembles a typical computer bus.) Byte multiplexing is particularly suited to lower speed devices with minimal device buffering. (IBM terminology) *See also* selector channel, multiplexer channel.

byte serial a method of data transmission where bits are transmitted in parallel as bytes and the bytes are transmitted serially. For example, the Centronics-style printer interface is byte-serial.

C

c   common symbol for speed of light in free space. $c = 3 \times 10^{10}$ cm/s.

C_{GD}   common notation for FET gate-to-drain capacitance.

C_{GS}   common notation for FET gate-to-source capacitance.

C-band   microwave frequency range, 3.95–5.85 Ghz.

C-element   a circuit used in an asynchronous as an interconnect circuit. The function of this circuit is to facilitate the handshaking communication protocol between two functional blocks.

cable   an assembly of insulated conductors, either buried or carried on poles (aerial cable).

cable limiter   a cable connector that contains a fuse. Cable limiters are used to protect individual conductors that are connected in parallel on one phase of a circuit.

cable tray   a specialized form of raceway used to hold insulated electric power cables in a building.

cache   an intermediate memory store having storage capacity and access times somewhere in between the general register set and main memory. The cache is usually invisible to the programmer, and its effectiveness comes from being able to exploit program locality to anticipate memory-access patterns and to hold closer to the CPU: most accesses to main memory can be satisfied by the cache, thus making main memory appear to be faster than it actually is.

A hit occurs when a reference can be satisfied by the cache; otherwise a miss occurs. The proportion of hits (relative to the total number of memory accesses) is the hit ratio of the cache, and the proportion of misses is the miss ratio. *See also* code cache, data cache, direct mapped cache, fully associative cache, set associative cache, and unified cache.

cache aliasing   a situation where two or more entries (typically from different virtual addresses) in a cache correspond to the same address(es) in main memory. Considered undesirable, as it may lead to a lack of consistency (coherence) when data is written back to main memory.

cache block   the number of bytes transferred as one piece when moving data between levels in the cache hierarchy or between main memory and cache. The term *line* is sometimes used instead of block. Typical block size is 16-128 bytes and typical cache size is 1-256 KB. The block size is chosen so as to optimize the relationship of the "cache miss ratio," the cache size, and the block transfer time.

cache coherence   the problem of keeping consistent the values of multiple copies of a single variable, residing either in main memory and cache in a uniprocessor, or in different caches in a multiprocessor computer. In a uniprocessor, the problem may arise if the I/O system reads and writes data into the main memory, causing the main memory and cache data to be inconsistent, or if there is aliasing. Old (stale) data could be output if the CPU has written a newer value in the cache, and this has not been transported to the memory. Also, if the I/O system has input a new value to main memory, new data would reside in main memory, but not in the cache.

cache hit   when the data referenced by the processor is already in the cache.

cache line   a block of data associated with a cache tag.

cache memory   *See* cache.

cache miss   a reference by the processor to a memory location currently not housed in the cache.

cache replacement   when a "cache miss" occurs, the block containing the accessed location must be loaded into the cache. If this is full, an "old" block must be expelled from the cache and replaced by the "new" block. The "cache replacement algorithm" decides which block should be replaced. An example of this is the "Least Recently Used (LRU)" algorithm, which replaces the block that has gone the longest time without being referenced.

cache synonym   *See* cache aliasing.

cache tag   a bit field associated with each block in the cache. It is used to determine where (and if) a referenced block resides in the cache. The tags are typically housed in a separate (and even faster) memory (the "tag directory") which is searched for in each memory reference. In this search, the high order bits of the memory address are associatively compared with the tags to determine the block location. The number of bits used in the tag depends on the cache block "mapping function" used: "Direct-mapped," "Fully associative," or the "Block-set-associative" mapped cache.

CAD   *See* computer-aided design.

cage-rotor induction motor   an induction motor whose rotor is occupied by copper or aluminum bars, known as rotor bars, instead of windings. Also commonly referred to as a squirrel-cage induction motor.

calculating board   a single-phase scale model of a power system that was used to calculate power flows before the advent of electronic computers.

calibration   the procedure of characterizing the equipment in place for a particular measurement set-up relative to some known quantity, usually a calibration standard traceable to the National Institute for Standards and Technology (NIST).

calibration kits   designed for use with vector network analyzers. With these kits you can make error-corrected measurements of devices by measuring known devices (standards) over the frequency range of interest. Calibration standards include shorts, open, sliding, and fixed loads.

calibration standards   a precision device used in the process of calibrating an EM measurement system. It can be a standard gain horn, an open, a short, a load, sphere, etc., used to characterize an RCS, antenna, or transmission line measurement system. Most calibration standards are provided with documentation that can be traced to a set of standards at the NIST.

call instruction   (1) command within a computer program that instructs the computer to go to a subroutine.

(2) an instruction used to enter a subroutine. When a call instruction executes, the current program counter is saved on stack, and the address of the subroutine (provided by the call instruction) is used as the new program counter.

calorimeter   a device used to determine particle energies by measuring the ionization of a particle shower in a heavy metal, usually iron and lead.

CAM   acronym for content-addressable memory or computer-aided manufacturing. *See* associative memory, computer-aided manufacturing.

CAMAC   acronym for computer automated monitor and control — an internationally accepted set of standards for electronic instrumentation, which specifies mechanical, electrical, and functional characteristics of the instrument modules.

camera   a device for acquiring an image, usually in a photographic or electronic form — in the latter case typically as a TV camera. Cameras may operate in optical, infra red or other wavelength bands.

camera calibration   a process in which certain camera parameters, or equivalently some quantities which are required for determination of the perspective projection on an image plane of a point in the 3-D world, are calculated by using the known correspondence between some points in the 3-D world and their images in the image plane.

camera model   (1) the representation of the geometric and physical features of a stereovision system, with relative references between the two camera coordinate systems, and absolute references to a fixed coordinate system.

(2) a mathematical model by which the perspective projection on an image plane of a point in the 3-D world can be determined.

can   slang for a pole-top distribution transformer.

candela (cd)   unit of measurement for luminous intensity (illuminating power in lumens/sr). The luminous intensity of 1/60 of 1 cm^2 of projected area of a blackbody radiator operating at the temperature of solidification of platinum (2046K). Historically, the unit of measurement for the light emitted by one flame of a specified make of candle.

candle   *See* candela.

candle power   *See* candela.

candlepower distribution   a curve, generally polar, representing the variation of luminous

intensity of a lamp or luminaire in a plane through the light center.

canned magnet   a magnet that is completely encased in its own vacuum jacket.

Canny edge detector   *See* Canny Operator.

Canny operator   an edge detector devised by John Canny as the optimal solution to a variational problem with three constraints. The general solution obtained numerically can be approximated in practical contexts by the first derivative of a Gaussian. Canny operator usually refers to the extension to two dimensions of this approximation: i.e., to use of a set of oriented operators whose orthogonal cross sections are a Gaussian and the derivative of a Gaussian. Its advantage is its capability for allowing edges and their orientations to be detected to sub-pixel accuracy. It uses a convolution with a Gaussian to reduce noise and a derivative to enhance edges in the resulting smoothed image. The two are combined into one step — a convolution with the derivative of a Gaussian. A hysteresis thresholding stage is included, to allow closed contours to remain closed. *See* infinite symmetric exponential filter.

CAP   *See* carrierless amplitude/phase modulation.

capability   an object that contains both a pointer to another object and a set of access permissions that specify the modes of access permitted to the associated object from a process that holds the capability.

capability curve   *See* capability diagram.

capability diagram   also called capability curve. Graphical representation of the complex power limits for safe operation of a synchronous machine. The vertical axis is average power P and the horizontal axis is reactive power Q. The region of allowable operation is determined by factors such as rotor thermal limit, stator thermal limit, rated power of prime mover (alternator operation), and stability torque limit.

capability list   a list of capabilities, usually associated with a process, defining a set of objects and the modes of access permitted to those objects. Computer systems have been designed to use capability lists to define the memory environment for process execution.

capacitance   the measure of the electrical size of a capacitor, in units of farads. Thus a capacitor with a large capacitance stores more electrons (coulombs of charge) at a given voltage than one with a smaller capacitance.

In a multiconductor system separated by nonconductive mediums, capacitance (C) is the proportionality constant between the charge (q) on each conductor and the voltage (V) between each conductor. The total equilibrium system charge is zero. Capacitance is dependent on conductor geometry, conductor spatial relationships, and the material properties surrounding the conductors.

Capacitors are constructed as two metal surfaces separated by a nonconducting electrolytic material. When a voltage is applied to the capacitor the electrical charge accumulates in the metals on either side of the nonconducting material, negative charge on one side and positive on the other. If this material is a fluid then the capacitor is electrolytic; otherwise, it is nonelectrolytic.

capacitance bridge   a circuit that includes two branches which form a balanced drive (two sinusoidal voltage sources connected in series with common point grounded) and two capacitances connected in series between free ends of the voltage sources. The detector of current (virtual ground of an operational amplifier is a suitable choice) is connected between the common point of the capacitors and ground. The circuit finds application in capacitive sensors.

capacitive reactance   the opposition offered to the flow of an alternating or pulsating current by capacitance measured in ohms.

capacitively coupled current   *See* capacitively coupled field.

capacitively coupled field   field applied to the affected limb by electrodes touching the skin (the current from the electrodes has both displacement and conduction components).

capacitor bank   (1) an assembly at one location of capacitors and all necessary accessories, such as switching equipment, protective equipment, and controls, required for a complete operating installation.

(2) a group of (typically 3) capacitors mounted on an electric power line for voltage boosting or power factor correction.

capacitor-start induction motor (CSIM)   a single-phase induction motor with a capacitor in series with its auxiliary winding, producing nearly a 90° phase difference between the main winding and the auxiliary winding currents at starting. This results in a high starting torque, so this motor is used for hard-to-start loads. The auxiliary winding and capacitor are removed from the circuit by a centrifugal switch as the machine approaches operating speed.

capacity miss   a category of cache misses denoting the case where the cache is not large enough to hold all blocks needed during execution of a program. *See also* conflict miss, cold start miss.

capacity region   for a multiple terminal communications system. The entire set of rate-vectors for which there exist channel codes such that the probability of making a decoding error can be made arbitrarily small. *See also* achievable rate region, multiple access channel.

capture effect   a phenomenon found in packet switched networks in which nonequal powers in packet radio networks using contention protocols lead to higher throughputs. In contention protocols used in packet radio networks, the transmitted packets are allowed to collide. If two packets collide and one is significantly stronger in power, this packet is more likely to be captured (detected) by the receiver.

capture range　the range of input frequencies over which the PLL can acquire phase lock.

capture register　internal register which, triggered by a specified internal or external signal, store or "capture" the contents of an internal timer or counter.

carbon brush　a block of carbon used to make an electrical contact to a rotating coil via the commutator of a DC machine or the slip rings of a synchronous machine.

carbon dioxide (CO_2)　linear gas molecule consisting of one carbon and two oxygen atoms, medium for an important class of lasers.

carbon dioxide laser　laser in which the ampliﬁying medium is carbon dioxide gas; efficient, powerful, and commercially important laser that is pumped and conﬁgured in many ways and has its principal output lines in the mid-infrared.

carbon resistor thermometer　a carbon resistor whose temperature sensitivity provides good temperature resolution.

carcinotron　a forward radial traveling wave amplifier in which microwave signals are fed to the radial slow wave structure.

card　a printed circuit board that can be plugged into a main board to enhance the functionality or memory of a computer.

card cage　mechanical device for holding circuit cards into a backplane.

cardinal series　the formula by which samples of a bandlimited signal are interpolated to form a continuous time signal.

cardinal vowel　according to English phonetician Daniel Jones, a vowel corresponding to one of the extreme positions of the vowel diagram.

carrier amplitude　amplitude of the radio frequency sinusoid used as a vehicle for transporting intelligence from the sending end of a communications link to the receiving end. For an AM, FM, or PM wave, the peak amplitude of the spectral component in the frequency domain about which symmetry exists. The carrier amplitude (as a function of time) contains a portion of the intelligence for angle modulation (*See* frequency modulation and phase modulation). In contrast, the carrier amplitude contains no information for AM or any of the SSB variations (*See* amplitude modulation and single sideband), but is merely used as a frequency marker.

carrier concentration　the number of mobile charge carriers per unit volume, positive (holes) or negative (electrons). In a semiconductor, both concentrations are present and are modifiable by externally applied electric fields.

carrier current communication　the use of electric lines to carry communication signals.

carrier frequency　in pulse-width-modulated (PWM) switching schemes, the switching frequency that establishes the frequency at which the converter switches are switched. In sine-triangle PWM, the carrier frequency is the frequency of the triangle waveform that the control or modulating signal is compared to.

carrier lifetime　the average duration an electron or a hole stays in a certain state.

carrier phase　the phase of a sinusoidal signal that is the carrier in a modulation scheme such as AM, FM, SSB, etc. The carrier may be defined in the form $A\cos(\omega_c t + \phi)$. The carrier is specified by the parameters A (amplitude), ω_c (carrier frequency), and ϕ (carrier phase).

carrier shift   the difference in frequency between the steady state, mark, and space in frequency shift keying (FSK) systems.

carrier signal   the RF signal in a communications system that has the modulating signal superimposed on it. This signal may have its frequency, amplitude, or phase varied to form a modulated signal. Without modulation it is a simple RF signal.

Many communication systems rely on the concept of sinusoidal amplitude modulation, in which a complex exponential signal $c(t)$ has its amplitude multiplied (modulated) by the information-bearing signal $x(t)$. This signal $x(t)$ is typically referred to as the modulating signal and the signal $c(t)$ as the carrier signal. The modulated signal $y(t)$ is then the product of these two signals.

$$y(t) = x(t)c(t)$$

carrier suppression   in SSB communications, the degree to which the carrier amplitude is reduced from its original value out of the modulator. (*See also* balanced modulator.) Carrier suppression is generally used as a method to significantly reduce the amount of unnecessary transmitted power, based upon the fact that no information is contained within the carrier amplitude in an AM waveform. It is sometimes desirable to only partly suppress the carrier, leaving what is termed a "pilot tone" at the carrier frequency.

carrier synchronization   a synchronization technique used in radio receivers. In all radio receivers some sort of carrier frequency synchronization is required; the phase synchronization is needed only if phase coherent demodulation is desired. Can be categorized as open and closed loop carrier synchronization.

carrier-sense multiple access (CSMA)   a random-access method of sharing a bus-type communications medium in which a potential user of the medium listens before beginning to transmit.

The channel sensing significantly reduces the probability of collisions. *Compare with* ALOHA.

carrier-to-interference ratio (CIR)   similar to signal-to-interference ratio but usually used in cellular communication systems where the carrier refers to the signal of interest and the interference refers to interference from other transmitters in the system. *See also* signal-to-interference ratio.

carrier-to-noise ratio   the ratio of the amplitude of the carrier signal to that of the noise in the IF bandwidth measured at any point in the receiver before any nonlinear process such as amplitude limiting or detection. The carrier to noise ratio is typically expressed in decibels.

carrierless amplitude/phase modulation (CAP)   an implementation of a quadrature amplitude modulation transmitter in which the passband in-phase and quadrature signals are generated directly via quadrature digital filters. A recent application for CAP is high-speed digital subscriber lines. *See also* quadrature amplitude modulation.

carry   overflow signal that occurs when the sum of the operands at the inputs of the adder equals the base. A binary adder, adding $1 + 1$ will produce a sum of 0 and carry of 1.

carry bit   *See* carry.

carry flag   *See* carry.

carry look-ahead adder   high-speed adder that uses extra combinational logic to generate all carries in an m-bit block in parallel. A method of generating the signals corresponding to the carries (borrows) in an addition (subtraction) circuit that does not require all the lower order carries to be determined; a high-speed carry.

Cartesian-based control   the system depicted in the figure. Notice that inverse kinematics is embedded into the feedback control loop. Due to the inverse kinematics calculations, Cartesian-based control is of greater computational complexity. Here X_d, $\dot{X}_d$, and $\ddot{X}_d$ denote position, velocity, and acceleration of the desired trajectory in Cartesian space. τ is a vector of generalized forces and q is a vector of generalized positions.

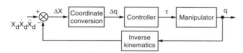

Cartesian-based control scheme.

Cartesian product   a mathematical operation on two sets. The Cartesian product of two sets, say A and B, denoted $A \times B$, is the set of all ordered pairs with the first element of each pair an element of A and the other an element of B. That is,

$$A \times B = \{(a, b) \mid a \in A \text{ and } b \in B\}$$

Cartesian space   *See* external space.

cartridge fuse   replaceable electrical safety device in which metal melts and interrupts the circuit when the current exceeds a preset limit in duration and magnitude.

cascade connection   a series connection of amplifier stages or networks in which the output of one feeds the input of the next.

cascade system   a 3-level system containing high-, intermediate-, and low-energy states. Resembling a cascade, these states are coupled, in that sequence, by two electromagnetic fields. *See also* cathodoluminescence.

cascode   a circuit technique in which the current output of the collector (drain) of a BJT (FET) is buffered by a common base (common gate) amplifier stage. The purpose is to increase the bandwidth and/or output resistance.

cascode amplifier   an amplifier consisting of a grounded-emitter input stage that drives a grounded-base output stage; advantages include high gain and low noise; widely used in television tuners. *See also* cascode.

CASE   *See* computer-aided software engineering.

castellation   recessed metallized feature on the edges of a chip carrier that interconnect conducting surfaces or planes within or on the chip carrier.

casual filter   a filter of which the transition from the passband to the stopband is gradual, not ideal. This filter is realizable.

catadioptric   an optical system made up of both refractive elements (lenses) and reflective elements (mirrors).

catastrophic code   a convolutional code in which a finite number of code symbol errors can cause an unlimited number of decoded symbol errors.

catastrophic encoder   a convolutional encoder with at least one loop in the state-transition diagram with zero accumulated code symbol weight, at least one nonzero information symbol and not visiting the zero state. After decoding, a finite number of (channel) errors can result in an infinite number of errors (catastrophic error propagation).

catastrophic error propagation   when the state diagram contains a zero distance path from some nonzero state back to the same state, the transmission of a 1 causes an infinite number of errors. *See also* catastrophic encoder.

catastrophic thermal failure   an immediate, thermally induced total loss of electronic function by a component or system.

catcher   a cavity resonator of a multicavity klystron proximate to the collector to catch microwave energy from the bunched electrons.

categoric input   a nonnumeric (symbolic) input, e.g., gender, color, which is usually fed to a network using one-out-of-N coding.

catenation   symbols strung together to form a larger sequence, as the characters in a word and the digits in a number.

cathode   the negative electrode of a device. *Contrast with* anode.

cathode ray tube (CRT)   a vacuum tube using cathode rays to generate a picture on a fluorescent screen. These cathode rays are in fact the electron beam deflected and modulated, which impinges on a phosphor screen to generate a picture according to a repetitive pattern refreshed at a frequency usually between 25 and 72 Hz.

cathodoluminescent   the property of luminescent crystals (phosphors) to emit visible light with bombarded electrons.

catoptric   an optical system made up of only reflective elements (mirrors).

CATV   *See* community-antenna television.

Cauchy distribution   the density function for a Cauchy disributed random variable X is:

$$f_X(x) = \frac{1}{1 + x^2}$$

Note that the moments for this random variable do not exist, and that the cumulative distribution function is not defined. *See* probability density function, moment.

Caurer filter   *See* elliptic filter.

causal system   a system whose output does not depend on future input; the output at time t may depend only on the input signal $\{f(\tau) : \tau \leq t\}$. For example, the voltage measured across a particular element in a passive electric circuit does not

depend upon future inputs applied to the circuit, and hence is a causal system.

If a system is not causal, then it is noncausal. An ideal filter which will filter in real time all frequencies present in a signal $f(t)$ requires knowledge of $\{f(\tau) : \tau > t\}$, and is an example of a noncausal system.

causality   a system $H : \mathcal{X}_e \to \mathcal{X}_e$, or equivalently, an operator that maps inputs from the extended space $\mathcal{X}_e$ into outputs from the same space where the output at time t is not a function of future inputs. This can be expressed using truncations as follows: A system H is causal if

$$[Hx(\cdot)]_T = [Hx_T(\cdot)]_T \qquad \forall x \in \mathcal{X}_e$$

See also extended spaces and truncation.

cavity   (1) a fully enclosed, hollow conductor, in which only time-harmonic electromagnetic fields of specific frequencies (i.e., resonant frequencies) exist. Each resonant frequency is identified by a collection of numbers in conjunction with a mode designator of the transverse electric, transverse magnetic, or transverse electromagnetic type.

(2) in optics, region of space that is partially or totally enclosed by reflecting boundaries and that therefore supports oscillation modes.

cavity dumping   fast removal of energy stored in a laser cavity by switching the effective transmission of an output coupling mirror from a low value to a high value.

cavity lifetime   one of several names used to indicate the time after which the energy density of an electromagnetic field distribution in a passive cavity maybe expected to fall to $1/e$ of its initial value; the name photon lifetime is also common.

cavity ratio (CR)   a number indicating cavity proportions calculated from length, width, and height. It is further defined into ceiling cavity ratio, floor cavity ratio, and room cavity ratio.

cavity short   a grounded metal rod connecting the body of an RF cavity. By grounding the cavity, it is kept from resonating.

Cayley–Hamilton theorem for 2-D general model   let T_{pq} be transition matrices defined by

$$ET_{pq} = \begin{cases} A_0 T_{-1,-1} + A_1 T_{0,-1} + A_2 T_{-1,0} \\ +I_n \quad \text{for } p = q = 0 \\ A_0 T_{p-1,q-1} + A_1 T_{p,q-1} \\ +A_2 T_{p-1,q} \\ \textit{for } p \neq 0 \quad \textit{and/or} \quad q \neq 0 \end{cases}$$

and

$$[Ez_1 z_2 - A_0 - A_1 z_1 - A_2 z_2]^{-1}$$
$$= \sum_{p=-n_1}^{\infty} \sum_{q=-n_2}^{\infty} T_{pq} z_1^{-(p+1)} z_2^{-(q+1)}$$

$$Ex_{i+1,j+1} = A_0 x_{ij} + A_1 x_{i+1,j} + A_2 x_{i,j+1}$$
$$+ B_0 u_{ij} + B_1 u_{i+1,j} + B_2 u_{i,j+1}$$

$i, j \in Z_+$ (the set of nonnegative integers) where $x_{i,j} \in R^n$ is the semistate vector, $u_{i,j} \in R^m$ is the input, and E, A_k, B_k ($k = 0, 1, 2$) real matrices with E possibly singular or rectangular. A pair (n_1, n_2) of positive integers n_1, n_2 such that $T_{pq} = 0$ for $p < -n_1$ and/or $q < -n_2$ is called the index of the model. Transition matrices T_{pq} of the generalized 2-D model satisfy

$$\sum_{p=0}^{n_1} \sum_{q=0}^{n_2} d_{pq} T_{p-k-1, q-t-1} = 0$$

for

$$\begin{cases} k < 0 \quad \text{and} \quad m_1 < k \leq 2n_1 - 1 \\ t < 0 \quad \text{and} \quad m_2 < t \leq 2n_2 - 1 \end{cases}$$

where d_{pq} are coefficients of the polynomial

$$\det [Ez_1 z_2 - A_0 - A_1 z_1 - A_2 z_2]$$
$$= \sum_{p=0}^{n_1} \sum_{q=0}^{n_2} d_{pq} z_1^p z_2^q$$

and m_1, m_2 are defined by the adjoint matrix

$$\text{adj } [Ez_1 z_2 - A_0 - A_1 z_1 - A_2 z_2]$$
$$= \sum_{i=0}^{m_1} \sum_{j=0}^{m_2} H_{ij} z_1^i z_2^j$$

$$(m_1 \leq n - 1, m_2 \leq n - 1)$$

Cayley–Hamilton theorem for 2-D Roesser model   let T_{ij} be transition matrix defined by

$$T_{ij} = \begin{cases} I \text{ (the identity matrix)} \\ \text{for } i, j = 0 \\ T_{10} := \begin{bmatrix} A_1 & A_2 \\ 0 & 0 \end{bmatrix}, T_{01} := \begin{bmatrix} 0 & 0 \\ A_3 & A_4 \end{bmatrix} \\ T_{10} T_{i-1,j} + T_{01} T_{i,j-1} \quad \text{for} \quad i, j \in Z_+ \\ 0 \text{ for } i < 0 \text{ or/and } j < 0 \end{cases}$$

(Z_+ is the set of nonnegative integers) of the 2-D of the Roesser model

$$\begin{bmatrix} x_{i+1,j}^h \\ x_{i,j+1}^v \end{bmatrix} = \begin{bmatrix} A_1 & A_2 \\ A_3 & A_4 \end{bmatrix} \begin{bmatrix} x_{ij}^h \\ x_{ij}^v \end{bmatrix} + \begin{bmatrix} B_1 \\ B_2 \end{bmatrix} u_{ij}$$

$i, j \in Z_+$ where $x_{ij}^h \in R^{n_1}$ and $x_{ij}^v \in R^{n_2}$ are the horizontal and vertical state vectors, respectively, $u_{ij} \in R^m$ is the input vector, and A_1, A_2, A_3, A_4, B_1, B_2 are real matrices. The transition matrices T_{ij} satisfy the equation

$$\sum_{i=0}^{n_1} \sum_{j=0}^{n_2} a_{ij} T_{i+h, j+k} = 0$$

for $h = 0, 1, \ldots$ and $k = 0, 1, \ldots$ where a_{ij} are coefficients of the 2-D characteristic polynomial

$$\det \begin{bmatrix} I_{n_1} z_1 - A_1 & -A_2 \\ -A_3 & I_{n_2} z_2 - A_4 \end{bmatrix}$$
$$= \sum_{i=0}^{n_1} \sum_{j=0}^{n_2} a_{ij} z_1^i z_2^j \quad (a_{n_1 n_2} = 1)$$

CB   *See* Citizen's band.

CBE   *See* chemical beam epitaxy.

CBR   *See* constant bit rate.

CCD   *See* charge-coupled device.

CCD memory   *See* charge-coupled-device memory.

CCDF   *See* complementary cumulative distribution function.

CCI   *See* cochannel interference.

CCIR   *See* International Radio Consultative Committee.

CCITT   Comité Consulatif International Télégraphique et Téléphonique.

CCITT two-dimensional   a modified relative element address designate scheme. The position of each changing element on the present line is coded with respect to the changing element on the reference line or the preceding changing element on the present line. The reference line lies immediately above the present line.

CCR   *See* cochannel reuse ratio.

CCS   *See* common channel signaling.

CCVT   *See* coupling capacitor voltage transformer.

CD   *See* compact disk, critical dimension.

cd   *See* candela.

CD-I   *See* compact disk-interactive.

CD-ROM   a read-only compact disk. *See also* compact disk.

CDF   *See* cumulative distribution function.

CDMA   *See* code division multiple access.

CdS   abbreviation for cadmium sulfide, a photoconductor with good visible light response.

CEL   *See* contrast enhancement layer.

CELL   *See* surface-emitting laser logic.

cell   (1) in mobile radio communications, the area serviced by one base station. One way of categorizing the cells is according to their size. Cell sizes may range from a few meters to many hundred kilometers. *See also* picocel, nanocel, nodal cell, microcell, macrocell, large cell, megacell, satellite cell.

(2) in ATM systems a small packet of fixed length. The CCITT chose a cell size of 53-bytes comprising a 5-byte header and 48-byte payload for their ATM network.

cell library   a collection of simple logic elements that have been designed in accordance with a specific set of design rules and fabrication processes. Interconnections of such logic elements are often used in semicustom design of more complex IC chips.

cell switching   means of switching data among the ports (inputs and outputs) of a switch such that the data is transferred in units of a fixed size.

cell-cycle-specific control   a type of control arising in scheduling optimal treatment protocols in the case when treated population is sensitive to the therapy only in chosen phases of its cell cycle. A mathematical model of population dynamics used to solve the problem of such control should be composed of subsystems sensitive and insensitive to the drug. It may be achieved by the use of compartmental models. Typical types of perturbations of the cell cycle considered as cell-cycle-specific control are as follows: cell arrest, cell killing, and alteration of the transit time. They could be applied for representation of cell synchronization, cell destruction, and cell recruitment from the specific phase.

cellular automaton   system designed from many discrete cells, usually assembled in one- or two-dimensional regular arrays, each of which is a standard finite state machine. Each cell may change its state only at fixed, regular intervals, and only in accordance with fixed rules that depend on cells' own values and the values of neighbors within a certain proximity (usually two- for one-dimensional, and four- for two-dimensional cellular automata). Cellular automata are a base of cellular computers; fine grain systems that are usually data-driven and used to implement neural networks, systolic arrays, and SIMD architectures.

cellular communications   traditionally, an outside-of-building radio telephone system that allows users to communicate from their car or from their portable telephone.

cellular manufacturing   grouping of parts by design and/or processing similarities such that the group (family) is manufactured on a subset of machines which constitute a cell necessary for the group's production.

cellular spectral efficiency   the cellular spectral efficiency of a system is defined as the sum of the maximum data rates that can be delivered to subscribers affiliated to all base stations in a re-use cluster of cells, occupying as small a physical area as possible. Mathematically, the cellular spectral efficiency, η, is defined as

$$\eta = \frac{\sum_{j=1}^{r} \sum_{i=1}^{K} R_{ij}}{B A_{cluster}} \text{bit/s/Hz/km}^2$$

where r denotes the number of cells in a re-use cluster, R_{ij} denotes the data rate measured in bits/s at some predefined BER available to subscriber I in cell j of the re-use cluster, B denotes the total bandwidth measured in hertz allocated to all cells

in the re-use cluster, and $A_{cluster}$ denotes the physical area, measured in square kilometers occupied by the re-use cluster.

CELP   *See* code excited linear prediction.

center frequency   (1) the frequency of maximum or minimum response for a bandpass or a bandstop filter, respectively; often taken as the geometric mean of the lower and upper cutoff frequencies.

(2) the frequency at the center of a spectrum display.

(3) the average frequency of the emitted wave after modulation by a sinusoidal signal.

(4) the frequency of a non-modulated wave. *See also* channel.

center of average   an approach of defuzzification that takes the weighted average of the centers of fuzzy sets with the weights equal to the firing strengths of the corresponding fuzzy sets.

center of gravity method   *See* centroid method.

center of projection   the point within a projector from which all the light rays appear to diverge; the point in a camera towards which all the light rays appear to converge before they cross the imaging plane or photographic plate.

central absolute moment   for random variable x, the pth central absolute moment is given by $E[|x - E[x]|^p]$. *See* central moment, absolute moment, expectation.

central limit theorem (CLT)   (1) a theorem that the distribution of the sum of independent and identically distributed random variables tends toward a Gaussian distribution as the number of individual random variables approaches infinity.

(2) in probability, the theorem that the density function of some function of n independent random variables tends towards a normal distribution as n tends to infinity, as long as the variances of

the variables are bounded: $0 < \sigma \leq v_i \leq \gamma < \infty$. Here σ and γ are positive constants, and v_i is the variance of the ith random variable. *See* Gaussian distribution.

central moment for random variable X the nth central moment is given by

$$E[(X - m)^2] = \int_{-\infty}^{\infty} (x - m)^2 f_X(x) dx$$

where $f_X(x)$ is the probability density function of X. *See* central absolute moment, absolute moment. expectation.

central processing unit (CPU) a part of a computer that performs the actual data processing operations and controls the whole computing system. It is subdivided into two major parts:

1. The arithmetic and logic unit (ALU), which performs all arithmetic, logic, and other processing operations,

2. The control unit (CU), which sequences the order of execution of instructions, fetches the instructions from memory, decodes the instructions, and issues control signals to all other parts of the computing system. These control signals activate the operations performed by the system.

centralized arbitration a bus arbitration scheme in which a central bus arbiter (typically housed in the CPU) accepts requests for and gives grants to any connected device, wishing to transmit data on the bus. The connected devices typically have different priorities for bus access, so if more than one device wants bus access simultaneously, the one with the highest priority will get it first. This prioritization is handled by the bus arbiter.

centrifugal switch a speed-sensitive switch operated by centrifugal force, mounted on the shaft of a motor, used to perform a circuit switching function. Used in single-phase induction motors to disconnect the starting winding when the motor approaches operating speed.

centripetal force force that is present during the robot motion. The force depends upon the square of the joint velocities of the robot and tends to reduce the power available from the actuators.

centroid (1) the center of a mass.

(2) description of the center of a particle beam profile.

(3) a region in the pattern space to which a remarkable number of patterns belong.

centroid defuzzification a defuzzification scheme that builds the weighted sum of the peak values of fuzzy subsets with respect to the firing degree of each fuzzy subset. Also called height defuzzification.

centroid method a widely used method of defuzzification whereby the centroid of the membership function of the fuzzy set is used as the defuzzified or crisp value. It is also known as the center of gravity method or the composite moments method.

centroidal profile a method for characterizing and analyzing the shape of an object having a well defined boundary. The centroid of the shape is first determined. Then a polar (r, θ) plot of the boundary is computed relative to this origin: this plot is the centroidal profile, and has the advantage of permitting template matching for a 2-D shape to be performed relatively efficiently as a 1-D process.

centrosymmetric medium a material that possesses a center of inversion symmetry. Of importance because, for example, second-order nonlinear optical processes are forbidden in such a material.

cepstrum inverse Fourier transform of the logarithm of the Fourier power spectrum of a signal. The complex cepstrum is the inverse Fourier transform of the complex logarithm of the Fourier transform of the signal.

ceramic ferrite a relatively inexpensive permanent magnet material with decent coercivity and low energy product that is composed of strontium or barium oxide and iron oxide. Also called hard ferrite.

cerebellar model articulation (CMAC) network a feedforward network developed originally as a model of the mammalian cerebellum. Several variants now exist, but basic operation involves the first layer of the network mapping the input into a higher-dimensional vector and a second layer forming the network output by means of a weighted sum of the first layer outputs. The weights can be trained using the LMS rule. Developed mainly for application in robotics, it has also been used in pattern recognition and signal processing.

Cerenkov counter a detector for charged particles. It consists essentially of a transparent medium such as a gas, which emits Cerenkov radiation when a charged particle passes through at a velocity greater than the velocity of light in the medium. The mass of a particle in a beam of known momentum can be determined with such a counter by measuring the characteristic angle at which the Cerenkov radiation is emitted.

Cerenkov radiation light emitted when a charged particle traverses a medium with a velocity greater than the velocity of light in the medium. The Cerenkov light is emitted in a cone centered on the particle trajectory. The opening angle of this cone depends on the velocity of the particle and on the velocity of light in the medium. The phenomenon involved is that of an electromagnetic shock wave and is the optical analogue of sonic boom. Cerenkov radiation provides an important tool for particle detection.

certainty equivalence principle a design method in which the uncertainties of process parameters are not considered. Found in self-tuning regulators where the controller parameters or the process parameters are estimated in real-time and are then used to design the controller as if they were equal to the true parameters. Although many estimation methods could provide estimates of parameter uncertainties, these are typically not used in the control design.

CFD *See* crossed field devices.

CFIE *See* combined field integral equation.

CGA *See* color graphics adapter.

chain code a method for coding thin contours or lines, for example in a bilevel picture, which encodes the direction of movement from one point to the next. For 8-connected contours, a three-bit code may be used at each point to indicate which of its eight neighbors is the succeeding point.

chain matrix *See* ABCD matrix.

chain parameters *See* ABCD parameters.

chain reaction a process in which high-energy neutrons emitted from fissile radioactive material are directed into more fissile material such that more neutrons are emitted. The process creates heat which is used to power thermal power plants.

chaining when the output stream of one arithmetic pipeline is fed directly into another arithmetic pipeline; used in vector computers to improve their performance.

chaining of fuzzy rules a reasoning strategy which searches the knowledge base and chain from rule to rule to form inferences and draw conclusions. In forward chaining, a chain of data-driven rules are evaluated for which the conditional parts are satisfied to arrive at the conclusion. Backward chaining is goal-driven in which subgoals are established, where necessary, through which a chain of rules are selected, eventually satisfying the goal.

103

chamfer distance   a digital distance based on a chamfer mask, which gives the distance between a pixel and those in its neighborhood; then the chamfer distance between two non-neighboring pixels (resp., voxels) is the smallest weighted length of a digital path joining them. The word "chamfer" comes from the fact that with such a distance a circle is in fact a polygon. The n-dimensional Manhattan and chessboard distances are chamfer distances; the Euclidean distance is not. In the 2–D plane, the best chamfer distances are given by the (3, 4) and (5, 7, 11) Chamfer masks: in the (3, 4) mask, a pixel is at distance 3 from its horizontal/vertical neighbors and at distance 4 from its diagonal neighbors, while in the (5, 7, 11) mask, it is at distance 5 from its horizontal/vertical neighbors, at distance 7 from its diagonal neighbors, and at distance 11 from its neighbors distant by 1 and 2 respectively along the two axes. *See* chessboard distance, Euclidean distance, Manhattan distance.

channel   (1) the medium along which data travel between the transmitter and receiver in a communication system. This could be a wire, coaxial cable, free space, etc. *See also* I/O channel.

(2) the conductivity path between the source and the drain of a field effect transistor.

(3) a single path for transmitting electrical signals. Example 1: The band of frequencies from 50 Hz to 15 KHz (Channel A) and 15 KHz to 75 KHz (Channel B) which frequency modulate the main carrier of an FM stereo transmitter. Example 2: A portion of the electromagnetic spectrum assigned for operation of a specific carrier from the FM broadcast band (88 to 108 MHz) of frequencies 200 KHz wide designated by the center frequency beginning at 88.1 MHz and continuing in successive steps to 107.9 MHz.

channel allocation   the act of allocating radio channels to cells, base stations, or cell sectors, in a radio network, also referred to as frequency allocation, or frequency planning. The allocation typically follows an algorithm that attempts to maximize the number of channels used per cell and minimize the interference in the network.

channel architecture   a computer system architecture in which I/O operations are handled by one or more separate processors known as channel subsystems. Each channel subsystem is itself made up of subchannels, in which control unit modules control individual I/O devices. Developed by IBM, and used primarily in mainframe systems, the channel architecture is capable of a very high volume of I/O operations.

channel capacity   a fundamental limit on the rate at which information can be reliably communicated through the channel. Also referred to as "Shannon capacity," after Claude Shannon, who first formulated the concept of channel capacity as part of the noisy channel coding theorem.

For an ideal bandlimited channel with additive white Gaussian noise, and an input average power constraint, the channel capacity is $C = 0.5 \log(1 + S/N)$ bit/Hz, where S/N is the received signal-to-noise ratio.

channel code   a set of codewords used to represent messages, introducing redundancy in order to provide protection against errors introduced by transmission over a channel. *See also* source code.

channel coding   the process of introducing controlled redundancy into an information sequence mainly to achieve reliable transmission over a noisy channel. Channel coding can be divided into the areas of block coding and trellis coding. Also called error control coding. *See also* block coding, trellis coding and convolutional coding.

channel command word   an "instruction" to an I/O channel. The commands consists of parameters (e.g., "operation," "data address," "count") giving the channel processor information on type of I/O operation requested (e.g., "read" or "write"), where the data is to be read or written, and the number of bytes involved in the data transfer.

In the IBM mainframe architecture there are six different types of channel control words: READ, READ BACKWARD, WRITE, CONTROL, SENSE, and JUMP.

channel control word   *See* channel command word.

channel encoder   a device that converts source-encoded digital information into an analog RF signal for transmission. The type of modulation used depends on the particular digital audio broadcasting (DAB) system, although most modulation techniques employ methods by which the transmitted signal can be made more resistant to frequency-selective signal fading and multipath distortion effects.

channel estimation   estimation of the radio channel parameters in the receiver. Typically delays, amplitudes, carrier phases, and direction-of-arrivals need to be estimated depending on the receiver configuration. Channel estimation is a modern way to look at receiver synchronization based mainly on feedback control loops, since in principle any method known to estimation theory can be applied to achieve synchronization of the receiver over an unknown radio channel.

channel I/O   an approach to I/O processing in which I/O operations are processed independent from the CPU by a channel system. *See also* channel architecture.

channel matched VQ   *See* channel optimized vector quantization.

channel measurement   *See* channel sounding.

channel modeling   the act of describing the effect of the (radio) channel on the transmitted signal in a form suitable for mathematical analysis, computer simulation or hardware simulation.

A channel model is a collection of channel characteristics essential to the performance of the communication system under study, organized in such a way that the basic performance trade-offs of the system can be analyzed or simulated without having to construct prototypes or carry out field tests.

channel optimized vector quantization (COVQ)   a combined source-channel code for block-based source coding (vector quantization) and block channel coding. A channel optimized vector quantizer can be designed using a modified version (taking channel induced distortion into account) of the generalized Lloyd algorithm). Also referred to as channel matched VQ. *See also* noisy channel vector quantization.

channel program   the set of channel control words that make up the instruction sequence that controls an I/O channel. *See also* channel control word.

channel reliability function   the rate function with infinitesimal error probability expressed by

$$E(R) = \begin{cases} \frac{1}{2}C_\infty & 0 \leq R \leq \frac{1}{4}C_\infty \\ \left(\sqrt{C_\infty} - \sqrt{R}\right)^2 & \frac{1}{4}C_\infty \leq R \leq C_\infty \end{cases}$$

for transmission of orthogonal or simplex signal over infinite bandwidth AWGN channel. C_∞ is the capacity of the infinite bandwidth white Gaussian noise channel, defined as

$$C_\infty = \frac{P_{av}}{N_o ln2} \text{bit/s}.$$

channel robust vector quantization   a vector quantizer that has been made robust against channel errors. *See also* noisy channel VQ.

channel robust VQ   *See* channel robust vector quantization.

channel sounding   the act of recording from a real, physical channel a selected set of characteristics describing the channel. This involves transmitting a known signal to the channel under study, and at the receiver processing the received signal to extract the required information, typically the impulse response or the transfer function of the channel.

channel spill   leakage of RF energy from a radio channel n into a radio channel $n \pm i$, $i \geq 1$, due to finite channel filter attenuation outside of the bandwidth of n.

channel step   *See* frequency synthesizer.

channel subsystem   the I/O processing component of a computer system conforming to the channel architecture model.

channel waveguide   a light guide that is either raised above or diffused into a substrate.

channel-to-case thermal resistance   the proportionality constant (denoted θ_{cc}) at thermal equilibrium between the temperature difference of the FET channel ($T_{channel}$) and a specified case surface (T_{case}) to the dissipated power in the channel (P_w), in units of $^\circ C/W$. The specified surface is usually the most effective in reducing the temperature. It includes the thermal resistance of the chip, die attach material (solder or adhesive), packaging and mounting medium, as applicable.

$$\theta_{cc} = \left. \frac{T_{channel} - T_{case}}{P_w} \right|_{equilibrium}$$

channelizer   a system that decomposes an RF signal into narrow-band output channels; term often applied to acousto-optic spectrum analyzers that are driven by RF frequency signals. *See also* acousto-optic spectrum analyzer.

chaos   (1) erratic and unpredictable dynamic behavior of a deterministic system that never repeats itself. Necessary conditions for a system to exhibit such behavior are that it be nonlinear and have at least three independent dynamic variables.

(2) in microelectronics, deterministic motion, in which the statistics are essentially those of a Gaussian random process.

chaotic behavior   a highly nonlinear state in which the observed behavior is very dependent on the precise conditions that initiated the behavior. The behavior can be repeated (i.e., it is not random), but a seemly insignificant change, such as voltage, current, noise, temperature, rise times, etc., will result in dramatically different results, leading to unpredictability. The behavior may be chaotic under all conditions, or it may be well behaved (linear to moderately nonlinear) until some parametric threshold is exceeded, at which time chaotic behavior is observed. In a mildly chaotic system, noticeable deviations resulting from small changes in the initial conditions may not appear for several cycles or for relatively long periods. In a highly chaotic system, the deviations are immediately apparent.

character   (1) letter, number or symbol as used on a computer keyboard.

(2) data type that represents an alphanumeric character as a group of bits, usually as an eight-bit byte.

character recognition   *See* optical character recognition.

character string   (1) a series of continuous bytes in memory, where each byte represents one character.

(2) data structure corresponding to ordered sequence of characters.

characteristic equation   the polynomial equation that results when the characteristic function is equated to zero. Its roots gives the singularities of the transfer function model, which in turn determine its transient behavior. Specifically, any root of the characteristic equation that has a negative real part indicates a stable decaying transient, while any root with a positive real part indicates an unstable growing transient. Any root with zero real part indicates a marginally stable transient that neither decays nor grows. The imaginary part of the root gives the frequency of oscillation of the transient signal. *See also* characteristic function.

characteristic function   (1) the name given to the denominator polynomial of a transfer function model. Through partial fraction expansion of a transfer function and subsequent inverse Laplace transformation, it is obvious that the characteristics of the system dynamics are defined by this function. For example, the transfer function

$$g(s) = \frac{9}{6 + 5s + s^2}$$

has characteristic function

$$\phi(s) = 6 + 5s + s^2 = (s + 2)(s + 3)$$

so its output response will contain terms like

$$y(t) = \alpha e^{-2t} + \beta e^{-3t} + \ldots$$

that are characteristic of the system itself. (Other terms in the response are attributed to the forcing input signal.) *See also* characteristic equation.

(2) a transformed probability density function,

$$\Phi_x(\omega) = E[\exp(j\omega^T x)]$$

useful in the analytic computation of higher order moments and convolutions of probability densities.

characteristic impedance   inherent property of a transmission line that defines the impedance that would be seen by a signal if the transmission line were infinitely long. If a signal source with a "source" or "reference" impedance equal to the

characteristic impedance is connected to the line there will be zero reflections.

characteristic loci   the plots of the eigenvalues of transfer function matrices, evaluated over a range of frequencies. These traces, which are parametrized by frequency, are shown on a single Nyquist plot and used to predict the closed loop stability of multiinput-multioutput systems, by application of the principle of the argument for complex variable functions. Unlike the Nyquist plots for single-input-single-output systems, an individual eigenvalue might not encircle the plane an integral number of times, yet the total encirclements of all the eigenvalues will be an integral number.

characteristic polynomial and equation of generalized 2-D model   the determinant

$$p(z_1, z_2) = \det[Ez_1z_2 - A_0 - A_1z_1 - A_2z_2]$$
$$= \sum_{i=0}^{n_1} \sum_{j=0}^{n_2} a_{ij} z_1^i z_2^j$$

$(n_1, n_2 \leq rank\ E)$ is called the 2-D characteristic polynomial of the generalized 2-D model

$$Ex_{i+1,j+1} = A_0 x_{ij} + A_1 x_{i+1,j} + A_2 x_{i,j+1}$$
$$+ B_0 u_{ij} + B_1 u_{i+1,j} + B_2 u_{i,j+1}$$

$i, j \in Z_+$ (the set of nonnegative integers) where $x_{ij} \in R^n$ is the semistate vector, $u_{ij} \in R^m$ is the input vector, and E, A_k, B_k ($k = 0, 1, 2$) are real matrices with E possibly singular or rectangular. $p(z_1, z_2) = 0$ is called the 2-D characteristic equation of the model.

characteristic polynomial assignment of 2-D Roesser model   consider the 2-D Roesser model

$$\begin{bmatrix} x_{i+1,j}^h \\ x_{i,j+1}^v \end{bmatrix} = \begin{bmatrix} A_1 & A_2 \\ A_3 & A_4 \end{bmatrix} \begin{bmatrix} x_{ij}^h \\ x_{ij}^v \end{bmatrix} + \begin{bmatrix} B_1 \\ B_2 \end{bmatrix} u_{ij}$$

$i, j \in Z_+$ (the set of nonnegative integers) with the state-feedback

$$u_{ij} = K \begin{bmatrix} x_{ij}^h \\ x_{ij}^v \end{bmatrix} + v_{ij}$$

where $x_{ij}^h \in R^{n_1}$, and $x_{ij}^v \in R^{n_2}$ are the horizontal and vertical state vectors, respectively, $u_{ij} \in R^m$ is the input vector, and $A_1, A_2, A_3, A_4, B_1, B_2$ are real matrices of the model,

$$K = [K_1, K_2] \in R^{m \times (n_1 + n_2)}$$

and $v_{ij} \in R^m$ is a new input vector. Given the model and a desired 2-D characteristic polynomial of the closed-loop system $p_c(z_1, z_2)$, find a gain feedback matrix K such that

$$\det \begin{bmatrix} I_{n_1}z_1 - A_1 - B_1K_1 & -A_2 - B_1K_2 \\ -A_3 - B_2K_1 & I_{n_2}z_2 - A_4 - B_2K_2 \end{bmatrix}$$

$$= p_c(z_1, z_2) = \sum_{i=0}^{n_1} \sum_{j=0}^{n_2} d_{ij} z_1^i z_2^j \ (d_{n_1 n_2} = 1)$$

characteristic polynomial of 2-D Fornasini–Marchesini model   the determinant

$$p(z_1, z_2) = \det[I_n z_1 z_2 - A_1 z_1 - A_{z_2} z_2]$$

$$= \sum_{i=0}^{n_1} \sum_{j=0}^{n_2} a_{ij} z_1^i z_2^j \ (a_{nn} = 1)$$

is called the 2-D characteristic polynomial of the 2-D Fornasini–Marchesini model

$$x_{i+1,j+1} = A_1 x_{i+1,j} + A_2 x_{i,j+1}$$
$$+ B_1 u_{i+1,j} + B_2 u_{i,j+1}$$

$i, j \in Z_+$ (the set of nonnegative integers) where $x_{ij} \in R^n$ is the local state vector, $u_{ij} \in R^m$ is the input vector, and A_k, B_k $(k = 1, 2)$ are real matrices.

$p(z_1, z_2) = 0$ is called the 2-D characteristic equation of the model.

characteristic polynomial of 2-D Roesser model   the determinant

$$p(z_1, z_2) = \det \begin{bmatrix} I_{n_1}z_1 - A_1 & -A_2 \\ -A_3 & I_{n_2}z_2 - A_4 \end{bmatrix}$$

$$= \sum_{i=0}^{n_1} \sum_{j=0}^{n_2} a_{ij} z_1^i z_2^j \ (a_{n_1 n_2} = 1)$$

is called the 2-D characteristic polynomial of the 2-D Roesser model

$$\begin{bmatrix} x_{i+1,j}^h \\ x_{i,j+1}^v \end{bmatrix} = \begin{bmatrix} A_1 & A_2 \\ A_3 & A_4 \end{bmatrix} \begin{bmatrix} x_{ij}^h \\ x_{ij}^v \end{bmatrix} + \begin{bmatrix} B_1 \\ B_2 \end{bmatrix} u_{ij}$$

$i, j \in Z_+$ (the set of nonnegative integers) where $x_{ij}^h \in R^{n_1}$, and $x_{ij}^v \in R^{n_2}$ are the horizontal and vertical state vectors, respectively, $u_{ij} \in R^m$ is the input vector, and $A_1, A_2, A_3, A_4, B_1, B_2$ are real matrices.

$p(z_1, z_2) = 0$ is called the 2-D characteristic equation of the model.

characterization   the process of calibrating test equipment, measuring, de-embedding and evaluating a component or circuit for DC RF and/or digital performance.

charge   a basic physical quantity that is a source of electromagnetic fields.

charge carrier   a unit of electrical charge that when moving, produces current flow. In a semiconductor two types of charge carriers exist: electrons and holes. Electrons carry unit negative charge and have an effective mass that is determined by the shape of the conduction band in energy-momentum space. The effective mass of an electron in a semiconductor is generally significantly less than an electron in free space. Holes have unit positive charge. Holes have an effective mass that is determined by the shape of the valence band in energy-momentum space. The effective mass of a hole is generally significantly larger than that for an electron. For this reason, electrons generally move much faster than holes when an electric field is applied to the semiconductor.

charge conservation   physical law (derived from Maxwell's equations) indicating that no change in the total charge within a certain volume can exist without the proper flow of charge (current) through that volume.

charge density   describes the distribution of charge along a line, on a surface or in a volume. May be discrete or continuous.

charge-coupled device (CCD)   a solid-state device used to record images. A CCD is a digital device which counts the photons that strike it by making use of the photoelectric effect. In a typical CCD array, a large number of such devices is collected into a 2-D grid. Each device corresponds to a single pixel, and the number of electrons in the device is linearly related to the brightness or intensity value at that point in the CCD.

charge-coupled device detector   a charge-coupled device (CCD) connected to photodetectors, where the photocharge is put into the CCD potential wells for transport and processing.

charge-coupled-device memory   large-capacity shift registers making use of charge-coupled devices (CCD), i.e., MOS devices in which data bits are stored dynamically as charge between a gate and the substrate. This forms a multigate MOS transistors with the source and drain terminals "stretched" apart, and a number of gate terminals in between. The first gate terminal (closest to the source) inserts bits (charge) into the register, and the following gates are controlled with overlapping clocks allowing the charge to move along the array. At the far (drain) end, the bit under the final gate terminal is detected as a change in current.

charge-spring model   *See* electron oscillator model.

charging current   that portion of an electric power line's current which goes to charge the capacitance of the line. The charging current is not available for power transmission.

chattering   fast switching. The term comes from the noise generated by the sustained rapid opening and closing of a switching element. *See also* discontinuous control.

Chattuck coil   a finely wound solenoid about a flexible, nonmagnetic core that is usually used in conjunction with a fluxmeter to measure magnetic potential between two points; a magnetic analog of a voltmeter.

CHDL   *See* computer hardware description language.

Chebyshev alignment   a common filter alignment characterized by ripples of equal amplitude within the pass-band and a steep rolloff in the vicinity of cutoff frequency.

Chebyshev filter   one of a class of commonly used low pass, high pass, band pass and band stop filters with an equiripple characteristic, designed to achieve relatively rapid rolloff rates near cutoff frequencies, at the expense of a loss of monotonicity in either the passbands or the stopbands. *See also* Butterworth filter.

checkerboarding   *See* fragmentation.

checkpoint   time in the history of execution at which a consistent version of the system's state is saved so that if a later event causes potential difficulties, the system can be restarted from the state that had been saved at the checkpoint. Checkpoints are important for the reliability of a distributed system, since timing problems or message loss can create a need to "backup" to a previous state that has to be consistent in order for the overall system to operate functionally.

checkpointing   method used in rollback techniques in which some subset of the system states (data, program, etc.) is saved at specific points (checkpoints), during the process execution, to be used for recovery if a fault is detected.

checksum   checksum is a value used to determine if a block of data has changed. The checksum is formed by adding all of the data values in the block together, and then finding the 2's complement of the sum. The checksum value is added to the end of the data block. When the data block is examined (possibly after being received over a serial line), the sum of the data values and checksum should be zero.

checksum character   in data communication and storage devices, an extra character is often added at the end of the data so that the total number of ones in a block, including the checksum character is even. The checksum character is used to detect errors within the data block.

chemical beam epitaxy (CBE)   a material growth technique that uses metal organic molecules in high vacuum growth chamber and a controlled chemical reaction on a heated substrate to grow a variety of II-VI, III-V, and group IV materials with atomic layer control. Used to create material structures for a variety of electronic and optical devices using quantum wells, heterostructures, and superlattices. This growth technique combined aspects of both MBE and MOCVD growth.

chemical laser   a laser in which the amplification results from one or more chemical reactions; potentially very powerful with principal output lines in the mid-infrared.

chemical sensor   the interface device for an instrumentation system that determines the concentration of a chemical substance.

chemical vapor deposition (CVD)   a process used in the manufacture of integrated circuits or optical fibers whereby a thin solid film of one material is deposited on the surface of another by using a radio frequency or other electrical energy source to dissociate a reactive gas.

chemically amplified resist   a type of photoresist, most commonly used for deep-UV lithography, which, upon post-exposure bake, will multiply the number of chemical reactions through the use of chemical catalysis.

chemiluminescence   light emitted as a result of a chemical reaction.

Chernobyl   typically refers to a fire at a nuclear power plant near Kiev in the Republic of the Ukraine.

chessboard distance   the distance between discrete points arising from the L^∞ norm. Given two discrete points $x = (x_1, \ldots, x_n)$, $y = (y_1, \ldots, y_n)$ on an n-dimensional integer lattice, the chessboard distance between x and y is $\max\{|x_1 - y_1|, \ldots, |x_n - y_n|\}$. So called because it equals the number of moves made by a King when going from one position to another in the game of chess. *See* norm.

chi-squared distribution   a probability distribution with n degrees of freedom and probability density function

$$f(x) = \frac{x^{\frac{n}{2}-1} e^{\frac{x}{2}} u(x)}{2^{\frac{n}{2}} \Gamma^{\frac{n}{2}}}$$

chip   (1) a small piece of semiconductor material upon which miniaturized electronic circuits can be built.

(2) an individual MMIC circuit or subsystem that is one of several identical chips that are produced after dicing up an MMIC wafer.

(3) in direct-sequence spread-spectrum transmission, the high bandwidth symbols, or pulses making up the signature sequence. They are used to spread the bandwidth of the data in frequency.

Usually the time duration of these pulses are many times smaller than that of the information symbols leading to significantly greater spreading of the signal bandwidth.

chip carrier a low-profile rectangular component package, usually square, whose semiconductor chip cavity or mounting area is a large fraction of the package size and whose external connections are usually on all four sides of the package.

chip chart this term is often used for the "gray scale" chart used in the process of aligning television camera systems. The gray scale provides logarithmic reflectance relationships.

chip select a control signal input to, e.g., a memory chip, used to make this particular chip "active" in reading or writing the data bus. Read or write is determined by another control input signal: the "R/W-signal." Typically, some of the high order bits from the CPU's address bus are decoded to form the chip select signals.

chip-to-chip optical interconnect optical interconnect in which the source and the detector are connected to electronic elements in two separate chips.

chirp the varying in time of a carrier frequency signal. *See also* chirp function.

chirp function a signal whose frequency varies monotonically with time, e.g., a linear chirp possesses a linear-frequency or a quadratic-phase variation.

chirp signal *See* chirp function.

chirping a shifting of the optical frequency often observed in modulated semiconductor lasers where the laser gain is modulated at high bandwidth; arises due to the later portions of the modulating signal seeing a different refractive index, or carrier density, than the earlier portions.

Cholesky decomposition a matrix-algebraic theorem that states that, for any positive definite square matrix $\mathbf{A}$, there exists a lower-left triangular matrix $\mathbf{G}$ such that $\mathbf{A} = \mathbf{G}\,\mathbf{G}^T$.

chopper *See* buck converter and DC chopper.

chopper-depth of modulation a marker normally associated with the monitoring of the depth of modulation of a television broadcast signal on a waveform monitor. The chopper reference is used to set the 0% modulation point relative to the video signal. The sync signal is typically at the 100% modulation level.

chroma the portion of the video signal defining the color information in the image. The chroma signal is defined by changes in the 3.579545 Mhz interlaced sinewave. Phase changes create changes in color, peak-to-peak changes in the sinewave alter the saturation of the color while changes in the DC level of the chroma signal alter the luminance (brightness).

chromatic aberration the failure of a lens to simultaneously focus all colors of light. It arises since the refractive index of a material depends on the wavelength of light.

chromatic color perceived color that possesses hue.

chromaticity the ratio of tune spread to momentum spread of the beam. Chromaticity affects the focusing and bending properties of magnets by making them sensitive to particle momentum. This results in focusing and bending dispersion of the beam in a manner analogous to an optical system.

chromaticity specification of color stimuli chromaticity coordinates relative to the R, G, B values correlated with hue and saturation.

chrominance (1) the color information in the video signal that is defined in terms of hue and saturation.

(2) the component of color which is independent of, and complementary to luminance; chrominance is 2-D: for example, it can be decomposed into hue and saturation. *See* hue, Intensity, luminance, saturation.

chronaxie the minimum duration of a unidirectional square-wave current needed to excite a nerve when the current magnitude is twice rheobase.

CIE *See* Commision International d'Eclairage.

CIE diagram the projection of the plane $(X + Y + Z) = 1$ onto the XY plane, where X, Y, Z are the respective tristimulus values as defined by the CIE (tristimulus values and Commision Internationale de l'Eclairage). The CIE diagram shows all of the visible chromaticity values and maps all colors with the same chromaticity but different value (luminances) onto the same points.

CIM *See* computer-integrated manufacturing.

CIR *See* carrier-to-interference ratio.

circle detection the location of circles in an image by a computer. Often accomplished with the Hough transform.

circle diagram (1) graphical representation of the operation of an induction machine. It is based on the approximate equivalent circuit and expresses stator and rotor current relations for all operating modes (motor, braking, generator) and all values of slip. Several variations of the diagram exist.
(2) graphical representation of the power flow through a transmission line. The maximum power flow through the line can be determined by the impedance of the line.

circuit a physical device consisting of an interconnection of elements, or a topological model of such a device. For example, an electric circuit may be constructed by interconnecting a resistor and a capacitor to a voltage source. A representation of this circuit is shown by the diagram in the figure.

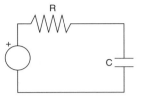

Circuit example.

circuit (STM) switching technology that provides a direct connection between two endpoints; data is transferred directly between the endpoints of a circuit without being stored in any intermediate nodes.

circuit breaker a device that makes and breaks the electrical contact between its input and output terminals. The circuit breaker is capable of clearing fault currents (tripping) as well as load currents. The circuit breaker consists of power contacts with arc clearing capability and associated control and auxiliary circuits for closing and tripping the breaker under the required conditions.

circuit protection devices or control measures used to safeguard electrical circuits from unsafe operating regions, such as overcurrents and overvoltages.

circuit switching a method of communication in which a physical circuit is established between two terminating equipments before communication begins to take place. This is analogous to an ordinary phone call.

circuit-set a closed path where all vertices are of degree 2, thus having no endpoints in the path.

circuit-switched service a telecommunications service, where the communications resource is retained for the whole period of communication. For example wired telephone services or mobile telephone services in the first- and second-generation systems can be classified as circuit-switched services.

circulant matrix a square $N \times N$ matrix $M = \{m_{i,j}\}$ such that $m_{i,j} = m_{(i+n)\bmod N, (j+n)\bmod N}$; that is, that each row of M equals the previous row rotated one element to the right. All circulant matrices are diagonalized by the discrete Fourier transform.

circular cavity a section of the circular waveguide closed at both ends by conducting plates.

circular convolution *See* periodic convolution.

circular mil the area of a circle which measures 0.001 inch in diameter.
Used to specify the cross-sectional area of a wire.

circular polarization a polarization state of a radiated electromagnetic field in which the tip of the electric field vector traces a circle as a function of time for a fixed position. The sense of rotation of the electric field vector is either right-hand or left-hand (clockwise or counter-clockwise).

circular register buffer a set of general purpose CPU registers organized to provide a large number of registers, which may be accessed a few at a time. The group of registers accessible at any particular time may be readily changed by incrementing or decrementing a pointer, with wraparound occurring from the highest numbered registers to the lowest numbered registers, hence the name circular register buffer. There is overlap between the groups of adjoining registers that are accessible when switching occurs. The overlapping registers can be used for passing arguments during subroutine calls and returns. The circular register buffer is a feature of the SPARC CPU architecture. In the SPARC CPU there are 256 registers, available 32 at a time, with an overlap of eight registers above and eight registers below the current group.

circular self-test path a BIST technique based on pseudorandom testing assured by arranging flip-flops of a circuit (during test) in a circular register in which each flip-flop output is ex-ored with some circuit signal and feeds the input of the subsequent flip-flop. This register simultaneously provides test pattern generation and test result compaction.

circularity measure the size invariant ratio of area divided by perimeter squared for small shapes, and much used as a preliminary discriminant or measure of shape: so-called because it is a maximum for circular objects.

circularly polarized light light composed of two orthogonal polarizations that are 90 degrees out of phase; the resultant light amplitude vector thus rotates about the direction of propagation at the optical frequency.

circulator a multiport nonreciprocal device that has the property that an electromagnetic wave incident in port 1 is coupled to port 2 only, an electromagnetic wave incident in port 2 is coupled into port 3 only and so on.

CIRF *See* cochannel interference reduction factor.

CISC processor *See* complex instruction set computer.

Citizen's band (cb) 40 channels where the carrier frequency is between 26.965 MHz and 27.405 MHz established by the FCC for short-distance personal or business communication.

city-block distance a distance measure between two real valued vectors $(x_1, x_2, \ldots, x_n)$

and $(y_1, y_2, \ldots, y_n)$ defined as

$$D_{city\ block} = \sum_{i=1}^{n} |x_i - y_i|$$

City-block distance is a special case of Minkowski distance when $\lambda = 1$. *See also* Minkowski distance.

cladding   the optical material that concentrically surrounds the fiber core and provides optical insulation and protection for the core. The refractive index of the cladding must be lower than that of the core material so that optical power is guided through the fiber by total internal reflection at the core-cladding boundary. *See also* total internal reflection, Snell's law.

clamping   the process of fixing either the minima or maxima of a voltage.

Clapp oscillator   an oscillator whose frequency is determined by a tuned parallel LC circuit with a split capacitance, i.e., two series capacitances, in the capacitive branch and an additional series tuning capacitance in the inductive branch. The Clapp oscillator is a variation of the Colpitts oscillator.

class   (1) in general, patterns are commonly discriminated into different categories according to certain properties they share. The categories in which a given set of patterns are partitioned are referred to as classes.
(2) in object orientation, is an entity that defines a set of objects which share the same attributes and processes.

class fuse   *See* U.L. classes.

class A amplifier   an amplifier in which the active device acts as a modulated current source biased midway between saturation and conduction cutoff. In a class A amplifier, as the amplitude of an applied sinusoidal signal is increased, the output will start to clip at both ends simultaneously. This is equivalent to a conduction angle of 360 degrees

as long as the output signal is not clipping, which is avoided. This term is often used to include any amplifier operating with signal levels low enough such that signal clipping is not present (i.e., small signal conditions).

class A-B amplifier   most current source amplifiers fall into this category, which includes all amplifiers biased somewhere between class A and class B. As the amplitude of a sinusoidal signal is increased, the output will start to cut off first. Further increases will cause clipping due to saturation. Thus the conduction angle is between 180 and 360 degrees, dependent on applied signal amplitude. Device saturation is usually avoided.

class B amplifier   an amplifier in which the active device acts as a modulated current source biased at conduction cutoff. In a class B amplifier, an applied sinusoidal signal will result in only half of the sinusoid being amplified, while the remaining half is cut off. Further increases in the signal amplitude will eventually cause the remaining half of the signal to saturate and clip, which is usually avoided. This is equivalent to a conduction angle of 180 degrees, regardless of signal amplitude.

class B-D amplifier   switched mode amplifier where the device is biased at cutoff, and the input signal is large enough to drive the amplifier into heavy saturation such that only a small percentage of time is spent in transition. The amplifier is literally switched between cutoff and saturation, and thus the saturation angle is a significant percentage of the conduction angle, which is 180 degrees. The unfiltered, broadband output current waveform of a class B-D amplifier resembles a stepped squarewave. It is important to note that only frequency related information (FM) is preserved in a class B-D amplifier, while all amplitude information (AM) is lost. Usually, class B-D power amplifiers are designed in a push-pull configuration to take advantage of both halves of a cycle.

class B-E amplifier transient switched mode amplifier where the device is biased at cutoff, the input signal is large enough to drive the amplifier into heavy saturation such that only a small percentage of time is spent in transition, and the design is such that during saturation the waveform is determined by the switch circuit transient response, while the waveform during cutoff is determined by the transient response to the entire circuit, including the load. The amplifier is literally switched between cutoff and saturation, the transient responses are well controlled, and thus the saturation angles approach the conduction angle, which is 180 degrees. The final tuned output current wave form of a class B-E amplifier resembles an ideal squarewave. It is important to note that only frequency related information (FM) is preserved in a class B-E amplifier, while all amplitude information (AM) is lost.

class C amplifier a current source amplifier biased beyond the conduction cutoff such that operation will not begin until the input signal reaches a specific amplitude, and results in less than half of an input sinusoid being amplified. If the signal amplitude is increased sufficiently, saturation and the associated clipping will occur. Thus the conduction angle is between 0 and 180 degrees, regardless of amplitude. Device saturation is usually avoided.

class D amplifier switched mode amplifier where the device is biased somewhere between class A and class B cutoff, and the input signal is large enough to drive the amplifier from cut-off to heavy saturation such that only a small percentage of time is spent in transition. The amplifier is literally switched between cutoff and saturation, and thus the saturation angle is a significant percentage of the conduction angle, which is 180 degrees. The unfiltered, broadband output current waveform of a class D amplifier resembles a stepped squarewave. It is important to note that only frequency related information (FM) is preserved in a class D amplifier, while all amplitude information

(AM) is lost. Usually, class D power amplifiers are designed in a push-pull configuration to take advantage of both halves of a cycle.

class E amplifier a transient switched mode amplifier where the device is biased somewhere between class A and class B cutoff, the input signal is large enough to drive the amplifier into heavy saturation such that only a small percentage of time is spent in transition, and the design is such that during saturation the waveform is determined by the switch circuit transient response, while the waveform during cutoff is determined by the transient response to the entire circuit, including the load. The amplifier is literally switched between cutoff and saturation, the transient responses are well controlled, and thus the saturation angles approach the conduction angle, which is 180 degrees. The final tuned output current waveform of a class E amplifier resembles an ideal squarewave. It is important to note that only frequency related information (FM) is preserved in a class E amplifier, while all amplitude information (AM) is lost.

class E-F amplifier a harmonic tuned or harmonic reaction amplifier (HRA) in which devices, biased for class B operation, are arranged in a push/pull configuration, and are utilized to inject each other with large harmonic currents in order to modulate the amplitude of the fundamental output current through the device, resulting in improved switching efficiency. The even order harmonics must be shorted at the output, while the odd order harmonics must be provided an open at the output.

class F amplifier a high-efficiency operation in amplifiers. The class F amplifier has a load impedance optimized not only for a fundamental wave but also for harmonic waves to improve efficiency. An efficiency of the class F amplifier is 100% under an ideal condition, where the optimum load impedance for even harmonic waves is short and that for odd harmonic waves is open.

class G amplifier   a frequency multiplying or harmonic amplifier biased somewhere between class A and class C, in which the input is tuned to the fundamental input frequency and the output is tuned to a frequency multiple of the input.

class H amplifier   frequency mixing amplifier biased somewhere between class A and class C, in which the inputs are tuned to the input frequencies and the output is tuned to either the sum or difference frequency.

class S amplifier   sampling or pulse width modulation amplifier in which a sampling circuit (or pulse width modulator), pulse amplifier and a low pass filter are cascaded. The input signal is sampled at a significantly higher rate than the input frequency (this requires a high frequency sampling signal), and the original signal is transformed into a constant amplitude pulse chain in which the pulse widths are proportional to the original signal's amplitude. The resulting pulse chain is amplified using any of the highly efficient switching methods desired. The output is then demodulated using a low pass filter, replicating the original signal. It is important to note that rapid variations in the input signal amplitude relative to the sampling signal will cause significant distortion or loss of information.

classified VQ   *See* classified vector quantization.

classified vector quantization (CVQ)   a vector quantization technique where different codebooks are developed based on image edge features. The codebook used to encode a particular block is determined by a classifier with differentiating capability between the types of features. A number of codebooks are developed each to encode blocks of pixels containing specific types of features. *See also* vector quantization.

classifier   a method of assigning an object to one of a number of predetermined classes.

clean cache block   a cache block (or "line") is clean if it is a copy of the information stored in memory. A clean block can be overwritten with another block without any need to save its state in memory.

clear   (1) to set the value of a storage location to zero (often used in the context of flip-flops or latches).
(2) clearing a bit (register) means writing a zero in a bit (register) location. Opposite to "set."

clearing time   the total time required to melt and clear, and thus totally open, a fuse-type overcurrent device.

cleaved coupled cavity   semiconductor laser configuration in which the amplifying region has been cleaved to introduce a mid-cavity reflecting boundary; added reflector is intended to improve mode-selectivity characteristics.

click noise   in a fading channel, the noise associated with a threshold crossing. In a fading radio channel situation, the moving user crosses the standing wave patterns in the propagation environment. As the user crosses the minima, the service quality temporarily downgrades and, in analogue systems, is noticeable as clicking.

climbers   two metal spikes, each of which is strapped to the inside of a line worker's legs, pointing down near the ankle. Plunged into the sides of a wooden utility pole, they provide purchase for the worker to scale the pole.

clipping   nonlinear distortion that occurs when the input to an amplifier exceeds the amplifier's linear range. The amplifier output saturates at its limit, giving a "clipped" appearance to the output waveform.

clock   (1) the oscillator circuit that generates a periodic synchronization signal.

(2) a circuit that produces a series of electrical pulses at regular intervals that can be used for timing or synchronization purposes.

clock cycle   one complete event of a synchronous system's timer, including both the high and low periods.

clock doubling   a technique in which the processor operates internally at double the external clock frequency.

clock duty cycle   the percentage of time that the electronic signal remains in the true or 1 state.

clock pulse   a digital signal that, via its rising edge or falling edge, triggers a digital circuit. Flip-flops and counters typically require clock pulses to change state.

clock recovery   in synchronous systems, the act of extracting the system clock signal from the received sequence of information symbols. *See also* symbol synchronization.

clock replacement algorithm   a page replacement algorithm described as follows: A circular list of page entries corresponding to the pages in the memory is formed. Each entry has a use bit which is set to a 1 when the corresponding page has been referenced. A pointer identifies a page entry. If the use bit of the page entry is set to a 1, the use bit is reset to a 0 and the pointer advances to the next entry. The process is repeated until an entry is found with its use bit already reset, which identifies the page to be replaced. The pointer advances to the next page entry for the next occasion that the algorithm is required. The word "clock" comes from viewing the pointer as an arm of a clock. Also known as a first-in-not-used-first-out replacement algorithm.

clock skew   the phenomenon where different parts of the circuit receive the same state of clock signal at different times because it travels in wires with different lengths. This skew of the signals causes a processing element to generate an erroneous output. Distribution of the clock by means of optical fibers, waveguides, a lens, or a hologram, eliminates clock skew.

clock speed   the rate at which the timing circuit in a synchronous system generates timing events.

closed convex set   a set of vectors C such that of $\mathbf{x}, \mathbf{y} \in C$ then $\lambda \mathbf{x} + (1-\lambda)\mathbf{y} \in C$ for all $0 \leq \lambda \leq 1$.

closed kinematic chain   in vision engineering, a sequence of links which forms a loop.

closed-loop control   control action achieved by a closed feedback loop, i.e., by measuring the degree to which actual system response conforms to desired system response and applying the difference to the system input to drive the system into conformance.

closed-loop DC motor acceleration   the use of sensors to provide feedback to the motor control circuit indicating the motor is actually accelerating before the starting resistors are removed from the armature circuit. Two popular methods to sense motor acceleration are CEMF coils, and current sensing coils.

closed-loop gain   the gain of an operational amplifier circuit with negative feedback applied (with the negative feedback loop "closed").

closed-loop optimal control   operation or structure of the controller with the decision mechanism which, under uncertainty, uses in the best possible way — with respect to a given *a priori* criterion — all information available to the controller; in particular, the closed-loop optimal control takes into account all future time instants at which new decisions will be made; the best known example of a closed-loop optimal control rule is the solution of the linear-quadratic-Gaussian problem (LQG problem).

closed-loop system any system having two separate paths inside it. The first path conducts the signal flow from the input of that system to the output of that same system. The second path conducts the signal flow from the output to the input of the system, thus establishing a feedback loop for the system. See the figure below for a general description of a closed-loop system. The

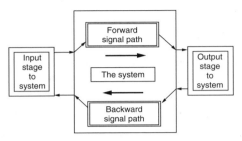

A closed-loop system.

forward and backward signal path construct the closed loop, which conduct the signal flow from the input stage to the output stage and then back to the input stage of that system.

closing a basic morphological operation. Given a structuring element B, the closing by B is the composition of the dilation *by B* followed by the erosion by B; it transforms X into $X \bullet B = (X \oplus B) \ominus B$. The closing by B is what one calls an algebraic closing; this means that: (*a*) it is a morphological filter; (*b*) it is extensive, in other words it can only increase an object. *See* dilation, erosion, morphological filter, structuring element.

closing/opening filter one of an important class of morphological filters. Let γ and φ be opening and closing operators respectively. The following operators can be obtained by composing γ and φ (i.e., by applying them in succession): $\gamma\varphi$, $\varphi\gamma$, $\gamma\varphi\gamma$, and $\varphi\gamma\varphi$. These are all morphological filters, and collectively they are called closing/opening filters or opening/closing filters. No further operators can be obtained by composing γ and φ. *See* closing, morphological filter, opening.

clothes pin slang for a wood or plastic clip used to secure a blanket to conductors.

CLT *See* central limit theorem.

cluster a group of data points on a space or a group of communicating computer machines. A cluster of computers on a local network can be installed to provide their service as a unique computer. This is frequently used for building large data storage and Web servers. In computer disks, a cluster consists of a fixed number of sectors. Each sector contains several bytes, for example 512.

cluster analysis in pattern recognition, the unsupervised analysis of samples in order to cluster them into classes based on (a) a distance metric and (b) a clustering algorithm. Typical algorithms minimize a cluster criterion (e.g., representation error) by grouping samples hierarchically or by iteratively reassigning samples to clusters. The K-means algorithm is an example of the latter. In the case of 2-D measurements, cluster analysis becomes a method of image segmentation.

clustering (1) any algorithm that creates the major clusters from a given set of patterns.
(2) a method of unsupervised learning that aims to discover useful structure in unlabeled data by grouping similar patterns.
See hierarchical clustering. *See also* distance measure, similarity measure.

clutter the name given to background signals which are currently irrelevant to a detection system; clutter is a form of structured noise. *See also* reverberation.

CMA *See* constant modulus algorithm.

CMAC network *See* cerebellar model articulation network.

CMMR *See* common mode rejection ration.

CMOS *See* complementary metal oxide semiconductor.

CO abbreviation for central switching office.

CO$_2$ *See* carbon dioxide.

CO$_2$ laser *See* carbon dioxide laser.

co-linear array a phased array of straight elements in which the axes of the elements lie along a common line. The elements are typically center fed half-wave dipoles or folded dipoles.

co-occurrence matrix an array of numbers that relates the measured statistical dependency of pixel pairs. Co-occurence matrices are used in image processing to identify the textural features of an image.

co-prime polynomials polynomials that have no common factors. For example, polynomials $(s^2 + 9s + 20)$ and $(s^2 + 7s + 6)$ are co-prime, while $(s^2 + 5s + 6)$ and $(s^2 + 9s + 14)$ are not, since they have a common $(s + 2)$ factor.

co-tunneling a cooperative process, whereby electrons can tunnel through two series connected tunneling barriers. In this process, the tunneling of an electron through one of the barriers causes a self-consistent shift of potential onto the second barrier, which results in the immediate inducing of a particle tunneling through this latter barrier.

coarticulation the transient process corresponding to the utterance of two phonemes. It is due to the movement of the articulatory organs between the different positions corresponding to the two phonemes.

coax *See* coaxial cable.

coaxial cable A transmission line formed by two concentric conductors separated by a dielectric designed to confine the fields and their energy in the medium between said conductors. It is often used in applications where signal interference between the cable and its surrounds must be kept to a minimum. Also called coax.

coaxial magnetron a radial magnetron where the anode and cathode are gradually transformed into a coaxial line.

cochannel interference (CCI) interference caused by radio transmitters operating on the same radio frequency as that of a particular wanted radio frequency signal.

cochannel interference reduction factor (CIRF) a key factor used to design a cellular system to avoid the cochannel interference.

cochannel reuse ratio (CRR) the reuse ratio between radio communication cells using the same radio channels.

cochannels radio channels occupying the same radio frequency allocation n.

cochlea a snail-shaped passage communicating with the middle ear via the round and oval window. Its operation consists of transducing the acoustical vibration to nerve impulses, subsequently processed in the brain.

Cockroft–Walton circuit a cascading voltage multiplier invented in 1932 by John Cockroft and Ernest Walton.

code (1) a technique for representing information in a form suitable for storage or transmission.
(2) a mapping from a set of messages into binary strings.

code acquisition the process of initial code synchronization (delay estimation) between the transmitter and receiver in a spread-spectrum system before the actual data transmission starts. It usually requires the transmission of a known sequence. *See also* code tracking.

code cache   a cache that only holds instructions of a program (not data). Code caches generally do not need a write policy, but see self-modifying code. Also called an instruction cache. *See also* cache.

code combining   an error control code technique in which several independently received estimates of the same codeword are combined with the codeword to form a new codeword of a lower rate code, thus providing more powerful error correcting capabilities. This is used in some retransmission protocols to increase throughput efficiency.

code converter   a device for changing codes from one form to another.

code division multiple access (CDMA)   a technique for providing multiple access to common channel resources in a communication system. CDMA is based on spread spectrum techniques where all users share all the channel resources. Multiple users are distinguished by assigning unique spreading codes to each user. Traditionally, individual detection is accomplished at the receiver through correlation or matched filtering.

code efficiency   the unitless ratio of the average amount of information per source symbol to the code rate, where the amount of information is determined in accordance with Shannon's definition of entropy. It is a fundamental measure of performance of a coding algorithm.

code excited linear prediction (CELP)   a class of linear predictive speech coding methods where the excitation is composed of sample vectors from VQ codebooks.

code hopping   the use of a new spreading code for each transmitted bit in a spread-spectrum system. *Compare with* frequency hopping.

code letter   *See* NEMA code letter.

code rate   in forward error control and line codes, the ratio of the source word length to the code word length, which is the average number of coded symbols used to represent each source symbol.

code segment   area in a process' virtual address space used to contain the program's instructions.

code tracking   the process of continuously keeping the code sequences in the receiver and transmitter in a spread-spectrum system synchronized during data transmission. *See also* code acquisition.

code V   a widely employed computer code for design of optical systems by Optical Research Associates.

codebook   a set of codevectors (or codewords) that represent the centroids of a given pattern probability distribution. *See also* vector quantization.

codebook design   a fundamental problem in vector quantization (VQ). The main question addressed by codebook design is how the codebook should be structured to allow for efficient searching and good performance. Several methods (tree-structured, product codes, M/RVQ, I/RVQ, G/SVQ, CVQ, FSVQ) for codebook design are employed to reduce computational costs low. *See also* vector quantization, tree-structured VQ.

codebook generation   a fundamental problem in vector quantization. Codebooks are typically generated by using a training set of images that are representative of the images to be encoded. The best training image to encode a single image is the image itself. This is called a local codebook. The main question addressed here is what codevectors should be included in the codebook. *See also* vector quantization.

codebook training   the act of designing a code-book for a source coding system. The LBG algorithm is often used to design the codebook for vector quantizers.

codec   word formed from encoder and decoder. A device that performs encoding and decoding of communications protocols.

coded modulation   an integrated modulation and coding approach for bandwidth-constrained channel where the redundancy introduced by the code is compensated by increasing the number of signals, for performance improvement without additional bandwidth or transmission power.

codeword   the channel symbol assigned by an encoder to a source symbol. Typically the codeword is a quantized scalar or vector.

coding   the process of programming, generating code in a specific language. The process of translating data from a representation form into a different one by using a set of rules or tables. *See also* ASCII code, EBCDIC, binary.

coding at primary rates for videoconferencing *See* image coding for videoconferencing.

coding gain   (1) the reduction in signal-to-noise ratio required for a specified error performance in a block or convolutional forward error control system as compared to an uncoded system with the same information rate, channel impairments, and modulation and demodulation techniques. In a trellis coded modulation system, it is the ratio of the squared free distance in the coded system to that of the uncoded system.

(2) the difference between the SNR/bit (dB) required for an uncoded and a coded system to attain the same arbitrary error probability. Depends on the code parameters and also on the SNR per bit.

coding of graphics   use of a representation scheme for graphics. Graphics coding is typically a two-level coding scheme. Both exact and approximate methods are applicable to this type of coding. Run-length coding, predictive coding, line-to-line predictive differential coding, and block coding are typical for graphic coding.

coding of line drawings   use of a representation scheme for line drawings. Line drawings are typically coded using chain codes where the vector joining two successive pixels are assigned a codeword. Higher efficiency is obtained by differential chain coding in which each pixel is represented by the difference between two successive absolute codes.

coding redundancy   redundancy.

coefficient of thermal expansion (CTE) mismatch   the difference between the coefficients of thermal expansion of two components, i.e., the difference in linear thermal expansion per unit change in temperature. (This term is not to be confused with thermal expansion mismatch).

coefficient of utilization (CU)   the ratio of the lumens reaching the working plane to the total lumens generated by the lamp. This factor takes into account the efficiency and distribution of the lumenaire, its mounting height, the room proportions, and the reflectances of the walls, ceiling, and floor.

coefficient sensitivity   let a transfer function be a ratio of polynomials

$$F(s) = \frac{N(s)}{D(s)} = \frac{a_0 + a_1 s + \ldots + a_m s^m}{d_0 + d_1 s + \ldots + d_n s^n}$$

in which the coefficients a_i and d_i are real and can be functions of an arbitrary circuit element x. For such an element x one may define the relative coefficient sensitivities as follows:

$$\mathbf{S}_x^{a_i} = \frac{\partial a_i}{\partial x} \frac{x}{a_i} \qquad \mathbf{S}_x^{d_i} = \frac{\partial d_i}{\partial x} \frac{x}{d_i}$$

The relationship between the function sensitivity and coefficient sensitivities can be established as well.

coercive field   when a ferroelectric material is cycled through the hysteresis loop the coercive field is the electric field value at which the polarization is zero. A material has a negative and a positive coercive field and these are usually, but not always, equal in magnitude to each other.

coercive force   the demagnetizing field applied to a permanent magnet that reduces its magnetic induction to zero; the x-intercept of the normal demagnetization curve. A commonly listed material property that indicates magnet performance in static conditions.

coercivity   *See* coercive force. *See also* intrinsic coercive force.

COGEN   *See* cogeneration.

cogeneration   (1) any of a number of energy generation systems in which two (or more) forms of energy are produced in forms practical for use or purchase by an end user. Typical systems produce electrical energy for sale to a utility and process steam for local space heating or other process uses. Cogeneration designs are generally adopted to increase the overall efficiency of a power generation process.

(2) typically, the production of heat energy, e.g., to heat buildings, as an adjunct to the production of electric power.

cognitive map   the cognitive map, introduced by R. Axelrod to study decision making processes, consists of points, or nodes, and directed links between the nodes. The nodes correspond to concepts. *See also* fuzzy cognitive map.

coherency   a property of a control area that has stiff interconnections between generators. Such an area thus may be described with the use of only a single frequency state variable.

coherence   (1) measure of the extent to which knowledge of a field at one point in space permits prediction of the field at another point.

(2) in an optical fiber.

(3) the coherence between two wide-sense stationary random processes is equal to the cross power spectrum divided by the square root of the product of the two auto-power spectra. The magnitude of the coherence so defined is thus between 0 and 1.

coherence bandwidth   the bandwidth over which the effect of communication channel can be assumed constant. Signals of bandwidth less than this can be transmitted without significant distortion.

coherence distance   *See* coherence length.

coherence length   distance over which the amplitude and phase of a wave can be predicted.

coherence time   the time over which the effect of communication channel can be assumed constant. Signals of duration less than this can be transmitted without significant distortion.

coherent   integration where magnitude and phase of received signals are preserved in summation.

coherent acousto-optical processor   acousto-optical (AO) signal processor where the light is amplitude-modulated by the acoustic wave in the AO device as opposed to intensity or power modulated.

coherent detection   detection technique in which the signal beam is mixed with a locally generated laser beam at the receiver. This results in improved receiver sensitivity and in improved receiver discrimination between closely spaced carriers.

coherent illumination   a type of illumination resulting from a point source of light that illuminates the mask with light from only one direction. This is more correctly called "spatially coherent illumination."

coherent light   light having a relatively long coherence length; laser light.

coherent optical communication   optical communication approaches where information is conveyed in the phase of the optical signal, therefore requiring that the phase of the optical sources be well controlled. *See also* optical communications.

coherent population trapping   a technique for creating a quantum mechanical coherence in a lambda system by a dissipative process. Ideally, the intermediate excited state in the lambda system decays rapidly compared to the other two states. The coherence arises because a particular linear combination of ground states is not coupled to the excited state. Atoms accumulate into this uncoupled superposition state by a process analogous to optical pumping, thereby creating a quantum mechanical coherence. This system often has counterintuitive properties, because it uses a dissipative process to create, rather than destroy, a quantum coherence.

coil   a conductor shaped to form a closed geometric path. Note that the coil will not be a closed conducting path unless the two ends of the coil are shorted together. Coils may have multiple turns, and may have various constructions including spool, preformed, and mush-wound. The coil may be wrapped around an iron core or an insulating form, or it may be self-supporting. A coil offers considerable opposition to AC current but very little to DC current.

coil pitch   *See* coil span.

coil side   that portion of a motor or generator winding that cuts (or is cut by) lines of magnetic flux and, thus, contributes to the production of torque and Faraday EMF in the winding.

coil span   the distance, measured either in number of coil slots or in spatial (mechanical)

degrees, between opposite sides of a winding of an electric machine. A full-span (full-pitch) winding is one in which the winding span equals the span between adjacent magnetic poles. Windings with span less than the distance between adjacent magnetic poles are called short-pitch, fractional-pitch, or chorded windings. Also called coil pitch.

cold plasma   a simplified model of the plasma state where the effects that depend on electron temperature are neglected. The particles are assumed to have no kinetic thermal motion of their own. The particles are at rest except for their induced velocities through the action of the self consistent electromagnetic fields.

cold reserve   the state of an idle thermal generating plant whose boilers and turbine are cold and must be brought up to operating temperature before power can be generated.

cold start   (1) a complete reloading of the system with no reassumption. All executed processes are lost.

(2) the starting of a computer system from a power-off condition.

(3) the state from which a thermal generation unit must be brought after being in cold reserve.

cold start miss   in a cache, a cold start miss occurs when a computer program is referencing a memory block for the first time, so the block has to be brought into the cache from main memory. Also called first reference misses or compulsory misses. When the cache is empty, all new memory block references are cold misses. *See also* capacity miss, conflict miss.

collapsible reel   a take-up reel used in line work which fits on the power-take-off of a line truck.

collector wall   the collector of a bipolar transistor is located below the surface of the substrate. The wall, or sidewall, is the vertical boundary of the collector that meets the substrate material.

The boundary usually forms a p-n junction that provides isolation from.

collet   a circular spring fingerstock connection element for a power vacuum tube.

colliding-pulse-modelocked (CPM) laser   a dye laser resonator design for producing femtosecond pulses; right and left travelling pulses collide in a thin intracavity absorber.

collimated   beam with nearly flat phase fronts and slow longitudinal variations of the transverse amplitude distribution.

collinear geometry acousto-optical tunable filter   acousto-optical tunable filter device where the acoustic and light waves propagate in the same direction. Also abbreviated collinear geometry AOTF.

collinear geometry AOTF   *See* collinear geometry acousto-optical tunable filter.

collision   (1) in a pipeline, a situation when two or more tasks attempt to use the same pipeline stage at the same time.

(2) in a hash table, when $n + 1$ different keys are mapped by a hashing function to the same table index (where n entries can be stored).

collision broadening   broadening of the spectral profile of an amplifying or absorbing transition due to inelastic or phase-interrupting elastic collisions.

collision vector   a binary number in which the ith bit is a 1 if submitting a task into the pipeline i cycles after a task will cause a collision.

color   (1) visual sensation associated with the wavelength or frequency content of an optical signal.

(2) representation a method of defining a signal or an image pixel value to be associated with a color index.

color blooming   phenomenon where the excess charge at a photo receptor can spread to neighboring receptors, and change their values in proportion to the overload. For RGB cameras, this effect can modify not only the luminance but also the chrominance of pixels. *See* color clipping, chrominance, luminance.

color burst   burst of eight to ten cycles of the 3.579545 MHz (3.58 MHz) chrominance subcarrier frequency that occurs during the horizontal blanking of the NTSC composite video signal. The color burst signal synchronizes the television receiver's color demodulator circuits.

color clipping   phenomenon where the intensity of the light on a photoreceptor exceeds some threshold, the receptor becomes saturated and its response is no longer linear, but limited to some bound. For RGB cameras, this effect can modify not only the luminance but also the chrominance of pixels. *See* color blooming, chrominance, luminance.

color coding   the process of identifying components' values and tolerances by means of a set of colored bands or dots.

color correction   in practical photometry it is known that the system used to measure luminescence will not possess the standard eye spectral response as specified by the 1931 International Commission on Illumination (CIE). The measurement "system" will undoubtedly consist of a photodetector, an optical filter and associated lenses. Unfortunately, system output is highly dependent on its spectral response.

color difference signals   the chrominance signal component that results from subtracting the luminance (Y) component from a primary color. The luminance signal corresponds to the changes in brightness as from a monochrome video signal. The three primary colors for color television signals are located in Maxwell's chromaticity diagram. Red is at a wavelength of 0.7 micrometer, green at 0.546 micrometer and blue at 0.436 micrometer. The luminance signal component results from the matrix addition of the primary colors. The matrix proportions are 30% red, 59% green, and 11% blue. Two color difference signals, (R-Y) and (B-Y) are sufficient to convey all the information necessary to reproduce full color at the TV receiver. The color difference signal (G-Y) can be determined by proper proportions of the (R-Y), (B-Y) and Y signals at the receiver.

color graphics adapter (CGA) a video adapter proposed by IBM in 1981. It is capable of emulating MDA. In graphic mode, it allows one to reach 640×200 (wide per high) pixels with 2 colors or 320×200 with 4 colors.

color image coding compression of color images is usually done by transforming RGB color space into a YC_1C_2 space, where Y represents luminance and C_1 and C_2 are color difference signals. The C_1 and C_2 signals are then subsampled, but coded with the same algorithm as the Y signal. Standard algorithms do not attempt to exploit correlations between the three signals.

color matching the process of mixing three fixed and independent primary colors so that an observer (trichromat) interprets the formulation as being the same as a specified but arbitrary color. In color television, the three primary colors are fixed at specific wavelengths bands λ_R, λ_G, and λ_B corresponding to colors red, green, and blue.

color preference index (CPI) measure appraising a light source for appreciative viewing of colored objects or for promoting an optimistic viewpoint by flattery.

color saturation a color with the dominant wavelength located at the periphery of Maxwell's chromaticity diagram. A fully saturated color is pure because it has not been contaminated by any other color or influence.

color signal the portion of a modulated signal that determines the colors of the intended output display.

color space the space $\mathcal{C}$ within which colors are represented in the image function $I : \mathcal{R}^2 \rightarrow \mathcal{C}$.

color temperature the color a black object becomes when it is heated. The standard color "white" occurs when a tungsten filament is heated to a temperature of 6800 degrees Kelvin. The temperature of 6800 K corresponds to a standard white raster as defined by the NTSC. The color temperature for white is useful for comparing color matching and color decoding among different displays that use different color phosphors. The standard "white" is obtained by mixing the 30% red, 59% green, and 11% blue color signals. Differences in the color saturation for the different phosphors found in television CRTs will modify the required proportions of red, green, and blue to produce the standard "white."

color-bar patterns a standard color-bar pattern for an NTSC video signal consisting of a composite video signal containing a 77% and a 100% white chip, and yellow, cyan, green, magenta, red, blue, and black chips. These patterns represent ideal color and luminance levels that can be input in a video system for setting levels and verifying system performance.

Colpitts oscillator a particular case of an LC-oscillator when X_1 and X_2 are capacitors (hence, $X_m = 0$), X_3 an inductance.

column decoder logic used in a direct-access memory (ROM or RAM) to select one of a number of rows from a given column address. *See also* two-dimensional memory organization.

column distance the minimum Hamming distance between sequences of a specified length encoded with the same convolutional code that differ in the first encoding interval.

column-access strobe   *See* two-dimensional memory organization.

comb filter   an electric wave filter that exhibits an amplitude versus frequency plot of periodically spaced pass bands interspersed with periodic stop bands. This plot resembles the teeth of an ordinary hair comb, from which the filter derives its name.

comb-line filter   filter consisting of parallel coupled transmission line resonators where all resonators are grounded on one side and capacitively loaded to ground on the other. Adjacent resonators are grounded on the same side. When fabricated as strip conductors in microstrip or stripline form, the metalized patterns have the appearance of a comb.

comb function   a function made of evenly spaced, equal amplitude time or frequency components (the Fourier transform of the comb function is another comb function). The comb function is useful for discretizing continuous signals and can be represented as the infinite sum of delta functions evenly spaced through time or frequency.

combination tone   various sum and difference frequency that are generated when two intense monochromatic fields interact with the same semi-classically described laser medium.

combinational lock   interconnections of memory-free digital elements.

combinational logic   a digital logic, in which external output signals of a device are totally dependent on the external input signals applied to the circuit.

combined cycle plant   a gas-turbine power plant in which the exhaust gases are used to heat water in a boiler to provide steam to run a turbo-generator.

combined field integral equation (CFIE)   a mathematical relationship obtained by combining the electric field integral equation (EFIE) and magnetic field integral equation (MFIE). It is normally used in electromagnetic scattering calculations from a conducting body to avoid non-physical interior resonances that appear by using either EFIE or MFIE alone.

combined source-channel coding   a general term for approaches to source-channel coding, where the source and channel codes are combined into one overall code. In the literature, the term is also used, more loosely, for approaches where (any kind of) joint optimization of the source and channel coding is utilized. Also commonly referred to as joint source-channel coding. *See also* channel optimized VQ.

combo trouble-shooter   in combined electric and gas utilities, a practice which is growing in popularity is the use of combo troubleshooters. The combo troubleshooter is cross trained in both electric and gas service practices. The cross functionality permits more efficient deployment of resources.

come-a-long   a ratcheted winch or block-and-tackle for pulling conductors into place.

Comité Consulatif International Télégraphique et Téléphonique (CCITT)   *International Consultative Committee for Telegraphy and Telephony*. This institution, based in Geneva, Switzerland, issues recommendations concerning all fields related to telecommunications.

command   (1) directives in natural language or symbolic notations entered by users to select computer programs or functions.

(2) instructions from the central processor unit (CPU) to controllers and other devices for execution.

(3) a CPU command, or a single instruction, ADD, LOAD, etc.

Commision International d'Eclairage (CIE) International standards body for lighting and color measurement. Known in English as the International Commission on Illumination.

commit   the phase of a transaction in which the new states are written to the global memory or database. The commit phase should not be started until it has been verified that performing the commit will not violate the system's consistency requirements. In most designs, the commit phase itself must be performed under more strict locking than the remainder of the transaction.

common base amplifier   a single transistor BJT amplifier in which the input signal is applied to the emitter terminal, the output is taken from the collector terminal, and the base terminal is connected to a constant voltage.

common centroid   a technique in the physical design of integrated circuits in which two transistors, which must be matched, are actually composed of multiple devices connected in parallel. By appropriately connecting the multiple devices, the effective center ("centroid") of the two transistors can be located at the same point, thus improving the matching in the presence of nonidealities in the integrated circuit fabrication process. *See also* cross-quad.

common channel signaling (CCS)   a technique for routing signaling information through a packet-switched network.

common collector amplifier   a single-transistor BJT amplifier in which the input signal is applied to the base terminal, the output is taken from the emitter terminal, and the collector terminal is connected to a constant voltage. Also referred to as an emitter follower, since the voltage gain of this configuration is close to unity (the emitter voltage "follows" the base voltage).

common drain amplifier   a single transistor FET amplifier in which the input signal is applied to the gate terminal, the output is taken from the source terminal, and the drain terminal is connected to a constant voltage. Also referred to as a source follower, since the voltage gain of this configuration is close to unity (the source voltage "follows" the gate voltage).

common emitter   a basic transistor amplifier stage whose emitter is common to both input and output loops. It amplifies voltage, current, and hence power.

common emitter amplifier   a single-transistor BJT amplifier in which the input signal is applied to the base terminal, the output is taken from the collector terminal, and the emitter terminal is connected to a constant voltage.

common gate amplifier   a single-transistor FET amplifier in which the input signal is applied to the source terminal, the output is taken from the drain terminal, and the gate terminal is connected to a constant voltage.

common mode gain   for a differential amplifier, the ratio of the output signal amplitude to the amplitude of a signal applied to both the amplifier input terminals (in common). For an ideal differential amplifier, the common mode gain would be zero; the deviation of a real differential amplifier from the ideal is characterized by the common mode rejection ratio (CMRR).

common mode noise   undesired electrical signals in lines that are equal in amplitude and phase with respect to a reference ground. Common mode voltages and currents can be generated by power electronic switching circuits and can interfere with control or other electronic equipment. Common mode currents will also sum into neutrals and grounding conductors, which may cause sensitive fault current detection relays to trip.

common mode rejection ratio (CMRR)   a measure of quality of an amplifier with differential inputs, defined as the ratio between the common-mode gain and the differential gain.

common source amplifier   a single-transistor FET amplifier in which the input signal is applied to the gate terminal, the output is taken from the drain terminal, and the source terminal is connected to a constant voltage.

common-channel interoffice signaling   the use of a special network, dedicated to signaling, to establish a path through a communication network, which is dedicated to the transfer of user information.

common-mode coupling   pick-up from an electromagnetic field that induces a change in potential on both signal leads of equal magnitude and phase relative to the ground reference potential.

communication link   a point-to-point communication system that typically involves a single information source and a single user. This is in contrast to a communications network, which usually involves many sources and many users.

communication theory   *See* information theory.

community-antenna television (CATV)   a television receiving and distribution system in which signals from television stations and sometimes FM stations are received by antennas, amplified, and then distributed to community subscribers via coaxial or fiber-optic cable. The system is known as cable TV.

commutating inductance   in switched circuits (converters, inverters, etc.), the inductance that is in series with the switching elements during the process of commutation from one topological state to another. This inductance results in noninstantaneous commutation due to the fact that current in an inductor cannot change instantaneously.

commutating pole   *See* interpole.

commutating winding   *See* interpole.

commutation   the process by which alternating current in the rotating coil of a DC machine is converted to unidirectional current. Commutation is accomplished via a set of stationary electrical contacts (brushes) sliding over multiple, shaft-mounted electrical contacts that turn with the machine rotor. The contacts are the connection points in a series-connected loop of the coils that make up the rotor winding. The brushes, sliding over these contacts, continually divide the loop into two parallel electrical paths between the brushes.

The brushes are positioned such that they make contact with those commutator segments that are connected to coils that are moving through a magnetic neutral point between poles of the machine's field flux. As a result, all coils making up one parallel path are always moving under a north magnetic pole, and the others are always moving under a south magnetic pole. The movement of the commutator contacts underneath the brushes automatically switches a coil from one path to the other as it moves from a north pole region to a south pole region. Since the coils in both paths move in the same direction, but through opposite flux regions, the voltages induced in the two paths are opposite. Consequently, the positive and negative ends of each path occur at the same points in the series loop, which are at the points where the brushes contact the commutator. The brush positions, thus, represent a unidirectional (or DC) connection to the rotating coil. *See also* commutator.

commutation angle   time in electrical degrees from the start to the completion of the commutation process. Also called overlap angle.

commutativity   a property of an operation; an operation is commutative if the result of the operation is not affected by any reordering of the operands of the operation. Additions and multiplication are commutative, whereas subtraction and division are not.

commutator   a cylindrical assembly of copper segments, insulated from each other, that make electrical contact with stationary brushes, to allow current to flow from the rotating armature windings of a DC machine to the external terminals of the machine. It also enables reversal of current in the armature winding. *See also* commutation.

commutator film   an oxide layer on the commutator surface, indicated by a dark color or a "film," that is required for proper commutator action and full loading of the machine. On a new DC machine commutator, or on a commutator that has just been stoned, there is no "film" on the commutator. It is advisable to refer to the manufacturer's technical manual for the proper procedure to "break in" the commutator and develop the film so the machine can be operated at rated conditions.

compact disk (CD)   a plastic substrate embossed with a pattern of pits that encode audio signals in digital format. The disk is coated with a metallic layer (to enhance its reflectivity) and read in a drive (CD player) that employs a focused laser beam and monitors fluctuations of the reflected intensity in order to detect the pits.

compact disk-interactive (CD-I)   a specification that describes methods for providing audio, video, graphics, text, and machine-executable code on a CD-ROM.

compact range   an electromagnetic measurement facility in which far-field conditions are achieved by the use of an offset parabolic reflector. The reflector is fed using a source antenna or other subreflector system located at its focus. The term "compact" range is used to describe the relative difference in its size compared to a true far-field range requiring a large separation distance between the source antenna and the device under test to achieve the same far-field conditions.

compactness measure   an alternative name for circularity measure.

compander   a point operation that logarithmically compresses a sample into fewer bits before

transmission. The inverse logarithmic function is used to expand the code to its original number of bits before converting it into an analog signal. Typically used in telecommunications systems to minimize bandwidth without degrading low-amplitude signals.

companding   a process designed to minimize the transmission bit rate of a signal by compressing it prior to transmission and expanding it upon reception. It is a rudimentary data compression technique that requires minimal processing.

companion matrix   the coefficient matrix in the state-equation representation of the network describable by a linear differential equation.

comparator   (1) a logic element that compares two binary numbers (A and B) to determine if $A = B$, $A < B$, or $A > B$. An exclusive NOR gate operates like a 1-bit comparator.

(2) a software tool that compares two computer programs, files, or sets of data to identify commonalities or differences. Typical objects of comparison are similar versions of source code, object code, data base files, or test results.

compare instruction   an instruction used to compare two values. The processor flags are updated as a result. For example, the instruction CMP AL,7 compares the contents of register AL with 7. The zero flag is set if AL equals 7. An internal subtraction is used to perform the comparison.

compartmental model   a dynamical model used in analysis of biomedical, pharmacokinetic, and ecological systems. The main idea in compartmental modeling is to "lump" in reality distributed system into a finite number of homogeneous, well-mixed subsystems called compartments or pools, which exchange materials with each other and with the environment. Usually the compartments are described by the first-order differential or difference equations, and in this sense they form a system of state equations. Although the most popular application of compartmental models is in modeling population dynamics and other biomedical phenomena, they may be also used to describe some engineering processes, e.g., distillation columns. The case of linear time-invariant compartments in which exchange rates are proportional to the state of the donor compartments may be treated by Laplace transforms and transfer function analysis.

compatibility (1) two different implementations of the same component whereby they may both be used in a system with no modification (often used in the context of new microprocessors running software compiled for older microprocessors).

(2) the capability of a functional unit to meet the requirements of a specified interface.

compensated pulsed alternator (CPA) *See* compulsator.

compensating winding a winding found in DC machines that is placed in the faces of the main field poles, and connected in series with the armature winding, to produce an mmf equal and opposite to the mmf of the armature, thereby reducing the effect of armature reaction.

compensation (1) operations employed in a control scheme to counteract dynamic lags or to modify the transformation between measured variables and controller output to produce prompt stable response.

(2) the alteration of the dynamic behavior of a process by the addition of system blocks. These are usually connected in cascade with the original process on either its input or its output variables, or both. *See also* compensator, pre-compensator and post-compensator.

compensator a system block added to an existing system (or process) to produce a combined transfer function that improves its performance when connected in a closed loop configuration.

See also compensation, pre-compensator and post-compensator.

compensatory behavior human dynamic behavior in which the operator's actions are conditioned primarily by the closed-loop man-machine system errors.

compensatory display for the simplest case, a display which shows only the difference between the desired input command and the system output.

compiler a program that translates a high level language program into an executable machine instruction program or other lower-level form such as assembly language. *See also* linker, assembler, interpreter, cross-assembler, cross-compiler.

complement (1) to swap 1's for 0's and 0's for 1's in a binary number.

(2) opposite form of a number system.

complement of a fuzzy set the members outside of a fuzzy set but within the universe of discourse. Represented by the symbol ¬.

Let A be a fuzzy set in the universe of discourse X with membership function $\mu_A(x)$, $x \in X$. The membership function of the complement of A, for all $x \in X$, is

$$\mu_{\neg A}(x) = 1 - \mu_A(x)$$

See also complement, fuzzy set, membership function.

complement operator the logical NOT operation. In a crisp (non-fuzzy) system, the complement of a set A is the set of the elements that are not members of A. The fuzzy complement represents the degree to which an element is not a member of the fuzzy set.

complementary arithmetic a method of performing integer arithmetic within a computer, in which negative numbers are represented in such a way that the arithmetic may be performed without regard to the sign of each number.

complementary cumulative distribution function (CCDF)   a function describing the probability $p(x)$ of achieving all outcomes in an experiment greater than x.

complementary metal oxide semiconductor (CMOS)   (1) refers to the process that combines n-channel and p-channel transistors on the same piece of silicon (complementary). The transistors are traditionally made of layers of metal, oxide, and semiconductor materials, though the metal layer is often replaced by polysilicon. There are a number of variations such as HCMOS, high-speed CMOS which scales down the elements compared to the standard MOS process and thus increases the speed and reduces the power consumption for each transistor in the CPU.

(2) A CMOS memory device used in computers to store information that must be available at startup. The information is maintained in the device by a small battery.

complete statistic   a sufficient statistic T where every real-valued function of T is zero with probability one whenever the mathematical expectation of that function of T is zero for all values of the parameter. In other words, if W is real-valued function, then T is complete if

$$E_\theta W(T) = 0 \ \forall \ \theta \in \Theta \Rightarrow P_\theta[W(T) = 0]$$
$$= 1 \ \forall \ \theta \in \Theta$$

completion unit   *See* retire unit.

complex amplitude   magnitude of a nearly harmonic function, complex to include phase deviations from a reference wave.

complex amplitude transmittance   transmittance of the complex amplitude, square root of the intensity transmission.

complex beam parameter   *See* beam parameter.

complex envelope   a low-pass complex valued signal used to represent a real band-pass signal. The complex envelope is obtained from the analytic signal with center frequency ω_c by multiplying the analytic signal by $e^{-j\omega_c t}$.

complex exponential signal   a signal of the form $x(t) = C^{j\omega t}$, where C is a constant and ω is the frequency in radians per second.

complex frequency   a complex number used to characterize exponential and damped sinusoidal motion in the same way that an ordinary frequency characterizes simple harmonic motion; designated by the constant s corresponding to a motion whose amplitude is given by Ae^{st}, where A is a constant and t is time, and $s = \sigma + j\omega$ where σ is the real part of s and ω is the imaginary part of s. ω is also known as the real angular frequency.

complex instruction set computer (CISC)   a processor with a large quantity of instructions, some of which may be quite complicated, as well as a large quantity of different addressing modes, instruction and data formats, and other attributes. The designation was put forth to distinguish CPUs such as those in the Motorola M68000 family and the Intel Pentium from another approach to CPU design that emphasized a simplified instruction set with fewer but possibly faster executing instructions, called RISC processors. One CISC processor, the Digital VAX, has over 300 instructions, 16 addressing modes, and its instruction formats may take up 1 to 51 bytes.

A CISC processor usually has a relatively complicated control unit. Most CISC processors are microprogrammed.

One of the benefits of a CISC is that the code tends to be very compact. When memory was an expensive commodity, this was a substantial benefit. Today, speed of execution rather than compactness of code is the dominant force.

131

See also microprogramming, reduced instruction set computer.

complex number a number consisting of a real part and an imaginary part, usually expressed in the form $a+bi$, where the "i" is used to distinguish that b represents the complex part of the number. i is mathematically defined as the positive root of -1.

complex power a complex number that represents electric power flow for an AC circuit. When expressed in rectangular form its real part is average power P in watts and its imaginary part is reactive power Q in reactive volt-amperes. When expressed in polar form its magnitude is apparent power S in volt-amperes and its angle is the power factor angle (the same angle as the impedance angle for a passive load). *See also* apparent power.

complex process (system) term used rather colloquially to denote controlled process (control system) possessing such characteristic features which, separately or jointly, allow to treat this process (system) as a complex entity; the features worth consideration are:

(1) The process is large in a physical sense — it occupies large space and there are large distances among its different elements,

(2) The model of the process is complicated and involves many variables, in particular control inputs,

(3) The process is composed of several interacting subprocesses and there are identifiable local objectives and local sets of decision variables,

(4) The control problem is seen as complicated due to the nature of the control objectives and the way by which free inputs are formed and influence the controlled process — even when the underlying physical process does not seem to be complex.

complex propagation constant propagation constant or wave number in a medium with gain or loss.

complex system *See* complex process.

complex transmittance the effect of a medium on both optical phase and amplitude of light traversing the medium.

complexity-constrained maximum-likelihood the maximum of the likelihood function given some quantifiable complexity constraint, M, i.e., $\max\{p(y|x, M)\}$. Breadth-first search algorithms can perform complexity-constrained maximum-likelihood detection on tree and trellis structured problems.

compliance matrix for the arm end point is defined formally by the following expression: $J_A K^{-1} J_A^T$ where J_A is an analytical Jacobian of the manipulator and K a positive definite matrix describing joint stiffness of the manipulator. Matrix K is invertible. Notice that the compliance matrix depends on the structure of the manipulator and changes with its position in Cartesian space. *See also* stiffness of a manipulator arm and analytical Jacobian.

compliant motion motion of the manipulator (robot) when it is in contact with its "environment," such as writing on a chalkboard or assembling parts.

component mounting site a location on a packaging and interconnecting structure, consisting of a land pattern and conductor fan-out to additional lands for testing or vias, used for mounting a single component.

composite a material usually consisting of a resin supporting fibers of a lightweight fabric that may be woven and treated in order to produce certain strength and/or electrical characteristics.

composite maximum method a method of defuzzification in which the defuzzified or crisp value is arrived at using the maximum value of the membership function of the fuzzy set.

composite moments method   *See* centroid method.

composite second order (CSO)   ratio of the power in the second-order distortion products to power in the carrier in a cable television channel.

composite sync   a synchronizing signal consisting of both horizontal and vertical sync information. Composite sync is used for providing synchronizing pulses to video equipment in the studio.

composite transform   a transform that can be factored into two or more transforms.

composite triple beat (CTB)   same as composite second order but for third-order distortion. *See* composite second order.

composite video   (1) a single video signal that contains luminance, color, and synchronization information. NTSC, PAL, and SECAM are all examples of composite video formats.

(2) the complete video signal. For B&W, it consists of the picture signal, blanking pulses, and synchronizing signals. For color, color synchronizing and color picture information are added. See figure.

compositional rule of inference   generalization of the notion of function. Let X and Y be two universes of discourse, A be a fuzzy set of X, and R is a fuzzy relation in $X \times Y$. The compositional rule of inference associates a fuzzy set B in Y to A in three steps:

(1) Cylindrical extension of A in $X \times Y$;

(2) Intersection of the cylindrical extension with R;

(3) B is the projection of the resulting fuzzy set on Y.

If we choose *intersection* as *triangular norm* and *union* as *triangular co-norm*, then we have the so-called *max-min composition* $B = A \circ R$, i.d.

$$\mu_B(y) = \bigvee_x \left[\mu_A(x) \bigwedge \mu_R(x, y) \right].$$

If we choose *algebric product* for *triangular norm* and *union* as *triangular co-norm*, then we have the so-called *max-product composition* $B = A \tilde{\circ} R$, i.d.

$$\mu_B(y) = \bigvee_x [\mu_A(x)\mu_R(x, y)].$$

The compositional rule of inference is the principal rationale behind *approximate reasoning*.

See also approximate reasoning, cylindrical extension of a fuzzy set, fuzzy relation, intersection of fuzzy sets, projection of a fuzzy set.

compound-connected DC machine   a direct current machine with two field windings in which one field winding is connected in series and one field winding is connected in parallel (shunt) with the armature winding. The shunt winding may be connected ahead of the series winding (long-shunt connection), or behind the series winding (short-shunt connection).

compound-rectifier exciter   a source of field current of a synchronous machine derived from the phase voltages and currents of the machine. The phase voltages and currents of the machine are fed through transformers, then rectified in order to provide DC quantities to the field winding. The components of the exciter are the transformers (voltage and current), rectifiers (including possible gate-circuitry), and power reactors; exclusive of all input control elements.

compression   (1) in information theory, the compact encoding (with a smaller number of bits) I_c of a digital image or signal I obtained by removing redundant or nonsignificant information, thus saving storage space or transmission time. Compression is termed lossless, if the transformation of I into I_c is reversible, otherwise it is termed lossy.

133

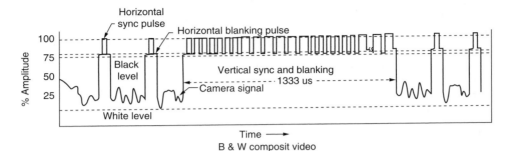

Composite video signal.

(2) in signal processing, at given bias levels and frequency, the ratio between the small signal power gain ($p_{outSS}/p_{incidentSS}$) under small signal conditions and the large signal power gain ($p_{outLS}/p_{incidentLS}$) at a given input power, expressed in decibels. As the input amplitude of a signal is increased, the output signal will eventually cutoff and/or clip due to saturation, resulting in compression. If the large signal is insufficiently large to cause cutout and/or clipping, then the compression will be at or near 0 dB.

$$G_{CR} = 10\log_{10}\left(p_{outSS}/p_{incidentSS}\right)$$
$$- 10\log_{10}\left(p_{outLS}/p_{incidentLS}\right)$$

compression coding   the lossy (irreversible) or loseless (reversible) process of reducing the amount of digital information required to represent a digital signal.

compression ratio   the ratio of the number of bits used to represent a signal before compression to that used after compression.

Compton laser   free-electron laser in which the amplification mechanism is considered to be Compton scattering.

compulsator   the compulsator (compensated pulsed alternator or CPA), is a specially designed rotating electrical alternator with a very low internal impedance that allows it to produce large,

repetitive pulses of current. These machines produce an alternating current output whose frequency is dependent upon the rotor speed and number of magnetic poles in the CPA. Typical output voltages of a CPA are 1,000–10,000 volts with output currents of up to 5,000,000 amperes and frequencies of 100–1,000 hertz.

compulsory miss   *See* cold start miss.

computational cut-off rate   *See* cut-off rate.

computational electromagnetics   the use of modern digital hardware to obtain solutions to Maxwell's equations and to visualize these solutions.

computational intelligence   *See* soft-computing.

computed tomography (CT)   *See* tomography.

computer   (1) an electronic, electromechanical, or purely mechanical device that accepts input, performs some computational operations on the input, and produces some output.

(2) functional unit that can perform substantial computations, including numerous arithmetic operations, or logic operations, without human intervention during a run.

(3) general or special-purpose programmable system that is able to execute programs automatically. It has one or more associated processing

units, memory, and peripheral equipment for input and output. Uses internal memory for storing programs and/or data.

computer architecture an image of a computing system as seen by a most sophisticated computer user and programmer. The above concept of a programmer refers to a person capable of programming in machine language, including the capability of writing a compiler. The architecture includes all registers accessible by any instruction (including the privileged instructions), the complete instruction set, all instruction and data formats, addressing modes, and other details that are necessary in order to write any program. This definition stems from the IBM program of generating the 360 system in the early 1960s. *Contrast with* computer organization. *See also* Flynn's taxonomy.

computer communication network collection of applications hosted on different machines and interconnected by an infrastructure that provides intercommunications.

computer generated hologram a hologram where the required complex amplitude and phase functions are generated by computer and written onto an optical medium.

computer hardware description language (CHDL) examples include VHDL and Verilog, current work in CHDL includes mainly languages for verification, and extensions of existing languages for system description and analog design. CHDL conferences are organized every year.

computer model a computer model of a device consists of a mathematical/logical model of the behavior of the device represented in the form of a computer program. A good computer model reproduces all the behaviors of the physical device in question, and can be confidently used to simulate the device in a variety of circumstances.

computer organization describes the details of the internal circuitry of the computer with sufficient detail to completely specifies the operation of the computer hardware. *Contrast with* computer architecture.

computer relay a protective relay that digitizes the current and/or voltage signals and uses a microprocessor to condition the digitized signal and implement the operating logic. *See* digital relay.

computer simulation a set of computer programs that allows one to model the important aspects of the behavior of the specific system under study. Simulation can aid the design process by, for example, allowing one to determine appropriate system design parameters or aid the analysis process by, for example, allowing one to estimate the end-to-end performance of the system under study.

computer torque control computed torque control is depicted in figure. The feedback controller sends its output through the inverse dynamic model. The feedback control law comprises and independent-joint PD controller with velocity reference, plus the desired acceleration. In the figure q_d, $\dot{q}_d$, and $\ddot{q}_d$ denote desired position, velocity, and acceleration vectors, respectively. q and $\dot{q}$ denote measured generalized position and velocity vectors. Finally, K_p and K_d are positive definite constant PD controller matrices.

computer vision *See* robot vision.

computer word data path of a computer (the size of virtual addresses);

(1) datum consisting of the number of bits that forms the fundamental registers, etc.;

(2) sequence of bits or characters that is stored, addressed, transmitted, and operated as a unit within a given computer. Computer words are one to eight bytes long, but can be longer for special applications.

computer-aided design (CAD) field of electrical engineering concerned with producing new

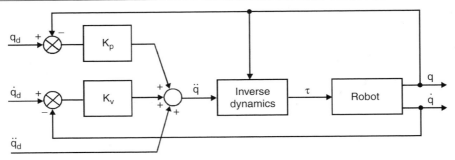

Computer torque control.

algorithms/programs which aid the designer in the complex tasks associated with designing and building an integrated circuit. There are many subfields of electrical CAD: simulation, synthesis, physical design, testing, packaging, and semiconductor process support.

computer-aided engineering (CAE) software tools for use by engineers.

computer-aided manufacturing (CAM) manufacturing of components and products when based heavily on automation and computer tools. *See also* computer-integrated manufacturing.

computer-aided software engineering (CASE) a computer application automating the development of graphic and documentation of application design.

computer-integrated manufacturing (CIM) manufacturing approach that makes substantial use of computers to control manufacturing processes across several manufacturing cells. *See also* computer-aided manufacturing.

concatenated code (1) a code that is constructed by a cascade of two or more codes, usually over different field sizes.

(2) the combination of two or more forward error control codes that achieve a level of performance with less complexity than a single coding stage would require. Serially concatenated coding systems commonly use two levels of codes,

with the inner code being a convolutional code and the outer code being a Reed–Solomon code. Parallel concatenated codes improve performance through parallel encoding and iterative serial decoding techniques. *See also* turbo code.

concentration gradient a difference in carrier concentration.

concentric resonator usually a symmetric laser resonator in which the mirror spacing is equal to twice the mirror curvature; mirrors have a common center of curvature.

concept formation the process of the incremental unsupervised acquisition of categories and their intentional descriptions.

The representative concept formation systems include EPAM, CYRUS, UNIMEM, COBWEB, and SGNN. *See also* self-generating neural network.

concurrency the notion of having multiple independent tasks available (tasks in this definition means any work to be done, not a formal computational entity).

concurrent processing having one logical machine (which may be a multiprocessor) execute two or more independent tasks simultaneously.

concurrent read and concurrent write (CRCW) shared memory model, in which concurrent reads and writes are allowed.

concurrent read and exclusive write (CREW) shared memory model, in which concurrent reads but only exclusive writes are allowed.

condenser lens lens system in an optical projection system that prepares light to illuminate the mask.

condition code internal flag used in the construction of CPUs. Many computers provide a mechanism for saving the characteristics of results of a particular calculation. Such characteristics as sign, zero result, carry or borrow, and overflow are typical of integer operations. The program may reference these flags to determine whether to branch or not.

condition code register register that contains the bits that are the condition codes for the CPU arithmetic or compare instructions.

condition variable a variable set as the result of some arithmetic or logical comparison.

conditionability of generalized 2-D model a mathematical relationship of interest in control systems.
The generalized 2-D model

$$Ex_{i+1,j+1} = A_0 x_{ij} + A_1 x_{i+1,j}$$
$$+ A_2 x_{i,j+1} + B_0 u_{ij}$$
$$+ B_1 u_{i+1,j} + B_2 u_{i,j+1}$$

is called conditionable if no two distinct solutions to the model for the same input sequence u_{ij} for $(i, j) \in [0, N_1] \times [0, N_2]$ coincide in all their boundary values (*See* boundary values of 2-D general model), where $x_{ij} \in R^n$ is the semistate vector $u_{ij} \in R^m$ is the input vector and E, A_k, B_k ($k = 0, 1, 2$) are real matrices with E possibly singular or rectangular.

conditional coding an approach to the solution of the problem of large code words and lookup tables in block coding. In this scheme one assumes that the receiver already knows the components $b_1, b_2, \ldots b_{N-1}$ of N-tuple b. Current component b_N can now be coded using this information. The assumption that there is statistical dependence between pixels is made.

conditional instruction an instruction that performs its function only if a certain condition is met. For example, the instruction JNZ TOP only jumps to TOP if the zero flag is clear (the "not zero" condition).

conditional statistic a statistic premised on the occurrence of some event. The probability of event E_1 given that E_2 has occurred is denoted by $p(E_1|E_2)$. *See* Bayes' rule.

conditionally addressed ROM read-only memory in which not every address can be used to access a valid word. Usually implemented from a PLA.

conductance (1) the reciprocal of resistance.
(2) a characteristic that describes the availability and the mobility of conduction electrons within a material. The values range from zero for a perfect insulator to infinity for a perfect conductor. The units are siemans.
(3) the ability of a substance to carry a thermodynamic flow, such as current, heat, energy, etc.

conducted emission an RF current propagated through an electrical conductor.

conducted noise unwanted electrical signals that can be generated by power electronic switching circuits. Conducted noise can travel through the circuit cables as common-mode or differential mode currents and can interfere with control circuits or other electronic equipment.

conduction angle the period during which a device is conducting, i.e., carrying current. While the device could be a switch or any other electrical element, such as inductor, phase coil, capacitor, resistor, this term is primarily applied to power electronic switching devices, which are gated to operate for some fraction of a power cycle.

The conduction angle is the sum of the positive transition angle (θ_+), the saturation angle (θ_{sat}), and the negative transition angle (θ_-). Hence, the conduction angle θ_{cond} is

$$\theta_{cond} = \theta_+ + \theta_{sat} + \theta_-$$

conduction band   the lowest energy band that is not completely occupied by electrons in a crystalline solid. *See also* valence band.

conduction current   the drift of electrons in a conductor (or of electrons and holes in a semiconductor).

conduction electron   a free electron in the conduction band of a semiconductor.

conductivity   (1) the reciprocal of resistivity.

(2) a measure of a material's ability to conduct electrical current. Conductivity σ is the ratio of the conduction current to the electric field in Ohm's Law:

$$J_c = \sigma E$$

conduit   a pipe through which an electrical cables are laid.

cone beam   term describing the shape of the beam formed with an X-ray source and beam restricter. Because the source is a finite distance from the target, beam divergence occurs because the photons in the beam are not travelling along truly parallel paths. Processing can be applied to correct for the beam shape during image reconstruction.

cone of protection   a method used to determine the extent of protection to surrounding structures afforded by a tall, grounded structure like a steel tower. Proposed prior to the "rolling ball" model, this method suggests that any structure which can fit within a right circular cone whose vertex is at the top of the tower will be protected from lightning strikes by that tower. The angle of the cone's vertex is a matter of some controversy. *See* rolling ball.

confidence interval   an interval around the estimator. The interval contains the unknown parameter with this probability.

confidence level   a probability that indicates the quality of an experiment.

configuration   operation in which a set of parameters is imposed for defining the operating conditions. The configuration of a personal computer regarding low-level features is frequently called set-up. At that level, the memory, the sequence of boot, the disk features, etc., are defined. The configuration of a computer also involves that of its operating system. For example, per MS-DOS see CONFIG.SYS and AUTOEXEC.BAT. The configuration of applicative software depends on the software under configuration itself.

confinement   condition according to which the amplitude of a beam falls to zero at large distances from the beam axis.

confinement condition   *See* confinement.

confinement diagram   diagram showing the values of the mirror curvatures of a two-mirror laser such that the electromagnetic modes are confined (satisfy the confinement condition); sometimes called a stability diagram because beam confinement can be associated with ray stability.

conflict miss   a cache miss category used to denote the case where, if the cache is direct-mapped or block-set-associative, too many blocks map to a set leading to that blocks can be expelled from the cache, even if the cache is not full, and later retrieved again. These are also called "collision misses." *See also* capacity miss.

conflict-free multiple access protocol class of multiple access protocols in which any transmission from a given user is guaranteed to have exclusive access to the channel. It will not be interfered with by a transmission from another user. One way of achieving this is to allocate the channel to the users either statically or dynamically. In static channel allocation schemes the channel can be divided into exclusive sub-channels in the time domain (TDMA), frequency domain (FDMA), code domain (CDMA), polarization domain (PDMA), or in the space domain (SDMA). In the dynamic channel allocation scheme, the channel is allocated on a demand basis using a reservation scheme or token passing.

conflicting goals objectives of several decision units in charge of given partitioned system, for example objectives of local decision units in a large-scale system, which would lead to mutually conflicting actions of the decision makers; a conflict may also arise between the objectives as perceived by a supremal unit and the local objectives; conflict between local and, eventually, between global and local objectives may or may not be alleviated by using the coordination instruments.

confocal parameter measure of the waist size of a Gaussian beam, 2 pi times waist spot size squared divided by wavelength; twice the Rayleigh length.

confocal resonator usually a symmetric laser resonator in which the mirror spacing is equal to the mirror curvature; mirrors have a common focal point.

conformal mapping a transformation $w = f(z)$ defined on a domain D with angle-preserving properties. The method of conformal mapping finds application, for example, in the quasi-static analysis of several transmission lines such as microstrips, coplanar waveguides, etc.

confusion matrix a matrix describing the likelihood of misclassification.

Let c_i, $i = 1, \ldots, n$ be the classes in which a given set of patterns can be partitioned. Let e_{c_i,c_j} be the percentage of patterns of class c_i that are erroneously recognized as patterns of class c_j. The matrix $E \doteq [e_{c_i,c_j}] \in \mathcal{R}^{n,n}$ is called a confusion matrix. Of course

$$\sum_{j=1}^{n} e_{c_i,c_j} = 1$$

congestion a state of a packet-based system where too many packets are present in the network and the overall performance degrades. To resolve the congestion, the system must employ some form of congestion control. *See also* preventive congestion control and reactive congestion control.

conical diffraction a scattering phenomenon in photorefractive crystals in which the scattered beam forms a cone of light. When a laser beam of finite transverse cross section passes through a photorefractive crystal, beam fanning often occurs. The hologram formed by the incident beam and the fanned light consists of a multitude of gratings because the fanned light spans a wide solid angle in space. When such a multitude of gratings is read out by a laser beam, only a subset of these gratings matches the Bragg condition with readout beam. The wave vectors of the Bragg-matched readout beams form hollow cones in momentum space. Therefore, conical diffractions are observed most of the time when fanning occurs. Conical diffraction is also often referred to as conical scatterings.

conical scattering *See* conical diffraction.

conjugate symmetric transform a property of a real-valued function that relates to its Fourier transform. If $x(t)$ is a real-valued function and its Fourier transform has the property that $X(-w) = X^*(w)$, where * denotes the complex conjugate. The transform $X(w)$ is said to be conjugate symmetric.

139

conjunction rule of inference   a rule of reasoning which states that if two propositions A and B are both individually true, then the combined proposition "A AND B" is also true.

connect/disconnect bus   *See* split-transaction.

connected component   a maximal-sized connected region. Also termed "blob."

connectedness   a graph or subgraph is said to be connected if there is at least one path between every pair of its vertices.

connection matrix   in a network of general topology, the connection matrix identifies how the circuit elements are connected together.

connection weight   in neural networks, within the processing element, an adaptive coefficient associated with an input connection. It is also referred to as synaptic efficacy.

connection-oriented service   a mode of packet switching in which a call is established prior to any information exchange taking place. This is analogous to an ordinary phone call, except that no physical resources need to be allocated.

connectionist model   one of many names given to the learning systems. The notion of learning systems has been developed in the fields of artificial intelligence, cybernetics and biology. In its most ambitious form, learning systems attempt to describe or mimic human learning ability. Attainment of this goal is still far away. The learning systems that have actually been implemented are simple systems that have strong relations to adaptive control. The learning systems are also known under the names of neural nets, parallel distributed processing models, etc. Examples of learning systems most commonly used are perceptron, Boltzmann machine, Hopfield network. An interesting feature of the neural networks is that they operate in parallel and that they can be implemented in

silicon. Using such circuits may be a new way to implement adaptive control systems.

connectionless service   a mode of packet switching in which packets are exchanged without first establishing a connection. Conceptually, this is very close to message switching, except that if the destination node is not active, then the packet is lost.

connectivity   specifies which sets are considered to be connected. Generally it is based on an adjacency relation between pixels (or voxels), so that a set X is connected if and only if for any $p, q \in X$ there is a sequence $p_0, \ldots, p_n$ ($n \geq 0$) such that $p = p_0$, $q = p_n$, and for each $k < n$, p_k is adjacent to p_{k+1}. *See* pixel adjacency, voxel adjacency.

connectivity check   a computerized procedure applied to a semiconductor chip's physical layout database which verifies the actual circuit on the chip is a correct implementation of the circuit described on the schematic diagrams of the chip.

consistency   a correctness criterion based on testing whether the result that is achieved by a set of operations being performed in parallel is identical to the result that would be obtained if the operations were performed sequentially without any overlap. Weaker tests have been proposed in order to trade hardware complexity for software responsibility and faster execution.

consistency of interests   situation in which there are several decision units with consistent goals. *Compare with* disagreement of interests.

consistency principle   a principle from possibility theory relating to the consistency between probability and possibility which states that the probability of an event is always at least as great as its possibility.

consistent estimator   an estimator whose value converges to the true parameter value as the sample size tends to infinity. If the convergence holds with probability 1, then the estimator is called strongly consistent or consistent with probability 1.

consistent goals objectives of several decision units in charge of a controlled partitioned system which, when followed, would lead jointly to overall optimal decisions (actions) of these units; independent decision makers contributing to common objectives, with consistent goals, form a team.

constant angular velocity normally used in disk storage units where the disk platters, rotate at a constant rotational speed. Because of this, and to have the same amount of data in each track, sectors on the inner tracks are more densely recorded than the outer tracks.

constant bit density on a disk, recording pattern in which the number of bits per unit distance is the same over all tracks.

constant bit rate (CBR) describes a traffic pattern in which the bits are sent at a fixed or constant rate. An 8-bit analog to digital converter sampling at 8 kilo-samples per second produces a CBR traffic stream with a bit rate of 64 kbps.

constant gain circle locus of input and output impedance points plots on the Smith chart that provide constant gain to an amplifier.

constant linear velocity used for example, in some optical disks where the platter rotates at different speed, depending on the relative position of the referenced track. This allows more data to be stored on the outer tracks than on the inner tracks. Because it takes time to vary the speed of rotation, the method is best suited for sequential rather than random access. *See also* constant angular velocity.

constant modulus algorithm (CMA) one of a number of algorithms (i.e., maximum ratio combining, Bartlett, Capon, and LMS) proposed in the literature for the adaptation of the weights associated with each radiating element and for combining signals received on radiating elements.

constant-current transformer Two-coil transformer with a moveable secondary coil used to provide constant output current to a variable load. Constant current is maintained by mounting both the primary and secondary coils on the center element of a shell-type core and allowing the secondary coil to move up and down with changes in demand for load current. Increasing current demand due to a reduction in load impedance causes the secondary coil to move away from the primary coil. Increasing the coil separation increases flux leakage and reduces the secondary output voltage. The reduced output voltage counteracts the demand for more current. Increases in load impedance reverse the process. Movement of the secondary coil is controlled automatically by attaching the secondary coil to a counterweight and pulley assembly and orienting the coil windings such that their flux directions oppose. Increases in secondary current increase the magnetic repulsion between the coils, which, aided by the counterweight, moves the secondary coil away from the primary. Reductions in secondary current produce the opposite effect.

constant-horse power drive a variable speed drive that is operating in a speed region where it is capable of delivering rated power. For DC machines, this region is above base speed and is achieved by field weakening. For AC induction motors, this region is above rated speed and is achieved by increasing the frequency of the applied voltage.

constant-torque drive a variable-speed drive that is operating in a speed region where it is capable of maintaining rated torque. For DC machines, this region is below base speed and is achieved by reducing the applied armature voltage. For AC induction motors, this region is below rated speed and is achieved by reducing the frequency of the applied voltage.

constitutive relation   describes the relation between the intensity vectors and the flux density vectors in a medium.

constitutive relationships   a set of equations that couple the electric-field intensity E, magnetic-field intensity H, the electric flux density D, and the magnetic flux density B to one another. For simple media, the constitutive relationships are $D = \epsilon E$, and $B = \mu H$, where ϵ and μ are the scalar permittivity and permeability of the medium, respectively.

constraining core   an internal supporting plane in a packaging and interconnecting structure, used to alter the structure's coefficient of thermal expansion.

constraint length   in convolutional codes, an indication of the number of source words that affect the value of each coded word. Two typical forms are:

(1) A code with constraint length K, in which the value of each coded word is affected by the present source word and up to $K - 1$ previous source words.

(2) The number of shifts over which a single message bit can influence the output of a convolutional encoder.

constraint propagation   artificial intelligence technique in which a hypothesis generates constraints that reduce the search space over the rest of the data. If no eventual contradiction is derived, then a "match" is achieved.

constructive algorithm   learning algorithm that commences with a small network and adds neural units as learning proceeds until the problem of interest is satisfactorily accommodated.

constructive solid geometry   method by which complex 3-D objects are defined as the combination of simpler solids.

contact head   *See* disk head.

contact potential   the internal voltage that exists across a p-n junction under thermal equilibrium conditions, when no external bias voltage is applied.

contact printing   a lithographic method whereby a photomask is placed in direct contact with a photoresist coated wafer and the pattern is transferred by exposing light through the photomask into the photoresist.

contactor   electromechanically actuated spring-loaded relay contacts normally used to control lights, heat, or other non-motor loads. In essence, it is an electromechanically operated switch that usually requires some form of pilot device for its actuation.

containment building   (1) a steel and concrete structure which encloses and isolates the radioactive portion of a nuclear power plant.

(2) a heavily re-inforced structure which surrounds the reactor and other radioactive portions of a nuclear power plant so as to contain radioactive gases or debris in the event of an explosion.

containment vessel   the heavy steel container which encloses the core of a pressurized-water reactor *cf* in a nuclear power plant.

content-addressable memory (CAM)   *See* associative memory.

contention   additional latency incurred as the result of multiple requestors needing access to a shared resource, which can only be used by one at a time.

contention protocol   class of a multiple access protocol where the users' transmissions are allowed to conflict when accessing the communication channel. The conflict is then resolved through the use of a static or dynamic conflict resolution protocol. Static resolution means that the conflict resolution is based on some preassigned priority. A static resolution can be probabilistic if the statistics of the probabilities are fixed. A common example is the p-persistent ALOHA protocol. The dynamic resolution allows for changing the parameters of the conflict resolution algorithm to reflect the traffic state of the system. A common example is the Ethernet protocol.

context   the privilege, protection and address-translation environment of instruction execution.

context switching   an operation that switches the CPU from one process to another, by saving all of the CPU registers for the first and replacing them with the CPU registers for the second.

context units   a set of memory units added to a feedforward network that receives information when an input is presented to the network and passes this information to the hidden layer when the next input is presented to the network.

contingency analysis   a plan for dealing with any of the probable faults which might befall a particular electric power system, the goal being to maintain power to the maximum number of customers and/or the most critical customers.

contingency list   in security analysis, a list, necessarily incomplete, of everything which could possibly go wrong in a section of an electric power system.

contingency ranking   the process of ranking the list of probable contingencies in order of severity.

contingency selection   the process of narrowing the list of probable contingencies, or disturbances, that can be further processed and studied to determine the extent of security violations in the system.

continuity equation   axiom that charge is a conserved quantity. In point-form, the continuity equation is stated as

$$-\frac{\partial \rho}{\partial t} = \nabla \cdot J,$$

where ρ is the charge density and J is the current density.

continuous duty   National Electrical Manufacturers Association (NEMA) classification describing an application in which a machine operates for long periods of time at relatively constant loads.

continuous Hopfield network   a Hopfield network with the same structure as the discrete version, the one difference being the replacement of the linear threshold units by neurons with sigmoidal characteristics. Any initial setting of the neuron outputs leads to a motion in the network's state space towards an attractor which, so long as the weights in the network are symmetric, is a fixed point. This allows the network to be employed for the solution of combinatorial optimization problems (its main application) by arranging the network's weights so that an optimal solution lies at a fixed point of the network's dynamics. *Compare with* discrete Hopfield network.

continuous rating   term often used to refer to the manufacturer's nameplate ratings for an electrical machine, which are the rated operating conditions guaranteed by the manufacturer for continuous-duty operation. *See also* continuous duty.

continuous signal   a continuous function of one or more independent variables such as time, that typically contains information about the behavior or nature of some phenomenon.

continuous spectrum   when an eigenvalue problem is defined over an infinite domain, the eigenvalues bunch together to form a continuum or a continuous spectrum. This concept is of fundamental importance for open waveguides either of electromagnetic or acoustic type.

continuous speech recognition   the process of recognizing speech pronounced naturally with no pauses between different words.

continuous system   See *incremental gain.*

continuous time signal   *See* signal.

continuous time system   a process that transforms continuous time input signals to continuous time output signals. *See also* system.

continuous tone image   coding a process that converts a digitized continuous tone image to a binary bit stream which has fewer bits than the original image for the purpose of efficient storage and transmission. *See* still image coding.

continuous wave (CW)   periodic and usually sinusoidal wave, in contrast to a pulsed or modulated wave.

continuous-valued logic similarity   by using the definition of the equivalence in continuous-valued logic, similarity between two variables $x = (x_0, \ldots, x_n)$ and $y = (y_0, \ldots, y_n)$, all components of which are continuous in the open interval $(0, 1)$, can be defined as

$$S_C(x, y) = \sqrt[\rho]{\sum_{i=1}^{n} (e\,(x_i, y_i))^\rho}$$

where ρ takes real value,

See equivalence in continuous-valued logic for definition of $e(x_i, y_i)$.

contour   (1) the edge that separates an object from other objects and the background. It must consist of one or several closed curves, one for the outer contour, and the others (if any) for the inner contours surrounding any holes. *See* contour filling, contour following, edge.

(2) following an operator which, starting from a contour point, follows the closed curve made by that contour. *See* contour.

contour filling   an object contour is generally built with an edge detector, but such a contour can be open, because some of its pieces, not recognized by the edge detector, may be missing. In order to close the contour, missing pieces can be added by an operator filling small holes in a contour. *See* contour, edge detection.

contrast   (1) a measure of the intensity difference (ratio) between an object and the image background.

(2) the difference in the perception of visual energy between picture white and picture black.

(3) enhancement alteration of the contrast in an image to yield more details or more information. *See* contrast, histogram stretching, histogram equalization.

contrast enhancement layer (CEL)   a highly bleachable coating on top of the photoresist that serves to enhance the contrast of an aerial image projected through it.

contrast rendition factor (CRF)   the ratio of visual task contrast with a given lighting environment to the contrast with sphere illumination.

contrast sensitivity   the responsiveness of the human visual system to low contrast patterns. In psychophysics, the threshold contrast is the minimum contrast needed to distinguish a pattern (such as a spatial sinusoid) from a uniform field of the same mean luminance, and the contrast sensitivity is the inverse of the threshold contrast. *See* human visual system.

control   intervention, by means of appropriate manipulated inputs, into the controlled process in the course of its operation; some form of observation of the actual controlled process behavior is usually being used by the controller.

control and status register (CSR) in a computer, the register that holds internal control and status information as a bit-wise array of Booleans.

control bus contains processor signals used to interface with all external circuitry, such as memory and I/O read/write signals, interrupt, and bus arbitration signals.

control channel the control channel used to transmit network control information. No user information is sent on this channel. *Compare with* traffic channel.

control chart plot of data over time indicating the fluctuation of the main statistical characteristics applied in statistical quality control. Control charts can be used to determine if a process is in a state of statistical control by examining past data and to determine control limits that would apply to future data in order to check if the process maintains in this state.

The individual observations are plotted against three lines. The center one represents an estimate of the process mean, standard deviation or other statistic, two others represent the lower control limit (LCL) and the upper control limit (UCL), respectively. If the control charts are being used for the first time, it is necessary to determine trial control limits. These limits should be revised if the points outside them are traced to a special cause which can be removed. The most frequently used control charts are $\bar{X}$, s (mean, standard deviation) and $\bar{X}$, R (mean, range) charts. Control charts are used to aid in identification of special causes of variation, reduction in product variability, and keeping good records.

control horizon end time of the control interval over which the operation of the control system is considered; if the control interval is infinite then the control horizon is also infinite.

control input *See* manipulated input.

control instruction machine instruction that controls the actions of a processor such as setting flags to enable specific modes of operation. Generally, control instructions do not perform computations. Sometimes control instructions include instructions that can effect sequential execution of a program, such as branch instructions.

control interval time interval over which the operation of the control system is considered; control interval can be finite or infinite. The notion of a control interval is essential when the controlled process features various accumulation phenomena.

control layer part of the controller responsible for performing tasks associated with a particular aspect of the control; particular control layer results from vertical decomposition of the controller into a multilayer control structure; the layers may differ in their function, or the control interval considered, or both. Typical examples of control layers would be regulation layer of a continuous industrial process (see direct control) and set point control layer of such process. *See also* optimizing control.

control line in a bus, a line used in a computer bus to administrate bus transfers. Examples are bus request (a device wants to transmit on the bus), or bus grant (the bus arbiter gives a device transmit access on the bus).

control memory a semiconductor memory (typically RAM or ROM) used to hold the control data in a microprogrammed CPU. This data is used to control the operation of the data path (e.g., the ALU, the data path busses, and the registers) in the CPU. If the control memory uses RAM, the CPU is said to be microprogrammable, which means that the CPU's instruction set can be altered by the user and the CPU can thus "emulate" the instruction set of another computer. Same as control store and micromemory.

control policy *See* control rule.

145

control problem   a design problem concerned with constructing a device called the controller whose goal is to force the controlled variable of the plant or process to behave in a desired manner. The elements of the control problem are control objective, a model of the process to be controlled, admissible controllers, and a means of evaluating the performance of a control strategy. *See also* controller, controlled variable.

control rod   an assembly of neutron-absorbing material, typically boron, which is extended into the core of a nuclear reactor to dampen the chain reaction.

control rule   decision mechanism (sequence of such mechanisms), used — within the considered control layer — to specify on-line the values of the control inputs; for example, the values of the manipulated inputs in case of the direct control layer or the set-point values in case of the optimizing control layer. Also known as control policy.

control scene   initial entity, a given "world," which is then partitioned into the controlled process and its (process) environment; control scene is the initial "world" which is of interest to a control engineer or a system analyst.

control store   *See* control memory.

control structure   essentially the same as the control system; this term is used when one wants to indicate that the controller is composed of several decision units, suitably interlinked; decision units of a control structure usually differ in their tasks, scope of authority and access to information; depending on the context one speaks of a control structure or of a decision structure; decentralized control, multilayer control, hierarchical control are examples of control structures.

control surface   any of the movable parts such as tabs, panels, or wings that control the depth of a submarine or the attitude of a flight vehicle moving through the atmosphere. For example, the yaw angle of an airplane is controlled by the rudder, the pitch angle by the elevators, while the roll angle by the ailerons. In fuzzy logic community, control surface may mean a plot of a typical fuzzy logic controller output as a function of its two inputs. The inputs to a typical fuzzy logic controller are the error between the desired and the actual plant output, and the change-in-error.

control system   (1) the entity comprising the controlled process and the controller. Control system is influenced by the environment of the process both through the free inputs to the process itself and through any current information concerning the behavior of these free inputs that is made available to the controller.

(2) an arrangement of interconnecting elements that interact and operate automatically to maintain a specific system condition or regulate a controlled variable in a prescribed manner.

control transformer   a step-down transformer used to provide power to the control portion of a power or motor circuit.

control under uncertainty   operation of the control system in a situation when there is a significant uncertainty regarding current or future values of the free inputs, which leads to uncertainty in the envisaged behavior of the controlled process and, eventually, there is uncertainty concerning the internal behavior of the controlled process; in particular when the forecasted scenarios of the future free input values tend to appear to be largely different from the actual future free input realization.

controllability   (1) the property of a system that ensures the existence of bounded control inputs to drive any arbitrary initial state to any arbitrary final state in finite time. For linear systems, an algebraic condition that involves system and input matrices can be used to test this property.

(2) the ability to establish the required test stimuli at each node in a circuit by setting appropriate values on the circuit inputs.

controllability at a given time   a characteristic of some dynamical systems. A linear dynamical system is said to be controllable at a given time if there exists a finite time t_1, such that it is controllable in a time interval $[t_0, t_1]$.

controllability condition for nonstationary discrete system   a condition found in some dynamical systems. A linear dynamical nonstationary discrete-time system is controllable in an interval $[k_0, k_1]$ if and only if the controllability matrix

$$\sum_{j=k_0}^{k_1-1} F(k_1, j+1) B(j) B^T(j) F^T(k_1, j+1)$$

is nonsingular. *See also* dynamical linear discrete time systems.

controllability in a fixed interval   a characteristic of some dynamical systems. A dynamical discrete system is said to be controllable in an interval $[k_0, k_1]$ if for any initial state $x(k_0) \in R^n$ and any vector $x_1 \in R^n$ there exists a sequence of admissible controls $u(k)$, $k = k_0, k_0+1, \ldots, k_1-2, k_1-1$ such that the corresponding trajectory of the dynamical system satisfies the condition

$$x(k_1, x(k_0), u) = x_1$$

controllability in a given time interval   a characteristic of some dynamical systems. A dynamical continuous system is said to be controllable in a given time interval $[t_0, t_1]$ if for any initial state

$$x(t_0) \in R^n$$

and any vector

$$x_1 \in R^n$$

there exists an admissible control

$$u \in L^2\left([t_0, t_1], R^m\right)$$

such that the corresponding trajectory of the dynamical system $x(t, x(t_0), u)$ satisfies the

following condition:

$$x(t_1, x(t_0), u) = x_1$$

controllability of nonstationary systems   a characteristic of some dynamical systems. A linear dynamical nonstationary system is controllable in time interval $[t_0, t_1]$ if and only if the $n \times n$ dimensional controllability matrix

$$W(t_0, t_1) = \int_{t_0}^{t_1} F(t_1, t) B(t) \\ \times B^T(t) F^T(t_1, t)\, dt$$

is nonsingular. *See also* dynamical linear continuous time system.

controllability of stationary systems   a characteristic of some dynamical systems. A linear dynamical stationary system is controllable if and only if

$$\text{rank} \, [B|AB|A^2B| \ldots |A^kB| \ldots |A^{n-1}B] = n$$

See also dynamical linear continuous time system.

controlled process   part of the control scene that can be influenced by the manipulated inputs set by the controller and the free inputs from the (process) environment; manipulated inputs will be set (adjusted) to realize specified control objectives. Control objectives can be stated in terms of constraints imposed on specified quantities, in terms of performance indices to be minimized or maximized, or in other form; free inputs to the process from its environment do not depend, by definition, on the process behavior.

controlled process model   model, usually stated in form of a set of differential or difference equations, describing the behavior of the controlled process as caused by its inputs; different models of the same controlled process may be used for various purposes, for steady-state control or for model-based predictive control.

controlled rectifier a rectifier that uses switching elements that have forward voltage blocking capability to allow a variable voltage DC output. *See also* thyristor.

controlled source a voltage or current source whose intensity is controlled by a circuit voltage or current elsewhere in the circuit. Also called dependent source.

controlled variable (1) the quantity, usually the output of a plant or process, that is being controlled for the purpose of the desired behavior, for example, transient response or steady-state response.

(2) variable associated with the behavior of the controlled process and such that one wants this variable either to follow a desired trajectory over a given time interval or to be kept at a prescribed constant value, i.e., at a specified set-point; introduction of a set of controlled variables is necessary to define a two-layer industrial controller with the regulation direct control layer and the set point optimizing control layer.

See also controller.

controller (1) the entity that enforces the desired behavior — as specified by the control objectives — of the controlled process by adjusting the manipulated inputs. The values of these inputs are either predetermined or decided upon (computed) using on-line, i.e., real time, decision mechanism of the controller — based on the currently available information. *See also* controlled variable.

(2) a device that generates the input to the plant or process. The role of the controller is to force the controlled variable of the plant or process to behave in a desired manner.

(3) a unit that directs the operation of a subsystem within a computer. For instance, a disk controller interprets data access commands from host computer (via a bus), and sends read/write, track seeking, and other control signals to the drive. During this time, the computer can perform other tasks, until the controller signals DATA READY for transfer via the CPU bus.

convection current a current in which electrons are released for movement outside of a material.

convective heat transfer the process by which a moving fluid transfers heat to or from a wetted surface.

convergence the condition when the electron beams from a multi-beam CRT meet at a single point. For example, the correct registration of the three beams in the color picture tube.

convergence in probability for some sequences of random numbers, the tendency to a single number.

To wit, for a sequence of numbers x_n, and a random variable x, if for all $\epsilon > 0$,

$$P\{| x_n - x |> \epsilon\} \to 0$$

for $n \to \infty$, then the sequence x_n tends to x in probability.

convergent state the equilibrium state of a dynamic system described by a first order vector differential equation is said to be convergent if there exists a $\delta = \delta(t_0)$, such that,

$$\| x(t_0) - x_e \| < \delta \Rightarrow \lim_{t \to \infty} x(t) = x_e$$

See also stable state.

converter a generic term used in the area of power electronics to describe a rectifier, inverter, or other power electronic device that transforms electrical power from one frequency and voltage to another.

convex fuzzy set (1) a fuzzy set that has a convex type of membership function.

(2) a fuzzy set in which all α-level sets are convex. *See also* α-level set.

convolution the mathematical operation needed to determine the response of a system from its stimulus signal and its weighting function. The convolution operation is denoted by the symbol "∗." The convolution of two continuous time signals $f_1(t)$ and $f_2(t)$ is defined by

$$f_1(t) * f_2(t) = \int_{-\infty}^{\infty} f_1(\tau) f_2(t - \tau) d\tau$$

$$= \int_0^t f_1(\tau) f_2(t - \tau) d\tau \quad \text{if} \quad f_1(t), f_2(t)$$

$$= 0, t < 0$$

The integral on the right-hand side of the above equation is called the convolution integral, and exists for all $t \geq 0$ if $f_1(t)$ and $f_2(t)$ are absolutely integrable for all $t > 0$. f_1 is the weighting function that characterizes the system dynamics in the time domain. It is equivalent to the response of the system when subjected to an input with the shape of a Dirac delta impulse function. Laplace transformation of the weighting function yields the transfer function model for the system.

The convolution of two discrete time signals $f_1[k]$ and $f_2[k]$ is defined by

$$f_1[k] * f_2[k] = \sum_{i=-\infty}^{\infty} f_1[i] f_2[k - i]$$

$$= \sum_{i=0}^{k} f_1[i] f_2[k - i] \text{ if } f_1[k], f_2[k] = 0, k < 0$$

The summation on the right-hand side of the above equation is called the convolution sum. Convolution is useful in computing the system output of LTIL systems.

convolution integral   *See* convolution.

convolutional code   (1) a code generated by passing the information sequence to be transmitted through a linear finite-state shift register and the coder memory couples the currently processed data with a few earlier data blocks. Thus the coder output depends on the earlier data blocks that have been processed by the coder.

(2) a channel code based on the trellis coding principle but with the encoder function (mapping) determined by a linear function (over a finite alphabet). The name "convolutional code" comes from the fact that the output sequence is a (finite alphabet) convolution between the input sequence and the impulse response of the encoder.

convolutional coding   a continuous error control coding technique in which consecutive information bits are combined algebraically to form new bit sequences to be transmitted. The coder is typically implemented with shift register elements. With each successive group of bits entering the shift register a new, larger set of bits are calculated for transmission based on current and previous bits. If for every k information bits shifted into the shift register, a sequence of n bits are calculated, the code rate is k/n. The length of the shift register used for storing information bits is known as the constraint length of the code. Typically, the longer the constraint length, the higher the code protection for a given code rate. *See also* block coding, error control coding.

cooling tower   a reinforced-concrete, spool-shaped structure in a thermal power plant in which condenser cooling water is itself cooled by convectively-driven air streams.

Cooper, Leon Niels   (1930–) Born: New York, New York

Best known for his work on superconductivity, resulting in a Nobel Prize he shared with John Bardeen and J. Robert Schrieffer. He postulated the idea that is now known as Cooper pairs, as part of the explanation of why some metals loose their conductivity at very low temperatures.

coordinate system   a system for defining the location of a point in space relative to some reference point and for defining a set of reference directions at each and every point in space.

coordinated rotation digital computer (CORDIC) algorithm for calculating trigonometric functions using only additions and shift operations.

coordinating unit *See* coordinator unit.

coordination process of influencing — by the coordinator — the local units in such a way as to make their behavior or decisions consistent with the objectives of the coordinator unit; coordination can be either iterative or periodic; iterative coordination plays an essential role in multilevel optimization. *See also* direct method, mixed method, price method coordination.

coordination by exception coordination performed only in unusual (emergency) situations, otherwise the control or decision making is designed to be fully decentralized.

coordination instrument any means by which the coordinator unit influences behavior or decisions of the subordinate (local) units; coordination instruments are often expressed in terms of vectors of values assumed by the coordination variables. In the case of price method, the coordination variables are the prices by which the interaction input and output variables are multiplied.

coordination strategy mechanism (algorithm) used to generate the values of the coordination instruments in the course of either iterative or periodic coordination; in case of iterative coordination, the convergence and the speed of convergence of the coordination strategy is of main concern; in case of periodic coordination, other issues are important — these can be stability of the coordination process as well as quantitative aspects of the controlled process behavior.

coordinator unit control (decision) agent in a hierarchical control structure, being in charge of decisions (control instruments) influencing operation of the local decision units; coordinator unit performs either iterative or periodic coordination of the local decisions; coordinator unit is often regarded as the supremal unit of the hierarchical control structure. Also called coordinating unit.

copolarized the plane wave whose polarization is the same as that of the reference plane wave (e.g., radiated wave from an antenna) is said to be copolarized (otherwise it is crosspolarized).

copper jacket timer a magnetic time-off timer that can be used in definite time DC motor acceleration starters and controllers. The copper jacket relay functions by slowing the dissipation of the magnetic field when the coil is turned off. After a certain amount of time the spring tension on the contactor overcomes the strength of the dissipating magnetic field — and causes the contacts to change state. Time delays with the copper jacket timer are adjusted by adding, or removing, permeable shims between the coil and copper jacket. The more shims in place the slower the magnetic field dissipates, hence the longer the time delay becomes.

copper loss electric loss due to the resistance in conductors, windings, brush contacts or joints, in electric machinery or circuits. Also referred to as $I^2 R$, the losses are manifested as heat.

coprime 2-D polynomial matrices *See* coprimeness of 2-D polynomial matrices.

coprime 2-D polynomials *See* coprimeness of 2-D polynomials.

coprimeness of 2-D polynomial matrices a mathematical relationship of interest in control systems.

A 2-D polynomial matrix $C(z_1, z_2)$ $(B(z_1, z_2))$ is called a right (left) divisior of $A(z_1, z_2)$ if there exists a matrix $B(z_1, z_2)$ $(C(z_1, z_2))$ such that

$$A(z_1, z_2) = B(z_1, z_2) C(z_1, z_2)$$

coprimeness of 2-D polynomials

A 2-D polynomial matrix $R(z_1, z_2)$ $(L(z_1, z_2))$ is called a common right (left) divisor of $A(z_1, z_2)$, and $B(z_1, z_2)$ if there exist two 2-D polynomial matrices $A_1(z_1, z_2)$, $B_1(z_1, z_2)$, $(A_2(z_1, z_2)$, $B_2(z_1, z_2))$ such that

$$A(z_1, z_2) = A_1(z_1, z_2) R(z_1, z_2)$$

and

$$B(z_1, z_2) = B_1(z_1, z_2) R(z_1, z_2)$$
$$(A(z_1, z_2) = L(z_1, z_2) A_2(z_1, z_2)$$

and

$$B(z_1, z_2) = L(z_1, z_2) B_2(z_1, z_2))$$

A 2-D polynomial matrix $P(z_1, z_2)$, $(Q(z_1, z_2))$ is called greatest common right (left) divisor of $A(z_1, z_2)$ and $B(z_1, z_2)$ if $P(z_1, z_2)$ $(Q(z_1, z_2))$ is a greatest right (left) divisor of $A(z_1, z_2)$ and $B(z_1, z_2)$ and any common right divisor of $A(z_1, z_2)$ and $B(z_1, z_2)$ is a right divisor of $P(z_1, z_2)$.

2-D polynomial matrices $A(z_1, z_2)$, $B(z_1, z_2)$ are called factor right (left) coprime if their greatest common right (left) divisor is a unimodular matrix $U(z_1, z_2)$ (nonzero $det\, U(z_1, z_2) \in R$).

2-D polynomial matrices $A \in R^{p \times m}[z_1, z_2]$, $B \in R^{p \times m}[z_1, z_2]$ $(p + q \geq m \geq 1)$ are called zero right coprime if there exists a pair (z_1, z_2) which is a zero of all $m \times m$ minors of the matrix $[\begin{smallmatrix} A \\ B \end{smallmatrix}]$. The minor coprimeness of two 2-D polynomial matrices implies their factor coprimeness.

coprimeness of 2-D polynomials   a mathematical relationship of interest in control systems.

A 2-D polynomial

$$p(z_1, z_2) = \sum_{i=0}^{n_2} a_i(z_1) z_2^i = \sum_{j=0}^{n_1} a_j(z_2) z_1^j$$

is called primitive if $a_i(z_1)$, $i = 1, 2, \ldots, n_2$ and $a_j(z_2)$, $j = 1, 2, \ldots, n_1$ are coprime (have

not a common factor). Two primitive 2-D polynomials are called factor coprime if their greatest common divisor is a constant. The primitive 2-D polynomials

$$a(z_1, z_2) = a_{n_1}(z_2)\, \bar{a}(z_1, z_2)$$
$$= a_{n_2}(z_1)\, \hat{a}(z_1, z_2)$$
$$b(z_1, z_2) = b_{m_1}(z_2)\, \bar{b}(z_1, z_2)$$
$$= b_{m_2}(z_1)\, \hat{b}(z_1, z_2)$$

$$\bar{a}(z_1, z_2) = z_1^{n_1} + \bar{a}_{n_1-1} z_1^{n_1-1} + \cdots$$
$$+ \bar{a}_1 z_1 + \bar{a}_0 \quad (\bar{a}_i = \bar{a}_i(z_2))$$

$$\hat{a}(z_1, z_2) = z_2^{n_2} + \hat{a}_{n_2-1} z_2^{n_2-1} + \cdots$$
$$+ \hat{a}_1 z_2 + \hat{a}_0 \quad (\hat{a}_i = \hat{a}_i(z_1))$$

$$\bar{b}(z_1, z_2) = z_1^{m_1} + \bar{b}_{m_1-1} z_1^{m_1-1} + \cdots$$
$$+ \bar{b}_1 z_1 + \bar{b}_0 \quad \left(\bar{b}_i = \bar{b}_i(z_2)\right)$$

$$\hat{b}(z_1, z_2) = z_2^{m_2} + \hat{b}_{m_2-1} z_2^{m_2-1} + \cdots$$
$$+ \hat{b}_1 z_2 + \hat{b}_0 \quad \left(\hat{b}_i = \hat{b}_i(z_1)\right)$$

are factor coprime if and only if

$$\det \left[\bar{B}^{n_1} + \bar{a}_{n_1-1} \bar{B}^{n_1-1} + \cdots \right.$$
$$\left. + \bar{a}_1 \bar{B} + \bar{a}_0 I_{m_1} \right] \neq 0$$

and

$$\det \left[\hat{B}^{n_2} + \hat{a}_{n_2-1} \hat{B}^{n_2-1} + \cdots \right.$$
$$\left. + \hat{a}_1 \hat{B} + \hat{a}_0 I_{m_2} \right] \neq 0$$

or

$$\det \left[\bar{A}^{m_1} + \bar{b}_{m_1-1} \bar{A}^{m_1-1} + \cdots \right.$$
$$\left. + \bar{b}_1 \bar{A} + \bar{b}_0 I_{n_1} \right] \neq 0$$

and

$$\det \left[\hat{A}^{m_2} + \hat{b}_{m_2-1} \hat{A}^{m_2-1} + \cdots \right.$$
$$\left. + \hat{b}_1 \hat{A} + \hat{b}_0 I_{n_2} \right] \neq 0$$

151

where

$$\bar{A} = \begin{bmatrix} 0 & 1 & 0 & \dots & 0 \\ 0 & 0 & 1 & \dots & 0 \\ \dots & \dots & \dots & \dots & \dots \\ 0 & 0 & 0 & \dots & 1 \\ -\bar{a}_0 & -\bar{a}_1 & -\bar{a}_2 & \dots & -\bar{a}_{n_1-1} \end{bmatrix}$$

$$\hat{A} = \begin{bmatrix} 0 & 1 & 0 & \dots & 0 \\ 0 & 0 & 1 & \dots & 0 \\ \dots & \dots & \dots & \dots & \dots \\ 0 & 0 & 0 & \dots & 1 \\ -\hat{a}_0 & -\hat{a}_1 & -\hat{a}_2 & \dots & -\hat{a}_{n_2-1} \end{bmatrix}$$

$$\bar{B} = \begin{bmatrix} 0 & 1 & 0 & \dots & 0 \\ 0 & 0 & 1 & \dots & 0 \\ \dots & \dots & \dots & \dots & \dots \\ 0 & 0 & 0 & \dots & 1 \\ -\bar{b}_0 & -\bar{b}_1 & -\bar{b}_2 & \dots & -\bar{b}_{m_1-1} \end{bmatrix}$$

$$\hat{B} = \begin{bmatrix} 0 & 1 & 0 & \dots & 0 \\ 0 & 0 & 1 & \dots & 0 \\ \dots & \dots & \dots & ,,, & \dots \\ 0 & 0 & 0 & \dots & 1 \\ -\hat{b}_0 & -\hat{b}_1 & -\hat{b}_2 & \dots & -\hat{b}_{m_2-1} \end{bmatrix}$$

coprocessor   a processor that is connected to a main processor and operates concurrently with the main processor, although under the control of the main processor. Coprocessors are usually special-purpose processing units, such as floating point, array, DSP, or graphics data processors.

copy-back   in cache systems an operation that is the same as write-back—a write operation to the cache that is not accompanied with a write operation to main memory. In copy-back, the data is written only to the block in the cache. This block is written to main memory only when it is replaced by another block. *See also* block replacement.

CORDIC   *See* coordinated rotation digital computer.

core   (1) the operating image of a process (sometimes used to refer to the part residing in physical memory), often written to disk if the program crashes (dumping core). Since magnetized ferrite rings (cores) were once used in main memory to store a single bit each. The name remained and now core memory means the same as main memory, although currently, main memory is chip-based. *See also* magnetic core memory.

(2) the ferromagnetic portion of a transformer or electric machine on which the coils are mounted. Typically made of laminated magnetic material, encircled by the windings, that provides a low reluctance path for magnetic flux.

(3) the central region of an optical fiber. The refractive index of the core must be higher than that of the cladding so that the optical power is guided through the fiber by total internal reflection at the core-cladding boundary. The core refractive index may be constant or may decrease with distance from the axis to the cladding. *See also* graded index and step index optical fiber.

(4) the section of a nuclear reactor in which the chain reaction is contained, comprising fuel rods, control rods, moderator, and coolant.

core lamination   *See* lamination.

core loss   loss in the ferromagnetic material comprising the core of an electric machine or transformer, composed of the sum of hysteresis losses and eddy current losses. These magnetic losses are caused by time varying fluxes in a ferromagnetic structure. Hysteresis losses are caused by friction in molecules as the dipoles in a structure change direction of alignment in response to an applied alternating voltage, while eddy current losses are resistive losses (I^2R), due to circulating currents in the core.

core memory   *See* magnetic core memory.

core-type transformer   a transformer in which the magnetic circuit upon which the windings are wound takes the form of a single ring. When the coils are placed on the core, they encircle the core. *See also* core.

coriolis forces forces/torques that depend upon the product of joint velocities.

corner cube antenna a traveling wave antenna typically four wavelengths long and placed 1.2 wavelengths from the apex of a 90-degree corner reflector often used in the submillimeter wave and terahertz frequency range for mixers and detectors.

corner detection the detection of corners, often with a view to locating objects from their corners, by a process of inference, in digital images.

corona a visible glow discharge which emanates from high-voltage conductors in regions of extremely high electric field intensity.

corona effect flow of electrical energy from a high-voltage conductor to the surrounding ionized air. This effect only becomes significant for potentials higher than 1000 V. This effect is characterized by a faint glow, a crackling noise and conversion of atmospheric oxygen to ozone.

corona loss the electric power lost in high voltage lines due to the radiation of energy by corona discharge. *See* corona.

corona resistance capacity of a material to bear the action of the corona effect. This capacity is particularly important for polymeric materials, which have to withstand the chemical degrading effect promoted by ozone generated by the corona effect. *See also* corona effect.

corona ring a toroidal metal ring connected to discontinuities on high-voltage conductors to reduce local field intensity and thus discourage the formation of corona discharge.

coronal projection a projection image formed on a plane parallel to the chest and perpendicular to the transverse and sagittal planes.

corporate feed also known as parallel feed, a method of feeding a phased array antenna in which the signal from a transmitter is successively split until there are enough feedlines to carry the signal to each array element. The name comes from the resemblance of the feed structure to the organizational chart of a corporation.

correction layer control layer of a multilayer controller, usually situated above the direct control layer and below the optimizing control layer, required to make such modifications of the decisions supplied by the optimizing layer — before these decisions are passed to the direct layer — that some specified objectives are met; for example, a correction layer of the industrial process controller may be responsible for such adjustment of a particular set point value that an important constraint is satisfied by the controlled process variables — in case when the optimizing layer is using inaccurate model of this controlled process.

correlation (1) the mathematical operation of comparing the behavior of two signals to determine how closely they are related. It is usefully applied to system identification when the input and output signals of a given system are compared by correlation to give the system transfer function model. The relevant equation is

$$\hat{g}(j\omega) = \frac{y(j\omega)u^*(j\omega)}{u(j\omega)u^*(j\omega)}$$

in terms of the FFTs of the input and output signals. The $*$ superscript indicates the complex conjugate. As the system output $y(t)$ generally contains terms in addition to those caused by the input $u(t)$, its Laplace transform is

$$y(s) = g(s) \times u(s) + d(s)$$

Thus the correlation of input and output can enhance the accuracy of the estimate for the system frequency response function $g(j\omega)$, since the correlation function then yields the following estimate of the actual process model:

$$\hat{g}(j\omega) = g(j\omega) + \frac{d(j\omega)u^*(j\omega)}{u(j\omega)u^*(j\omega)}$$

Clearly, the effect of the disturbance $d(t)$ is readily reduced if the input is chosen either to be large in amplitude, compared to the disturbance, or to be pseudo-random so that it does not correlate with the disturbance and the second numerator becomes zero, by definition.

(2) the temporal or spatial function or sequence resulting from integrating lagged products of two signals, i.e., the products of a signal with the time-inverted version of a second signal at various relative delays with respect to each other. If the two signals are identical, a maximum correlation output amplitude results at full overlap. *See also* convolution.

correlation bandwidth   the frequency for which the autocorrelation of the transfer function reaches a given threshold, for example $x\%$ of the central value.

correlation coefficient   a measure of the ability to predict one random variable x as the linear function of another y. The correlation coefficient

$$\rho = \frac{E[xy] - E[x]E[y]}{\sigma_x \sigma_y}$$

satisfies $-1 \leq \rho \leq 1$, where $|\rho| = 1$ implies a deterministic linear relationship between x and y, and $\rho = 0$ implies lack of correlation. *See* correlation.

correlation detector   the optimum structure for detecting coherent signals in the presence of additive white Gaussian noise.

correlation function   a mathematical description that describes the common relationship between two processes. For a single time dependent quantity, the autocorrelation function describes the loss of information in the process description of the quantity. *See* autocorrelation function.

correlation peak detector   a photodetector array that detects and outputs only the peak in a spatial correlation function.

correlation receiver   a receiver utilizing the autocorrelation or cross-correlation properties of the transmitted signal to detect the desired information from the received signal. A correlation receiver is theoretically equivalent to a matched filter receiver. Correlation receivers are often used in spread spectrum systems and channel measurement systems.

correlation sidelobes   the correlation function other than the peak; particularly where signals are designed to give a sharply peaked correlation function.

correlation similarity   the (unnormalized) correlation of two real vectors $x = (x_0, \ldots, x_n)$ and $(y_0, \ldots, y_n)$ is defined as their inner product:

$$C = \sum_{i=1}^{n} x_i y_i.$$

correlator   a circuit that calculates the correlation function. *See also* correlation function.

correspondence problem   the problem of matching points in one image with their corresponding points in a second image.

corresponding point   a point, in a set of images representing a different view of the same scene, onto which the same physical point in the real world is projected.

corrugated horn   a horn with grooves on its inner walls. The grooves create the same boundary conditions on all four walls of the horn. This results in the elimination of spurious diffraction at the edges of the horn aperture.

coset leader   each possible error pattern of the code word representing the left-most column of the standard array.

cosine modulated filter bank a filter bank with each of its analysis filters and synthesis filters being the modulation of a low-pass prototype filter by a cosine function. In the frequency domain, it is equivalent to shifting the low-pass prototype filter by different frequencies to form a bank of band-pass filters covering the entire frequency band.

cosine roll-off filter a filter that has an impulse response which satisfies the Nyquist I criterion for zero intersymbol interference. The filter has the following transfer function.

$$H(\omega) = \begin{cases} 1, & |\omega| \leq \frac{\pi(1-\alpha)}{T} \\ \cos^2 \frac{T}{4\alpha} \left(|\omega| - \frac{\pi(1-\alpha)}{T} \right), & \frac{\pi(1-\alpha)}{T} < |\omega| < \frac{\pi(1+\alpha)}{T} \\ 0, & |\omega| \geq \frac{\pi(1+\alpha)}{T} \end{cases}$$

where T is the transmitted symbol period and α is a parameter that is known as the excess bandwidth ($0 \leq \alpha \leq 1$). The case $\alpha = 0$ yields the ideal low-pass filter, and the case $\alpha = 1$ yields a filter referred to as the raised cosine filter.

cosine transform a transform that consists of a set of basis functions, which are cosine functions. Usually referred to as discrete cosine transform. *See* discrete cosine transform.

cost function a nonnegative scalar function that represents the cost incurred by an inaccurate value of the estimate. Also called penalty function.

Costas loop a carrier synchronization loop in a digital communications receiver that uses a quadrature phase detector in place of a conventional square-law device.

cotree the complement of a tree in a network.

Cotton–Mouton effect second-order anisotropic reciprocal magnetooptic effect that causes a linearly polarized incident light to transmit through as an elliptically polarized output light wave when the propagation direction of the incident light is perpendicular to the direction of the applied magnetization of the magnetooptic medium. It is also known as magnetic linear birefringence.

Coulomb blockade the situation in which a particle has insufficient thermal energy to allow the necessary energy exchange during a tunneling process. Hence, the bias supply must supply energy to the electron to account for the stored energy change in tunneling, which requires $V > e/2C$, where C is the capacitance ($eV \ll kBT$).

Coulomb force electric force exerted on an electrically charged body, which is proportional to the amount of the charge and the electric field strength in which the charged body is placed.

Coulomb's Law the force of repulsion/ attraction between two like/unlike charges of electricity concentrated at two points in an isotropic medium is proportional to the product of their magnitudes and inversely proportional to the square of the distance between them and to the dielectric constant of the medium.

Coulomb, Charles (1763–1806) Born: Angouleme, France

Best known for his study of electric charge and magnetism resulting in Coulomb's Law, as well as his studies in friction. Coulomb also invented the torsion balance in 1777. He used this device in many experiments. Coulomb began his career in the military, but resigned when the French Revolution began. His experience as a military engineer involved him in a wide variety of different projects. It also gave him time to continue his own experimental work. Coulomb's law states that the force between two charges is proportional to the product of the charges and inversely proportional to the square of the distance between the two charges. Coulomb is honored by having his name used as the unit of electric charge, the coulomb.

counter (1) a variable or hardware register that contains a value that is always incremented or decremented by a fixed amount, and always in the same direction (usually incremented by one, but not always).

(2) a simple Moore finite state machine that counts input clock pulses. It can be wired or enabled to count up and/or down, and in various codes.

counter-EMF   a voltage developed in an electrical winding by Faraday's Law that opposes the source voltage, thus limiting the current in the winding.

counter-EMF starter   a type of DC-motor starter that reduces the resistance in the starting circuit as the voltage across the armature rises.

counter-propagation learning   *See* hierarchical feature map.

counter-rotating field theory   mathematical theory in which a magnetic field with stationary spatial direction but sinusoidally varying magnitude is decomposed into two constant-magnitude fluxes rotating in opposite directions. The angular velocity, ω, of the rotating fluxes equals the time-domain angular frequency of the stationary flux, and the magnitude of the rotating fluxes is half the magnitude of the stationary flux. Mathematically, this is described by the following equation:

$$B\cos\theta\sin\omega t = 0.5B\sin\omega t - \theta$$
$$+ 0.5B\sin\omega t + \theta$$

where the left-hand term represents the time-varying flux magnitude along the spatial direction θ, and the right-hand terms represent forward and backward rotating fields of constant magnitude rotating about θ. The following diagram illustrates this relationship. Resolution of the stationary flux into two rotating fluxes allows induction machine performance to be analyzed using standard rotating field theories. The total machine performance is determined from the net result of both rotating fluxes. Also called double revolving-field theory.

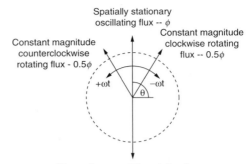

The vector sum of the rotating fluxes produces the stationary oscillating flux

Counter-rotating field concept.

counter-torque   torque developed in opposition to the rotation of a machine. It is produced as load current flows in the presence of and perpendicular to magnetic flux in a machine that is generating electric power.

counterpoise   a ground wire buried beneath an overhead line to lower footing impedance.

counterpoise ground   buried conductor routed under transmission lines designed to achieve low earth electrode resistance.

coupled inductor   two inductors within a switching converter, which have similar voltage waveforms, are wound on the same magnetic core to steer the ripple from one winding to another. A typical example is a Ćuk converter with an isolation transformer where the input and output inductors are coupled with the transformer to provide zero ripple at the input and the output. *See also* Ćuk converter.

coupled line filter   a type of microstrip or stripline filter that is composed of parallel transmission lines. Bandwidth is controlled by adjusting the transmission line spacing. Wider bandwidths are obtained by tighter coupling. A two-port circuit is formed by terminating two of the four ports in either open or short circuits, which leaves ten possible combinations. Different combinations are used to synthesize low-pass, bandpass, all pass, and all stop frequency responses.

coupled lines   the electromagnetic field of two unshielded transmission lines in proximity can interact with each other to form coupled lines. Usually three conductors are needed. Examples of coupled lines are coupled microstrip lines and coupled striplines.

coupled Riccati equations   a set of differential or algebraic matrix equations of the Riccati type arising in the jump linear quadratic problem which should be solved simultaneously. The set of coupled differential Riccati equations has a form

$$\dot{K}(i,t) + A(i)'K(i,t) + K(i,t)A(i)$$
$$\sum_{j=1}^{s} q_{ij} K(j,t) + Q(i)$$
$$- K(i,t)B(i)R(i)^{-1}B(i)'K(i,t) = 0$$

for $i \in \mathbf{S} = \{1, 2, \ldots, s\}$, $t \in [0, T]$ and terminal condition $K(i,T) = 0$ where $A(i), B(i), Q(i), R(i)$ are, respectively, system and weighting, state and input matrices in the mode i, and q_{ij} is the transition rate from mode i to j. If the system is stochastically stabilizable and the pairs $(A(i), \sqrt{Q(i)})$ are observable for all i where $\sqrt{Q(i)} = C(i)$ such that $C(i)'C(i) = Q(i)$, then the solution $K(i,t)$ is finite for each i and

$$\lim_{T-t \to \infty} K(i,t) = \hat{K}(i)$$

where $\hat{K}(i)$ is the set of solutions to the coupled algebraic Riccati equations:

$$A(i)'\hat{K}(i) + \hat{K}(i)A(i) + \sum_{j=1}^{s} q_{ij}\hat{K}(j) + Q(i)$$
$$- \hat{K}(i)B(i)R(i)^{-1}B(i)'\hat{K}(i) = 0$$

Coupled Riccati equations are also used in N-person linear quadratic games. In this case $i =$

$1, 2, \ldots, N$ is the number of the player and the set of coupled differential matrix equations has the following form:

$$\dot{K}(i,t) + A'K(i,t) + K(i,t)A + Q(i)$$
$$- K(i,t) \sum_{j=1}^{j=N} BR(j)^{-1}B'K(j,t) = 0$$

for $i \in \{1, 2, \ldots, N\}$, $t \in [0, T]$ and terminal condition $K(i,T) = 0$ where A, B are, respectively, state and input matrices in the state equation, $Q(i), R(i)$ are weighting state and input matrices for the i-th player, and its solution defines the gain in a Nash equilibrium in the game.

coupled wave equations   a special case of Maxwell's equations describing the propagation of two interacting electromagnetic fields in a nonlinear material. For example, in four wave mixing or optical phase conjugation, the probe and conjugate fields often obey coupled wave equations.

coupler   (1) a passive, wavelength-insensitive, fiber optic component that combines all inputs and distributes them to the outputs with a defined splitting ratio. The most common types are termed fused taper and dichroic.

(2) a device inserted into an optical fiber communications link that is used to insert additional optical energy or signals onto the link and/or tap off some of the energy or signal on the link to another optical fiber. A coupler may also be used to combine power from two or more input fibers into one output fiber or to distribute the combined input power to a number of output fibers. Couplers are made in $1 \times N$, $N \times N$, and $N \times M$ (input) $\times$ (output) configurations where N and M are the number of fibers. The power combining and/or splitting ratios can vary from 50%/50% (a 3 dB coupler) in a symmetric 2×2 coupler to $> 99\%/ < 1\%$ in an asymmetric coupler.

coupling capacitor voltage transformer (CCVT)   a potential transformer that uses the impedance of a small capacitance to reduce the power line voltage to measureable levels.

coupling factor   in a coupler, the ratio of power in the coupled port to that applied at the input port.

coupling efficiency   the efficiency to which a signal can be coupled from one transmission line or resonator to another.

coupling field   a region of interaction where energy is transferred between systems. The energy may be transferred between systems of a similar nature (i.e., electrical–electrical) or between systems of a different nature (i.e., electrical–mechanical).

coupling loss   measure of the power dissipated as a result of coupling a signal from one transmission line or resonator to another.

covariance   the expectation of the product of two mean-removed random quantities:

$$C_{fg}(t_1, t_2) = E\left[f(t_1)g(t_2)^T\right]$$
$$-E[f(t_1)]E[g(t_2)^T]$$

See also autocovariance, correlation, variance.

covert channel   a mechanism by means of which information can be communicated indirectly despite design features that prevent direct communication. Many covert channels utilize side effects of operations, such as effects on timing or scheduling that can be seen from programs or processes that are not supposed to be direct destinations of any communications from a sender that in fact performs operations that emit signals by affecting timing or scheduling within the system.

COVQ   *See* channel optimized vector quantization.

cow magnet   a long, pill-shaped alnico magnet that is placed in the second stomach of a cow to capture ferrous debris and prevent it from passing through the digestive system.

CPA   acronym for compensated pulsed alternator. *See* compulsator.

CPI   *See* color preference index or cycles per instruction.

CPM laser   *See* colliding pulse modelocked laser.

CPU   *See* central processing unit.

CPU time   the time that is required to complete a sequence of instructions. It equals to the (cycle time) × (number of instructions) × (cycles per instruction).

CR   *See* cavity ratio.

Cramer-Rao bound   a lower bound on the estimation error covariance for unbiased estimators. In particular, the estimation error covariance $\Lambda_e(x)$, which is a function of the unknown quantity x to be estimated, must satisfy $\Lambda_e(x) \geq I(x)$, where I is the Fisher information matrix. If an estimator achieves the Cramer-Rao bound with equality for all x, it is efficient; if an efficient estimator exists, it is the maximum likelihood estimator. *See* bias, maximum likelihood estimation.

crash   an incorrect operation of computer software or hardware leading to temporary or permanent loss of part of the data.

CRC   *See* cyclic redundancy check.

CRC character   a type of error detection code commonly used on disk and tape storage devices. Data stored on a device using CRC has an additional character added to the end of the data that makes it possible to detect and correct some types of errors that occur when reading the data back.

CRC-code   code that employs cyclic redundancy checking. *See* cyclic redundancy check.

CRCW   *See* concurrent read and concurrent write.

credit assignment problem   during neural network learning, the problem of determining how to apportion credit (blame) to individual components for network behavior that is appropriate (inappropriate) to the task being learned.

crest factor   the ratio of the peak value of a signal to its RMS value.

CREW   *See* concurrent read and exclusive write.

CRF   *See* contrast rendition factor.

crisp set   in fuzzy logic and approximate reasoning, this term applies to classical (nonfuzzy Boolean) sets that have distinct and sharply defined membership boundaries. *See also* fuzzy set.

critical   refers to the state of a chain reaction which is self-sustaining but which produces just enough free neutrons to compensate for those lost to the moderator and leakage.

critical angle   the incidence angle, defined by Snell's law, where the incident wave is totally reflected at the interface of two different dielectric media.

critical band   broadly used to refer to psychoacoustic phenomena of limited frequency resolution in the cochlea. More specifically, the concept of critical bands evolved in experiments on the audibility of a tone in noise of varying bandwidth, centered around the frequency of the tone. Increasing the noise bandwidth beyond a certain critical value has little effect on the audibility of the tone.

critical clearing angle   (1) following a balanced three-phase fault at the stator terminals of a synchronous machine, the maximum value of the angular position of the rotor prior to the removal (clearing) of the fault such that the rotor will obtain synchronous speed without slipping poles following the removal (clearing) of the fault. The corresponding time for the rotor to achieve this angle is specified as the critical clearing time.

(2) the largest allowable angular deviation from synchronism that may be borne by a power system such that the system remains stable: the edge of instability.

critical damping   the least amount of damping such that the system does not freely oscillate. For a characteristic equation of the form: $s^2 + 2\zeta\omega_n s + \omega_n^2$, where $\zeta = 1.0$; the system is critically damped if the roots of the characteristic equation are repeated and real.

critical dimension (CD)   the size (width) of a feature printed in resist, measured at a specific height above the substrate.

critical frequency   the rate of picture presentation, as in a video system or motion picture display, above which the presented image ceases to give the appearance of flickering. The critical frequency changes as a function of luminance, being higher for higher luminance.

critical path   a signal path from a primary input pin to a primary output pin with the longest delay time in a logic block.

critical point   *See* equilibrium point.

critical race   a change in two input variables that results in an unpredictable output value for a bistable device.

critical region   a set of instructions for a process that access data shared with other processes. Only one process may execute the critical region at a time.

critical section   *See* critical region.

critically sampled   sampling at the Nyquist frequency.

Crosby direct FM transmitter   after its inventor, Murray Crosby. Also known as the "serrasoid modulator." Direct frequency modulation (FM) of an inductor/capacitor (LC) oscillator is essentially straightforward: One of the frequency determining elements value is varied in accordance with the baseband information.

cross-polarization   the field component orthogonal to the desired radiated field component (co-polar component).

cross chrominance   NTSC video artifact that causes luminance information to be present in the decoded chroma signal (luminance crosses into chrominance). Cross chrominance is a result of mixing high frequency luminance information with the chrominance information in the composite video signal. An example of cross luminance is the rainbow pattern observed when tweed or a herringbone pattern appears in a TV scene.

cross color   *See* cross chrominance.

cross luma   *See* cross luminance.

cross luminance   NTSC video artifact that causes chrominance information present in the decoded luma signal (chrominance crossed into luminance). Cross luminance has the appearance similar to a zipper caused by the color subcarrier. Television receivers that use a line comb filter to separate luma and chroma signals have cross-luminance components appearing on sharp horizontal edges. Television receivers that use a bandpass filter for luma–chroma separation have cross-chrominance components appearing on sharp vertical edges.

cross modulation   an undesired intermodulation of an electromagnetic carrier wave by another electromagnetic carrier wave that is either physically adjacent to it or near it in terms of its radio frequency.

cross polar discrimination   the ratio of co-polar to cross-polar received field strengths in a depolarizing medium. Usually expressed in decibels.

cross polar isolation   the ratio between the field received by an antenna with one polarization from co-polar and cross-polar transmissions. This expresses the receiver's ability to detect one signal when dual-polarized signals are transmitted through a depolarizing medium.

cross power spectrum   for two jointly wide sense stationary random processes the Fourier transform of their cross-correlation.

cross spectra   computation of the energy in the frequency distribution of two different electrical signals.

cross-assembler   a computer program that translates assembly language into machine code for a target machine different from the one on which the cross-assembler runs.

cross-compiler   a computer program that translates a source code into machine or assembly code for a target machine different from the one on which the cross-compiler runs.

cross-correlation   a measure of the correlation or similarity of two signals. For random processes $x(t)$ and $y(t)$, the cross-correlation is given by: $R_{xy}(t_1, t_2) = Ex(t_1)y(t_2)$. *See* wide sense stationary.

cross-correlation function   a function describing the degree of similarity between two signals, as a function of time-shift between the signals. *See also* correlation function.

cross-field theory   a conceptual way to envision the operation of a single-phase induction motor.

The rotor current is assumed to produce a magnetic field electrically and spatially orthogonal to the field produced by the main stator winding, thus contributing to a rotating magnetic field.

cross-quad   the simplest form of the common centroid concept, in which two matched transistors are formed by connecting diagonally opposite devices in a two-by-two array of transistors.

crossarm   a transverse, generally insulated member mounted horizontally on a utility pole. It carries insulators and allows wide spacing of overhead conductors.

crossarm brace   a brace, often insulated, which keeps a crossarm from rotating on its attachment bolts.

crossbar switch   a structure that allows N units to communicate directly with each other, point to point. Which pairs are connected depends on how the switch is configured at that point in time. Crossbars are usually implemented for small (8 or less) numbers of nodes, but not always.

crossed field devices (CFD)   radial or linear forward traveling wave amplifier or backward wave oscillator where radial or transverse DC accelerating electric fields are perpendicular to axial or longitudinal DC magnetic fields, respectively.

crossing number   the even number obtained by adding the number of changes in binary value on going once around a particular pixel location; crossing number is useful for finding skeletons of shapes and for helping with their analysis: it is a measure of the number of 'spokes' of an object emanating from a specified location.

crossmodulation   modulation of a desired carrier by an undesired interfering signal due to interaction of the two signals in one or more non-linear elements.

crossover frequency   the frequency where the magnitude of the open-loop gain is 1.

crossover point   a crossover point of a fuzzy set A is the point in the universe of discourse whose membership in A is 0.5.

crosspoint   a point at which two overlaping neighboring fuzzy sets have the same membership grade.

crosspoint level   the membership grade of a fuzzy set at a crosspoint.

crosstalk   (1) undesired coupling (resistive, capacitive, or inductive) from one signal wire or transmission line to a colocated signal wire or transmission line.

(2) capacitative interference between two parallel transmission lines, in which a signal on one line affects a signal on the other line.

crowbar   a triggered, shunt device that diverts the stored energy of the beam power supply.

CRR   *See* cochannel reuse ratio.

CRT   *See* cathode ray tube.

crystal filter reference amplifier   a generic phrase that refers to a cascade of amplifying stages coupled together with the aid of crystal filter networks. Various topologies are available for the connection of these crystals (lattice, bridged, hybrid, and ladder). The outstanding feature of such an arrangement is that the amplitude versus frequency and phase versus frequency characteristics are essentially determined by the crystal arrangements.

crystal oscillator   an electronic circuit used for generating a sinusoidal waveform whose frequency is determined by a piezoelectric crystal device.

crystalline phase in crystalline materials the constituent atoms are arranged in regular geometric ways; for instance in the cubic phase the atoms occupy the corners of a cube (edge dimensions $\approx$ 2–15 Å for typical oxides).

CSI *See* current-source inverter.

CSIM *See* capacitor-start induction motor.

CSMA *See* carrier-sense multiple access.

CSMA/CD CSMA with collision detection. A modification of the CSMA protocol, where a user continues to monitor (sense) the channel after it initiates its transmission to determine if other terminals have also initiated transmissions. The terminal aborts the transmission if it detects a collision. This protocol is used in the local area network specified in the standard IEEE 802.3, which defines a local area computer network commonly referred to as the Ethernet. *See also* carrier-sense multiple access.

CSMA-CA CSMA with collision avoidance, a modification of the CSMA multiple access protocol to make it suitable for use in a radio environment. Prior to the transmission of a message, a source and destination node undertake an exchange of short messages to ensure that the destination node is idle and capable of receiving a transmission from the source node. *See also* carrier-sense multiple access.

CSO *See* composite second order.

CSR *See* control and status register.

CT *See* current transformer.

CT2 (cordless telephone, 2nd generation). A digital cordless communication system developed in the United Kingdom, which operates in the 864-868 MHz band and can provide both private and public cordless telephony services. CT2-based public telephony services are often described as Telepoint services (q.v.). CT2 systems have 40 radio channels, each 100 kHz wide. Each radio channel provides a duplex traffic channel using simple "ping-pong" TDD. The carrier bit rate is 72 Kb/s, and CT2 terminals have a peak transmit power of 10 mW.

CTB *See* composite triple beat.

CTE *See* coefficient of thermal expansion mismatch.

CU *See* coefficient of utilization.

cubic voxel *See* voxel.

cubical quad antenna a parasitic array antenna in which the elements are made of square loops of wire with a perimeter of one wavelength. These antennas are typically used in the HF or VHF range and usually have only two or three elements, resulting in a structure that fills a roughly cubical volume.

Ćuk converter named after its inventor, Slobodan Ćuk, this device can be viewed as a boost converter followed by a buck converter with topologic simplification. A capacitor is used to transfer energy from the input to the output. The output voltage v_o is related to the input voltage v_i by $v_o = v_i d/(1-d)$ and it can be controlled by varying the duty ratio d. Note that the output voltage is opposite polarity to the input.

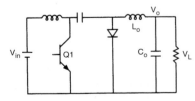

Ćuk converter.

cumulative distribution function (CDF) for a random variable $\mathbf{x}$, the probability that $\mathbf{x}$ is less than or equal to some value x, denoted $F_{\mathbf{x}}(x)$. $F_{\mathbf{x}}-\infty$ is zero and increases monotonically to $F_{\mathbf{x}}(+\infty) = 1$. For a continuous probability density function $p(x)$, the CDF is $F_{\mathbf{x}}(x) = \int_{-\infty}^{x} p(t)\,dt$. The CDF is used in image processing to carry out histogram equalization. *See* histogram equalization, probability density function.

cumulatively compounded a compound-wound DC machine in which the flux produced by the MMF of the shunt field winding and the flux produced by the MMF of the series field winding are in the same direction.

Curie temperature the temperature above which a material ceases to be magnetic; at this temperature a ferromagnetic material becomes paramagnetic.

current the flow of charge, measured in amperes (1 ampere = 1 coulomb/s).

current density vector field the field (commonly denoted J) that is related to the electric field intensity vector field by the conductivity of the medium that the fields are located in. One of the quantities found on the right side of Ampere's Law. The units are (amperes/square meter).

current distribution factor in economic dispatch studies, the proportion of a power line's total current which is contributed by a particular generating plant.

current fed inverter an inverter in which the current is switched instead of the voltage to create AC current from a DC source. A large inductor is used to maintain a nearly constant current on the DC side which is then directed to the load in an alternating fashion by the use of switching elements.

current feedback op-amp an op-amp in which the output voltage is controlled by the current in the input stage, rather than the voltage. The advantage of the current feedback op-amp is that (to first order) its closed-loop performance is not subject to the gain-bandwidth tradeoff that affects voltage feedback op-amps.

current gain short-circuit current gain that helps describe the physical operation of the transistor. The current gain is the dimensionless ratio of the peak RF output current ($I_{out.pk}$) to the peak incident RF current ($I_{incident.pk}$). Hence, current gain, G_I is

$$G_I = \frac{I_{out.pk}}{I_{incident.pk}}$$

current limiting the output current is limited to a preset level even under a shorted output condition. This can be accomplished by reducing the output voltage to prevent the current limit from being exceeded.

current limiting fuse a fuse that limits the level of fault current from that which is available. It operates by developing a substantial voltage across the fuse following the melting of the fuse element.

current mirror a configuration of two matched transistors in which the output is a current that is ideally equal to the input current. In the case of a BJT current mirror, the collector of the first transistor is forced to carry the input current. This establishes a corresponding base-emitter voltage, which is applied to the second transistor. If the two devices are matched, then the collector current of the second transistor will equal that of the first transistor, thus "mirroring" the input current. This is a commonly used configuration in integrated circuits, which can take advantage of the inherent matching available by fabricating the two transistors in close proximity to each other.

current regulator a device used to control the magnitude and phase of the current in DC, AC or other electrical variable speed drives. May use different control strategies like hysteresis current control or ramp comparison current control.

current source amplifier   the most common group of amplifiers is made up of current sourcing amplifiers, in which the active device acts as a modulated current source. All class A, B, A-B, and C amplifiers fit into this general group. Parameters such as device characteristics, quiescent bias point, RF load line, and amplitude and waveform of the applied signal should be included with the class definition, thus defining the major contributors to the physical actions taking place in one of these amplifiers.

current source inverter   *See* current fed inverter.

current source region   region of the I-V curve(s) of a device in which the output current is relatively constant (slope near zero) for changing output voltage.

current transducer   a device used to measure current in a variety of applications including variable speed drives. May give out a current proportional to the measured current or a voltage proportional to the measured current. Electrical isolation may be obtained by using a current transformer (cannot be used at DC) or a hall effect transducer that can be used down to DC.

current transformer (CT)   (1) a transformer that is employed to provide a secondary current proportional to primary current flowing. The primary "winding" is often created by passing the system conductor or bus bar through an opening in the device and the secondary is typically rated at a standard value to match common meters and display units. Current transformers are used in current measurement, protective relays, and power metering applications. The load (meter) on a CT should never be removed without first shorting the

secondary of the CT, otherwise dangerous voltage levels may result when the load is removed.

(2) a device which measures the instantaneous current through a conductor of an electric power line and transmits a signal proportional to this current to the power system's instrumentation.

current unit   a protective relay that monitors the magnitude of a power transmission line's current flow.

current withstand rating   the current withstand rating of a device is the maximum short term current that can flow in the device without causing damage. *See* ampacity.

current-controlled multivibrator   a multivibrator with oscillation period controlled by the current that is charging and discharging a timing capacitor. Most of the circuits are based on two symmetric controlled current sources connected by a timing capacitor, and a steering circuit (connected between power supply voltage and this capacitor), which includes two switches. The switches take turns providing the conducting path for the current sources, and the current in the timing capacitor changes direction each half of the period. If the current sources are controlled by an external voltage, the circuit becomes a voltage-controlled multivibrator. Current- and voltage controlled multivibrators are a vital part of phase-locked loops and frequency synthesizers.

current-limiting device   when operating within its current-limiting range, a current-limiting device is open in 0.25 cycle or less and limits the maximum short-circuit current to a magnitude substantially less than the short-circuit current available at the fault point.

current-mode control   the inductor current is sensed to control the pulse width. The switch is turned on by a constant frequency clock and turned off when the inductor current reaches the control reference. This method produces subharmonic oscillations if the steady-state duty ratio is greater than or equal to 0.5. An artificial stabilizing ramp is necessary to achieve stable control over the full range of duty ratios. When the slope of the ramp is equal to the falling slope of the inductor current, one cycle response is achieved. (i.e., the input perturbations are rejected, and the current reference is followed within one cycle).

current-source inverter (CSI) an inverter with a DC current input. Current-source inverters are most commonly used in very high power AC motor drives.

cursor (1) the symbol on a computer screen that indicates the location on the screen that subsequent input will affect.

(2) a movable, visible mark used to indicate a position of interest on a display surface.

curvature (1) property of a planar curve: the limit, for arc length going to zero, of the ratio between the change in tangent angle and the change in arc length.

(2) a geometric property that describes the degree that a surface or a curve is bent. The curvature of a curve is the magnitude of the rate of change of the unit tangent vector with respect to arc length. The curvature of a surface is given in terms of a metric tensor which embodies two principal (planar) curvatures, κ_1 and κ_2. The curvature of a surface is sometimes characterized by the so-called Gaussian curvature $\kappa = \kappa_1 \kappa_2$.

curvature function a function which gives curvature values at different locations of a curve or surface.

cut-off frequency the minimum frequency at which a waveguide mode will propagate energy with little or no attenuation.

cut-off rate a measure of reliability of a channel. At information rates below the cutoff rate, the probability of error averaged over the ensemble of randomly selected codes can be forced to

zero exponentially with the block length (for block codes), or constraint length (trellis codes). The cut-off rate is less than or equal to the channel capacity. Also known as the computational cut-off rate, since for trellis codes, it is at this rate that the expected number of computations required to perform sequential decoding becomes unbounded. *See also* capacity.

cut-off wavelength the wavelength above which a fiber exhibits single mode operation (and below which exhibits multimode operation).

cutoff frequency *See* cut-off frequency.

cutoff rate *See* cut-off rate.

cut-out a pole-mounted device which contains a high-voltage fuse integral with a disconnect switch. It is used to protect and disconnect primary circuits in distribution work.

cutset a minimal subnetwork, the removal of which cuts the original network into two connected pieces.

CVD *See* chemical vapor deposition.

CVQ *See* classified vector quantization.

CW *See* continuous wave.

cybernetics (1) the field of control and communication theory in general, without specific restriction to any area of application or investigation.

(2) the behavior and design of mechanisms, organisms, and/or organizations that receive and generate information and respond to it in order to obtain a desired result.

cycle *See* clock cycle.

cycle stealing an arrangement in which a DMA controller or I/O channel, in order to use the I/O bus, causes the CPU to temporarily suspend its use of the bus. The CPU is said to hesitate. *See also* direct memory access.

cycle time   time required to complete one clock cycle (usually measured in nanoseconds).

cycles per instruction (CPI)   a performance measurement used to judge the efficiency of a particular design.

cyclic access   in devices such as magnetic and optical disks (and older bubble memories) that store data on rotating media, the property that any individual piece of data can be accessed once during each cycle of the media.

cyclic code   a linear block code where every cyclic shift of a codeword is another codeword in the same code. This property is an outcome of the significant algebraic structure that underlies these codes.

cyclic redundancy check (CRC)   (1) is used to detect transmission errors over communication channels in digital systems. When using the CRC technique, the modulo-2 sum of the message bits is calculated after grouping them in a special way and then appended to the transmitted message and used by the receiver to determine if the message was received correctly. The number of bits used in the CRC-sum is typically 8 or 16, depending on the length of the message and the desired error-detection capability.

A value calculated from a block of data to be stored or transmitted with the data block as a check item. CRCs are generated using a shift register with feedback and are described by the length of the register and the feedback terms used.

(2) a coding scheme for detecting errors in messages. Given a polynomial $C(x)$ of order $c - 1$, and an m-bit message $M(x)$ represented as a polynomial of order $m - 1$, calculate $R(x) = (x^c M(x))/C(x)$, then construct the transmitted message $T(x) = (x^c M(x)) \oplus R(x)$ which is exactly divisible by $C(x)$.

Supposing $E(x)$ represents any errors introduced during transmission, the receiver calculates $(T(x) + E(x))/C(x)$. If this is not zero, $E(x)$ is non-zero and the received message is erroneous. If this is zero, either no errors were introduced or $E(x)$ is such that the CRC is not strong enough to detect the error. A suitable choice of $C(x)$ will make this unlikely.

cycloconverter   a frequency changing converter that synthesizes lower frequency sine waves from portions of a polyphase set of higher frequency sine waves. For example, a three-phase, 400 Hz input could be used to create a 60 Hz supply of any desired phase order.

cyclotron frequency   a frequency of electron oscillation in a cyclotron. A frequency of circular motion of an electron under magnetic fields applied perpendicularly to the plane of the circular motion.

cylinder   a stack of tracks where these tracks are at a constant radial position (the same track number) on a disk or disk pack.

cylindrical extension of a fuzzy set   let A be a fuzzy set in a Cartesian product space X^i, and X^n be another Cartesian product space including X^i, then the cylindrical extension of A in X^n is a fuzzy set in X^n, denoted as *cext* $(A; X^n)$, with membership function defined equal to membership in X^i. *See also* fuzzy set, membership function, projection of a fuzzy set.

cylindrical lens   a lens that has curvature, or optical focusing power, in one direction only.

cylindrical microstrip line   structure that consists of an angular strip on a cylindrical ground plane separated by a dielectric substrate. The cylindrical conductor acts as a ground plane, the angular strip guides the electromagnetic wave along the longitudinal direction.

cylindrical stripline   the angular strip in the homogeneous media between the two cylindrical conductors. The inner and outer conductors act as ground planes, the angular strip guides the electromagnetic wave along the longitudinal direction.

cylindrical wave   an electromagnetic wave in which each wavefront (surface of constant phase) forms a cylinder of infinite length and propagates in toward or away from the axis of the cylinder. A uniform cylindrical wave has the same amplitude over an entire wavefront; a nonuniform cylindrical wave has varying amplitude.

cylndrical-rotor machine   a synchronous machine with a cylindrical rotor containing a distributed field winding and an essentially uniform air-gap. This design is limited to two and four pole machines (3600 and 1800 rpm at 60 Hz) and is usually used in large generators. *See also* salient-pole machine.

D

d axis   *See* direct axis.

D flip-flop   a basic sequential logic circuit, also known as bistable, whose output assumes the value (0 or 1) at its D input when the device is clocked. Hence it can be used as a single bit memory device, or a unit delay.

D'Arsenval meter   a permanent-magnet moving-coil instrument with a horseshoe form magnet. It measures direct current only.

D-TDMA   *See* dynamic time division multiple access.

D/A   *See* digital-to-analog converter.

DAC   *See* digital-to-analog converter.

Dahlin's method   a synthesis procedure for finding a digital control law that will produce a defined closed loop behavior for a given process, when it is controlled in a feedback configuration. The synthesis equation is

$$k(z) = \frac{(1 - e^{-T/\lambda})z^{-N-1}}{1 - e^{-T/\lambda}z^{-1} - (1 - e^{-T/\lambda})z^{-N-1}}$$
$$\times \frac{1}{g(z)}$$

where $k(z)$ is the pulse transfer function for the synthesized controller, T is the sampling time of the digital control loop, λ is the time constant specified for the closed loop system, NT is the process deadtime, and $g(z)$ is the pulse transfer function for the process.

daisy chain   (1) a type of connection when devices are connected in series.

(2) a hardware configuration where a signal passes through several devices. A signal will be passed through a device if that device is not requesting service, or not passed through if the device is requesting service.

(3) an interrupt-prioritizing scheme in which the interrupt acknowledge signal from the CPU is connected in series through all devices. A shared interrupt-request line connects all devices to the CPU with a single common line. When one (or several) devices activates its request line, the CPU will (after some delay) respond with an acknowledge to the first device. If this device did request an interrupt, it will be serviced by the CPU. However, if the device did not request an interrupt, the acknowledge is just passed through to the next device in the daisy chain. This process is repeated until the acknowledge signal has passed through all the connected devices on this chain. The scheme implements prioritized service of interrupts by the way the devices are electrically connected in the daisy chain: the closer a device is to the CPU, the higher its priority.

A more general case exists where several daisy chains are used to form priority groups, where each chain has a unique priority. The CPU will service interrupts starting with the daisy chain having the highest priority. In this scheme, any device may be connected to more than one priority group (chain), using the interrupt priority level appropriate for the particular service needed at this moment.

DAMA protocol   *See* demand assign multiple access protocol.

damage   the failure pattern of an electronic or mechanical product.

damped sinusoid   a sinusoidal signal multiplied by a decaying. For example, $x(t) = 10e^{-3t}cos(10t)$.

damper winding   an uninsulated winding, embedded in the pole shoes of a synchronous machine, that includes several copper bars short-circuited by conducting rings at the ends, used to reduce speed fluctuation in the machine by developing an induction-type torque that opposes any change in speed.

damping   a characteristic built into electrical circuits and mechanical systems that prevents rapid or excessive corrections that may lead to instability or oscillatory conditions.

damping coefficient   electrical torque component in phase with the rotor speed.

damping factor   a measure of the ability of the PLL to track an input signal step. Usually used to indicate the amount of overshoot present in the output to a step perturbation in the input.

damping ratio   the ratio of the real part of the resonant frequency to the undamped resonant frequency, in a second-order system.

dark current   a noise source in photodetectors, corresponding to undesired output signals in the absence of light.

darkfield performance   the amount of light that is blocked in a light valve projector when a totally dark scene is being displayed. This parameter is critical to good contrast ratio.

Darlington bipolar junction transistor   a combination of two bipolar junction transistors (BJT) where the emitter current of one transistor drives the base of the second transistor. The arrangement reduces the current required from the base driver circuit, and the effective current gain of the combination is approximately the product of individual gains. The configuration can be made from two discrete transistors or can be obtained as a single integrated device.

dart leader   in lightning, a continuously moving leader lowering charge preceding a return stroke subsequent to the first. A dart leader typically propagates down the residual channel of the previous stroke.

dashpot timer   the dashpot timer is a fluid time-on timer that can be used in definite time DC motor acceleration starters and controllers. The dashpot timer functions where a magnetic field forces a piston to move within a cylinder when the coil is energized. The movement of the piston is limited by fluid passing through an orifice on the piston. The amount of fluid passing through the orifice is controlled by a throttle value, which in turn determines the amount of time delay. After the fluid pressure equalizes across the piston, movement stops and contacts change state. The fluid can either be air (pneumatic dashpot) or oil (hydraulic dashpot). When the timer is deenergized, a check valve allows the pressure to equalize across the piston rapidly, thereby causing the contacts to change state "immediately."

data   any information, represented in binary, that a computer receives, processes, or outputs.

data access fault   a fault, signaled in the processor, related to an abnormal condition detected during data operand fetch or store.

data acquisition   (1) method used for capturing information and converting it to a form suitable for computer use.
(2) process of measuring real-world quantities and bringing the information into a computer for storage or processing.

data bottleneck   a computer calculation in which the speed of calculation is limited by the rate at which data is presented to the processor rather than by the intrinsic speed of the processor itself. Ultra high speed parallel processors are very frequently limited in this way.

data buffer   *See* buffer.

data bus set of wires or tracks on a printed circuit or integrated circuit that carry binary data, normally one byte at a time.

data cache a small, fast memory that holds data operands (not instructions) that may be reused by the processor. Typical data cache sizes currently range from 8 kilobytes to 8 megabytes. *See* cache.

data communications equipment (DCE) a device (such as a modem) that establishes, maintains, and terminates a session on a network.

data compression theorem Claude Shannon's theorem, presenting a bound to the optimally achievable compression in (lossless) source coding. *See also* Shannon's source coding theorem.

data dependency the normal situation in which the data that an instruction uses or produces depends upon the data used or produced by other instructions such that the instructions must be executed in a specific order to obtain the desired results.

data detection in communications, a method to extract the transmitted bits from the received signal.

data flow architecture a computer architecture that operates by having source operands trigger the issue and execution of each operation, without relying on the traditional, sequential von Neumann style of fetching and issuing instructions.

data fusion analysis of data from multiple sources — a process for which neural networks are particularly suited.

data logger a special-purpose processor that gathers and stores information for later transfer to another machine for further processing.

data path the internal bus via which the processor ships data, for example, from the functional units to the register file, and vice versa.

data pipeline a mechanism for feeding a stream of data to a processing unit. Data is pipelined so that the unit processing the data does not have to wait for the individual data elements.

data preprocessing the processing of data before it is employed in network training. The usual aim is to reduce the dimensionality of the data by feature extraction.

data processing inequality information theoretic inequality, a consequence of which is that no amount of signal processing on a signal can increase the amount of information obtained from that signal. Formally stated, for a Markov chain $X \to Y \to Z$,

$$I(X; Z) \leq I(X; Y)$$

The condition for equality is that $I(X; Y|Z) = 0$, i.e., $X \to Z \to Y$ is a Markov chain.

data reduction coding system any algorithm or process that reduces the amount of digital information required to represent a digital signal.

data register a CPU register that may be used as an accumulator or a buffer register or as index registers in some processors. In processors of the Motorola M68000 family, data registers are separate from address registers in the CPU.

data segment the portion of a process' virtual address space allocated to storing and accessing the program data (BSS and heap, and may include the stack, depending on the definition).

data stripe storage methodology where data is spread over several disks in a disk array. This is done in order to increase the throughput in disk accesses. However, latency is not necessarily improved. *See also* disk array.

data structure a particular way of organizing a group of data, usually optimized for efficient storage, fast search, fast retrieval, and/or fast modification.

data tablet   a device consisting of a surface, usually flat, and incorporating means for selecting a specific location on the surface of the device and transmitting the coordinates of this location to a computer or other data processing unit that can use this information for moving a cursor on the screen of the display unit.

data terminal equipment (DTE)   a device, such as a subscriber's computer, Exchange workstation, or Exchange central system, that controls data flow to and from a computing system. It serves as a data source, data sink, or both, and provides for the data communication control function according to protocols. Each DTE has an address that is a 12-digit number uniquely identifying the subscriber's connection to the network. DTE term is usually used when referring to the RS-232C serial communications standard, which defines the two end devices of the communications channel: the DTE and the DCE (data communications equipment). The DCE is usually a modem and the DTE is a UART chip of the computer.

data-oriented methodology   an application development methodology that considers data the focus of activities because data are more stable than processes.

database computer   a special hardware and software configuration aimed primarily at handling large databases and answering complex queries.

dataflow computer   a form of computer in which instructions are executed when the operands that the instructions require become available rather than being selected in sequence by a program counter as in a traditional von Neumann computer. More than one processor is present to execute the instructions simultaneously when possible.

Daubechies wavelets   a class of compactly supported orthogonal and biorthogonal wavelets, first proposed by Ingrid Daubechies, that can be obtained by imposing sufficient conditions on wavelet filters.

daughter board   a computer board that provides auxiliary functions, but is not the main board (motherboard) of a computer system (and is usually attached to the motherboard).

dB   *See* decibel

dBc   ratio of the signal power (p) to a reference signal power (p_{ref}), usually the modulation carrier signal, expressed in decibels referenced to a carrier (dBc). Thus a harmonic signal that is 1/100th of the power in a desired fundamental signal is at -20 dBc.

$$P_{dBd} = 10\log_{10}\left(\frac{p}{p_{ref}}\right)$$

dBm   Power ratio in decibels referenced to 1 milliwatt.

DBS   *See* direct broadcast satellite.

DBS receiver   electronic assembly that accepts as input a microwave signal from a satellite, containing transmitted TV signals "modulated" onto the signal. The receiver first amplifies the low-level signal, then processes the signal by first converting it to a lower "IF" frequency and then demodulating the signal to separate the TV signals from the microwave carrier signal. A basic way of looking at the relationship of the microwave carrier to the TV signal is to think of the carrier signal as an envelope with a message letter inside. The message letter is the TV signal. Demodulation is the process of carefully removing the message from the envelope (carrier). The noise figure of receiver is a measure of the amount of noise that will be added to the signal (carrier and TV signal) by the receiver. If the receiver adds too much noise, the result will be a snowy picture on the TV screen.

dBW   ratio of the signal power in watts (p_{watts}) to a 1 W reference power, expressed in decibels referenced to 1 W (dBW). Thus 1 watt signal power is equal to 0 dBW, and -30 dBW is equal to 0 dBm.

$$P_{dBW} = 10\log_{10}\left(\frac{p_{watts}}{1 \cdot Watt}\right)$$

DC   direct current. *See* DC current, DC voltage.

DC block   A circuit simulation component that behaves like a capacitor of infinite value.

DC chopper   a DC to DC converter that reduces the voltage level by delivering pulses of constant voltage to the load. The average output is equal to the input times the duty cycle of the switching element.

DC circuit   electrical networks in which the voltage polarity and directions of current flow remain fixed. Thus such networks contain direct currents as opposed to alternating currents, thereby giving rise to the term.

DC current   constant current with no variation over time. This can be considered in general terms as an alternating current (AC) with a frequency of variation of zero, or a zero frequency signal. For microwave systems, DC currents are provided by batteries or AC/DC converters required to "bias" transistors to a region of operation where they will either amplify, mix or frequency translate, or generate (oscillators) microwave energy.

DC drain conductance   for an FET device under DC bias, the slope of the output drain to source current (I_{DS}) versus output drain to source voltage (V_{DS}) for a fixed gate to source voltage (V_{GS}),

expressed in siemens.

$$g_d = \left.\frac{\partial I_{DS}}{\partial V_{DS}}\right|_{V_{GS} = \text{constant}}$$

DC generator commutator exciter   a source of energy for the field winding of a synchronous machine derived from a direct current generator. The direct current generator may be driven by an external motor, a prime mover, or by the shaft of the synchronous machine.

DC input power   the total DC or bias power dissipated in a circuit, which is usually dependent on signal amplitudes, expressed in watts. This may include input bias, bias filtering, regulators, control circuits, switching power supplies and any other circuitry required by the actual circuit. These considerations should be explicitly specified, as they will affect how efficiency calculations are performed.

DC link   the coupling between the rectifier and inverter in a variable speed AC drive.

DC link capacitor   a device used on the output of a rectifier to create an approximately constant DC voltage for the input to the inverter of a variable speed AC drive.

DC link inductor   an inductor used on the output of a controlled rectifier in AC current source drives to provide filtering of the input current to the current source inverter. If used in conjunction with a capacitor, then it is used as a filter in voltage source drives.

DC load flow   a fast method of estimating power flows in an electric power system in which the problem is reduced to a DC circuit, with line impedances modeled as resistances and all generator bus voltages presumed to remain at their nominal values.

DC machine   an electromechanical (rotating) machine in which the field winding is on the stator and carries DC current, and the armature winding is on the rotor. The current and voltage in the armature winding is actually AC, but it is rectified by the commutator and brushes.

DC motor   a motor that operates from a DC power supply. Most DC motors have a field winding on the stator of the machine that creates a DC magnetic field in the airgap. The armature winding is located on the rotor of the machine and the DC supply is inverted by the commutator and brushes to provide an alternating current in the armature windings.

DC motor drive   a converter designed to control the speed of DC motors. Controlled rectifiers are generally used and provide a variable DC voltage from a fixed AC voltage. Alternatively, a chopper, or DC–DC converter, can be employed to provide a variable DC voltage from a fixed DC voltage.

DC offset current   the exponentially decaying current component that flows immediately following a fault inception. DC offset is the result of circuit inductance, and is a function of the point in the voltage wave where the fault begins. The offset for a given fault can range from no offset to fully offset (where the instantaneous current peak equals the full peak–peak value of the AC current).

DC restoration   reinsertion of lost DC level information into a signal after using AC signal coupling; in television applications, the DC component of a composite video signal represents the average brightness of the picture. After AC coupling of the composite video, the DC level includes the average luminance signal plus the fixed average of the sync and lanking signals, causing picture level racking errors. For a positive video signal, the average value of mostly white scenes will be slightly lower than it should be; for mostly dark scenes, the DC average could become negative (due to the sync and blanking signals) when it should be slightly positive. Clamping circuits restore the DC average of the video by establishing the DC level of the sync tips.

DC servo drive   a feedback, speed control drive system used for position control. Servos are used for applications such as robotic actuators, disk drives, and machine tools.

DC test   tests that measure a static parameter, for example, leakage current.

DC to RF conversion efficiency   dimensionless ratio of RF power delivered to the load (p_{out}) versus total DC input power dissipated in the amplifier (p_{DC}). With the DC to RF conversion efficiency given my η_{DC} we have

$$\eta_{DC} = \frac{p_{out}}{p_{DC}}$$

DC transconductance   for an FET device under DC bias, the slope of the output drain to source current (I_{DS}) versus input gate to source voltage (V_{GS}) for a fixed drain to source voltage (V_{DS}), expressed in siemens. Given by g_m we have

$$g_m = \left.\frac{\partial I_{DS}}{\partial V_{GS}}\right|_{V_{DS}\,=\,\text{constant}}$$

DC voltage   constant voltage with no variation over time. This can be considered in general terms as an alternating current (AC) with a frequency of variation of zero, or a zero frequency signal. For microwave systems, DC voltages are provided by batteries or AC/DC converters required to "bias" transistors to a region of operation where they will either amplify, mix or frequency translate, or generate (oscillators) microwave energy.

DC–AC inverter   *See* inverter.

DC–DC converter   a switching circuit that converts direct current (DC) of one voltage level to direct current (DC) of another voltage level. A typical DC–DC converter includes switches, a low pass filter (to attenuate the switching frequency ripple), and a load. The size of magnetic components and capacitors can be reduced and bandwidth can be increased when operating at high frequency. Most DC–DC converters are pulse-width modulated (PWM), while resonant or quasi-resonant types are found in some applications. Commonly used topologies include the buck converter, boost converter, buckboost converter, and Ćuk converter. Isolation can be achieved by insertion of a high frequency transformer.

DC-free code   *See* balanced code.

DCA   *See* dynamic channel assignment.

DCE   *See* data communications equipment or distributed computing environment.

DCS-1800   digital communication system–1800 MHz. A micro-cell version of GSM that operates at a lower transmitter power, higher frequency band, and has a larger spectrum allocation than GSM, but in most other respects is identical to GSM. The DCS-1800 spectrum allocation is 1710–1785 MHz for the up-link and 1805–1880 MHz for the down-link (i.e., 2×75 MHz). The peak transmit power for portable DCS-1800 terminals is 1 W.

DCT   *See* discrete cosine transform.

de Haas–Shubnikov oscillation   *See* Shubnikov–de Haas oscillation.

de-assert   to return an enabling signal to its inactive state.

de-emphasis   refers to the receiving process of correcting the amplitude of certain signal components that have been "pre-emphasized" prior to transmission in order to improve signal-to-noise (S/N) ratio. In commercial FM broadcast receivers, de-emphasis is accomplished with a simple resistor capacitor lowpass filter that represents the inverse transfer function characteristic of the pre-emphasis network situated at the transmitter. See the figure for a de-emphasis network. *See* pre-emphasis.

de-regulation   the removal of some government controls on public utilities, generally including the unbundling of certain services, the dismantling of vertically-integrated utilities, and the introduction of competition among various utility companies for customer services.

dead band   (1) the portion of the operating range of a control device or transducer over which there is no change in output.

(2) referring to an automatic controller behavior, a range of values of the controlled variable in which no corrective action occurs. This type of controller behavior is responsible for the time lag, called dead zone lag, which can cause instability of the controlled system if other conditions are present.

Also known as dead zone.

dead end   an installation in which an electric power line terminates at a pole or tower, typically for purposes of structural stabilty.

dead man   (1) a stand on which to rest a utility pole when setting the pole by hand.

(2) a buried log used as a guy anchor.

dead-end shoe   a fixture for securing a wire or strain insulator to a utility pole.

dead tank breaker   a power circuit breaker where the tank holding the interrupting chamber is at ground potential. Oil circuit breakers, for example, are typically dead tank breakers.

dead zone   *See* dead band.

deadbeat 2-D observer   a system described by the equations

$$z_{i+1,j+1} = F_1 z_{i+1,j} + F_2 z_{i,j+1} + G_1 u_{i+1,j}$$
$$+ G_2 u_{i,j+1} + H_1 y_{i+1,j} + H_2 y_{i,j+1}$$
$$\hat{x}_{i,j} = L z_{i,j} + K y_{i,j}$$

$i, j \in Z_+$ (the set of nonnegative integers) is called a full-order deadbeat observer of the second generalized Fornasini–Marchesini 2-D model

$$Ex_{i+1,j+1} = A_1 x_{i+1,j} + A_2 x_{i,j+1}$$
$$+ B_1 u_{i+1,j} + B_2 u_{i,j+1}$$
$$y_{i,j} = C x_{i,j} + D u_{i,j}$$

$i, j \in Z_+$ if there exists finite integers M, N such that $\hat{x}_{i,j} = x_{i,j}$ for $i > M$, $j > N$, and any u_{ij}, y_{ij} and boundary conditions x_{i0} for $i \in Z_+$ and x_{0j} for $j \in Z_+$ where $z_{ij} \in R^n$ is the local state vector of the observer at the point (i, j), $u_{ij} \in R^m$ is the input, $y_{i,j} \in R^p$ is the output, and $x_{i,j} \in R^n$ is the local semistate vector of the model, F_1, F_2, G_1, G_2, H_1, H_2, L, K, E, A_1, A_2, B_1, B_2, C, D are real matrices of appropriate dimensions with E possibly singular or rectangular, Z_+ is the set of nonnegative integers. In a similar way, a full-order asymptotic observer can be defined for other types of the generalized 2-D models.

deadbeat control of 2-D linear systems   given the 2-D Roesser model

$$\begin{bmatrix} x^h_{i+1,j} \\ x^v_{i,j+1} \end{bmatrix} = \begin{bmatrix} A_1 & A_2 \\ A_3 & A_4 \end{bmatrix} \begin{bmatrix} x^h_{ij} \\ x^v_{ij} \end{bmatrix} + \begin{bmatrix} B_1 \\ B_2 \end{bmatrix} u_{ij}$$

$i, j \in Z_+$ (the set of nonnegative integers)

$$y_{ij} = [\, C_1 \ C_2 \,] \begin{bmatrix} x^h_{ij} \\ x^v_{ij} \end{bmatrix}$$

with boundary conditions $x^h_{0j} = 0$ for $j \geq N_2$ and $x^v_{i0} = 0$ for $i \geq N_1$, find an input vector sequence

$$u_{ij} = \begin{cases} \neq 0 & \text{for} \quad 0 \leq i \leq N_1, 0 \leq j \leq N_2 \\ = 0 & \text{for} \quad i > N_1 \text{ and } j > N_2 \end{cases}$$

such that the output vector $y_{ij} = 0$ for all $i > N_1$ and $j > N_2$ where $x^h_{ij} \in R^n_1$ and $x^v_{ij} \in R^n_2$ are the horizontal and vertical state vectors, respectively, and A_1, A_2, A_3, A_4, C_1, C_2 are real matrices.

deadlock   a condition when a set of processes using shared resources or communicating with each other are permanently blocked.

deadtime   the time that elapses between the instant that a system input is perturbed and the time that its output starts to respond to that input.

debug   to remove errors from hardware or software. *See also* bug.

debug port   the facility to switch the processor from run mode into probe mode to access its debug and general registers.

debugger   (1) a program that allows interactive analysis of a running program, by allowing the user to pause execution of the running program and examine its variables and path of execution at any point.
(2) program that aids in debugging.

debugging   (1) locating and correcting errors in a circuit or a computer program.
(2) determining the exact nature and location of a program error, and fixing the error.

debuncher   a radio frequency cavity phased so that particles at the leading edge of a bunch of beam particles (higher momentum particles) are decelerated while the trailing particles are accelerated, thereby reducing the range of momenta in the beam.

Debye material   a dispersive dielectric medium characterized by a complex-valued frequency domain susceptibility function with one or more real poles. Water is an example of such a material.

Debye media   *See* Debye material.

decade   synonymous with power of ten. In context, a tenfold change in frequency.

decade bandwidth   10 : 1 bandwidth ratio (the high-end frequency is ten times the low-end frequency). The band edges are usually defined as the highest and lowest frequencies within a contiguous band of interest at which the loss equals

L_{Amax}, the maximum attenuation loss across the band.

decay   a transformation in which an atom, nucleus, or subatomic particle changes into two or more objects whose total rest energy is less than the rest energy of the original object.

decay heat   the fraction of the total energy obtained from a nuclear fission reaction which is produced by delayed neutrons and by the secondary decay of fission daughters.

decay length   the average distance a species of a particle at a given energy travels before decaying.

decay time   in the absence of any pump or other excitation mechanisms, the time after which the number of atoms in a particular state falls to $1/E$ of its initial value.

decentralized control   a structure of large-scale control systems based on system decomposition onto interconnected subsystems in order to simplify control design. Decentralized control systems are usually designed in the form of local feedback controllers and are chosen to fit information structure constraints imposed by the decomposition. To ensure robustness with respect to interconnections between subsystems, the local controllers should be robust and/or coordination should ensure robustness of the overall system. *See also* decomposition.

decibel (dB)   a unit of measure that describes the ratio between two quantities in terms of a base 10 logarithm. For example, the ratio between the power level at the input and output of an amplifier is called the power gain and may be expressed in decibels as follows:

$$G(dB) = 10 \log_{10}(P_{out}/P_{in})$$

Terms such as dBm, dBuV, dBW indicate that the decibel measurement was made relative to an established standard. A common power measure reference is 0 dBm, which is defined to be 1 mW (milliwatt, 0.001 W). A common voltage reference is 1 μV (1 microvolt).

decimal   from the number system that has base 10 and employs 10 digits.

decimation   an operation which removes samples with certain indexes from a discrete-time signal and then re-indexes the remaining samples. Most frequently, decimation refers to keeping every nth sample of a signal. Also known as downsampling.

decision boundary   a boundary in feature space which separates regions with different interpretations or classes; e.g., the boundary separating two adjacent regions characterizing the handwritten characters 'E' and 'F'. In practice, the regions associated with neighboring classes overlap; consequently most decision boundaries lead to some erroneous classifications, so an error criterion is used to select the "best" boundary. *See* classifier, Bayesian classifier.

decision directed   the use of previously detected information bits in an estimator, detector, or adaption algorithm in an adaptive filter. Usually improves performance compared to nondecision directed counterparts, but introduces potential problems with error propagation (erroneous bit feed back).

decision level   the boundary between ranges in a scalar quantizer. On one side of the decision level, input values are quantized to one representative level; on the other side, input values are quantized to a different representative level.

decision mechanism   rules and principles by which the information available to a given decision unit (control unit) is processed and transformed into a decision; typical decision mechanisms within control systems are fixed decision rule and optimization-based mechanism. Decision mechanisms can assume a hierarchical form — one may then talk about a hierarchical decision structure.

decision support system   a system whose purpose is to seek to identity and solve problems.

decision tree analysis   decomposing a problem into alternatives represented by branches where nodes (branch intersections) represent a decision point or chance event having probabilistic outcome. Analysis consists of calculating expected values associated with the chain of events leading to the various outcomes.

decision-directed adapatation   a method for adapting a digital equalizer in which decisions at the output of the equalizer are used to guide the adaptive algorithm. *See also* adaptive algorithm, LMS.

declaration   phase or statement in which a new variable is requested; the declaration of a variable is based on the definition of its type or class. This phase leads to a instantiation of the variable.

decode cycle   the period of time during which the processor examines an instruction to determine how the instruction should be processed. This is the part of the fetch-decode-execute cycle for all machine instruction.

decode history table   a form of branch history table accessed after the instruction is decoded and when the branch address is available so that the table need only store a Boolean bit for each branch instruction to indicate whether this address should be used.

decoder   (1) a logic circuit with N inputs and 2^N outputs, one and only one of which is asserted to indicate the numerical value of the N input lines read as a binary number.

(2) a device for converting coded information to a form that can be understood by a person or machine.

decoder source   the coded signal input to the decoder. In information theory the decoder source is modelled as a random process.

decoding   (1) the act of retrieving the original information from an encrypted (coded) data transmission.

(2) the process of producing a single output signal from each input of a group of input signals.

(3) the operation of the decoder. The inverse mapping from coded symbols to reconstructed data samples. Decoding is the inverse of encoding, insofar as this is possible.

decomposition   (1) an operation performed on a complex system whose purpose is to separate its constituent parts or subsystems in order to simplify the analysis or design procedures.

(1) For large-scale systems, decomposition is performed by neglecting links interconnecting subsystems. It is followed by design of local control systems on the base of local objectives and coordination, which enables reaching global goals. *See also* decentralized control.

(2) For optimization algorithms, decomposition is reached by resolving the objective function or constraints into smaller parts, for example, by partitioning the matrix of constrains in linear programs followed by the solution of a number of low dimensional linear programs and coordination by Lagrange multipliers.

(3) For uncertainties, decomposition is performed to make their model trackable, for example, by dividing them into matched and mismatched parts. *See also* matching conditions.

(4) For linear time-invariant systems in state form, Kalman's decomposition is a transformation of state matrix in a way that indicates its controllable and observable, controllable and unobservable, uncontrollable and observable, and

uncontrollable and unobservable parts. *See also* controllability.

(5) For matrix transfer functions, decomposition transforms it to the form composed of specific blocks used in chosen design procedure. *See also* H infinity design.

(2) in fuzzy systems, if the fuzzy relation R is a composition of the fuzzy relations A and B, then the determination of A given B and R is referred to as a decomposition of R with respect to B.

decorrelation   the act of removing or reducing correlation between random variables. For random vectors a discrete linear transform is often used to reduce the correlation between the vector components.

decorrelator   a linear code-division multiple-access receiver, which takes the data vector output from a bank of filters matched to the desired users' spreading sequences, and multiplies by a matrix which is the inverse of the spreading-sequence correlation matrix.

decoupled load flow   a load-flow study in which certain simplifying assumptions permit an accelerated solution.

decoupling problem for 2-D linear systems given the 2-D Roesser model

$$\begin{bmatrix} x^h_{i+1,j} \\ x^v_{i,j+1} \end{bmatrix} = \begin{bmatrix} A_1 & A_2 \\ A_3 & A_4 \end{bmatrix} \begin{bmatrix} x^h_{ij} \\ x^v_{ij} \end{bmatrix} + \begin{bmatrix} B_1 \\ B_2 \end{bmatrix} u_{ij}$$

$$y_{ij} = [\, C_1 \ C_2 \,] \begin{bmatrix} x^h_{ij} \\ x^v_{ij} \end{bmatrix}$$

$i, j \in Z_+$ (the set of nonnegative integers), find matrices $K = [K_1, K_2]$ and $H \,(\det H \neq 0)$ of the control law

$$u_{ij} = K \begin{bmatrix} x^h_{ij} \\ x^v_{ij} \end{bmatrix} + H v_{ij} \quad i, j \in Z_+$$

such that the transfer function matrix of the closed-loop system

$$T_c(z_1, z_2) = [\, C_1 \ C_2 \,]$$
$$\times \begin{bmatrix} I_{n_1} z_1 - A_1 - B_1 - K_1 \\ -A_3 - B_2 - K_1 \end{bmatrix}$$
$$\begin{bmatrix} -A_2 - B_1 K_2 \\ I_{n_2} z_2 - A_4 - B_2 K_2 \end{bmatrix}$$
$$\times \begin{bmatrix} B_1 \\ B_2 \end{bmatrix} H$$

is diagonal and nonsingular, where $x^h_{ij} \in R^{n_1}$ and $x^v_{ij} \in R^{n_2}$ are horizontal and vertical state vectors, respectively, and $u_{ij} \in R^m$ is the input vector, $v_{ij} \in R^m$ is a new input vector, $y_{ij} \in R^p$ is the output vector, and $A_1, A_2, A_3, A_4, B_1, B_2, C_1, C_2$, are given real matrices.

decrement   to reduce the value of a variable or content of a register by a fixed amount, normally one.

decryption   a process, implemented in hardware or software, for reconstructing data previously coded by using a cryptography algorithm, that is encrypted data. These algorithms are typically based on keywords or codes.

DECT   *See* digital European cordless telephone.

deep-bar rotor   a squirrel-cage induction motor rotor in which the rotor bars are deep and narrow, to make the effective resistance, and therefore the torque, higher at starting.

deep-UV lithography   lithography using light of a wavelength in the range of about 150 nm to 300 nm, with about 250 nm being the most common.

default   the value or status that is assumed unless otherwise specified.

deferred addressing   *See* indirect addressing.

deferred evaluation   a scheduling policy under which a task is not scheduled until it has been determined that the results of that task are required by another task that is in execution and cannot proceed without these results.

definite time DC motor acceleration   when DC motors accelerate during their starting sequence, starting resistors are removed from the armature circuit in steps. In definite time DC motor acceleration (also referred to as open loop DC motor acceleration), the starting resistors are removed in definite time increments, whether the motor is actually accelerating or not.

definite-purpose motor   any motor design, listed and offered in standard ratings with standard operating characteristics, with special mechanical features for use under service conditions other than usual or for use on a particular type of application.

definition   phase or statement in which a new type, or a class, or a frame for variables is defined. The definition of typed constant is also typically allowed.

deflector   any of a number of optical devices that change the direction of an optical beam, using mechanisms such as diffraction, mechanical mirror motion, and refraction.

defocus   the distance, measured along the optical axis (i.e., perpendicular to the plane of the best focus) between the position of a resist-coated wafer and the position if the wafer were at best focus.

defocusing quadrupole magnet   a quadrupole magnet that focuses beam in the vertical plane and defocuses in the horizontal plane.

DeForest, Lee   (1873–1961) Born: Council Bluffs, Iowa.

Best known for his contributions to the development of radio communications. DeForest's greatest invention was called the audion triode. This vacuum tube was based on an earlier patented tube developed by John A. Fleming. This tube, which was both an amplifier and a rectifier, allowed the development of radios, radar, television, and some early computers. DeForest's life was noted for controversial, and often poor, business decisions.

deformable mirror device   a type of device for light modulation, especially spatial light modulation, employing micromechanical structures, such as cantilevered mirrors or mirrors with torsional motions, to deflect incident light rays.

defuzzification   the process of transforming a fuzzy set into a crisp set or a real-valued number.

defuzzifier   a fuzzy system that produces a crisp (non fuzzy) output from the results of the fuzzy inference engine. The most used defuzzifiers are

(1) maximum defuzzifier that selects as its output the value of y for which the membership of the output membership function $\mu_B(y)$ is maximum;

(2) centroid defuzzifier determines the center of gravity (centroid), $\overline{y}$ of B, and uses this value as the output of the fuzzy inference system.

See also fuzzy inference engine, fuzzy inference system.

degenerate common emitter   a combination of the common-emitter and emitter-follower stages with a very well-defined gain.

degenerate four-wave mixing   a four-wave mixing process in which all of the interacting waves have the same frequency. In certain geometrical arrangements, this process leads to optical phase conjugation and in addition can be for certain types of optical information processing.

degenerate modes   two modes with different field structures having the same cutoff frequency in a waveguide or the same resonant frequency in a cavity.

degenerate two-wave mixing   a special case of two wave mixing in which the two beams are of exactly the same frequency. In two-wave mixing, if the two laser beams are of the same frequency, a stationary interference intensity pattern is formed. This leads to a stationary volume refractive index grating. Such a kind of two-wave mixing is referred to as degenerate two-wave mixing.

degradation   situation in which a signal has been corrupted by noise, blurred by some point-spread function or distorted in some other fashion.

degree of membership   the degree to which a variable's value belongs in a fuzzy set. The degree of membership varies from 0 (no membership) to 1 (complete membership).

degree of mobility   each prismatic or revolute joint has one degree of freedom and provides the mechanical structure with a single degree of mobility.

degree of visual angle   the angle subtended by an object of a given width a given distance away from the viewer.

degrees of freedom   the number of independent position variables that have to be specified in order to locate all parts of the mechanism is defined as a number of degrees of freedom. Therefore, the degrees of freedom is defined as the minimal number of position variables necessary for completely specifying the configuration of the mechanism.

delay   (1) the time required for a signal to propagate along a wire.

(2) the difference in the absolute angles between a point on a wavefront at the device output and the corresponding point on the incident input wavefront, expressed in seconds or degrees. Delay

can exceed 360 degrees. Given by t_d, we have

$$t_d = \theta_{out} - \theta_{in}$$

delay angle   *See* firing angle.

delay locked loop   *See* delay-locked loop.

delay power spectrum   a function characterizing the spread of average received power as a function of delay. Can be obtained from the scattering function by integrating over the Doppler shift variable. *See also* scattering function, multipath propagation.

delay profile   *See* delay power spectrum.

delay range   the difference in arrival times between the first and last significant component of the impulse response of a wideband communication channel. Also known as the total excess delay.

delay resolution   the capability, measured in units of delay (seconds), of a signal used for channel measurement to resolve received signal components which arrive with different delays. If two signal components arrive at the receiver with a delay separation less than the delay resolution, they will be observed as one signal, superimposed on each other. The actual value of the delay resolution depends on the criterion by which two signal components are defined to be resolved. An approximate measure is given by the inverse of the channel (or signal) bandwidth. *See also* multipath propagation.

delay slot   in a pipelined processor, a time slot following a branch instruction. An instruction issued within this slot is executed regardless of whether the branch condition is met, so it may appear that the program is executing instructions out of order. Delay slots can be filled (by compilers) by rearranging the program steps, but when this is not possible, they are filled with "no-operation" instructions.

delay spread   a measure of the time through which the duration of a transmitted signal is extended by dispersion in a wideband communication channel. Usually measured as the RMS delay spread, i.e., the second moment of the time-averaged channel impulse response.

delay-line   a transmission line of the appropriate length to result in a specific time delay. As an example, a line at 100 MHz that is 90 degrees long (one-quarter wavelength) will exhibit a time delay of 2 ns.

delay-locked loop (DLL)   (1) a pseudo-noise (PN) sequence tracking loop typically used in receivers for spread spectrum signals. The tracking loop has two branches that correlate the received signal with two shifts of a locally generated PN sequence: an advanced and a retarded time shift of the phase of the signal being tracked.
    (2) a technique for symbol synchronization based on time-shifted and reversed correlation functions of the desired symbol waveform, which results in a control function with an s-shape (termed an s-curve). The control function is used in a feedback loop similar to a PLL to adjust the timing of receiver clock used in sampling the received signal. DLLs are used, e.g., in spread-spectrum receivers to maintain chip synchronization.

delayed AGC   *See* delayed automatic gain control.

delayed automatic gain control   automatic gain control in which the control mechanism becomes effective only when the input signal is larger than a predetermined level.

delayed branch instruction   a form of conditional branch instruction in which one branch is executed irrespective of the outcome of the branch.

Then the branch takes effect. Used to reduce the branch penalty.

delayed neutrons   neutrons emitted by fission daughters after some time delay.

delta connection   a three-phase power source or load in which the elements are connected in series and are thus represented on a schematic diagram as a triangular configuration.

delta function   in discrete-time, the function given by

$$\delta[n] = \begin{cases} 1, & n = 0 \\ 0, & n \neq 0 \end{cases}$$

In continuous-time, the (Dirac) delta "function" is not a function at all and, although not mathematically rigororous, is a tremendously important concept throughout signal and system theory. The Dirac delta "function" is defined as

$$\delta[t] = \begin{cases} \infty, & t = 0 \\ 0, & t \neq 0 \end{cases}$$

such that $\int f(t)\delta(t - t_o)dt = f(t_o)$. It is of relevance in defining the impulse response of a system.

delta gun   cathode ray tube (CRT) electron gun structure has the red, green, and blue electron gun components configured in the shape of an equilateral triangle; the structure provides the smallest CRT neck size and has the smallest deflection yoke diameter, but requires color registration (color convergence) correction in both the horizontal and the vertical CRT face. *See* cathode ray tube.

delta modulation   a special case of differential pulse code modulation (DPCM) where the digital code-out represents the change, or slope, of the analog input signal, rather than the absolute value of the analog input signal. A "1" indicates a rising slope of the input signal. A "0" indicates a falling slope of the input signal. The sampling rate is dependent on the derivative of the signal, since a rapidly changing signal would require a rapid sampling rate for acceptable performance.

delta modulation control   a pulse-time modulation method transplanted from the delta modulation in signal processing. The difference between the control reference and the switched signal is integrated and fed to a Schmitt trigger. When the integrated value reaches a predefined upper bound the switch is turned off. When the integrated value reaches a predefined lower bound the switch is turned on.

delta rule   a supervised learning algorithm, based upon gradient descent, that was developed for application to single-layer networks of linear threshold units. For each input pattern x, the weight wij connecting input xj to unit i is adjusted according to $Dwij = h(ti - wi \cdot x)xj$, where wi is the vector of weights wij, ti is the target output for unit i, and h is a positive constant. This rule is also known as the Widrow–Hoff rule and the LMS algorithm.

delta–delta transformer   a three-phase transformer connection formed by connecting three single-phase transformers in which the windings on both the primary and the secondary sides are connected in series to form a closed path.

delta-wye transform   a transformation between delta and wye connections.

delta–wye transformer   a three-phase transformer connection formed by connecting three single-phase transformers in which the primary windings are connected in series to form a closed path while one end of each of the secondary windings is connected to a common point (the neutral).

demagnetization   the act of removing a device from being in a magnetic state, i.e., rearranging the atomic magnetic domains in a disoriented fashion.

demagnetizing field   the magnetic field produced by divergences in the magnetization of a magnetic sample.

demand assign multiple access protocol   telephone signalling mechanism in which the access is established for the duration of a call.

demand fetch   in a cache memory, the name given to fetching a line from the memory into the cache on a cache miss when it is requested and not before.

demand meter   an electric meter which shows both the energy used and the peak power demand in a given period.

demand paging   condition where each page in virtual memory is loaded into main memory, after first being referenced by the processor (i.e., not in advanced). The first reference to each page will thus always cause a "page fault" (page not in main memory). After these initial page faults, most of the pages needed by the program are in main memory.

demodulation   the process by which a modulated signal is recovered back to its original form. It is the general process of extracting the information-bearing signal from another signal. Modulation is the general process of embedding an information-bearing signal into a second carrier signal. An important objective in modulation is to produce a signal whose frequency range is suitable for transmission over the communication channel to be used. *See also* modulation.

DeMorgan's theorem   a formula for finding the complement of a Boolean expression. It has two forms:

(1) $\overline{A \vee B} = \overline{A} \wedge \overline{B}$
(2) $\overline{A \wedge B} = \overline{A} \vee \overline{B}$

183

where A and B are Boolean variables and $\wedge$ represents logical AND and $\vee$ represents logical OR and the overbar represents the logical complement.

demultiplexer   a logic circuit with K inputs and I controls which steers the K inputs to one set of 2^I sets of output lines. *Compare with* multiplexer.

demultiplexing   the inverse operation of multiplexing which enables the transmission of two or more signals on the same circuit or communication channel.

Denavit–Hartenberg notation   a system that describes the translational and rotational relationships between adjacent links. The D-H representation results in 4×4 homogeneous transformation matrix representing each link's coordinate system at the joint with respect to the previous link's coordinate system. The D-H representation of a rigid link depends on four geometric parameters associated with each link. Every coordinate frame is assigned according to the three rules:

(1) The z_{i-1} axis lies along the axis of motion of the ith joint.

(2) The z_i axis is normal to the z_{i-1} axis, and pointing away from it.

(3) The y_i axis completes the right-handed coordinate system as required.

Referring to the figure, the four D-H parameters are defined as follows:

- q_i is the joint angle from the x_{i-1} axis to the x_i axis about the z_{i-1} axis (using the right-hand rule),
- d_i is the distance from the origin of the $(i-1)$-th coordinate frame to the intersection of the z_{i-1} axis with the x_i axis along z_{i-1} axis,
- a_i is the offset distance from the intersection of the z_{i-1} axis with the x_i axis to the origin of the ith frame along the x_i axis (in another words it is the shortest distance between the z_{i-1} and z_i axes),

- α_i is the offset angle from the z_{i-1} to the z_i axis about the x_i axis (using the right-hand rule).

For a revolute joint d_i, a_i, and α_i are called the *link parameters* or *joint parameters* and remain constant. q_i is called the *joint variable*. For a prismatic joint, q_i, a_i, and α_i are the link parameters and remain constant, while d_i is the joint variable. The D-H transformation matrix for adjacent coordinate frames has the following form:

$$^{i-1}A_i = \begin{bmatrix} \cos q_i & -\cos \alpha_i \sin q_i & \sin \alpha_i \sin q_i & a_i \cos q_i \\ \sin q_i & \cos \alpha_i \cos q_i & -\sin \alpha_i \cos q_i & a_i \sin q_i \\ 0 & \sin \alpha_i & \cos \alpha_i & d_i \\ 0 & 0 & 0 & 1 \end{bmatrix}$$

denormalized number   nonzero number whose leading significand bit is zero and whose exponent has a fixed value. These numbers lie in the range between the smallest normalized number and zero.

density estimation   statistical methods for estimating the probability density from a given set of examples.

density function (DF)   an alternative name for probability density function (PDF).

density matrix   representation for the wave functions of quantum mechanics in terms of binary products of eigenfunction expansion amplitudes; with ensemble averaging the density matrix representation is convenient for phenomenological inclusion of relaxation processes.

density matrix formalism of quantum mechanics   a mathematical formulation of the theory of quantum mechanics more general than those based on a description in terms of a wavefunction or a state vector, because it can treat situations in which the state of the system is not precisely known. The density matrix formalism is often used in laser physics and in nonlinear optics, for example, under situations in which collisional dephasing effects are important.

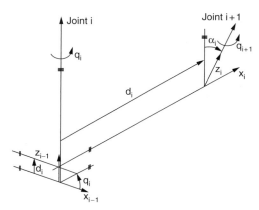

The Denavit–Hartenberg parameters.

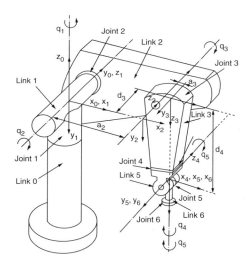

A PUMA 560 robot arm illustrating joints, links, and the D-H parameters.

dependability system feature that combines such concepts as reliability, safety, maintainability, performance, and testability.

dependency a logical constraint between two operations based on information flowing among their source and/or destination operands; the constraint imposes an ordering on the order of execution of (at least) portions of the operations. For example, if the first operation in a sequential program produces a result that is an operand of the second operation of the program, that second operation cannot be performed until the first operation has been completed, since its operand value will not be available earlier.

dependent source *See* controlled source.

depleted uranium uranium in which the proportion of fissile U-235 has been reduced below useful levels.

depletion layer space charge region or layer adjacent to a semiconductor junction where the majority carrier concentration has been depleted or reduced below the thermal equilibrium value.

depletion mode an FET that is on when zero volts bias is applied from gate to source.

depolarization (1) the change of the polarization state of a wave propagating through an anisotropic medium.

(2) phenomenon that occurs when a wave becomes partially or completely unpolarized.

(3) changing the original polarization of a propagating wave into a different type.

depolarizing scattering change in the polarization of the light due to strong scattering.

deposited multi-chip module (MCM-D) a multi-chip module built using the deposition and thin-film lithography techniques that are similar to those used in integrated circuit manufacturing.

depth in computer vision, the distance to a surface, as perceived subjectively by the observer.

Also, the number of bits with which each pixel is represented in a digital image.

depth map   a map of depth in a scene corresponding to each pair of coordinates in an image of the scene.

depth of field   the range of depths over which objects in the field of vision are in acceptable focus.

depth of focus (DOF)   the total range of focus that can be tolerated; that is, the range of focus that keeps the resulting printed feature within a variety of specifications (such as critical dimension, photoresist profile shape, and exposure latitude).

depth of penetration   distance inside a material interface that an impinging wave has attenuated by $1/e$ where $e = 2.7183$ (Euler's constant).

depth-first search   a search strategy for tree or trellis search where processing is performed depth first, i.e., a particular path is processed through the depth of the tree/trellis as long as it fulfills a certain threshold criterion (e.g., based on the Fano metric). If a path fails the threshold test, a new path is considered. Also known as sequential search.

derating factor   the fraction (or percent) of nominal rating to which a specified quantity must be reduced due to unusual operating conditions. Examples of conditions that may require application of a derating factor are high altitude, high ambient temperature, frequent motor starting, and "plugging" operation of a motor.

de-regulation   the removal of some government controls on public utilities, generally including the unbundling of certain services, the dismantling of vertically-integrated utilities, and the introduction of competition among various utility companies for customer services.

derivative control   control scheme whereby the actuator drive signal is proportional to the time derivative of the difference between the input (desired output) and the measured actual output.

descrambling   the process of restoring a transmitted signal to its original form commonly used in CATV systems. The positive-trap, negative-trap, and baseband are common methods of scrambling requiring such a means to restore the signal to its original form.

descriptor   an object describing an area of space within memory. A descriptor contains information about the origin and length of the area.

desensitization   a reduction in a device output signal power due to one or more additional signals that compress the device output, expressed as a negative ratio of the desensed signal output power to the output power without the additional signals, in decibels.

$$D = -10 \log_{10} \left(\frac{P_{desensed}}{P_{undesensed}} \right)$$

design for testability   designing a semiconductor component such that it is easier to feed it a set of test vectors that guarantees, or provides sufficient reassurance, that the component was manufactured (and designed) correctly.

design model   a mathematical model that is used to design a controller. The design model may be obtained by simplifying the truth model of the process. The truth model is also called the simulation model. The truth model is usually too complicated for controller design purposes. The controller performance is tested using the truth model. *See also* truth model.

designed distance   the guaranteed minimum distance of a BCH forward error control code that is designed to correct up to t errors.

destination operand   where the results of an instruction are stored, e.g., the instruction MOV

AL, 7 uses AL as the destination operand (7 is the source operand).

destructive physical analysis (DPA)   devices are opened and analyzed for process integrity and workmanship.

destructive read   reading process in which the information is lost from memory after being read out.

detectability   the property of a system concerning the existence of a stabilizing output injection. For linear time-invariant systems, it is characterized by the observability (see the definition) of the unstable modes.

detective quantum efficiency (DQE)   of a photodetector, the ratio of its quantum efficiency (Q_e) to its power noise factor (k), i.e., $DQE = 3DQ_e/k$. This is a general relationship that applies to electrical, photographic, and biological (e.g., the eye) types of photodetectors.

detector   (1) a device that converts RF input signals to a corresponding DC output signal.

(2) in optics, a circuit, usually containing a diode that converts the intensity of light into an electrical signal. Used in a variety of applications including power sensing, leveling, and modulation envelope reproduction. *See also* photodetector.

deterministic control of uncertain systems   an approach to the control of uncertain systems that is effective over a specified range of the system parameter variations. These methods do not require on-line identification of the values of the system parameters to yield the desired robust performance. No statistical information about uncertain parameters is required. The two main approaches are variable structure sliding mode control and Lyapunov theory approach.

deuteron   the nucleus of the deuterium atom (an isotope of the hydrogen) consisting of a proton and a neutron.

developed power   the power converted from electrical to mechanical form in an electric motor. The developed power of a motor can be calculated from the developed torque and motor speed.

developed torque   the torque created by an electric motor including torque required to overcome the friction and windage losses of the motor. This value will be higher than the shaft torque, which is actually delivered to the load.

development   the process by which a liquid, called the developer, selectively dissolves a resist as a function of the exposure energy that the resist has received.

deviation ratio   the allowable frequency deviation for an FM signal divided by the highest modulating frequency.

device   a hardware entity that exists outside of the motherboard, and is accessed through device drivers. Devices often relate to I/O (floppy drives, keyboards, etc.).

deviation   a measure of the dispersion among the elements in a set of data. Sometimes, a deviation is defined as failure to meet the specified critical limits.

device controller   (1) a device used to connect a peripheral device to the main computer; sometimes called a peripheral controller.

(2) software subroutine used to communicate with an I/O device.

device driver   program that controls an input/output device, usually providing a standard form of interface to the programs that utilize the device. Device drivers perform the basic functions of device operation.

device register   register in an I/O device that may be read or written by the processor to determine status, effect control, or transfer data.

device scaling to increase device speed and circuit density as well as reduce power consumption, all three dimensions of transistors are reduced and the impurity concentrations increased by interrelated scale factors.

dewar a vacuum insulated, cryogenic radiation insulated, low conduction heat loss vessel for cryogenic fluids. A dewar is used for keeping material, chiefly liquids, cold or hot. The common thermos bottle is an example of a dewar. In the accelerator business dewars are often used to store large quantities of cryogenic liquids.

DF *See* dissipation factor.

DFB laser *See* distributed feedback laser.

DFT *See* discrete Fourier transform.

Dhrystone *See* Dhrystone benchmark.

Dhrystone benchmark synthetic benchmark program consisting of a representative instruction mix used to test the performance of a computer. Does not compute anything in particular. Another synthetic benchmark is the Whetstone benchmark.

DIAC a two-terminal AC device that, once gated on by sufficient forward voltage, permits the flow of current until reverse biased. It is often used as a trigger device to drive the gate of other power electronic devices.

Diac's delta function *See* delta function.

diagnostic (1) one of a set of tests to run through a system that determines whether the system is functioning correctly.

(2) pertaining to the detection and isolation of faults or failures. For example, a diagnostic message, a diagnostic manual.

diagonal clipping distortion that occurs in an AM demodulator (usually associated with diode detection), where the capacitor discharge time constant is set too long for the detector to accurately follow fast changes in the AM signal envelope. Sometimes referred to as "failure to follow distortion," diagonal clipping can also occur in AM modulators when the intelligence bandwidth exceeds that of the modulator.

diagonal dominance a measure of the amount of interaction that exists between variables in a multi-input–multi-output (MIMO) system. It is quantified by Gershgorin circles or bands that are often plotted on an inverse Nyquist array (or INA) diagram that shows the frequency response matrix of the system in a graphical form. Its practical significance relates to the fact that a diagonally dominant system can be controlled by multiple single variable controllers, whereas a nondominant process might require sophisticated and costly multivariable techniques for effective control. *See also* inverse Nyquist array.

diamagnetic materials with magnetization directed opposite to the magnetizing field, so that the permeability is less than one; metallic bismuth is an example.

die an individual MMIC circuit or subsystem that is one of several identical chips that are produced after dicing up an MMIC wafer.

dielectric (1) a medium that exhibits negligible or no electrical conductivity and thus acts as a good electrical insulator.

(2) a medium characterized by zero conductivity, unity relative permeability, and a relative permittivity greater than one. Also known as an insulator.

Dielectries are usually used to separate two conducting bodies such as to form a capacitor.

dielectric constant (1) a quantity that describes how a material stores and dissipates electrical energy.

(2) ratio of the electrical capacity of a condenser, which has a given material as the dielectric, to the capacity of an identical condenser, but with air as the dielectric.

(3) permittivity of a medium normalized to the permittivity of free space; a measure of the response of a dielectric to an applied electric field.

(4) an electric property of an insulator or semiconducting material, which describes how differently electric fields will behave inside of the material as compared to air. As an example, $e_r = 12.9$ for GaAs as compared to $e_r = 1$ for air. In integrated circuits, an effective dielectric constant (e_{eff}) is used, since the electric fields supported by the signals traveling through the conductors on the circuit flow through both air and the insulator or semiconductor simultaneously.

dielectric discontinuity   interface between two media with different dielectric permittivity properties.

dielectric medium   medium that is polarizable but relatively nonconducting.

dielectric resonator   an unmetallized dielectric object of high dielectric constant and high quality factor that can function as an energy storage device.

dielectric resonator antenna (DRA)   an antenna where a dielectric resonator is used as the radiation element.

dielectric resonator [stabled] oscillator (DRO)   a dielectric resonator is a cylindrically shaped piece of material, or "puck," that has the properties of having low-loss resonant frequencies that are determined primarily by the size of the cylinder. Placing a dielectric resonator near a microstrip line can form a resonant circuit that will frequency stabilize a voltage-controlled oscillator.

dielectric slug tuner   system of two movable dielectric pieces of material placed on a transmission line for the purpose of matching a wide range of load impedances by means of placing the dielectrics in proper positions.

dielectric step discontinuity   the junction between different dielectric waveguides.

dielectric waveguide   a waveguide that relies on differences in permittivity among two or more materials to guide electromagnetic energy without the need for ground planes or metallic strips. Such guides of rectangular, circular, elliptical, and other cross sections are made of dielectric materials and used for transmitting signals. Transmission is accomplished by the total internal reflection mechanism inside the waveguide.

difference amplifier   *See* differential amplifier.

difference engine   a mechanical calculator developed by Babbage in 1823.

difference equation   the mathematical model of a LTIL discrete time system. *See also* discrete time system, LTIL system.

difference of Gaussian filter   a bandpass filter whose point spread function is the difference of two isotropic Gaussians with different variances. The result is a "Mexican hat" shape similar to the Laplacian of a Gaussian. Various physiological sensors, including some filters in early vision, appear to have DOG point spread functions. See Marr-Hildreth operator

difference-frequency generation   a second-order nonlinear optical process in which two input beams are applied to a nonlinear optical material and an output is produced at the difference of the frequencies of the two input beams.

difference-mode signal   if two arbitrary signals v_1 and v_2 are applied to the inputs of a differential amplifier, then the common-mode signal is the arithmetic average of the two signals. That is,

$$(v_1 + v_2)/2$$

differential amplifier   an amplifier intended to respond only to the difference between its input voltages, while rejecting any signal common to both inputs.

The differential amplifier is designed such that the difference between the two inputs is amplified (high differential gain), while the signals appearing at either individual input (referenced to ground potential) sees a very low gain (low common-mode gain, usually loss). The differential amplifier is usually used as the first component at the receiving end of a communications link using twisted pair cable (either shielded or unshielded) as the transmission medium. This provides a method to reject any common-mode noise induced onto the twisted pair transmission line, including common-mode noise falling within the useful bandwidth of the communications link. The figure of merit for the differential amplifier is its common mode rejection ratio (CMRR), computed by dividing the differential-mode gain by the common-mode gain.

differential coding   a coding scheme that codes the differences between samples. *See* predictive coding.

differential entropy   the entropy of a continuous random variable. For a random variable X, with probability density function $f(x)$ on the support set $\mathcal{S}$, the differential entropy $h(X)$ is defined as

$$h(X) = - \int_{\mathcal{S}} f(x) \log f(x)$$

provided the integral exists. Also written $h(f)$, emphasizing the sole dependence upon the density. *See also* entropy, relative entropy, mutual information.

differential gain   the amplification factor of a circuit that is proportional to the difference of

two input signals. The differential gain may be expressed in percentage form by multiplying the above amplification factor by 100, or in decibels by multiplying the common logarithm of the differential gain by 20.

differential inclusion   a multivalued differential equation,

$$\dot{\mathbf{x}} \in \mathbf{F}(t, \mathbf{x}) \, ,$$

where $\mathbf{F}(t, \mathbf{x})$ is a nonempty set of velocity vectors at $\mathbf{x} \in \mathbb{R}^n$ for each time t on some time interval. The set $\mathbf{F}(t, \mathbf{x})$ can be viewed as the set of all possible "velocities" $\dot{\mathbf{x}}(t)$ of a dynamical system modeled by the multivalued, or multifunction, differential equation. A solution $\mathbf{x}(t)$ is an absolutely continuous function on some time interval whose velocity vector $\dot{\mathbf{x}}$ lies in the set $\mathbf{F}(t, \mathbf{x})$ for almost all t. *See also* Filippov method.

differential kinematics   equation $v = J(q)\dot{q}$ can be interpreted as the differential kinematics mapping relating the n components of the joint velocity vector to the $r \leq m$ components of the velocity vector v of concern for the specific task. Here n denotes number of degrees of mobility of the structure, m is the number of operational space variables, and r is the number of operational space variables necessary to specify a given task. *See also* geometric Jacobian.

differential length vector   the vector sum of the differential length changes in each of the three coordinate directions along a given curve.

differential mode gain   for a differential amplifier, the ratio of the output signal amplitude to the amplitude of the difference signal between the amplifier input terminals.

differential pair   a two-transistor BJT (FET) amplifier in which a differential input signal is applied to the base (gate) terminals of the two transistors, the output is taken differentially from the collector (drain) terminals, and the emitter (source) terminals are connected together to a constant current source. Also known as an emitter-coupled pair (BJT) or source-coupled pair (FET). This configuration is often used as the basis of the differential input stage in voltage feedback op-amps.

differential pair oscillator   a device used instead of a transistor in any LC-oscillator. Two distinct advantages result from employing the differential pair as the active element. The first is that the output signal may be taken at the collector of transistor that is external to the oscillator feedback loop, and second is that, if a tuned circuit is used as a load, the distortion of the output signal is much less than it would be for a single transistor oscillator. The second advantage follows from the fact that the differential pair collector currents do not include even harmonic components and, in addition, the amplitudes of existing high-order harmonics are smaller than they are for a single transistor.

differential peak detector   a circuit commonly used for the demodulation of FM signals; it utilizes two peak detectors, a differential amplifier and a frequency selective circuit. Also known as a balanced peak detector.

differential protection   a protective relaying scheme in which the currents entering and leaving the protected line or device are compared.

differential protection unit   a protective unit based on the difference of currents flowing in and out of a protected zone.

differential pulse code modulation (DPCM)   (1) a class of methods for pulse code modulation (or scalar quantization) where (linear) prediction is used in order to utilize the temporal redundancy in the source signal to enhance performance. Also referred to as predictive PCM or predictive SQ. *See also* pulse-code modulation, scalar quantization, adaptive differential pulse-code modulation.

(2) in image processing, a lossy predictive coding scheme. In this scheme m pixels in a causal neighborhood of the current pixel is used to estimate (predict) the current pixel's value. The basic components of the predictive coder comprises predictor, quantizer, and code assigner.

differential relay   a differential relay is a protective relay that measures current going into a device from all sources by means of a network of paralleled current transformers. Ideally, the operational current is zero for normal conditions, and rises to a high value (proportional to fault current) when a fault comes on inside the differential zone. Differential relays are commonly applied in bus protection, transformer protection, generator protection, and large motor protection.

differential volume element   in a given coordinate system, the product of the differential length changes in each of the three coordinate directions.

differential-mode coupling   pick-up from an electromagnetic field that induces a change in potential on both signal leads of equal magnitude but opposite phase relative to the ground reference potential.

differentially compounded   a compound machine in which the flux produced by the MMF of the shunt field winding and the flux produced by the MMF of the series field winding oppose each other. Most often obtained by incorrectly connecting the machine, the differentially compounded machine may demonstrate very erratic behavior.

diffraction   (1) distortion of an electromagnetic wave due to the proximity of a boundary or aperture.

(2) a bending or scattering of electromagnetic waves. Basically a redistribution within a wavefront when it passes near the edge of an opaque object.

(3) the propagation of light in the presence of boundaries. It is the property of light that causes the wavefront to bend as it passes an edge.

differentiator   a filter that performs a differentiation of the signal. Since convolution and differentiation are both linear operations, they can be performed in either order.

$$(f * g)'(x) = f'(x) * g(x) = f(x) * g'(x).$$

Thus, instead of filtering a signal and then differentiating the result, differentiating the filter and applying it to the signal has the same effect. This filter is called a differentiator. A low-pass filter is commonly differentiated and used as a differentiator.

diffracted beam   diffraction that takes place when the wavelength of an incident beam is short compared to the interaction distance. Particles exhibit wave like characteristics in their passage through matter. In striking a target the incident beam scatters off nucleons. The scattered waves then combine according to the superposition principle and the peak of this scattered wave is called the diffracted beam.

diffraction angle   angle corresponding approximately to the rate of spreading of an electromagnetic wave that has been transmitted through an aperture; with Gaussian beams the far field half angle for a radius equal to the spot size.

diffraction coefficient   in the Geometric Theory of Diffraction, the coefficient that is proportional to the contribution to the scattered field due to the fringe currents near an edge or corner of a scattering target.

diffraction efficiency of Bragg cell   ratio of the intensity of the principal diffracted beam to the intensity of the undiffracted beam.

diffraction grating   an array of reflecting or transmitting lines that mutually enhance the effects of diffraction.

diffraction loss   loss from an electromagnetic beam due to finite aperture effects.

diffraction tomography   generalization of computerized tomography incorporating scattering effects.

diffuse density   signal that has uniform energy density, meaning that the energy flux is equal in all parts of a given region.

diffuse intensity   the energy scattered in all directions out of the forward or specular directions. Sometimes also called incoherent component of the intensity.

diffuse multipath   the result of multipath propagation observed as overlapping signal components, due to delay differences of multipath components being less than the delay resolution of the signal. Observable in the delay power spectrum as a continuous distribution of power over delay. *See also* multipath propagation, delay power spectrum.

diffuse scattering   the component of the scattering from a rough surface that is not in the specular direction. It is caused by reflections from local surfaces oriented in planes different from that of the mean surface. *See also* specular scattering.

diffuse transmittance   a transmitted signal that has uniform energy density.

diffusion   a region of a semiconductor into which a very high concentration of impurity has been diffused in order to substantially increase the majority carrier concentration in that region.

diffusion pump   second stage of the vacuum system. Hot oil showers the particles in a vacuum and creates a better vacuum. After a mechanical (roughing) pump is used to remove about 99.99% of the air in the beam tube, the remaining air can then be removed by a diffusion pump, down to about $1E^{-9}$ torr.

diffusion under field (DUF)   a local thin layer of semiconductor with a very high carrier concentration located under and in contact with the collector of a vertical bipolar transistor to provide a low-resistivity connection to it.

diffusive scattering   when the photon mean free path is much smaller than the scatterer dimensions and then the energy is scattered uniformly in all directions.

DigiCipher HDTV system   a high-definition television (HDTV) digital transmission television system proposed to the FCC by the American Television Alliance composed of General Instruments and Massachusetts Institute of Technology. The DigiCipher HDTV proposal submitted to the FCC in August of 1991 was the first system to provide an all digital television system that promised spectrum compatibility with the existing television channel allocation. The system used quadrature amplitude modulation (16-QAM) digital transmission at a 4.88 MHz symbol rate.

digital   circuits or systems that employ two valued (binary) signals denoted by the digits 0 and 1. Normally binary 1 is used to indicate high/true and binary 0 to indicate low/false (Positive Logic).

digital cellular radio   cellular radio product designed to transmit its signals digitally.

digital communications   communication techniques that employ binary bits to encode information.

digital European cordless telephone (DECT)   a digital microcell system operating in the 1.88–1.90 GHz band designed to provide high-capacity wireless voice and data services indoors and outdoors in small local networks. DECT systems use a TDMA/FDMA multiple access scheme, with 10 radio frequency carriers, each 1728 kHz wide, divided into 2×12 time slots. This provides a total of 120 duplex traffic channels, with duplexing via TDD. The carrier bit rate is 1152 kb/s. DECT portable terminals have a peak transmit power of 250 mW.

digital filter   the computational process or algorithm by which a sampled signal or sequence of numbers (acting as an input) is transformed into a second sequence of numbers termed the output signal. The computational process may be that of low-pass filtering (smoothing), bandpass filtering, highpass filtering, interpolation, the generation of derivatives, etc.

digital halftone   halftone technique based on patterns of same size dots designed to simulate a shade of gray between white paper and full colorant coverage.

digital image   (1) an array of numbers representing the spatial distribution of energy in a scene obtained by a process of sampling and quantization.

(2) a discrete-value function $\hat{I}(k, l) = I(k\Delta x, l\Delta y)$ obtained by sampling, at equispaced positions, the continuous function $I(x, y)$ which measures image intensity at position x, y of the image plane. I can be single-valued for monochrome images or m-valued (usually $m = 3$) for color images.

digital modulation signal generator   an RF signal generator capable of providing signals with digital modulation formats such as Gaussian minimum shift keying (GMSK), Pi/4 differential quadrature phase shift keying (DQPSK), and code division multiple access (CDMA).

digital optical computing   optical computing that deals with binary number operations, logic gates, and other efforts to eventually build a general-purpose digital optical computer. In digital optical computing, new optical devices are sought to replace elements in an electronic computer. The digital optical computer may be primarily based on already known computer architectures and algorithms.

digital optics   optical systems that handle digital data.

digital relay   a relay in which decisions are made by a digital computer, typically a microprocessor system.

digital serial processing   processing of more than one but not all bits in one clock cycle. If the digit size is W_1 and the word length is W, then the word is processed in W/W_1 clock cycles. If $W_1 = 1$, then the system is referred to as a bit-serial, and if $W_1 = W$, then the system is referred to as a bit-parallel system. In general, the digit size W_1 need not be a divisor of the word length W, since the least and most significant bits of consecutive words can be overlapped and processed in the same clock cycle.

digital signal processor (DSP)   microprocessor specifically designed for processing digital signals. DSPs are typically well suited to perform multiplications and additions in chain, even in floating point. They are less suitable for managing interrupts and large amounts of memory. For reaching high performance, a neat division between memory for data and memory for programs is adopted, with the constraint of having a high number of pins.

digital simulator   a simulator that allows the user to check the function and perform a timing analysis of a digital system.

digital subscriber line (DSL)   in telephony, a digital connection between a customer premise and a central switching office (CO) using twisted-pair (copper) as the transmission medium. Although DSLs were originally introduced for narrowband ISDN applications (144 kbps), recent

enhancements of DSLs (definitions follow) now support a broader range of higher-rate services.

digital sum variation   a measure of the maximum possible imbalance in a line coded sequence. Definitions vary; a common definition is the total number of running digital sum values that can arise in the encoded sequence.

digital tachometer   a device with a sensor that senses pulses from a rotating axis and converts them to digital output calibrated in rotations per minute (rpm).

digital video interactive (DVI)   a compression/decompression scheme used in digital audio and video.

digital voltmeter (DVM)   a modern solid state device capable of measuring voltage and displaying the value in digitized form. The term is also used loosely for the digital multimeter, which can also measure current and resistance.

digital–optical computing   that branch of optical computing that involves the development of optical techniques to perform digital computations.

digital-to-analog converter (DAC, D/A)   a device that changes a digital signal to an analog signal of corresponding magnitude.

digital-video effects   application of digital technology to manipulate video information for production, to compress video data, to transmit video signals, and to process or transform video signals for various display systems.

The original analog video is digitized for application to computer-type circuits that can produce effects such as video mixing or overlay; editing of video signals; compression of video data; synchronizing video systems; signal transformation and timebase correction for various display formats and signal conversion back to analog form. The primary video artifacts observed from the digital-video effects result from either digitizing to an insufficient number of bits per picture element (pixel) or from too few samples for the video block processing algorithms. Using fewer than 8 bits/pixel for each color component will cause poor signal-to-noise ratios and produce scenes that have the appearance of a poster or cartoon. The poster appearance results from contours that are too abrupt and from colors that are not smoothly blended. Similarly, algorithms that process large blocks of pixels tend to reduce resolution and produce blocks with color shifts.

digitization   a process applied to a continuous quantity that samples the quantity first, say, in time or spatial domain, and then quantizes the sampled value. For instance, a continuous-time signal can be first sampled and then quantized to form a digital signal, which has been discretized in both time and magnitude.

digitize   the action of converting information from analog to digital form.

digitizer   *See* data tablet.

dilation equation   the equation

$$\pi(t) = \sum_{n=-\infty}^{\infty} a(n)\pi(2t - n)$$

with $\pi(t)$ being scaling function and $a(n)$ being the coefficients. It states the fact that, in multiresolution analysis, a scaling space is contained in a scaling space with finer scale.

dilation   a fundamental operation in mathematical morphology. given a structuring element B, the dilation by B is the operator transforming X into the Minkowski sum $X \oplus B$, which is defined as follows:

(1) if both X and B are subsets of a space E, $X \oplus B = \{x + b \mid x \in X, b \in B\}$.

(2) if X is a gray-level image on a space E and B is a subset of E, for every $p \in E$ we have $(X \oplus B)(p) = \sup_{b \in B} X(p - b)$.

(3) if both X and B are gray-level images on a space E, for every $p \in E$ we have $(X \oplus B)(p) = \sup_{h \in E} [X(p - h) + B(h)]$.

(4) with the convention $\infty - \infty = -\infty$ when $X(p - h), B(h) = \pm\infty$. (In the two items above, $X(q)$ designates the gray-level of the point $q \in E$ in the gray-level image X.) *See* erosion, structuring element.

Dill parameters   three parameters, named A, B, and C, that are used in the Dill exposure model for photoresists. A and B represent the bleachable and nonbleachable absorption coefficients of the resist, respectively, and C represents the first-order kinetic rate constant of the exposure reaction. (Named for Frederick Dill, the first to publish this model.) Also called the ABC parameters.

dilution transformer   one of several hedges, (e.g., somewhat, quite, rather, and sort of) that dilute the characteristics of a fuzzy set.

diminished radix complement   form of the complement representation of negative numbers. In the binary system, the radix complement is called the 2s complement and the diminished radix complement is called the 1s complement.

diode   a two-terminal device that permits the flow of electric current in only one direction.

Diodes are most often constructed by abutting n-type and p-type regions of a semiconductor, that has significantly higher electrical conductivity in one direction (forward-biased) than the other (reverse-biased).

Diode devices may be specially designed for low-power, high switching speed applications (signal diodes) or higher-power applications (rectifier diodes).

diode detector   a device that by use of rectification and the use of inherent nonlinearity separates a modulating signal from its carrier.

195

diode gun Plumbicon   a Plumbicon tube with an electron gun that operates with positive voltage applied to G_1 with respect to the cathode. The diode gun principle provides a finer beam spot size and lower beam temperature. This results in higher resolution and improved lag performance compared to triode gun tubes. The diode gun also provides a much higher current reserve for highlight handling when used in conjunction with a dynamic beam control circuit.

diode laser   laser in which the amplification takes place in an electrically pumped semiconducting medium. Also known as a semiconductor laser or a heterojunction laser.

diode rectifier   a circuit in which the output voltage is fixed by the circuit parameters and the load. The direction of power flow is not reversible. An example of a single-phase diode-bridge rectifier with a capacitor filter is shown. Note that the diodes are on only for a short duration, while the rectified line voltage is greater than the capacitor voltage.

dioptric   an optical system made up of only refractive elements (lenses).

dip   *See* sag.

DIP   *See* dual in-line package.

DIP switch   set of micro-switches (on/off or deviators) that are compliant with the DIP for the position of their pin (connections); thus, they can be installed in standard sockets for integrated circuits.

dipole   *See* magnetic dipole.

dipole antenna   a straight wire, with two arms, of oppositely pulsating charges, typically $\lambda/2$ or $\lambda/4$.

Dirac delta function   Delta function.

Dirac's delta function   not actually a function, Dirac's delta function is defined functionally by its property of "choosing" a single value of the integrand when integrated:

$$f(r_0) = \int f(r)\delta(r - r_0)dr$$

when the volume of integration includes the point r_0.

direct addressing   address of the operand (data upon which the instruction operates) of the instruction is included as part of the instruction.

direct axis (d axis)   the magnetic axis of the rotor field winding of a synchronous machine. The axis between the poles of a DC machine.

direct axis magnetizing (armature) reactance   a reactance that represents all the inductive effects of the d-axis stator current of a synchronous machine, except for that due to the stator winding leakage reactance. In Park's d-axis equivalent circuit of the synchronous machine, this reactance is the only element through which both the stator and rotor currents flow. Its value may be determined by subtracting the stator winding leakage reactance from the steady-state value of the d-axis operational impedance or from the geometric and material data of the machine.

direct axis synchronous reactance   the sum of the stator winding leakage reactance and the direct-axis magnetizing (armature) reactance of a synchronous machine. This represents the balanced steady-state value of the direct-axis operational impedance of the synchronous machine, and thus characterizes the equivalent reactance of the machine during steady-state operation.

direct axis transient reactance   a value that characterizes the equivalent reactance of the d-axis windings of the synchronous machine between the initial time following a system disturbance (subtransient interval) and the steady state. This reactance cannot be directly mathematically related to the d-axis operational impedance. However, in models in which the rotor windings are represented as lumped parameter circuits, the d-axis transient reactance is expressed in closed form as the sum of the stator winding leakage reactance, and the parallel combination of the d-axis magnetizing reactance and the field winding leakage reactance.

direct broadcast satellite (DBS)   refers to TV signal transmission and distribution from a base station up to a satellite, and then down to consumers who have suitable satellite receiving antennas and down-converter receivers.

$$F(dB) = 10 \log_{10}(SNR_{in}/SNR_{out})$$

direct burial   (1) the practice of burying a specially-armored power or communications cable in a ditch without the use of a surrounding conduit.

(2) a term applied to any cable which is meant for direct burial.

direct control   bottom, first, control layer of a multilayer controller, directly responsible for adjusting the manipulated inputs to the controlled process; typical example of direct control is the regulation layer of an industrial control system, where the manipulated inputs are used to make the controlled variables follow the desired set point values.

direct control layer   *See* direct control.

direct converter   a frequency converter that converts an RF signal to a baseband signal directly in receivers. It converts a baseband signal to an RF signal directly in transmitters.

direct current machine   a DC machine is an electromechanical dynamo that either converts direct current electrical power into mechanical

power (DC motor), or converts mechanical power into direct current electrical power (DC generator). Some DC machines are designed to perform either of these functions, depending on the energy source to the dynamo.

direct current motor   a rotation machine energized by DC electrical energy and used to convert electrical energy to mechanical energy.

direct digital synthesizer   an oscillator that generates sinusoidal wave by digital calculation and digital to analog conversion. It can generate an arbitrary frequency signal in a fine channel step.

direct drive   a drive in which no gear reducer is used.

direct drive robot   a mechanical arm where all or part of the active arm joints are actuated with the direct drive. Due to the fact that many actuators are best suited to relatively high speeds and low torques, a speed reduction system is required. Gears are the most common elements used for reduction. A robot with a gear mechanism is called a geared robot. Gears are located at different joints; therefore, usually geared robots have a transmission system, which is needed to transfer the motion from the actuator to the joint.

direct dynamics   the direct dynamics problem consists of determining the joint generalized accelerations $\ddot{q}(t)$ assuming that joint generalized forces $\tau(k)$, joint positions $q(t)$, and joint velocities $\dot{q}(t)$ are known. Solution of the direct dynamics leads to the dynamic simulation which is defined as follows: for a given set of joint generalized forces and initial values of the joint positions and velocities integrate the equations of motion in order to find a set of joint accelerations, velocities, and positions. The dynamics simulation is very useful for manipulator simulation.

197

direct fuzzy control   the use of fuzzy control directly in the inner control loop of a feedback control system.

direct kinematics   direct kinematics (or forward kinematics) for an arbitrary manipulator and given the joint variables vector $q(t) = [q_1(t), q_2(t), \ldots, q_n(t)]^T$ and the geometric link parameters finds the position and orientation of the end-effector of the manipulator with respect to a reference coordinate frame. The direct kinematics problem can be solved by successive multiplication of the D-H transformation matrices from the base of the manipulator towards its end-effector. More precisely the kinematics can be represented mathematically as a continuous map assigning to every joint position (configuration of the manipulator) a corresponding position and orientation of the end-effector. $k : Q \rightarrow E$ where Q and E are called, respectively, the inertial (joint) space and the external space of the manipulator. Suppose that a manipulator has r-revolute and $(n - r)$ prismatic joints. Then the internal space is defined as $Q = T^r \times R^{n-r}$, where T^r denotes the r dimensional torus, and R^{n-r} $(n - r)$ dimensional space of real numbers.

direct mapped   *See* direct mapped cache.

direct mapped cache   a cache where each main memory (MM) block is mapped directly to a specific cache block. Since the cache is much smaller than the MM, several MM blocks map to the same cache block.

If, for example, the cache can hold 128 blocks, MM block k will map onto cache block k modulo 128. Because several MM blocks map onto the same cache block, contention may arise for that position. This is resolved by allowing the new block to overwrite the old one, making the replacement algorithm very trivial in this case.

In its implementation, a high-speed random access memory is used in which each cache line and the most significant bits of its main memory address (the tag) are held together in the cache at a location given by the least significant bits of the memory address (the index). After the cache line is selected by its index, the tag is compared with the most significant bits of the required memory address to find whether the line is the required line and to access the line.

direct memory access (DMA)   used in a computer system when transferring blocks of information between I/O devices (e.g., disk memory) to/from the main memory with minimal intervention from the CPU. A "DMA controller" is used and can, after initiation by the CPU, take control of the address, control, and data busses. The CPU initiates the DMA controller with parameters such as the start address of the block in main memory, number of bytes to be transferred, and the type of transfer requested (read or write). The transfer is then completely handled by the DMA controller, and the CPU is typically notified by an interrupt when the transfer service is completed. While the DMA transfer is in progress, the CPU can continue executing the program doing other things. However, as this may cause access conflicts of the busses between the CPU and the DMA controller, a "memory bus controller" handles prioritized bus requests from these units. The highest priority is given to the DMA transfer, since this normally involves synchronous data transfer that cannot wait (e.g., a disk or tape drive). Since the CPU normally originates the majority of memory access cycles, the DMA control is considered as "stealing" bus cycles from the CPU. For this reason, this technique is normally referred to as "cycle stealing."

direct methanol fuel cell (DMFC)   a liquid-or–vapor fed proton exchange membrane fuel cell operating on a methanol/water mix and air. *See also* akaline fuel cell, proton exchange membrane fuel cell, phosphoric acid fuel cell, solid oxide fuel cell.

direct method   Lyapunov's second method of investigating the stability of dynamical systems.

The method is called a direct method because no knowledge of the solution of the differential equations modeling a dynamical system is required when investigating the stability of an equilibrium solution. *See also* equilibrium solution and stability.

direct method coordination   coordination by the direct method amounts to iterating the coordination variables (direct coordination instruments) defined as the interaction inputs and outputs and, if needed, any other variables that, when fixed, provide for independence of the local decision problems. The results of these problems are used by the coordinator to check whether its objectives are satisfied — if not, then the direct coordinating instruments are changed (iterated), etc.

direct modulation   modulation of the optical intensity output from a semiconductor diode laser by direct modulation of the bias current.

direct scattering theory   predicts the distribution of scattered intensity from knowledge of the structure of the inhomogeneous medium.

direct semiconductor   a semiconductor whose band gaps are between electronic states with the same momentum, thus allowing optical transitions to occur between them.

direct sequence   refers to a technique for digital spread spectrum modulation. The data symbol is modulated by a higher-rate pseudo-random sequence. The resulting signal is thus at a higher rate, or equivalently, occupies a larger bandwidth. This is denoted direct sequence spreading.

direct stroke   (1) a lightning strike which terminates on a piece of equipment

(2) an approach to the design of overhead electric transmission lines which assumes that only direct strokes to the line will be disruptive to a power system.

direct write lithography   a lithography method whereby the pattern is written directly on the wafer without the use of a mask.

direct-access storage   storage in which an item can be accessed without having to first access all other items that precede it in the storage; however, sequential access may be required to a few preceding systems. An example is a disk, in which blocks may be accessed independently, but access to a location within a block is preceded by access to earlier locations in the block. *See also* sequential-access storage.

direct-axis subtransient open-circuit time constant   a constant that characterizes the initial decay of transients in the d-axis variables of the synchronous machine with the stator windings open-circuited. The interval characterized is that immediately following a disturbance, during which the effects of all amortisseur windings are considered. A detailed (derived) closed-form expression for the subtransient open-circuit time constant of a machine with a single d-axis amortisseur winding is obtained by taking the reciprocal of the smallest root of the denominator of the d-axis operational impedance. An approximate (standard) value is often used, in which it is assumed the field winding resistance is very small and the detailed expression simplified.

direct-axis subtransient reactance   the high-frequency asymptote of the d-axis operational impedance of a synchronous machine. This value characterizes the equivalent reactance of the d axis of the machine during the initial time following a system disturbance. In models in which the rotor windings are represented as lumped parameter circuits, the d-axis subtransient reactance is expressed in closed form as the sum of the stator winding leakage reactance, and the parallel combination of the d-axis magnetizing reactance and the d-axis rotor leakage reactances.

direct-axis subtransient short-circuit time constant   a constant that characterizes the initial decay of transients in the d-axis variables of the synchronous machine with the stator windings short-circuited. The interval characterized is that immediately following a disturbance, during which the effects of amortisseur windings are considered. A detailed (derived) closed-form expression for the subtransient short-circuit time constant of a machine with a single d-axis amortisseur winding is obtained by taking the reciprocal of the largest root of the numerator of the d-axis operational impedance. An approximate (standard) value is often used, in which it is assumed the field winding resistance is small and the detailed expression simplified.

direct-axis transient open-circuit time constant a constant that characterizes the decay of transients in the d-axis variables of the synchronous machine with the stator windings open-circuited. The interval characterized is that following the subtransient interval, but prior to steady-state, during which the effects of the amortisseur windings are small (possibly negligible). A detailed (derived) closed-form expression for the transient open-circuit time constant of a machine with a single d-axis amortisseur winding is obtained by taking the reciprocal of the smallest root of the denominator of the d-axis operational impedance. An approximate (standard) value is often used, in which it is assumed the amortisseur winding resistance is infinite and the detailed expression simplified.

direct-axis transient short-circuit time constant a constant that characterizes the decay of transients in the d-axis variables of the synchronous machine with the stator windings short-circuited. The interval characterized is that following the subtransient interval, but prior to steady-state, in which the effects of the amortisseur windings are small (possibly negligible). A detailed (derived) closed-form expression for the short-circuit transient time constant is obtained by taking the reciprocal of the smallest root of the numerator of the d-axis operational impedance.

An approximate (standard) value is often used, in which it is assumed the amortisseur winding resistance is infinite and the detailed expression simplified.

direction cosine similarity between two variables $x = (x_0, \ldots, x_n)$ and $y = (y_0, \ldots, y_n)$, it is defined as

$$\cos \theta = \frac{(x, y)}{\| x \| \, \| y \|}$$

where (x, y) is the inner product of x and y and $\| x \|$ is the Euclidean norm of x.

direction line a curve to which the given field is tangential at every point on the curve. Also called stream line or flux line.

direction of arrival *See* angle of arrival.

directional coupler a passive, 3 or 4 port device used to sample a portion of the forward (incident) signal or the reverse (reflected) signal, or both (dual directional coupler) in an RF, microwave, or mmW circuit. Directional couplers are usually described in terms of coupling factor and directivity. The coupling factor describes what fraction of the incident (or reflected) power appears at the desired coupled port. Directivity describes the fraction of power coupled to the same port due to reverse power in the main arm of the coupler.

directional overcurrent relay an overcurrent relay that operates only for overcurrents flowing in the tripping direction. Direction sensing is typically done with respect to a voltage or current signal, which is not affected by fault location.

directional power relay a protective relay that operates for power flow in a given direction. Applications are in cases where normal power flow is in one direction, including anti-motoring protection on a turbine-generator and fault backfeed protection on parallel step-down transformers.

directivity the maximum ratio of an antenna's ability to focus or receive power in a given direction relative to a standard; the standard is usually an isotropic radiator or a dipole. Only depends on the radiation pattern shape and does not include the efficiency of the antenna.

directory a table used in the directory method to maintain cache coherence in multiprocessors. Contains entries identifying the caches that hold copies of memory locations.

directory look-aside table (DLT) *See* translation look-aside buffer.

Dirichlet conditions a set of conditions guaranteeing that a signal $x(t)$ will be equal to the N-term Fourier approximation of $x(t)$ as $N \to \infty$ except at isolated values of t for which $x(t)$ is discontinous. The conditions are that

(1) the signal $x(t)$ must be absolutely integrable,

(2) $x(t)$ must have a finite number of maxima and minima during a period, and

(3) $x(t)$ must have a finite number of (finite) discontinuities in any finite interval of time.

dirty bit a status bit used to indicate if a block (e.g., cache block, page, etc.) at some level of the memory hierarchy has been modified (written) since it was first loaded in. When the block is to be replaced with another block, the dirty bit is first checked to see whether the block has been modified. If it has, the block is written back to the next lower level. Otherwise, the block is not written back.

dirty page a page in memory that has been altered since last loaded into main memory. *See also* dirty bit.

disable action that renders a device incapable of performing its function; the opposite of enable.

disagreement of interests situation in which there are several decision units with conflicting goals. *Compare with* consistency of interests.

disassembler a computer program that can take an executable image and convert it back into assembly code.

disc *See* magnetic disk. Also spelled "disk."

DISC Cerenkov counter a differential isochronous self-collimating Cerenkov counter. This device is used to identify particles over a wide range of masses and can also be used to give an independent calibration of the average momentum of a beam line.

disco contraction of "distribution company," a firm which owns the electric distribution network in a service area but neither generates nor transmits electric power.

disconnect switch a manually operated switching device used to disconnect circuit conductors and their associated load(s) their source of electrical power.

discontinuity effect an appropriate equivalent circuit model for discontinuities that have a tendency to disturb the electric and magnetic fields in their vicinity.

discontinuity manifold *See* sliding surface.

discontinuous control the control law that is allowed to vary discontinuously to account for sudden switching. An example of a discontinuous controller is

$$u = -U\text{sign}(s(e)) \,,$$

where e is a control error and s is a function of e, where

$$\text{sign}(s) = \begin{cases} -1 & \text{if } s < 0 \\ 1 & \text{if } s > 0. \end{cases}$$

201

discrete cosine transform (DCT) a popular format for video compression. The spatial signal is expanded in a cosine series, where the higher frequencies represent increased video resolution.

The forward 2-D DCT of an $n \times n$ block is defined as

$$F(u, v) = \frac{4C(u)C(v)}{n^2} \sum_{j=0}^{n-1} \sum_{k=0}^{n-1} f(j, k)$$

$$\cos\left(\frac{(2j + 1)u\pi}{2n}\right) \cos\left(\frac{(2k + 1)v\pi}{2n}\right)$$

and the inverse is defined as

$$f(j, k) = \sum_{j=0}^{n-1} \sum_{k=0}^{n-1} C(u)C(v)F(u, v)$$

$$\cos\left(\frac{(2j + 1)u\pi}{2n}\right) \cos\left(\frac{(2k + 1)v\pi}{2n}\right)$$

where $C(w) = \frac{1}{\sqrt{2}}$ for $w = 0$ and 1 for $w = 1, 2, \ldots, n - 1$.

There is a family of DCTs, of which the DCT-II described above is the one commonly used. These other types of DCT, specifically the DCT-IV, are sometimes used in calculating fast transforms.

The $N = 8$ element DCT is particularly important for image data compression and is central to the JPEG and MPEG standards. As a matrix, the 8-element DCT is

For images that exhibit high pixel to pixel correlation, the DCT is indistinguishable from the Karhunen–Loeve transform.

discrete data channel the concatenation of all communication system elements between and including the modulator and demodulator.

discrete Fourier transform (DFT) the sum of complex exponentials representing a sampled sequence. This transform is obtained to represent a reasonable approximation of a signal for which only a finite sample exists. Defined as

$$X(x) = \sum_{n=0}^{N-1} x(n)e^{-j(2\pi/N)nk}$$

where $x(n)$ represents a sequence of finite samples of a signal; N is the number of samples in the sequence.

discrete fuzzy set a fuzzy set that includes only those sample points of a continuous variable.

discrete Hadamard transform *See* Hadamard transform.

discrete Hopfield network a single layer, fully connected network that stores (usually bipolar) patterns by setting its weight values w_{ij} equal to the (i, j) entry in the sum of the outer products of the patterns. The network can be used as an associative memory so long as the number of stored patterns is less than about 14% of the number of neural elements. *Compare with* continuous Hopfield network.

$$\begin{pmatrix} 0.3536 & 0.3536 & 0.3536 & 0.3536 & 0.3536 & 0.3536 & 0.3536 & 0.3536 \\ 0.4904 & 0.4157 & 0.2778 & 0.0975 & -0.0975 & -0.2778 & -0.4157 & -0.4904 \\ 0.4619 & 0.1913 & -0.1913 & -0.4619 & -0.4619 & -0.1913 & 0.1913 & 0.4619 \\ 0.4517 & -0.0975 & -0.4904 & -0.2778 & 0.2778 & 0.4904 & 0.0975 & -0.4517 \\ 0.3536 & -0.3536 & -0.3536 & 0.3536 & 0.3536 & -0.3536 & -0.3536 & 0.3536 \\ 0.2778 & -0.4904 & 0.0975 & 0.4157 & -0.4157 & -0.0975 & 0.4904 & -0.2778 \\ 0.1913 & -0.4619 & 0.4619 & -0.1913 & -0.1913 & 0.4619 & -0.4619 & 0.1913 \\ 0.0975 & -0.2778 & 0.4157 & -0.4904 & 0.4904 & -0.4157 & 0.2778 & -0.0975 \end{pmatrix}$$

discrete multipath   the result of multipath propagation observed as clearly separable, discrete signal components, seen in the delay power spectrum as a set of discrete peaks at various delays. *See also* multipath propagation, delay power spectrum, specular reflection.

discrete network   an electronic network composed of separate, i.e., individual, components.

discrete sine transform (DST)   a unitary transform mapping N samples $g(n)$ to N coefficients $G(k)$ according to: $G(k) = \sqrt{\frac{2}{N+1}} \sum_{k=0}^{N-1} g(n) \sin\left(\frac{nk\pi}{N+1}\right)$ with inverse $g(n) = \sqrt{\frac{2}{N+1}} \sum_{k=0}^{N-1} G(k) \sin(\frac{nk\pi}{N+1})$.

discrete spectrum   the eigenvalues of a differential equation with real coefficients and finite boundary conditions form a discrete spectrum. By extension, also the modes of closed waveguides originate a discrete spectrum.

discrete time Fourier series   representation of a periodic sequence x_n with period N by the sum of a series of harmonically related complex exponential sequences: $x_n = \frac{1}{N} \sum_{N-1}^{k=0} X_K e^{j2\pi knN}$. The x_k are the Fourier series coefficients, obtained by $x_k = \sum_{N-1}^{n=0} x_n e^{-j2\pi knN}$.

discrete time signal   a signal represented by samples at discrete moments of time (usually regularly spaced). The samples may take values from a continuous range, so the term is usually used to differentiate a sampled analog signal from a digital signal which is quantized. *See* signal.

discrete time system   a process that transforms discrete time input signals to discrete time output signals.

discrete time white Gaussian noise   noise samples modeled as independent and identically distributed Gaussian random variables.

discrete Walsh-Hadamard transform   *See* Hadamard transform.

discrete wavelet transform (DWT)   a computation procedure that calculates the coefficients of the wavelet series expansion for a given finite discrete signal.

discriminator   a circuit whose output voltage varies in magnitude and polarity in direct proportion to the difference between the input voltage and a standard signal. A discriminator that converts phase deviations at the input to a linearly proportionate variation in output voltage is called a phase discriminator and is used in FM detection.

disk   *See* magnetic disk. Also spelled "disc."

disk actuator   a mechanical device that moves the disk arms over the disk surface(s) in order to position the read/write head(s) over the correct disk track.

disk arm   a mechanical assembly that positions the head over the correct track for reading or writing a disk device. The arm is not movable on a fixed-head disk, but is on a moving-head disk.

disk array   a number of disks grouped together, acting together as a single logical disk. By this, multiple I/O requests can be serviced in parallel, or that the bandwith of several disks can be harnessed together to service a single logical I/O request.

disk cache   a buffer memory area in main memory used to hold blocks of data from disk storage. The cache can hide much of the rotational and seek latencies in disk accesses because a complete data block (disk sector) is read or written together. Disk caches are normally managed by the machine's operating system, unlike a processor cache, which is managed by hardware.

disk capacitor   a small single-layer ceramic capacitor with a dielectric insulator consisting of conductively silvered opposing surfaces.

disk controller   unit that carries out the actions required for the proper operations of a disk unit.

disk drive   assembly consisting of electronics and mechanical components, to control disk and disk-head movement and to exchange data, control, and status signals with an input/ouput module, as required for the proper reading or writing of data to or from a disk. The head disk assembly plus all the associated electronics.

disk format   the (system-dependent) manner in which a track of a disk is partitioned so as to indicate for each sector: the identity, the start and end, synchronization information, error-checking information, etc. A disk must be formatted before any initial writing can take place.

disk head   read/write head used in a disk drive. Such a head may be fixed-gap, in which the head is positioned at a fixed distance from the disk surface; contact-head, in which the head is always in contact with the surface; and aerodynamic, in which the head rests lightly on the surface when the disk is motionless but floats a small distance above when the tape is rotating. Typically, contact heads are used in floppy disks, and aerodynamic head are used in Winchester disks.

In earlier systems, "fixed-head disks" having one read/write head per track were used in some disk systems, so the seek time was eliminated. However, since modern disks have hundreds of tracks per surface, placing a head at every track is no longer considered an economical solution.

disk latency   time between positioning a read/write head over a track of data and when the beginning of the track of data passes under the head.

disk operating system (DOS)   a set of procedures, services, and commands that can be used by the computer user for managing its resources with a special attention to disks managing. The most famous DOS for personal computers is the MS-DOS (Microsoft-DOS). It is a mono-task and mono-user operating system.

disk pack   a stack of disk platters that can be removed for off-line storage.

disk platter   metal disks covered with a magnetic material for recording information.

disk scheduling   algorithms used to reduce the total mechanical delays in disk accesses, as seen by a queue of simultaneous I/O requests. E.g., if a "shortest-seek-time-first" scheduling algorithm is used, seek times can be reduced. That is, among the queue of pending I/O requests, the one next serviced is the one requiring the shortest seek time from the current location of the read/write head. The disk scheduling algorithm is run by the computer operating system.

disk sector   the smallest unit that can be read or written on a track; adjacent sectors are separated by a gap. A typical track has 10–100 sectors.

disk spindle   a stack of disk platters.

disk striping   the notion of interleaving data across multiple disks at a fine grain, so that needed data can be accessed from all the disks simultaneously, thus providing much higher effective bandwidth.

disk track   connectric circle over which a read/write head moves; adjacent tracks are separated by a gap. A typical disk has hundreds to thousands of tracks.

diskette   a floppy disk is a flexible plastic diskette coated with magnetic material. It is a smaller, simpler, and cheaper form of disk storage media than the hard disk, and also easily removed for transportation.

disocclusion the uncovering of an object. *See* occlusion.

disparity in binocular vision, the relative difference of position of an object with respect to a background between the left and right images. It is usually measured in minutes of arc.

dispatch the determination of the power output of each plant in an electric power system.

dispersed generation problem in economic dispatch calculations, the task of accounting for the production capacity of co-generation or dispersed sources such as photovoltaic or wind generation plants.

dispersion (1) a characteristic of a medium in which the propagation velocity of a wave varies as a function of signal frequency.

As a pulse propagates through an optical fiber, its chromatic components will spread out or "disperse" in time. This phenomenon limits the distance between optical regenerators in fiber communication systems. There are four sources of dispersion: modal dispersion, material dispersion, waveguide dispersion, and nonlinear dispersion.

(2) the variation of the index of refraction of a material as a function of wavelength.

dispersion compensator a device that compensates for the accumulated chromatic dispersion in a fiber optic transmission system. Three main schemes exist: fiber grating dispersion compensators, dispersion compensating fiber, and phase conjugation or mid-span spectral inversion.

dispersion diagram a plot of propagation constant versus frequency.

dispersion of authority a situation in which the decisions, for example, decisions concerned with the manipulated inputs to the controlled process, are distributed between several (control) decision units; dispersion of authority may result either from natural, legislative, or other reasons, or may appear due to design of the controller; decentralized control, in particular, is based on the dispersion of authority between the local decision units. Also known as dispersion of control.

dispersion of control *See* dispersion of authority.

dispersion shifted fiber single mode optical fibers with zero dispersion in the 1550 nm telecommunications window. Prior to 1985, single mode optical fibers were designed to have zero dispersion in the 1310 nm telecommunications window.

dispersive medium (1) a medium for which the permittivity or the permeability (or both) are frequency dependent.

(2) in optics, medium in which the index of refraction varies significantly with frequency.

displaced frame difference (DFD) the difference between a given digital image frame and its estimate obtained by using the motion compensation technique. It is useful in image (sequence) data compression and motion estimation.

displacement current a field quantity that describes the completion of a circuit when a conducting path is not present.

displacement parameter complex parameter representing the displacement of the amplitude and phase centers of a Gaussian beam from the axis of an optical system.

display a device that provides a visual non-permanent display of system input and or output. Common display technology includes CRT (cathode ray tube), LED (light emitting diode), PDP (plasma display panel), EL (electroluminescense) and LCD (liquid crystal display).

dissipated power   the power dissipated as a heat, which is defined by subtracting an RF output power from a DC input power.

dissipation   the phenomenon associated with the attenuation of a propagating wave in a medium with material losses.

dissipation factor (DF)   the ratio of the effective series resistance of a capacitor to its reactance at a specified frequency measured in percentage. Also known as loss tangent.

dissipation power   (1) ratio of real power (in phase power) to reaction power (shifted $90°$ out of phase).

(2) the ratio of the imaginary to real parts of the complex permittivity, expressed as a dimensionless ratio.

dissipative half-space   used for analysis of complex systems, it is constructed by dividing the infinite space in two by some convenient fashion. The resulting two half-spaces can both be infinite, or one finite and one infinite. The resulting half-space is then filled with a dissipative material (other than a perfect insulator or conductor).

dissipator   a form of air terminal which is meant to prevent lightning strikes to a structure by reducing the surface charge on the earths's surface in the immediate region of the structure.

distance   *See* chamfer distance, chessboard distance, Euclidean distance, Hamming distance, Hausdorff distance, inter-feature distance, Mahalanobis distance, Manhattan distance.

distance between symbol strings   a measure of the difference between two symbol strings. The most frequently used distance measures between two strings include Hamming distance, edit or Levenshtein distance, and maximum posterior probability distance. *See also* Hamming distance, edit distance, maximum posterior probability distance.

distance measure   a function $d(x, y)$ defined on a metric space, such that

$$d(x, y) \geq 0, \text{ where } = \text{ holds iff } x = y,$$
$$d(x, y) = d(y, x), \text{ and}$$
$$d(x, y) \leq d(s, z) + d(z, y).$$

See also similarity measure.

distance profile   for convolutional codes, the minimum Hamming weight of all sequences of a specific length emerging from the zero state. A distance profile for one code is superior to that of another if

(1) all values of the distance profiles of the two codes up to a certain depth p (lower than the constraint length) are equal; and

(2) the superior distance profile code has higher values of the distance profile for all depths above the given depth p.

distance protection   relaying principle based upon estimating fault location (distance) and providing a response based upon the distance to the fault.

distance relay   *See* impedance relay.

distance resolution   delay resolution mapped into the spatial domain, measured in units of distance (meters). Each unit of time delay corresponds to a unit of distance d traveled by the radio wave through the equation

$$d = c\tau,$$

where τ is the delay and c is the speed of light. Distance resolution gives the smallest difference in path length resolved by a signal. Is also a measure of the spatial resolution capability of a signal (or measurement system).

distance transform   (1) a function $\mathcal{D} : \mathcal{R}^n \to \mathcal{R}$ that assigns to each point of a set $\mathcal{S}$ a value representing the distance between such a point and the complement set of $\mathcal{S}$.

(2) a map of all the pixels in a shape showing the closest distance of each point in the shape from the background; also, an image in which the distance maps of all the shapes in the image are indicated. *See* distance.

distortion   (1) addition of an unwanted component to an electronic signal.

(2) undesired change in a signal's amplitude and/or phase as the result of it passing through an active nonlinear circuit. Numerous figures of merit have been adopted to describe various aspects of signal distortion.

distortion cross modulation   nonlinear distortion of a system or carrier signal characterized by the appearance in the output of frequencies equal to the sum and difference of the desired carrier's frequency and the unwanted cross modulating carrier's frequency. Although harmonic components are also caused to be present, they are not usually a part of measurements of this effect.

distortion factor   *See* total harmonic disturbance.

distortion-rate theory   *See* rate-distortion theory.

distributed amplifier   composed of two main artificial transmission-line sections consisting of series inductors and shunt capacitors, which are usually supplied by the FET transistor. Excellent bandwidth performers are obtainable and the amplifier can be designed as wideband low-noise amplifiers and are relatively easy to simulate and fabricate.

distributed amplifier   the input and output capacitance of the active devices can be absorbed into distributed circuits (i.e., transmission lines) to obtain a very broad bandwidth.

distributed antenna   typically consists of a set of discrete radiators fed by a common cable from a single signal source.

distributed arbitration   a scheme used for bus arbitration where multiple bus masters can access the bus. Arbitration is not done centrally (by a bus arbiter), but instead done in a distributed fashion. A mechanism to detect when more than one master tries to transmit on the bus is included. When this happens, one (or all) stops transmitting and will reattempt the transmission after a short (e.g., random) time delay.

Compare with centralized arbitration.

distributed computing   an environment in which multiple computers are networked together and the resources from more than one computer are available to a user. *See also* distributed computing environment.

distributed computing environment (DCE) an industry-standard, comprehensive, and integrated set of services that supports the development, use, and maintenance of distributed computing technologies. DCE is independent of the operating systems and network types. It provides interoperability and portability across heterogeneous platforms, and provides security services to protect and control access to data. DCE also provides services that make it easy to find distributed resources; for instance, directory service, a DCE component, is a central repository for information about resources in a distributed system. Distributed time service (DTS), another DCE component, provides a way to synchronize the times on different hosts in a distributed system. DCE gives a model for organizing widely scattered users, services, and data. It runs on all major computing platforms and is designed to support distributed applications in heterogeneous hardware and software environments. Particularly important for the World Wide Web and security of distributed objects.

distributed computing system   a system whose different parts can run on different processors.

distributed control   a control technique whereby portions of a single control process are located in two or more places.

distributed element   a circuit element in which dimension is not negligible relative to the wavelength. The characteristics of a distributed element depend upon a dimension such as a length.

distributed feedback laser (DFB)   a laser source where the optical feedback is distributed throughout the length of the gain medium. Feedback then occurs through Bragg diffraction, and the laser operates in only one optical mode.

distributed generation   small power plants at or near loads and scattered throughout the service area.

distributed memory   denotes a multiprocessor system where main memory is distributed across the processors, as opposed to being equally accessible to all. Each processor has its own local main memory (positioned physically "close"), and access to other processors' memory takes place through passing of messages over a bus. The term "loosely coupled" could also be used to describe this type of multiprocessor architecture to contrast it from shared-memory architectures, which could be denoted as "strongly coupled."

distributed memory architecture   a multiprocessor architecture in which physical memory is distributed among the processing nodes, as opposed to being in a central location, equidistant from all processors.

distributed refresh   in a DRAM, carrying out refresh operations one at a time, at regular intervals. Requires that all rows be refreshed in a time less that the time before which any given row needs to be refreshed. *See also* burst refresh.

distributed sample scrambling   a modification of the reset scrambling technique in which information regarding the state of the scrambling sequence generator is embedded into the encoded sequence in a distributed fashion for purposes of synchronizing the descrambling sequence generator.

distribution   (1) the possibility to execute different parts of a system on different processors.

(2) that class of electric power system work which is concerned with the distribution of electric power within a load area such as a residential or commercial area, or within an industrial installation. The distribution circuit extends from the local substation and terminates at the customer's meter.

distribution function   cumulative distribution function.

distribution management system (DMS)   a system that helps manage the status of the distribution network, crews and their work flow, system safety, and abnormal conditions.

distribution switchboard   a switchboard used in the distribution system, typically within a building.

distribution transformer   a transformer designed for use on a power distribution system (typically 2.4 kV to 34.5 kV) to supply electrical power to a load at the proper utilization voltage.

disturbance   a sudden change or a sequence of changes in the components or the formation of a power system. Also called fault.

disturbance decoupling of generalized 2-D linear systems   given the second generalized 2-D Fornasini–Marchesini model with disturbances

$$Ex_{i+1,j+1} = A_1x_{i+1,j} + A_2x_{i,j+1}$$
$$+ B_1u_{i+1,j} + B_2u_{i,j+1}$$
$$+ H_1z_{i+1,j} + H_2z_{i,j+1}$$
$$y_{ij} = Cx_{ij}$$

$i, j \in Z_+$ (the set of nonnegative integers), find a state-feedback $u_{ij} = K x_{ij}$ such that the output y_{ij} is not affected by the disturbances for $i, j \in Z_+$ where $x_{ij} \in R^n$, $u_{ij} \in R^m$, $y_{ij} \in R^p$, $z_{ij} \in R^q$ are semistate vector, input vector, output vector, and disturbance vector, respectively, and E, A_k, B_k, H_k $(k = 1, 2,)$ are real matrices with E possibly singular or rectangular.

dithering   computer technique allowing the display and printing of gray-level images on devices having a small number of available colors (generally two, as in black-and-white CRTs and printers). The two main approaches are matrix dithering, where a matrix of black and white dots is associated with each gray-level, and error-diffusion dithering, where each gray-level pixel in turn is assigned an available color, and the error is spread to its unprocessed neighbors (as in the Floyd-Steinberg method). *See* halftone.

divergence   the angle that the trajectory of each particle makes with the beam axis. Accelerator systems always try to reduce beam divergence.

divergence theorem   consider a volume V bounded by a surface S and an outer normal n. Let also $\mathbf{F}$ be a vector function which, together with its partial derivatives, is continuous at all points of V and S. The divergence theorem states that

$$\int_V div\ \mathbf{F} d\mathbf{x} = \int_S \mathbf{F} \cdot \mathbf{n} dS.$$

The above equation may also be used to define the divergence.

diversity combining   a communication technique that combines the signals received through different, possibly independent channels. The most common methods of combining are maximum-ratio combining, equal gain combining, selection combining, and switched combining.

diversity frequency   a method for increasing the reliability of digital communications in which

multiple copies of the signal, or other types of redundant information, are transmitted. Frequency diversity implies that the received signal occupies a much wider bandwidth than the minimum bandwidth needed to carry the information.

diversity path   a form of diversity in which multiple copies of the signal are created via different paths from the transmitter to receiver.

diversity selection   a form of diversity reception in which the receiver selects the strongest signal among the copies received. The weaker signals are simply ignored.

divide by zero   error occurring when a division per zero is operated. This case is mathematically undefined. In many cases, this problem is detected directly at level of microprocessor that activates an exception and leaving true a status flag. The exception can be managed for recovering the error and avoiding the interruption of the program execution. Also known as divide per zero.

divide per zero   *See* divide by zero.

divider   functional unit consisting of circuits that implement either integer or floating-point division.

DLL   *See* delay-locked loop.

DLT   directory look-aside table. *See also* translation look-aside buffer.

DMA   *See* direct memory access.

DMFC   *See* direct methanol fuel cell.

DMS   *See* distribution management system.

DOA   direction of arrival. *See also* angle of arrival.

DOF   *See* depth of focus.

domain module or area of execution that is to be kept isolated from other domains; a domain may have special properties that define the nature of the isolation and the limits on communication to and from the domain. In the context of secure system design, a domain may be an execution of a process that has specific security attributes.

dominant mode the mode of a waveguide having the lowest cutoff frequency or of a cavity having the lowest resonant frequency.

don't care a function that can be taken either as a minterm or a maxterm at the convenience of the user.

donor an impurity in a semiconductor that donates a free electron to the conduction band.

door *See* lid.

dopant an impurity substance (such as phosphorus or boron) added in very small controlled quantities to a semiconductor base material (such as silicon or gallium arsenide), thereby changing the material conduction characteristics by modifying electron and/or hole concentration. A donor dopant is one that gives rise to electrons and an acceptor dopant is one that gives rise to holes.

doping the process of introducing impurity atoms into pure silicon to change its electrical properties. The impurities may be either n-type (introducing an additional conducting electron) or p-type (introducing the absence of a conducting electron, also called a "hole").

Doppler broadened lineshape function spectral function that results from Doppler shifts caused by the velocity distribution of atoms or molecules in a gas; a Gaussian function for a Maxwellian velocity distribution.

Doppler broadening broadening of a spectral line due to Doppler shifts caused by the random motion of atoms or molecules in a vapor.

Doppler effect *See* Doppler shift.

Doppler filter a filter used to resolve targets from each other and from extraneous returns from other objects (called clutter) by filtering in the velocity or Doppler domain. So-called because the Doppler effect causes frequency shifts proportional to velocity variations in tracked objects.

Doppler frequency a shift in frequency of the returned power from a target as a result of the target's motion relative to the illuminating source.

Doppler linewidth characteristic width of a Doppler-broadened spectral line; usually the full width at half maximum when the line is Gaussian.

Doppler power spectrum a function characterizing the spread of average received power as a function of Doppler shift. Can be obtained from the scattering function by integrating over the delay variable. *See also* scattering function, multipath propagation.

Doppler radar radar-based technique used in measuring the velocity of a moving target or wind by measuring the Doppler shift (Doppler effect).

Doppler shift a frequency shift in a received signal caused by time-variant transmission delay, or equivalently time-variant propagation path length. This in turn is caused by movement of terminals with respect to each other, or by movement of reflecting objects. In optics, frequency shifts imposed on laser beams such as when used in laser radar or when diffracted by an acoustic wave.

The Doppler shift depends on the frequency of the signal and the angle of arrival of the signal relative to the direction of movement of the receiver. For a signal consisting of a range of frequencies, each frequency component will experience a different shift. Hence the received signal will have a different bandwidth than the transmitted signal (Doppler dispersion).

For a tone (continuous-wave) signal of frequency f, the Doppler shift f_D observed on a single propagation path of changing length is given by

$$f_D = \frac{fv}{c},$$

where v is the rate (in m/s) of path length change, and c is the speed of light.

Also known as *Doppler effect*.

Doppler spread   the increase of bandwidth of a signal due to doppler shifting of multipath energy arriving at a mobile receiver from differing directions.

Doppler-free spectroscopy   a spectroscopic technique in which two or more light fields are used to produce a spectral line that is not Doppler shifted or Doppler broadened. For example, in the case of lambda systems or vee systems, this requires that the electromagnetic fields co-propagate, whereas for cascade systems, the fields must counterpropagate.

DOS   *See* disk operating system.

dose to clear   the amount of exposure energy required to just clear the resist in a large clear area for a given process.

dosimeter   an instrument used for measuring or evaluating the absorbed dose of radiation. It may depend on the measurement of ionization for its operation or may simply involve the darkening of a piece of photographic film ("film badge").

dot pitch   the center-to-center distance between adjacent green phosphor dots of the red, green, blue triad in a color display.

dot-matrix printer   a printer that produces readable characters by imprinting a large number of very small dots.

double bridge   a Wheatstone bridge modification designed to increase the precision of measurements for low-value resistors. To avoid the error due to resistance of the connection (called yoke) between the unknown resistor and the standard resistor, the ends of the unknown resistor and the standard resistor are connected to the balance detector (usually a galvanometer for this type of bridge) via two small resistors, the ratio of which is the same as the ratio of resistors in the ratio arms. Then the yoke resistance is eliminated from the balance condition and the unknown resistor can be found using the same formula as in an ordinary Wheatstone bridge.

dots per inch (DPI)   a measure of the density of line-printer plots in dots per inch.

double buffering   used in terminal-to-computer communication where a number of remote terminals are connected to a single computer through a "multiplexer." This unit connects n low-speed bit-serial transmission lines onto a single high-speed bit-serial line (running n times faster) using STDM (synchronous time division multiplex) of the connection. In the multiplexer, each low-speed line is connected to a "one-character buffer," converting the received low-speed bitstream to a high-speed bitstream using two shift-registers (buffers). The first one receives the low-speed line character bits (8 bits), clocked by the low-speed "receive clock." When a complete character has been received, it is moved to the second shift-register, where it is stored until it can be shifted out on the high-speed line on the appropriate time slot. Meanwhile, the next character is assembled in the first shift-register. For full-duplex operation (simultaneous two-way communication), a similar structure is needed for the opposite (computer-to-terminal) connection.

double conversion   the process where an incoming RF signal is mixed with a local oscillator (LO) signal to produce the first intermediate frequency (IF). This IF is then mixed with a second fixed LO to produce a second IF signal

double conversion receiver   a receiver that uses two heterodyne operations before detection generating two intermediate frequencies, first intermediate frequency (IF) and second IF, respectively. *See also* intermediate frequency.

double heterostructure dye laser   laser in which the amplification takes place in a dye; a broadly tunable laser.

double lambda system   a quantum mechanical system composed of two interacting lambda systems that share common lower states. A double lambda system is also a double vee system. Double lambda systems can be used to construct resonant closed loop interactions in which some linear combination of the phases of the four fields is constant in time.

double line to ground fault   *See* double phase to ground fault.

double phase ground fault   a fault with two transmission lines being connected to the ground.

double revolving-field theory   *See* counter rotating-field theory.

double sideband   a modulation scheme resulting in a spectrum consisting of the carrier frequency, one signal that is the sum of the carrier and the modulating signal, and one signal that is the difference between the carrier and the modulating signal.

double tuned   a circuit, amplifier, or other device having a response that is the same as two single-tuned circuits.

double word   data block that contains twice the number of bits as the machine word size in a microprocessor.

double-cage rotor   *See* dual-cage rotor.

double-frequency recording   *See* magnetic recording code.

double-line contingency   a malfunction of a power system which involves the simultaneous failure of two transmission lines.

double-sided assembly   a packaging and interconnecting structure with components mounted on both the primary and secondary sides.

double-sided disk   a disk in which both sides of a platter are covered with magnetic material and used for storing information.

doublet   a system of two quadrupole magnets in close proximity, and with opposite polarity, used to simultaneously constrain the beam size in two dimensions at some point downstream. Doublet (Quadrupole) is a beam optical system consisting of two quadrupoles of opposite sign, which provides net particle focusing in all planes.

doubly-fed induction motor   an induction motor with a wound-rotor. The rotor and stator windings are connected to separate sources of electric energy. The machine can be used as a generator to provide frequency conversion or precise speed control in the motor-mode.

downconductor   the cable which connects an air terminal or lightning rod to ground. *See* air terminal.

download   to bring data from a remote source to local storage.

down-sampling   decimation.

downstream   in power distribution work, the direction in which power flows, i.e., towards the load.

DPA   *See* destructive physical analysis.

DPCM *See* differential pulse-code modulation.

DPCM encoding *See* differential pulse-code modulation.

DPI dots per inch.

DPLL *See* dual phase-locked loop.

DQE *See* detective quantum efficiency.

DRA *See* dielectric resonator antenna.

drain terminal of an FET (usually identical in structure to the source) to which electrons flow. Electrons in the FET channel flow down the drain, and current flow is defined as the negative direction of electron movement, since electrons are negative. In p-channel FETs, current flows from source to drain. In n-channel FETs, current flows from drain to source. The drain is usually considered to be the metal contact at the surface of the die.

drain conductance the increase in drain current when the magnitude of the applied FET drain-to-source voltage is increased. Mathematically, the derivative of drain current with respect to drain voltage.

drain saturation current the drain-to-source current flow through the JFET under the conditions that $V_{GS} = 0$ and $|V_{DS}| > |V_P|$ such that the JFET is operating in the active or saturated region.

drain-source leakage the current flowing in the channel of a MOSFET when its gate and source are shorted together. The magnitude of the leakage current is strongly influenced by the applied drain-source voltage, the gate length and substrate doping concentration.

drain-to-source voltage (VDS) potential difference between the FET drain and source

terminals, this voltage determines the device operational region and limits the output power. For an n-channel device this voltage is normally positive, and negative for a p-channel device. Magnitudes usually range up to as high as 10 V for a low noise device and much higher for power devices.

DRAM *See* dynamic random access memory.

drift (1) movement of free carriers in a semiconductor due to the electric field.
(2) the relatively unimpaired fluctuation of adaptive filter coefficients in the direction of least sensitivity along the eigenvector corresponding to the minimum eigenvalue is known as coefficient drift. Coefficient drift can be a problem if this eigenvalue gets very small, as coefficients can drift out of the allowed region. Also important as random, often temperature-induced, fluctuations in the output levels of DC amplifiers.

drift chamber a series chambers used to detect particle trajectories. They are similar to multi-wire proportional chambers, except the wire spacing is increased. The correlation between the position of an ionized track produced by a charged particle and the time of appearance of an electric pulse at the wire is used to measure the distance of the trajectory from the wires.

drift space space where electrons move due only to their inertia.

drip-proof machine a machine with ventilating openings constructed in such away that drops of liquid or solid particles falling on it, at an angle less than 15 degrees from the vertical, can enter the machine neither directly nor by striking on it, run along a horizontal or inwardly inclined surface.

drive circuit a circuit that produces gate trigger pulses, of desired level and timing, to turn on and off active switches (just turn on for natural-commutated switches) in switching circuits.

driving-point admittance   the admittance measured at the antenna terminals when the antenna is in free space (not loaded).

DRO   *See* dielectric resonator [stabled] oscillator.

dropout   equipment misoperation due to an interruption, noise, or sag.

dropout compensation   signal compensation provided when a loss of information from magnetic tape defects occurs during VCR tape playback. Signal dropouts result from oxide or dust interfering with the tape-head to tape contact when the VCR is in the record or the playback operational mode. A horizontal delay line stores the luminance signal of the previous line that is used to replace the lost video information. The signal dropout is detected as a loss of the tape playback. FM signal activates an electronic switch. The switch substitutes the stored delay line contents into the video path to compensate for the lost video from the tape playback.

dropout current   the current at which a magnetically-operated device will revert to its deenergized position.

dropout voltage   the voltage level where proper equipment operation is hindered.

Drude material   a frequency-dependent dielectric whose complex permittivity is described by an equation with two poles, one of them at $w = 0$. A collisional plasma is an example of such a material duration–bandwidth reciprocity relation.

Drude media   *See* Drude material.

drum memory   an old form of backing memory. Similar to magnetic disks in operation, but here the magnetic film is deposited on the surface of a drum, instead of a disk.

dry-type transformer   a transformer that is cooled by circulating air or gas through or around the transformer housing.

DSB   *See* double sideband modulation.

DS-CDMA   direct sequence code division multiple access. *See also* direct sequence and code division multiple access.

DSL   *See* digital subscriber line.

DSP   *See* digital signal processor.

DTE   *See* data terminal equipment.

DTW   *See* dynamic time warping.

dual control   an adaptive control that performs two different functions at the same time: one that applies probing signals to learn more about the system dynamics and the other that tries to keep the output at a desired value.

dual functions   mathematical relationship between two sets used to represent a signal. If a signal $s(t)$ can be represented by a complete set $\{\pi_n(t)\}$, i.e., $s(t) = \sum_n a_n \pi(t)_n$, there must exist a dual set $\{\hat{\pi}_n(t)\}$ such that the expansion coefficients $a_n = <s(t), \hat{\pi}_n(t)>$. $\pi_n(t)$ and $\hat{\pi}_n(t)$ are called dual functions. As a special case, if the set $\{\pi_n(t)\}$ is orthonormal, then it is its own dual.

dual in-line package (DIP)   a standard case for packaging integrated circuits. The package terminates in two straight, parallel rows of pins or lead wires. This standard has been more recently substituted by surface-mount standards. *Compare with* single in-line package.

dual mode filters   filters realized by using two resonances inside the same cavity, hence allowing saving of space, volume, and weight.

dual phase-locked loop (DPLL)   programmable, low-jitter, low-power, and high-performance devices. DPLLs are capable of synthesizing two low-jitter clocks with user-selected, industry-standard frequencies, phased-locked to the system reference timing. They accept a wide range of popular telecom and networking input frequencies and can be programmed to generate a range of output frequencies. Used for wide area network (WAN) and ISDN applications.

dual port memory   memory system that has two access paths; one path is usually used by the CPU and the other by I/O devices.

dual-cage rotor   a three-phase induction motor rotor with two separate squirrel cage windings, that give the effect of varying rotor resistance. The outer cage has high resistance to obtain high starting torque, while the inner cage has low resistance to reduce losses at full load.

dual-element fuse   a fuse constructed with two different types of fusible elements in series. One element is designed to melt very quickly in the presence of fault current. The other is designed to melt after a time delay when exposed to overload conditions. The fusible elements are somewhat similar in operation to the thermal and magnetic elements of an inverse-time circuit breaker. Dual-element and time-delay are often used interchangeably.

duality property of Fourier transform   the property that results from the symmetry between the Fourier transform synthesis and analysis equations. To illustrate, let $f(w)$ be the Fourier transform of the time function $g(t)$ $(f(w) = F\{g(t)\})$. Substituting t for w in f gives the time function $f(t)$. Applying a Fourier transform to $f(t)$ gives the function $2\pi g(-w)$, a frequency domain function similar to the original time function $g(t)$ $(g(-w) = \frac{1}{2\pi} F\{f(t)\})$.

DUF   *See* diffusion under field.

dump   an area of steel and dense concrete into which unwanted particle beam can be steered so that its energy can be dissipated in a safe and controlled manner. The dump resistor is switched into the magnet/capacitor circuit in order to dissipate the stored energy in the magnets/capacitors.

duoplasmatron   a type of ion-producing source that develops protons by extracting positive ions from an arc struck in hydrogen gas.

duplex   a method of winding the armature of a commutated electric machine such that the number of parallel electrical paths between brushes is double that provided by a simplex winding. Duplex windings are constructed by placing consecutive coils in alternate coil slots and continuing the winding twice around the rotor, filling the empty slots on the second pass. The result is two complete, identical windings between brush positions rather than the one winding that is produced when coils are placed in adjacent slots. *See also* simplex, multiplex, reentrancy.

duplex channel   two-way simultaneous (and independent) data communication, e.g., between a computer and remote terminals.

duplex ultrasound   simultaneous display of speed versus time for a chosen region and the two-dimensional B-mode image.

duplication with complementary logic   fault detection based on circuit duplication and comparison. One module is designed using positive logic and the other module uses negative logic. This assures detecting common mode faults.

dust cover   a cover to protect the terminals of a pad-mount transformer.

dust-ignition-proof machine   a machine designed with a casing or specialized enclosure to safely contain any internal ignition or flammable substances or components, and prevent them from igniting external flammables such as explosive gases, vapors, and dust particles.

DUT   device under test

duty cycle   (1) the ratio of the turn-on time of a semiconductor switch to the sum of the turn-on and turn-off times;

$$\text{Turn-on}/(\text{Turn-on} + \text{Turn-off})$$

(2) the mode of operation that an electric machine is classified, in consideration of thermal limits, e.g., continuous operation, intermittent operation.

Also known as duty ratio.

duty ratio   *See* duty cycle.

DVI   *See* digital video interactive.

DVM   See digital volt meter.

DX-center   stands for defect unknown(X)/complex. Name originally given to defects that caused persistent photoconductivity (PPC) in GaAs. The term is now commonly applied to similar defects in both III-V and II-VI compound semiconductors. The presently accepted interpretation of the microscopic nature of the DX-center is a large deformation associated with a negatively charged substitutional impurity.

dyadic decomposition   wavelet decomposition in which the dilation is a dyadic number, i.e., 2^j with j being the decomposition level; this corresponds to a octave frequency band division in the frequency domain.

dyadic Green's functions   Green's functions that relate together vector field quantities.

dyed resist   a photoresist with an added non-photosensitive chemical that absorbs light at the exposing wavelength.

dynamic allocation   allocation of memory space that is determined during program execution. Dynamic allocation can be used to designate space on the stack to store objects whose lifetimes match the execution interval of a subroutine; this allocation is performed upon entry to the subroutine.

dynamic beam control   a circuit in a camera designed to instantaneously increase beam current to handle highlights in a scene.

dynamic brake   the braking operation of a machine by extracting electrical energy and then dissipating it in a resistor.

dynamic braking   since AC motors do not have separate field excitation, dynamic braking is accomplished by continuing to excite the motor from the drive. This causes a generative current to the drives DC intermediate bus circuit. The dynamic brake resistors are then placed across the DC bus to dissipate the power returned. The brake resistor is usually switched by a transistor or other power switch controlled by the drive.

dynamic channel assignment (DCA)   a technique of assigning or reassigning radio channels in a communications system in order to respond to current interference or propagation conditions. Such assignment or reassignment may be made before or during radio transmission. DCA is typically implemented in a distributed fashion, with mobile terminals and/or fixed cell sites taking measurements to determine local interference levels, and then applying an algorithm to choose the best radio channel on which to initiate or reassign communication.

dynamic convergence   the process of making the electron beam fall on a specified surface during scanning, as in the case of a CRT. This process is necessary for all three-gun CRTs that do not have in-line video displays or CRTs.

dynamic linking   deferring the determination of the association between a symbol used in a program module and the object to which it refers until that object must be accessed (i.e., during program execution).

dynamic load line   graphical plot showing the instantaneous relationship between voltage across and current through a transistor when driven by an input AC signal.

dynamic matching   *See* variational similarity.

dynamic memory allocation   the run-time assignment of small units of memory to an active program. Used typically to support growing structures such as lists.

dynamic path reconnect   used in IBM's high-end computer systems to allow a "subchannel" to change its channel path each time it cycles through a disconnect/reconnect with a given device. This enables it to be assigned to another available path, rather than just wait for the currently allocated path to become free.

dynamic programming   introduced by Bellman, one of the best known methods for solving the optimal control problems. A recursive method to compute the optimal control as a function of the state, dynamic programming is used in multistage systems by working backward from the final stage.

Dynamic programming is based on The Optimality Principle. The Principle says that optimal control strategy has the feature that regardless of initial state and initial decision, decision in the next step must form an optimal control strategy with respect to the final state of the previous decisions. This principle allows us to find an optimal strategy in a numerical way. The principle serves to limit the number of potentially optimal control strategies that must be investigated. It also implies that optimal control strategies must be determined by working backward from the final stage.

dynamic random access memory (DRAM) a semiconductor memory using one capacitor and one access transistor per cell (bit). The information is stored dynamically on a small charge on the cell capacitance, and can be read or written through the "access transistor" in the cell. Since the charge will slowly leak away (through semiconductor junctions), the cells need to be "refreshed" once every few milliseconds. This is typically done using on-chip circuitry. DRAMs have very high storage density, but are slower than SRAMs (static RAMs). *See also* burst refresh, distributed refresh.

dynamic range   refers to the range of input signal amplitudes over which an electronic device will operate within a set of specified parameters. Usually expressed in decibels. In a communications receiver, the upper end of the dynamic range is determined by the largest tolerable input signal, while the lower end is set by the receiver's sensitivity. The sensitivity is the minimum discernible signal for a specific signal-to-noise ratio (SNR). *See also* signal-to-noise ratio.

dynamic reconfiguration   changes of optical paths from sources to detectors which are instantly controllable. Paths of optical signals are controlled and changed by an optical crossbar switch that is usually a spatial light modulator.

dynamic scattering   procedure to study the change of state of atoms and molecules by analyzing the frequency shift and fluctuations of scattered light.

dynamic scheduling   (1) creating the execution schedule of instructions at run-time by the hardware, which provides a different schedule than strict program order (i.e., a processor issues instructions to functional units out of program order). The processor can dynamically issue an instruction as soon as all its operands are available and the required execution unit is not busy. Thus, an instruction is not delayed by a stalled previous instruction unless it needs the results of that previous instruction.

(2) changing the software program schedule dynamically depending on data or operating conditions.

(3) automatic adjustment of the multiprocessing program at run time that reflects the actual number of CPUs available presently. For instance, a DO loop with 100 iterations is automatically scheduled as 2 blocks with 50 iterations on a two-processor system, as 10 blocks with 10 iterations on a ten-processor system, and as one block on a single-processor machine. This enables one to run multiprocessor programs on single-processor computers.

dynamic simulation   *See* direct dynamics.

dynamic stability   a measure of a power system to return to a pre-disturbance steady-state condition following a disturbance.

dynamic system   *See* static system.

dynamic time division multiple access (D-TDMA)   time division multiple access scheme in which the channels are assigned dynamically. *See also* time division multiple access.

dynamic time warping (DTW)   a recognition technique based on nonlinear time alignment of unknown utterances with reference templates.

dynamic time warping   in problems of temporal pattern recognition, each exemplar can be regarded as a sequence of vectors. The process of pattern matching requires carrying out an optimal alignment of the vectors composing the sequences so as to minimize a proper distance. For example, in automatic speech recognition, the problem of isolated word recognition requires producing an optimal alignment between the incoming word to be classified and a reference template. Let

$$A = \begin{bmatrix} a_1, \dots, a_M \end{bmatrix}$$
$$B = \begin{bmatrix} b_1, \dots, b_N \end{bmatrix}$$

be two sequences of vectors ($a_i, b_i \in \mathcal{R}^p$) that must be aligned optimally. Formally, determining the optimal alignment consists of finding a warping function

$$C = c_1, \dots, c_K$$

where $c(k) = [i_k, j_k]$, such that the distance

$$d(A, B) = \sum_{k=1}^{K} (a_{i(k)} - b_{j(k)})^2$$

is minimum. The optimization must take place under the following conditions:

(1) *monotonic condition*

$$i(k) \geq i(k-1) \quad \text{and} \quad j(k) \geq j(k-1)$$

(2) *Boundary conditions*

$$i(1) = k(1) = 1$$
$$i(K) = M$$
$$j(K) = N$$

(3) *non-skip condition* $i(k) - i(k-1) \leq 1$ and $j(k) - j(k-1) \leq 1$

(4) *efficiency condition*

$$|i(k) - j(k)| < Q$$

The solution of this problem can be obtained by Belmann's dynamic programming. The algorithm that produces the optimal template alignment is referred to as dynamical time warping (DTW).

dynamical linear nonstationary continuous-time finite-dimensional system   a system described by the linear ordinary differential state-equation

$$x'(t) = A(t)x(t) + B(t)u(t)$$

and the linear algebraic output equation

$$y(t) = C(t)x(t) + D(t)u(t)$$

where

$$x(t) \in R^n$$

is the state vector,

$$u(t) \in R^m$$

is the input vector,

$$y(t) \in R^q$$

is the output vector.

$$u \in L^2_{loc}([t_0, \infty), R^m)$$

is an admissible control, $A(t)$ is $n \times n$ dimensional matrix, with piecewise-continuous elements, $B(t)$ is $n \times m$ dimensional matrix, with piecewise-continuous elements, $C(t)$ is $q \times n$ dimensional matrix with piecewise-continuous elements, $D(t)$ is $q \times m$ dimensional matrix with piecewise-continuous elements. The solution of the state equation has the form

$$x(t, x(t_0), u) = F(t, t_0)x(t_0)$$
$$+ \int_{t_0}^{t} F(t, s)B(s)u(s)ds$$

where $F(t, s)$ is $n \times n$ dimensional transition matrix for a dynamical system.

dynamical linear nonstationary discrete-time finite-dimensional system   a system described by the linear difference state equation

$$x(k + 1) = A(k)x(k) + B(k)u(k) \quad (1)$$

and the linear algebraic output equation

$$y(k) = C(k)x(k) + D(k)u(k)$$

where $x(k) \in R^n$ is the state vector, $u(k) \in R^m$ is a control vector, $y(k) \in R^q$ is an output vector, and $A(k)$, $B(k)$, $C(k)$, and $D(k)$ are matrices of appropriate dimensions with variable coefficients.

Solution of the difference state equation (1) has the form

$$x(k, x(k_0), u) = F(k, k_0)x(k_0)$$
$$+ \sum_{j=k_0}^{j=k-1} F(k, j + 1)B(j)u(j)$$

where $F(k, j)$ is $n \times n$ dimensional transition matrix defined for all

$$k \geq j$$

in the following manner:

$$F(k, k) = I_{nxn} \quad \text{for} \quad k \in Z$$
$$F(k, j) = F(k, j + 1)A(j)$$
$$= A(k - 1)A(k - 2) \ldots A(j + 1)A(j)$$

for $k > j$.

dynamical linear stationary continuous-time finite-dimensional system   a system described by the linear differential state-equation

$$x'(t) = Ax(t) + Bu(t) \quad (1)$$

and the linear algebraic output equation

$$y(t) = Cx(t) + Du(t)$$

where

$$x(t) \in R^n$$

is the state vector,

$$u(t) \in R^m$$

is the input vector,

$$y(t) \in R^q$$

is the output vector, A, B, C, and D are constant matrices of appropriate dimensions. The transition matrix of (1) has the form $F(t, s) = e^{A(t-s)}$.

dynamical linear stationary discrete-time finite-dimensional system   a system described by the linear difference state equation

$$x(k + 1) = Ax(k) + Bu(k) \quad (1)$$

and the linear algebraic output equation

$$y(k) = Cx(k) + Du(k)$$

where $x(k) \in R^n$ is a state vector, $u(k) \in R^m$ is a control vector, $y(k) \in R^q$ is an output vector, and A, B, C, and D are constant matrices of appropriate dimensions. The transition matrix of (1) has the form $F(k, j) = A^{k-j}$.

dynamical systems with delays   a system described by the linear state equation

$$x'(t) = A_0 x(t) + A_1 x(t - h) + Bu(t) \quad (1)$$

where $x(t)$ is n-dimensional vector, $u(t)$ is m-dimensional control vector, and A_0, A_1, and B are constant matrices of appropriate dimensions and $h > 0$ is a constant delay. For a given admissible control and initial data, the above differential equation (1) with deviating argument has a unique solution derived by the method of steps.

The state space for dynamical system (1), $X = W_1^{(2)}([-h, 0], R^n)$, is infinite-dimensional Sobolev space of absolutely continuous functions defined on $[-h, 0]$ with values in R^n and with square integrable derivatives.

Linear unbounded operator connected with the dynamical system (1) generates the solution in the state space X and has infinite number of eigenvalues each of finite multiplicity. The corresponding eigenfunctions may form the basis in the infinite-dimensional state space X.

It should be stressed that it is possible to consider other types of linear dynamical systems with delays, namely systems with multiple delays, systems with delays in the control or neutral dynamical systems with delayed derivative.

dynamo   a term used to describe any of a variety of rotating machines that convert mechanical to electrical energy, or less commonly, electrical to mechanical energy. Dynamos typically consist of a stationary structure, called the stator, supporting a rotating element called the rotor. Energy conversion occurs via Faraday induction. A field winding (or in some smaller machines, permanent magnets) is mounted on one of the mechanical structures and produces a magnetic flux. An armature winding is mounted on the other structure, and rotation of the rotor produces relative motion between the field flux and the coils of an armature winding, inducing a Faraday voltage in the armature coil. This Faraday induced voltage is the source of electrical energy at the dynamo output.

dynamometer   a rotating device used to measure the steady-state torque and power output of rotating machines. Dynamometers generally provide precise control of the load torque applied to a test machine, and power output is determined through precise speed measurements.

E

E modes the wave solutions with the zero magnetic field component along the direction of propagation. Also known as transverse magnetic (TM) modes.

E plane in measuring an antenna's radiation pattern, the plane that contains the current in the element and therefore the electric field intensity vector field. This plane is perpendicular to the H plane cut.

E-beam excitation *See* electron beam excitation.

E-plane sectoral horn a horn antenna where the aperture is formed by flaring the walls in the direction of the E plane. The H-plane dimension is left unchanged from that of the waveguide feed.

early stopping a technique applied to network training that is aimed at assuring good generalization performance. Training on a finite set of data eventually leads to overfitting, and this can be avoided by periodically assessing generalization performance on a set of test data. As training proceeds, network performance on both the training set and the test set gradually improves, but eventually performance on the test set begins to deteriorate, indicating that training should be stopped. The set of network weights that give the best performance on the test data should be employed in the trained network.

early vision the set of (mainly perceptual) processes occurring at an early stage of the vision process, typically at the retinal level.

EAROM electrically alterable programmable read-only memory. *See also* electronically erasable programmable read-only memory.

earth electrode system a network of electrically interconnected rods, plates, mats, or grids installed for the purpose of establishing a low-resistance contact with earth. The design objective for resistance to earth of this subsystem should not exceed 10 Ω.

Earth station the interface point for communications to and from a satellite. An earth station (also known as a hub) consists of an antenna and transmit and receive subsystems.

earth wire an overhead wire which is maintained at ground potential for purposes of lightning shielding and system grounding.

earthing *See* grounding.

EBCDIC *See* extended binary-coded-decimal interchange code.

ebullient heat transfer the heat transfer process associated with the formation and release of vapor bubbles on a heated surface.

ECC *See* error-correcting code.

ECG *See* electrocardiography.

echo cancellation technique that removes the unwanted effects that result from impedance mismatch. Echo cancellation frequently makes use of adaptive filtering algorithms which can be realized by digital filters.

echo canceller in telephony, a filter that removes echoes caused by leakage of the transmitted signal through a hybrid. The hybrid is the interface between two wire (local) and four wire (long-distance) facilities, and separates inbound and outbound signals. Echo cancellers are used in analog telephony to cancel echoes from signals that traverse the entire round-trip connection, and in full-duplex voiceband modems, where the echo is due to leakage through the near-end hybrid.

echo width in a 2-D scattering problem, the width needed to capture the exact amount of incident power, that when the scattering body is replaced by a cylinder, and the cylinder radiates the captured power, the amount of power received at a specified point is the same as that received if the scattering body is not replaced. Echo width is the 2-D analog of radar cross section in 3-D. The units of echo width are meters.

Eckert, John Presper (1919–1995) Born: Bryn Mawr, Pennsylvania

Best known as one of the designers of ENIAC (Electronic Numerical Integrator And Calculator), an early computer. Like many computer pioneers during World War II, Eckert was looking for more efficient ways of calculating trajectory tables for artillery and ranging systems for radar. Eckert graduated from the University of Pennsylvania and remained there to work with John Mauchly. Eckert and Mauchly later formed a company and continued to develop and refine their machine. Eckert eventually sold the company to what would become the Sperry Rand Corporation. Here he produced UNIVAC I (Universal Automatic Computer), one of the first commercially successful computers. The chief improvement of this machine over its predecessors was in its use of a stored memory.

ECL *See* emitter-coupled logic.

ecliptic plane of earth's orbit around the sun.

economic dispatch a generation scheme in which units are utilized such that the greatest profit is generated for the utility.

economic interchange an arrangement between interconnected electrical power systems whereby a system can meet its load demand by buying power from one or more of the other systems in the interconnection group.

ectopic beat a heart beat that originates from other than the normal site.

eddy current a circulating current in magnetic materials that is produced as a result of time-varying flux passing through a metallic magnetic material.

eddy current brake a braking device in which energy is dissipated as heat by generating eddy currents.

eddy current drive a magnetic drive coupled by eddy currents induced in an electrically conducting member by a rotating permanent magnet, resulting in a torque that is linearly proportional to the slip speed.

eddy current loss the energy wasted in sustaining undesirable eddy currents in an electrical conductor.

EDFA *See* erbium doped fiber amplifier.

edge a substantial change, over a small distance, in the values of an image's pixels — typically in the gray level values. Edges can be curved or straight and are important because they are often the boundaries between objects in an image.

edge condition an electric/magnetic field perpendicular to a dielectric/metallic edge shows a singular behavior referred to as the edge condition.

edge coupled microstrip lines the microstrip lines that have the same ground plane and are on the same dielectric substrate and are parallel to each other. The coupling is mainly due to the fringing fields at the edges of the lines.

edge detection the location of edges in an image by computer. *See* Canny operator, gradient edge detector, gradient filter, Marr-Hildreth operator, Sobel operator, straight edge detection.

edge elements   the basis functions that are associated with the edges of the discretizing elements (such as triangles, tetrahedrals, etc.) in a numerical method such as finite element method.

edge enhancement   a type of image processing operation where edges are enhanced in contrast, such as by passing only the high spatial frequencies in an image.

edge guide   two conductor transmission lines in which one of the conductors is a thin sheet of width substantially larger than the gap to the second conductor. The guided wave is largest at the edge or boundary of the thin sheet conductor, standing across to the second conductor, and propagating in the direction of the thin sheet conductor edge or boundary.

edge-sensitive   pertaining to a bistable device that uses the edge of a positive or negative pulse applied to the control input, to latch, capture, or store the value indicated by the data inputs.

edge-triggered   *See* edge-sensitive.

Edison, Thomas Alva   (1847–1931) Born: Milan, Ohio

Best known as the holder of 1069 patents secured during his lifetime. Among these were patents for the phonograph and the incandescent filament lamp. Edison was largely self-taught. His early interest in communication devices stemmed from his employment as a telegraph operator. He used the profits from the sale of his first invention, a "stock ticker," to set up a lab in Newark, New Jersey. Always a shrewd commercial developer, he followed the invention of the light bulb with work on developing efficient generators to power these bulbs. Edison is considered the archetypal American inventive genius.

Edison Electric Institute   a trade group of investor-owned public electric utilities in the USA.

edit distance   the edit distance between two strings A and B is defined as

$$D_{edit} = \min\{a + b + c\}$$

where B is obtained from A with a replacements, b insertions, and c deletions. There is an infinite number of combinations $\{a, b, c\}$ to achieve this. One of the ways to find the minimum from these is dynamic programming. Edit distance is also called Levenshtein distance.

EEG   *See* electroencephalography.

EEPROM   *See* electrically erasable programmable read-only memory.

effective address   (1) the computed address of a memory operation.

(2) the final actual address used in a program. It is usually 32 or 64 bits wide.

The effective address is created from the relative address within a segment (that is, relative to the base of a segment), which has had applied all address modification specified in the instruction word. Depending on the configuration of the memory management unit, the effective address may be different from the real address used by a program, i.e., the address in RAM, ROM, or I/O space where the operation occurs.

(3) when a memory location is referenced by a machine instruction, the actual memory address specified by the addressing mode is called the effective address. For example, if an instruction uses an indirect mode of addressing, the effective address is to be found in the register or memory location whose address appears in the instruction.

effective aperture   a figure of merit for an antenna giving the equivalent aperture size required to intercept from a field of uniform power density an amount of power equal to that received by the antenna placed in the same field. At a given wavelength, the effective aperture is directly proportional to the gain.

effective dielectric constant   (1) simple and single dielectric constant used to describe a complicated configuration of media with a variety of dielectric constants in an equivalent model.

(2) the resulting computational effects of having two dielectric materials in a microstrip transmission line.

effective isotropic radiated power   the product of the total radiated power by the directive gain of the antenna.

effective isotropically radiated power (EIRP) in antenna theory, the amount of power needed by an isotropic radiator to produce the same radiation intensity at a receiver as the original antenna in the main beam direction. EIRP, which is expressed in decibel-meters or decibel-watts, can be calculated by multiplying power supplied to an antenna by its directive gain in the desired direction. *See also* effective radiated power.

effective length   the ratio of the voltage induced across an antenna terminating impedance divided by the incident field strength.

effective mass   an approach whereby a particular response is described using classical equations by defining an effective mass whose value differs from the actual mass. For example, an electron in a lattice responds differently to applied fields than would a free electron or a classical particle.

effective permittivity   a simple and single permeability constant used to describe a complicated configuration of media with a variety of permeability constants in an equivalent model.

effective radiated power (ERP)   the effective power output from an antenna in a specified direction, including transmitter output power, transmission line loss, and antenna power gain.

efficiency   (1) the ratio of the input power to the output power. It is a figure of merit for the energy cost effectiveness of a device.

(2) in antennas, the ratio of the power radiated to the input power. This term is sometimes defined with the mismatch loss $(1 - \gamma^2)$ included in the total efficiency of the antenna; other times, it is omitted from the calculation.

efficient estimator   an unbiased estimator which achieves the Cramer-Rao bound. *See* Cramer-Rao bound.

EFIE   *See* electric field integral equation.

EGA   *See* enhanced graphics adapter.

EIA   Electronics Industry Association.

eigenfunction   the name given to an eigenvector when the eigenvectors arise as solutions of particular types of integral equation. *See* eigensystem, eigenvector.

eigenfunction expansion   a method used to expand a given field in terms of eigenfunctions. It is particularly used in modal analysis of waveguide discontinuities.

eigenstate   a linear combination of quantum mechanical basis states that is constant in time. A quantum mechanical system starting in an eigenstate will remain unchanged in time except for an overall phase. The phase varies as the product of the eigenvalue and time. Quantum mechanical eigenstates are analogous to normal modes of coupled oscillator systems in classical mechanics.

eigensystem   a system where the output of a system is the input function multiplied by a constant. *See* eigenfunction, eigenvalue, eigenvector.

eigenvalue   the multiplicative scalar associated with an eigenfunction or an eigenvector. For example, if $Ax = \lambda x$, then λ is the eigenvalue associated with eigenvector x. If A is a covariance matrix, then λ represents the variance of one of the principal components of A. *See* eigensystem, eigenvector, principal component.

eigenvector for a linear system A, any vector x whose direction is unchanged when operated upon by A. *See* eigensystem, eigenvalue.

eigenvalue assignment a technique that, given a set of desired eigenvalues, has the objective to compute a constant state feedback gain K such that the closed-loop state equation has precisely these eigenvalues.

eight connected *See* pixel adjacency.

eighteen connected *See* voxel adjacency.

Einstein coefficients coefficients introduced by Einstein to represent the spontaneous decay rate, A, and the stimulated emission or absorption coefficient, B, in the presence of an electromagnetic energy density or intensity.

EIRP *See* effective isotropically radiated power.

EISA *See* extended industry standard architecture.

elastic scattering (1) a scattering process in which energy is conserved.

(2) the state in which there is no energy transfer between the light wave and the scattering medium, and hence no frequency change.

elasto-optic effect the change in refractive index of a medium due to mechanical forces. Also known as the photoelastic effect.

elastomer a polymer material that can undergo reversible shape changes, expanding or contracting under influence of an applied electric field; often used for mechanical light modulation devices.

elbow in URD work, the termination of a buried cable where it attaches to the distribution transformer.

electret a material similar to ferroelectrics, but charges are macroscopically separated and thus are not structural. In some cases the net charge in the electrets is not zero, for instance when an implantation process was used to embed the charge.

electric charge a basic physical quantity that is a source of electromagnetic fields. The units of electric charge are coulombs.

electric charge density the fundamental, macroscopic source of the electromagnetic field that quantifies the average number of discrete electric charges per unit volume. SI units are coulombs per cubic meter.

electric current density a source vector in electromagnetics that quantifies the amount of electric charge crossing some cross-sectional area per unit time. The direction of the electric current density is in the direction of electric charge motion. SI units are amperes per square meter.

electric field in a region of space, if a test charge q experiences a force F then the region is said to be characterized by an electric field of intensity E given by

$$E = \frac{F}{q}$$

electric field integral equation (EFIE) the integral equation based on the boundary condition of the total electric field. It is often used in moment method description of microstrip circuits and metallic antennas. *See also* moment method.

electric field intensity a force field that is a measure of the magnitude and direction of the force imparted upon a discrete charge normalized to the discrete charge's value. Depends on material characteristics. The units are volts per meter.

electric flux density basic electromagnetic field quantity used to describe the effects of permeable matter to the electric field; it is expressed in SI units of coulombs per square meter.

electric furnace a method of smelting metals and applying high heat for industrial processes which makes use of the heat from an electric arc struck between (typically) carbon electrodes.

electric permittivity tensor relationship between the electric field vector and the electric displacement vector in a medium with no hysteresis; displacement divided by the electric field in scalar media.

electric polarization vector an auxiliary vector in electromagnetics that accounts for the creation of atomic dipoles in a dielectric material due to an applied electric field. Macroscopically, the electric polarization vector is equal to the average number of electric dipole moments per unit volume. Mathematically, $P = D - \epsilon_0 E$, where D is the electric flux density, E is the electric field intensity, and ϵ_0 is the free space permittivity. SI units are coulombs per square meter.

electric susceptibility tensor relationship between the electric field vector and the electric polarization vector in a medium with no hysteresis. It is the polarization divided by the permittivity of free space and the electric field in scalar media.

electric vector potential a vector function that is used to derive solutions for electric and magnetic fields.

electric vehicle (1) a vehicle powered by an electrical energy storage device such as batteries.

(2) (EV) a vehicle (typically a car or truck) enabled by high-efficiency electric motors and controllers and powered by alternative energy sources, such as fuel cells. Electric vehicles are zero-emission vehicles. *See* regenerative braking, hybrid electric vehicle.

electrical breakdown *See* breakdown.

electrical degrees a convenient way of representing the distance around the circumference of a machine with two poles spanning the entire $360°$ of the circumference,

$$\text{Electrical Degree} = \text{Pairs of poles} \times \text{Mechanical Degree}$$

electrical network a collection of interconnected electrical devices.

electrical tree a microscopic cracking pattern which forms in the insulation of electric power cables which are not exposed to water *See* tree, water tree.

electrically alterable read-only memory (EAROM) a PROM device that can be erased electronically. More costly than the EPROM (erasable PROM) device, which must be erased using ultra-violet light.

electrically erasable programmable read-only memory (EEPROM) a nonvolatile semiconductor memory, it is used at the place of PROM or EPROM. They can be programed and erased by electrical means. *See also* electrically programmable read-only memory.

electro-acoustic smart material one of several materials that have self-adaptive characteristics in their acoustic behavior (such as transmission, reflection, and absorption of acoustical energy) in response to an external stimulus applied as a function of the sensed acoustical response.

electro-optic coefficient a parameter that describes the change in refractive index in a medium with application of an electric field. The linear electro-optic coefficient describes media where the index change is linearly related to applied field; the quadratic electro-optic coefficient describes media where the index change is related to the square of the applied field, i.e., the Kerr effect.

electro-optic device   any device that uses the electro-optic effect, such as for optical beam modulation and deflection.

electro-optic effect   an optical phenomenon in certain types of crystals in which the application of an applied electric field results in a change in the dielectric tensor, or equivalently a change in the refractive index of the medium. In the linear electro-optic effect, the resulting change in refractive index is proportional to the electric field strength; in the quadratic electro-optic effect, the resulting change in refractive index is proportional to the square of the electric field strength. The linear electric-optic effect was first studied by F. Pockels in 1893.

electro-optic modulator   a device that uses the electro-optic effect to alter the temporal and spatial character of a light beam.

electro-optic scanner   a device that uses the electro-optic effect to scan a light beam across a range of angles.

electro-optic smart materials   materials in which optical properties are changed self-adaptively with an external electric stimulus proportional to the sensed optical characteristics.

electro-optic switch   a device that uses the electro-optic effect to switch a light beam on or off.

electro-optic wave retarder   a device that will retard the phase of an optical beam by a fixed amount using the electro-optic effect; in particular, the retardation of one polarization component of the light beam using the electro-optic effect in a birefringent material.

electroacoustics   science of interfacing between acoustical waves and corresponding electrical signals. This includes the engineering of transducers (e.g., loudspeakers and microphones), but also parts of the psychology of hearing, following the notion that it is not necessary to present to the ear signal components that cannot be perceived.

electrocardiography (ECG)   the device (electrocardiograph) or the output (electrocardiogram) depicting the body surface recording of the electrical activity of the heart.

electroencephalography (EEG)   recordings of the electrical potentials produced by the brain.

electroluminescence   a general term for optical emission resulting from the passage of electric current.

electrolyte   current-conducting solution between two electrodes or plates of a capacitor, at least one of which is covered by a dielectric.

electrolytic capacitor   a capacitor solution between two electrodes or plates of a capacitor, at least one of which is covered by a dielectric.

electromagnet   a magnet that employs an electric current in a coil to produce a magnetic field.

electromagnetic compatibility (EMC)   the ability of a system or equipment to operate within design tolerances in its intended environment, with adjacent systems and equipment, and with itself, so that the effect of any electromagnetic disturbances produced by the systems or equipment is reduced.

electromagnetic energy   energy contained in electromagnetic fields and associated polarizable and magnetizable media.

electromagnetic environmental effects   encompasses all electromagnetic disciplines, including electromagnetic compatibility (EMC); electromagnetic interference (EMI); electromagnetic vulnerability (EMV); electromagnetic pulse (EMP); radiation hazard (RADHAZ) (hazard of electromagnetic radiation to personnel, ordnance, and fuels (HERP, HERO, HERF)); lightning, p-static; electrostatic discharge (ESD), and emission control (EMCON).

electromagnetic interference (EMI) (1) any electromagnetic disturbance that interrupts, obstructs, or otherwise degrades or limits the effective performance of electronics/electrical equipment. It can be induced intentionally, as in some forms of electronic warfare, or unintentionally, as a result of a spurious emissions and responses, intermodulation products, and the like. Additionally, EMI may be caused by atmospheric phenomena, such as lightning and precipitation static and non-telecommunication equipment, such as vehicles and industry machinery.

(2) unwanted high-frequency electrical signals, also known as radio frequency interference (RFI), which can be generated by power electronic circuits switching at high frequencies. The signals can be transmitted by conduction along cables (450 kHz to 30 MHz) or by radiation (30 MHz to 40 GHz) and can interfere with control or other electronic equipment.

electromagnetic interference filter a filter used to reduce or eliminate the electromagnetic interference (EMI) generated by the harmonic current injected back onto the input power bus by switching circuits. The harmonic current is caused by the switch action that generates switch frequency ripple, voltage and current spikes, and high-frequency ringing. Generally called an EMI filter.

electromagnetic pulse (EMP) a large impulsive-type electromagnetic wave generated by nuclear or chemical explosions.

electromagnetic radiation an electromagnetic wave created by the acceleration or deceleration of charge.

electromagnetic smart materials materials such as shielding materials, radar-absorbing materials (RAMs), and electromagnetic surface materials, in all of which some electromagnetic properties can be adaptively controlled by means of an external stimulus dictated by the sensed electromagnetic response.

electromagnetic spectrum the frequency and wavelength of electromagnetic radiation. We have the following classification reported in the figure, while the microwave frequency band designations is reported in the table.

electromagnetic susceptibility a device's failure to perform appropriately if there is an electromagnetic disturbance.

electromagnetic torque the torque produced in a machine by the interaction of the magnetic fields and/or by the varying reluctance principle where the field attempts to maximize its intensity in a machine during electromechanical energy conversion.

electromagnetic vulnerability (EMV) the inability of a device, equipment, or system to perform without degradation when subjected to electromagnetic environment of a specified power level and frequency range.

electromagnetic wave wave in which the electric and magnetic variables are solutions of the Maxwell–Heaviside equations.

electromagnetic wave propagation the phenomenon of electromagnetic energy propagating in the form of waves of the coupled electric and magnetic field intensity vectors.

electromagnetically induced transparency a technique to render optically dense media transparent by using a long-lived quantum mechanical coherence. Depending on the quantum mechanical system, electromagnetically induced transparency can be viewed as a special case of effects such as

Microwave frequency band designations

| Frequency | Wavelength | IEEE Radar Band Designation | |
		Old	New
1–2 GHz	30–15 cm	L	D
2–3 GHz	15–10 cm	S	E
3–4 GHz	10–7.5 cm	S	F
4–6 GHz	7.5–5 cm	C	G
6–8 GHz	5–3.75 cm	C	H
8–10 GHz	3.75–3 cm	X	I
10–12.4 GHz	3–2.42 cm	X	J
12.4–18 GHz	2.42–1.67 cm	Ku	J
18–20 GHz	1.67–1.5 cm	K	J
20–26.5 GHz	1.5–1.13 cm	K	K
26.5–40 GHz	1.13 cm–7.5 mm	Ka	K
40–300 GHz	7.5–1.0 mm	mm	

Fano interference or coherent population trapping applied to optically dense media.

electromagnetics   the study of the effect of electric charges at rest and in motion.

electromechanical equation   a basic non-linear equation that governs the rotational dynamics of a synchronous machine in stability studies. The equation is given by

$$2H d^2\delta / \omega_s dt^2$$

where H is a constant defined as the ratio of the kinetic energy in megajoules at synchronous speed to the machine rating in MVA, ω_s is the synchronous speed in rads per second, and δ is the load angle expressed in electrical degrees. Various forms of the swing equation serve to determine stability of a machine within a power system. Solution of the swing equation yields the load angle δ as a function of time. Examination of all the swing curves (plot of δ w.r. to time) shows whether the machines will remain in synchronism after a disturbance. *See also* swing equation.

electromechanical relay   a protective relay that uses electrical, magnetic, and mechanical circuits to implement the operating logic.

electron band   a range or band of energies in which there is a continuum (rather than a discrete set as in, for example, the hydrogen atom) of allowed quantum mechanical states partially or fully occupied by electrons. It is the continuous nature of these states that permits them to respond almost classically to an applied electric field.

electron beam excitation   electron impact excitation due to a free electron beam rather than, for example, the conduction electrons in a gas discharge; permits high-pressure excitation without arcing.

electron beam lithography   lithography performed by exposing resist with a focused beam of electrons.

electron beam welding   a welding process that produces coalescence of metals with the heat obtained from a concentrated beam composed primarily of high-velocity electrons impinging on the surfaces to be joined.

electron collision frequency   the average number of collisions per second an electron has with heavy particles in a medium such as plasma.

electron impact excitation   excitation of an atom or molecule resulting from collision by an electron.

electron multiplication   the phenomenon where a high-energy electron strikes a surface and causes additional electrons to be emitted from the surface. Energy from the incident electron transfers to the other electrons to cause this. The result is electron gain proportional to the incident electron energy.

electron oscillator model   simplified classical model for an atomic or molecular medium in which the charges are assumed to be bound together by springs rather than quantum mechanical potentials; provides good qualitative explanation for absorption and dispersion.

electron plasma   a plasma medium in which electrons are the mobile charge carriers and the ions form the stationary compensating positive charge background.

electron wave   the wave described by Bloch function solutions to the problem of an electron in a periodic lattice of ions.

electron-beam lithography   refers to a lithographic (or photographic) process in which the exposure energy is provided by the energy carried by a beam of focused electrons rather than by photons (light).

electronic bottleneck   the factor limiting the speed and capacity of a fiber optic communication network is ultimately the link between photons of light and the electronics required to transmit or receive and process them. In order to exploit the full bandwidth capacity of a single-mode optical fiber an optical network should minimize the number of optical to electrical conversions and instead process the signals in the optical domain.

electronic brake   a power electronic system designed to decelerate a motor. For an induction machine, the brake supplies DC current to the stator winding, producing a stationary magnetic field, negative slip, and braking torque.

electronic motor starter   starter in which solid state devices provide reduced voltage to the motor for starting, thus limiting the starting current.

electronic nonlinear response   the nonlinear optical response resulting from the motion of bound electrons. It is characterized by moderately large response and very short (several fs) response times.

electronic overload device   an overload device that employs an electronic circuit to sense motor voltage and current for the purpose of providing precise motor overload protection. *See also* overload heater, overload relay.

electronic switch   an electronic circuit that controls analog signals with digitary (binary) signals.

electronic transition   alteration in the electronic structure of a material such that one electron temporarily changes its energy level through the absorption or emission of energy.

electronic warfare   contention for the control of the electromagnetic (EM) spectrum, to allow active and passive EM sensing and communications while denying the same ability to adversaries. Includes deceptive EM techniques.

electronically commutated machine   a DC machine with rotor-mounted permanent magnets, and concentrated (square) stator windings. The machine fundamentally behaves like a DC machine with linear speed and torque characteristics, and hence is also called a brushless DC machine. The machine is distinguished from conventional DC machines in its substitution of a six-switch inverter for brushes and copper commutator bars and in its need for the rotor position information in real time.

electronically erasable programmable read-only memory (EEPROM) a term used to denote a programmable read-only memory where the cells are electronically both written and erased. Also known as electrically alterable read-only memory. *See also* electronically programmable read-only memory.

electronically programmable read-only memory (EPROM) Programmable read-only memory that is electronically written but requires ultraviolet light for erasure.

electroplastic effect plastic deformation of metals with the application of high-density electric current.

electroplastic smart material material with smart properties of elastic deformation changes proportional to a controlled electric current applied in proportion to the sensed deformation.

electrorheological property property exhibited by some fluids that are capable of altering their flow characteristics depending on an externally applied electric field.

electrorheological smart fluid fluid with smart flow characteristics dictated to change self-adaptively by means of an electric field applied in proportion to the sensed flow parameters.

electrorheological smart material material with smart properties of elastic deformation changes proportional to a controlled electric current applied in proportion to the sensed deformation.

electroslag welding a welding process that produces coalescence of metals with molten slag that melts the filler metal and the surfaces of the parts to be joined.

electrostatic discharge (ESD) the discharge of a body through a conducting path between two pins of an IC. Circuits located at the inputs and outputs of ICs protect the internal devices from ESD events.

electrostatic precipitator a method of extracting dust from stack gases or ventilating systems in which ions are laid on the dust particles by high-voltage electrodes and then attracted electrostatically into a trap.

electrostatic voltmeter a voltmeter, typically used for voltages in the kilovolt range, in which the pointer is moved by the electrostatic attraction of a pair of metal plates across which the voltage to be measured is applied.

electrostriction the tendency of materials to become compressed in the presence of an applied electric field. The change in density is proportion to the square of the electric field strength. This process leads to an increase in the refractive index of the material, describable by $\delta n = n_2 I$, where n_2 is the (positive) coefficient of the nonlinear refractive index and I is the intensity of the field in units of power per unit area. For condensed matter, a typical value of n_2 is 10^{-20} m^2/W.

element factor in antenna theory, that part of the radiation pattern that is governed by the geometrical shape of the antenna that constrains the current.

Elias' upper bound for any (n, k) block code, the minimum distance is bounded asymptotically as

$$\frac{d_{\min}}{n} \le 2A(1 - A)$$

where the parameter A is related to the code rate through the equation

$$\frac{k}{n} = 1 + A \log_2 A + (1 - A) \log_2(1 - A)$$

$$0 \le A \le \frac{1}{2}$$

ellipse detection   the detection of ellipses in digital images, often with a view to locating elliptical objects or those containing ellipses; ellipse detection is also important for the location of circular features on real objects following orthographic projection or perspective projection.

ellipsometry   measurement of the changes of light polarization produced by scattering.

elliptic   *See* elliptic filter.

elliptic filter   (1) filter with an equal ripple passband frequency response, but in which the stop band also exhibits an equal ripple (peaks) stopband response. Also known as a Caurer filter.

(2) member of a class of low pass, high pass, band pass and band stop filters with an equi-ripple characteristic, designed to achieve optimally rapid rolloff rates near cutoff frequencies at the expense of a loss of monotonicity in both the passbands and the stopbands. For example, an elliptic low pass filter design is equiripple in the passband and stopband, and has a squared magnitude response of the form

$$|H(jw)|^2 = \frac{1}{1 + \varepsilon^2 U_N^2(w/w_c)},$$

where $U_N(w)$ is a Jacobian elliptic function. *See also* Butterworth filter, Chebychev filter.

elliptical polarization   the polarization state of a radiated electromagnetic field in which the tip of the electric field vector traces an ellipse as a function of time for a fixed position. The sense of rotation of the electric field vector is either right-hand or left-hand (clockwise or counterclockwise). Circular polarization and linear polarization are special cases of elliptical polarization.

embedded computer   (1) a computing machine contained in a device whose purpose is not to be a computer. For example, the computers in automobiles and household appliances are embedded computers.

(2) a device, consisting of a microprocessor, firmware (often in EPROM), and/or FPGAs/EPLDs, which is dedicated to specific functions, and becomes an inseparable component of a device or system, in contrast to devices that are controlled by stand-alone computers. Embedded computers use embedded software, which integrates an operating system with specific drivers and application software. Their design often requires special software–hardware codesign methods for speed, low power, low cost, high testability or other special requirements.

(3) software that is part of a larger system and performs some of the requirements of that system; e.g., software used in an aircraft or rapid transit system. Such software does not provide an interface with the user. *See* firmware.

embedded computer system   *See* embedded computer, embedded system.

embedded passives   3-D packaging solution, consisting of embedding passive elements into the mounting substrate, for increasing the packaging efficiency.

embedded system   software that is part of a larger system and performs some of the requirements of that system; e.g., software used in an aircraft or rapid transit system. Such software does not provide an interface with the user. *See also* firmware.

EMC   *See* electromagnetic compatibility.

EMI   *See* electromagnetic interference.

EMI filter   *See* electromagnetic interference filter.

emission credit   a scheme in the USA for the control of nitrous oxide and sulfur dioxide emissions from industrial plants. A utility can purchase from the government the right to emit a certain quantity of pollutant. If it does not emit this amount, the credit can be sold on the market to another plant which may need to emit some pollutant. The resale value of the pollution credit forms a financial incentive for a facility to emit as little pollution as possible.

emissivity the fraction of the power incident on a material that is reradiated after being absorbed by the material. For a material in thermal equilibrium, the emissivity is equal to the absorptivity.

emitter follower *See* common collector amplifier.

emitter-coupled logic (ECL) a very high speed bipolar transistor logic circuit family.

emitter-coupled pair *See* differential pair.

EMP *See* electromagnetic pulse.

empirical distribution function *See* histogram.

empirical model mathematical model based on curve-fitting specific mathematical functions to measured data, rather than on device physics. Empirical models generally have a low to midrange modeling valuation coefficient.

EMTP the Electro-Magnetic Transient Program, a computer program which simulates an electric power system such that its response to disturbances may be accurately predicted.

emulate executing a program compiled to one instruction set on a microprocessor that uses an incompatible instruction set, by translating the incompatible instructions while the program is running.

emulation a model that accepts the same inputs and produces the same outputs as a given system. To imitate one system with another. *Contrast with* simulation.

emulation mode state describing the time during which a microprocessor is performing emulation.

emulator (1) the firmware that simulates a given machine architecture.

(2) a device, computer program, or system that accepts the same inputs and produces the same outputs as a given system. *Compare with* simulator.

EMV *See* electromagnetic vulnerability.

encapsulation property of a program that describes the complete integration of data with legal process relating to the data.

enclosure a box, cabinet, wall, fence, barrier, or other means designed to protect personnel from accidentally contacting energized electrical parts and to protect the electrical parts from physical damage.

encoder (1) a device that directly creates a digital signal based on an analog value.

(2) a logic circuit with 2^N inputs and N outputs, the outputs indicating the number of the one input line that is asserted.

(3) a device used to obtain positional information. The encoder uses a disk mounted on the shaft of a rotating machine and a light source. The combination provides lights pulses that can be decoded to provide the angular position of the motor. *See also* absolute and incremental encoder.

encoding (1) the act of placing information to be transmitted in a form that can be transmitted over a particular medium and will be recognizable by the receiver. *See also* coding.

(2) in computing systems, to represent various pieces of information by a defined sequence of the binary digits 0 and 1, called bits. To apply the rules of a code.

encryption the transformation employed to transform information to be transmitted (plaintext) into a format that is unintelligible (ciphertext or a cryptogram). The ciphertext can then be transmitted via a communication channel without revealing the contents of the plaintext. This is achieved by means of an encryption key. A system for performing encryption is also known as a cipher. The information to be encrypted is referred to as plaintext, and the encrypted message resulting from encryption is referred to as ciphertext. The intended receiver of the ciphertext also has the encryption key, and by having both the ciphertext and the encryption key available, the original plaintext can be recovered. *See also* encryption key, block cipher, stream cipher, public key cryptography, private key cryptography.

encryption key a codeword used for decryption of ciphertext into plaintext in encryption systems. Ciphertext can then be transmitted via a communication channel without revealing the contents of the plaintext. The plaintext can only be recovered by someone in possession of the encryption key. *See also* encryption, block cipher, stream cipher.

end bell the cap that forms the end of the stator housing for an electric machine with a cylindrical frame.

end-around carry technique used in one's complement arithmetic, in which a carryover of the result of an addition or subtraction beyond the leftmost bit during addition (or subtraction) is "wrapped around" and added to the result.

end-effector a part found at the free end of the chain of the links which form a manipulator. Position and orientation of the manipulator are referred to the end-effector. The frame attached the end-effector is known as end-effector frame. The end-effector can be a gripper or it can be attached to the end-effector. The orthonormal end-effector frame consists of three unit vectors a-approach (it approaches the object) o-orientation (which is normal to the sliding plane between the fingers of the gripper), and n-normal vector to two others so that the frame (n, o, a) is right-handed.

end-fire the pattern factor is maximum in the E plane (for a dipole antenna along the z axis, this is the plane where $\theta = 0$ or 180 degrees).

end-fire array a linear phased array antenna in which the amplitudes and phases of the element excitations are adjusted such that the direction of maximum radiation is along the axis passing through the center of the array elements.

endian an adjective used with a qualifier to indicate how long values are constructed from smaller values that reside in memory. A "big-endian" machine constructs long values by placing the smaller parts from the lower addresses at the left end of the value (the "big" end).

endoscope remote imaging system used by physicians to examine a patient's internal organs. The cylindrical rod is placed either through an existing opening in the body or through a small incision. The rod contains an illuminating fiber optic bundle and an optical system to deliver the image to a video monitor. In a conventional endoscope, the image is transmitted through a series of lenses to a camera outside of the patient. In an electronic endoscope the image is collected by a CCD camera positioned inside the patient.

endpoint detection the process of isolating spoken words, typically used subsequently for word recognition.

ends a crimped-type wire connector.

energy that which does work or is capable of doing work. In electrical systems, it is generally a reference to electrical energy measured in kilowatt hours.

energy band continuous interval of energy levels that are allowed in the periodic potential field of the crystalline lattice.

energy banking pertaining to the maintenance of a thermal unit on hot reserve.

energy compaction in a transformation, the concentration of the input signal energy into a relatively small part of the transformed signal. A linear transform of a random vector compacts the signal energy when the energy or variance of a small number of transform coefficients is large relative to the variance of the other coefficients.

energy conservation the conservation of energy between the input and output of a system; i.e., the energy of the output signal is equal to that of the input signal to within a constant factor. A unitary or orthogonal transform conserves energy in that the energy or magnitude of the output vector is equal to that of the input vector.

energy-density spectrum for a continuous time signal $x(t)$ the function $|X(w)|^2$, where $X(w)$ is the Fourier Transform of x. The function $|X(w)|^2$ is referred to as an energy-density spectrum because $\Delta |X(w_0)|^2$ is proportional to the energy in the signal x in a bandwidth Δ around the frequency w_0. Analogously, a discrete time signal $x[n]$ has an energy-density spectrum $|X(\Omega)|^2$.

energy gap the width of the energy interval between the top of the valence band and the bottom of the conduction band.

energy level one of the specific values possible for the energy of an electron in an atom or molecule.

energy product the product of the magnetic flux density, B and magnetic field intensity, H at any operating point on the normal demagnetization curve, indicating the energy delivered by the magnet. The maximum energy product is commonly used to designate varying grades of materials.

energy relaxation time the characteristic time for energy loss to scattering processes.

energy signal a continuous time signal $f(t)$ is an energy signal if

$$\int_{-\infty}^{\infty} f(t)^2 dt < \infty.$$

For example, the signal $f(t) = e^{-t}, t \geq 0$ is an energy signal. The signal $f(t) = 1$ is not an energy signal. The definition has practical significance if the signal f is a voltage, or current, applied to an electric circuit, as shown in the figure. The energy dissipated by the circuit is given by

$$\frac{1}{R} \int_{-\infty}^{\infty} v(t)^2 dt.$$

A discrete time signal is an energy signal if

$$\sum_{k=-\infty}^{\infty} f(k)^2 < \infty.$$

enhanced backscattering peak of the angular distribution of scattered intensity in the retrore-flection direction, due to multiple scattering of light in dense volumes or very rough boundaries.

enhanced graphics adaptor (EGA) a video adapter proposed by IBM in 1984. It is capable of emulating CGA and MDA. It can reach 43 lines with 80 columns. In graphic mode, it can reach 640×350 pixels (wide per high) with 16 colors selected from a pallet of 64.

enhanced small device interface *See* enhanced small disk interface.

enhanced small disk interface (ESDI) mass storage device interface similar to MFM and RLL except that the clock recovery circuits are in the peripheral device rather than in the controller. Originally designed by Maxtor.

enhancement improvement of signal quality without reference to a model of signal degradation restoration, image enhancement.

enhancement mode an FET that is off when zero volts bias is applied from gate to source.

ENI *See* equivalent noise current.

ENR *See* excess noise ratio noise.

enriched uranium uranium which contains 1.5% to 5% of fissile U-235 and is thus suitable for use in power reactors.

enrichment the process of increasing the ratio of one isotope of a chemical element, e.g. fissile uranium, to another, less desireable isotope, e.g. non-fissile uranium.

ensemble processor a parallel processor consisting of a number of processing elements, memory modules, and input/output devices under a single control unit. It has no interconnection network to provide interprocessor or processor-memory communications.

entity a software process that implements a part of a protocol in a computer communication network.

entity relationship diagram a diagram that describes the important entities in a system and the ways in which they are interrelated.

entropy information theoretic quantity representing the amount of uncertainty in guessing the value of a random variable. For a discrete random variable X, with probability mass function $p(x)$ defined on the support set S, the entropy, $H(X)$ is defined as

$$H(X) = -\sum_{x \in S} p(x) \log p(x)$$

Also written $H(p)$, emphasizing the sole dependence upon the mass function. *See also* differential entropy, relative entropy, mutual information.

entropy coding a term generally used as equivalent to lossless source coding. The name comes from the fact that lossless source coding can compress data at a rate arbitrarily close to the entropy of the source.

entropy estimation noiseless source coding theorem states that the bit rate can be made arbitrarily close to the entropy of the source that generated the image. Entropy estimation is the process of characterizing the source using a certain model and then finding the entropy with respect to that model. The major challenge here is to approximate the source structure as close as possible while keeping the complexity of the model and the number of parameters to a minimum.

entry mask bit pattern associated with a subroutine entry point to define which processor registers will be used within the subroutine and, therefore, which should be saved upon entry to the subroutine. Some processor designs perform this state saving during the execution of the instruction that calls a subroutine.

entry point instruction that is the first instruction in a subroutine.

ENV *See* equivalent noise voltage.

envelope the imaginary waveform produced by connecting the peak values of a modulated carrier wave. For AM, the amplitude of the carrier sinusoid is a function of time

$$e(t) = E(t) \sin(\omega_c + \theta)$$

where $E(t)$ is a function of the intelligence. In this case, the envelope of the resulting double sideband AM waveform, $e(t)$, is a scaled representation of the intelligence.

envelope delay the time it takes for a specific reference point on the envelope of a modulated wave to propagate between two points in a circuit or transmission system. In a time invariant system, the envelope delay is the derivative (i.e., rate of change) of phase (in radians) with respect to angular frequency (also in radians) as the envelope of the signal passes between two points in the system. Also referred to as time delay.

envelope detection   a device which produces an output waveform that is proportional to the real envelope of its input; i.e., when the input is $A(t)cos(2\pi f_c t + \theta(t))$, the output is $v(t) = CA(t)$ with C a constant.

envelope detector   the optimum structure for detecting a modulated sinusoid with random phase in the presence of additive white Gaussian noise.

environment   a set of objects outside the system, a change in whose attributes affects, and is affected by, the behavior of the system.

environmental dispatch   a generation scheme in which units are committed so as to minimize disturbance to the natural environment.

epipolar line   in stereo vision, the intersection of an image plane with an epipolar plane, which is the plane identified by a point P and the centers of projection of the two cameras.

epitaxial layer   a doped layer of semiconducting material grown on the surface of a prepared semiconductor substrate. Various methods are utilized, resulting in a very thin and dimensionally well controlled active layer for fabricating semiconductor devices. "Epi" parts generally are more uniform and perform better than ion implanted devices, but are also more expensive to fabricate.

EPROM   *See* erasable programmable read-only memory.

equal area criterion   a criterion used often in power system studies whenever the stability of a single machine system needs to be determined without actually solving the swing equation. It is a direct approach. *See also* electromechanical equation or swing equation.

equal gain combining   a method of diversity combining in which the outputs of several communication channels are first co-phased and then summed. The resulting signal has increased mean power and reduced fade depth. *See also* angle diversity, antenna diversity.

equal ripple   in-band power gain or power loss vs. frequency response in which the minimum dips are equal in power and maximum peaks are equal in power.

equalization   a method used in communication systems to compensate for the channel distortion introduced during signal transmission.

equalizer   a device used at the receiver (typically a digital filter) that attempts to suppress or cancel intersymbol interference.

equalizer adaptive   an equalizer that automatically adjusts its parameters to suppress or cancel intersymbol interference. In the case of a digital filter, the parameters are the filter coefficients, which are adapted to minimize a cost criterion (e.g., mean-squared error).

equalizer decision-feedback   a nonlinear digital filter that consists of a linear prefilter $P(z)$ for suppressing precursor intersymbol interference, and a feedback filter $F(z)$ for suppressing postcursor intersymbol interference.

equalizer minimum mean squared error (MMSE)   a digital equalizer (filter) in which the coefficients are selected to minimize the mean-squared error between the transmitted symbols and the filter outputs.

equalizer transversal   an equalizer that is implemented as a finite impulse response digital filter. Also called a "tapped-delay line."

equalizer zero-forcing   an equalizer in which the parameters are adjusted to eliminate intersymbol interference (at the expense of enhancing the noise).

equalizing pulse interval   series of 12 sync pulses inserted in the vertical blanking interval of each field of the NTSC composite video signal. The equalizing pulse makes a transition between the composite signal blanking level and the sync level at one-half horizontal line intervals. The duration at the sync level is 3.575 ± 0.425 percent of the horizontal line time (0.45 to 0.5 times the horizontal sync). There are three horizontal lines of equalizing pulses (6 pulses) preceding the vertical synchronization signal and 3 horizontal lines of equalizing pulses (6 pulses) following the vertical synchronization interval. The equalizing pulse interval repeats every field of 262.5 horizontal lines.

equalizing pulses   in an encoded video signal, a series of $2\times$ line frequency pulses occurring during vertical blanking, before and after the vertical synchronizing pulse. Different numbers of equalizing pulses are inserted into different fields to ensure that each field begins and ends at the right time to produce proper interlace. The $2\times$ line rate also serves to maintain horizontal synchronization during vertical blanking.

equiband   a complex method of chrominance signal demodulation, that does not require dissimilar filters and delay equilization; it most frequently utilizes a bandwidth of 800 KHz.

equilibrium manifold   *See* sliding surface.

equilibrium point   for a continuous-time system $dx/dt = f(t, x)$, a constant solution x^* such that $f(t, x^*) = 0$. For a discrete-time system

$x(k + 1) = f(k, x(k))$, a constant solution x^* such that $x^* = f(k, x^*)$.

equilibrium solution   consider a dynamic system described by a first-order vector differential equation of the form

$$\dot{x}(t) = f(x(t), t, u(t))$$

If there exists a vector x_e such that

$$f(x_e, t, 0) = 0 \quad \forall t$$

then whenever $x(t) = x_e$, the system is said to be in equilibrium and x_e is called an equilibrium state.

equilibrium state   *See* equilibrium solution.

equipment hazard   a possible source of peril, danger, risk, or difficulty.

equiripple filter   a filter designed using an iterative algorithm, for example Remez exchange algorithm, to minimize the maximal deviation from the desired magnitude frequency response in both passband and stopband for a given filter length. It has equal amplitude ripples within the passband and stopband. Chebyshev filter.

equivalence in continuous-valued logic   similar to equivalence in conventional logic equivalence in continuous-valued logic between two variables x and y, which are continuous in the open interval (0, 1), can be defined as

$$e(x, y) = \max\{\min(x, y), \min((1-x), (1-y))\}.$$

See equivalence in logic.

equivalence in logic   for two Boolean variables x and y, defined as

$$(x \equiv y) = (\bar{x} \wedge \bar{y}) \vee (x \wedge y)$$

equivalence theorem an electromagnetic theorem: If the tangential magnetic and electric fields are known everywhere on some closed surface S, then these fields may be replaced with equivalent electric and magnetic surface currents, respectively. These equivalent currents will produce the same field structure exterior to S as the original fields and the null field internal to S.

equivalent circuit a combination of electric circuit elements chosen to represent the performance of a machine or device by establishing the same relationships for voltage, current, and power.

equivalent control an algorithm used to determine a system's dynamics when restricted to a sliding surface. The method entails combining the solution to an algebraic equation involving the time derivative of the function describing the sliding surface and the dynamical system's model. *See also* variable structure system and sliding mode control.

equivalent current a theoretical current used to obtain the scattered field from a surface or discontinuity. The equivalent current is formulated to represent the actual physical currents so as to result in an equivalent scattered field.

equivalent impedance the impedance of the windings of an electromagnetic machine reflected to one side (component) of the machine. For example, in a transformer, the equivalent impedance consists of the combined leakage reactances and resistances of the primary and secondary windings.

equivalent noise current (ENI) a noise current source that is effectively in parallel with either the noninverting input terminal (ENI^+) or the inverting input terminal (ENI^-) and represents the total noise contributed by the op amp if either input terminal is open circuited.

equivalent noise temperature an alternative way of describing the noise properties of two port networks.

equivalent noise voltage (ENV) a noise voltage source that is effectively in series with either the inverting or noninverting input terminal of the op amp and represents the total noise contributed by the op amp if the inputs were shorted.

equivalent reactance the reactance of the windings of an electromagnetic machine reflected to one side (component) of the machine. *See also* equivalent impedance.

equivalent resistance the resistance of the windings of an electromagnetic machine reflected to one side (component) of the machine. *See also* equivalent impedance.

equivalent source fictitious source used in the equivalence theorem.

equivalent sphere illumination (ESI) the level of sphere illumination that would produce task visibility equivalent to that produced by a specific lighting environment.

equivalent system dynamics a dynamical system model resulting from substituting the *equivalent control* into the plant's modeling equation. The equivalent system's trajectory is confined to a surface that is parallel to the sliding surface if the system's initial condition is off the sliding surface. If the initial condition is on the sliding surface, then the equivalent system's trajectory will stay on the sliding surface.

erasable optical disk a magneto-optical disk that can be both read/written and erased. A thermo-magneto process is used for recording and erasure of information. The recording process uses e.g., the "laser power modulation" or the "magnetic field modulation" technique.

erasable programmable read-only memory (EPROM) a nonvolatile chip memory, it is used in the place of PROM. EPROMs present a glass on the case that allows one to see the chip.

They can be erased by exposing the chip at the ultraviolet light for typically 20 minutes. Once erased they can be reprogrammed. The programming has to be performed by using a special algorithm and a supplementary V_{pp}. Also called UVPROM. *See also* electronically erasable programmable read-only memory.

erasure   in a forward error control system, a position in the demodulated sequence where the symbol value is unknown. Depending on the quality of the received signal when no decision is made on a particular bit, the demodulator inserts an erasure in the demodulated data. Using the redundancy in the transmitted data, the decoder attempts to fill in the positions where erasures occurred. *See* binary erase channel.

erbium doped fiber amplifier (EDFA)   an optical amplifier based on the rare-earth erbium that can amplify optical signals in the 1550 nm telecommunications window. The useful gain region extends from approximately 1530 nm to 1565 nm.

EREW   *See* exclusive reads and exclusive writes.

ergodic process   a process that has all possible ergodic properties.

For process $X(t)$ when the time average

$$\lim_{T \to \infty} (1/T) \int_{-T/2}^{T/2} X(t)dt$$

exists and equals the corresponding expected value $E[X(t)]$, it is said that $X(t)$ is ergodic in the mean. There are ergodic properties associated with the mean, autocorrelation, and power spectral density as well as all finite-order joint moments.

ergodicity   stochastic processes for which ensemble averages can be replaced by temporal averages over a single realization are said to be ergodic. For a stochastic process to be ergodic,

a single realization must in the course of time take on configurations closely resembling the entire ensemble of processes. Stationary filtered white noise is considered ergodic, while the sinusoidal process $A \cos(\omega t + \phi)$ with random variables A and ϕ is not.

Erlang B   a formula (or mathematical model) used to calculate call blocking probability in a telephone network and in particular in cellular networks. This formula was initially derived by A. K. Erlang in 1917, a Danish pioneer of the mathematical modeling of telephone traffic, and is based on the assumption that blocked calls are forever lost to the network. *See also* Erlang C.

Erlang C   similar to Erlang B, the formula is based on a traffic model where the call arrival process is modeled as a Poisson process, the call duration is of variable length and modeled as having an exponential distribution. The system is assumed to have a queue with infinite size that buffers arriving calls when all the channels in the switch are occupied. The model is based on the assumption that blocked calls are placed in the queue.

Erlang capacity   maximum number of users in the system which leads to the maximum allowable blocking probability (for example, 2%). *See also* blocking probability.

erosion   an important basic operation in mathematical morphology. Given a structuring element B, the erosion by B is the operator transforming X into the Minkowski difference $X \ominus B$, which is defined as follows:

(1) if both X and B are subsets of a space E, $X \ominus B = \{z \in E | \forall b \in B, \ z + b \in X\}$

(2) if X is a gray-level image on a space E and B is a subset of E, for every $p \in E$ we have $(X \ominus B)(p) = \inf_{b \in B} X(p + b)$

(3) if both X and B are gray-level images on a space E, for every $p \in E$ we have $(X \ominus B)(p) = \inf_{h \in E}[X(p + h) - B(h)]$ with the convention $\infty - \infty = +\infty$ when $X(p + h), B(h) = \pm\infty$.

In (2) and (3), $X(p)$ designates the gray-level of the point $p \in E$ in the gray-level image X. *See* dilation, structuring element.

ERP *See* effective radiated power.

error (1) manifestation of a fault at logical level. For example, a physical short or break may result in logical error of stuck-at-0 or stuck-at-1 state of some signal in the considered circuit.

(2) a discrepancy between a computed, observed, or measured value or condition and the true, specified, or theoretically correct value or condition. *See* anomaly, bug, defect, exception.

error control coding *See* channel coding.

error-correcting code (ECC) code used when communicating data information in and between computer systems to ensure correct data transfer. An error correcting code has enough redundancy (i.e., extra information bits) in it to allow for the reconstruction of the original data, after some of its bits have been the subject of error in the transmission. The number of erroneous bits that can be reconstructed by the receiver using this code depends on the Hamming distance between the transmitted codewords. *See also* error detecting code.

error-correction the mechanism by which a receiving circuit is able to correct errors that have occurred in an encoded transmission. *See also* Hamming code, error-correcting code.

error correction capability of a code is bounded by the minimum distance and for an (n, k) block code, it is given by $t = [(d_{min}-1)/2]$, where $[x]$ denotes the largest integer contained in x.

error detecting code code used when communicating data information in and between computer systems to ensure correct data transfer. An error detecting code has enough redundancy (i.e.,

extra information bits) in it to allow for the detection of the original data, after some of its bits have been the subject of error in the transmission. The number of erroneous bits that can be detected by the receiver using this code depends on the Hamming distance between the transmitted codewords. *See also* error correcting code.

error detection the process of detecting if one or more errors have occurred during a transmission of information. Channel codes are suitable for this purpose. The family of CRC-codes are an example of channel codes specially designed for error detection. *See also* error detecting code.

error detection capability the capability of a code to detect error is bounded by the minimum distance, and for an (n, k) block code, it is given by $d_{min} - 1$.

error extension the multiplication of errors that might occur during the decoding of a line coded sequence, or during the decoding of a forward error control coded sequence when the number of symbol errors exceeds the error correction capability of the code.

error function mathematical function over some interval in which a calculated result is compared to a known quantity (usually data) by utilizing a difference quantity to determine how well the mathematical function replicates the known quantity over that interval. Most error functions utilize an area or least squares difference function.

error latency length of time between the occurrence of an error and the appearance of the resulting failure.

error recovery process of regaining operational status and restoring system integrity after the occurrence of an error with the use of special hardware and software facilities.

error state diagram a diagram that illustrates all possible error events with respect to an assumed reference event, e.g., an error sequence compared to a reference sequence, both produced by a finite state machine. All possible sequences that diverge and later merge with the reference sequence are accounted for in the diagram. The transfer function of the error state diagram can be used for performance evaluation of detection/decoding algorithms working on noise outputs from a finite state machine.

ESD   *See* electrostatic discharge.

ESDI   *See* enhanced small disk interface.

ESI   *See* equivalent sphere illumination.

ESPRIT   acronym for estimation of signal parameters via rotational invariance techniques. A subspace-based estimation technique based on two identical, displaced sensor arrays.

estimation   in adaptive control, a key role is played by on-line determination of process parameters. A recursive parameter estimator is present in many adaptive control schema such as self-tuning regulator (explicitly) or model-reference adaptive controller (implicitly).

Usually, parameters estimation is viewed in the broader context of system identification formed by selection of model structure, experiment design, parameters estimation, and validation. In adaptive control, the parameters vary continuously, and it is necessary to estimate them recursively. A basic technique used for parameters estimation is the Least Squares Method. This method is particularly useful if the model has the property of being linear in the parameters.

estimator   any function of the sample points. Hence, an estimator is a random variable. To be useful, an estimator must have a good relationship to some unknown quality that we are trying to determine by experiment.

etching   a reactive process where material is removed from a semiconductor device or printed circuit board. Usually a photosensitive material is exposed through a photomask, and either a wet chemical process or a dry plasma process is used to selectively remove material to leave a particular pattern behind after the etch process is completed.

Ethernet   a standard for interconnecting devices on a local area network (LAN).

Euclidean distance   a distance measure between two real valued vectors $(x_1, x_2, \ldots, x_n)$ and $(y_1, y_2, \ldots, y_n)$ defined as

$$D_{Euclidean} = \sqrt{\sum_{i=1}^{n}(x_i - y_i)^2}$$

Euclidean distance is the special case of Minkowski distance when $\lambda = 2$. *See also* Minkowski distance.

Euler number   a topological invariant of an object having an orientable surface. Assuming that the surface is endowed with the structure of a graph with vertices, edges, and faces (where two neighboring faces have in common either a vertex or an edge with its two end-vertices, their interiors being disjoint): the Euler number is $V - E + F$, where V, E, and F are respectively the number of vertices, edges and faces; this number $V - E + F$ does not depend on the choice of the subdivision into vertices, edges, and faces. For a bounded 2–D object in a Euclidean or digital plane, the Euler number is equal to the number of connected components of that object, minus the number of holes in it. For 2–D binary digital figures on a bounded grid, the Euler number can easily be computed by counting the number of occurrences of some local configurations of on and off pixels. Also called genus.

eureka   in a multiprocessor system, a coordination (synchronization) operation generating a completion signal that is logically ORed among all processors participating in an asynchronously parallel action. The interpretation and name come from its use in systems that delegate various portions of a search space to different processing modules. A processor executes the eureka operation when it has found the desired object or value; this can serve as a signal for the others to abort their attempts to find a solution.

eutectic   alloy composition with minimum melting temperature at the intersection of two solubility curves.

eutectic alloy overload device   an overload device that employs a melting alloy as the actuating element. *See also* overload heater, overload relay.

EUV lithography   lithography using light of a wavelength in the range of about 5 to 50 nm, with about 13 nm being the most common. Also called soft X-ray lithography.

EV   *See* electric vehicle.

even function   a real-valued function $x(t)$ in which $x(-t) = -x(t)$ for all values of t. *Compare with* odd function.

even mode impedance   characteristic impedance of a transmission line when a single and certain even mode exists on it.

even order response   a circuit gain or insertion loss versus frequency response in which there are an even number of peaks in the ripple pattern, due to an even number of paired elements in the circuit. Even order circuits exhibit a peak for each element pair and a loss equal to L_{amax} (maximum attenuation loss across the band) at DC (low pass) or at ω_0 (band pass).

even-mode characteristic impedance   characteristic impedance of a circuit due to an even-mode current or voltage excitation. Often applied in the context of a transmission line coupler where the even-mode excitation consists of applying equal amplitude voltages or currents of identical phase on two conductors. The resulting impedance under this excitation is defined as the even mode characteristic impedance.

even-order response   circuit gain or insertion loss versus frequency response in which there are an even number of peaks in the ripple pattern, due to an even number of paired elements in the circuit. Even-order circuits exhibit a peak for each element pair and a loss equal to L_{amax} (maximum attenuation loss across the band) at DC (low pass) or at ω_o (band pass).

even signal   a signal that has even symmetry. If $x(t)$ is an even signal, it satisfies the condition $x(t) = x(-t)$. *See* odd signal.

event   (1) a specific instance taken from some sample space, normally with an associated probability or probability density; also commonly an idealized infinitesimal point in (x, y, z, t) space at which some occurrence is taken to happen. For example, a flash of light at time t_0 at position (x_0, y_0, z_0) — event (x_0, y_0, z_0, t_0) — will lead at a later time t_1 to a wave of light passing a point (x_1, y_1, z_1) — namely event (x_1, y_1, z_1, t_1).

(2) a nonsequential change in the sequencing of macroinstructions in a computer. Events can be caused by a variety of factors such as external or internal interrupts (traps) or branch statements.

event table   table listing all events, and their corresponding effects as well as reactions to them. Exception conditions/responses table is a special type of event table.

exact absolute controllability   a dynamical system where the attainable set K_∞ is equal to the whole state space $W_1^{(2)}([-h, 0], R^n)$.

exact coding   coding methods that reproduce the picture at the receiver without any loss. This method is also called information-lossless or exact coding techniques. Four methods of exact coding are run-length coding, predictive coding, line-to-line predictive differential coding, block coding. However, several coding schemes use these in a hybrid manner.

exact controllability of infinite dimensional system an infinite dimensional system where the attainable set K_∞ is equal to the whole state space X.

exception (1) an unusual condition arising during program execution that causes the processor to signal an exception. This signal activates a special exception handler that is designed to handle only this special condition. Division by zero is one exception condition. Some vendors use the term "trap" to denote the same thing.

(2) an event that causes suspension of normal program execution. Types include addressing exception, data exception, operation exception, overflow exception, protection exception, underflow exception.

exception handler a special block of system software code that reacts when a specific type of exception occurs. If the exception is for an error that the program can recover from, the program can recover from the error and resume executing after the exception handler has executed. If the programmer does not provide a handler for a given exception, a built-in system exception handler will usually be called, which will result in terminating the process that caused the exception. Finally, the reaction to exception can be halting of the system. As an example, a bus error handler is the system software responsible for handling bus error exceptions.

excess delay the arrival times of a component of the impulse response of a wideband communication channel relative to the first arriving component. Hence the total excess delay, the difference in arrival time between the first and last significant components.

excess loss the ratio of the actual propagation loss between two antennas to the free space loss for two antennas separated by the same distance in a vacuum. Usually expressed in decibels.

excess noise (1) thermal noise in excess or exceeding thermal noise at 290°K. The excess noise ratio (ENR) is the ratio of the excess noise to the noise at 290°K, expressed in decibels.

(2) Noise in excess of the thermal noise (n_e), which is a function of the device, frequency and bias current, also known as $1/f$, $1/f_a$, flicker or popcorn noise. This noise rapidly dies off such that above a few megahertz it becomes insignificant. It is a major concern in generating phase noise in oscillators.

$$ENR = 10 \log_{10} \left(\frac{T - 290°K}{290°K} \right)$$
$$n_e = f(\text{device}, 1/f^a, I_{bias})$$

excess noise ratio noise (ENR) a noise source used in noise figure measurements. ENR is the ratio of the source's noise power when it is on to the noise power when it is off. The ENR values are entered and stored in the noise figure meter to calibrate a measurement.

excess-N representation method of representing floating point numbers, in which a positive or negative exponent is stored as a positive integer by adding the value N to it. For example, in excess-128 representation, an exponent of -40 would be stored as the value 88.

exchangeable disk *See* removable disk.

excimer a molecule formed by the excited state of one atom and one or more other atoms that remains a molecule only as long as the excited state lifetime.

excimer laser laser using a gas or gases to create an excited dimer (e.g., KrF), usually resulting in pulsed deep-UV radiation.

excitation population of excited states of a laser medium at the expense of some energy source.

excitation system   the DC voltage source and its accompanying control and protection systems connected to the synchronous generator rotor.

exciter   a DC source that supplies the field current to produce a magnetic flux in an electric machine. Often it may be a small DC generator, placed on the same shaft of the electrical machine.

exciting current   the current drawn by a transformer primary with its secondary open circuited. It is the vector sum of the core loss current I_c and the magnetizing branch current I_m. The exciting current I_e is also the current measured in the open circuit test on a transformer. The exciting current is calculated as the ratio of the primary induced EMF and the impedance of the tank circuit. On load, it is equal to the difference between the primary and reflected secondary currents of the transformer.

exciton laser   laser (or laser-like system) in which the amplified field consists of excitons rather than electromagnetic waves or photons.

exclusive OR   Boolean binary operator typically used for comparing the status of two variables or signals. Sometimes written "XOR." The truth table for $\oplus \equiv X$ XOR Y is as follows:

X	Y	$X \oplus Y$
F	F	F
F	T	T
T	F	T
T	T	F

exclusive reads and exclusive writes (EREW)   shared memory model, in which only exclusive reads and exclusive writes are allowed.

execution cycle   sequence of operations necessary to execute an instruction.

execution time   amount of time it takes a computation (whether an instruction or an entire program) to complete, from beginning to end. Time during which an instruction or a program is executed. The portion of one machine cycle needed by a CPU's supervisory-control unit to execute an instruction.

execution unit   in modern CPU implementations, the module in which actual instruction execution takes place. There may be a number of execution units of different types within a single CPU, including integer processing units, floating point processing units, load/store units, and branch processing units.

exhaustive search   a search through all possibilities before deciding on what action to take. For example, for the maximum-likelihood (exhaustive search) detection of a sequence of k bits, all 2^k possible bit sequences are considered and the one with the largest likelihood is selected.

expanded memory   expanded memory specification, EMS, was born for adding memory to PCs (the so-called LIM-EMS). PCs were limited in memory to 640 kb even if the 8088/8086 CPU's limit is 1 Mb. Thus, in order to overcome this limit, the additional memory was added by using a paging mechanism: up to four windows of 16 kb of memory included into the 640 kb to seen up to 8 Mb of memory divided into pages of 16 kb. To this end, special memory boards were built. Currently, the MS-DOS is still limited to 640 kb, but the new microprocessors can address even several gigabytes over the first megabyte. Thus, to maintain the compatibility with the previous version and the adoption of the MS-DOS, the presence of expanded memory is simulated by means of specific drivers.

expectation   the integral of a function with respect to some probability measure. If $f(\cdot)$ is a deterministic function and x is random, governed by probability density $p(x)$, then

$$E[f(x)] = \int_{-\infty}^{\infty} f(x)p(x)dx.$$

expected value of a random variable   ensemble average value of a random variable that is given by integrating the random variable after scaling by its probability density function (weighted average) over the entire range.

expert system   a computer program that emulates a human expert in a well-bounded domain of knowledge.

explicit value   a value associated with the bit string according to the rule defined by the number representation system being used.

explosion-proof machine   National Electrical Manufacturers Association (NEMA) classification describing an electrical machine that is totally enclosed and whose enclosure is designed to withstand an internal explosion of a specified gas or vapor that may accumulate within the enclosure. The specification also requires that the design prevent ignition of the specified gas or vapor surrounding the enclosure due to sparks, flashes, or explosions of the specified gas within the enclosure.

exponent   (1) the field within a floating-point format that determines the power to which the mantissa should be raised.

(2) a shorthand notation for representing repeated multiplication of the same base. 2^4 is exponential notation to multiply two by itself four times: $2^2 = 2 \cdot 2 \cdot 2 \cdot 2 = 16$. 4 is called the exponent, indicating how many times the number 2, called the base, is used as a factor.

(3) the component of a binary floating-point number that signifies the integer power of two by which the significand is multiplied in determining the value of the represented number.

exponential distribution   a probability density function having the following exponential behavior:

$$ f(x) = \begin{cases} \lambda e^{-\lambda x}, & x \geq 0 \\ 0 & x < 0 \end{cases}. $$

where $\lambda > 0$. This distribution can describe a number of physical phenomena, such as the time for a radioactive nucleus to decay, or the time for a component to fail. *See also* probability density function, Cauchy distribution, Gaussian distribution.

exponential stability   (1) the property of an asymptotically stable equilibrium solution that guarantees an exponentially decreasing (to zero) norm in time of the difference between the solution and the equilibrium point.

(2) a special case of uniform asymptotic stability of an equilibrium point of $\dot{\mathbf{x}} = \mathbf{f}(t, \mathbf{x})$.

exposure   the process of subjecting a resist to light energy (or electron energy in the case of electron beam lithography) for the purpose of causing chemical change in the resist.

exposure dose   *See* exposure energy.

exposure energy   the amount of energy (per unit area) that the photoresist is subjected to upon exposure by a lithographic exposure system. For optical lithography, it is equal to the light intensity times the exposure time. Also called exposure dose.

exposure field   the area of a wafer that is exposed at one time by the exposure tool.

express feeder   a feeder to which laterals are connected only at some distance from the substation. These thus traverse areas fed by other feeders and are used to supply concentrated loads or new subdivisions. *See* feeder, lateral.

expulsion fuse   a fuse used on primary distribution lines which extinguishes the arc that results when it blows by explosively ejecting the fuse wire from its enclosure.

expulsion tube arrester   a gapped lightning arrester which establishes the power-follow arc in a tube lined with a substance which generates a sufficient quantity of gas when heated to blow out the arc. *See* power follow, lightning arrester.

expurgated code a code constructed from another code by deleting one or more codewords from the original code.

extended binary-coded-decimal interchange code (EBCDIC) character code developed by IBM and used in mainframe computers. It is closely related to the Hollerith code for punched cards.

extended code a code constructed from another code by adding additional symbols to each codeword. Thus an (n, k) original code becomes an $(n + 1, k)$ code after the adding of one redundant symbol.

extended industry standard architecture (EISA) a bus architecture designed for PCs using an Intel 80386, 80486, or Pentium microprocessor. EISA buses are 32 bits wide and support multiprocessing. The EISA bus was designed by IBM competitors to compete with micro channel architecture (MCA). EISA and MCA are not compatible with each other, the principal difference between EISA and MCA is that EISA is backward compatible with the ISA bus (also called the AT bus), while MCA is not. Therefore, computers with an EISA bus can use new EISA expansion cards as well as old AT expansion cards, while computers with an MCA bus can use only MCA expansion cards.

extended Kalman filter state estimation method based on linearization (see the definition) of nonlinear system and measurement equations about the current state estimate. The forms of both the filter gain and estimation error covariance equations are similar to those in the Kalman filter (see the definition). However, due to the linearization about the current state estimate, both equations are dependent on the current state estimate and cannot be calculated off-line.

extended memory in PCs, the memory located over the first megabyte of memory. This kind of memory is seen in the PC as continuous. Operating systems such as Windows NT, LINUX, and OS/2 are capable of working in protected mode, and thus at 32 bits. When MS-DOS is adopted to allow the exploitation of this memory as EMS, special software drivers have to be used. This overcomes the limit of 640 kb imposed by the real-mode and adopted by MS-DOS.

extended source a (light) source in which rays are emitted from a large source area. *Compare with* point source.

extended space for a space of functions $\mathcal{X}$, another space of functions, denoted $\mathcal{X}_e$, defined as the space of all functions whose truncation belongs to $\mathcal{X}$. *See also* truncation.

extended storage *See* solid state disk.

extension principle a basic identity for extending the domain of nonfuzzy or crisp relations to their fuzzy counterparts.

external cavity klystron a klystron device in which the resonant cavities are located outside the vacuum envelope of the tube.

external event event occurring outside the CPU and I/O modules of a computer system that results in a CPU interrupt. Examples include power fail interrupts, interval timer interrupts, and operator intervention interrupts.

external fragmentation in segmentation, leaving small unusable areas of main memory that can occur after transferring segments into and out of the memory.

external interrupt a signal requesting attention that is generated outside of the CPU.

external memory secondary memory of a computer.

external modulation   modulation of the optical intensity using an optical intensity modulator to modulate a constant power laser.

external space   the space of allowable positions and orientations of the end-effector of the manipulator. In general the external space, with the special Euclidean group $SE(3) \cong SO(3) \times R^3$, consists of rotations ($SO(3)$) and translations (R^3) in R^3. The external space is called as Cartesian space, test-oriented space, or operational space by many roboticians. Some people mean by Cartesian space the space in which the position of a point is given with three numbers, and in which the orientation of a body is given by three numbers. This is because in order to position and orient a body in Cartesian space we need 6 coordinates. In general orientation is specified by 3×3 orientation matrix that forms orthonormal vectors. Orientation can be specified in terms of a minimal representation describing the rotation of the end-effector frame with respect to the reference frame, e.g., Euler angles or rotation about an arbitrary axis and equivalent angle. Then manipulator position and orientation is $m \times 1$ vector, with $m \leq n$ (n denotes number of degrees of freedom). Using minimal representation of the orientation matrix, direct kinematics equation can be written in the following form: $x = g[\begin{smallmatrix}\phi\\p\end{smallmatrix}] = k(q)$ where ϕ is a set of vectors for minimal representation and p denotes three position coordinates.

external stability   stability concepts related to the input–output behavior of the system.

externally vented machine   classification describing an electrical machine constructed with an open frame in which ventilation air is forced through the machine by blower(s) mounted outside the machine enclosure.

extinction angle   time in electrical degrees from the instant the current in a valve reaches zero (end of conduction) to the time the valve voltage changes sign and becomes positive.

extinction cross section   the sum of the scattering and the absorption cross sections.

extrapolation   one of several methods to estimate the values of a sequence $r(k)$ for lags $|k| > p$ from the given values of $r(k)$ for $|k| \leq p$.

extrinsic   associated with the outside or exterior. In devices and device modeling, extrinsic refers to that part of the device or model associated with the passive structures that provide interconnects and contacts to other components, but are still considered a part of the device.

extrinsic fiber optic sensor   a fiber optic sensor where the fiber delivers light to and from a sensing element external to the fiber. Chemical sensors are an example where the sensing element exhibits a change in optical property such as absorption, fluorescence or phosphorescence upon detection of the species to be measured.

F

f_H common notation for higher band edge frequency in hertz.

f_L common notation for lower band edge frequency in hertz.

Fabry–Perot etalon interferometer consisting of two highly reflecting flat or spherical mirrors; only resonant frequencies are transmitted.

Fabry–Perot laser a laser source where the gain medium is placed within a Fabry–Perot cavity, which provides feedback into the laser medium. Several simultaneous lasing modes are supported in such cavities.

Fabry–Perot resonator any open (as opposed to a cavity) resonator, usually the assembly of two parallel plates resembling the optical Fabry–Perot interferometer. *See also* standing-wave resonator.

Fabry–Perot structure Fabry–Perot etalon or interferometer that has an optically nonlinear medium in its cavity.

facsimile the process of making an exact copy of a document through scanning of the subject copy, electronic transmission of the resultant signals modulated by the subject copy, and making a record copy at a remote location.

facsimile encoding a bilevel coding method applied to the encoding and transmission of documents. Facsimile systems may include support for grayscale image coding too, which is described under still image coding.

FACTS *See* flexible AC transmission system.

fading channel signal fluctuation caused by multiple propagation paths over a radio channel is called fading. May be categorized as fast fading or slow fading, depending on the rate of fading with respect to the information symbol rate. May also be categorized as frequency selective or frequency nonselective, depending on the transfer function of the radio channel. *See also* fading margin, fading Rayleigh, fading Rician.

fading margin the margin by which the average signal-to-noise ratio (SNR) in a radio communications link is over-designed, in order to compensate for variations in the short-term SNR that occur due to fading of the signal. The signal fading is typically due to multipath propagation, which arises due to the presence of multiple reflectors in the radio link. Utilizing diversity schemes and transmitter power control are typically employed in order to reduce the fading margin required in a radio system.

fading rate the rate at which the received signal level crosses the median signal level in a downward direction (i.e., with a negative slope). It is usually expressed in fades per second or fades per minute, depending on the actual rate of fading.

fading Rayleigh in mobile wireless communications, wide fluctuations in received signal strength (e.g., swings of 30 – 40 dB) and phase caused by scattering of the transmitted signal off of surrounding objects. The scattering induces a Gaussian distribution on the in-phase and quadrature signal components, so that the received signal envelope has a Rayleigh distribution. *See also* Rayleigh distribution.

fading Rician similar to Rayleigh fading, the only difference being that a direct line-of-sight component is present in the received signal in addition to the scattered signal. The received signal envelope has a Rician distribution. *See also* Rice distribution.

fail safe   pertaining to a circuit, for a set of faults, if and only if for any fault in this set and for every valid input code either the output is correct or assumes some defined safe state.

fail-stop processor   a processor that does not perform incorrect computation in the event of a fault. Self-checking logic is often used to approximate fail-stop processing.

failure   manifestation of an error at system level. It relates to execution of wrong actions, nonexecution of correct actions, performance degradation, etc.

failure mechanism   a physical or chemical defect that results in partial degradation or complete failure of a product.

fairness   (1) the concept of providing equivalent or near-equivalent access to a shared resource for all requestors.

(2) a fair policy requires that tasks, threads, or processes are allowed access to a resource for which they compete.

(3) the degree to which a scheduling or allocation policy is equitable and nondiscriminatory in granting requests among processes competing for access to limited system resources such as memory, CPU, or network bandwidth.

fall time   (1) in digital electronics, the period of duration of the transition of a digital signal from a stable high-voltage level to a stable low-voltage level.

(2) in optics, the time interval for the falling edge of an optical pulse to transition from 90% to 10% of the pulse amplitude. Alternatively, values of 80% and 20% may be used.

falling edge   (1) the region of a waveform when the wave goes from its high state to its low state.

(2) the high-to-low transition in voltage of a time-varying digital signal.

false color   the replacement of a color in a colored image by a different color, usually not present in the original image. Used to highlight regions or distinguish pixels of similar colors. *See* pseudo-color.

false sharing   the situation when more than one processor accesses different parts of the same line in their caches but not the same data words within the line. This can cause significant performance degradation because cache coherence protocols consider the line as the smallest unit to be transferred or invalidated.

fan-in   (1) the number of inputs to a module. This is usually used in connection with logic gates.

(2) multiple inputs of a channel. If the channel is a bus, only an input is allowed at a time. When N light beams are combined with an optical element, only $1/N$ of the power of each beam finds its way into the combined beam. Large fan-in is quite impractical for VLSI, because it uses a unique discrete channel for each input and current flows are limited by transistor capacitance.

fan-out   (1) the limit to the number of loading inputs that can be reliably driven by a driving device's output.

(2) multiple outputs of a bus. A signal is distributed into multiple channels. With a fan-out of N, each channel receives only $1/N$ of the light power. Large fan-out is quite impractical for VLSI, because it uses a unique discrete channel for each output and current flows are limited by transistor capacitance.

fan beam reconstruction   reconstruction of a computed tomography image from projections created by a point source that emits a fan- or wedge-shaped beam of radiation. Fan beam reconstruction enables data to be gathered much more quickly than by using a linear beam to produce parallel projections. *See* computed tomography, image reconstruction, projection, radon transform.

Fano algorithm   a sequential decoding algorithm for decoding of trellis codes.

Fano mode   a bound nonradiative surface mode that propagates along an interface and decays in a nonoscillatory manner in a direction perpendicular to the propagation. Such a mode occurs when one of the media is a plasma medium and has a negative dielectric function. A metal at optical frequency is an example of such a medium.

Fano's inequality   information theoretic inequality bounding the probability of incorrectly guessing the value of one random variable based on observation of another. If P_e is the probability of incorrectly guessing a random variable $X \in \mathcal{X}$, based upon observation of the random variable Y, then

$$H\left(P_e\right) + P_e \log(|\mathcal{X} - 1) \geq H(X|Y)$$

Named after its discoverer, R. M. Fano (1952). Used in proving the weak converse to the channel coding theorem (Shannon's second theorem).

Fano's limit   theoretical limit relating the achievable gain and bandwidth of a given passive lossless matching network when terminated in an arbitrary load impedance.

far field   (1) that region of space in which the electric field and magnetic field components of an electromagnetic wave are related by the impedance of free space. The far field is generally considered to begin no closer to the NIER source than a distance of several wavelengths or several times the antenna aperture.

(2) that region of the field of a certain source where the angular field distribution depends in a known way from the distance of the source. Generally, in free space, if we consider as a source an antenna with maximum overall dimension D, assumed large compared to the wavelength, the far-field region is commonly taken to exist at distances greater than $2D^2/\lambda$ from the antenna, λ

being the wavelength. Also called the Fraunhofer region.

far pointer   a pointer to a far segment. In 80 × 86 architecture, a far pointer specifies the segment address and the offset.

far-field pattern   graph or chart representing the absolute or normalized antenna gain as a function of angle (typically azimuth or elevation) and used to describe the directional properties of an antenna in the far field (Fraunhofer region).

far-infrared (FIR)   spectral region often considered to range from about 10 to 100 micrometers.

farad   the basic unit of measure in capacitors. A capacitor charged to 1 volt with a charge of 1 coulomb (1 ampere flowing for 1 second) has a capacitance of 1 farad.

Faraday effect   the rotation of the plane of polarization of a high-frequency signal (microwave RF, optical field) in the presence of a magnetic field.

Faraday rotation   (1) rotation in the direction of polarization experienced by a wave traveling through an anisotropic medium. Important examples of media in which the phenomenon occurs include the earth's ionosphere and ferrites biased by a static magnetic field.

(2) depolarization caused in a plasma (e.g., the ionosphere) resulting from interaction between the ions of the plasma and the magnetic field of the wave.

(3) rotation in the polarization vector experienced by a wave after it propagates through a gyromagnetic medium.

Faraday rotator   a magneto-optical device that changes the orientation plane of polarized light when it passes parallel to a magnetic field through a substance with pronounced absorption lines.

Faraday shield   an electrostatic (E field) shield made up of a conductive or partially conductive material or grid. A Faraday cage or screen room is effective for protecting inside equipment from outside radiated RF energies.

Faraday's law   one of Maxwell's equations that describes the fundamental relationship between induced voltage and a time-varying magnetic field. For a conducting coil, the induced voltage is proportional to the time rate of change in the magnetic flux linking the coil. This change may be produced either by actual variation of field strength or by relative motion between coil and field. *See also* Maxwell's equations.

Faraday, Michael   (1791–1867) Born: Newington, Surrey, England

Best known as the greatest experimental physicist of the 19th century. It was Faraday who invented the electric motor, generator, and transformer, and first described electromagnetic induction and the laws of electrolysis. Faraday had no formal schooling, although he attended many lectures. The most inspirational of these lectures were by the famed chemist Sir Humphry Davy. Faraday became Davy's assistant and thus began an extraordinary career as an experimentalist. Faraday's contributions are recognized by the use of his name as the unit of electrical capacitance, the farad in the SI system, and the Faraday constant in electrolysis.

farm   *See* processor farm.

fast Fourier transform (FFT)   a computational technique that reduces the number of mathematical operations in the evaluation of the discrete Fourier transform (DFT) to $N \log_2 N$.

fast neutrons   neutrons emitted from fission reactions which travel at velocities higher than the average velocity of atoms under random thermal motion.

fast packet networks   networks in which packets are transferred by switching at the frame layer rather than the packet layer. Such networks are sometimes called frame relay networks. Current thinking views frame relay as a service, rather than transmission, technology.

fast reactor   a reactor which maintains a critical chain reaction with fast neutrons and which does not require a moderator.

father wavelet   the scaling function in the coarsest resolution in wavelet analysis. *See* mother wavelet.

fault   (1) in hardware, a physical defect or imperfection of hardware. Typical circuit faults are shorts, opens in conductor, defects in silicon, etc. *See also* disturbance.

(2) in software, the manifestation of an error.

fault avoidance   a technique used to prevent or limit fault occurrence (for example, with signal shielding, fan-out limitation, and power dissipation decrease).

fault confinement   technique that limits the spread of fault effects to some area of the system and prevents propagation of these effects to other areas.

fault coverage   the measure of test quality expressed as the percentage of detected faults.

fault detection   the process of locating distortions or other deviations from the ideal, typically during the process of automated visual inspection, e.g., in products undergoing manufacture.

fault indicator   a small indicating unit equipped with a permanent magnet and pivoting pointer which is hung on a transmission line suffering intermittent faults of unknown origin. After a fault occurs, fault indicators are inspected. Each shows the presence and direction of a fault, thus allowing the defect to be located.

fault kva   fault kilovolt-amps (kva) is the fault level expressed in terms of volt-amps rather than amps. One advantage of using volt-amps rather than amps is that the same flow is experienced on both sides of a transformer when expressed in volt-amps, while the flow changes due to the transformer turns ratio when it is expressed in amps. Volt-amps for a three-phase fault are expressed as $1.73 \times$ rms line-line voltage $\times$ rms symmetrical fault current. Volt-amps for a single phase fault are defined as $1.73 \times$ rms line-line voltage $\times$ rms symmetrical current in the faulted phase.

fault latency   the length of time between the occurrence of a fault and the appearance of an error.

fault masking   a technique that hides the effects of faults with the use of redundant circuitry or information.

fault mva   fault megavolt-amps (mva) is the fault level expressed in terms of volt-amps rather than amps. One advantage of using volt-amps rather than amps is the same flow is experienced on both sides of a transformer when expressed in volt-amps, while the flow changes due to the transformer turns ratio when it is expressed in amps. Volt-amps for a three-phase fault are expressed as $1.73 \times$ rms line-line voltage $\times$ rms symmetrical fault current. Volt-amps for a single phase fault are defined as $1.73 \times$ rms line-line voltage$\times$ rms symmetrical current in the faulted phase.

fault prevention   any technique or process that attempts to eliminate the possibility of having a failure occur in a hardware device or software routine.

fault resistance   the resistance that occurs at the point of fault due to voltage drop across an arc or due to other resistance in the fault path.

fault secure   pertaining to a circuit, with respect to a set of faults, if and only if for any fault in this set, and any valid input code the output is a non-code or correct code (the output is never an invalid code). The circuit is considered to operate properly if the output is a code word.

fault simulation   an empirical method used to determine how faults affect the operation of the circuit and how much testing is required to obtain the desired fault coverage.

fault tolerance   correct execution of a specified function in a circuit (system), provided by redundancy despite faults. The redundancy provides the information needed to negate the effects of faults.

fault tree   the identification and analysis of conditions and factors that cause or contribute to the occurrence of a defined undesirable event, usually one that significantly affects system performance, economy, safety, or other required characteristics.

fault-tolerant control system   a system that exhibits stability and acceptable performance in the presence of component faults (failures) or large changes in the system that resemble failures.

fax   abbreviation for a machine that makes and transmits facsimiles. *See* facsimile.

FCA   *See* fixed channel assignment.

FCEV   *See* fuel cell electric vehicle.

FCM   *See* fuzzy cognitive map or fuzzy c-means.

FCT   *See* field controlled thyristor.

FDD   *See* frequency division duplex.

FDDI   *See* fiber distributed data interface.

FDM   *See* frequency division multiplexing.

FDMA   *See* frequency division multiple access.

FDTD   *See* finite difference time domain.

feature   a measurable characteristic of an object in an image. Simple examples would be area, perimeter, and convexity. More complex features use vectors; examples include moments, Fourier descriptors, projections, and histogram based features. Features are frequently used to recognize classes of object, and sets of simple features can be collected into a vector for this purpose. Using both area and perimeter, for instance, one can quickly distinguish between a circle and a triangle. May also refer to a characteristic of a whole image: such a feature could then be used in image database analysis. *See* object/feature.

feature detection   the detection of smaller features within an image with a view to inferring the presence of objects. This type of process is cognate to pattern recognition. Typically, it is used to locate products ready for inspection or to locate faults during inspection. A feature can be detected by finding points having optimal response to a given combination of local operations such as convolutions or morphological operators. *See* object detection.

feature extraction   a method of transforming raw data, which can have very high dimensionality, into a lower dimensional representation that still contains the important features of the data.

feature map   a fixed geometrical structure (often two dimensional) for unsupervised learning that maps the input patterns to different output units in the structure so that similar input patterns always trigger nearby output units topographically. *See also* self-organizing system, self-organizing algorithm.

feature measurement   the measurement of features, with the aim of recognition or inspection to determine whether products are within acceptable tolerances.

feature orientation   measurement of the orientation of features, either as part of the recognition process or as part of an inspection or image measurement process.

feature recognition   the process of locating features and determining what types of features they are, either directly or indirectly through the location of sub-features followed by suitable inference procedures. Typically, inference is carried out by application of Hough transforms or association graphs.

feature size   the characteristic size of electronic components on a die.

FEC   *See* forward error correction.

feedback   (1) signal or data that is sent back to a commanding unit from a control process output for use as input in subsequent operations. In a closed-loop system, it is the part of the system that brings back information on the process condition under control.

(2) the provision of a path from the output to the input of a system, such that the output may be made a function of both the input and the previous outputs of the system.

(3) the technique of sampling the output of an amplifier and using that information to modify the amplifier input signal. A portion of the output is "fed back" to the input. Positive feedback occurs when the output is added to the input; negative feedback occurs when the output is subtracted from the input. Negative feedback, invented by communications engineer Harold Black in 1928, usually results in a gain–bandwidth tradeoff: decreasing and stabilizing the amplifier gain, while increasing the bandwidth. According to Norbert Wiener, feedback is a method of controlling a system by reinserting into it the results of its past performance.

feedback amplifier   a circuit configuration of amplifiers that has a feedback path. A negative feedback configuration is commonly used in amplifiers for its stable performance, where a portion of the output signal is added to the input signal at 180 degrees out of phase. There are two types of basic feedback configurations: parallel feedback and a series feedback.

feedback control   the regulation of a response variable of a system in a desired manner using measurements of that variable in the generation of the strategy of manipulation of the controlling variable.

feedback decoding   a majority-logic decoding method for decoding of convolutional codes.

feedback linearization   a method of using feedback to cancel out nonlinearities in a dynamical system model so that the resulting closed-loop system model is linear. The method uses tools from differential geometry. *See also* Lie derivatives.

feedback oscillator   electronic circuits designed to provide specific signals of desired waveshape. The feedback oscillator should be envisioned as two basic network subsystems: the active portion (amplifier) and the passive (or feedback, FB) portion. The output of one subsystem is connected to the input of the other and vice versa. The passive network is lossy but provides the important function of establishing the necessary electrical phase shift to establish oscillation at the desired frequency. *See also* Barkhausen criterion.

feeder   overhead lines or cables that are used to distribute the load to the customers. They interconnect the distribution substations with the loads.

feeder circuit   an electrical circuit designed to deliver power from the service equipment or separately derived system to the branch circuit panelboard(s) on a facility. For large systems, there may be more than one level of feeder circuits. *See also* branch circuit.

feedforward amplifier   a circuit configuration of low distortion amplifiers where a distorted voltage extracted from the output signal is fed again to the output port so as to cancel the distorted voltage included in the output signal.

feedforward control   a form of compensation in which the measurement of a disturbance is used to take preventative control action before the effect of the disturbance is noted at the system output (and then only compensated by feedback control action). A typical feedforward control configuration is shown in the figure. The controller is designed to have the same effect as the cascade combination of the process and process D.

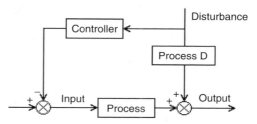

A typical feedforward control configuration.

feedforward network   *See* feedforward neural network.

feedforward neural network   one of two primary classifications of neural networks (the other is recurrent). In a feedforward neural network the **x**-vector input to the single functional layer of processing elements (this input typically occurs via a layer of input units) leads to the **y**-vector output in a single feedforward pass. An example of the feedforward neural network is a multilayer perceptron.

female connector   a connector presenting receptacles for the insertion of the corresponding male connector that presents pins.

fenestration   any opening or arrangement of openings (normally filled with media for control) for the admission of daylight.

frequency independent antenna   antenna with very large bandwidths. Upper frequency can be about 40 times higher than the lower frequency. Examples: spiral and log periodic antennas.

Ferranti, Sebastian Ziani de   (1864–1930) Born: Liverpool, Lancashire, England

Best known for developing systems of high-voltage AC power systems. The generating and transmission systems he designed still form the basis for most modern power systems. As a principal in the London Electric Supply Corporation, Ltd., Ferranti demonstrated that high-voltage AC current could be distributed and then stepped down for use in a more efficient and economical system than the smaller DC current systems then operating. Ferranti invented a number of other devices and systems as a consultant and in his own company. He was named president of the Institute of Electrical Engineers in 1911.

ferrite   a term applied to a large group of ceramic ferromagnetic materials usually consisting of oxides of magnesium, iron, and manganese. Ferrites are characterized by permeability values in the thousands and are used for RF transformers and high Q coils.

ferrite beads   small toroids made of ferrites which are slipped over a conductor in order to suppress RF currents. The beads act as RF chokes at high frequencies.

ferrite core   a magnetic core made up of ferrite (compressed powdered ferrromagnetic) material, having high resistivity and low eddy current loss.

ferrite core memory   *See* magnetic core memory.

ferrite material   a material that has very low conductivity (σ) and very large permeability (μ).

Its properties can be altered when an external magnetic field is applied. It is used in ferrite loaded loop antennas, for example, to increase the flux through the loop antenna.

ferroelectric material   a polar dielectric in which the crystallographic orientation of the internal dipole moment can be changed by the application of an electric field.

ferrofluid   iron based solution employed in voice coil/pole piece gap improving magnetic flux and power handling capacity.

ferromagnetic   materials in which internal magnetic moments spontaneously line up parallel to each other to form domains, resulting in permeabilities considerably higher than unity (in practice, 1.1 or more); examples include iron, nickel, and cobalt.

ferroresonance   a resonant phenomenon involving inductance that varies with saturation. It can occur in a system through the interaction of the system capacitance with the inductance of, for example, that of an open-circuited transformer. Ferroresonance resembles, to some extent, the normal resonance that occurs wherever L-C circuits are encountered. If the capacitance is appreciable, ferroresonance can be sustaining or result in a limited over voltage enough to damage the cable or the transformer itself.

ferroresonant transformer   a transformer that is designed to operate as a tuned circuit by resonating at a particular frequency.

Fessenden, Reginald Aubrey   (1866–1932) Born: East Bolton, Quebec, Canada

Best known as a radio pioneer who described the principle of amplitude modulation and the heterodyne effect. Fessenden became the chief scientist at Edison's laboratory and then spent two years with Westinghouse. He later held teaching posts at Purdue University and Western University of

Pennsylvania (now the U. of Pittsburgh.) During his career, Fessenden was the holder of over 500 patents — only Edison had more.

FET   *See* field-effect transistor.

fetch cycle   the period of time during which an instruction is retrieved from memory and sent to the CPU. This is the part of the fetch–decode–execute cycle for all machine instruction.

fetch on demand   *See* fetch policy.

fetch on miss   *See* fetch policy.

fetch policy   policy to determine when a block should be moved from one level of a hierarchical memory into the next level closer to the CPU.

There are two main types of fetch policies: "fetch on miss" or "demand fetch policy" brings in an object when the object is not found in the top-level memory and is required; "prefetch" or "anticipatory fetch policy" brings in an object before it is required, using the principle of locality. With a "fetch on miss" policy, the process requiring the objects must wait frequently when the objects it requires are not in the top-level memory. A "prefetch" policy may minimize the wait time, but it has the possibility of bringing in objects that are never going to be used. It also can replace useful objects in the top-level memory with objects that are not going to be used. *See also* cache and virtual memory. The prefetching may bring data directly into the relevant memory level, or it may bring it into an intermediate buffer.

fetch-and-add instruction   for a multiprocessor, an instruction that reads the content of a shared memory location and then adds a constant specified in the instruction, all in one indivisible operation. Can be used to handle multiprocessor synchronizations.

fetch–execute cycle   the sequence of steps that implement each instruction in a computer instruction set. A particular instruction is executed by executing the steps of its specific fetch–execute cycle. The fetch part of the cycle retrieves the instruction to be executed from memory. The execution part of the cycle performs the actual task specified by the instruction. Typically, the steps in a fetch–execute cycle are made up of various combinations of only three operations:

(1) the movement of data between various registers in the machine,

(2) the addition of the contents of two registers or the contents of a register plus a constant with the results stored in a register, and, less frequently,

(3) shift or rotate operations upon the data in a register.

fetching   the process of reading instructions from a stored program for execution.

FFT   *See* fast Fourier transform.

FH-CDMA   frequency hopping code division multiple access. *See* frequency hopping and code division multiple access.

FIB   *See* focused ion beam.

fiber Bragg grating   a distributed Bragg reflector written by ultraviolet light in the core of a photosensitive optical fiber. Multiple weak Fresnel reflections coherently add in phase to produce a strong reflection over a well defined narrow band of wavelengths.

fiber cladding   the region of an optical fiber having a lower index of refraction than the core region, to allow confinement of light in the core.

fiber distributed data interface (FDDI)   an American National Standards Institute standard for 100 megabits per second fiber-optic local-area networks. Incorporates token processing and supports circuit-switched voice and packetized data. For its physical medium, it uses fiber optic cable, in a dual counter-rotating ring architecture.

fiber optic bundle   an optical system to deliver the image to a video monitor. In a conventional endoscope, the image is transmitted through a series of lenses to a camera outside of the patient. In an electronic endoscope the image is collected by a CCD camera positioned inside the patient.

fiber optic sensor   a sensor employing an optical fiber to measure chemical composition or a physical parameter, such as temperature, pressure, strain, vibration, rotation or electromagnetic fields. Light is launched down a fiber to a transducer or sensing element that alters the properties of the light in response to the parameter being measured. The altered light is returned back down a fiber to a detector.

fiber Raman amplifier   provides amplification of signals through stimulated Raman scattering in silica fibers. Raman amplification differs from stimulated emission in that in stimulated emission, a photon of the same frequency as the incident photon is emitted where as in the case of Raman amplification, the incident pump photon looses energy to create another photon at a lower frequency.

fiber-optic cable   a glass fiber cable that conducts light signals and can be used in token ring local area networks and metropolitan area networks. Fiber optics can provide higher data rates than coaxial cable. They are also immune to electrical interference.

fiber-optic interconnect   interconnect that uses an optical fiber to connect a source to a detector. An optical fiber is used for implementing a bus. The merits are large bandwidth and high speed of propagation.

fidelity   a qualitative term used to describe how closely the output amplitude of a device faithfully reproduces that of its input. Faithful reproduction here refers to the preservation of those characteristics of the input signal (e.g., amplitude, frequency, and/or phase shift) essential to proper operation of the device under question. Waveform fidelity degradation is most often characterized by a change in signal shape (time domain), or a change in the relative scaling of its frequency components as the signal propagates through the circuit or system (frequency domain). *See also* amplitude linearity.

field   (1) the member of an electrical machine that provides the main magnetic flux, which then interacts with the armature causing the desired machine operation (i.e., motor or generator).

(2) a description of how a physical quantity varies as a function of position and possibly time. *See also* electric field, finite field.

field circuit   a set of windings that produces a magnetic field so that the electromagnetics induction can take place in electric machines.

field controlled thyristor (FCT)   a thyristor controlled by change in the magnitude of the field current.

field current control   a method of controlling the speed of a DC motor by varying the field resistance, thus producing a change in the field current.

field discharge resistor   a resistor used to dissipate the energy stored in the inductance of a field winding. It may be a standard power resistor that is connected across the winding just prior to opening the supply switch, or a permanent non-linear resistive device that has high resistance at normal voltage but low resistance when voltage rises at switching.

field loss protection   a fault-tolerant scheme used in electric motors. Some DC motor control circuits provide field loss protection in the event the motor loses its shunt field. Under a loss of field, DC motors may overspeed causing equipment damage and/or personal injury. In a motor controller that has field loss protection, a sensor determines when the shunt field has lost current flow, then secures the motor before an overspeed condition occurs.

field memory   video memory required to store the number of picture elements for one vertical scan (field) of video information of an interlaced scanned system.

The memory storage in bits is computed by multiplying the number of video samples made per horizontal line times the number of horizontal lines per field (vertical scan) times the number bits per sample. A sample consists of the information necessary to reproduce the color information.

Storage requirements can be minimized by sampling the color video information consisting of the luminance (Y) and the two color difference signals, (R - Y) and (B - Y). The color signal bandwidth is less than the luminance bandwidth that can be used to reduce the field memory storage requirements. Four samples of the luminance (Y) signal is combined with two samples of the (R - Y) signal and two samples of the (B - Y) signal. The preceding video sampling technique is designated as 4:2:2 sampling and reduces the field memory size by one-third. Field memory for NTSC video sampled at 4 times the color subcarrier frequency at 8 bits/pixel would require 3.822 megabits of RAM when 4:2:2 sampling is used.

field mill   a measuring instrument for electric fields consisting of a rotary array of blades which revolves near a capacitive pickup. The AC variation of the pickup's output is proportional to the electric field in the region near the instrument in the direction of the rotor axis.

field orientation   a term related to the vector control of currents and voltages that enables direct control of machine torque in AC machines. Vector control refers to the control of both the amplitude and phase angle, and the control results in a desired spatial orientation of the electromagnetic fields in the machine.

field oxide   an insulating silicon oxide layer used in integrated circuits to electrically isolate components.

field programmable gate array (FPGA)   (1) a programmable logic device that consists of a matrix of programmable cells embedded in a programmable routing mesh. The combined programming of the cell functions and routing network define the function of the device.

(2) a gate array with a programmable multi-level logic network. Reprogrammability of FPGAs make them generic hardware and allow them to be reprogrammed to serve many different applications. FPGAs consist of SRAMS, gates, latches, and programmable interconnects.

(3) a reconfigurable electronic device that has all of the advantages of one-time programmable devices such as PALs and PLAs, except that the FPGA can be reprogrammed in place.

field propagator   the analytical description of how electromagnetic fields are related to the sources that cause them. Common field propagators in electromagnetics are the defining Maxwell equations that lead to differential equation models, Green's functions that produce integral equation models, optical propagators that lead to optics models, and multipole expansions that lead to modal models.

field rate   the rate at which a field of video is generated. The field rate for NTSC television is 59.94 fields per second consisting of alternating even and odd fields. Each field consists of 262.5 lines of video, generating the even or odd numbered lines of video.

field reversing   a method of achieving a reversal of rotation of a DC motor by reversing the field flux.

field singularity   *See* edge condition.

field strength   in general terms the magnitude of the electric field vector (in volts per meter) or the magnitude of the magnetic field vector (in ampere-turns per meter). As used in the field of EMC/EMI, the term is applied only to measurements made in the far field and is abbreviated as FS. For measurements made in the near field, the term electric field strength (EFS) or magnetic field strength (MFS) is used, according to whether the resultant electric or magnetic field, respectively, is measured.

field weakening   a method of achieving speed increase in DC motors by reducing the field flux (increasing field circuit resistance).

field-by-field alignment   a method of alignment whereby the mask is aligned to the wafer for each exposure field (as opposed to global alignment).

field-effect transistor (FET)   a majority-carrier device that behaves like a bipolar transistor with the important difference that the gate has a very high input impedance and therefore draws no current.

An active device with three terminals — gate, source, and drain — in the active (amplifier) mode of operation, the drain current is related to the gate-source voltage. The relationship is usually approximated by a square law, but there are significant deviations from the square law depending on factors such as device geometry. The FET can also be used as a switch, with the gate-source voltage controlling the "on/off" state of the conducting channel between source and drain terminals. The input resistance at the gate is extremely high (usually of order tens of megaohms) and the gate current is negligibly small (usually of order picoamperes or less).

There are various families of FETs, including MOSFETs and JFETs. Within each family, there are two types of FET, n-channel and p-channel (named for the sign of the majority carriers that form the current conducting path between source and drain).

Some FETs also have a fourth terminal, the "substrate" or "body" terminal. The p-n junctions between the substrate and the drain and source terminals should be reverse-biased to insure proper device operation.

field-oriented control   speed control of an induction motor obtained by varying the magnitude and orientation of the airgap magnetic field. This is also referred to as vector control and requires sensing of the rotor position. Vector controllers allow the induction motor to operate very much like a DC motor, including development of rated torque at zero speed.

FIFO   *See* first-in-first-out.

FIFO memory   commonly known as a queue. It is a structure where objects are taken out of the structure in the order they were put in. Compare this with a LIFO memory or stack. A FIFO is useful for buffering data in an asynchronous transmission where the sender and receiver are not synchronized: the sender places data objects in the FIFO memory, while the receiver collects the objects from it.

figure of merit   performance evaluation measure for the various target and equipment parameters of a sonar system. It is a subset of the broader sonar performance given by the sonar equations, which includes reverberation effects.

file format   the structure of the computer file in which an image is stored. Often the format consists of a fixed-size header followed by the pixel values written from the top to the bottom row and within a row from the left to the right column. However, it is also common to compress the image. *See* graphics interchange format, header, image compression, tagged image file format (TIFF).

Filippov method   a definition of a solution to a system of first-order differential equations with discontinuous right-hand side,

$$\dot{\mathbf{x}} = \mathbf{f}(t, \mathbf{x}),$$

proposed by A. F. Filippov. A vector function $\mathbf{x}(t)$ defined on the interval $[t_1, t_2]$ is a solution to the above system of differential equations in the sense of Filippov, if it is absolutely continuous and for almost all $t \in [t_1, t_2]$,

$$\dot{\mathbf{x}}(t) \in \mathbf{F}(t, \mathbf{x}),$$

where $\mathbf{F}(t, \mathbf{x})$ is an appropriately constructed convex set. *See also* differential inclusion.

fill-in when solving a set of sparse linear equations using Gaussian elimination, it is possible for a zero location to become nonzero. This new nonzero is termed a fill-in.

FILO *See* first-in-last-out.

filter (1) a network, usually composed of inductors and capacitors (for lumped circuit), or transmission lines of varying length and characteristic impedance (for distributed circuit), that passes AC signals over a certain frequency range while blocking signals at other frequencies. A bandpass filter passes signals over a specified range (flow to fhi), and rejects frequencies outside this range. For example, for a DBS receiver that is to receive satellite transmitted microwave signals in a frequency range of 11 GHz to 12 GHz, a band-pass filter (BPF) would allow signals in this frequency range to pass through with minimum signal loss, while blocking all other frequencies. A low-pass filter (LPF) would allow signals to pass with minimum signal loss as long as their frequency was less than a certain "cutoff frequency" above which significant signal blocking occurs.

(2) an operator that transforms image intensity $I\mathbf{x}$ of pixel $\mathbf{x}$ into a different intensity $\hat{I}\mathbf{x}$, depending on the values of a set of (usually neighboring) pixels (which may or may not include $\mathbf{x}$). Filtering is performed to enhance significant features of an image or to remove nonsignificant ones or noise.

filtered backprojection an algorithm for image reconstruction from projections. In the filtering part of the algorithm, the projections are measured, their Fourier transforms computed, and the transforms are multiplied (filtered) by a weighting function. In the backprojection part, the inverse Fourier transforms of the weighted projections are computed and summed to yield the reconstructed image. Filtered backprojection is the reconstruction algorithm currently used by almost all commercial computed tomography scanners. *See also* Fourier transform, image reconstruction, projection, radon transform, reconstruction, tomography.

filter bank a set of filters consisting of a bank of analysis filters and a bank of synthesis filters. The analysis filters decompose input signal spectra into a number of directly adjacent frequency bands for further processing, and the synthesis filters recombine the signal spectra from different frequency bands.

filtering (1) an estimation procedure in which the present value of the state vector (see the definition) is estimated based on the data available up to the present time.

(2) the process of eliminating object, signal or image components which do not match up to some pre-specified criterion, as in the case of removing specific types of noise from signals. More generally, the application of an operator (typically a linear convolution) to a signal.

fin efficiency a thermal characteristic of an extended surface that relates the heat transfer ability of the additional area to that of the base area.

final test electrical test performed after assembly to separate "good" devices from "bad."

finesse measure of the quality of a Fabry-Perot interferometer; free spectral range divided by linewidth (full width at half maximum).

finger stick an insulated stick like a hot-stick used to actuate a disconnect-switch atop a pole.

finite difference method a numerical technique for solving a differential equation wherein the differential equation is replaced by a finite difference equation that relates the value of the solution at a point to the values at neighboring points.

finite difference time domain (FDTD) a numerical technique for the solution of electromagnetic wave problems that involves the mapping of the Maxwell equations onto a finite difference mesh and then following the time evolution of an initial value problem. This technique is widely used to investigate the performance of a complex RF structures.

finite differences a method used to numerically solve partial differential equations by replacing the derivatives with finite increments.

finite element a numerical technique for the solution of boundary value problems that involves the replacement of the set of differential equations describing the problem under consideration with a corresponding set of integral equations. The area or volume of the problem is then subdivided with simple shapes such as triangles and an approximation to the desired solution with free parameters is written for each subregion and the resulting set of equations is minimized to find the final solution. This approach is useful for solving a variety of problems on complex geometries.

finite-extent sequence the discrete-time signals with finite duration. The finite-extent sequence $\{x(n)\}$ is zero for all values of n outside a finite interval.

finite field a finite set of elements and two operations, usually addition and multiplication, that satisfy a number of specific algebraic properties. In honor of the pioneering work by Evariste Galois, finite fields are often called Galois fields and denoted $GF(q)$, where q is the number of elements in the field. Finite fields exist for all q which are prime or the power of a prime.

finite impulse response (FIR) filter any filter having an impulse response which is nonzero for only a finite period of time (therefore having a frequency response consisting only of zeros, no poles). For example, every moving average process can be written as the output of a FIR filter driven by white noise. Impulse response function, moving average, infinite impulse response (IIR) filter.

finite state machine (FSM) a mathematical model that is defined in discrete time and has a finite number of possible states it can reside in. At each time instance, an input, x, is accepted and an output, y, and a transition from the current state, S_c, to a new state, S_n, are generated based on separate functions of the input and the current state. A finite state machine can be uniquely defined by a set of possible states, S, an output function, $y = f(x, S_c)$, and a transition function, $S_n = g(x, S_c)$. An FSM describes many different concepts in communications such as convolutional coding/decoding, CPM modulation, ISI channels, CDMA transmission, shift-register sequence generation, data transmission and computer protocols. Also known as finite state automata (FSA), state machine.

finite state VQ (FSVQ) a vector quantizer with memory. FSVQ form a subset of the general class of recursive vector quantization. The next state is determined by the current state S_n together with the previous channel symbol u_n by some mapping function.

$$S_{n+1} = f(u_n, S_n), n = 0, 1, \ldots$$

This also obeys the minimum distortion property

$$\alpha(\mathbf{x}, s) = \overset{-1}{\min} d(\mathbf{x}, \beta(u, s))$$

with a finite state $S = [\alpha_1, \alpha_2, \ldots, \alpha_k]$, such that the state S_n can only take on values in S. The states can be called by names in generality.

finite wordlength effect   any perturbation of a digital filter output due to the use of finite precision arithmetic in implementing the filter calculations. Also called quantization effects.

FIR   *See* far-infrared.

firing angle   time in electrical degrees from the instant the valve voltage is positive to the application of firing pulse to the valve (start of conduction). Also called delay angle.

firm power   an amount of electric power intended to be available at all times to a commercial customer, regardless of system conditions.

firm real-time   *See* firm real-time system.

firm real-time system   a real-time system that can fail to meet one or more deadlines without system failure. *Compare with* soft real-time, hard real-time.

firmware   software that cannot be modified by the end user.

first difference   for a sequence $\{x(n)\}$ the sequence obtained by simply subtracting its $(n-1)$th element from its nth element, i.e.,

$$y(n) = x(n) - x(n - 1)$$

first metal   *See* first metal layer.

first metal layer   the first metal layer applied in a fabrication process. The second layer applied would be the second metal layer and so on. Also known as first metal, first metallization, and metal 1.

first metallization   *See* first metal layer.

first order hold (FOH)   for a signal $f(k)$, the sequence of straight lines connecting the sample points of $f(k)$. It interpolates the values between two adjacent samples $f(k)$ and $f(k + 1)$ using a linear approximation given by

$$x(t) = x(kT_s) + \frac{t - kT_s}{T_s} \left(x((k + 1)T_s) - x(kT_s) \right)$$

first order system   the system that can be described by a linear first-order difference equation. The output of the first-order system $y(n)$ is equal to a linear combination of the past output value $y(n - 1)$ and the input value $x(n)$, i.e.,

$$y(n) = \alpha x(n) + \beta y(n - 1)$$

first-fit memory allocation   a memory allocation algorithm used for variable-size units (e.g., segments). The "hole" selected is the first one that will fit the unit to be loaded. This hole is then broken up into two pieces: one for the process and one for the unused memory, except in the unlikely case of an exact fit, there is no unused memory.

first-in-first-out (FIFO)   a queuing discipline whereby the entries in a queue are removed in the same order as that in which they joined the queue.

first-in-last-out (FILO)   a queuing rule whereby the first entries are removed in the opposite order as that in which they joined the queue. This is typical of Stack structures and equivalent to last-in-first-out (LIFO).

first-swing stability   criterion to determine transient stability by use of the swing equation. The rotor angle immediately following a severe disturbance usually increases. The criterion states that if the rotor angle swings back and decreases a short time after the disturbance, then the system is first-swing stable.

Fisher information   a quantitative measurement of the ability to estimate a specific set of parameters. The Fisher information $J(\theta)$ is

fissile material

defined by

$$J(\theta) = E_\theta \left(\frac{\partial \ln f_\theta(y)}{\partial \theta} \right)^2 = -E_\theta \left(\frac{\partial^2 \ln f_\theta(y)}{\partial^2 \theta} \right)$$

where Y is a N-dimensional vector indexed by a vector of parameters θ. *See also* Cramer-Rao bound.

fissile material   an isotope which has a significant probabilitty of undergoing nuclear fission, e.g., U235, plutonium-239, thorium-232, and enriched uranium.

fission   the nuclear reaction in which a single heavy nucleus is split into two or more lighter nucleii called "daughter" products and emit highly energetic sub-atomic particles plus energy in the process.

fixed channel assignment (FCA)   a technique of assigning radio channels in a communications system in a fixed and predetermined way in accordance with predicted rather than actual interference and propagation conditions. Such assignments are not changed during radio transmission.

fixed losses   that component of the copper losses in DC shunt, short-shunt, and long-shunt machines' field circuit, that does not vary with change in the load current. With a fixed field power supply, it is an accepted industry agreement to not consider the losses in the field circuit rheostat in computing the efficiency and hence consider the field losses as fixed losses.

fixed point   *See* equilibrium point.

fixed reference D/A converter   the analog output is proportional to a fixed (nonvarying) reference signal.

fixed resolution hierarchy   an image processing scheme in which the original and reconstructed image are of the same size. Pixel values are refined as one moves from level to level. This is primarily used for progressive transmission. Tree-structured VQ and transform-based hierarchical coding are two of the fixed resolution hierarchies.

fixed shunt   *See* shunt capacitor.

fixed station (FS)   that part of a radio communications system that is permanently located in a given geographical location. Another name for the base station in a cellular radio telephone network.

fixed termination   a broadband termination of a specific nominal impedance, usually 50 ohms, used to terminate the signal path of a transmission line. Fixed terminations are used in the calibration of network analyzers and general microwave measurements.

fixed-gap head   *See* disk head.

fixed-head disk   a disk in which one read/write head unit is placed at every track position. This eliminates the need for positioning the head radially over the correct track, thus eliminating the "seek" delay time. Rarely used today because modern disks consist of hundreds of tracks per disk surface, making it economically infeasible to place a head unit at every track.

fixed-length instruction   the machine language instructions for a computer all have the same number of bits.

fixed-point processor   a processor capable of operating on scaled integer and fractional data values.

fixed-point register   a digital storage element used to manipulate data in a fixed-point representation system whereby each bit indicates an unscaled or unshifted binary value. Common encoding schemes utilized in fixed-point registers include Unsigned Binary, Sign-Magnitude, and Binary Coded Decimal representations.

fixed-point representation   (1) a number representation in which the radix point is assumed to be located in a fixed position, yielding either an integer or a fraction as the interpretation of the internal machine representation. *Contrast with* floating-point representation.

(2) a method of representing numbers as integers with an understood slope and origin. To convert a fixed-point representation to its value, multiply the representation by the slope (called the DELTA in Ada) and add the origin value. Fixed-point representation provides fast computations for data items that can be adequately represented by this method.

FLA   *See* full load amperage.

flag   (1) a bit used to set or reset some condition or state in assembly language or machine language. For instance, the inheritance flag, and the interrupt flag. As an example, each maskable interrupt is enabled and disabled by a local mask bit. An interrupt is enabled when its local mask bit is set. When an interrupt's trigger event occurs, the processor sets the interrupt's flag bit.

(2) a variable that is set to a prescribed state, often "true" or "false," based on the results of a process or the occurrence of a specified condition. Same as indicator.

flag register   (1) a register that holds a special type of flag.

(2) a CPU register that holds the control and status bits for the processor. Typically, bits in the flag register indicate whether a numeric carry or overflow has occurred, as well as the masking of interrupts and other exception conditions.

flash EEPROM   *See* flash memory.

flash memory   a family of single-transistor cell EPPROMs. Cell sizes are about half that of a two-transistor EEPROM, an important economic consideration. Bulk erasure of a large portion of the memory array is required.

The mechanism for erasing the memory is easier and faster than that needed for EEPROM. This allowed their adoption for making memory banks on PCMCIA for replacing hard disks into portable computers. Recently, also used for storing BIOS on PC main boards. In this way, the upgrading of BIOS can be made by software without opening the computer by non-specifically skilled people.

flashover   arcing between segments of the commutator of a DC machine. Flashover may occur due to distortion of the airgap flux as a result of heavy overloads, rapidly changing loads, or operation with a weak main field.

flat address   a continuous address specification by means of a unique number. Operating systems such as Windows NT, OS/2 and Mac OS use such a type of address. MS-DOS adopts the real-mode that adopts a segmented memory.

flat fading   *See* frequency nonselective fading.

flat pack   a component with two straight rows of leads (normally on 0.050-in centers) that are parallel to the component body.

flat panel display   a very thin display screen used in portable computers. Nearly all modern flat-panel displays use liquid crystal display technologies. Most LCD screens are backlit to make them easier to read in bright environments.

flat topping   a distortion mechanism in an amplifying stage in which the stage is unable to faithfully reproduce the positive peaks of the output signal. Reasons for such problems include poor regulation of the plate voltage supply.

flat voltage start   the usual initial assumption made when beginning a power-flow study. All voltages are assumed to be 1.0 p.u., and all angles are assumed to be zero.

flat-compounded   characteristic of certain compound-wound DC generator designs in which the output voltage is maintained essentially constant over the entire range of load currents.

Fleming, John Ambrose   (1849–1945) Born: Lancaster, Lancashire, England

Best known as the inventor of the thermionic diode valve, also known as the Fleming valve. This valve was later adapted by Lee DeForest into a form of vacuum tube he called the Audion Triode. Fleming was an avid experimentalist as well as a solid theoretician. He began his studies under James Clerk Maxwell and ended up consulting for Edison, Swan, and Ferranti, as well as Marconi. Thus Fleming can be considered one of the earliest pioneers in the field of radio and television electronics.

Fletcher-Powell algorithm   an iterative algorithm of Fletcher and Powell (1963) for finding a local minimum of the approximation error. In its original form, the Fletcher-Powell algorithm performs an unconstrained minimization.

flexible AC transmission system   a transmission scheme in which each power line is maintained at its optimal impedance, generally by means of thyristor-controlled series compensators.

flexible link   opposite to rigid link, is subject to elastic deformations of its structure. Basic assumption is that its model is described by the so-called Euler–Bernoulli beam equations of motion in which of second or higher order in the deformation variables are neglected.

flexible manipulator   member of a class of robot manipulators with flexible links. *See* flexible link.

flexible manufacturing systems (FMS)   manufacturing systems that deal with high-level distributed data processing and automated material flow using computer-controlled machine tools, robots, assembly robots and automated material handling, etc., with the aim of combining the benefits of a highly productive transfer line, and a highly flexible job shop.

flexure node   standing wave pattern in loudspeaker membrane caused by applied energy.

flicker   (1) the apparent variation of lighting luminance over time to an observer.

(2) the apparent visual interruptions produced in a TV picture when the field (one half of a TV frame) frequency is insufficient to completely synchronize the visual images. Occurs when the field rate is too low and the biologic characteristic of the human eye known as the persistence of vision does not give the illusion of continuous motion. Flicker is eliminated if the field rate exceeds 50 to 60 Hz.

(3) repetitive sags and swells in the electric service voltage, often accompanied by periodic harmonic distortion.

flicker fusion   the perception by the human visual system of rapidly varying lights (flicker) as being steady (fused). Flicker fusion is why florescent lights and scenes in movies and television appear to have constant illumination.

flicker noise   occurs in solid-state devices and vacuum tubes. Its power varies inversely with frequency; therefore, it is also called $1/f$ noise.

flip-chip for microwave   a packaging interconnection, in which the chips are bonded with contact pads, face down by solder bump connection. The solder bumps support the weight of the chip and control the height of the joint. It is a very good approach for high-frequency packages.

flip-flop   a basic digital device capable of storing one bit of information (1 or 0). In an "edge-triggered" flip/flop, information is stored in the device at the transition (positive or negative) of a clock-signal. A flip/flop is typically constructed using two "latches" in series. The first one (the "master") opens and closes by one clock-phase, and the second (the "slave") opens and closes using another (non-overlapping) phase. A flip/flop may be implemented using a static or dynamic logic style. In the latter case, the size is only half of the static version, but the information is lost (due to leakage) after some time, needing "refresh" (or storing new data) for proper operation.

float switch   a switch that is operated by a fluid level in a tank or process channel.

floating point characteristic   part of a number that represents the exponent.

floating point operation   arithmetic operation (e.g., add or multiply) involving floating point numbers (i.e., numbers with decimal points).

floating-point operations per second (FLOP) a measure of processing speed, as in megaFLOPs or gigaFLOPs.

floating-point register   a register that holds a value that is interpreted as being in floating-point format.

floating-point representation   in floating-point notation, a number is represented as a fractional part times a selected base (radix) raised to a power. This is the counterpart of scientific notation used in digital systems. The decimal equivalent of the floating-point value can be written $N.n = f \times r^e$ where f is the fraction or mantissa and e is a positive or negative integer called the exponent. *Contrast with* fixed-point representation.

floating-point unit   a circuit that performs floating point computations, which is generally addition, subtraction, multiplication, or division.

FLOP   *See* floating-point operations per second.

floppy disk   a flexible plastic disk coated with magnetic material. Enclosed in a cardboard jacket having an opening where the read/write head comes into contact with the diskette. A hole in the center of the floppy allows a spindle mechanism in the disk drive to position and rotate the diskette. A floppy disk is a smaller, simpler, and cheaper form of disk storage media, and is also easily removed for transportation.

Floquet mode   a solution to Maxwell's equations that can be supported by an infinite, periodic structure.

Floquet's theorem   a basic theorem underlying the theory of wave propagation in periodic structures.

flow dependency   *See* true data dependency.

flow diagram   *See* flowchart.

flowchart   a traditional graphic representation of an algorithm or a program, in using named functional blocks (rectangles), decision evaluators (diamonds), and I/O symbols (paper, disk) interconnected by directional arrows which indicate the flow of processing. Also called flow diagram.

flower pot   a cover for the bushing of a pad-mount transformer.

fluidized bed combustion   a method of solid-fuel combustion in which the fuel, usually coal, is pulverized and mixed with a ballasting substance and burned on a bed of pressurized air. If the ballasting agent is crushed limestone, sulfur from the coal is absorbed and carried out as solid ash.

fluorescence   emission of light from an electronically excited state that was produced by absorption of radiation with a wavelength shorter than the emitted light. Fluorescence emission is a quantum mechanically allowed transition between electronic levels of the same spin state, resulting in emission of light with a very short lifetime, typically nanoseconds.

fluorescent lamp   typically a lamp made by exciting a low pressure discharge in mercury vapor and other gases; mercury, when excited in the discharge, predominantly emits 257 nm radiation (ultraviolet) which is absorbed by a phosphor on the inside wall of the lamp tube; the phosphor fluoresces, emitting a white light spectrum in the visible.

fluoroscopy   a mounted fluorescent screen on which the internal structure or parts of an optically opaque object may be viewed as shadows formed by the transmission of X-rays through the object.

flush   the act of clearing out all actions being processed in a pipeline structure. This may be achieved by aborting all of those actions, or by refusing to issue new actions to the pipeline until those present in the pipeline have left the pipeline because their processing has been completed.

flux   (1) lines that indicate the intensity and direction of a field. Intensity is usually represented by the density of the lines.
(2) a measure of the intensity of free neutron activity in a fission reaction, closely related to power, the product of neutron density and neutron velocity, e.g., neutrons per square cm per second.

flux density   lines of magnetic flux per unit area, measured in tesla; $1T = 1Wb/m^2$.

flux line   *See* direction line.

flux linkage   quantity that indicates the amount of flux associated with a coil. Flux linkage is denoted by the symbol λ and expressed in Webers (Wb) or Weber-Turns (Wb-t). For a single turn coil, flux linkage is the same as the flux. Flux linkages of an N turn coil are $N\phi$ Wb-t.

fluxmeter   an instrument that measures the change in magnetic flux within a coil by integrating the induced voltage with respect to time.

flyback converter   the isolated version of the buck-boost converter. In this case, the transformer is also an inductor that stores energy when the transistor is on and releases energy to the output when the transistor is off. The primary and the secondary of the transformer conduct alternately. The advantage of this topology is that through the addition of a second winding on the input inductor, i.e., to form a transformer, electrical isolation is achieved. This type of transformer is called a flyback transformer or flyback inductor. *See also* buck-boost transformer.

flyback inductor   *See* flyback converter.

flyback transformer   *See* flyback converter.

flying head disk   a disk storage device that uses a read/write head unit "flying" over (i.e., very closely above) the disk surface on a thin air bearing. Used e.g., in "Winchester" disks (a sealed "hard disk"). This is in contrast to, e.g., floppy disks, where the head unit is actually in physical contact with the disk when reading or writing. *See also* disk head.

Flynn's taxonomy   a classification system that organizes computer processor types as either single-instruction stream or multiple-instruction stream and either single-data stream or multiple-data stream. The four resultant types of computer processors are known as SISD, MISD, SIMD, and MIMD (see table below). Due to Michael J. Flynn (1966).

flywheel   a heavy wheel placed on the shaft of an electrical machine for storing kinetic energy. A flywheel may be used to help damp speed transients or, to help deliver energy to impact loads such as a punch press.

FM   *See* frequency modulation.

FMS   *See* flexible manufacturing systems.

FM sound   a carrier wave whose instantaneous frequency is varied by an amount proportional to the instantaneous amplitude of the sound input modulating signal.

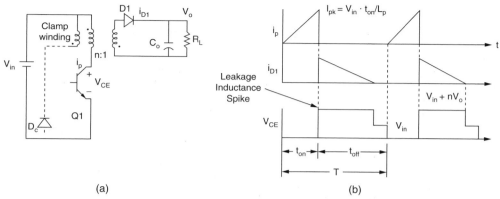

Flyback converter.

Flynn's Classification for Computer Architectures

	Single-Data Stream	Multiple-Data Stream
Single-Instruction Stream	von Neumann architecture/uniprocessors RISC	Systolic processors Wavefront processors
Multiple-Instruction Stream	Pipelined architectures Very long instruction word processors	Dataflow processors Transputers

focal length   distance from a lens or mirror at which an input family of parallel light rays will be brought to a focus.

focus   the position of the plane of best focus of the optical system relative to some reference plane, such as the top surface of the resist, measured along the optical axis (i.e., perpendicular to the plane of best focus).

focus of attention   the center of the region detected by a visual attention mechanism, which is the first stage of a pattern recognition system that is aimed to detect regions where a target object is likely to be found.

focus of expansion   the point in an image from which feature points appear to be diverging when the camera is moving forwards or objects are moving towards the camera.

focus-exposure matrix   the variation of linewidth (and possibly other parameters) as a function of both focus and exposure energy. The data is typically plotted as linewidth versus focus for different exposure energies, called the Bossung plot.

focused ion beam (FIB)   a lithography technique similar to electron beam lithography in which a stream of charged ions is raster-scanned to produce an image in a resist.

focuser   a focusing electrode for an electron steam in a vacuum tube.

fold-over distortion   the names given to the phenomena associated with the generation of an erroneously created signal in the sampling process are called "fold-over" distortion or aliasing. Sampling a signal exhibits many of the same properties that a "mixer" circuit in RF communications possesses. The mathematical relationship for a mixer

269

circuit and a sampling circuit is expressed by the trigonometric identity

$$\sin A \cdot \sin B = .5\cos(A - B) - .5\cos(A + B)$$

From the equation it is evident that if the A radian frequency is not twice the B radian frequency, then the $A - B$ term will produce a signal whose radian frequency is less than B's. This created signal will appear within the original frequency bandwidth. These error signals can be minimized by incorporating an anti-aliasing filter (i.e., a low-pass filter) on the input to the sample-and-hold circuit. The anti-aliasing filter bandlimits the input frequencies so that fold-over distortion or aliasing is minimized.

folded sideband   an FM method used frequently in video tape recording (VTR) systems. The term refers to the fact that there is no corresponding upper sideband to the one existing sideband.

folding   the technique of mapping many tasks to a single processor.

foldover   a form of distortion in a digital communications system which occurs when the minimum sampling rate is less that two times the analog input signal frequency.

footing impedance   the electrical impedance between a steel tower and distant earth.

foot-candle   the unit of illuminance when the foot is taken as the unit of length. It is the illuminance on a surface one square foot in area on which there is a uniformly distributed flux of one lumen.

force manipulability ellipsoid   a model that characterizes the end-effector forces and/or torques, F, that can be generated with the given set of joint forces and/or torques, τ, with the manipulator in a given posture. Mathematically assuming that the joint forces and/or torques, satisfy the equation $\tau^T \tau = 1$, the ellipsoid (taking

into account kineto-static duality) can be written as follows: $F^T(JJ^T)F = 1$. Here, J denote geometric Jacobian of the manipulator. *See also* kineto-static duality.

forced commutation   the use of external circuitry to artificially force a current zero, thus allowing a diode or thryistor to turn off.

forced interruption   an interruption in electric supply caused by human error, inappropriate equipment operation, or resulting from situations in which a device is quickly taken out of service by automatic or manual switching operations.

forced outage   the unscheduled interruption of electric power to a portion of a power system due to equipment failure, weather conditions, or other mishaps.

forced outage rate   a measure of performance usually applied to generation units. It is the ratio of equipment down-time vs. the total time that the unit is available for operation.

forced system   a dynamic system is said to be forced if it is excited by a nonzero external source.

foreground   (1) a process that is currently the process interacting with the user in a shell, as opposed to background processes, which get suspended when they require input from the user.

(2) a term describing a Unix command that is using your terminal for input and output, so that you will not get another Unix prompt until it is finished.

(3) the foreground of a window is a single color or a pattern. The foreground text or graphics is displayed against the background of the window.

fork   (1) the action of a process creating another (child) process without ending itself.

(2) in Unix, a process that can create a new copy of itself (a child) that has its own existence until either it terminates (or is killed) or its parent terminates. The prepare handler is called before

the processing of the fork subroutine commences. The parent handler is called after the processing of the fork subroutine completes in the parent process. The child handler is called after the processing of the fork subroutine completes in the child process.

form   term used to indicate the structure and dimensions of a multiterm equation without details within component terms.

formants   the main frequencies that result from typical short-time spectral analysis of vowels. Depending on the application, two or three formants are commonly extracted.

forming gas   a mixture of hydrogen and nitrogen (typically 5% hydrogen and 95% nitrogen),

typically used in the post-metal annealing process in wafer fabrication.

forward channel   also referred to as forward link, it is the radio link (channel) in a cellular or microcellular network where the base station is the transmitter and the mobile terminal is the receiver. In the case of cellular networks, the term is synonymous with down-link. The mobile to base station link is known as the reverse link or up-link.

forward converter   an isolated version of a buck converter. The primary and the secondary of the transformer conduct at the same time. The transformer does not store energy. An additional winding with a diode is often used to reset the magnetizing energy when the transistor is off.

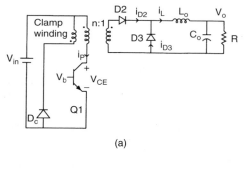

(a)

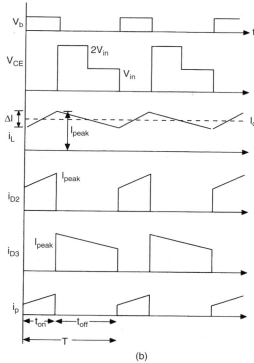

(b)

Forward converter.

271

forward error correction (FEC) an error control system used for simplex channels (only a forward channel) where the channel code is used to determine the most likely transmitted sequence of information symbols. *Compare with* automatic repeat request.

forward error recovery a technique (also called roll-forward) of continuing processing by skipping faulty states (applicable to some real-time systems in which occasional missed or wrong responses are tolerable).

forward voltage the voltage across the device when the anode is positive with respect to the cathode.

forward wave in a traveling wave tube, microwaves propagate in the same direction with electrons in the electron beam. Electromagnetic waves travel in the positive direction of the coordinate system.

forward–backward procedure an efficient algorithm for computing the probability of an observation sequence from a hidden Markov model.

forwarding (1) sending a just-computed value to potential consumers directly, without requiring a write followed by a read.

(2) to immediately provide the result of the previous instruction to the current instruction, at the same time that the result is written to the register file. Also called bypass.

Foster's reactance theorem states that the driving-point impedance of a network composed of purely capacitive and inductive reactances is an odd rational function of frequency.

Foster–Seeley discriminator similar to the balanced slope detector. A circuit used to convert an incoming FM signal to a related AM signal to provide a demodulation process.

foundry for semiconductor manufacturing, a vendor providing wafer processing.

four connected *See* pixel adjacency.

four-level laser laser in which the most important transitions involve only four energy states; usually refers to a laser in which the lower level of the laser transition is separated from the ground state by much more than the thermal energy kT (contrast with three-level laser).

four-point starter a manual motor starter that requires a fourth terminal for the holding coil. Because of its independent holding coil circuit, it is possible to vary the current in the field circuit independently of the holding coil circuit. The disadvantage is that the motor starter holding relay will not drop out with loss of the field; however, proper overcurrent protection should shut down the motor in the event of field loss.

four-quadrant operation (1) a signed representation of electrical or mechanical variables in the phase plane in order to situate the different modes for energy transfer. This term can be used both for power electronics and electrical machines. For electrical variables, the four-quadrant operation is defined by the voltage–current (or current–voltage) characteristic with the two variables expressed as instantaneous or mean values. For mechanical variables, the term is defined by the torque–speed (or speed–torque) characteristic with the same time-domain representation as previously explained. The energy transfer is defined with electrical power in the voltage–current curve and with mechanical power in the torque–speed curve. The four-quadrant operation is related to reversible power in electromechanical systems. For example, in the case of power electronics, the four-quadrant operation is defined only with electrical variables to visualize the way of energy transfer in a static converter. For electrical machines, the same operation can be defined with electrical variables at the input (motor) or at the output (generator) and also with mechanical variables at the output (motor) or at the input (generator).

(2) the four combinations of forward/reverse rotation and forward/reverse torque of which a regenerative drive is capable. These are: motoring: forward rotation/forward torque; regeneration: forward rotation/reverse torque; motor: reverse rotation/reverse torque; and regeneration: reverse rotation/forward torque.

four-wave mixing a nonlinear optical phenomenon in which four optical beams interact inside nonlinear media or photorefractive crystals. When four beams of coherent electromagnetic radiation intersect inside a nonlinear or photorefractive medium, they will, in general, form six interference patterns and induce six volume refractive index gratings in the medium. The presence of the index gratings will affect the propagation of these four beams. This may lead to energy coupling. The coupling of the four optical beams is referred to as four-wave mixing. In one of the most useful four-wave mixing configurations, the four beams form two pairs of counterpropagating beams. In this particular configuration, some of the refractive index gratings are identical in their grating wavevectors. This leads to the generation of phase conjugate waves. Four-wave mixing is a convenient method for the generation of phase conjugated waves.

four-way interleaved splitting a resource into four separate units that may be accessed in parallel for the same request (usually in the context of memory banks).

Fourier amplitude the amplitude angle (modulo 2π) taken by the Fourier transform. Specifically, given the Fourier transform $F(f) = A(f)(u)e^{i\,\Phi(f)(u)}$, then $A(f)$ represents the amplitude. *See* Fourier transform.

Fourier binary filter a filter placed in the Fourier plane of an optical system constructed with only two amplitude or two phase values.

Fourier filter a filter, or mask, placed in the Fourier transform plane.

Fourier integral the integral which yields the Fourier transform of an absolutely integrable function f over n-dimensional Euclidean space:

$$F(f)(\mathbf{u}) = \int f(\mathbf{x})\,e^{-i\,\mathbf{u}\cdot\mathbf{x}}d\mathbf{x}$$

where the frequency $\mathbf{u}$ is expressed in radians. *See* Fourier transform.

Fourier optics optical systems that utilize the exact Fourier transforming properties of a lens.

Fourier optics relay lens a lens system that produces the exact Fourier transform of an image. Two such relay lens will reproduce an image without any phase curvature.

Fourier phase the phase angle (modulo 2π) taken by the Fourier transform. Specifically, given the Fourier transform $F(f) = A(f)(u)e^{i\,\Phi(f)(u)}$, then $\Phi(f)$ represents the phase. *See* Fourier transform.

Fourier phase congruence for a 1-D real-valued signal f and a point p, the Fourier phase that the signal f would have if the origin were shifted to p; in other words, it is the Fourier phase of f translated by $-p$. The congruence between the phases at p for the various frequencies — in other words the degree by which those phases at p are close to each other — can be measured by $f(p)$ where $f(p)^2 + H(f)(p)^2$, where $H(f)$ is the Hilbert transform of f. *See* Fourier amplitude/phase, Hilbert transform.

Fourier plane a plane in an optical system where the exact Fourier transform of an input image is generated.

Fourier series Let $f(t)$ be a continuous time periodic signal with fundamental period T such that

$$f(t) = \sum_{n=-\infty}^{\infty} c_n e^{j\omega_0 t}, \; -\infty < t < \infty$$

where ω_0 is the fundamental frequency, c_0 is a real number, and c_n, $n \neq 0$ are complex numbers. This representation of $f(t)$ is called the *exponential Fourier series* of $f(t)$. The coefficients c_n are called the Fourier coefficients and are given by

$$c_n = \frac{1}{T} \int_T f(t) e^{-j\omega_0 t} dt, \; n = 0, \pm 1, \pm 2, \ldots$$

The function $f(t)$ can also be expressed as

$$f(t) = c_0 + \sum_{n=1}^{\infty} (a_n \cos(n\omega_0 t) + b_n \sin(n\omega_0 t))$$

with Fourier coefficients a_n and b_n determined from

$$a_n = \frac{2}{T} \int_T f(t) \cos(n\omega_0 t \, dt), \; n = 1, 2, 3, \ldots$$

$$b_n = \frac{2}{T} \int_T f(t) \sin(n\omega_0 t \, dt), \; n = 1, 2, 3, \ldots$$

This representation of $f(t)$ is called the *trigonometric Fourier series* of $f(t)$. A signal $f(t)$ has a Fourier series if it satisfies Dirichlet conditions, given by (i) $f(t)$ is absolutely integrable over any period, (ii) $f(t)$ is piecewise continuous over any period, and (iii) $\frac{d}{dt} f(t)$ is piecewise continuous over any period. Virtually all periodic signals used in engineering applications have a Fourier series.

Fourier transform　a linear mathematical transform from the domain of time or space functions to the frequency domain. The discrete version of the transform (DFT) can be implemented with a particularly efficient algorithm (fast Fourier transform or FFT). The discrete Fourier transform of a digital image represents the image as a linear combination of complex exponentials.

The Fourier transform of a continuous time period signal $f(t)$ is given by

$$F(\omega) = \int_{-\infty}^{\infty} f(t) e^{-j\omega t} dt, \; -\infty < \omega < \infty .$$

If the signal $f(t)$ is absolutely integrable and is well behaved, then its Fourier transform exists. For example, the rectangular pulse signal

$$f(t) = \begin{cases} 1, & -\frac{T}{2} \leq t < \frac{T}{2} \\ 0, & \text{otherwise} \end{cases}$$

has Fourier transform $F(w) = \frac{2}{\omega} \sin \frac{\omega T}{2}$.

The inverse Fourier transform of a signal is given by

$$f(t) = \frac{1}{2\pi} \int_{-\infty}^{\infty} F(\omega) e^{j\omega t} d\omega$$

See also optical Fourier transform, two-dimensional Fourier transform.

Fourier, Jean Baptiste Joseph　(1768–1830) Born: Auxerre, France

Best known for the development of new mathematical tools. The Fourier series, which describes complex periodic functions, and the Fourier integral theorem, which allows complex equations to be broken into simpler trigonometric equations for easier solution, are named in honor of their discoverer. Fourier was an assistant lecturer under the great mathematicians Joseph Lagrange and Gaspard Monge at the Ècole Polytechnique in Paris. He also served in a number of positions in Napoleon's government, eventually resigning in protest. His contributions to the field of electrical engineering are well recognized.

Fowler–Nordheim oscillation　oscillations in internal field emission tunneling current due to quantum interference.

Fox and Li method　iterative diffraction integral method for calculating the electromagnetic modes of a laser resonator.

FPC coefficients the parameters of the linear system that performs the linear prediction of a given stochastic process.

FPGA *See* field-programmable gate array.

FQO *See* four-quadrant operation

fractal from the Latin *fractus*, meaning broken or irregular, a fractal is a rough geometric shape that is self-similar over multiple scales, i.e., its parts are approximate copies of larger parts, or ultimately of the whole. Magnifying a fractal produces more detail. Fractals are good models of many natural phenomena such as coast lines, clouds, plant growth, and lightning as well as artificial items such as commodity prices and local-area network traffic. *See also* fractal coding.

fractal coding the use of fractals to compress images. Regions of the image are represented by fractals, which can then be encoded very compactly. Fractal coding provides lossy compression that is independent of resolution and can have extremely high compression ratios, but it is not yet practical for general images. Some of the methods are covered by patents. *See also* compression, compression ratio, fractal, image compression, image resolution, lossy compression, model based image coding, spatial resolution, vector image.

fractional arithmetic mean radian bandwidth dimensionless ratio of the bandwidth ($bw = \omega_H - \omega_L$) to the arithmetic radian center frequency (ω_{oa}). The band edges are usually defined as the highest (ω_H) and lowest (ω_L) frequencies within a contiguous band of interest at which the loss equals L_{Amax}, the maximum attenuation loss across the band.

$$bw_a = \frac{\omega_H - \omega_L}{\omega_{oa}} = \frac{2\,(\omega_H - \omega_L)}{\omega_H + \omega_L}$$

fractional Brownian motion a nonstationary generalization of Brownian motion; it is a zero-mean Gaussian process having the following covariance:

$$E\,[B_H(t)B_H(s)] = \sigma^2 2(|t|^{2H} + |s|^{2H} - |t-s|^{2H})$$

This process has interesting self-similarity and spectral properties; it is a fractal with dimension $D = 2 - H$. *See* Brownian motion.

fractional discrimination coding a preprocessing scheme where the images are processed with fractional discrimination functions and then coded. This method of coding enables efficient local feature based encoding of pictures while preserving global features. This is a type of contextual coding scheme.

fractional geometric mean radian bandwidth dimensionless ratio of the bandwidth ($bw = \omega_H - \omega_L$) to the geometric radian center frequency (ω_{oa}). The band edges are usually defined as the highest (ω_H) and lowest (ω_L) frequencies within a contiguous band of interest at which the loss equals L_{Amax}, the maximum attenuation loss across the band.

$$bw_a = \frac{\omega_H - \omega_L}{\omega_{oa}} = \frac{(\omega_H - \omega_L)}{\sqrt{\omega_H + \omega_L}}$$

fractional horsepower National Electrical Manufacturers Association (NEMA) classification describing any "motor built in a frame smaller than that having a continuous rating of 1 horsepower, open type, at 1700 to 1800 rpm."

fractional rate loss for frame-by-frame transmitted convolutionally encoded data, the fraction of overhead (compared to the size of the frame) needed to put the encoder into a known state.

fragmentation waste of memory space when allocating memory segments for processes. Internal fragmentation occurs when memory blocks are rounded up to fix block sizes, e.g., allocated in sizes of power of 2 only. E.g., if 35 K of data is allocated a 64 K block, the difference ($64 - 35 = 29K$) is wasted. External fragmentation occurs between allocated segments, as a result of allocating different sized segments for processes entering and leaving memory. This latter fragmentation is also called checkerboarding.

frame   (1) a set of four vectors giving position and orientation information.

(2) the basic element of a video sequence. The standard frame rate for TV standards is 25 frames/s (European standards, e.g., PAL and SECAM) or 30 frames/s (U.S. and Japanese standards, e.g., NTSC).

(3) in paging systems, a memory block whose size equals the size of a page. Frames are allocated space according to aligned boundaries, meaning that the last bits of the address of the first location in the frame will end with n zeros (binary), where n is the exponent in the page size. Allocating frames for pages makes it easy to translate addresses and to choose a frame for an incoming page (since all frames are equivalent).

(4) time interval in a communication system over which the system performs some periodic function. Such functions can be multiple access functions (e.g., TDMA multiple access frame) or speech-processing functions (e.g., speech coding frame, interleaving frame, or error control coding frame).

frame grabber   a device which is attached to an electronic camera and which freezes and stores images digitally, often in gray-scale or color format, typically in one or three 8-bit bytes per pixel respectively.

frame memory   video memory required to store the number of picture elements for one complete frame of electronically scanned video information. The memory storage in bits can be computed by multiplying the number of video samples made per horizontal line, times the number of horizontal

lines per field (vertical scan), times the number bits per sample, times the number of fields/frame. A sample consists of the information necessary to reproduce the color information.

The NTSC television system consists of two interlaced fields per frame. Storage requirements are usually minimized by sampling the color video information consisting of the luminance (Y) and the two color difference signals, ($R - Y$) and ($B - Y$). The color signal bandwidth is less than the luminance bandwidth that can be used to reduce the field memory storage requirements. Four samples of the luminance (Y) signal is combined with two samples of the ($R - Y$) signal and two samples of the ($B - Y$) signal. The preceding video sampling techniques is designated as 4:2:2 sampling and reduces the field memory size by one third.

The number of memory bits that are required to store one NTSC frame is two times the number of bits required to store one NTSC field. Field memory for NTSC video sampled at 4 times the color subcarrier frequency at 8 bits/pixel would require 7.644 megabits of RAM when 4:2:2 sampling is used.

frame processing   image processing applied to frames or sequences of frames.

frame rate   in analog and digital video, the rate per second at which each frame or image is refreshed. The NTSC (National Television Systems Committee) rate for analog television is 30 frames per second.

frame store   a device which stores images digitally, often in gray-scale or color format, typically in one or three 8-bit bytes per pixel respectively; a frame store often incorporates, or is used in conjunction with, a frame grabber.

frame synchronization   a method to obtain rough timing synchronization between transmitted and received frames. The degree of synchronization obtained depends on the level of frame synchronization (super-frame, hyper-frame). The level of frame synchronism also determines the actual method and the place where the synchronism takes place.

frame synchronizer (1) a time-base corrector effective over one frame time. Video data may be discarded or repeated in one-frame increments; in the absence of an input, the last received frame is repeated, thus producing a freeze-frame display.

(2) a device that stores video information (perhaps digitally) to reduce the undesirable visual effect caused by switching nonsynchronous sources.

Franklin, Benjamin (1706–1790) Born: Boston, Massachusetts

Best known as the greatest American statesman, scientist, and entrepreneur of his day. He was the person who described electric charge, introduced the terms "positive" and "negative" as descriptive of charges, and described the fundamental nature of lightening. Franklin received no formal training in the sciences. His printing business gave him the financial security necessary to carry out numerous fundamental experiments on the nature of electricity. The results of these experiments were later given theoretical validation by Michael Faraday. Franklin's political contributions to the United States were also extraordinary. Amazingly, this additional responsibility did not deter Franklin from numerous other experiments and inventions in a variety of fields.

Fraunhofer region *See* far field.

free distance the minimum Hamming distance between two convolutionally encoded sequences that represent different valid paths through the same code trellis. The free distance equals the maximum column distance and the limiting value of the distance profile.

free electron laser laser in which the active medium consists of electrons that are subject to electric and magnetic fields but are not associated with atoms or molecules.

free input a quantity influencing the controlled process from outside, generated within the process environment, and being beyond direct or indirect influence of the controller; a free input is defined by a continuous trajectory over given time interval or by a sequence of values at given time instants.

free input forecasting process of predicting — at a given time — future values of the free input; these future values can be predicted in different forms: single future free input trajectory (scenario) over a specified time interval (forecasting interval); bunch of possible scenarios over specified time interval, (or intervals) — eventually with attached weights; stochastic model valid at the considered forecasting time and other. Free input forecasting can be essential for making most of the decisions concerning current values of the manipulated inputs.

free input model model, usually stated in form of a set of differential or difference equations, describing the behavior of the free inputs; for example, ARMA process is often used as the free input model. Usually free input model is driven by a white noise stochastic process or by a sequence of set bounded — but otherwise time-independent — quantities.

free running frequency the frequency at which an oscillator operates in the absence of any synchronizing input signal. The frequency is usually determined by a time constant from externally connected resistance and capacitance.

free space loss the propagation loss experienced between two isotropic antennas in a vacuum resulting from spreading of power over spherical wavefronts centered at the transmitting antenna and from the finite aperture of the receiving antenna. The free space loss increases proportional to the square of the distance from the antenna and the square of the frequency.

free system *See* unforced systems.

free-page list   a linked-list of information records pointing to "holes" (i.e., free page frames) in main memory.

free-space interconnect   interconnect using optical elements such as lenses, gratings, and holograms, in which optical signals may cross each other in the space. The main advantages are high interconnection density and parallelism, and dynamic reconfiguration.

free-wheeling diode   a diode connected in parallel with the load of a half-wave rectifier to prevent the return of energy from the load to the source. Due to the stored energy in the inductive load, the current must continue after the source voltage becomes negative. The free-wheeling diode provides a path for the current to circulate and allows the diode or SCR in the rectifier to turn off.

Fremdhold integral equation   a linear integral equation wherein the limits of integration are fixed.

frequency   the repetition rate of a periodic signal used to represent or process a communication signal. Frequency is expressed in units of hertz (Hz). 1 Hz represents one cycle per second, 1 MHz represents one million cycles per second, and 1 GHz represents one billion cycles per second.

frequency chirp   a monotonic change in optical frequency with time; often used for laser radar ranging in analogy with conventional radar and for ultrashort pulse generation via pulse compression or autocorrelation. *See also* chirp signal.

frequency compensation   the modification of the amplitude–frequency response of an amplifier to broaden the bandwidth or to make the response more nearly uniform over the existing bandwidth.

frequency converter   an equipment or circuit that converts an RF signal to an intermediate (IF) signal in receivers. It converts an IF signal to an RF signal in transmitters.

frequency correlation function   a function characterizing the similarity of a received signal with respect to a shift in frequency.

frequency deviation   in a frequency modulated system, the number of hertz the carrier is varied during the modulation process.

frequency distortion   caused by the presence of energy storage elements in an amplifier circuit. Different frequency components have different amplifications, resulting in frequency distortion, and the distortion is specified by a frequency response curve.

frequency division duplex (FDD)   a technique based on the allocation of two separate frequency bands for the transmission in both directions in a link. FDD typically requires a guard band between the two frequency bands in order to eliminate interference between the transmitter and receiver at a terminal. In a full duplex communication system, information flows simultaneously in both directions between two points.

frequency division multiple access (FDMA)   a multiple-access technique based on assigning each user a unique frequency band upon which transmission takes place. *See also* time division multiple access, code division multiple access.

frequency division multiplexing (FDM)   refers to the multiplexing of signals by shifting each signal to a different frequency band. *See also* frequency division multiple access.

frequency domain   representation of a signal by frequency components, such as its Fourier transform.

frequency domain sampling   a procedure that is a dual of the time-domain sampling theorem, whereby a time-limited signal can be reconstructed from frequency- domain samples.

frequency domain storage   an optical data storage technique in which individual bits are stored by saturating or bleaching spectral holes into an inhomogeneously broadened absorption spectrum. Usually, this hole burning is accomplished using electromagnetic fields of discrete frequencies. The maximum frequency domain storage density is given by the ratio of inhomogeneous width to homogeneous width of the absorption spectrum.

frequency hopping   a system where the carrier frequency used for modulation is changed according to some predetermined hopping pattern. The receiver hops according to the same pattern in order to retrieve the information. If the rate of change is lower or equal to the data rate, the process is known as slow frequency hopping, otherwise as fast. Frequency hopping is used for counteracting intentional interference, frequency selective fading, and as a multiple access technique (FH-CDMA).

frequency modulation (FM)   angle modulation in which the instantaneous frequency of a sine-wave carrier is caused to depart from the carrier frequency by an amount proportional to the instantaneous value of the modulating wave.

frequency nonselective channel   a radio channel that has a uniform transfer function over the bandwidth of information signal. In such a channel, fading leads to reduced SNR at the receiver, but does not lead to distortion of the symbol shape and hence intersymbol interference.

frequency pulling   shift of the frequency of a laser oscillation mode from the empty cavity or nondispersed cavity frequency toward the center frequency of the amplifying transition.

frequency regulation   the change in the frequency of an unloaded generator with respect to its frequency in a fully-loaded state. Typically applied to small, isolated power systems such as emergency power units.

frequency relay   a protective relay which monitors the frequency of the electric power system.

frequency resolution   a measure of the ability of a system to resolve different frequencies in a signal. As the frequency resolution increases, more finely-spaced frequency components can be resolved. The time resolution of a system is roughly inversely proportional to the frequency resolution; the uncertainty principle places a lower bound on the time-frequency resolution product.

frequency response   consider a system with transfer function given by

$$H(s) = \frac{Y(s)}{F(s)}.$$

The term *frequency response* is used to denote $H(j\omega)$, (commonly written $H(\omega)$). The frequency response is often used in describing the steady-state (stable) system response to a sinusoidal input. For example, in the case of a stable continuous-time system with input signal $f(t) = A\cos(\omega t)$, the steady-state output signal is given by

$$y_{ss}(t) = A|H(j\omega)\cos(\omega t + \angle H(j\omega).$$

The term $|H(j\omega)|$, $-\infty < \omega < \infty$, is called the *magnitude response*. The term $\angle H(j\omega)$, $-\infty < \omega < \infty$, is called the *phase response*. The magnitude and phase responses are typically represented in the form of a *Bode plot*. *See also* Bode plot, transfer function.

frequency reuse   a way to increase the effective bandwidth of a satellite system when available spectrum is limited. Dual polarizations and multiple beams pointing to different earth regions may utilize the same frequencies as long as, for example, the gain of one beam or polarization in the directions of the other beams or polarization (and vice versa) is low enough. Isolations of 27–35 dB are typical for reuse systems.

frequency reuse cluster a group of cells in a cellular communications network wherein each frequency channel allocated to the network is used precisely once (i.e., in one cell of the group). The size of the frequency reuse cluster is one of the major factors that determines the spectral efficiency and ultimately the capacity of a cellular network. For a given channel bandwidth, the smaller the cluster size, the higher the network capacity.

frequency selective filter a filter that passes signals undistorted in one or a set of frequency bands and attenuate or totally eliminate signals in the remaining frequency bands.

frequency selective surface (FSS) filter made of two-dimensional periodic arrays of apertures or metallic patches of various shapes. Several layers can be used to obtain a structure with a set of desired spectral properties.

frequency shift keying (FSK) (1) an encoding method where different bits of information are represented by various frequencies; used for spread-spectrum signal encoding for security and reduced interchannel interference.

(2) a digital modulation technique in which each group of successive source bits determines the frequency of a transmitted sinusoid.

frequency space the transformed space of the Fourier transform.

frequency synchronization the process of adjusting the frequency of one source so that it exactly matches that of another source: more specifically, so that n periods of one frequency are exactly equal to m periods of the other frequency, for integral n and m. *See* phase-locked loop.

frequency synthesizer an oscillator that produces sinusoidal wave with arbitrary frequency.

In common cases, generated frequencies are allocated with a frequency spacing called the channel step.

frequency variation a change in the electric supply frequency.

frequency-modulation recording *See* magnetic recording code.

fresh fuel nuclear fuel which has never participated in a nuclear reaction and is thus only slightly radioactive.

Fresnel region the region in space around an antenna at which the fields have both transverse and radial components and the antenna pattern is dependent on the distance from the antenna. The Fresnel, or near-field, region is typically taken to be $r < 2D^2/\lambda$, where r is the distance from the antenna, D is the maximum dimension of the antenna, and λ is the wavelength.

Fresnel zone an indicator of the significant volume of space occupied by a radio wave propagating along a line-of-sight path between the transmitter and receiver. At an arbitrary point which is at distance d_1 from the transmitter and at distance d_2 from the receiver, along the axis joining the transmitter and the receiver, a radio wave with wavelength λ occupies a volume, which at that point between the transmitter and the receiver, has a radius which is given by the radius of the first Fresnel zone. The radius of the first Fresnel zone is given by

$$R = \sqrt{\frac{\lambda d_1 d_2}{d_1 + d_2}}$$

From this equation one can see that the significant volume occupied by a radio wave is essentially ellipsoidal. In order for a transmitter and receiver to communicate successfully, it is important for this volume to be kept free from any obstruction, especially at microwave frequencies.

fricative   a phoneme pronounced when constricting the vocal tract so as the air flow becomes turbulent and the corresponding frequency spectrum looks like a broadband noise. The utterances with frication can occur with or without phonation. When phonation also occurs, the presence of formants can be detected in the spectrum.

Friis transmission formula   a formula used to compute the received signal power in a radio communication link given the amount of power transmitted, the effective apertures or gains of the transmit and receive antennas, the distance between the antennas, and the wavelength.

fringing   the portion of the flux at the air gap in a magnetic circuit that does not follow the shortest path between the poles.

frit   a relatively low softening point material of glass composition.

Fritchman model N-state Markoff model describing N error states of a channel.

frog   the top of a tower.

front end   (1) an initial processing unit that provides a user interface and/or reformats input data for subsequent computations on a special-purpose or high-performance back end processor.
    (2) the portion of the compiler that does machine-independent analysis.

front of a motor   the end of a motor that is opposite to the major coupling or driving pulley.

front porch   video blanking level duration of approximately 1.27 microseconds contained within the horizontal blanking interval of the composite NTSC signal. The front porch duration is 2% of the total horizontal line time, starting at the end of a horizontal line of video signals and ending at the beginning of the horizontal line sync signal.

frontside bus   the term used to describe the main bus connecting the processor to the main memory (see also backside bus).

frustrated total reflection   *See* attenuated total reflection.

FS   *See* fixed station.

FSA   finite state automata. *See* finite state machine.

FSK   *See* frequency shift keying.

FSM   *See* finite state machine.

FSS   *See* frequency selective surface.

FSVQ   *See* finite state vector quantizer.

fuel cell   an electrochemical device that continuously converts chemical energy into electric energy (and heat) for as long as fuel and oxidant are supplied. Fuel cells are related to, but different from, batteries. For example, a fuel cell does not need to be recharged. And, when hydrogen is used as fuel, the fuel cell generates only power and drinking water. *See* alkaline fuel cell, direct methanol fuel cell, molten carbonate fuel cell, proton exchange membrane fuel cell, solid oxide fuel cell.

fuel cell electric vehicle (FCEV)   *See* electric vehicle.

fuel cycle   (1) the operating time of a nuclear reator between shutdowns for refueling.
    (2) the life cycle of nuclear fuel, e.g., raw materials, to fresh fuel, to irradiated fuel, to spent fuel, to recycling, and finally to residual waste.

fuel pool   a large vat of water used to store both fresh and spent nuclear fuel at the reactor site.

fuel rod   long, thin canister, typically made of zirconium or other metal, which contains the radioactive fuel in a nuclear reactor core.

fulgurite   a vitrified tube of fused sand created by power fault arcs or lightning strikes to sandy soil.

full adder   a combinational logic circuit that produces a two-bit sum of three one-bit binary numbers.

full bridge amplifier   a class-D amplifier based on a full-bridge inverter configuration.

full load amperage (FLA)   a value, found on the nameplate of an induction motor, indicating the expected current drawn by the motor when operating at rated voltage and load. The current measured at the input of an electrical apparatus which has a rated variable at the output. The full-load amperage is also defined as a value that permits the system to operate in a safe condition if it is equal to the rated value. The full-load amperage is equal to the rated value if and only if the voltage is at its rated value. The full load is defined with the electrical power in the case of transformers, generators, or power electronics converters and it is defined with the mechanical power for electrical motors.

full permutation symmetry   *See* nonlinear susceptibility, symmetries of functions of a lens.

full rank Jacobian   a Jacobian matrix where $\dim(R(J)) = r$, $\dim(N(J)) = n - r$. At a singular configuration of the Jacobian, one has $\dim(R(J)) + \dim(N(J)) = n$ independently of the rank of the matrix J. *See* null space of the Jacobian and range of Jacobian.

full-load speed   the speed of a motor that produces rated power when operating at rated voltage and, for AC motors, frequency.

full-load torque   the torque of a motor that is producing rated power at rated speed.

full-wave analysis   the rigorous computation of electromagnetic fields without approximations (apart for numerical discretization) is often referred to as full-wave analysis.

full-wave control   both the positive and negative half cycle of the waveforms are controlled.

full-wave rectifier   a device that passes positive polarity portions of a signal and reverses negative polarity portions of an AC signal. Ideally, for a sinusoidal input $v_i(t) = V_m \cos(\omega t)$, the output of an ideal full-wave rectifier is $v_o(t) = |V_m \cos(\omega t)|$.

full width at half max (FWHM)   the width of a distribution function (i.e., the horizontal distance between two points having the same vertical coordinate) defined at a vertical position halfway between the maximum and minimum. This value exists for certain functions that are of interest as possible point spread functions (e.g., Gaussian, Moffat function) and gives a rough idea of the amount of blur to be expected.

fully associative cache   a cache block mapping technique where a main memory block can reside in any cache block position. A number of bits from the address received from the CPU is compared in parallel with the cache tags to see if the block is present in the cache. Although costly to implement, and impractical for large cache sizes, it is the most flexible mapping method. *See also* cache, direct mapped cache, and associativity.

function   a programming construct that creates its own frame on a stack, accepting arguments, performing some computation, and returning a result.

function sensitivity   in a circuit described by the function $F(s, x)$ where s is a complex variable, and x is a passive or active element in the circuit realization of the function, the equation

$$\mathbf{S}_x^{F(s,x)} = \frac{\partial F(s,x)}{\partial x}\frac{x}{F(s,x)}$$

Under conditions of sinusoidal steady state, when $s = j\omega$, the function $F(j\omega, x)$ is described by the sensitivity

$$\mathbf{S}_x^{F(j\omega,x)} = \text{Re } \mathbf{S}_x^{F(j\omega,x)} + j \text{ Im } \mathbf{S}_x^{F(j\omega,x)}$$
$$= \mathbf{S}_x^{|F(j\omega,x)|} + j \frac{\partial \arg F(j\omega, x)}{\partial x/x}$$

Thus, the real part of the function sensitivity gives the relative change in the magnitude response, and the imaginary part gives the change in the phase response, both with respect to a normalized element change.

function test   a check for correct device operation generally by truth table verification.

function-oriented design   a design methodology based on a functional viewpoint of a system starting with a high-level view and progressively refining it into a more detailed design.

functional block   ASICs designed using this methodology are more compact than either gate arrays or standard cells because the blocks can perform much more complex functions than do simple logic gates.

functional decomposition   the division of processes into modules.

functional redundancy checking   an error detection technique based on checking bus activity generated by a processor (master) with the use of another slave processor which monitors this activity.

functional unit   a module, typically within a processor, that performs a specific limited set of functions. The adder is a functional unit.

fundamental component   the first order component of a Fourier series of a periodic waveform.

fundamental frequency   for a continuous time periodic signal, $f(t)$ with fundamental period T (seconds), $\omega_0 = 2\pi/T$ (rad/sec). *See also* periodic signal.

fundamental mode   (1) operating mode of a circuit that allows only one input to change at a time.

(2) lowest order mode of a laser cavity; usually refers to a Gaussian transverse field distribution without regard to the longitudinal mode order.

fuse   an overcurrent device that employs one or more fusible elements in series. The fusible element(s) are designed to melt and interrupt the circuit when current above a threshold value flows in the circuit.

fuse coordination   the process of matching the fuse or circuit breaker interruption capability to overload current and short-circuit current and insuring that the protective device closest to a fault opens first so as to minimize the service interruption.

fuse cutout   a primary distribution voltage level fuse that employs a replaceable fusible link and provides a means of disconnect. The interrupting rating of a fuse cutout can be somewhat lower than that of a power fuse, however.

fuse link   used in nonerasable programmable memory devices. Each bit in the memory device is represented by a separate fuse link. During programming, the link is either "blown" or left intact to reflect the value of the bit. *See also* fusible link ROM.

fuse reducer   an adapter that allows fuses to be installed in fuseclips designed for larger fuses.

fuse saving   the practice of tripping distribution line reclosers or circuit breakers on a fast trip to

283

beat sectionalizing fuses that are protecting laterals. The following reclose will restore all load. If the fault remains following the fast trip(s), subsequent slow trip(s) will allow the fuse to operate on a permanent fault on the lateral. Fuse saving decreases the customer outage rate but causes more sags to customers in a fault situation.

fused disconnect   a disconnect switch that also employs fuse(s) for the purpose of overcurrent protection.

fusible link ROM   read-only memory using fuse links to represent binary data. *See also* fuse link.

fusion   a nuclear reaction in which two light nuclei are combined into a heavier nucleus with a release of energy. Fusion power reactors have been proposed but as yet never successfully constructed.

Futurebus   a bus specification standardized by the IEEE, originally defined for CPU-memory data transfers.

fuzz stick   a short hot-stick (insulated pole).

fuzzification   a procedure of transforming a crisp set or a real-valued number into a fuzzy set.

fuzzifier   a fuzzy system that transforms a crisp (nonfuzzy) input value in a fuzzy set. The most used fuzzifier is the *singleton fuzzifier,* which interprets a crisp point as a fuzzy singleton. It is normally used in fuzzy control systems. *See also* fuzzy inference system, fuzzy singleton.

fuzziness   the degree or extent of imprecision that is naturally associated with a property, process or concept.

fuzzy adaptive control   adaptive control involving fuzzy logic concepts or fuzzy control involving adaptation. Examples of such control are self-organizing fuzzy control, model reference fuzzy adaptive control, and fuzzy self-tuning control.

fuzzy aggregation network   artificial neural network that can be trained to produce a compact set of fuzzy rules with conjunctive and disjunctive antecedents. An aggregation operator is used in each node.

fuzzy algorithm   an ordered set of fuzzy instructions which upon execution yield an approximate solution to a specified problem. Examples of fuzzy algorithms include fuzzy c-mean clustering, fuzzy-rule-based classification, etc.

fuzzy AND   *See* triangular norm.

fuzzy approximation   an action or process of using a fuzzy system to approximate a nonlinear function or mapping.

fuzzy associative memory   a look-up table constructed from a fuzzy IF-THEN rule base and inference mechanism to define a relationship between the input and output of a fuzzy controller, or a fuzzy system. *See also* fuzzy inference system.

fuzzy automata   based on the concept of fuzzy sets, a class of fuzzy automata is formulated similar to Mealy's formulation of finite automata.

fuzzy basis functions   a set of fuzzy membership functions or their combinations in a fuzzy system that form a basis of a function space, normally for function approximation.

fuzzy behavioral algorithm   a relational algorithm that is used for the specific purpose of approximate description of the behavior of a system.

fuzzy C   the C programming language incorporating fuzzy quantities and fuzzy logic operations.

fuzzy c-means (FCM) a fuzzy version of the commonly used K-means clustering algorithm. The main feature of FCM is that a type of membership function is utilized in computing a distance measure. Also called fuzzy ISODATA.

fuzzy clustering a method used to cluster data into subsets based on a distance or similarity measure that incorporates fuzzy membership functions.

fuzzy cognitive map (FCM) nonlinear dynamical systems whose state trajectories are constrained to reside in the unit hypercube $[0, 1]^n$. The components of the state vector of an FCM stand for fuzzy sets or events that occur to some degree and may model events, actions, goals, etc. The trajectory of an FCM can be viewed as an inference process. For example, if an FCM's trajectory starting from an initial condition $\mathbf{x}(0)$ converges to an equilibrium state $\mathbf{A}$, then this can be interpreted as the FCM providing an answer to a question "What if $\mathbf{x}(0)$ happens?" In this sense it can be interpreted that the FCM stores the rule "IF $\mathbf{x}(0)$ THEN the equilibrium $\mathbf{A}$." FCMs were introduced by B. Kosko.

fuzzy complement the complement of a fuzzy set A is understood as NOT (A).

fuzzy concentration the concentration of a fuzzy set produces a reduction by squaring the membership function of that fuzzy set.

fuzzy conditional statement an IF-THEN statement of which either the antecedents or consequent(s) or both may be labels of fuzzy sets. Also know as fuzzy rule, linguistic fuzzy model, linguistic rules. *See* fuzzy IF-THEN rule.

fuzzy control the application of fuzzy logic and fuzzy inference rules utilizing knowledge elicited from human experts in generating control decisions for the control of processes. More specifically, a means of expressing an operator's knowledge of controlling a process with a set of fuzzy IF-THEN rules and linguistic variables.

fuzzy control system stability stability of a fuzzy control system that normally includes a plant to be controlled and a fuzzy controller.

fuzzy controller devices for implementing fuzzy control, normally including the following components: a fuzzifier to transform a crisp real-valued number to a fuzzy set; a fuzzy rule base to define IF-THEN control rules; a fuzzy inference engine to combine the IF-THEN rules; and a defuzzifier to transform a fuzzy set to a crisp real-valued number.

fuzzy controller design a process of determining a fuzzy controller including a fuzzifier, a fuzzy rule base, a fuzzy inference engine, and a defuzzifier.

fuzzy decision rule a decision rule with fuzzified antecedents.

fuzzy decision tree a decision tree with fuzzy decision functions.

fuzzy decision tree algorithm an algorithm to generate a fuzzy decision tree, such as the branch-bound-backtrack algorithm.

fuzzy decisional algorithm a fuzzy algorithm that serves to provide an approximate description of a strategy or decision rule.

fuzzy definitional algorithm a finite set of possible fuzzy instructions that define a fuzzy set in terms of other fuzzy sets (and possibly itself, i.e., recursively).

fuzzy digital topology an extension of the topological concepts of connectedness and surroundness in a digital picture using fuzzy subsets.

fuzzy dilation   an operator that increases the degree of belief in each object of a fuzzy set by taking the square root of the membership function.

fuzzy dynamic model   model involving fuzzy logic concepts and system dynamics. A typical example is a hierarchical model with a higher level of fuzzy inference rules and a lower level of analytical linear dynamic equations.

fuzzy dynamic system   system involving fuzzy logic concepts and system dynamics. A typical example is a fuzzy control system with a dynamic plant and a fuzzy controller.

fuzzy entropy   the entropy of a fuzzy set is a functional to measure the degree of fuzziness of a fuzzy set based on Shannon's function. It has been applied to provide a quantitative measure of ambiguity to the problems of gray-tone image enhancement.

fuzzy estimation   an action or process of deducing state variables from given input and output measurements in a fuzzy dynamic state space model.

fuzzy expert system   a rule-based system for approximate reasoning in which rules have fuzzy conditions and their triggering are driven by fuzzy matching with fuzzy facts. *See also* fuzzy inference system.

fuzzy filter   a filter involving fuzzy logic concepts. A typical example is a filter constructed from a number of local filters through fuzzy membership functions.

fuzzy gain scheduling   gain scheduling involving fuzzy logic concepts. A typical example is the use of a number of fuzzy IF-THEN rules to adjust a controller gain.

fuzzy generational algorithm   a fuzzy algorithm serves to generate rather than define a fuzzy set. Possible applications of generational algorithms include generation of handwritten characters and generation of speech.

fuzzy geometry   a way to describe an image by fuzzy subsets.

fuzzy global control   a fuzzy control for an overall plant or a control based on a fuzzy global model that consists of a number of local models smoothly connected through a set of membership functions.

fuzzy global model   a fuzzy dynamic model consisting of a number of local linear models or simple nonlinear models smoothly connected through a set of nonlinear membership functions.

fuzzy grammar   an attributed grammar that uses fuzzy primitives as the syntactic units of a language.

fuzzy H-infinity control   H-infinity control involving fuzzy logic concepts or fuzzy control to achieve an H-infinity controller performance index.

fuzzy H-infinity filter   an H-infinity filter involving fuzzy logic concepts or a fuzzy filter to achieve an H-infinity filter performance index.

fuzzy hierarchical systems   fuzzy systems of hierarchical structures. A typical example is a two-level fuzzy system with a higher level of fuzzy inference rules and lower level of analytical linear models.

fuzzy identification   a process of determining a fuzzy system or a fuzzy model. A typical example is identification of fuzzy dynamic models consisting of determination of the number of fuzzy space partitions, determination of membership functions, and determination of parameters of local dynamic models.

fuzzy IF-THEN rule   rule of the form

$$\text{IF } x \text{ is } A \text{ THEN } y \text{ is } B$$

where A and B are linguistic values defined by fuzzy sets on universe of discourse X and Y respectively (abbreviated as $A \longrightarrow B$). The statement "*x is A*" is called the *antecedent* or *premise,* while "*y is B*" is called the *consequence* or *conclusion.* A fuzzy IF-THEN rule can be defined as a binary *fuzzy relation.* The most common definition of a fuzzy rule $A \longrightarrow B$ is as *A coupled with B,* i.e.,

$$R = A \longrightarrow B = A \times B,$$

or

$$\mu_{A \times B}(x, y) = \mu_A(x) \star \mu_B(y),$$

where $\star$ is a triangular norm.

See also fuzzy relation, fuzzy set, linguistic variable, triangular norm.

fuzzy implication   *See* fuzzy IF-THEN rule.

fuzzy inference   a fuzzy logic principle of combining fuzzy IF-THEN rules in a fuzzy rule base into a mapping from a fuzzy set in the input universe of discourse to a fuzzy set in the output universe of discourse. A typical example is a composition inference.

fuzzy inference engine   a device or component carrying out the operation of fuzzy inference, that is, combining fuzzy IF-THEN rules in a fuzzy rule base into a mapping from a fuzzy set in the input universe of discourse to a fuzzy set in the output universe of discourse. *See also* approximate reasoning, fuzzy inference system, fuzzy rule.

fuzzy inference system   a computing framework based on fuzzy set theory, fuzzy IF-THEN rules, and approximate reasoning. There are two principal types of fuzzy inference systems:

(1) Fuzzy inference systems mapping fuzzy sets into fuzzy sets (pure fuzzy inference systems) that are composed of a knowledge base containing the definitions of the fuzzy sets and the database of fuzzy rules provided by experts; and a fuzzy inference engine that performs the fuzzy inferences.

(2) Fuzzy inference systems performing nonlinear mapping from crisp (nonfuzzy) input data to crisp (nonfuzzy) output data. In the case of a Mamdani fuzzy system, in addition to a knowledge base and a fuzzy inference engine, there is a fuzzifier that represents real-valued inputs as fuzzy sets, and a defuzzifier that transforms the output set to a real value. In the case of a Sugeno fuzzy system, special fuzzy rules are used, giving a crisp (nonfuzzy) conclusion, and the output of the system is given by the sum of those crisp conclusions, weighted on the activation of the premises of rules. Some fuzzy systems of this type hold the universal function approximation property.

See also defuzzifier, fuzzifier, fuzzy set, fuzzy IF-THEN rules, fuzzy reasoning, Sugeno fuzzy rule, universal function approximation property.

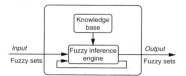

Pure fuzzy inference system.

fuzzy input–output model   input–output models involving fuzzy logic concepts. A typical example is a fuzzy dynamic model consisting of a number of local linear transfer functions connected by a set of nonlinear membership functions.

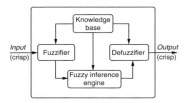

Mamdani fuzzy system.

fuzzy integral   an aggregation operator used to integrate multiattribute fuzzy information. It is a functional defined by using fuzzy measures, which corresponds to probability expectations. Two commonly used fuzzy integrals are Sugeno integral and Choquet integral.

fuzzy intensification   an operator that increases the membership function of a fuzzy set above the crossover point and decreases that of a fuzzy set below the crossover point.

fuzzy intersection   the fuzzy intersection is interpreted as "A AND B," which takes the minimum value of the two membership functions.

fuzzy ISODATA   *See* fuzzy c-means.

fuzzy local control   fuzzy control in a sense of local region of plant operation range, which is a key feature of fuzzy control. A typical example is a local control based on a local fuzzy dynamic model.

fuzzy local model   a fuzzy model in a sense of local region of plant operation range. A typical example is a fuzzy model of Takagi–Sugeno–Kang type.

fuzzy local system   fuzzy system in a sense of local region of plant operation range. A typical example is a fuzzy system of Takagi–Sugeno–Kang type.

fuzzy logic   introduced by Zadeh (1973), gives us a system of logic that deals with fuzzy quantities, the kind of information humans manipulate that is generally imprecise and uncertain. Fuzzy logic's main characteristics is the robustness of its interpolative reasoning mechanism. *See also* approximate reasoning, linguistic variable, modifier, generalized modus ponens.

fuzzy logic controller   controller based on a fuzzy logic system. *See also* fuzzy logic system.

fuzzy logic system   a system of logic that deals with fuzzy quantities, that are unprecise and uncertain.

fuzzy measure   a subjective scale used to express the grade of fuzziness similar to the way probability measure is used to express the degree of randomness. An example of fuzzy measure is g_λ-fuzzy measure proposed by Sugeno.

fuzzy membership function   function of characterizing fuzzy sets in a universe of discourse which takes values in the interval [0, 1]. Typical membership functions include triangular, trapezoid, pseudo-trapezoid, and Gaussian functions.

fuzzy minus   the minus operation applied to a fuzzy set to give an intermidiate effect of dilation. *See also* fuzzy plus.

fuzzy model   *See also* fuzzy inference system.

fuzzy model-based control   fuzzy control based on various kinds of mathematical or linguistic models. A typical example is control of dynamic systems based on dynamic fuzzy models, which consist of a number of local linear models smoothly connected through a set of nonlinear membership functions.

fuzzy modeling   combination of available mathematical description of the system dynamics with its linguistic description in terms of IF-THEN rules. In the early stages of fuzzy logic control, fuzzy modeling meant just a linguistic description in terms of IF-THEN rules of the dynamics of the plant and the control objective. Typical examples of fuzzy models in control application includes Mamdani model, Takagi–Sugeno–Kang model, and fuzzy dynamic model.

fuzzy neural control a control system that incorporates fuzzy logic and fuzzy inference rules together with artificial neural networks.

fuzzy neural network artificial neural network for processing fuzzy quantities or variables with some or all of the following features: inputs are fuzzy quantities; outputs are fuzzy quantities; weights are fuzzy quantities; or the neurons perform their functions using fuzzy arithmetic.

fuzzy neuron a McCulloch–Pitts neuron with excitatory and inhibitory inputs represented as degrees between 0 and 1; output is a degree to which it is fired.

fuzzy nonlinear control nonlinear control involving fuzzy logic concepts or fuzzy control with application to nonlinear systems.

fuzzy number a convex fuzzy set of the real line such that
  1. it exists exactly one point of the real line with membership 1 to the fuzzy set;
  2. its membership function is piecewise continuous.
  In fuzzy set theory, crisp (nonfuzzy) numbers are modeled as fuzzy singletons.
  See also convex fuzzy set, fuzzy singletons.

fuzzy observer a device to estimate the states of a dynamic system, involving fuzzy logic concepts. A typical example is an observer constructed from a number of local observers through fuzzy membership functions.

fuzzy operator logical operator used on fuzzy sets for fuzzy reasoning. Examples are the complement (NOT), union (OR), and the intersection (AND).

fuzzy optimal control optimal control involving fuzzy logic concepts or fuzzy control to achieve an optimal control performance index.

fuzzy optimal filter optimal filter involving fuzzy logic concepts or fuzzy filter to achieve an optimal filter performance index.

fuzzy OR *See* triangular co-norm.

fuzzy output feedback fuzzy control based on feedback of a plant output. This is closely related to fuzzy dynamic models.

fuzzy parameter estimation a method that uses fuzzy interpolation and fuzzy extrapolation to estimate fuzzy grades in a fuzzy search domain based on a few cluster center-grade pairs. An application of this method is to estimate mining deposits.

fuzzy partition partition of a plant operating space based on fuzzy logic concepts. A typical example is a partition of a state space by overlapping subspaces which are characterized by a set of fuzzy membership functions.

fuzzy pattern matching a pattern matching technique that applies fuzzy logic to deal with ambiguous or fuzzy features of noisy point or line patterns.

fuzzy plus the operation of plus applied to a fuzzy set to give an intermediate effect of concentration. *See also* fuzzy minus.

fuzzy PROLOG the PROLOG programming language incorporating fuzzy quantities and fuzzy logic operations.

fuzzy proposition a proposition in which the truth or falsity is a matter of degree.

fuzzy reasoning approximate reasoning based on fuzzy quantities and fuzzy rules. *See also* approximate reasoning.

fuzzy relation a fuzzy set representing the degree of association between the elements of two or more universes of discourse. *See also* fuzzy sets.

fuzzy relational algorithm a fuzzy algorithm that serves to describe a relation or relations between fuzzy variables.

fuzzy relational matrix a matrix whose elements are membership values of the corresponding pairs belonging to a fuzzy relation.

fuzzy relaxation a relaxation technique with fuzzy membership functions applied.

fuzzy restriction a fuzzy relation that places an elastic constraint on the values that a variable may take.

fuzzy rule *See* fuzzy IF-THEN rule.

fuzzy rule bank a collection of all fuzzy rules that are arranged in N-dimensional maps.

fuzzy rule base a set of fuzzy IF-THEN rules. It is a central component of a fuzzy system and defines major functions of a fuzzy system.

fuzzy rule minimization a technique to simplify the antecedent and consequent parts of rules and to reduce the total number of rules.

fuzzy rule-based system system based on fuzzy IF-THEN rules, or another name for fuzzy systems. *See also* fuzzy inference system.

fuzzy rules of operation a system of relational assignment equations for the representation of the meaning of a fuzzy proposition.

fuzzy self-tuning self-tuning based on fuzzy logic concepts or fuzzy systems with some kind of self-tuning strategy or adaptation.

fuzzy set introduced by L. Zadeh (1965). A fuzzy set A in a universe of discourse X is characterized by a *membership function* μ_A which maps each element of X to the interval [0, 1]. A fuzzy set may be viewed as a generalization of a classical (crisp) set whose membership function only takes two values, zero or unity. *See also* crisp set, membership function.

fuzzy singleton a fuzzy set of membership value equal to 1 at a single real-valued point and 0 at all other points in the universe of discourse. *See also* membership function, fuzzy set.

fuzzy sliding mode control a combination of available mathematical description of the system dynamics with its linguistic description in terms of IF-THEN rules. In the early stages of fuzzy logic control, fuzzy modeling meant just a linguistic description in terms of IF-THEN rules of the dynamics of the plant and the control objective. *See also* TSK fuzzy model and fuzzy system.

fuzzy space the region containing the fuzzy sets created by set theoretic operations, as well as the consequent sets produced by approximate reasoning mechanisms.

fuzzy state feedback fuzzy control based on feedback of plant states.

fuzzy state space model fuzzy dynamic models in state space format. A typical example is a set of local linear state space models smoothly connected through a set of nonlinear membership functions.

fuzzy system a fuzzy system is a set of IF-THEN rules that maps the input space, say X, into the output space Y. Thus, the fuzzy system approximates a given function $F : X \rightarrow Y$ by covering the function's graph with patches, where each patch corresponds to an IF-THEN rule. The patches that overlap are averaged. *See also* fuzzy inference system.

fuzzy thresholding algorithm a thresholding algorithm with a threshold selected based on fuzzy measure.

fuzzy tuning the tuning of a control system using fuzzy logic and fuzzy rules.

fuzzy union an operator that take the maximum value of the membership grades.

FWHM *See* full width at half max.

G

G (giga)   a prefix indicating a quantity of 10^9. For instance, a gigabyte (GB) of storage is $1,000,000,000$ (typically implemented as 2^30) bytes.

G_{CR} common notation for compression. *See* compression.

g_d common notation for DC drain conductance.

G_I common notation for current gain. G_I is dimensionless.

g_m common notation for DC transconductance.

G_P common notation for power gain in decibels.

G_T common notation for transducer gain in amperes/volt.

G_V common notation for voltage gain. G_V is dimensionless.

G-line   a line of the mercury spectrum corresponding to a wavelength of about 436 nm.

GaAlSb/InGaAlAsSb   nearly lattice matched semiconductor heterostructure system capable of "staggered" band lineups in which the electrons and holes congregate in separate layers. Unique transport properties are utilized in optical devices and tunnel structures.

GaAs   periodic table symbol for gallium arsenide. *See* gallium arsenide.

GaAs laser   a semiconductor laser for wavelengths in the near infrared. The active medium is a gallium-arsenide semiconductor alloy.

GaAs/AlGaAs   most commonly grown semiconductor epitaxial heterostructure due to its lattice match, common anion, and existing technology base of GaAs devices.

Gabor transform   a short-time Fourier transform in which the window function is the Gaussian function.

gain   (1) the ratio of the output variable of a device to its input variable. For calculation purposes, the dimensionality of the gain is simply the unit of the output variable divided by the unit of the input variable. The gain of a device is a dimensionless value only when the electrical units of both the input and output variables are the same (e.g., voltage gain, current gain, power gain, etc.). In this case, a gain greater than one indicates an increase from input to output, while a value for gain less than one is indicative of a decrease (or attenuation). The overall gain of several cascaded components is found by multiplying the individual gains of each component in the system. Gain is often expressed in decibels to facilitate calculation of cascaded gains in a system. *See also* decibel.

(2) the ratio of the radiation intensity of a particular antenna to that of an isotropic radiator, in the same direction and at the same distance.

gain circles   circles of constant gain plotted on the Smith chart that can be used to graphically impedance match a device to achieve a desired gain. The circles are generated by plotting on the Smith chart the solution for the source reflection coefficient, Γ_s, or load reflection coefficient, Γ_L, in the transducer gain equation for a fixed value of

$$G = \frac{|S_{21}|^2 \left(1 - |\Gamma_S|^2\right)\left(1 - |\Gamma_L|^2\right)}{|\left(1 - S_{11}\Gamma_S\right)\left(1 - S_{22}\Gamma_L\right) - S_{12}S_{21}\Gamma_L\Gamma_S|^2}$$

for a fixed value of G_t, where S_{11}, S_{22}, S_{12}, and S_{21} are the scattering parameters for the device.

gain coefficient   factor multiplying distance in a formula for the exponential amplitude or intensity growth of a wave in an unsaturated amplifying medium.

gain compensation   this deals with the assumption in motion estimation in interframe coding that illumination is spatially and temporally uniform. Under these assumptions, the monochrome intensities $b(\mathbf{z}t)$ and $b(\mathbf{z}t - \tau)$ of two consecutive frames are related by $b(\mathbf{z}t)$ and $b(\mathbf{z} - \mathbf{D}, t - \tau)$ where τ is the time between two frames, $\mathbf{D}$ is the two-dimensional translation vector. Often this assumption about uniform illumination is not correct. In some situations, a multiplicative factor called *gain* is used to change the intensity. This is called gain compensation. This results in estimating $\mathbf{D}$ using gradient-type algorithms to minimize the square of the prediction error.

gain focusing   focusing or collimation of an electromagnetic beam by the profile of the gain; gain guiding.

gain guided laser diode   electrically pumped semiconductor laser in which the mode fields are confined in the transverse direction by the profile of the gain.

gain medium   medium for which an output electromagnetic wave has more power than the corresponding input, essential for laser operation.

gain ripple   difference between the maximum gain ($G_{max} = 10 \log_{10} [p_{outmax}/p_{incident}]$) and the minimum gain across the band ($G_{min} = 10 \log_{10} [p_{outmin}/p_{incident}]$), expressed in decibels. The band edges are usually defined as the highest and lowest frequencies within a contiguous band of interest at which the gain equals the minimum required passband gain.

$$G_{ripple} = G_{max} - G_{min}$$
$$= 10 \log_{10} \left(\frac{p_{outmax}}{p_{incident}} \right)$$
$$- 10 \log_{10} \left(\frac{p_{outmin}}{p_{incident}} \right)$$

gain saturation   reduction in gain that occurs when the intensity of a laser field depletes the population inversion.

gain switching   rapid turn-on of the gain in a laser oscillator for the purpose of producing a large output pulse.

gain–bandwidth product   for amplifiers based on a voltage-feedback op-amp, increasing the closed-loop gain causes a proportional decrease to the closed-loop bandwidth; thus the product of the two is a constant. This constant is called the gain–bandwidth product, and is a useful figure-of-merit for the performance of the op-amp.

gain-guided laser   laser in which the mode fields are confined in the transverse direction by the profile of the gain.

gain-shape vector quantization (GSVQ)   *See* shape-gain vector quantization.

gain-transfer measurement method   common antenna gain measurement scheme in which the absolute gain of the antenna under test is determined by measuring its gain relative to a gain standard (i.e., antenna with accurately known gain).

GaInN/AlInN   rapidly evolving semiconductor heterostructure system with ability to emit light in the green and blue regions of the spectrum for long-lifetime LEDs and lasers.

Galerkin's method   in an integral equation technique used to solve a numerical electromagnetic problem, the method in which the expansion and testing functions are the same.

gallium arsenide (GaAs)   a composite widely used in the fabrication of active elements.

gallium phosphide (GaP)   (1) a semiconductor for high-speed electronics and that is part of the family of III-V compounds for semiconductor lasers.

(2) an acousto-optic material with good acousto-optic figure of merit and with low acoustic attenuation for high-bandwidth acousto-optic cells.

galloping   a low-frequency vibration of electric power lines caused by wind.

Galois field   a finite field with q elements denoted as $GF(q)$, important in the study of cyclic codes. *See also* finite field.

Galvani, Luigi   (1737–1798) Born: Bologna, Italy

Best known for the use of his name in the description of zinc plating, or "galvanizing" metals. Galvani was an anatomist who studied the flow of electricity in frog's legs when metal contacts were applied. Many of Galvani's ideas were later proved incorrect by Alessandro Volta. However, his pioneering work was honored by the reference to Galvanic current as an early name for electric current. Andre Ampere also honored him by naming his current measurement apparatus a galvanometer.

gamma camera   a device that uses a scintillation crystal to detect gamma photons for radionuclide imaging. Also known as a scintillation camera.

gamma correction   inside a CRT, the luminance of a pixel is represented by a voltage V; however the luminance of the ray produced by that voltage is not proportional to V, but rather to V^γ, where γ is a device-dependent constant between 2 and 2.5. It is thus necessary to compensate for this effect by transforming the luminance L into $L^{1/\gamma}$; this is called gamma correction. Some CCD cameras implement gamma correction directly on the signal that they generate. Some computer monitors allow modification of the factor γ or gamma correction through software.

gamma rays   electromagnetic radiation of very high energy (greater than 30 keV) emitted after nuclear reactions or by a radioactive atom when its nucleus is left in an excited state after emission of alpha or beta particles.

gamma-ray laser   laser producing its output in the gamma-ray region of the spectrum, often considered to be any wavelength below about 0.1 angstrom; graser, not yet demonstrated.

ganged   operation of a multiple phase device with all phases operated simultaneously.

GaP   *See* gallium phosphide.

gap filler   a low-power transmitter that boosts the strength of transmitted DAB RF signals in areas that normally would be shadowed due to terrain obstruction. Gap fillers can operate on the same frequency as DAB transmissions or on alternate channels that can be located by DAB receivers using automatic switching.

gapless arrester   a lightning arrester which is distinguished from a gapped arrester by having a continuous conductive path between the conductor and ground.

gapped arrester   a lightning arrester whose conducting path contains a gas- or air-filled spark gap which must be broken down by the lightning impulse voltage.

garbage   an object or a set of objects that can no longer be accessed, typically because all pointers that direct accesses to the object or set have been eliminated.

garbage collection   the process by which the memory a program no longer uses is automatically reclaimed and reused. Relying on garbage collection to manage memory simplifies the interfaces between program components. Two main methods used to implement garbage collection are reference counting and mark-and-sweep.

garbage collector   a software run-time system component that periodically scans dynamically allocated storage and reclaims allocated storage that is no longer in use (garbage).

gas cable   an electric power cable which is pressurized with an insulating gas, typically sulfur hexaflouride, for its primary insulation.

gas capacitor   a capacitor whose dielectric is composed of a high-pressure gas, often nitrogen or an electronegative gas such as sulfur hexaflouride.

gas circuit breaker   a circuit breaker in which the arc between the contacts is extinguished by immersion in or a blast of an electronegative gas.

gas discharge excitation   electron impact excitation or pumping of a laser medium occurring as a result of collisions between the lasing atoms or molecules and the discharge electrons.

gas dynamic excitation   excitation or pumping of a laser medium occurring as a result of heating followed by rapid expansion and cooling of a gaseous laser medium.

gas insulated switchgear   circuit breakers and switches whose primary insulation is compressed gas.

gas laser   laser in which the amplifying medium is a gas.

gas metal arc welding (GMAW)   a welding process that produces coalescence of metals by heating them with an arc between a consumable filler metal electrode and the parts to be joined. The process is used with shielding gas and without the application of pressure.

gas substation   an electric power substation in which the conductors are insulated from each other and the ground by a high-pressure gas, generally sulfur hexaflouride.

gas tungsten arc welding (GTAW)   a welding process that produces coalescence of metals by heating them with an arc between a nonconsumable tungsten electrode and the parts to be joined. The process is used with shielding gas and without the application of pressure. Filler may or may not be used.

gate   (1) a logical or physical entity that performs one logical operation, such as AND, NOT, or OR.

(2) the terminal of a FET which controls the flow of electrons from source to drain. It is usually considered to be the metal contact at the surface of the die. The gate is usually so thin and narrow that if any appreciable current is allowed to flow, it will rapidly heat up and self-destruct due to I-R loss. This same resistance is a continuing problem in low noise devices, and has resulted in the creation of numerous methods to alter the gate structure and reduce this effect.

gate array   a semicustom integrated circuit (IC) consisting of a regular arrangement of gates that are interconnected through one or more layers of metal to provide custom functions. Generally, gate arrays are preprocessed up to the first interconnect level so they can be quickly processed with final metal to meet customer's specified function.

gate length   the length in microns of that portion of the FET channel, as measured in the direction of channel current flow, that can be pinched off by application of the proper control bias. Gate lengths are usually on the order of 0.1 to 2 microns. The f_{max} of a FET is inversely proportional to the gate length. If the length is very short (less than 0.05 microns for GaAs), then quantum effects play a role in device operation.

gate turn off device   (1) any power semiconductor switching device which can turned off with a signal to its gate.

(2) a thyristor capable of gate-turn-off operation.

gate turnoff thyristor   a thyristor that can be turned off with a negative gate current while conducting current in the forward direction.

gate width   the width in microns of that portion of a FET channel associated with a single gate, as measured in the direction orthogonal to channel current flow, that can be pinched off by application of the proper control bias. A single FET structure may include many subcells placed in parallel with each other, resulting in a much higher gate periphery for more power (i.e., 4 individual FETs, each with a gate of 100 microns, may be paralleled together to form a 4 × 100 micron FET with a gate periphery of 400 microns).

gate-to-drain breakdown voltage   the breakdown voltage of the reverse biased gate-to-drain junction of FET. Usually it is specified at a predetermined value of current per millimeter of gate periphery.

gate-to-drain capacitance   the capacitance between the gate and drain terminals of a FET. It is formed primarily by the capacitance of the FET's physical interconnect structure (extrinsic fixed capacitance) and by the depleted region capacitance (intrinsic capacitance), which is a function of voltage and temperature.

gate-to-source breakdown voltage   the breakdown voltage of the reverse-biased gate-to-source junction of FET. Usually it is specified at a predetermined value of current per millimeter of gate periphery.

gate-to-source capacitance   the capacitance between the gate and source terminals of a FET. It is formed primarily by the capacitance of the FET's physical interconnect structure (extrinsic fixed capacitance) and by the depleted region capacitance (intrinsic capacitance), which is a function of voltage and temperature.

gate-to-source voltage   the potential difference between the FET gate and source terminals, this voltage controls the channel from saturation to pinchoff. This voltage is normally negative for an n-channel FET, and positive for a p-channel FET. However, either of these devices can be operated in a slightly enhanced mode, allowing low-level excursions of the opposite polarity.

gateway   *See* router.

Gauss' law   fundamental law of electromagnetic field that states that the total electric/magnetic flux through a closed surface is equal to the total electric/magnetic charge enclosed.

Gauss' theorem   *See* divergence theorem.

Gaussian aperture   aperture in which the transmission profile is a Gaussian function of radius.

Gaussian approximation   an approximation used in statistics where the distribution of a sum of random variables is approximated by the Gaussian distribution. Such an approximation is based on the central limit theorem of probability.

Gaussian beam   electromagnetic beam solution of the paraxial wave equation in which the field has spherical phase fronts and is a Gaussian function of distance from the beam axis.

Gaussian distribution   a probability density function characterized by a mean μ and covariance Σ:

$$f(x) = (2\pi)^{-N/2} |\Sigma|^{-1/2} \, e^{-(x-\mu)^T \Sigma^{-1}(x-\mu)/2},$$

where $|\Sigma|$ represents the determinant of Σ and N represents the dimensionality of x. If x is scalar then the above function simplifies considerably to its more familiar form:

$$f(x) = \frac{1}{\sqrt{2\pi}\,\sigma} e^{-\frac{(x-\mu)^2}{2\sigma^2}}.$$

The Gaussian distribution is tremendously important in modeling signals, images, and noise, due

297

to its convenient analytic properties and due to the central limit theorem probability density function, mean, covariance. *See* Cauchy distribution, exponential distribution.

Gaussian elimination   the standard direct method for solving a set of linear equations. It is termed direct because it does not involve iterative solutions. Variations of this scheme are used in most circuit simulators.

Gaussian mirror   mirror in which the reflection profile is a Gaussian function of radius.

Gaussian noise   a noise process that has a Gaussian distribution for the measured value at any time instant.

Gaussian process   (1) a random process where the joint distribution of a set of random variables $X_1, X_2, \ldots, X_n$ determined as values of the process at the points $t_1, t_2, \ldots, t_n$ is an n-variate Gaussian distribution for all sets of points $t_1, t_2, \ldots, t_n$, and all values of the integer n.
(2) a random (stochastic) process $x(t)$ is Gaussian if the random variables $x(t_1), x(t_2), \ldots, x(t_n)$ are jointly Gaussian for any n.

Gaussian pulse   pulse in which the field is a Gaussian function of time from the pulse maximum.

Gaussian sphere   unit sphere, with an associated spherical coordinate system, normally used to represent orientations.

gaussmeter   an instrument to measure the flux density due to a magnetic field.

Ge   *See* germanium.

geared robot   an arbitrary robot equipped with gears is called a geared robot. *See* direct drive robot.

genco   a contraction of "generating company," which is a company which generates electric power but does not engage in transmission or distribution activities.

gender   an adapter presenting two male or two female connectors for reversing the type of cable connector. Connectors can be of the same type or not.

general response formula for 2-D Roesser model   the solution to the 2-D Roesser model

$$\begin{bmatrix} x_{i+1,j}^h \\ x_{i,j+1}^v \end{bmatrix} = \begin{bmatrix} A_1 & A_2 \\ A_3 & A_4 \end{bmatrix} \begin{bmatrix} x_{ij}^h \\ x_{ij}^v \end{bmatrix} + \begin{bmatrix} B_1 \\ B_2 \end{bmatrix} u_{ij} \quad (1a)$$

$i, j \in Z_+$ (the set of nonnegative integers)

$$y_{ij} = [C_1, C_2] \begin{bmatrix} x_{ij}^h \\ x_{ij}^v \end{bmatrix} + D u_{ij} \quad (1b)$$

with boundary conditions $x_{0j}^h, j \in Z_+$ and $x_{i0}^v, i \in Z_+$ is given by

$$\begin{bmatrix} x_{ij}^h \\ x_{ij}^v \end{bmatrix} = \sum_{p=1}^{i} T_{i-p,j} \begin{bmatrix} 0 \\ x_{p0}^v \end{bmatrix}$$

$$+ \sum_{q=1}^{j} T_{i,j-q} \begin{bmatrix} x_{0q}^h \\ 0 \end{bmatrix} \quad (2)$$

$$+ \sum_{p=0}^{i} \sum_{q=0}^{j} \left(T_{i-p-1,j-q} \begin{bmatrix} B_1 \\ 0 \end{bmatrix} \right.$$

$$\left. + T_{i-p,j-q-1} \begin{bmatrix} 0 \\ B_2 \end{bmatrix} \right) u_{pq}$$

where $x_{ij}^h \in R^{n_1}$ and $x_{ij}^v \in R^{n_2}$ are the horizontal and vertical state vectors, $u_{ij} \in R^m$ is the input vector, $y_{ij} \in R^p$ is the output vector, $A_1, A_2, A_3, A_4, B_1, B_2, C_1, C_2, D$ are given real matrices, and the transition matrix T_{pq} is defined by

$$
T_{pq} = \begin{cases}
I_n & \text{for } p = q = 0 \\
\begin{bmatrix} A_1 & A_2 \\ 0 & 0 \end{bmatrix} & \text{for } p = 1, q = 0 \\
\begin{bmatrix} 0 & 0 \\ A_3 & A_4 \end{bmatrix} & \text{for } p = 0, q = 1 \text{ and} \\
T_{10}T_{i-1,j} + T_{01}T_{i,j-1} & \text{for} \\
\quad i, j \in Z_+ \ (i + j > 0) \\
0 & \text{for } p < 0 \text{ or/and } q < 0
\end{cases}
$$

Substitution of (2) into (1b) yields the response formula.

general response formula for generalized 2-D model the solution x_{ij} to the generalized 2-D model

$$
Ex_{i+1,j+1} = A_0 x_{ij} + A_1 x_{i+1,j} + A_2 x_{i,j+1}
$$
$$
+ B_0 u_{ij} + B_1 u_{i+1,j}
$$
$$
+ B_2 u_{i,j+1} \quad \text{(1a)}
$$
$$
y_{ij} = C x_{ij} + D u_{ij} \quad \text{(1b)}
$$

$i, j \in Z_+$ (the set of nonnegative integers) with admissible boundary conditions x_{i0}, $i \in Z_+$ and x_{0j}, $j \in Z_+$ is given by

$$
x_{ij} = \sum_{p=1}^{i+n_1} \sum_{q=1}^{j+n_2} \left(T_{i-p-1,j-q-1} B_0 \right.
$$
$$
+ T_{i-p,j-q-1} B_1 + T_{i-p-1,j-q} B_2 \big) u_{pq}
$$
$$
+ \sum_{p=1}^{i+n_1} \left(T_{i-p,j-1} [A_1, B_1] \right.
$$
$$
+ T_{i-p-1,j-1} [A_0, B_0] \big) \begin{bmatrix} x_{p0} \\ u_{p0} \end{bmatrix}
$$
$$
+ \sum_{q=1}^{i+n_2} \left(T_{i-1,j-q} [A_2, B_2] \right.
$$
$$
+ T_{i-1,j-q-1} [A_0, B_0] \big) \begin{bmatrix} x_{0q} \\ u_{0q} \end{bmatrix}
$$
$$
+ T_{i-1,j-1} [A_0, B_0] + \begin{bmatrix} x_{00} \\ u_{00} \end{bmatrix} \quad \text{(2)}
$$

where $x_{i,j} \in R^n$ is the semistate local vector, $u_{i,j} \in R^m$ is the input vector, E, A_k, B_k ($k = 0, 1, 2$) C, D are given real matrices with E possibly singular and the transition matrices T_{pq} are defined by

$$
ET_{pq} = \begin{cases}
A_0 T_{-1,-1} + A_1 T_{0,-1} \\
+ A_2 T_{-1,0} + I & \text{for } p = q = 0 \\
A_0 T_{p-1,q-1} + A_1 T_{p,q-1} \\
+ A_2 T_{p-1,q} \\
\quad \text{for } p \neq 0 \text{ and/or } q \neq 0
\end{cases}
$$

and

$$
[Ez_1 z_2 - A_0 - A_1 z_1 - A_2 z_2]^{-1}
$$
$$
= \sum_{p=-n_1}^{\infty} \sum_{q=-n_2}^{\infty} T_{pq} z_1^{-(p+1)} z_2^{-(q+1)}
$$

pair (n_1, n_2) of positive integers n_1, n_2 such that $T_{pq} = 0$ for $p < -n_1$ and/or $q < -n_2$ is called the index of the model. Substitution of (2) into (1b) yields the response formula.

general-purpose motor term often used to describe National Electrical Manufacturers Association (NEMA) class B, and less often class A, induction motors. General-purpose motors are those typically used when relatively low starting currents, low slip, good speed regulation, moderate starting torque, and high efficiency are the predominant concerns.

general-purpose register a register that is not assigned to a specific purpose, such as holding condition codes or a stack pointer, but that may be used to hold any sort of value. General-purpose registers are typically not equipped with any dedicated logic to operate on the data stored in the register.

general-use switch a manually operated switching device designed for general use. It is designed to interrupt rated current at rated voltage.

generalization the process of inferring rules and taking decisions after a learning phase has taken place. The process is supposed to take place on data not used during learning.

generalized cone   data structure for volumetric representations, generated by sweeping an arbitrarily-shaped cross section along a 3-D line called the "generalized cone axis."

generalized delta rule   the weight update rule employed in the backpropagation algorithm.

generalized Lloyd algorithm (GLA)   a generalization of the Lloyd (or Lloyd–Max) algorithm for scalar quantizer design to optimal design of vector quantizers. *See also* K-means algorithm.

generalized modus ponens   generalization of the classical *modus ponens* based on the compositional rule of inference.

Let be a fuzzy rule $A \longrightarrow B$, that can be interpreted as a fuzzy relation R, and a fuzzy set A', then the compositional rule of inference maps a fuzzy set

$$B' = A' \circ R = A' \circ (A \longrightarrow B)$$

that can be interpreted as

premise1:	x is A'
(fact)	
premise2 :	if x is A then y is B
(fuzzy rule)	
consequence :	y is B'
(conclusion)	

For example, if we have the fuzzy rule *"If the tomato is red then it is ripe,"* and we know *"The tomato is more or less red"* (fact), the generalized modus ponens can infer *"The tomato is more or less ripe."*

See also compositional rule of inference, fuzzy inference system, fuzzy relation, linguist variable, modifier.

generator   in electrical systems, any of a variety of electromechanical devices that convert mechanical power into electrical power, typically via Faraday induction effects between moving and stationary current-carrying coils and/or magnets. Electrostatic generators use mechanical motion to physically separate stationary charges to produce a large electrostatic potential between two electrodes.

generator coherency   a group of generators where the rotor angles swing in synchronism with one another following a disturbance. Usually, generators in close electrical proximity and at some distance from the fault tend to be coherent.

generator differential relay   a generator differential relay is a differential relay specifically designed for protection of electric power generators. Variations include allowances for split-phase winding machines.

generator inertia constant   a term proportional to the combined moment of inertia of the turbine-generator mass.

generator matrix   a matrix used to describe the mapping from source word to code word in a linear forward error control code. The mapping is described through multiplication of the source word by this matrix using element-wise finite field arithmetic. A linear code is completely specified by its generator matrix.

generator polynomial   uniquely specifies a cyclic code and has degree equal to the number of the parity bits in the code. For an (n, k) cyclic code, it is the only code word polynomial of minimum degree $(n - k)$.

genetic algorithm   an optimization technique that searches for parameter values by mimicking natural selection and the laws of genetics. A genetic algorithm takes a set of solutions to a problem and measures the "goodness" of those solutions. It then discards the "bad" solutions and keeps the "good" solutions. Next, one or more genetic operators, such as mutation and crossover, are applied to the set of solutions. The "goodness" metric is applied again and the algorithm iterates until all solutions meet certain criteria or a specific number of iterations has been completed optimization.

genlock a shortened term for "generator lock," meaning that one sync generator system is locked to another. When video systems are genlocked, sync and burst information is the same for both systems.

geodesic lens a lens composed of circularity symmetric spherical depression (or dimples) in the surface of a thin-film waveguide. In this type of lens, light waves are confined in the waveguide and follow the longer curved path through the lens region. Waves propagating near the center of the lens travel a longer path than waves near the edge, modifying the wave front so that focusing can occur.

geometric distribution a discrete probability density function of a random variable $\mathbf{x}$ that has the form $p\{\mathbf{x} = k\} = p(1-p)^k k = 0, 1, 2, \ldots$ this is the probability that k independent trials, each with probability of success p and failure $1 - p$, fail before one succeeds. *See* probability density function.

geometric Jacobian the Jacobian (or more precisely the Jacobian matrix, but roboticists usually shorten it to simply Jacobian) is a mapping from velocities in joint space (generalized velocities) to velocities in Cartesian space. This mapping is written in the form $v = J(q)\dot{q}$ where v is a six-dimensional vector of linear and angular velocities of the end-effector and $\dot{q}$ is a vector of generalized velocities $\dot{q}(t) = [\dot{q}_1(t), \dot{q}_2(t), \ldots, \dot{q}_n(t)]^T$. The Jacobian has dimensions $6 \times n$. All points in which the Jacobian is not invertible are called singularities of the mechanism or singularities. The Jacobian is defined for an arbitrary manipulator and depends on its geometric parameters (more precisely, Denavit–Hartenberg parameters). *See also* analytical Jacobian.

geometric radian center frequency the logarithmic radian center frequency, it is the logarithmic mid-point between the higher (ω_H) and lower (ω_L) band edges, expressed in radians/second. The band edges are usually defined as the highest and lowest frequencies within a contiguous band of interest at which the loss equals L_{Amax}, the maximum attenuation loss across the band.

$$\omega_{OG} = \sqrt{\omega_H \cdot \omega_L}$$

or

$$\log_{10}(\omega_{OG}) = \frac{\log_{10}(\omega_H) + \log_{10}(\omega_L)}{2}$$

geometric theory of diffraction a correction to geometrical optics that includes diffracted fields due to corners or edges and accounts for energy diffracted into the shadow region.

geometric transformation transforms the pixel co-ordinates of an image to effect a change in the spatial relationships of elements in the image. The change often takes the form of a stretching or warping of the image.

geometrical optics a high-frequency technique using ray tracing for tubes of rays to determine incident, reflected, and refracted fields. Most useful in real media when the wave amplitude varies slowly compared to the wavelength.

geosynchronous orbit an orbit 22,753 miles above the earth in which an object will orbit the earth once every 24 hours above the equator and will appear to be stationary from the earth's surface.

geothermal energy thermal energy in the form of hot water and steam in the earth's crust.

germanium (Ge) an optical material for construction of components and systems at infrared wavelengths, especially in the 8 to 14 micron region. Also, an acousto-optic material for infrared wavelengths.

Gershgorin circle   measures the relative size of the off-diagonal elements in a transfer function matrix at a given frequency. When evaluated over a range of frequencies, these circles sweep out a band, centered on the diagonal elements, that is used in the prediction of closed loop stability for multi-input–multi-output systems. This theory is based on Gerhsgorin's theory on the bounds of matrix eigenvalues. *See also* diagonal dominance and inverse Nyquist array.

GFCI   *See* ground-fault circuit interrupter.

GFI   *See* ground fault interrupter.

GFLOP   *See* gigaflop.

ghosting   the formation of an image in which a ghost image (i.e., a similar, fainter, slightly displaced image) appears superimposed on the intended image; ghosts result from a form of crosstalk, and typically arise by radio transmission via an alternative path occurring as a result of random reflection.

giant magnetoresistance effect   huge change in electrical resistance with relatively small change in magnetic field exhibited by multilayers of many combinations of metallic materials. Used in the next generation of hard disk memory.

giant pulse   large output pulse from a laser oscillator that results when the cavity losses are quickly raised from a high value holding a pumped laser below threshold to low value bringing the laser above threshold (Q-switching or loss-switching) or when the gain occurs in the form of a short pulse (gain switching).

Gibbs phenomenon   the rippling phenomenon in the reconstruction of a function around a discontinuity based on the Fourier series of the function. Specifically, given a discontinuous periodic function f which is square-integrable on its period, the Fourier series of f converges to f in the square mean sense. When the periodic function corresponding to the n first terms of that Fourier series is built, then ripples appear at the vicinity of each discontinuity of f. When n tends to infinity, these ripples tend to zero in the square mean sense, but their maximum amplitude does not tend to zero; this produces an overshoot by a factor of 1.1789797 and an undershoot by a factor of 0.9028233. A similar effect arises for a square-integrable non-periodic function reconstructed from its Fourier transform. The phenomenon causes ringing around edges in low bit rate image coding. *See* Fourier integral, Fourier series, norm.

Gibbs random field   a class of random fields described by normalized exponentials of potential functions over cliques. Let $S = s_1, s_2, \ldots$ be points on a lattice, and let $\omega \in \Omega$ be a sample of a random field on this lattice. Points on the lattice interact with other lattice points in a local, predefined manner; that is, the energy of the lattice is given by the sum of the energies of locally-interacting sets of points in the lattice: each such set of points is known as a clique $C \in \mathcal{C}$. Specifically, the energy of clique C in sample ω is $V_C(\omega)$, where V is a potential function depending only upon the points in C. Then the total energy of a sample ω is given by

$$E(\omega) = \sum_{C \in \mathcal{C}} V_C(\omega).$$

With this definition, a probability measure π is a Gibbs random field if

$$\pi(\omega) = \frac{1}{Z} e^{-E(\omega)/T}$$

where $Z = \sum_{\omega} e^{-E(\omega)/T}$ is a normalization constant.

Gibson mix   an analysis of computer machine language instructions that concluded that approximately 1/4 of the instructions accounted for 3/4 of the instructions executed on a computer.

GIF   *See* graphics interchange format.

gigaflop (GFLOP)   1000 million floating point operations per second.

Gilbert cell   a four-transistor configuration combining the differential pair and current mirror concepts. With appropriate signal conditioning at the input and output terminals, the cell can be used for many analog signal processing applications such as analog multiplication.

Gilbert, William   (1544–1603) Born: Colchester, Essex, England

Best remembered as an early investigator into electric charge and magnetism. He is also considered by many to be the inventor of the modern scientific method. This is due to his rigorous experimental methodology, and the detailed records he kept on his investigations. Isaac Newton and Francis Bacon both acknowledged his contributions in this regard. Gilbert was, by training, a physician and held the post of royal physician in the courts of Elizabeth I and James I.

Givens transformation   a transformation, proposed by Givens, that transforms a general matrix to a triangular form. The Givens transformations $G_{ij\theta}$ are functions of three parameters. Transforming a vector b, $a = G_{ij\theta}b$; then $a_n = b_n$ for all $n \neq i$, $n \neq j$, and the two-vector $[a_i a_j]$ is equal to the rotation of vector $[b_i b_j]$ by an angle θ in the plane. Givens rotations can be used to successively set elements of a matrix to zero by an appropriate selection of $ij\theta$.

GKS   *See* graphical kernel system.

GLA   *See* generalized Lloyd algorithm.

glass laser   laser in which the host medium for doping with laser atoms is a glass.

glitch   (1) an incorrect state of a signal that lasts a short time compared to the clock period of the circuit. The use of "glitch" in describing power systems is generally avoided. *See also* hazard.

(2) slang for a transient that causes equipment crashes, loss of data, or data errors.

global alignment   a method of alignment where the mask is aligned globally to the whole wafer (as opposed to field-by-field alignment).

global interconnection   interconnection in which every source is connected to all detectors and every detector is connected to all sources. Global interconnection is easily implementable using optics because, unlike electrons, photons do not interact with each other, and an optical system is inherently a parallel processor.

global memory   in a multiprocessor system, memory that is accessible to all processors. *See also* local memory, distributed memory.

global minimum   a point at which a function attains its lowest value over the domain of its arguments.

global observability of generalized 2-D model   the generalized 2-D model

$$Ex_{i+1,j+1} = A_1 x_{i+1,j} + A_2 x_{i,j+1}$$
$$+ B_1 u_{i+1,j} + B_2 u_{i,j+1}$$
$$y_{ij} = C x_{ij}$$

is called globally observable if any of its global semistates

$$X(q) := \{x_{i-q,-i}, i = \ldots, -1, 0, 1, \ldots;$$
$$q = 0, 1, \ldots\}$$

can be calculated using future outputs and inputs of the model, i.e.,

$$U(k) := \{u_{i+k,-i}, i = \ldots, -1, 0, 1, \ldots;$$
$$k = 0, 1, \ldots\}$$
$$Y(k) := \{y_{i+k,-i}, i = \ldots, -1, 0, 1, \ldots;$$
$$k = 0, 1, \ldots\} \text{ for } k \geq q$$

global positioning system (GPS) system of 18 primary satellites in medium earth orbit, distributed so that at least four are simultaneously visible from each point on the globe; typically used in timing and positioning applications.

global reconstructibility of generalized 2-D model the generalized 2-D model

$$Ex_{i+1,j+1} = A_1 x_{i+1,j} + A_2 x_{i,j+1}$$
$$+ B_1 u_{i+1,j} + B_2 u_{i,j+1}$$
$$y_{ij} = C x_{ij}$$

is called globally reconstructible if any of its global semistates

$$X(q) := \{x_{i-q,-i}, i = \ldots, -1, 0, 1, \ldots;$$
$$q = 0, 1, \ldots\}$$

may be calculated using past outputs and inputs of the model, i.e.,

$$U(k) := \{u_{i+k,-i}, i = \ldots, -1, 0, 1, \ldots;$$
$$k = 0, 1, \ldots\}$$
$$Y(k) := \{y_{i+k,-i}, i = \ldots, -1, 0, 1, \ldots;$$
$$k = 0, 1, \ldots\} \text{ for } k \leq q$$

globally asymptotically stable equilibrium an asymptotically stable equilibrium (see the definition) with a region of attraction (see the definition) equal to $\Re^n$ (n-dimensional real Euclidean space).

globally asymptotically stable state *See* asymptotically stable in the large.

glove in power line work, this refers to conductors which are energized with voltages low enough to be safely contacted by workers wearing suitable rubber gloves, to the range of voltages at which the practice is allowed, and to the gloves themselves.

GMAW *See* gas metal arc welding.

GMSK *See* minimum-shift keying Gaussian.

goat head the top of a tower.

Gold book *See* IEEE Color Books.

Gold sequences a set of spreading sequences developed by the coding theorist R. Gold in 1966 that are typically used in a multiple access system utilizing direct sequence spread spectrum. The set of Gold sequences is the solution of a problem in sequence design, where the criteria is the minimization of the maximum cross-correlation between any two sequences in the set under all possible cyclic shifts of one sequence relative to the other. *See also* cross-correlation.

Gompertz dynamics the simplest nonlinear, with sigmoidal growth function, model of population dynamics. The unperturbed growth of population is described by the first-order differential equation

$$\dot{x} = gx ln \left(\frac{x_{\max}}{x} \right)$$

where x is an average number of individuals in the population with maximal value (the so-called plateau population) x_{max} and g is a positive growth parameter. The Gompertz equation may be simplified using the dimensionless scaling

$$y = ln \left(\frac{x_{\max}}{x} \right)$$

which leads to a linear equation of the form

$$\dot{y} = -gy$$

The simplest way to introduce perturbations to the dynamics represented by affine control actions is to add a bilinear term into the Gompertz equation

$$\dot{x} = g \ln \left(\frac{x_{\max}}{x} \right) - kux$$

where u is a control variable and k represents a perturbed population loss (or growth) parameter. By applying the dimensionless scaling, the equation could be once more transformed into the linear one with respect to variable y. The unperturbed Gompertz dynamics could also be represented by two first-order differential linear equations

$$\dot{x} = q(t)x; \ \dot{q} = -gq; \ x(0) = x_0; \ q(0) = q_0$$

with $x_0 e^{-\frac{q_0}{g}} = x_{\max}$.

Gordon–Haus limit expresses the maximum performance of a soliton-based communication system as a bitrate-length product.

GOS *See* grade of service.

governor a device connected to a rotating machine by which the speed-regulating system is automatically adjusted to maintain constant speed under various load conditions.

governor power flow the inherent response of prime movers or governors to a change in the operating condition of the power system in an attempt to balance the power equation. All generators participate at some level in the change, with larger generators picking up relatively larger amounts.

GPS *See* global positioning system.

grade of service (GOS) a probabilistic measure of service or equipment availability in a telecommunications network, expressed as the probability of a particular service (e.g., a completed telephone connection) being denied at request time.

graded index medium in which the index of refraction varies as a function of position; usually

refers to variations transverse to the direction of propagation, and quadratic transverse variations are especially important.

graded index fiber an optical fiber where the core refractive index varies from a maximum at its center to a minimum value at the core-cladding interface. The profile is typically designed so that all modes of a multimode fiber propagate at approximately the same velocity. *See also* step index fiber.

graded index lens cylindrical optical elements with a refractive-index profile across the cross section of the element to provide the transfer; lens in which the transmitted waves are focused or defocused by transverse variations of the index of refraction rather than by transverse variations of the thickness.

graded index optical fiber an optical fiber with a refractive index in the core that decreases monotonically from the fiber axis to the interface between the core and cladding. The index usually goes from a higher value in the center of the core to a lower value of refractive index at the core/cladding interface. *See also* graded index profile.

graded index optics optical elements that use a refractive-index profile across the cross-section of the element, typically a cylinder, to provide particular transfer functions as for lenses; the profile is generally produced by ion implantation. Also called GRIN optics.

graded index profile an index of refraction in an optical fiber core that decreases with distance from the core axis out to the core/cladding boundary. The variation of the index of refraction with the radial distance from the fiber axis can be given approximately as a power law.

gradient a vector function denoted by ∇f or **grad** f, where f is a continuous, differentiable scalar function. For a 2-D function $f(x, y)$, the gradient is

$$\nabla f = \begin{bmatrix} \frac{\partial f}{\partial x} \\ \frac{\partial f}{\partial y} \end{bmatrix}$$

The magnitude of this gradient is

$$|\nabla f| = \sqrt{\left(\frac{\partial f}{\partial x} + \frac{\partial f}{\partial y} \right)^2}$$

In image processing, the term gradient often refers to the magnitude of the gradient. *See* Sobel operator.

gradient descent   a method for finding the minimum of a multidimensional function $f(x)$. The technique starts at some point and advances towards the minimum by iteratively moving in a direction opposite to that of the gradient:

$$x_{i+1} = x_i - \alpha \frac{\partial f}{\partial x}$$

where α is some scalar, usually set empirically. *See* gradient, optimization.

gradient edge detector   an edge detector that defines an edge to be present at a pixel only if the magnitude of the gradient at that pixel is greater than some threshold. *See* edge, edge detection, gradient, Sobel operator.

gradient index optics   optical components, e.g., optical fibers, within which the refractive index changes gradually between two extremes.

gradient space   a 2-D representation of the orientation of a 3-D surface in space, in which the two components of a point $P \equiv (p, q)$ are the first-order partial derivatives of a surface of the form $z = f(x, y)$.

grammar   an ordered 4-tuple $(G = T, V, P, S)$, where T is said to be the set of terminals, V is the set of variables or non-terminals, P a set of production rules and $S \in V$ is called the start symbol. Terminals are the symbols with which the strings of the language are made. For example, T may be the English alphabet, the ASCII character set, or the set $\{0, 1\}$. Non-terminals are symbols that are replaced by a string of zero or more terminals and non-terminals. The production rules specify which strings can be used to replace non-terminals. The symbol S is the first symbol with which every production starts. All the strings that can be generated from S using rules in P are said to be in $L(G)$, the language of G. For example, let G be the grammar (T, V, P, S) such that

$$T = \{a, b\}$$
$$V = \{S\}$$

where S is the start symbol and

$$P = \{S \rightarrow aSb, S \rightarrow \varepsilon\}$$

where ε is the empty string. The language $L(G)$ is the set of all strings of the form $a^n b^n$. The $\rightarrow$ symbol signifies that S can be replaced with whatever follows the arrow.

Gram-Schmidt orthogonalization   a recursive procedure for whitening (decorrelating) a sequence of random vectors. *See* whitening filter.

grant signal   a control signal on a bus that gives permission to control the bus to another module.

graph   a couple $G = (E, V)$ where V is a set of nodes and $E \subseteq V \times V$ is a set of edges. Graphs are widely used in modeling networks, circuits, and software.

graph search   an optimization technique used to find the minimum cost path from a starting point to a goal point, through a graph of interconnected nodes. Each link between nodes has an associated path cost, which must be selected based on the problem of interest. *See* optimization.

graphics   the discipline dealing with the generation of artificial images by a computer. Its two main aspects are geometric modelling, whose subject is the computational representation of the geometry and topology of objects and scenes, and rendering, which studies the generation of images from the interaction of light and objects. Graphics has been generalized to the synthesis of animated image sequences, in which case one speaks of computer animation. Also called computer graphics, image synthesis.

graphic adapter   an adapter for interfacing the computer toward a monitor. *See* MDA, CGA, EGA, VGA.

graphic controller   *See* graphic adapter.

graphical kernel system (GKS)   a standard for computer graphics recognized by both ANSI and ISO. GKS defines the manipulation of graphic objects, including their visualization, print, etc. All manipulation is performed by regarding the graphic adoption as a independent device driver. In this way, GKS applications can be executed on different kinds of graphics adapters.

Graphics Interchange Format (GIF)   a popular image-file format that compresses the image with LZW coding. file format, image compression, Lempel-Ziv-Welch (LZW) coding. The Graphics Interchange Format© is the Copyright property of CompuServe Incorporated. GIF(sm) is a Service Mark property of CompuServe Incorporated.

grating   *See* diffraction grating.

grating lobe   a lobe in an antenna radiation pattern whose peak amplitude is equal to that of the main lobe.

grating spectrometer   an instrument that provides a spectral decomposition of a optical source using a diffraction grating to give spatial dispersion of wavelengths.

gravitational torque   torque that depends upon the position of the robot in the gravitational field.

Gray book   *See* IEEE Color Books.

Gray code   a code in which each of a sequence of code words differs by one bit from the preceding one, and the assigned value of each code word is one more than that of the preceding one. Such a code avoids glitches (i.e., sharp momentary unwanted spikes) when, in an electromechanical system, the sensors giving the code words are imperfectly aligned. One possible three-bit Gray code, a three bit binary code for comparison purposes, and the decimal equivalent value are given in the table below:

gray level   the individual numerical value corresponding to a particular degree of brightness in a digital image, often on an 8-bit gray scale consisting of 256 gray levels stretching from pure black to pure white.

gray level co-occurrence   a means of measuring texture and other brightness variations in digital images by generating matrices which tabulate the frequencies with which different gray levels co-exist at different distance vectors from each other.

Gray code

Decimal value	0	1	2	3	4	5	6	7
Gray code representation	000	001	011	010	110	111	101	100
Binary representation	000	001	010	011	100	101	110	111

gray level saturation   the restriction of image gray scale so that intensities above a certain level become fixed at the white level corresponding to the highest numerical value available within the current storage capacity, typically one byte, of each pixel.

gray scale   (1) an optical pattern in discrete steps between light and dark.

(2) intrinsic modulation property of the marking technology that enables dots of either different size or intensity to be printed.

(3) in digital images, a method used to represent varying levels of brightness. Typically, gray images use 8-bit gray level for the gray scale so that for each picture element, 0 represents pure black and 255 represent pure white with other gray levels in between these extremes.

grayscale   *See* gray scale.

Green book   *See* IEEE Color Books.

Green's function   the function that satisfies a given differential equation having as source term a Dirac delta function. *See* delta function.

Green's theorems   consider a closed, regular, surface S bounding a volume V where the two scalar functions ϕ and ψ, continuous together with their first and second derivatives throughout V and on the surface S, are defined. Green's first identity states

$$\int_V \nabla \psi \cdot \nabla \phi \, dV + \int_V \psi \nabla^2 \phi \, dV = \int_S \psi \frac{\partial \phi}{\partial n} \, dS$$

while Green's second identity takes the form

$$\int_V \left(\psi \nabla^2 \phi - \phi \nabla^2 \psi \right) dV$$
$$= \int_S \psi \frac{\partial \phi}{\partial n} \, dS - \int_S \phi \frac{\partial \psi}{\partial n} \, dS$$

which is frequently referred to as Green's theorem.

grey level   *See* gray level.

grey scale   *See* gray scale.

grid   refers to the regular array of vertical and horizontal wires used for interconnecting the chip.

grid array   a technique for combining the output of amplifiers or oscillators in space by using a two-dimensional spatial array of elements placed on a uniformly spaced grid.

GRIN optics   *See* graded index optics.

grip   a twisted wire tie which secures a wire to an insulator or other fixture.

grip teeth   a set of jaws which secures a wire to a hoist or come-a-long *cf* so that it can be pulled into place.

gripple   *See* gain ripple.

grooved media   on an optical disk, the embossment of the disk surface with grooves such that the disk tracks are either the grooves themselves or the regions between the groves.

ground   (1) an earth-connected electrical conducting connection that may be designed or non-intentionally created.

(2) the electrical "zero" state, used as the reference voltage in computer systems.

ground bounce   a transient variation in the potential of the ground terminal of a logic device caused by variations in the supply current acting on the ground impedance of the circuit as seen by the device. Usually caused by simultaneous turnon of the pullup and pulldown sections of totem-pole outputs.

ground bounce noise   ground bounce occurs when a large number of semiconductor circuit components are mounted on a common semiconductor chip substrate, so that they are imperfectly insulated from each other. In normal operation the substrate should act as an insulator; however, during certain unusual fluctuations in signal levels, the systems power and ground connections can experience fluctuations, which affect the performance of each component in a random way that has the characteristics of noise, much like capacitive coupling.

ground current the current that flows in a power system in a loop involving earth and (in some usages) other paths apart from the three phases.

ground fault interrupter a protective device used in commercial and residential wiring which monitors equipment connected to an electrical outlet and shuts off the power when a ground fault in the equipment is detected.

ground fault neutralizer an inductor connected between the neutral of Y windings of a generator or transformer and ground. It is tuned to the machine's capacitance so as to minimize ground fault current.

ground lamp indicator lamp on electrical distribution switchboards that darkens when a ground condition exists on one (or more) of the busses.

ground loop an undesired conductive path between two conductive bodies in a radial grounding system that are connected to a common ground.

ground plane a perfectly or highly conducting half space beneath an antenna. Also, an unetched layer of metal on a printed circuit board over which microstriplines and printed antennas are formed.

ground rod a metallic, rod-type electrode designed to be driven into the earth. It serves as an earth connection for grounding purposes. Other types of earth electrodes include buried plates, rings, and grids. For buildings, its primary function is to keep the entire grounding system at earth potential.

ground wave a vertically polarized TEM wave propagating close to the ground. It is one of the three modes of propagation (ground, sky, and space waves).

ground-fault circuit interrupter (GFCI) a device designed to detect ground-fault current above a threshold value (several milliamperes) and then interrupt the source of electrical power by opening a circuit breaker or a set of contacts. GFCIs are designed for personnel protection and are generally available in the form of circuit breakers and receptacles.

ground-signal-ground (GSG) (1) configuration of the contact tips of a coplanar microwave probe. A single signal contact is positioned in the center of two parallel ground plane contacts.

(2) a network of wires buried in the earth around a electric power transmission line tower to reduce footing impedance.

grounded system an electrical distribution system in which one of the normal current-carrying conductors, often the neutral, is intentionally grounded.

grounding *See* ground.

grounding transformer a transformer connected to an otherwise ungrounded three-phase system for the purpose of providing a path for ground current flow. Zig-zag transformers and grounded wye-delta transformers can be used as grounding transformers.

group code a recording method used for 6250 bpi (bits per inch) magnetic tapes.

group decision support system a special decision support system used in support of groups of decision makers that might be geographically dispersed.

group delay   the derivative of the radian phase with respect to frequency.

group detection   a special strategy for multiuser detection in a multiple access system. The users get divided into a number of detection groups according to an appropriate criterion. The users in the designated first group are detected using a predetermined detection technique; the corresponding multiple access interference generated by this group of users is estimated, and then subtracted from the received signal. This detection and cancellation is continued successively until all users in all groups are detected.

group velocity   (1) a quantity proved to indicate (in most dispersive media) the speed of energy propagation of wavepackets.
(2) in an optical fiber, the speed at which a signal superimposed on an optical wave propagates.

Groupe Special Mobile (GSM)   a digital cellular communications network standard developed in Europe in the 1980s and now implemented in many countries throughout the world. The acronym originally denoted Groupe Special Mobile, after the French-based group that developed the system. Now the acronym denotes Global System for Mobile. GSM is one of a group of systems that are usually referred to as second-generation cellular systems. The system uses 200 kHz channels, GMSK modulation and frequency hopping. GSM transmits digitally encoded speech at a rate of approximately 13 kbps and makes extensive use of channel error-correcting codes.

GSG   *See* ground-signal-ground.

GSM   *See* Global Special Mobile.

GSVQ   gain-shape vector quantization. *See* shape-gain vector quantization.

GTAW   *See* gas tungsten arc welding.

GTO   *See* gate turnoff thyristor.

guaranteed cost control   a methodology of robust controller design whose objective is to ensure that the value of the performance index for the real system represented by the family of models with uncertainty changing within a given set is guaranteed to be not greater than a given value. Guaranteed cost control techniques are usually applied to nominally linear systems and quadratic cost functionals.

guard band   a design technique for a color CRT intended to improve the purity performance of the device by making the lighted area of the screen smaller than the theoretical tangency condition of the device geometry.

guard digits   those extra digits of the significand in a floating-point operand that must be retained in order to allow correct normalization and rounding of the result's significand.

guarded machine   classification describing an electrical machine constructed with an open frame in which rotating components or live electrical parts are guarded by screens. Screen openings are restricted so that a rod with a diameter of 0.75 inches cannot contact any part within 4 inches of the guarded opening, and a rod with a diameter of 0.5 inches cannot contact any part more than 4 inches inside the guarded opening.

guide wavelength   the distance over which the fields of propagating modes repeat themselves in a waveguide.

guided scrambling   an extension of the self-synchronizing scrambling line code procedure in which the source bit stream is augmented with appropriately valued bits prior to scrambling in order to ensure that the scrambled sequence contains good line code properties.

guided wave interconnect a means for connecting electrical systems such as computers and computer circuit boards optically, using optical fibers and optical circuits of channel waveguides.

guy a wire which extends at an angle from a utility pole to the ground in order to brace the pole against toppling due to unbalanced forces from the utility lines it supports.

guy anchor one of several appliances used to hold a guy under tension in the earth.

guy guard a metal or plastic plate or tube mounted on a guy wire at ground level to increase the visibility of the wire and decrease its ability to cut into a person or object pressed against it.

H

H   *See* horizontal.

H_∞ design   a group of robust controller design methods based on the methodology of the Hardy space H_∞ consisting of all complex-valued functions of complex variable that are analytic and bounded in the open right half-plane. The least bound that may be imposed on this function is its H_∞ norm. Since the open right half-plane may be replaced by the imaginary axis $j\omega$, H_∞ methods provide a direct generalization of the classical frequency domain approach to control system design.

A standard problem is to design a controller that ensures the internal stability of the closed-loop system and minimizes the H_∞ norm of the transfer function between the inputs (reference signals, disturbances) and errors. Since this transfer function is equal to the sensitivity function, such design results in optimal sensitivity. The standard problem is then transformed into an equivalent model-matching problem with a fixed, possibly unstable transfer function derived from the plant and a "free parameter" stable compensator to be chosen. The compensator is found by minimization of the supremum over all frequencies of the modeling error. Finally, an optimal (or suboptimal) controller is synthesized based on the found optimal (or suboptimal) solution to the model-matching problem. To meet specific dynamic objectives the transfer functions are modified by pre- and postfilters in the form of frequency dependent weighting functions. Although the primary problem is formulated in frequency domain, it may be solved both by input–output and state space techniques. In the former case the algorithms are based on spectral and inner–outer factorizations and approximation theorems for complex functions.

In the latter case the problem may be attacked by linear-quadratic game theoretic approach resulting in a set of Riccati equations. H_∞ (H infinity) methods may be used in robust stabilization, robust performance design, disturbance attenuation, optimal tracking, model following, optimal sensitivity design, etc.

H infinity design   *See* H_∞ design.

H modes   the wave solutions with zero electric field component in the direction of propagation. Also known as transverse electric (TE) modes.

H parameters   characterizes a microwave network with an arbitrary number of ports by relating the total voltages and currents at the ports.

H-D curve   *See* Hurter-Driffield curve.

H-mode   *See* tranverse electric wave.

H-plane   in measuring an antenna's radiation pattern, the plane that is perpendicular to the current in the element and therefore contains the magnetic field intensity vector field. This plane is perpendicular to the electric field (E) plane cut.

H-plane sectoral horn   a horn antenna where the aperture is formed by flaring the walls in the direction of the H-plane. The electric field (E) plane dimension is left unchanged from that of the waveguide feed.

H-tree   a popular clock distribution tree topologically that resembles the H shape. It introduces the least amount of clock skew compared to other distribution topologies.

Haas effect   states that the first sound heard will mask subsequent short delay arriving sounds, the combination appearing as a louder source. Also called law of the first wavefront.

Haar transform   unitary transform mapping N samples g(n) to N coefficients $G(k)$ in a way that corresponds to repeated two-point averaging and two-point differencing. The 2×2 Haar transform is

$$H_2 = \frac{1}{\sqrt{2}} \begin{pmatrix} 1 & 1 \\ 1 & -1 \end{pmatrix}$$

The scaling factor allows the same matrix to be used for the inverse transform. The 4×4 Haar transform can be interpreted as follows: first apply the 2×2 transform to two independent pairs of samples; then apply the 2×2 transform to the two average coefficients just computed. Larger Haar transforms are constructed by continuing this process recursively. The Haar transform yields coefficients equal to the subband values generated by dyadic decomposition with the Haar wavelet. This transform has achieved rather less use than the other transforms in this family, such as the discrete cosine, Fourier, and Hadamard transforms.

Haar wavelet   the orthonormal wavelet pair $(\frac{1}{\sqrt{2}}, \frac{1}{\sqrt{2}})$, $(\frac{1}{\sqrt{2}}, \frac{-1}{\sqrt{2}})$. Analysis and synthesis pairs are identical. This is the most compact wavelet pair. Dyadic subband decomposition with the Haar wavelets yields coefficients equal to those from the Haar transform.

hacker   a person who explores computer and communication systems, usually for intellectual challenge, commonly applied to those who try to circumvent security barriers (crackers).

Hadamard matrix   an $n \times n$ matrix H with elements ± 1 is a Hadamard matrix of order n if $HH^T = nI$, i.e., the rows are all mutually orthogonal, as are the columns. Hadamard matrices can only exist for $n = 1, 2$ or n an integer multiple of 4. Hadamard matrices of order 2^i can be constructed by the recursion

$$H_1 = (1)$$
$$H_{2n} = \begin{pmatrix} H_n & H_n \\ H_n & -H_n \end{pmatrix}$$

These are known as the Sylvester matrices. Because of their properties, Hadamard matrices find application in the theory of error control codes, and code division multiple access. Named after Jacques Salomon Hadamard (1865–1963). *See also* orthogonal, Walsh cover.

Hadamard transform   (1) *See* Walsh–Hadamard transform.

(2) a unitary transform mapping N samples $g(n)$ to N coefficients $G(k)$ according to the transform matrix H_N, where

$$H_2 = \frac{1}{\sqrt{2}} \begin{pmatrix} 1 & 1 \\ 1 & -1 \end{pmatrix}$$

and larger arrays are formed by the recursive definition

$$H_N = \frac{1}{\sqrt{2}} \begin{pmatrix} H_{\frac{N}{2}} & H_{\frac{N}{2}} \\ H_{\frac{N}{2}} & -H_{\frac{N}{2}} \end{pmatrix}$$

The inverse transform is identical. The Hadamard transform was formerly used for data compression because its entries are all 1 or -1, allowing computation without multiplications. In this context it is now superseded by the discrete cosine transform.

half adder   a logic circuit that produces the sum and carry outputs for two input signals. A half adder has no carry input.

half bridge amplifier   a class-D amplifier based on a half-bridge inverter configuration. Not suitable for amplification of DC to low-frequency signals because the capacitor leg cannot provide unidirectional current.

half subtracter   a logic circuit that provides the difference and borrow outputs for two input signals. A half subtracter has no borrow input.

half-band filter   a filter whose even-indexed coefficients are all zeros except the one in the filter center.

half-height point   a point at which the membership grade is equal to 0.5.

half-life   the average time needed for half the nuclei of a radioactive element to decay.

half-power bandwidth   the frequency at which a lowpass filter amplitude response falls to $1\sqrt{2}$ of the DC response is known as the half-power point. The two-sided half-power bandwidth is twice the half-power point.

half-wave rectifier   a device that passes positive polarity portions of a signal and blocks negative polarity portions of an AC signal. Ideally, for a sinusoidal input $v_i(t) = V_m \cos(\omega t)$, the output equals the input while the input is positive and is zero while the input is negative.

half-wave symmetry   a periodic function that satisfies $x(t) = -x(t - T/2)$.

half-wave voltage   the voltage required to produce an amount of refractive index change in a medium that will retard the phase of a traversing optical wave.

halftone   technique of simulating continuous tones by varying the amount of area covered by the colorant. Typically accomplished by varying the size of the printed dots in relation to the desired intensity.

Hall effect   the phenomenon whereby charge carriers are displaced perpendicularly to their drift velocity when current flows in the presence of a magnetic field. The resulting shift in carriers inside the conductor or semiconductor produces a transverse Hall voltage that is proportional to the strength of the magnetic field (for constant current).

Hall measurement   a semiconductor characterization that uses a crossed electric and magnetic field to yield information on the conductivity of a test sample.

Hall probe   a device with a Hall effect transducer to sense a magnetic field.

halogen   one of the halide atoms known for being highly reactive.

halt instruction   an instruction (typically privileged) that causes a microprocessor to stop execution.

Hamming code   an encoding of binary numbers that permit error detection and correction first discovered by Richard Hamming at Bell Laboratories.

A Hamming code is a perfect code with code word length $n = 2^m - 1$ and source word length $k = n - m$ for any $m > 2$. A Hamming code can correct an error involving a single bit in the binary number. It can also detect an error involving two bits. *See also* parity.

Hamming distance   the number of digit positions in which the corresponding digits of two binary words of the same length differ. The minimum distance of a code is the smallest Hamming distance between any pair of code words. For example, if the sequences are 1010110 and 1001010, then the Hamming distance is 3.

Hamming net   a pattern recognition network that has a set of prototype patterns stored in its weights. A given input pattern is identified with the prototype whose Hamming distance from the input pattern is least.

Hamming weight   the number of nonzero symbols in a given sequence of symbols.

Hamming window   a tapering function used to truncate functions spectrally or in the time domain (i.e., in frequency it is a specifically formed low pass filter). It is defined as: $w(x) = 0.54 + 0.46 * \cos(2 * \pi * x/x_o)$, $|x| < x_o/2$. The Hamming window sacrifices filter sharpness at the cutoff frequency for a smoother stop band behavior.

hand   *See* end-effector.

hand-held computer   a small lightweight computer that performs functions such as electronic mail, handwriting recognition for taking notes, and holding addresses and appointments. Also called a "palm-top."

hand-off   in a cellular system, the process by which the mobile terminal switches from communicating with one base station to a communication with neighboring base station as the mobile travels through the different radio cells.

handline   a rope used to pull tools and equipment from an assistant on the ground to a worker atop a utility pole.

handover   *See* hand-off.

handshaking   I/O protocol in which a device wishing to initiate a transfer first tests the readiness of the other device, which then responds accordingly. The transfer takes place only when both devices are ready.

handwritten character recognition   the process of recognizing handwritten characters that are clearly separated.

Hankel transform   the 2-D Fourier transform of a function with circular symmetry and arises in the analysis of optical systems. The (zero-order) Hankel transform $F(\rho)$ of a function $f(r)$ for $r \geq 0$ is $F(\rho) = 2\pi \int_0^\infty rf(r)J_0(2\pi\rho r)dr$ and the (zero-order) inverse Hankel transform is $f(r) = 2\pi \int_0^\infty \rho F(\rho)J_0(2\pi\rho r)d\rho$ where $J_0(r)$ is the zero-order Bessel function of the first kind, i.e., $J_0(r) = \frac{1}{\pi} \int_0^\pi \cos(r\sin\theta)d\theta$. *See* Fourier transform.

Hanning window   a raised cosine window function whose impulse response is

$$w_n = \begin{pmatrix} 0.5 + 0.5\cos(n\pi/N) & |n| < N \\ 0 & |n| \geq N \end{pmatrix}.$$

hard bake   the process of heating the wafer after development of the resist in order to harden the resist patterns in preparation for subsequent pattern transfer. Also called postbake.

hard bug   a name for a crimped copper wire connector.

hard contaminant   a contaminant or foreign object which is at least partly opaque to X-radiation: typical hard contaminants are pieces of metal, glass or stone.

hard decision   demodulation that outputs a q-ary value for each demodulated symbol in a sequence of q-ary symbols. *See also* soft decision.

hard disk   a rigid magnetic disk used for storing data. A typically nonremovable collection of one or more metallic disks covered by a magnetic material that allows the recording of computer data. The hard disk spins about its spindle while an electromagnetic head on a movable arm stays close to the disk's surface to read from or write to the disk. Each disk is read and written on both above and below. N disks are read/written by using $2N$ heads. The information is stored by cylinders, circular segments of the collection of the disks. Cylinders are divided in sectors as a pie. The mean time to access data is typically close to 10 msec.

Generally, hard disks are the backing memory in a hierarchical memory. *See also* floppy disk.

hard fault   *See* permanent fault.

hard ferrite   *See* ceramic ferrite.

hard magnetic material   a ferromagnetic material that retains its magnetization when the magnetizing field is removed; a magnetic material with significant coercivity.

hard real-time   *See* hard real-time system.

hard real-time system   a real-time system in which missing even one deadline results in system failure. *Compare with* soft real-time, firm real-time.

hard X-ray having sufficient photon energy to penetrate "hard" material; usually more than about 15 keV.

hard-decision decoding decoding of encoded information given symbol decisions of the individually coded message symbols. *Compare with* soft-decision decoding.

hardware computer constructs that have a physical manifestation.

hardware accelerator a piece of hardware dedicated to performing a particular function (such as image convolution or matrix-vector products) which would otherwise be performed in software. Although much less flexible, dedicated hardware implementations can give significant speed improvements over software, and are especially useful for real-time applications. *See* real-time system.

hardware interrupt an interrupt generated by a hardware device, for example, keyboard, the DMA, PIC, the serial adapter, the printer adapter, etc. Other hardware interrupts can be generated by the control unit or by the ALU, for example, for the presence of a division per zero, for attempting to execute an unknown instruction. This last class of hardware interrupts is called internal exception.

hardware noise radio frequency emissions due to arcing of utility lines at defective connectors.

harmonic (1) the name associated with a number used to denote the frequency components that exist in a certain fourier series representation for a certain function of time $f(t)$. The Fourier series representation is given

$$f(t) = \sum_{n=-\infty}^{\infty} F_n * [\cos{(n\omega_0 t)} + j \sin{(n\omega_0 t)}]$$

where $n = 0, 1, 2, 3, \ldots$, $\omega_o = 2\pi/T$, $j = \sqrt{-1}$, T is the period of the function $f(t)$, F_n

is the coefficient of the Fourier series for a certain value of n:

1st harmonic has $F_n = F_1$ and

$$[\cos{(1\omega_0 t)} + j \sin{(1\omega_0 t)}]$$

2nd harmonic has $F_n = F_2$ and

$$[\cos{(2\omega_0 t)} + j \sin{(2\omega_0 t)}]$$

3rd harmonic has $F_n = F_3$ and

$$[\cos{(3\omega_0 t)} + j \sin{(3\omega_0 t)}]$$

etc.

(2) sinusoidal component of a periodic waveform that has a frequency equal to an integer multiple of the basic frequency (or fundamental frequency). Thus the third harmonic of a power system voltage in the U.S. has a frequency of 3×60, or 180 Hz. For electric systems powered by sinusoidal sources, harmonics are introduced by nonlinear devices such as saturated iron cores and power electronic devices.

harmonic amplifier a type of amplifier that utilizes various forms of harmonic and mixing actions. These amplifiers may pump up the fundamental by increasing the switching efficiency of the active device. Others may actually be used as frequency multipliers or frequency converters (mixers). All class F, G, and H amplifiers fit into this general group. Parameters such as device characteristics, quiescent bias point, RF load line, significant harmonic and/or mixing frequencies, and amplitude and waveform of the applied signal(s) should be included with the class definition, thus defining the major contributors to the physical actions taking place in one of these amplifiers.

harmonic analysis the branch of mathematics dealing with the decomposition of signal functions as a linear combination of basis functions which represent "waves" of various frequencies. When the basis functions are sines and cosines each with a frequency that is an integer multiple of the signal's frequency, we have trigonometric harmonic analysis; in other words classical Fourier analysis, which provides the amplitudes and phases of the constituent sinusoids. (Fourier transform.) With other basis functions, for example wavelets, we have non-trigonometric harmonic analysis (wavelet, wavelet transform). Abstract harmonic analysis studies the generalization of Fourier analysis to abstract spaces.

harmonic balance technique   one of several techniques for analyzing nonlinear circuits. The nonlinear circuit is divided into two portions of linear and nonlinear elements, and a portion of linear elements is calculated in a frequency domain and a portion of nonlinear elements is calculated in a time domain, respectively. The calculated voltages or currents at connecting nodes of these portions are balanced by using Fourier transforming or inverse Fourier transforming.

harmonic component   a Fourier component of order greater than one of a periodic waveform.

harmonic content   the internally generated, harmonically related spectral output from a device or circuit. Harmonic energy is that energy that is at exact multiples of the fundamental frequency, generated by the nonlinearities within the device or circuit acting on the fundamental frequency.

harmonic converter   found in a microwave receiver, this component uses the technique of harmonic mixing to convert the RF signal to a lower IF frequency for further processing. Harmonic converters can be used as part of a vector network analyzer.

harmonic distortion   caused by the nonlinear transfer characteristics of a device or circuit. When a sinusoidal signal of a single frequency (the fundamental frequency) is applied at the input of a nonlinear circuit, the output contains frequency components that are integer multiples of the fundamental frequency (harmonics). The resulting distortion is called harmonic distortion.

harmonic frequency   integral multiples of fundamental frequency. For example, for a 60-Hz supply, the harmonic frequencies are 120, 180, 240, 300, ...

harmonic generation   in nonlinear optics, the process in which a laser beam interacts with a material system to produce new frequency components at integer multiples of the frequency of the incident beam. Under carefully controlled circumstances, the lower-order harmonics (e.g., second and third) can be generated with high ($> 50\%$) efficiency. Under different circumstances, harmonics as high as the 30th can be generated.

harmonic load-pull measurement   a measurement method where transfer characteristics of a device at the fundamental frequency can be measured by electrically changing the load impedance at harmonic frequencies.

harmonic orthogonal   set the set of functions $e^{j\omega t}$. It is called harmonic because each basis function is a harmonic of a certain frequency and because the inner product between any two functions is zero:

$$\int_{-\infty}^{+\infty} e^{j\omega_1 t} e^{j\omega_2 t} dt = 0, \omega_1 \neq \omega_2$$

$$\int_{-\infty}^{+\infty} e^{j\omega t_1} e^{j\omega t_2} d\omega = 0, t_1 \neq t_2$$

harmonic tuning   the process of tuning an amplifier circuit to a frequency that is an integral multiple of the fundamental frequency at which the circuit would normally operate.

harmonically pumped mixer   mixer where the intermediate frequency (IF) signal is at a frequency which is the sum or difference of the RF and an integer multiple (usually two) of the LO (local oscillator) frequency.

Hartley oscillator   a particular case of LC-oscillators when $X_1 + X_2 + 2X_m$ is realized as a single tapped coil, and X_3 is a capacitor. Well suited for variable-frequency operation by varying a single capacitor.

Hartley oscillators are usually not used at VHFs of higher frequencies. Similarly, the circuit is avoided at very low audio frequencies. It is important to distinguish the Hartley oscillator from the Armstrong topology. In the Armstrong oscillator, no ohmic connection exists between the two inductors. Instead, coupling is entirely magnetic.

Harvard architecture   a computer design feature where there are two separate memory units: one for instructions and the other for data only. The name originates from an early computer development project at Harvard University in the late 1940s. *Compare with* Princeton architecture.

hash table   a table storing a mapping function whose domain is sparsely used and that is accessed by indices that are computed from the search field ("key") using a many–one mapping (called a hash function). Hash tables are used for many memory and name mapping functions, such as symbol tables in assemblers and compilers.

hashed page table   a page table where the translation of each virtual page number is stored in a position determined by a hash function applied to the virtual page number. This technique is used to reduce the size of page tables.

hashing   the act of translating a search key into a table index using a many–one mapping. *See also* hash table.

Hausdorff distance   an important distance measure, used in fractal geometry, among other places. Given a distance function d defined on a Euclidean space E, one derives from it the Hausdorff distance H_d on the family of all compact (i.e., bounded and topologically closed) subsets of E; for any two compact subsets K, L of E, $H_d(K, L)$ is the least $r \geq 0$ such that each one of K, L is contained in the other's dilation by a closed ball of radius r, that is: $K \subseteq \bigcup_{p \in L} B_r(p)$ and $L \subseteq \bigcup_{p \in K} B_r(p)$, where $B_r(p) = \{q \in E \,|\, d(p, q) \leq r\}$

Hayes-compatible modem   refers to a modem when it is capable of responding at the commands of modems made by Hayes Microcomputer Products. The Hayes set of commands represents a sort of standard for modems.

haystack response   bandpass frequency response characterized by flat midband response with sloping sides.

hazard   a momentary output error that occurs in a logic circuit because of input signal propagation along different delay paths in the circuit.

hazardous location   a classification system used to define locations that are susceptible to fire and explosion hazards associated with normal electrical arcing. A class I hazardous location contains a flammable concentration of flammable vapors. A class II hazardous location contains a combustible concentration of combustible dusts. A class III hazardous location contains an ignitable concentration of ignitable fibers.

hazardous outage   an outage which has been assessed to have potential life threatening consequences. The criteria for a hazardous outage varies considerably from one utility to another. In some cases arcing lines is sufficient for a hazardous outage. Other utilities would consider at least fire or explosion to qualify an outage as hazardous.

haze tails   when self-converging deflection yokes are used on a CRT, overfocusing occurs on the vertical center line of the beam spot. Around the periphery of the screen, this produces a hazy area above and below the spot, referred to as haze tails. This degrades picture contrast.

HBMT *See* hybrid bipolar MOS thyristor.

HBT *See* heterojunction bipolar transistor.

HDA *See* head disk assembly.

HDSL *See* high-speed digital subscriber line.

HDTV *See* high definition television.

head an electromagnet that produces switchable magnetic fields to read and record bit streams on a platter's track.

head disk assembly (HDA) collection of platters, heads, arms, and acutators, plus the air-tight enclosing of a magnetic disk system.

head-medium gap the distance between the read–write head and the disk in magnetic or optical disk memory devices.

header (1) a data structure containing control information that is placed at the head of a datagram; when placed at the end of a datagram this is referred to as a trailer. Addressing information and checksum are examples of the control information contained in a header.

(2) a section of an image file, usually of fixed size and occurring at the start of the file, that contains information about the image, such as the number of rows and columns and the size of each pixel.

heap data storage structure that accepts items of various sizes and is not ordered. Contrast with stack.

heat sink aluminum mountings of various shapes and sizes used for cooling power semiconductor devices. Heat sinks can be cooled by either

natural convection or a fan, and heat dissipation can be improved by a coating of black oxide or if the heat sink is made with fins.

heater *See* overload heater.

Heaviside characteristic an activation function according to which the neuron output takes on the value unity when its weighted input exceeds (or equals) the neural threshold value and zero otherwise.

Heaviside, Oliver (1850–1925) Born: London, England

Best known for his theoretical work in electrical engineering. Much of his work is contained in his three-volume work called *Electromagnetic Theory*. The final volume was published in 1912. Heaviside extended and improved the works of Hamilton and Maxwell and deduced the concepts of capacitance, impedance, and inductance. Heaviside was self-taught and irascible. At first, much of his work was dismissed as unorthodox or too theoretical to be of practical value. Heaviside is best known in physics for his correct prediction of the existence of the ionosphere.

heavy water water in which a heavy isotope of hydrogen substitutes for the hydrogen atoms.

Hebbian algorithm in general, a method of updating the synaptic weight of a neuron w_i using the product of the value of the ith input neuron, x_i, with the output value of the neuron y. A simple example is: $w_i(n+1) = w_i(n) + \alpha y(n)x_i(n)$ where n represents the nth iteration and α is a learning-rate parameter.

Hebbian learning a method of modifying synaptic weights such that the strength of a synaptic weight is increased if both the presynaptic neuron and postsynaptic neuron are active simultaneously (synchronously) and decreased if the activity is asynchronous. In artificial neural networks, the synaptic weight is modified by a function of the correlation between the two activities.

Hebb's principle   to update the weights of the simple feedforward neural network (e.g., perceptron) a very simple idea, called Hebb's principle, can be used. The principle states the following: Apply a given pattern to the inputs and clamp the outputs to the desired response, then increase the weights between nodes that are simultaneously excited. This principle was formulated by Hebb in 1949, in an attempt to model neural networks. Mathematically it can be expressed as follows:

$$w_{ij}(t+1) = w_{ij} + \gamma u_i(t) \left(y_j^0(t) - y_j(t) \right)$$

where y_j^0 is the desired response and y_j is the response of the network. By regarding the weights as parameters, one may treat the above formula as a gradient method for parameters estimation.

hedge   a special linguistic term by which other linguistic terms are modified. Examples are very, fairly, more or less, and quite.

height defuzzification   *See* centroid defuzzification.

heirarchical coding   coding where image data are encoded to take care of different resolutions and scales of the image. Additional data is transmitted from the coder to refine the image search. *See also* progressive transmission.

heirarchical interpolation   a technique of forming image pyramids. In this method, pixels are interpolated to higher levels in the hierarchy. Thus, only the interpolated pixels need a residual in order to be reconstructed exactly, since the subsampled pixels are already correct.

helical antenna   a wire antenna that is helical in shape. Typically, the helical antenna is fed by a coaxial cable, the outer conductor of which is connected to a ground plane and the inner conductor of which protrudes through the ground plane and

is shaped in the form of a helix. Since a helical antenna emits an elliptically polarized wave when operating in axial mode, it is often used in satellite communication applications.

helical beam tube   a backward wave amplifier based on interaction between a forward helical electron beam launched into a waveguide and backward traveling microwave in the waveguide.

helical scan tape   a magnetic tape in which recording is carried out on a diagonal to the tape, by a head that spins faster than the tape movement. Popular for VCRs, camcorders, etc., as it improves recording density and tape speed.

helium neon laser   neon laser in which excitation of the neon atoms occurs primarily by collisions with electron-impact-excited helium atoms; the first gas laser, and especially important for its CW visible oscillation lines.

Helmholtz coils   a pair of coaxial coils of the same diameter that are spaced one radius apart. They are used to generate a field of uniform strength or to measure the magnetic moment of a magnet.

Helmholtz equation   a partial differential equation mathematically described by

$$(\nabla^2 + k^2)\phi = 3Df,$$

where ∇^2 is the Laplacian, k is the wavenumber, f is the forcing function, and ϕ is the equation's solution.

HEMT   *See* high electron mobility transistor.

Henry, Joseph   (1797–1878) Born: Albany, New York

Best known as the first Director (1846) of the Smithsonian Institution, and President of the National Academy of the Sciences. Henry was largely self-taught, but his early experiments garnered him sufficient recognition to become a Professor of Natural Philosophy at New Jersey

College (now Princeton). Henry's early experiments resulted in the development of a practical electric motor and a relay later quite important in telegraphy.

hermetic seal   a seal that is such that the object is gas-tight (usually a rate of less than 1×10^{-6} cc/s of helium).

Hermite form of 2-D polynomial matrix   denote by $F^{m \times n}(z_1)[z_2]$ ($F^{m \times n}[z_1][z_2]$) the set of $m \times n$ polynomial matrices in z_2 with coefficients in the field $F(z_1)$ (polynomial coefficients in z_1). 2-D polynomial matrix $A(z_1, z_2) \in F^{m \times n}[z_1, z_2]$ of full rank has Hermite form with respect to $F^{m \times n}[z_1][z_2]$ if

$$
A_H(z_1, z_2) = \begin{cases}
\begin{bmatrix}
a_{11} & a_{12} & \dots & a_{1n} \\
0 & a_{22} & \dots & a_{2n} \\
\dots & \dots & \dots & \dots \\
0 & 0 & \dots & a_{nn} \\
0 & 0 & \dots & 0 \\
\dots & \dots & \dots & \dots \\
0 & 0 & \dots & 0
\end{bmatrix} & \text{if } m > n \\[2pt]
\begin{bmatrix}
a_{11} & a_{12} & \dots & a_{1n} \\
0 & a_{22} & \dots & a_{2n} \\
\dots & \dots & \dots & \dots \\
0 & 0 & \dots & a_{nn}
\end{bmatrix} & \text{if } m = n \\[2pt]
\begin{bmatrix}
a_{11} & a_{12} & \dots & a_{1m} & \dots & a_{1n} \\
0 & a_{22} & \dots & a_{2m} & \dots & a_{2n} \\
\dots & \dots & \dots & \dots & \dots & \dots \\
0 & 0 & \dots & a_{mm} & \dots & a_{mn}
\end{bmatrix} & \\
\quad \text{if } m < n
\end{cases}
$$

where $\deg_{z_2} a_{ii} > \deg_{z_2} a_{ki}$ for $k \neq i$ ($\deg_{z_2}$ denotes the degree with respect to z_2). In a similar way, the Hirmite form of $A(z_1, z_2)$ with respect to $F[z_1][z_2]$) can be defined. $A(z_1, z_2)$ can be reduced to its Hermite form $A_H(z_1, z_2)$ by the use of elementary row operations or equivalently by premultiplication by suitable unimodular matrix $U(z_1, z_2)$ (det $U(z_1, z_2) \in F(z_1)$), i.e., $A_H(z_1, z_2) = U(z_1, z_2)A(z_1, z_2)$. See for example, T. Kaczorek, *Two-Dimensional Linear Systems*, Springer-Verlag, Berlin, 1985.

Hermite Gaussian beam   electromagnetic beam solution of the paraxial wave equation in which the field is a product of a Hermite-polynomial and a Gaussian function of distance from the beam axis.

Hermitian matrix   a square matrix that equals its conjugate transpose.

hertz   a measure of frequency in which the number of hertz measures the number of occurrences (of whatever is being measured) per second.

Hertz dipole   a straight, infinitesimally short and infinitesimally thin conducting filament with uniform current distribution. The current amplitude is usually assumed to vary sinusoidally in time.

Hertz, Heinrich Rudolf   (1857–1894) Born: Hamburg, Germany

Best known as the person who discovered radio waves. Hertz had several professorships including posts at Berlin, Kiel, Karlsruhe, and Bonn. Hertz's work was both theoretical and experimental. He succeeded in proving important elements of Maxwell's equations, and was able to demonstrate that radio waves could be generated. He died early from a degenerative bone disease. We honor his name by referring to the hertz as a unit of frequency.

Hertzian dipole   *See* Hertz dipole.

Hertzian potential   potentials used for the computation of electromagnetic field. By using the Hertzian vector potentials of electric and magnetic type Π_e, Π_h, respectively, the fields are expressed as

$$
\mathbf{E} = k^2 \Pi_e + \nabla\nabla \cdot \Pi_e - j\omega\mu\nabla \times \Pi_h
$$
$$
\mathbf{H} = j\omega\varepsilon\nabla \times \Pi_e + k^2\Pi_h + \nabla\nabla \cdot \Pi_h
$$

where ϵ is the dielectric permittivity, μ the magnetic permeability, and $k = \omega\sqrt{\epsilon\mu}$, with ω denoting the angular frequency.

heterodyne detection method of measuring the frequency content of an optical signal in which that signal is combined with a known CW reference signal from a local oscillator by means of a square-law detector; sometimes provides a more direct and sensitive representation of the frequency content of a signal than homodyne detection.

heterodyne receiver a receiver where the low-level RF signal is mixed with a local oscillator signal to produce an intermediate frequency (IF). The IF frequency is usually between 10 to 100 MHz. It can be amplified with a low noise amplifier. Heterodyne receivers have much better sensitivity and noise characteristics than direct detection receivers.

heterogeneous having dissimilar components in a system; in the context of computers, having different types or classes of machines in a multi-processor or multicomputer system.

heterojunction a junction between two crystals of different bulk composition, typically referring to semiconductor–semiconductor interfaces. The prototype example is AlAs/GaAs.

heterojunction bipolar transistor (HBT) proposed by Shockley (U.S. Patent 2569347).

heterojunction field effect transistor (HFET) a field effect transistor that uses a heterojunction in the channel parallel to the current flow direction to improve the carrier transport by separating the dopant from conduction region. Other names for this device include MODFET, (MOdulation Doped Field Effect Transistor and HEMT, High Electron Mobility Transistor.

heterojunction laser *See* diode laser.

heterostructure refers to a composite structure in which two dissimilar materials are chemically joined at a common interface. Examples are GaAs-AlGaAs, metal on semiconductor, oxide on semiconductor, etc.

hexadecimal notation expressing numbers in base-16 format. *See also* hexadecimal number system.

hexadecimal number system the numbering system that uses base-16, commonly used in computer systems. The digits 10–15 are generally represented by the letters A–F.

Heuristic special problem-specific knowledge that can be used to help direct a search efficiently towards an appropriate, though possibly suboptimal, solution. *See also* Heuristic search.

heuristic search a search that uses problem-specific knowledge in the form of heuristics in an effort to reduce the size of the search space or speed convergence towards a solution. Heuristic searches sacrifice the guarantee of an optimal solution for reduced search time.

HEV *See* hybrid electric vehicle.

hexagonal pixel See pixel.

HFET *See* heterojunction field effect transistor.

HgCdTe semiconductor alloy important as the active photoconductive or photovoltaic element in mid- to long-wavelength infrared detectors.

HgTe/CdTe heterostructure combination with properties similar to the HgCdTe alloy, but with added tailorability from the ability to modify properties such as band gap layer by layer.

hi-bi fiber optical fiber with high linear birefringence. Light launched on one of the fiber's polarization axes will maintain its polarization state as it propagates along the length of the fiber.

hidden layer a layer of neurons in a multi-layer perceptron network which is intermediate between the output layer and the input layer.

hidden Markov model (HMM) a discrete-time, discrete-space dynamical system governed by a Markov chain that emits a sequence of observable outputs, usually one output (observation) for each state in a trajectory of such states. More formally, a HMM is a 5-tuple $(\Omega_X, \Omega_O, A, B, \pi)$, where Ω_X is a finite set of possible states, Ω_O is a finite set of possible observations, A is a set of transition probabilities, B is a set of observation probabilities, and π is an initial state distribution.

hidden station problem problem related to multiple access protocols in mobile environment where two mobile stations may not be able to detect each other's transmission, leading to the possibility of a transmission collision.

hidden unit a neural unit in a network that has no direct connection to either the network inputs or to the network outputs.

hierarchical clustering a clustering technique that generates a nested category structure in which pairs or sets of items or categories are successively linked until every item in the data set is connected. Hierarchical clustering can be achieved agglomeratively or divisively. The cluster structure generated by a hierarchical agglomerative clustering method is often a dendrogram, but it can be also as complex as any acyclic directed graph. All concept formation systems are hierarchical clustering systems, including self-generating neural networks. *See also* clustering, concept formation, self-generating neural network.

hierarchical coding coding a signal at several resolutions and in order of increasing resolution. In an hierarchical image coder an image is coded at several different sizes and in order of increasing size. Typically, the smaller sized images are used to encode the larger size images to obtain better compression.

hierarchical control operation or structure of the control system, where the controller, or the given control layer, is composed of several local decision units coordinated (*See* coordination) by a supremal unit (coordinator unit); hierarchical control in the above narrow sense is tantamount to the multilevel control, yet the term hierarchical control is also used in broader meaning — covering both multilevel and multilayer control as well as integrated multilayer–multilevel control structures occurring in large industrial and environmental applications.

hierarchical feature map a hybrid learning network structure also called counter-propagation network that consists of a unsupervised (hidden) layer and a supervised layer. The first layer uses the typical (instar) competitive learning, the output will be sent to the second layer, and an outstar supervised learning is performed to produce the result. *See also* counter propagation learning, outstar training.

Kohonen calls a feature map that learns hierarchical relations as a "hierarchical feature map."

hierarchical memory also known as multilevel memory. The organization of memory in several levels of varying speed and capacity such that the faster memories lie close to the processor and slower memories lie further away from the processor. Faster memories are expensive, and therefore are small. The memories that lie close the processor store the current instruction and data set of the processor. When an object is not found in the memories close to the processor, it is fetched from the lower levels of the memory hierarchy. The top levels in the hierarchy are registers and caches. And the lowest level in the hierarchy is the backing memory, which is usually a disk.

high byte the most significant byte of a multibyte numeric representation.

high definition television (HDTV) a set of standards for broadcast television that incorporate all-digital transmission, greatly improved audio and video quality, audio and video compression, and integration of data with television audio and video.

high electron mobility transistor (HEMT) a depletion-mode field-effect transistor in which the channel is a two-dimensional electron gas at a heterointerface. Other names: MODFET, TEGFET, and 2DEGFET.

high frequency emphasis filter a band-limited filter whose frequency response increases with frequency. Typically such a filter would have unity gain or less at low frequencies.

high order interleaving in memory interleaving, using the most significant address bits to select the memory module and least significant address bits to select the location within the memory module.

high phase order (HPO) polyphase systems that contain 6, 9, or more phases — rather than the standard three-phase system. HPO systems may be used to provide a means of transmitting more electrical power down existing right-of-ways than three-phase systems, without an increase in transmission voltage, and without an increase in EMF levels at the edge of the right-of-way.

high rupturing capacity (HRC) a term used to denote fuses having a high interrupting rating. Most low-voltage HRC-type fuses have an interrupting rating of 200 kA RMS symmetrical.

high side pertains to the portion of a circuit connected to the higher voltage winding of a power transformer.

high state a logic signal level that has a higher electrical potential (voltage) than the other logic state. For example, the high state of TTL is defined as being greater than or equal to 2.0 V.

high-impedance state the third state of tri-state logic, where the gate does not drive or load the output line. *See* tri-state circuit.

high-level transmitter a transmitter in which the modulation process takes place at a point where the power level is at or near the output power.

high-level vision the highest stage of vision, leading to the full extraction of the 3-D information and to its exploitation for the description and understanding of the scene.

high-loss resonator a resonator having a high value of loss (usually diffraction loss) per round trip; unstable resonator.

high-pass filter (1) a filter that has a transfer function, or frequency response, whose values are small for frequencies lower than some intermediate frequency.

(2) a filter whose impulse response is a high-pass signal.

high-pass signal a signal whose Fourier transform, or power spectral density, is a small value for all frequency components that are less than some intermediate frequency value. As a formal definition, if $X(\omega)$ is the Fourier transform of the signal $x(t)$, then $X(\omega) = 0$ for $|\omega| < B$, for some $B > 0$. *Compare with* band-pass signal.

high-performance metal oxide semiconductor (HMOS) technology a variation of the MOS process that scales down devices to achieve higher speed and thus lower power consumption in integrated circuits.

high-rate quantization theory a theory developed by Bennet, Schutzenberger, Zador and others. The theory analyzes and predicts the performance of (infinitely) fine quantization (high-rate quantization).

325

high-resistance grounded system an electrical distribution system in which the neutral is intentionally grounded through a high resistance. The high-resistance grounded wye system is an alternative to solidly grounded and ungrounded systems. High-resistance grounding will limit ground fault current to a few amperes, thus removing the potential for arcing damage inherent in solidly grounded systems.

high-speed carry in an arithmetic logic unit, a carry signal that is generated separately from the generation of the result and is therefore faster; a lookahead carry.

high-speed digital subscriber line (HDSL) a digital subscriber line (DSL) in which two twisted-pairs provide a rate of 1.544 Mbps (T1 rate) in both directions (full-duplex). Each twisted-pair provides a full-duplex connection of 784 kbps. (This is larger than 1.544/2 Mbps due to the duplication of a synchronization bit).

high-speed metal oxide semiconductor (HMOS) technology *See* high-performance metal oxide semiconductor technology.

high-stability diode gun Plumbicon a diode gun Plumbicon tube with electrostatic focus and magnetic deflection, which uses a high-stability electrode structure evaporated and bonded to the tube envelope.

high-voltage DC (HVDC) transmission transmission of electric power (at typically 500–1500 kV) using DC rather than AC. This can be desirable for several reasons:

(1) For economic reasons when a large amount of power is to be transmitted over a long distance, i.e., 300–400 miles, or via underwater cables;

(2) For the connection of asynchronous AC systems; and

(3) Improved transient stability and dynamic damping of the electrical system oscillations.

higher-order mode mode that spatially varies faster than the fundamental in a cavity, or waveguide.

higher-order unit a neural unit whose input connections provide not the usual weighted linear sum of the input variables but rather a weighted polynomial sum of those variables.

highway *See* bus.

Hilbert space an inner product space with the additional property that a certain norm defined on the space makes it a complete metric space. An inner product space is a linear space on which an inner product is defined, while completeness means that there are no 'missing' vectors arbitrarily close to but not included in the space.

Hilbert transform (1) a transform that relates the real and imaginary parts of a complex quantity, such as the index of refraction.

(2) for f, a function of one real variable, the transform $H(f)$ defined as follows: $H(f)(x) = \lim_{\varepsilon \to 0} \frac{1}{\pi} \int_{|t| \geq \varepsilon} \frac{f(x-t)}{t} dt$. An alternative formula, which coincides with the previous one, except possibly on a zero-measure set of points of discontinuity of f is: $H(f)(x) = \lim_{\varepsilon \to 0} \frac{1}{\pi} \int_{-\infty}^{+\infty} f(x - t) \frac{t}{t^2 + \varepsilon^2} dt$. when f is square — integrable, the Fourier transforms $F(f)$ and $F(H(f))$ if f and $H(f)$ are related by: $F(H(f))(u) = -i \, \text{sign}(u) F(f)(u)$ almost everywhere; thus for positive frequencies, the Fourier phase is shifted by $-\pi/2$; in particular, f and $H(f)$ have the same L^2 norm. *See* Fourier amplitude/phase, Fourier transform.

histogram (1) a plot of the frequency of occurrence of the gray levels in an image.

(2) the frequency distribution of a set of numbers. In image processing, the distribution of the gray levels in an image, typically plotted as the number or percentage of pixels at a certain gray level vs. the gray levels. If the ordinate is the ratio of the number of pixels at a gray level to the total

number of pixels, the histogram is an approximation of the probability density function of the gray levels. Probability density function.

histogram equalization   a technique for computing an image gray level transformation to redistribute the pixel intensity values and "flatten" the pixel intensity histogram. Histogram equalization can be used to enhance the contrast of the image. Also called histogram leveling, histogram flattening. Contrast enhancement, histogram.

histogram modeling   (1) making the histogram of an image have another shape, for example, the flat line produced by histogram equalization. This procedure is usually not automatic, i.e., the shape is usually specified by a person. Also called histogram modification or histogram specification. *See* histogram equalization.

(2) the fitting of a function to a histogram in order to obtain an analytical expression that approximates the histogram.

histogram sliding   the addition of a constant to all pixels in an image to brighten or darken the image. A positive constant makes the histogram slide (translate) to the right; a negative constant makes it slide to the left. *See* histogram.

histogram stretching   expansion or contraction of the gray-level histogram of an image in order to enhance contrast. Usually performed by multiplying or dividing all pixels by some constant value. *See* Contrast enhancement, histogram.

hit   the notion of searching for an address in a level of the memory hierarchy and finding the address mapped (and the data present) in that level of the hierarchy. Often applies to cache memory hierarchies.

hit rate   the percentage of references to a cache that find the data word requested in the cache. Also

the probability of finding the data and given by

$$\frac{\text{number of times required word found in cache}}{\text{total number references}}$$

Also known as hit ratio.

hit ratio   *See* hit rate.

hit–miss ratio   in cache memory, the ratio of memory access requests that are successfully fulfilled within the cache to those that require access to standard memory or to an auxiliary cache.

hit–miss transform   a class of transforms in image processing that locate objects larger than some specific size, S_1, and smaller than some size S_2.

HL7   a data communications protocol for interfacing components of a hospital information system.

HMIC   *See* hybrid microwave integrated circuit.

HMM   *See* hidden Markov model.

HMOS   *See* high-performance metal oxide semiconductor technology.

hold signal   an input signal to a processor indicating that a DMA device is requesting the use of the bus. Also called bus request.

hold time   the time required for the data input(s) to be held stable prior to (or after) the control input changes to latch, capture, or store the value indicated by the data inputs.

hold-up time   the time duration that the output voltage is maintained, within specification, during an interruption of the input power. Typical hold-up time is one to three line cycles.

hole   fictitious positive charge representing the motion of electrons in the valence band of a

327

semiconductor; the number of holes equals the number of unoccupied quantum states in the valence band.

hole burning   localized reduction in gain of a laser amplifier due to saturation by an intense signal; localization may be in space or frequency.

hole mobility   *See* spatial hole burning, spectral hole burning.

Hollerith card   a punched card used for data storage. Developed by Herman Hollerith for use in the 1890 U.S. census, these cards were widely used for early computer data storage. Now obsolete.

Hollerith code   an encoding of data where the data is represented by a pattern of punched holes in a card.

Hollerith, Herman   (1860–1929) Born: Buffalo, New York

Best known for his development of a tabulating machine using punched cards for entering data. This machine was used to tabulate the 1890 U.S. census. The success of this machine spurred a number of other countries to adopt this system. The company Hollerith formed to develop and market his ideas became one of several companies that merged to form IBM, International Business Machines. Hollerith's initial interest in automatic data processing probably stems from his work as an assistant to several census preparers at Columbia University. The success of Hollerith's work gives him claim to the title of the father of automated data processing.

hollow beam   an electron beam with hollow core.

hollow cathode   a negative electrode formed using a section of hollow tube; used in low pressure discharges.

hologram   medium that, when illuminated optically, provides a three-dimensional image of stored information, sometimes called holograph.

holograph   *See* hologram.

holographic data storage   a technique used to store multiple images, or 2-D arrays of digital information, as multiplexed holograms in an optically sensitive material. Multiplexing techniques include angle, frequency, and spatial position. Angle and frequency multiplexing techniques are based on Bragg selectivity where angle multiplexing favors transmission geometries and frequency multiplexing favors reflection geometries.

holographic interconnect   free-space interconnect that uses holograms to control optical paths from sources to detectors. The hologram provides the interconnection by forming the intentionally distorted image of the source array on the detector array. A hologram can be viewed as a combination of gratings. Each grating directs a light beam passing through it to a new direction. The hologram can be displayed by a spatial light modulator to achieve dynamic reconfiguration.

holography   the science of making and reading holograms.

homodyne detection   method of measuring the frequency content of an optical signal in which that signal is combined with itself by means of a square-law detector; simpler but sometimes less informative and sensitive representation of the frequency content of a signal than heterodyne detection.

homogeneous   having all nodes or machines in a multiprocessor or multicomputer be of the same type or class.

homogeneous broadening   spectral broadening of a transition in a laser medium due to irreversible dephasing processes like spontaneous emission, collisions, photon interactions, or spin exchange.

homogeneous coordinates a 2-D point (x, y) with a third coordinate point (x, y, W), where W is a constant, typically 1. Points represented in homogeneous coordinates allow for translation, rotation and scaling to be applied as multiplications of transformation matrices and row vectors. Without homogeneous coordinates, translations are treated as additions while scaling and rotation are treated as multiplications.

homogeneous linear estimator an estimator that is a homogeneous linear function of the data.

homogeneous medium optical medium in which none of the properties (e.g., gain, index of refraction, etc.) are functions of position.

homogeneous solution a system of linear constant-coefficient differential equations has a complete solution that consists of the sum of a particular solution and a homogenous solution. The homogenous solution satisfies the original differential equation with the input set to zero. Analogous definitions exist for difference equations.

homogeneous transformation matrix a 4×4 matrix that maps a position vector expressed in homogeneous coordinates from one coordinate system to another coordinate system (reference coordinate frame). A homogeneous transformation matrix has the form

$$T = \begin{bmatrix} R_{3\times3} & p_{3\times1} \\ 0_{1\times3} & 1 \end{bmatrix} = \begin{bmatrix} n_x & o_x & a_x & p_x \\ n_y & o_y & a_y & p_y \\ n_z & o_z & a_z & p_z \\ 0 & 0 & 0 & 1 \end{bmatrix}$$

where $R_{3\times3}$ represents the rotation matrix, which consists of three unit vectors n (normal), o (orientation), a (approach). The orientation matrix is an orthonormal matrix. $p_{3\times1}$ represents the position vector of the origin of the rotated coordinate

system with respect to the reference system. $0_{1\times3}$ is 1×3 zero vector.

homojunction a junction between regions of the same bulk material that differs in the concentration of dopants. The typical example is the n-p diode.

homomorphic filter an image enhancement technique based upon the illumination-reflectance model. The homomorphic filter assumes the image function $f(x, y)$ is the combination of two functions $i(x, y)$ and $r(x, y)$. By taking the natural log of the images the components are separated and can be processed separately; The reflectance of an image usually contains high frequency components while the illumination component tends to vary slowly (the low frequencies); thus, the filter applied to the logarithm image should affect the low and high frequency components differently. The most common use for the filter is to enhance images by compressing the brightness range while increasing contrast; the filter applied in this case is a circularly symmetric high pass filter with the stop band magnitude <1 and pass band magnitude >1.

homomorphic signal processing the processing of signals from a nonadditive model under a tranformation which renders the model additive. For example, a logarithmic transformation may be used to transform a multiplicative noise model into an additive noise model.

homopolar generator an electromagnetic generator in which the magnetic flux passes in the same direction from one magnetic member to the other over the whole of a single air gap area. Such generators have been built to supply very large pulsed currents.

homopolar machine *See* homopolar generator.

homopolar magnetic bearing a magnetic bearing in which the rotating member always experiences a magnetic field of the same polarity.

329

homothetic   a copy of a signal that retains the original shape, but is scaled in size. For a signal $x(t)$, the homothetic $x_\lambda(t) = x(\lambda t)$ retains the shape of $x(t)$ but is scaled along the t axis as a function of the parameter λ. Typically used in association with structuring elements in mathematical morphology.

homotopy method   a technique for solving nonlinear algebraic equations $F(x) = 0$ based on higher-dimensional function embedding and curve tracing. The idea is to construct a parameterized function such that at one parameter value, say $\lambda = \lambda_0$, the system of equations is easy to solve or has one or more known solutions, and at another parameter value, say $\lambda = \lambda_f$, the system of equations is identical to that of the system of interest, $F(x) = 0$. A homotopy method may then be interpreted as geometric curve following through solution space from the known solutions of the 'easy' problem to the unknown solutions of $F(x) = 0$.

Hopfield memory   *See* Hopfield model.

Hopfield model   a neural algorithm capable of recognizing an incomplete input. Also known as Hopfield memory.

Hopfield suggested that an incomplete input can be recognized in an iterative process, in which the input is gradually recognized in every cycle of the iteration. The iteration is completed when the input finally matches with a stored memory. The Hopfield model is a sort of associative memory. A hologram can also be directly used as associative memory. The main difference of the nonneural holographic associative memory and the Hopfield model is as follows. The direct holographic associative memory is one step and its signal-to-noise ratio depends on the incompleteness of the input that cannot be improved. The Hopfield model is an iterative process involving a nonlinear operation, such as thresholding in which the signal-to-noise ratio of the input can be improved gradually during the iterative process. A large number of optical systems have been proposed to implement the Hopfield model, including the first optical neural networks. Those optical implementations are primarily based on optical matrix-vector or tensor-matrix multiplication.

Hopfield network   a recurrent, associative neural network with n processing elements. Each of the processing elements receives inputs from all the others. The input that a processing element receives from itself is ignored. All of the processing elements output signals are bipolar. The network has an energy function associated with it; whenever a processing element changes state, this energy function always decreases. Starting at some initial position, the system's state vector simply moves downhill on the network's energy surface until it reaches a local minimum of the energy function. This convergence process is guaranteed to be completed in a fixed number of steps. *See also* continuous Hopfield network, discrete Hopfield network.

Hopper, Grace Murray   (1906–1992) Born: New York, New York

Best known as the author of the first compiler. Hopper began her career as a mathematics professor at Vassar College. During WW II she volunteered for service and was assigned to work at Harvard with Howard Aiken on the Mark I computer. She later joined J. Presper Eckert and John Mauchly working on the UNIVAC computer. It was at this time she wrote the first compiler. Her compiler and her views on computer programming significantly influenced the development of the first "English-like" business computer language, COBOL.

hopping sequence   *See* frequency hopping.

horizontal (H)   in television signals, H may refer to any of the following: the horizontal period or rate, horizontal line of video information, or horizontal sync pulse.

horizontal microinstruction   a microinstruction made up of all the possible microcommands available in a given CPU. In practice, some encoding is provided to reduce the length of the instruction.

horizontal polarization   a term used to identify the position of the electric field vector of a linearly polarized antenna or propagating EM wave relative to a local reference, usually the ground or horizon. A horizontally polarized EM wave is one with its electric field vector aligned parallel to the local horizontal.

horizontal rate   the rate at which a line of video is drawn; for NTSC television, the horizontal frequency is 15,734.264 Hz for a rate of 63.5 μs/line (the same as line rate).

horizontal scanning   in radar systems, the rotation of the antenna parallel to the horizon. In video display systems, the process of examining or displaying an image with multiple horizontal lines.

horizontal sync pulse   a blacker-than-black signal level contained within the horizontal blanking interval of the composite NTSC signal. The horizontal sync pulse duration is 7.5 ± 0.5 percent of the horizontal line time and an amplitude of 25% of the peak-to-peak video signal. The horizontal sync pulse frequency is 15750 Hz monochrome and 15734 Hz for color NTSC television formats. The NTSC modified the monochrome horizontal frequency slightly to interleave the color subcarrier and to provide an integer relationship of 286 between the horizontal rate and the 4.5 MHz sound carrier.

During the equalizing pulse interval, the horizontal sync pulse rate is doubled and the pulse width is halved. Accurate horizontal synchronizing the video display is preserved throughout the NTSC frame by measuring the horizontal line time between the blanking level to the sync level transitions.

horn antenna   an aperture antenna formed by a waveguide that has been flared out on one end.

horn gap   a V-shaped spark gap which provides a method of extinguishing a power-follow arc by allowing the arc to climb the sides of the V until it is too long to be maintained.

horsepower-rated switch   a manually operated switching device designed for motor circuit applications. It is designed to interrupt the rated overload current at rated voltage of a motor with a horsepower rating that is less than or equal to the horsepower rating of the switch.

hose   *See* line hose.

host   a computer that is the one responsible for performing a certain computation or function.

hot   an energized conductor.

hot electron   an electron in the conduction band of a semiconductor having a superthermal kinetic energy.

hot electron bolometer   a superconducting resistor structure that uses rapid heating and cooling of electrons by an RF field to produce a time varying resistance, useful as a mixer element at submillimeter and THz frequencies.

hot electron transistor   a transistor, usually fabricated from heterostructure materials, that uses electron transport over an energy step to obtain high electron velocities and hopefully high speed operation.

hot line work   work performed on energized electric power lines. *See* glove, hot-stick, barehand.

hot reserve   the state of an idle thermal generating plant whose boilers and turbines are hot and can thus be quickly brought into service.

hot restart   reassumption, without loss, of all operations in the system from the point of detected fault.

hot standby   a technique for achieving fault tolerance that requires having a backup computer running in parallel with the primary computer so that the backup may take over, if the primary should fail.

hot stick   an insulated pole used by line workers to make connections to and otherwise manipulare energized overhead conductors.

hot tap   a clamp, applied with a hot stick that connects a branch circuit to an existing conductor and typically applied while the system is energized.

hot wire   an energized conductor, particularly as opposed to a neutral or ground wire, both of which are typically maintained at ground potential

hot-carrier diode   *See* Schottky barrier diode.

Hotelling transform   a transformation whose basis vectors are the eigenvectors of the covariance matrix of the random vectors. *See* Karhunen—Loève transformation.

Hough transform   a transform which transforms image features and presents them in a suitable form as votes in a parameter space, which may then be analyzed to locate peaks and thereby infer the presence of desired arrangements of features in the original image space: typically, Hough transforms are used to locate specific types of object or shape in the original image. Hough transform detection schemes are especially robust.

Householder transformation   a matrix Q that maps each vector to its reflection through a defined hyperplane; specifically,

$$Q = I - 2\frac{uu^T}{u^T u}$$

reflects through the plane having normal vector u.

HPA   acronym for high-power amplifier.

HPO   *See* high phase order.

HRC   *See* high rupturing capacity.

HSI/HSV System   a system whereby color is represented by hue, saturation, and intensity or value; hence the acronym. This system tends to be more intuitive for users than the other two since it is similar to an artist's tint, shade and tone. The HSV hexcone is created in a cylindrical coordinate system, where hue is the angle around the vertical axis, value is the height along the vertical axis and saturation is the perpendicular distance from the vertical axis. *See* saturation.

HTC   *See* hydro-thermal coordination problem.

HTGR   acronym for high temperature gas-cooled reactor.

hue   one of the characteristics that distinguishes one color from another. Hue defines color on the basis of its position in the spectrum (red, blue, green, yellow, etc.). Hue is one of the three characteristics of television color. Hue is often referred to as tint. In NTSC and PAL video signals, the hue information at any particular point in the picture is conveyed by the corresponding instantaneous phase of the active video subcarrier.

hue saturation intensity (HSI)   a color model based on the specification of hue (H), saturation (S), and Intensity (I). A useful and convenient property of this model is the fact that intensity is separate from the color components. The hue and saturation components (together referred to as chromaticity) relate closely to color perception in humans, however the HSI intensities are linear and do not correspond to those observed by the eye. The conversion from RGB to HSI goes as

$$H = \frac{\arccos\left(0.5\left((R-G)+(R-B)\right)\right)}{\sqrt{(R-G)^2+(G-B)*(R-B)}}$$
$$S = 1 - (3/(R+G+B)) * (\min\{R,G,B\})$$
$$I = (R+G+B)/3$$

See Also RGB, color spaces.

Huffman coding   a variable length coding scheme whose codewords are generated from the probability distribution of the source. Decoding a Huffman codeword corresponds to traversing an unbalanced binary tree according to the value (0 or 1) of each bit in the word; the leaves of the tree are the source symbols, with the most probable ones being the shortest distance from the root of the tree. Huffman coding achieves an average code rate equal to the source entropy if and only if all the probabilities are negative powers of 2. In general it achieves less compression than arithmetic coding but is easier to implement.

Huffman encoding   *See* Huffman coding.

hum bars   horizontal black and white bars that extend over the entire TV picture and usually drift slowly through it. Hum bars are caused by an interfering power line frequency or one of its harmonics.

human visual system (HVS)   the collection of mechanisms in humans which process and interpret the visual world. These mechanisms include the eye, the retina, the optic nerve, the visual cortex and other parts of the brain.

hunting   a mechanical oscillation in the speed of a synchronous machine due to changes in the load. Damper windings are used to reduce the hunting by providing a torque that opposes the change in speed.

Hurter–Driffield (H-D) curve   the standard form of the H-D or contrast curve is a plot of the relative thickness of resist remaining after exposure and development of a large clear area as a function of log-exposure energy. The theoretical H-D curve is a plot of log-development rate versus log-exposure energy. (Hurter and Driffield are the two scientists who first used a related curve in 1890.)

Hurwitz matrix   a square matrix with real elements and whose all eigenvalues have negative real parts.

Hurwitz polynomial   a polynomial $p(s) = a_n s^n + a_{n-1} s^{n-1} + a_{n-2} s^{n-2} + \cdots + a_0$, where all of the a_i's are either real or complex numbers, is said to be Hurwitz if all of its roots have negative real part.

Huygen's principle   a principle stating that each point of a wave front can be considered to be a source of a secondary spherical wave that combines with the other secondary spherical waves to form a new wave front. Huygen's principle is often used as an approximate technique for solving diffraction and antenna radiation problems. *See also* equivalence theorem, equivalent source.

HVC color coding   technique whereby the HVS contrast sensitivity functions for different wavelengths are taken into account for compressing the color planes independently with different sets of parameters.

HVDC transmission   *See* high-voltage DC transmission.

hybrid bipolar MOS thyristor (HBMT)   like its transistor cousin, it is an advanced device designed to take advantage of the unique characteristics of both bipolar devices (high current capability with low forward voltage drop) and MOS device (high switching speeds with low gate drive requirements).

hybrid circuit   a circuit based on at least two different technologies. For instance, a circuit built by using solid state circuits and tubes.

hybrid computer   a computer based on at least two different technologies. For instance, a computer presenting both digital and analog circuits and signals.

hybrid coupler   generally a four port circuit, which has an electrical response such that a signal applied to port one is divided equally in amplitude between ports three and four. The phase of the signals at ports three and four are either 0, 90, or 180 degrees apart, dependent on the type of hybrid coupler. Port two is completely isolated, and no signal appears at this port.

hybrid electric vehicle (HEV)   an automotive vehicle that uses an electric motor in combination with an internal combustion engine. *See* electric vehicle, regenerative braking.

hybrid magnetic bearing   an active magnetic bearing that also incorporates permanent magnets, thus reducing energy consumption and improving performance.

hybrid microwave integrated circuit (HMIC)   a planar assembly that combines different circuit functions formed by strip or microstrip transmission lines printed onto a dielectric substrate and incorporating discrete semiconductor solid state devices and passive distributed or lumped circuit elements, interconnected with wire bonds.

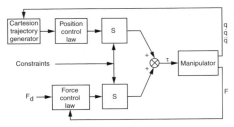

The hybrid position/force controller.

hybrid mode   a solution to Maxwell's equations for a particular structure that is neither TE nor TM.

hybrid numerical method   a method that makes use of two or more different techniques (e.g., mode matching and finite differences) in order to efficiently solve a given boundary value problem. *See also* finite differences, mode matching.

hybrid parameters   circuit parameters used especially in transistor models; when referring to a two-port, they relate the voltage at port 1 with its current, and the voltage at port 2 with the current of port 2 with its voltage, and the current at port 1 according to the equation

$$V_1 = h_{11}I_1 + h_{12}V_2$$
$$I_2 = h_{21}I_1 + h_{22}V_2$$

hybrid position/force controller   a general hybrid position/force controller is presented in the figure. The hybrid controller consists of two sub-controllers. One is a position controller and the second is a force controller. Trajectory generator inputs a Cartesian path in terms of position, velocity, and acceleration. F_d is a desired external force associated with the task in Cartesian space. Matrix S is a selection matrix with ones and zeros on the diagonal. Where a one is present in S, a zero is present in S', and position control is in effect. It is assumed that joint positions, velocities, and accelerations are measured. Next they are transformed via kinematical computations to Cartesian positions, velocities, and accelerations. Desired and actual parameters are compared in position controller. Desired and measured forces are compared in force controller. Finally, manipulator is controlled at the joint force and/or torque level.

hybrid redundancy   a technique combining active and passive redundancy. Passive redundancy prevents generation of error results, and active redundancy improves fault tolerance capabilities by replacing faulty hardware with spare elements.

hybrid stepper motor   a stepper motor that combines the rotor design characteristics of variable-reluctance and permanent magnet stepper motors. Hybrid stepper motor rotors consist of an axially magnetized cylindrical permanent magnet capped on each pole by toothed, soft iron caps. Teeth on the caps are displaced with respect to each other to provide stepping control. Hybrid stepper motors combine the higher torque capability of permanent magnet motors with the higher step resolution of variable-reluctance motors.

hydroelectric generator   large, three-phase synchronous alternator powered by a water-driven turbine. *See also* generator.

hydro-thermal coordination   the practice of manipulating water levels in the reservoirs of a power system's hydroelectric plants with respect to the generation levels of the thermal plants in the system with the objective of minimizing generation costs, as well as satisfying waterway considerations such as water conservation, flood control capacity, recreation, and environmental requirements.

hydrophone   receiving sensors that convert sound energy into electrical or optical energy (analogous to underwater microphones).

hydropower   conversion of potential energy of water into electricity using generators coupled to impulse or reaction water turbines.

hyperpolarizability   a measure of the nonlinear optical response of an atom or molecule. The hyperpolarizability of order $n\gamma(n)$ is defined by the relation $p(n) = C\gamma(n)E^n$, where E is the electric field strength of the applied laser field and $p(n)$ is the nth order contribution to the dipole moment per molecule. C is a coefficient of order unity whose definition is different among different workers.

hysteresis   (1) the phenomenon that the magnetic state of a substance is dependent upon its magnetic history, so that its magnetization for an

increasing magnetizing force differs from that for a decreasing magnetizing force.

(2) the characteristic of magnetic materials that causes the trajectory of the flux density vs. field intensity curve as the intensity is increased to be different from that when the intensity is decreased, giving rise to a loss, which is proportional to the area enclosed by the two trajectories.

hysteresis brake   a braking device utilizing hysteresis to provide a constant braking torque irrespective of slip speed.

hysteresis control   a time-optimal feedback control method in which the control variable reaches a reference value in the shortest possible time and then stays within a prescribed hysteresis band around the set point through manipulation of the system state between two configurations. The actual variable is compared with the reference value, and if the error exceeds the hysteresis band, then the control input is changed such that the control variable is forced to decrease. On the other hand, if the actual variable falls below the hysteresis band then the control input is changed such that the control variable increases in magnitude.

hysteresis curve   a graph describing the relationship between the magnetic flux density and the magnetic field intensity in a (usually ferromagnetic) material.

hysteresis drive   *See* hysteresis torque coupling.

hysteresis loss   the energy loss due to hysteresis in a magnetic material subjected to a varying magnetic field.

hysteresis motor   any of a variety of single-phase AC motors that use the hysteresis properties of hard magnetic materials to develop torque. Stator windings of a hysteresis motor can be of any design that produces a rotating flux within the machine. Motion of the rotating flux over the rotor magnetizes the hard magnetic material on the rotor; however, the hysteresis characteristics of the material cause the alignment of magnet flux to lag the rotating stator flux. This misalignment produces rotor torque. Because of the nature of the torque production, hysteresis motors operate at synchronous speed and have a constant torque characteristic, which permits them to synchronize any load that they can accelerate.

hysteresis torque coupling   a magnetic drive in which the magnetizing stator magnet drives a rotor of hysteresis material through the complete hysteresis cycle once per rotation, resulting in a constant torque characteristic irrespective of relative speed.

Hz   hertz. *See* frequency.

I

I and Q   the I and Q signals are used in video transmission to generate a chroma phasor or vector. By varying the amplitude of the I and Q signals, all colors in the visible spectrum can be generated and transmitted. The I and Q signals are derived from the RGB signals where I = .6R − 0.28G − 0.32B, Q = .21R − .52G + .31B.

I$_{DC}$ DC current in amperes.

I$_{IN_p}$ peak input current in amperes.

I$_{OUT_p}$ peak output current in amperes.

I$_{DS}$ FET channel current.

I&P   *See* interconnections and packaging.

I-V characteristics   the charts describing the current through a diode as a function of its voltage bias.

I/O   input/ouput. Operations or devices that provide data to or accept data from a computer.

I/O bandwidth   the data transfer rate into or out of a computer system. Measured in bits or bytes per second. The rate depends on the medium used to transfer the data as well as the architecture of the system. In some instances the bandwidth average rate is given and in others the maximum rate is given. *See also* I/O throughput.

I/O buffer   a temporary storage area where input and output are held. Having I/O buffers frees a processor to perform other tasks while the I/O is being done. Data transfer rates of the processor and an I/O device are, in general, different.

The I/O buffer makes this difference transparent to both ends. *See also* peripheral buffer.

I/O bus   a data path connecting input or output devices to the computer.

I/O card   a device used to connect a peripheral device to the main computer; sometimes called an I/O controller, I/O interface, or peripheral controller.

I/O channel   *See* input/output channel, subchannel I/O.

I/O command   one of the elementary types of commands that a computer can execute. I/O commands are typically used to initiate I/O operations, sense completion of commands, and transfer data. Not all computers have I/O commands. *See also* memory mapped I/O.

I/O controller   (1) a device used to connect a peripheral device to the main computer; sometimes called an I/O card.

(2) the software subroutine used to communicate with an I/O device. *See also* I/O routine.

I/O device   a physical mechanism attached to a computer that can be used to store output from the computer, provide input to the computer, or do both.

I/O interface   a device used to connect a peripheral device to the main computer; sometimes called an I/O card or I/O controller.

I/O interrupt   a signal sent from an input or output device to a processor that the status of the I/O device has changed. An interrupt usually causes the computer to transfer program control to the software subroutine that is responsible for controlling the device. *See also* interrupt-driven I/O.

I/O port   (1) a register designed specifically for data input–output purposes.

(2) the place from which input and output occurs in a computer. Examples include printer port, serial port, and SCSI port.

I/O processor an input/output processor with a unique instruction set, dedicated to performing I/O operations exclusively, thus alleviating the burden off the CPU. It usually has a separate local bus for I/O operations data traffic, thus permitting the CPU to access memory on the main system bus without interruption. For example, the Intel 8086 CPU has an 8089 I/O processor associated with it. Both can operate simultaneously, in parallel.

I/O register a special storage location used specifically for communicating with input/output devices.

I/O routine a function responsible for handling I/O and transferring data between the memory and an I/O device. *See also* I/O controller.

I/O system the entire set of input/output constructs, including the I/O devices, device drivers, and the I/O bus.

I/O throughput the rate of data transfer between a computer system and I/O devices. Mainly determined by the speed of the I/O bus or channel. Throughput is typically measured in bits/second or bytes/second. In some instances, the throughput average rate is given and in others the maximum rate is given. *See also* I/O bandwidth.

I/O trunk *See* I/O bus.

I/O unit the equipment and controls necessary for a computer to interact with a human operator or to access mass storage devices or to communicate with other computer systems over communication facilities.

I-line a line of the mercury spectrum corresponding to a wavelength of about 365 nm.

IA *See* index assignment.

IAEA International Atomic Energy Agency, an organization which monitors nuclear materials and energy.

IC *See* integrated circuit.

ICE *See* in-circuit emulator.

ideal filter a system that completely rejects sinusoidal inputs of the form $x(t)4 = A\cos\omega t$, $-\infty < t < \infty$, for ω within a certain frequency range, and does not attenuate sinusoidal inputs whose frequencies are outside this range. There are four basic types of ideal filters: low-pass, high-pass, band-pass, and bandstop.

ideal operating amplifier an op amp having infinite gain from input to output, with infinite input resistance and zero output resistance and insensitive to the frequency of the signal. An ideal op amp is useful in first-order analysis of circuits.

ideal transformer a transformer with zero winding resistance and a lossless, infinite permeability core resulting in a transformer efficiency of 100 percent. Infinite permeability would result in zero exciting current and no leakage flux. For an ideal transformer, the ratio of the voltages on the primary and secondary sides would be exactly the same as the ratio of turns in the windings, while the ratio of currents would be the inverse of the turns ratio.

ideality factor the factor determining the deviation from the ideal diode characteristic $m = 1$. At small and large currents $m \approx 2$.

idempotent in an operator, if applying it twice gives the same result as applying it only once; mathematically speaking, if ψ is the operator and X the object to which it is applied, this means that $\psi(\psi(X)) = \psi(X)$.

idler wave   the additional, often unwanted, output wave produced by an optical parametric oscillator designed to generate a signal wave.

IEC   *See* International Electro-technical Commission.

IEEE   *See* Institute of Electrical and Electronics Engineers.

IEEE Color Books   a series of seven books related to industrial and commercial power systems containing recommended IEEE/ANSI standards and practices. The books are color-coded as to subject as follows:

Gray: Power systems in commercial buildings

Green: Grounding industrial and commercial power systems

Brown: Power system analysis

Gold: Design of reliable industrial and commercial power systems

Orange: Emergency and standby power for industrial and commercial power systems

Red: Electric distribution practice in industrial plants

Buff: Protection and coordination of industrial and commercial power systems.

IEEE float encoding   *See* NaN.

IEEE representation   an encoding convention for the representation of floating-point values within computing systems.

IEEE Standard 754   the specification that defines a standard set of formats for the representation of floating-point numbers.

IF   *See* intermediate frequency.

IF amplifier   an amplifier having high and controllable gain with a sharp band-pass characteristic centered around the intermediate frequency (IF).

IGBT   *See* insulated gate bipolar junction transistor.

ignitron   a high-voltage mercury switch. The device is found in modulators used to dump the capacitor bank voltage in the event of a PA crowbar. An ignitron passes electrical current to a pool of liquid mercury at ground potential.

IID   *See* independent and identically distributed.

II-VI semiconductor   binary semiconductor made from elements in the periodic table two columns to the "left" and "right" of silicon, e.g., CdTe, ZnS, etc.

III-V semiconductor   a binary semiconductor made from elements from columns III and V in the periodic table.

illumination   the effect of a visible radiation flux received on a given surface. Illumination is measured by the illuminance, which is the luminous flux received by surface unit, usually expressed in lux. One lux equals 1 lumen/m^2.

illumination system   the light source and optical system designed to illuminate the mask for the purpose of forming an image on the wafer.

ill-posed problem   a problem whose solution may not exist, may not be unique, or may depend discontinuously on the data. A problem is well-posed if it can be shown that its solution exists, is unique, and varies continuously with perturbation of the data. If any of the three conditions does not hold, the problem is ill-posed.

image acquisition   the conversion of information into an image. Acquisition is the first stage of an image processing system and involves converting the input signal into a more amenable form (such as an electrical signal), and sampling and quantizing this signal to produce the pixels in the image. Hardware, such as lenses, sensors and transducers are particularly important in image acquisition. *See* analog-to-digital converter, digitization, image, pixel, quantization, sample.

image analysis the extraction of information from an image by a computer. The results are usually numerical rather than pictorial and the information is often complex, typically including the recognition and interpretation of objects in or features of the image.

image classification (1) the division by a computer of objects in an image into classes or groups. Each class contains objects with similar features.

(2) the division of the images themselves into groups according to content.

image coding the usual reasons for coding messages lie in the areas of data compression and data security. However, the main motivation for recoding images is to achieve the high rates of data compression required to cope with the storage and transmission of prodigious amounts of image data. *See* still image coding, video coding.

image coding for videoconferencing this is also called coding at primary rates for videoconferencing. In North America and Japan, one of the most plentiful is the DSI rate of 1.544 Mb/s. In Europe, the corresponding rate is 2.044 Mb/s. The main difference is that in videoconferencing the video camera is fixed and picture data is produced only if there are moving objects in the scene. Typically, the amount of moving data is much smaller than the stationary area.

image compression representation of still and moving images using fewer bits than the original representation. *See* still image coding and video coding for methods.

image compression or coding the process of reducing the number of binary digits or bits required to represent the image.

image correlation the calculation of the correlation function where the two signal inputs are images.

image degradation blurring of an image wavefront due to scattering as it passes an inhomogeneous medium, e.g., astronomical images due to atmospheric scattering.

image dissector a nonstorage photoemission device used as a light detector in early television camera tubes. Electron streams, emitted by light striking a photocathode film, are focused on a plane containing a small aperture into a photomultiplier structure. Scanning is accomplished by magnetically deflecting the electron stream across the aperture.

image enhanced mixer mixer that operates on the principle of terminating the image frequency component at an optimal impedance level that improves the conversion loss of the mixer. The optimal impedance termination allows the image product to remix with the LO frequency component such that the down-converted signal is in phase with the IF component and thus increases its magnitude. The image spectral component is at a frequency that is either the sum or difference of the LO and IF components.

image enhancement processing an image to improve its appearance, or to make it more suited to human or machine analysis. Enhancement is based not on a model of degradation (for which, restoration) but on qualitative or subjective goals such as removing noise and increasing sharpness. Edge emphasis and smoothing are examples of enhancement techniques.

image feature an attribute of a block of image pixels.

image frequency rejection ratio ratio of output for the image frequency below that for the desired signal when both image and desired signal levels at the input are the same.

image frequency response output generated in a superheterodyne receiver due to an undesired signal with a frequency spaced by twice the IF from desired frequency.

image fusion   the merger of images taken with different imaging modalities or with different types of the same modality, for example, different spectral bands. *See* multispectral image.

image guide   a type of dielectric waveguide based on a dielectric rod waveguide with a conducting plate placed at the plane of symmetry.

image hierarchy   a way to organize the image data in order of importance. Each level corresponds to a reconstructed image at a particular resolution or level of quality. Generally, they are classified into fixed and variable resolution hierarchies.

image impedance   the input impedance into a passive 2-port device when that device is terminated in an infinite chain of identical devices connected to both ports such that each port is connected to the same port (i.e., port 1 connects to port 1 and port 2 connects to port 2 throughout the infinite chain, giving a perfect match). The impedance looking into port 1 is Z_{I11}, and into port two is Z_{I22}. For a uniform transmission line, the image impedances are equal to each other and also equal to the characteristic impedance of the line.

image information   stored digitally in a computer, often representing a natural scene or other physical phenomenon, and presented as a picture. The brightness at each spot in the picture is represented by a number. The numbers may be stored explicitly as pixels or implicitly in the form of equations. *See* bitmapped image, image acquisition, image processing, imaging modality, pixel, vector image.

image processing   the manipulation of digital images on a computer, usually done automatically (solely by the computer) but also semi-automatically (the computer with the aid of a person). The goal of the processing may include compressing the image, making it look better, extracting information or measurements from it or getting rid of degradations. *See also* image acquisition, image analysis, image classification, image compression, image enhancement, image recognition, image restoration.

image recognition   the identification and interpretation of objects in an image. Image recognition is a higher-level process that requires more intelligence than other parts of image processing such as image enhancement. Artificial intelligence, neural nets or fuzzy logic are often used. *See* artificial intelligence, artificial neural network, fuzzy logic, classifier, image classification, image understanding, pattern recognition.

image reconstruction   (1) the process of obtaining an image from nonimage data that characterizes that image.

(2) the conversion of a digital image to a continuous one suitable for display. analog-to-digital converter, interpolation.

(3) in tomography, the reconstruction of a spatial image from its projections. *See* computed tomography, Radon transform.

image registration   the spatial alignment of a pair of images, which may involve correcting for differences in translation, rotation, scale, deformation or perspective. Image registration is especially important in binocular vision.

image rejection filter   a filter usually placed before a mixer to preselect the frequency, which when "mixed" with the local oscillator frequency, will down-convert to the desired intermediate frequency. In the absence of an image rejection filter, there are two RF frequencies, which when mixed with the local oscillator frequency f_{lo}, will down-convert to the intermediate frequency f_{if}. Specifically, these are given by $f_{lo} + f_{if}$ and $f_{lo} - f_{if}$. For example, if the local oscillator frequency is

10 GHz and the local oscillator frequency is 1 GHz, then signals at both 11 GHz and 9 GHz will produce an intermediate frequency signal of 1 GHz without an image rejection filter.

image resolution   a measure of the amount of detail that can be present in an image. Spatial resolution refers to the amount a scene's area that one pixel represents, with a smaller area being a higher resolution. Brightness resolution refers to the amount of luminance or intensity that each gray level represents, with a smaller amount being a higher resolution.

image restoration   the removal of degradations in an image in order to recover the original image. Some typical image degradations and possible causes are: blurring (intervening atmosphere or media, optical abberrations, motion), random noise (photodetectors, electronics, film grain), periodic noise (electronics, vibrations) and geometric distortions (optical abberrations, angle of image capture). *See* degradation, motion compensation, noise, noise suppression.

image reversal   a chemical process by which a positive photoresist is made to behave like a negative photoresist.

image segmentation   the division of an image into distinct regions, which are later classified and recognized. The regions are usually separated by edges; pixels within a region often have similar gray levels. *See* edge detection, image classification, image recognition.

image smoothing   the reduction of abrupt changes in the gray levels of an image. Usually performed by replacing a pixel by the average of pixels in some region around it. Also called image averaging. *See also* smoothing.

image theory   a solution technique for determining the fields radiated by sources in the presence of material boundaries; typically the boundary is between a dielectric and a perfect electric conductor. The equivalence theorem is used to generate an equivalent problem in a homogeneous medium with image sources that provide the same boundary conditions as the original problem.

image transform   the conversion of the image from the spatial domain to another domain, such as the Fourier domain. It is often easier and more effective to carry out image processing such as image enhancement or image compression in the transformed domain. *See* block transform, cosine transform, discrete Fourier transform, discrete Hadamard transform, discrete Hadamard transform, discrete cosine transform, discrete sine transform, discrete wavelet transform (DWT), distance transform, Fourier transform, Haar transform, Hadamard transform, Hankel transform, Hilbert transform, Hotelling transform, Hough transform, Karhunen-Loeve transform, Medial axis transform, Radon transform, separable image transform, sine transform, slope transform, top hat transform, transform, Walsh transform, Wavelet transform.

image understanding   the interpretation by a computer of the contents of an image. The process seeks to emulate people's ability to intelligently extract information from or make conclusions about the scene in an image. Also called image interpretation.

imaginary power   *See* reactive power.

imaging modalities   the general physical quantity that the pixels in an image represent; the type of energy an image processing system converts during image acquisition. The most common modality is visible light, but other modalities include invisible light (infrared or x-ray), sound (ultrasound) and magnetism (nuclear magnetic resonance). *See* computed tomography, fluoroscopy, image acquisition, magnetic resonance imaging, multispectral image, nuclear magnetic resonance, positron emission tomography, synthetic aperture radar, tomography, ultrasound, x-ray image.

IMC   *See* internal model control.

immediate addressing   an addressing mode where the operand is specified in the instruction itself. The address field in the instruction holds the data required for the operation.

immediate operand   a data item contained as a literal within an instruction.

immersed flow   a flow of electrons emitted from an electron gun exposed to the focusing magnetic fields.

immittance   a response function for which one variable is a voltage and the other a current. Immittance is a general term for both impedance and admittance and is generally used where the distinction is irrelevant.

immunity to a distrubance   an equipment or systems capability to operate if an electromagnetic disturbance occurs.

impact excitation   excitation of an atom or molecule resulting from collision by another particle such as an electron, proton, or neutron.

IMPATT diode   acronym for impact avalanche and transit time diode. Negative resistance device used at high frequencies used to generate microwave power. Typically used in microwave cavity oscillators.

impedance (Z)   (1) electrical property of a network that measures its ability to conduct electrical AC current for a given AC voltage. Impedance is defined as the ratio of the AC voltage divided by the AC current at a given point in the network. In general, impedance has two parts: a real (resistive) part and an imaginary (inductive or capacitive "reactive") part. Unless the circuit is purely resistive (made up of resistors only), the value of impedance will change with frequency.

(2) in an antenna, usually defined at the input to an antenna, the impedance is the ratio of the applied (or induced) voltage to the current flowing into (or out of) the antenna input. More generally, it is defined as the ratio of the electric field to the magnetic field.

impedance inverter   circuit whose input impedance, when terminated with a load, is inversely proportional to said load impedance — typically implemented with a quarter-wave transmission line.

impedance matching   one of the main design activities in microwave circuit design. An impedance matching network is made up of a combination of lumped elements (resistors, capacitors, and inductors), or distributed elements (transmission lines of varying characteristic impedance and length). Impedance matching networks transform network impedance from one value to another. For example, on the input to a low noise transistor, the impedance of an incoming 75 ohm transmission line would be transformed by the input matching network to the impedance Z_{opt}, required to achieve the minimum noise figure of the transistor. The Smith chart is a tool commonly used by microwave engineers to aid with impedance matching.

impedance matching network   a combination of reactive elements connected between a circuit and a load that transforms the load impedance into another impedance value to achieve improved performance such as maximum power transfer, reduced reflections, or optimum noise performance.

impedance matrix   the matrix formed in the method of moments by expanding the electric current in a series of basis functions, expressing the electric field in terms of the current, and taking the inner product with a set of testing functions.

The resulting matrix equation is of the form

$$[V_m] = [Z_{mn}][I_n]$$

where $[V_m]$ represents the known excitation, $[Z_{mn}]$ is the impedance matrix, and $[I_n]$ is the unknown current.

impedance measurement   measurement of the impedance (ratio of voltage to current) at a circuit port.

impedance parameters   circuit parameters relating the voltages across the ports of a multiport to the port currents.

impedance relay   a protective relay that senses the operational impedance at a location, i.e., the ratio of voltage to current at any given time. During fault conditions on the protected line, the impedance relay will sense the impedance (distance in ohms) between the location of the relay and the fault. Typical impedance relay characteristics are mho and reactance. Impedance relays are widely used in sensing phase faults on transmission lines. Ground impedance relays are available that measure the distance to a single phase to ground fault using a modified technique. Also known as distance relay.

impedance scaling   modifying a filter circuit to change from a normalized set of element values to other usually more practical element values by multiplying all impedance by a constant equal to the ratio of the desired (scaled) impedance to the normalized impedance.

impedance standard substrate (ISS)   a substrate, usually made of alumina, containing thin film gold calibration standards used in the calibration of vector network analyzers used in conjunction with microwave coplanar probes.

impedance transformation   *See* matching.

implicant   a first-order implicant is a pair of logically adjacent minterms. A second-order implicant is a set of logically adjacent first-order implicants, and so on.

implication   a rule of reasoning which states that the truth of a proposition or condition implies the truth of another. A common implication rule is of the form "If A then B."

implied addressing   a form of addressing where the register or memory address is not specified within the instruction but is assumed.

imprecise interrupt   an implementation of the interrupt mechanism in which instructions that have started may not have completed before the interrupt takes place, and insufficient information is stored to allow the processor to restart after the interrupt in exactly the same state. This can cause problems, especially if the source of the interrupt is an arithmetic exception. *See also* precise interrupt.

imprecision   a sense of vagueness where the actual value of a parameter can assume the specified value to within a finite tolerance limit.

impressed current   a current generated from an independent source. Often used to represent antennas.

improper modes   in open waveguides the eigenfunctions relative to the continuous spectrum, which are defined over an infinite interval, are often referred to as improper modes.

impulse   a unit pulse. *See also* implusive transient.

impulse breakdown   a test of electrical insulation in which lightning or switching impulses are applied.

impulse function   a function $I(n)$ whose value is non-zero at sample position $n = 0$, and zero elsewhere. The unit impulse function $\delta(n)$ is unity at sample position $n = 0$ and zero elsewhere. A shifted impulse is also termed an impulse. *See* Dirac delta function.

impulse generator   (1) an electronic device delivering single pulses of various shapes, preferably square.

(2) a high-voltage trigger generator.

impulse noise   a noise process with infrequent, but very large, noise spikes; it is also known as shot noise or salt and pepper noise. The phrase "impulsive noise" is frequently used to characterize a noise process as being fundamentally different from Gaussian white noise, in being derived from a probability density function with very heavy (long) tails. Applying impulsive noise to an image leaves most pixels unaffected, but with some pixels very bright or dark.

impulse response   the output of a linear time-invariant system when the input is a pulse of short time duration. The system can be entirely characterized by the impulse response.

In the case of a continuous time system with input $f(t)$, the impulse signal $\delta(t)$ is defined as

$(i) \quad \delta(t) = 0, \quad t \neq 0$

$(ii) \displaystyle\int_{-\epsilon}^{\epsilon} \delta(t)dt = 1, \quad \text{for any } \epsilon > 0,$

and the impulse response is the zero state system response to an input $f(t) = \delta(t)$. In the case of a discrete time system with input $f[k]$, the impulse signal is defined as

$(i) \quad \delta[k] = 1, \quad k = 0$
$(ii) \quad \delta[k] = 0, \quad k \neq 0,$

and the impulse response is the zero state system response to input $f[k] = \delta[k]$.

impulse train   a signal that consists of an infinitely long series of Dirac delta functions with period T: $\sum_{n=-\infty}^{\infty} \delta(t - nT)$. This signal can be used to sample a continuous-time waveform.

impulsive transient   a rapid frequency variation of voltage or current during steady-state operation in which the polarity is mostly unidirectional.

in-circuit emulator (ICE)   a device that replaces the processor and provides the functions of the processor plus testing and debugging functions.

in-line gun   a CRT electron gun structure that has the red, green, and blue electron gun components aligned in a horizontal plane. The in-line gun structure requires color registration (color convergence) correction in the vertical CRT face plate axis only.

in-order issue   the situation in which instructions are sent to be executed in the order that the instructions appear in the program.

in-phase signal   in quadrature modulation, the signal component that multiplies $\cos 2\pi f_c t$, where f_c is the carrier frequency.

INA   *See* inverse Nyquist array.

incandescent lamp   a lamp made by heating a metal filament in vacuum; not a burning candle.

incident power   power in an electromagnetic wave that is traveling in an incident direction.

incoherent illumination   a type of illumination resulting from an infinitely large source of light that illuminates the mask with light from all possible directions. This is more correctly called spatially incoherent illumination.

incoherent light   light in which detailed knowledge of the fields at one point in space does not permit prediction of the fields at another point.

increment   to add a constant value (usually 1) to a variable or a register. Pointers to memory are usually incremented by the size of the data item pointed to.

incremental encoder   similar to an absolute encoder, except there are several radial lines drawn around the disc, so that pulses of light are produced on the light detector as the disc rotates. Thus incremental information is obtained relative to the starting position. A larger number of lines allows higher resolution, with one light detector. A once per rev counter may be added with a second light detector. *See also* encoder.

incremental gain   a system $H : \mathcal{X}_e \to \mathcal{X}_e$ is said to be Lipschitz continuous, or simply continuous, if there exists a constant $\Gamma(H) < \infty$, called incremental gain, such that

$$\Gamma(H) = \sup \frac{\|(Hx_1)_T - (Hx_2)_T\|}{\|(x_1)_T - (x_2)_T\|}$$

where the supremum is taken over all $x_1, x_2 \in \mathcal{X}_e$ and all real numbers T for which $x_{1T} \neq x_{2T}$. *See also* extended space, gain.

incremental model   small-sign differential (incremental) semiconductor diode equivalent RC circuit of a diode, biased in a DC operating point.

incremental passivity   a system $H : X_e \to X_e$ is said to be incrementally passive if

$$\langle Hx_1 - Hx_2, x_1 - x_2 \rangle_T \geq 0 \quad \forall x_1, x_2 \in \mathcal{X}_e$$

See also extended space, inner product space, and passivity.

incremental resistance   the slope dV/dI of the I-V characteristic of a circuit element.

incremental strict passivity   a system H: $X_e \to X_e$ is said to be incrementally strictly passive if there exist δ such that

$$\langle Hx_1 - Hx_2, x_1 - x_2 \rangle_T \geq$$
$$\delta \langle x_1 - x_2, x_1 - x_2 \rangle_T \quad \forall x_1, x_2 \in \mathcal{X}_e$$

See also extended space, inner product space, passivity, strict passivity, and incremental passivity.

incrementally linear system   a system that has a linear response to changes in the input, i.e., the difference in the outputs is a linear (additive and homogenous) function of the difference in the inputs. *See* linear system.

independence   a complete absence of any dependence between statistical quantities. In terms of probability density functions (PDFs), a set of random quantities are independent if their joint PDF equals the product of their marginal PDFs:

$$p_{x_1, x_2, \ldots}(x_1, x_2, \ldots) = p_{x_1}(x_1) \cdot p_{x_2}(x_2) \ldots$$

Independence implies uncorrelatedness. *See* correlation, probability density function.

independent and identically distributed (IID)   a term to describe a number of random variables, each of which exhibits identical statistical characteristics, but acts completely independently, such that the state or output of one random variable has no influence upon the state or output of any other.

independent event   event with the property that it gives no information about the occurrence of the other events.

independent identically distributed process   a random process $x[i]$, where $x[i]$ and $x[j]$ are independent for $i \neq j$, and where the probability distribution $p(x[i])$ for each element of the process is not a function of i. *See* independence, probability density function, random process.

independent increments process   a random process $x(t)$, where the process increments over non-overlapping periods are independent. That is, for $t_i < s_i, s_i \leq t_{i+1}$, then

$$(x(s_1) - x(t_1)), (x(s_2) - x(t_2)), \ldots$$

are all independent. *See* independence, random process.

independent joint control control of a single joint of a robot while all the other joints are fixed.

index that part of memory address used to access the locations in the cache, generally the next most significant bits of the address after the tag.

index assignment (IA) each reproduction vector in a (vector) quantizer is represented by an index that is transmitted or stored. The index assignment problem is the problem of assigning indices to reproduction vectors, in such a way that errors in the transmission (or storage) influence the reproduction fidelity as little as possible.

index grating through the photorefractive effect. This volume index grating is referred to as a photorefractive grating. The photo-induced refractive index grating can also be erased by the illumination of another beam of light. *See also* index-guided laser diode.

index of fuzziness an index to measure the degree of ambiguity of a fuzzy set by computing the distance between the fuzzy set and its nearest ordinary set.

index of refraction (1) a parameter of a medium equal to the ratio of the velocity of propagation in free space to the velocity of propagation in the medium. It is numerically equal to the square root of the product of the relative permittivity and relative permeability of the medium.
(2) in optics, a complex quantity describing the optical transmission properties of a medium; the real part corresponds to the ratio of speed of light in vacuum to the speed of light in the medium, while the imaginary part corresponds to attenuation of the light by the medium.

index register register used to index a data structure in memory; the starting address of the structure will be stored in a base register. Used in indexed addressing.

index-guided laser electrically pumped semiconductor laser in which the mode fields are confined in the transverse direction by the profile of the index of refraction.

index-guided laser diode *See* index-guided laser.

indexed addressing an addressing mode in which an index value is added to a base address to determine the location (effective address) of an operand or instruction in memory. Typically, the base address designates the beginning location of a data structure in memory such as a table or array and the index value indicates a particular location in the structure.

indexing *See* indexed addressing.

indicator *See* flag.

indirect addressing an addressing mode in which an index value is added to a base address to determine the location (effective address) of an operand or instruction in memory. Typically, the base address designates the beginning location of a data structure in memory such as a table or array and the index value indicates a particular location in the structure. Also known as deferred addressing.

indirect weld parameters (IWP) a collection of parameters that establish the welding equipment setpoint values. Examples include voltage, current, travel, speed, electrode feed rate, travel angle, electrode geometry, focused spot size, and beam power.

indiscernibility relation in Pawlak's information system $S = (U, A)$ whose universe U has n members denoted x, y, z, ... and the set A consisting of m attributes $\mathbf{a}_j$, the indiscernibility relation on U with respect to the set $B \subseteq A$ defines the partition U/B of U into equivalence classes such that $x, y \in U$ belong to the same class of U/B if for every attribute $\mathbf{a} \in B$,

$$\mathbf{a}(x) = \mathbf{a}(y).$$

In terms of negotiations, the partition U/B may be interpreted as the partition of the inverse U into blocks of negotiators that have the same opinion on all of the issues of B.

induced emission   enhanced emission of electromagnetic radiation due to the presence of radiation at the same frequency; also called stimulated emission.

induced voltage   voltage produced by a time-varying magnetic flux linkage. *See also* Faraday's Law.

inductance   a parameter that describes the ability of a device to store magnetic flux. The units are henrys per meter.

induction furnace   a method of smelting or heating metals with eddy currents induced by a high-frequency coil surrounding the crucible.

induction generator   an induction machine operated as a generator. If the machine is connected to an AC system and is driven at greater than synchronous speed, the machine can convert mechanical energy to electrical form. The induction generator requires a source of reactive power.

induction machine   classification of any of a variety of electrical machines in which an AC current in the stator coils is used to produce a rotating magnetic flux that, by Faraday action, induces an AC voltage in a set of coils (the induction coils) on the machine's rotor. The rotor coils are shorted to cause a second AC current to flow in the rotor coils, which produces, in turn, a second rotating flux. The interaction of the rotor- and stator-produced fluxes creates torque.

induction motor   *See* induction machine.

induction regulator   *See* induction voltage transformer.

induction theorem   states that if the incident tangential magnetic and electric fields are known everywhere on some closed surface, then these fields may be replaced with equivalent electric and magnetic surface currents.

induction voltage transformer   specially constructed transformer with a rotating primary coil that is used to provide voltage regulation on individual power circuits. The secondary of an induction regulator is mounted on the stationary shell of cylindrical core, and the primary is mounted on a movable, center rotor. In the neutral position, the magnetic axes of the primary and secondary coils are oriented 90 degrees to each other, reducing the magnetic coupling to zero. In this position, energizing the primary does not induce voltage in the secondary; however, rotating the primary coil in either direction from the neutral position creates mutual flux linkage and causes a secondary voltage to appear. Rotation in one direction causes secondary voltage to be in phase with the primary; rotation in the opposite direction causes secondary voltage to be out of phase with the primary. Voltage regulation is provided by connecting the primary coil across the line to be regulated and connecting the secondary coil in series with the load. By positioning the primary coil based on load demand in the line, secondary voltage can be used to adjust line voltage either up or down. Induction regulators are also equipped with a short-circuited coil mounted on the primary in spatial quadrature with the primary coil. In the neutral position, this coil has maximum coupling with the secondary coil, which minimizes the inductive reactance in the load line due to the secondary coil.

inductive coupling   a means of transferring electrical energy from one part (area) of a circuit to another part without requiring any ohmic (wire) connection. Instead, magnetic flux linkages couple two inductors (coils). The coils must be in close proximity in order to establish sufficient mutual inductance.

inductive discontinuity   a type of discontinuity that exhibits an inductive, or quasi-inductive, behavior. As an example, magnetic plane iris in metallic rectangular waveguides produces this type of response.

inductor   a two-terminal electrical element that satisfies a prescribed algebraic relationship in the flux-current $(\phi - I)$ plane.

industrial inspection   the use of computer vision in manufacturing, for example, to look for defects, measure distances and sizes, and evaluate quality. *See* computer vision, inspection system.

industrial robot   a mechanical structure or manipulator that consists of a sequence of rigid body (links) connected by means of articulation joints. A manipulator is characterized by an arm, a wrist, and an end-effector. Motion of the manipulator is imposed by actuators through actuation of the joints. Control and supervision of manipulator motion is performed by a control system. Manipulator is equipped with sensors that measure the status of the manipulator and the environment.

inelastic scattering   when there is energy exchange between the impinging wave and the medium. This also produces frequency shifts.

inertia constant   the energy stored in an electric machine running at synchronous speed and given in megajoules per megavoltampere of machine rating.

inertial confinement   a method of initiating and containing the plasma of a fusion reaction by illuminating a pellet of solid heavy hydrogen with powerful laser beams from all directions simultaneously.

inference engine   a reasoning mechanism for manipulating the encoded information and rules in a knowledge base to form inferences and draw conclusions.

infinite bus   an electrical supply with such large capacity that its voltage (and frequency, if AC) may be assumed constant, independent of load conditions. If a machine's capacity is small relative to the electric supply system to which it is connected, it may be assumed to be operating on an infinite bus.

infinite impulse response (IIR) filter   any filter having an impulse response of infinite length, therefore having a frequency response with at least one pole. In direct contrast with finite impulse response (FIR) filters, IIR filters employ feedback, which leads to an impulse response of infinite extent; consequently a causal IIR filter will depend on all previous inputs. As a simple example, every autoregressive process can be expressed as the output of an IIR filter driven by white noise. *See* finite impulse response filter, autoregressive process.

infinite symmetric exponential filter (ISEF) one type of optimal edge detecter. In two dimensions, the filter is

$$f(x, y) = a \times e^{-p(|x|+|y|)}$$

The resulting edge detector has excellent localization and signal-to-noise ratio. *Compare with* Canny edge detector.

infinite-dimensional dynamical system   linear stationary infinite-dimensional dynamical system is described by the following abstract ordinary differential state equation:

$$x'(t) = Ax(t) + Bu(t)$$

where $x(t)$ is the state vector that belongs to infinite-dimensional Banach space X, $u(t)$ is the input vector that belongs to infinite-dimensional Banach space U, A is a linear generally unbounded operator that is a generator of a strongly continuous semigroup of linear bounded operators $S(t) : X \to X$, for $t > 0$, $B : u \to x$ is a linear bounded operator. Solution of the state equation has the form

$$x(t, x(0), u) = S(t)x(0) + \int_0^t S(t - s)Bu(s)ds$$

information   a mathematical model of the amount of surprise contained in a message. For a discrete random source with a finite number of possible symbols or messages the information associated with the symbol x_k is $I_k = -\log_2(p_k)$ where p_k is the probability of the symbol x_k. The expected value of the information of the symbols is the first order entropy of the source.

information gain   for an attribute in a set of objects to be classified is a measure of the importance of the attribute to the classification. The information gain G_i of the ith attribute A_i of a set of n objects S in the classification is defined as

$$G_i = I(S) - E_i$$

where $I(S)$ is the expected information (or entropy) for the classification and E_i is the expected information required for the value of A_i to be known. $I(S)$ is defined as

$$I(S) = -\sum_{c=1}^{N_c} \frac{n_c}{n} \log_2 \frac{n_c}{n}$$

where N_c is the total number of classes in the classification, and n_c is the number of objects in the cth class C_c. E_i is defined as

$$E_i = \sum_{k=1}^{N_i} \frac{n_{ik}}{n} I(S_{ik})$$

where S_{ik} is the subset of S in which A_i of all objects takes its kth value, N_i is the number of values A_i can take, n_{ik} is the number of objects in

S_{ik}, and the information required in S_{ik} is

$$I(S_{ik}) = -\sum_{c=1}^{N_c} \frac{n_{ikc}}{n_{ik}} \log_2 \frac{n_{ikc}}{n_{ik}}$$

where n_{ikc} is the number of objects in S_{ik} belonging to class C_c.

information theory   theory relating the information content of a message to its representation for transmission through electronic media. This subject includes the theory of coding, and also topics such as entropy, modulation, rate distortion theory, redundancy, sampling.

infrared detector   semiconductor or other material with a measurable change, usually, but not always, in electrical conductivity, when exposed to infrared light.

infrared (IR)   invisible electromagnetic radiation having wavelengths longer than those of red light; often considered to range from about 0.7 micrometers to 100 micrometers.

infrared laser   laser producing its output in the infrared region of the spectrum.

InGaAs/InAlAs/InP   lattice matched semiconductor heterostructure system used for
   (1) optical fiber communications sources and detectors, and
   (2) high-speed electronic circuits.

inheritance   in object orientation, the possibility for different data types to share the same code.

inhibit   to prevent an action from taking place; a signal that prevents some action from occurring. For example, the READY signal on a memory bus for a read operation may inhibit further processing until the data item has become available, at which time the READY signal is released and the processor continues.

inhibit sense multiple access (ISMA) protocol   multiple access protocol attempting to solve the hidden station problem. In ISMA, the base station transmits an "idle" signal at the beginning of a free slot. Each user has to monitor this signal before it attempts to access the base station. ISMA is sometimes called busy tone multiple access (BTMA). *See also* busy tone multiple access.

inhomogeneous broadening   spectral broadening of a transition in a laser medium due to intrinsic differences between the constituent atoms or molecules.

inhomogeneous medium   medium whose constitutive parameters are functions of position. For example, in optical media, the properties of gain, index of refraction, etc., might be functions of position.

initial rest   a system satisfies an initial rest assumption if the output of the system is zero until the input of the system becomes nonzero $(x(t) = 0, t \leq t_0 \Rightarrow y(t) = 0, t \leq t_0)$. Systems described by linear constant-coefficient differential or difference equations are causal, linear, and time invariant if at initial rest.

initialize   (1) to place a hardware system in a known state, for example, at power up.

(2) to store the correct beginning data in a data item, for example, filling an array with zero values before it is used.

injection electroluminescence   refers to the case where the electroluminescence the injection of carriers across a p-n junction.

injection laser diode   semiconductor laser in which amplification is achieved by injection of holes or electrons into the active region.

injection locking   two oscillators can be injection-locked together by providing small coupling between them. In that case, the phase relation between the two oscillators is going to be constant over time.

inner product space   over the field F, $\mathcal{H}$ is an inner product space if it is a linear space together with a function $\langle , \rangle : \mathcal{H} \times \mathcal{H} \to F$, such that

(1) $\langle x, y \rangle = \overline{\langle y, x \rangle}$, $\forall x, y \in \mathcal{H}$.
(2) $\langle \lambda x + \mu y, z \rangle = \lambda \langle x, z \rangle + \mu \langle y, z \rangle$,
    $\forall x, y, z \in \mathcal{H}, \forall \lambda, \mu \in F$.
(3) $\langle x, x \rangle \geq 0$.
(4) $\langle x, x \rangle = 0$ if and only if $x = 0$.

The field F could be either the real or the complex numbers.

input   data conveyed to a computer system from the outside world. The data is conveyed through some input device.

input admittance   the inverse of input impedance, measured in mhos.

input backoff   *See* backoff.

input buffer   a temporary storage area where input is held. An input buffer is necessary when the data transfer rate from an input device is different from the rate at which a computer system can accept data. Having an input buffer frees the system to perform other tasks while waiting for input.

input current shaping   a technique to force the line current to assume the same shape as the line voltage of a rectifier in order to achieve low harmonic distortion and high power factor correction.

input device   a peripheral unit connected to a computer system, and used for transferring data to the system from the outside world. Examples of input devices include keyboard, mouse, and light pen.

input impedance   (1) when a voltage is applied to a conducting material, a current will flow through the material. The ratio of the voltage to the current is known as the input impedance and is a complex number with magnitude measured in units of ohms. *See also* Ohm's law.

(2) impedance seen when looking into the input terminals of an antenna.

input layer   a layer of neurons in a network that receives inputs from outside the network. In feedforward networks, the set of weights connected directly to the input neurons is often also referred to as the input layer.

input neuron   a neuron in a network that receives inputs from outside the network. In feedforward networks, the set of weights connected directly to the input neurons is often also referred to as the input layer.

input register   a register to buffer transfers to the memory from the I/O bus. The transfer of information from an input device takes three steps:

(1) from the device to its interface logic,

(2) from the interface logic to the input register via the I/O bus, and

(3) from the input register to the memory.

input return loss   negative ratio of the reflected power to the incident power at the input port of a device, referenced to the source or system impedance, expressed in decibels. The negative sign results from the term "loss." Thus, an input return loss of 20dB results when 1/100th of the incident power is reflected.

input routine   (1) a function responsible for handling input and transferring the input to memory.

(2) the software subroutine used to communicate with an input device. *See also* I/O controller and I/O routine.

input stability circle   circular region shown graphically on a Smith chart that defines the range of input impedance terminations for which an active device or circuit remains unconditionally stable, and those regions for which it is potentially unstable, at a particular frequency.

input vector   a vector formed by the input variables to a network.

input-output curve   a plot of the rate of fuel consumption vs. the power output for a thermal generating plant.

input–output equivalence of generalized 2-D system matrices   the 2-D polynomial matrix

$$\begin{bmatrix} P & -Q \\ C & D \end{bmatrix}$$

$$P := E z_1 z_2 - A_0 - A_1 z_1 - A_2 z_2$$
$$Q := B_0 + B_1 z_1 + B_2 z_2$$

is called the system matrix of the generalized 2-D model

$$E x_{i+1,j+1} = A_0 x_{ij} + A_1 x_{i+1,j}$$
$$+ A_2 x_{i,j+1} + B_0 u_{ij}$$
$$+ B_1 u_{i+1,j} + B_2 u_{i,j+1}$$
$$y_{ij} = C x_{ij} + D u_{ij}$$

$i, j \in Z_+$ (the set of nonnegative integers) where $x_{ij} \in R^n$ is the semistate local vector, $u_{ij} \in R^m$ is the input vector, $y_{ij} \in R^p$ is the output vector and E, A_k, B_k ($k = 0, 1, 2$), C, D are real matrices with E possibly singular or rectangular. Two system matrices $S_i := [\begin{smallmatrix} P_i & -Q_i \\ C_i & D_i \end{smallmatrix}]$ for $i = 1, 2$ are called input–output equivalent if $C_1 P_1^{-1} Q_1 + D_1 = C_2 P_2^{-1} Q_2 + D_2$.

input-output stability   system condition where bounded outputs are obtained as a response to bounded inputs. *See also* bounded function.

input/output channel (I/O channel)   input/output subsystem capable of executing a program, relieving the CPU of input/output-related tasks. The channel has the ability to execute read and write instructions as well as simple arithmetic and sequencing instructions that give it complete control over input/output operations. (IBM terminology) *See also* direct memory access, selector channel, multiplexer channel.

input/output port   a form of register designed specifically for data input/output purposes in a bus-oriented system.

INR   *See* interference to noise ratio.

inrush current   the transient current drawn by an electrical apparatus when it is suddenly connected to a power source. The inrush current may be larger in magnitude than the steady-state full-load current. The transient response is short in time and the electrical equipment generally supports the inrush current, provided it does not happen frequently. For a single transient, the thermal limit of the equipment is not reached, but if it is switched on and switched off several times within a short period, the temperature can rise very quickly. In case of transformers, the inrush current is not sinusoidal even if the voltage is due to the hysteresis of the ferromagnetic core.

insertion loss   (1) worst-case loss of the device across the stated frequency range. The loss due to the insertion of the unit in series with a signal path.

(2) transmission loss of an RF or microwave component or system, typically measured in decibels.

insolation   incident solar radiation.

inspection system   an image processing system, usually automatic or semi-automatic, that performs industrial inspection.

instantaneous   the range of 0.5 to 30 cycles of the supply frequency.

instantaneous contact   the contacts of a contactor or relay that open or close with no time delay.

instantaneous frequency   (1) the time rate of change of the angle of an angle-modulated wave.

(2) the frequency of radiation at some chosen instant of time; the rate of change of phase in radians per second, divided by 2π.

instantaneous overcurrent relay   an overcurrent relay that operates with no intentional delay following sensing of a power frequency overcurrent, i.e., a current above its set point.

instantaneous power   power in an AC or modulated signal at a given instant in time.

instantaneous-trip circuit breaker   essentially an inverse-time circuit breaker with the thermal element removed. It will only trip magnetically in response to short-circuit currents. Thus, it is often referred to by other names, such as magnetic circuit breaker and magnetic-only circuit breaker.

instar   *See* instar configuration.

instar configuration   a term used for a neuron fed by a set of inputs through synaptic weights. An instar neuron fires whenever a specific input pattern is applied. Therefore, an instar performs pattern recognition. *See also* instar training. *Compare with* outstar configuration.

instar training   an instar is trained to respond to a specific input vector and to no other. The weights are updated according to

$$w_i(t+1) = w_i(t) + \mu(x_i - w_i(t))$$

where x_i is the ith input, $w_i(t)$ the weight from input x_i, μ is the training rate starting from about 0.1 and gradually reduced during the training. *See also* instar configuration. *Compare with* outstar training.

Institute of Electrical and Electronics Engineers (IEEE)   a professional organization of electrical engineers and computer scientists. The world's largest professional organization for engineers.

instruction   specification of a collection of operations that may be treated as an atomic entity with a guarantee of no dependencies between these operations.

instruction access fault   a fault, signaled in the processor, related to abnormal instruction fetches.

instruction cache   *See* code cache.

instruction counter   the memory register within a computer that maintains the location of the next instruction that will be fetched for execution. This register is also called a program counter.

instruction decoder   the part of a processor that takes instructions as input and produces control signals to the processor registers as output. All processors must perform this function; some perform it in several steps, with part of the decoding performed before instruction issue and part after.

instruction decoder unit   in modern CPU implementations, the module that receives an instruction from the instruction fetch unit, identifies the type of instruction from the opcode, assembles the complete instruction with its operands, and sends the instruction to the appropriate execution unit, or to an instruction pool to await execution.

instruction fetch unit   in modern CPU implementations, the module that fetches instructions from memory, usually in conjunction with a bus interface unit, and prepares them for subsequent decoding and execution by one or more execution units.

instruction field   a portion of an instruction word that contains a specified value, such as a register address, a 16-bit immediate value, or an operand code.

instruction format   the specification of the number and size of all possible instruction fields in an instruction-set architecture.

instruction issue   the sending of an instruction to functional units for execution.

instruction pipeline   a structure that breaks the execution of instruction up into multiple phases, and executes separate instructions in each phase simultaneously.

instruction pointer   another name for program counter, the processor register holding the address of the next instruction to be executed.

instruction pool   in modern CPU implementations, a holding area in which instructions that have been fetched by an instruction fetch unit await access to an execution unit.

instruction prefix   a field within a program instruction word used for some special purpose. Found only rarely. The Intel X86 architecture occasionally uses instruction prefixes to override certain CPU addressing conventions.

instruction reordering   a technique in which the CPU executes instructions in an order different from that specified by the program, with the purpose of increasing the overall execution speed of the CPU.

instruction repertoire   *See* instruction set.

instruction scheduling   the relocation of independent instructions, which is done to maximize instruction-level parallelism (and/or minimize instruction stalls).

instruction set   the instruction set of a processor is the collection of all the machine-language instructions available to the programmer. Also known as instruction repertoire.

instruction window for an out-of-order issue mechanism, a buffer holding a group of instructions being considered for issue to execution units. Instructions are issued from the instruction window when dependencies have been resolved.

instruction-level parallelism the concept of executing two or more instructions in parallel (generally, instructions taken from a sequential, not parallel, stream of instructions).

instrument transformer a two-winding transformer designed and optimized for metering applications. The essential features are accurate input to output ratios for the measured parameter and minimal burden (or load) on the circuit being tested. *See also* current transformer, voltage transformer.

insulated gate bipolar junction transistor (IGBT) a hybrid electronic switch that has the high input impedance and high speed characteristic of a MOSFET with the conductivity characteristic (low saturation voltage) of a bipolar transistor.

insulation class specification of insulating material. Primarily determined by the maximum temperature that the material can withstand. The class of insulating system used in a transformer or electrical machine is a key factor in determining its rated load.

insulation coordination the practice of selecting the insulation of a power system such that breakdowns from overvoltages (such as from a lightning strike) will occur at points where the least harm will result.

insulation resistance measured value of electrical resistance of the insulation of a product or device.

insulator a device designed to separate and prevent the flow of current between conductors. Properties of the dielectric (insulating) material

and geometry of the insulator determine maximum voltage and temperature ratings.

insulator string a chain of two or more strain insulators that are coupled together to increase the total insulation level of the assembly.

integer a fixed-point, whole number value that is usually represented by the word size of a given machine.

integer unit in modern CPU implementations, a type of execution unit designed specifically for the execution of integer-type instructions.

integral control control scheme whereby the signal that drives the actuator equals the time integral of the difference between the input/desired output and the measured actual output.

integral equation a type of equation where the unknown function appears under an integral. When this is the only place that the unknown appears, the integral equation is commonly called a first-kind equation, while if the unknown also appears outside the integral, it is a second-kind integral equation.

One-dimensional examples are

$$y(s) = \lambda \int_a^b K(s,t)x(t)dt$$

and

$$x(s) = y(s) + \lambda \int_a^b K(s,t)x(t)dt$$

where λ is a possibly complex scalar parameter and $y(s)$ is known (often called the driving term); $x(t)$ is the unknown function and $K(s,t)$ is the known kernel. The above are linear Fredholm integral equations of the first and second kind, respectively.

integral horsepower motor a motor built in a frame as large as or larger than that of a motor of open construction having a continuous rating of one horsepower at 1700–1800 rpm.

integrated AOTF   acousto-optical tunable filter device using integrated wave guide optics and surface acoustic waves (SAWs) for acousto-optical interaction.

integrated circuit (IC)   (1) an assembly of miniature electronic components simultaneously produced in batch processing, on or within a single substrate, that performs an electronic circuit function.
(2) many transistors, resistors, capacitors, etc., fabricated and connected together to make a circuit on one monolithic slab of semiconductor material.

integrated optics   also known as guided-wave optics; generally, describes the devices that couple to fiber optics, but does not include the fibers.

integrated project support environment   a software engineering environment supporting all the stages of the software process from initial feasibility studies to operation and maintenance.

integrated services digital network (ISDN)   a network that provides end-to-end digital connectivity to support a wide range of services, including voice and nonvoice services, to which users have access by a limited set of standard multipurpose user–network interfaces.

integrating A/D   device that takes an analog input signal and integrates it over time to produce a digital signal that represents the area under the curve, or the integral.

integrator   for an input of $x(t)$ and an output of $\int_{-\infty}^{t} x(\tau)d\tau$, is one of the basic signal processing building blocks. Integrators can be implemented using operational amplifiers.

integrity (of data)   (1) a belief in the truth of the information represented by a set of data.

(2) a condition stating that the information in a set of data does satisfy a set of logical constraints (the integrity constraints).

intelligence   the aggregated and processed information about the environment, including potential adversaries, available to commanders and their staff.

intelligent control   a sensory-interactive control structure incorporating cognitive characteristics that can include artificial intelligence techniques and contain knowledge-based constructs to emulate learning behavior with an overall capacity for performance and/or parameter adaptation. Intelligent control techniques include adaptation, learning, fuzzy logic, neural networks, genetic algorithm, as well as their various combinations.

intelligent material   *See* smart material.

intelligibility   objective quantitative measure of speech perception.

intensifying transformer   one of a class of hedges — for example very, extremely — which intensifies the characteristics of a fuzzy set.

intensity   a measure of the strength of a light field. Two different definitions of intensity are commonly encountered. Especially in the discipline of radiometry, intensity is usually defined to be power per unit solid angle. Alternatively, especially in laser physics intensity is defined to be power per unit area. The intended definition can usually be deduced from context or from dimensional analysis.

intensity average   spatially or temporally averaged value of the intensity.

intensity modulation   alteration of the power level of an optical beam, usually to impose information onto the beam.

intensity optical same as optical power per unit area; time-average value of the Poynting vector $E \times H$ measured in watts per meter squared, sometimes also representable as the energy density times the speed of light.

intensity-dependent refractive index the property of many materials that the measured value of the refractive index depends on the intensity of the light field used to perform the measurement. Typically, the refractive index changes by an amount describable by $\delta n = n_2 I$, where n_2 is the coefficient of the nonlinear refractive index and I is the intensity of the field in units of power per unit area.

inter-area mode oscillation this mode of oscillation is characterized by the swinging of many machines in one part of the system against machines in other parts. These oscillations are caused by the presence of weak ties between two or more areas.

inter-cell interference the interference caused by the transmitters in a cellular communications system (base station or mobile terminal) on a receiver of interest where the interfering transmitters are located on cells other than the receiver cell. *Compare with* intra-cell interference.

inter-feature distance the distance between a pair of feature points in an image, often used to help identify the object, or (in conjunction with other measures of this type) to infer the presence of the object.

inter-record gap the space between two records of data stored on a magnetic medium such as a tape. This space helps prevent interference between the two records. It can also contain markers marking the beginning and end of the records.

interaction region a region where an electron beam and microwaves transfer energy to each other.

interaural attribute attribute of ear input signals (e.g., localization in the horizontal plane) that depend on differences between, or ratios of measures of, two ear input signals.

interconnect the physical manifestation of an interconnection network.

interconnection density number of interconnection channels per unit area or unit volume. The interconnection density of optical interconnects is high because unlike electrons, light beams can cross each other without any change.

interconnection network the combination of switches and wires that allows multiple processors in a multiprocessor or multicomputer to communicate.

interconnections and packaging (I&P) elements of the electrical signal transmission path from the driver chip to the receiver chip. Various elements that make up I&P are chip wirebonds and package pins, circuit boards, connectors, motherboards, and cables.

interdigital capacitor a capacitor consisting of two pieces of electrodes that have multiple fingers. Each finger of these electrodes is adjacently arranged in turn with a narrow gap. It utilizes the gap capacitance between adjacent fingers.

interdigital filter filter consisting of parallel coupled transmission line resonators where each resonator is grounded on one side and is either open-circuited or capacitively loaded to ground on the other side. Adjacent resonators are grounded on the opposite ends.

interface the set of rules specified for communicating with a defined entity.

interface state for a semiconductor device such as a transistor, a region where trapping of charge

357

carriers can occur at the interface between regions of differing electrical properties.

interface trap   energy level at the insulator/semiconductor interface that can trap electrons or holes. The occupation of these traps follow Fermi statistics.

interference   (1) a process in which two waves interact. Interference occurs when the complex properties of the wave are combined instead of simply their two amplitudes. The wave interference can be constructive when they both have the same phase, or destructive when they have opposite phases.

(2) a disturbance on a communication system caused by the presence of a signal, or small number of signals, of man made origin. The interference may be due to the presence of signals external (other systems) or internal (e.g., multiple access interference) to the system of interest. *Compare with* background noise.

(3) an interaction whereby a process that is supposed to be independent (and isolated) from another process does indeed perform some action that has an effect visible from the second process. Interference can be caused by direct writing to the state of the second process or by indirect effects, such as by monopolizing the use of a system module.

(4) any external electrical or electromagnetic disturbance which causes an undesired response or degradation of the desired signal.

interference cancellation   a signal processing technique where an interfering signal is estimated, regenerated based on the estimate, and then canceled from the received signal, leaving a potential interference-free, desired signal. Examples are multiple access interference cancellation, co-channel interference cancellation, echo cancellation.

interference channel   a multiple transmitter, multiple receiver communications system in which each received signal is a (possibly non-deterministic) function of two or more transmitted signals. *See also* broadcast channel, multiple access channel.

interference to noise ratio (INR)   the total interference measured at a receiver divided by the receiver noise power. The INR is a useful measure for comparing the interference distributions between different types of radio systems.

interframe   referring to the use of more than a single image frame (e.g. an algorithm which differences successive image frames or detects frame- to-frame motion).

interframe coding   image coding schemes that utilize the temporal correlation between image frames.

interframe coding 3-D   in situations where motion picture film or television images need to be coded, three-dimensional transform coding is used. In this approach, the frames are coded L at a time and partioned into $L \times L \times L$ blocks per pixel prior to transformation.

interframe hybrid transform coding $L \times L$ blocks of pixels are first 2-D-transformed. The resulting coefficients are then DPCM (interframe) coded and transmitted to the receiver. This is then reversed at the receiver for reconstruction. In the interframe hybrid transform coding, the DC components are determined exclusively by the amount of spatial detail in the picture.

interframe prediction   this technique uses a combination of pixels from the present field as well as previous fields for prediction. For scenes with low detail and small motion, field difference prediction appears to perform better than frame difference prediction.

interharmonic component   the frequency component that is not an integer value of the periodic supply waveform.

interior permanent magnet machine a permanent magnet machine in which the permanent magnets are buried in the rotor iron. With this construction the airgap is small, allowing the airgap flux to be reduced by current control. This allows an operating mode in which the top speed of the motor can be as much as five times the base speed, and is one of the main features of this motor.

interlaced *See* interlaced scanning.

interlaced scanning a system of television and video scanning whereby the downward rate of travel of the scanning electron beam is increased by sending every other line whereby field 1 (odd field) is superimposed on field 2 (even field) to create one picture frame. Interlaced scanning is used to reduce flicker.

interleaved memory a memory system consisting of several memory modules (or banks) with an assignment of addresses that makes consecutive accesses fall into different modules. This increases the effective memory bandwidth, as several memory requests can be satisfied (concurrently by several modules) in the same time as it would take to satisfy one memory request. For example, a simple arrangement (sequential interleaving), which favors sequential access, is to assign address k to memory bank k mod N, where N is the number of modules employed: the reference stride is 1, so consecutive requests are to different bands.

There are other schemes for assigning addresses to banks, all of which schemes aim to (for nonunit strides) give the same performance as with a stride of 1. These include skewed addressing, which is similar to sequential interleaving but with a fixed displacement (the skew) added to the chosen module number; dynamic skewed addressing, in which the skew is variable and determined at run-time; pseudo-random skewing, in which the module number is chosen on a pseudo-random basis; prime-number interleaving, in which the number of modules employed is prime relative to the degree of interleaving; and superinterleaving, in which the number of modules exceeds the degree of interleaving.

interleaved storage the notion of breaking storage into multiple pieces that may be accessed simultaneously by the same request, increasing the bandwidth from the storage.

interleaving (1) regular permutation of symbols usually used to split bursts of errors caused by channels with memory (e.g., fading channels) making a coding scheme designed for memoryless channels more effective. The two main categories are *block interleaving* and *convolutional interleaving*. The inverse process is called deinterleaving.

(2) a characteristic of a memory subsystem such that successive memory addresses refer to separately controlled sections or banks of memory. Sequential accesses to interleaved memory will activate the banks in turn, thus allowing higher data transfer rates.

interlock the mechanism that stalls a pipeline while a result needed in the pipeline is being produced.

intermediate frequency (IF) (1) the frequency produced when two signals are combined in a mixer circuit.

(2) an operating frequency of an internal stage used in a superheterodyne system. This frequency is converted from radio frequency by using a local oscillator and a mixer for receivers. From this frequency, radio frequency is converted by using a local oscillator and a mixer for transmitters.

intermediate-level vision the set of visual processes related to the perception/description of surfaces and of their relationships in a scene.

intermittent fault a fault that appears, disappears, and then reappears in a repeated manner.

intermodal distortion distortion in the temporal shape of an optical signal transmitted through

359

an optical fiber caused by the differential time delays between the numerous modes propagating in a multimode fiber.

intermodulation   the heterodyning (or mixing) of at least two separate and distinct electrical signals within a nonlinear system. For two waves mixing together, the resultant intermodulation products are given by

$$F_{IM} = PF_1 + QF_2$$

where F_{IM} is the resultant intermodulation product frequency, P and Q are integers (zero, positive, or negative), and the order, R, of the intermod is given by

$$R = \mid P \mid + \mid Q \mid$$

intermodulation distortion   in optical devices, nonlinear transfer functions that result from the interaction of multiple signal components among each other. Strong signal components lead to levels of distortion that limit the low end of a device dynamic range.

intermodulation intercept point   a figure of merit for intermodulation product suppression, measured in decibels.

internal bus   a bus used to connect internal components of a computer system, such as processors and memory devices.

internal forwarding   a mechanism in a pipeline that allows results from one pipeline stage to be sent directly back to one or more waiting pipeline stages. The technique can reduce stalls in the pipeline.

internal fragmentation   in paging, the effect of unused space at the end of a page. On average the last page of a program is likely to be 50% full. Internal fragmentation increases if the page size is increased.

internal memory   main memory of a computer.

internal model control (IMC)   a feedback control structure in which the output from a model of the process is subtracted from the process output to form the feedback signal. Using this loop

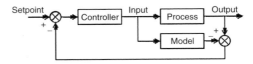

Internal model control structure.

configuration as a theoretical construct, the design of closed loop controllers (which is a difficult nonlinear mathematical problem) becomes equivalent to the design of an open loop controller (which is a simpler linear problem). IMC is a special case of Q-parametrization or Youla parametrization of all stabilizing controllers, and is closely related to the Smith Predictor invented in the late 1950s. Unstable processes need to be stabilized first by an inner loop before application of this theory. Practical designs for processes subjected to input disturbances should be checked for internal performance. *See also* inverse response compensator.

internal model principle   in process control, one of the most common problems is regulation. This implies the importance of considering the attenuation of disturbances that act on the process. One of the many ways of dealing with disturbances is called internal model principle.

Using the internal model principle, the pole placement procedure can be modified to take disturbances into account. In many cases, the disturbances have known characteristics, which can be captured by assuming that the disturbance in the process model is generated by some dynamical system. Using internal model principle implies that a model of the disturbance is included in the controller.

internal performance   defines the performance of a feedback system to stimuli that enter the loop at internal points. Naive designs may produce automatic control loops that have good performance in the traditional sense, yet exhibit poor internal performance. As an example, the process modeled by the transfer function

$$g(s) = \frac{4}{1 + 7s}$$

when controlled in a unity feedback configuration by the digital control law

$$k(z) = \frac{0.794(z - 0.842)}{z - 1}$$

illustrates clearly the difference between the traditional concept of performance and that of internal performance. The control loop has good performance in the traditional sense, as its response to a setpoint change (at time $t = 0$) and an output disturbance (at $t = 10$) is much faster than the open loop system, yet its internal performance is poor since it rejects input disturbances sluggishly (after $t = 20$). The concept is particularly important for processes subjected to input disturbances in which the closed loop is faster or more damped than the open loop process itself. *See also* internal stability.

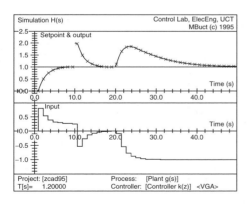

Illustration of the difference between performance and internal performance.

internal photoemission emission of energetically excited electrons or holes over a energy barrier into its contacting collector by a process in which the emitting electrons or holes are optically excited from the ground state of the contacting emitter.

internal quantum efficiency the product of injection efficiency and radiative efficiency corresponds to the ratio of power radiated from the junction to electrical power supplied.

internal stability defines the stability of a system to signals that enter the loop at all possible points. The difference between the traditional concept of stability and the modern concept of internal stability is illustrated by the open loop control configuration, which is internally unstable yet might easily have been classified as stable under the traditional definition. The relevant equations are

$$y = g(s)k(s) \times e + g(s) \times v$$
$$= \frac{1}{s + 1} \times e + \frac{1}{s - 1} \times v$$

where blocks $g(s)$ and $k(s)$ are transfer functions, signals e and v are inputs, signal u is an internal variable, and signal y is the output. Clearly the relationship between y and e is stable, while that between y and v remains unstable. The traditional definition of stability would have missed the latter phenomena. Exactly the same effect is observed when the above system is put into closed loop. The concept of internal stability relates closely to those of controllability and observability in state space theory of dynamic systems. *See also* internal performance.

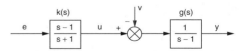

Illustration of internal stability.

International Electro-technical Commission (IEC) the international standards and conformity assessment body for all fields of electrotechnology.

International Radio Consultative Committee (CCIR) one of two international committees that exist for the purpose of carrying out studies of technical and other problems related to the interworkings of their respective national telecommunication systems to provide a worldwide telecommunications network. It operates under the auspices of the International Telecommunication Union (ITU). After each plenary assembly, the ITU publishes recommendations that deal with point-to-point radio relay systems. A purpose of those recommendations is to make the performance of such systems compatible with metallic line systems that follow the CCITT (The International Telegraph and Telephone Consultative Committee) recommendations.

internet a network formed by the interconnection of networks.

interpixel redundancy the tendency of pixels that are near each other in time or space redundancy to have highly correlated gray levels or color values. Reducing interpixel redundancy is one way of compressing images. *See* image compression.

interpolation (1) the process of finding a value between two known values on a chart or graph.
(2) the reconstruction of missing signal samples based on observed samples. Frequently, the conversion of a discrete signal to a discrete signal on a finer time step or (in the limit) to a continuous signal.

interpolative coding coding schemes that involve interpolation.

interpolative vector quantization (IRVQ) technique in which a subsample of the interpolated

original image is used to form the predicted image, and then a residual image is formed based upon this prediction. This approach reduces blocking artifacts by using a smooth prediction image.

interpole a set of small poles located midway between the main poles of a DC machine, containing a winding connected in series with the armature circuit. The interpole improves commutation by neutralizing the flux distortion in the neutral plane caused by armature reaction.

interpreter a computer program that translates and immediately performs intended operations of the source statements of a high-level language program.

interprocess communication (IPC) the transfer of information between two cooperating programs. Communication may take the form of signal (the arrival of an event) or the transfer of data.

interrupt an input to a processor that signals the occurrence of an asynchronous event. The processor's response to an interrupt is to save the current machine state and execute a predefined subprogram. The subprogram restores the machine state on exit, and the processor continues in the original program.

interrupt descriptor table used in protected mode to store the gate descriptors for interrupts and exceptions.

interrupt disable (1) a state in which interrupts to the CPU are held, but not processed. In most systems it is possible to disable interrupts selectively, so that certain types of interrupts are processed, while others are disabled and held for later processing.
(2) the operation that changes the interrupt state of the CPU from enabled to disabled.

interrupt enable (1) a state in which interrupts to the CPU can be processed.

(2) the operation that changes the interrupt state of the CPU from disabled to enabled.

interrupt handler a predefined subprogram that is executed when an interrupt occurs. The handler may perform input or output, save data, update pointers, or notify other processes of the event. The handler must return to the interrupted program with the machine state unchanged.

interrupt line a wire carrying a signal to notify the processor of an external event that requires attention.

interrupt mask a bit or a set of bits that enables or disables the interrupt line to be transmitted at the interrupt detector circuit (inside or outside the CPU). The mechanisms of masking are typically implemented into interrupt controllers.

interrupt priority a value or a special setting that specifies the precedence to serve the corresponding interrupt with respect to other interrupt signals.

interrupt service the steps that make up the operation by which the CPU processes an interrupt. Briefly, the CPU suspends execution of its current program and branches to a special program known as an interrupt handler or interrupt routine to take appropriate action. Upon completion of the interrupt service, the CPU will take one of a number of actions, depending on circumstances: it can

(1) return to the previous task if conditions permit;

(2) process other pending interrupts, or

(3) request further action from the operating system.

interrupt-driven I/O an input/output (I/O) scheme where the processor instructs an I/O device to handle I/O and proceeds to execute other tasks; when the I/O is complete, the I/O device will interrupt the processor to inform completion. An interrupt-driven I/O is more efficient than a programmed I/O where the processor busy-waits until the I/O is complete.

interruptible load a load, typically of a commercial customer, which by contract may be interrupted by the utility for purposes of system stability.

interrupting capacity *See* interrupting rating.

interrupting rating for a circuit breaker, fuse, or switch, the maximum fault level that the device can safely interrupt. The interrupting rating can be expressed in terms of amps or volt-amps.

interruption *See* sustained interruption, momentary interruption, temporary interruption.

intersection of fuzzy sets a logic operation, corresponding to the logical AND operation, forming the conjunction of two sets. In a crisp (non-fuzzy) system, the intersection of two sets contains the elements which are common to both. In fuzzy logic, the intersection of two sets is the set with a membership function which is the smaller of the two.

Denoted $\cap$, fuzzy intersection is more rigorously defined as follows. Let A and B be two fuzzy sets in the universe of discourse X with membership functions $\mu_A(x)$ and $\mu_B(x)$, $x \in X$. The membership function of the intersection $A \cap B$, for all $x \in X$, is

$$\mu_{A \cap B}(x) = \min\{\mu_A(x), \mu_B(x)\}$$

See also fuzzy set, membership function.

intersubband laser a type of quantum well laser in which the lasing transition couples states that are accessible to a single type of carrier, (i.e., either electrons or holes).

intersymbol interference (ISI) the distortion caused by the overlap (in time) of adjacent symbols.

intersymbol interference postcursor intersymbol interference caused by the causal part of the channel impulse response $h(t)$ (i.e., $h(t)$, $t > 0$). For each transmitted symbol, postcursor intersymbol interference is contributed from preceding transmitted symbols.

intertie another name for tie line.

interval polynomial a polynomial

$$p(s, \mathbf{a}) = a_0 + a_1 s + a_2 s^2 + \cdots + a_n s^n$$

with the uncertain coefficient vector

$$\mathbf{a} = [a_0, a_1, a_2, \ldots, a_n]$$

that may take values from the $n + 1$ dimensional rectangle

$$\mathbf{A} = \left\{ \mathbf{a} | a_i \in \left[a_i^-; a_i^+ \right], i = 0, 1, 2, \ldots, n \right\}$$

intra-cell interference interference arising in a cellular communication system that due to other transmitters (base stations or mobile terminals) in the same cell. *Compare with* inter-cell interference.

intra-chip optical interconnect optical interconnect in which the source and the detector are connected to electronic processing elements in a single chip.

intrafield predictor for television a technique used to deal with problems caused in TV signal prediction. In particular, problems due to lack of correlation because of the presence of chrominance components. One method of overcoming such difficulties is to sample the composite signal at integral multiples of subcarrier frequencies. Predictors can be designed to match the bimodal spectrum of the composite signal to predict the baseband luminance and modulated chromiance with one predictor.

intraframe coding image coding schemes that are based on an intraframe restriction; that is, to separately code each image in a sequence.

intraframe image video processing method that operates within a single frame without reference to other frames. *See* interframe.

intramodal distortion distortion in the temporal shape of an optical signal transmitted through an optical fiber caused by the wavelength dispersion of a propagating mode due to the finite spread of wavelengths in the optical source used to transmit the signal.

intrinsic term associated with the inside or interior. In devices and device modeling, intrinsic refers to that part of the device or model associated with the active semiconductor structures that control device operation, or provide the desired functions.

intrinsic coercive force the demagnetizing field required to reduce the intrinsic induction to zero; the x-intercept of the intrinsic demagnetization curve. This quantity is used to gage the field required to magnetize a material and its ability to resist demagnetization.

intrinsic demagnetization curve the second quadrant portion of the hysteresis loop generated when intrinsic induction (B_i) is plotted against applied field (H), which is mathematically related to the normal curve; most often used to determine the effects of demagnetizing (or magnetizing) fields.

intrinsic fiber optic sensor a fiber optic sensor where a property of the fiber itself responds to the measured parameter. Examples include microbend and interferometric sensors.

intrinsic image an image onto which physical properties of the imaged scene (e.g., range, orientation, reflectance) are mapped.

intrinsic impedence (1) the impedance presented when a source is open-circuited.

(2) a characteristic parameter associated with a medium that is the ratio of the magnitudes of the transverse components of the electric field intensity and magnetic field intensity for a wave propagating in a given direction. It has units of ohms.

intrinsic induction the vector difference between the magnetic induction and the applied external magnetic field, or the magnetic field established by the magnetic material itself.

intrinsic permutation symmetry *See* symmetries of nonlinear susceptibility.

intrinsic semiconductor a semiconductor material in which the number of holes equals the number of electrons.

invalid entry an entry in a table, register or module that contains data that is not consistent with the global state. Invalidity is used in cache coherence mechanisms.

invariant subspace of generalized 2-D model a linear subspace V of the local state space X that satisfies certain conditions. It is called an (E, A, B)-invariant subspace of the generalized 2-D model

$$Ex_{i+1,j+1} = A_1 x_{i+1,j} + A_2 x_{i,j+1}$$
$$+ B_1 u_{i+1,j} + B_2 u_{i,j+1}$$

$i, j \in Z_+$ (the set of nonnegative integers) if

$$\begin{bmatrix} A_1 \\ A_2 \end{bmatrix} V \subset EV \times EV + Im \begin{bmatrix} B_1 \\ B_2 \end{bmatrix}$$

in product space $X \times X$ where $x_{ij} \in R^n$ is the local semistate vector, $u_{ij} \in R^m$ is the input vector, and E, A_k, B_k $(k = 1, 2)$ are real matrices. There exists a state feedback $u_{ij} = K x_{ij}$, $i, j \in Z_+$ such that

$$\left(\begin{bmatrix} A_1 \\ A_2 \end{bmatrix} + \begin{bmatrix} B_1 \\ B_2 \end{bmatrix} K \right) V \subset EV \times EV$$

if and only if V is an (E, A, B)-invariant subspace of the model.

inverse Chebychev filter filter that exhibits an inverse Chebychev response in the stop-band region. That is, in the stop-band region of the filter, the gain of the filter as a function of frequency is equal ripple in magnitude. This is in contrast with many other types of filters, which in the filter stop-band region exhibit a gain response that is monotonic with frequency.

inverse dynamics the inverse dynamics problem consists of determining the joint generalized forces $\tau(t)$ (acting at the joints of the manipulator) needed to generate the motion specified by the joint generalized position $q(t)$, velocities $\dot{q}(t)$, and accelerations $\ddot{q}(t)$. The inverse dynamics solution is used in model based control algorithms for manipulators.

inverse dynamics linearizing control inverse dynamics linearizing control is an operational space control scheme that uses a feedback control signal that leads to the system of double integrators. In this case, nonlinear dynamics of the manipulator is compensated and the inverse dynamics linearizing control allows full trajectory tracking in the operational space.

inverse filter for a linear time invariant (LTI) filter, the filter which, when cascaded with it produces white noise output for a white noise input. The inverse filter of an LTI filter with impulse response $h(n)$ has impulse response $a(n)$, where $h(n) * a(n) = \delta(n)$. Alternatively, in the frequency domain, the inverse filter $A(z)$ satisfies the relation $A(z)H(z) = 1$. The inverse filter $A(z)$ is said to whiten the filter $h(n)$ to produce $\delta(n)$.

inverse Fourier transform *See* Fourier transform.

inverse homogeneous transformation matrix in general, the inverse of the homogeneous transformation matrix describes the reference co-ordinate frame with respect to the transformed frame and has the form

$$T^{-1} = \begin{bmatrix} n_x & n_y & n_z & -p \cdot n \\ o_x & o_y & o_z & -p \cdot o \\ a_x & a_y & a_z & -p \cdot a \\ 0 & 0 & 1 & 1 \end{bmatrix}$$

where "·" denotes a scalar product between two vectors.

inverse kinematics problem for a desired position and orientation of the end-effector of the manipulator and the geometric link parameters with respect to the reference coordinate system, calculates a set of desired joint variables vectors. The inverse kinematics problem has usually multiple solutions. In general, the inverse kinematics can be solved using algebraic, interactive, or geometric methods.

inverse Laplace transform *See* Laplace transform.

inverse Nyquist array (INA) an array of Nyquist plots of the elements of the inverse frequency response matrix $\hat{Q}(j\omega)$ of a multi-input–multi-output transfer function model $Q(s)$ of an open loop multivariable system. Each diagonal graph can be viewed as a single-input–single-output inverse Nyquist diagram from which the stability of the multivariable system can be deduced, under certain conditions relating to diagonal dominance. A typical INA diagram, for a MIMO system with two inputs and two outputs, clearly shows the individual polar plots and the Gershgorin circles forming a band with canters along the diagonal elements. Rosenbrock, H. H., *Computer-Aided Control System Design,* Academic Press, London, 1974. *See also* Gershgorin circle.

inverse Nyquist plot a graphical representation of the complex function

$$\hat{q}(jw)$$

which is the inverse of the frequency response model of the open loop transfer function system $q(s)$ that is connected in a unity feedback control loop. The stability of the resulting closed loop system is deduced by the principle of the argument applied to this complex function.

inverse problem essentially, the problem of inverting a "forward" system. More specifically, suppose we have some system $S(x)$ parameterized in terms of x, and suppose we can define observations

$$y = O(S(x))$$

Then the "forward" problem is the deduction of y (or its statistics) based on knowledge of x. The "inverse" problem is the deduction of x (or its statistics) based on knowledge of y (and possibly given prior statistics for x).

inverse response compensator a feedback control configuration in which the dynamics of the process are canceled by a cascade precompensator that includes a model of the process. The block diagram for an inverse response compensator is shown in the figure. Other configura-

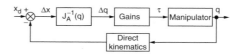

Block diagram for an inverse response compensator.

tions based on the model are also possible. For example, the inner control loop might easily have been a state space control law, with or without asymptotic setpoint tracking.

inverse scattering theory yields physical parameters of the inhomogeneous medium from knowledge of the scattered intensity distribution.

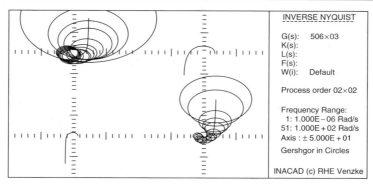

Typical inverse Nyquist array diagram.

inverse system the inverse of a system S_1 that transforms input signals x into output signals y is the system S_2 that, when placed in series with S_1, results in a larger cascade that leaves the input signal unchanged (x in, x out). *See* invertible system.

inverse translation buffer (ITB) a device to translate a real address into its virtual addresses, to handle synonyms. Also called a reverse translation buffer and an inverse mapper.

inverse-time circuit breaker a circuit breaker in which the allowed current and time are inversely proportional. It contains a thermal element and a magnetic element in series. The thermal element is designed to trip as a result of heating over time in response to overload currents, while the magnetic element is designed to trip magnetically, with no intentional time delay, in response to short-circuit currents. Also called a thermal-magnetic circuit breaker.

inverse wavelet transform a computation procedure that calculates the function using the coefficients of the wavelet series expansion of that function.

inversion when a positive (negative) voltage is applied between a conductor and a p-type (n-type) semiconductor separated by a thin dielectric layer,
the majority carrier holes (electrons) are repelled and minority carrier electrons (holes) are trapped at the surface.

inversion layer an equilibrium layer of minority carriers at a semiconductor surface or interface.

inversion population usually the density of atoms or molecules in the higher state of a laser transition minus the density in the lower state.

inversion symmetry the study of how the physical properties of a material change under the hypothetical inversion of the coordinate of each particle through the origin of the coordinate system. *See also* centrosymmetric medium.

inversionless laser a laser that operates without the need for a population inversion. In general, there are two types of inversionless lasers: those that have a hidden inversion in some eigenstate basis and those that do not. Raman and Mollow lasers can be included in the category of hidden inversion. Lasers in the category of no hidden inversion are possible because it is polarization, not population, that produces gain. These lasers generally require at least three quantum mechanical states (for example, lambda and vee systems) and make simultaneous use of Raman-like gain as well as induced transparency of the type associated with coherent population trapping or Fano interference.

Oscillators based on parametric gain, for example optical parametric oscillators or phase conjugate oscillators, are generally not considered inversionless lasers.

inverted magnetron   a radial magnetron in which the anode structure is inside and the cathode structure is outside.

inverted page table   the number of entries in a conventional page table in a virtual memory system is the number of pages in the virtual address space. In order to reduce the size of the page table, some systems use an inverted page table that has only as many entries as there are pages in the physical memory, and a hashing function to map virtual to physical addresses in nearly constant time.

inverter   (1) switching circuit that converts direct current to alternating current. An inverter can be classified as a natural-commutation type or a self-commutation type. The output AC of a natural-commutated inverter is synchronized to the AC line. This type of inverter is a thyristor rectifier in a reversed order. The output AC of a self-commutated inverter is independent, i.e., both the frequency and amplitude may be controlled. Commonly used self-commutated inverters include square-wave inverters and PWM inverters.

(2) a physical or logical gate that changes a signal from high to low, or low to high.

invertibility of generalized 2-D linear system
let

$$T = T(z_1, z_2)$$
$$= C\,[Ez_1, z_2 - A_0 - A_1z_1 - A_2z_2]^{-1}$$
$$\times (B_0 + B_1z_1 + B_2z_2) + D$$

be the transfer matrix of the generalized 2-D model

$$Ex_{i+1,j+1} = A_0x_{ij} + A_1x_{i+1,j} + A_2x_{i,j+1}$$
$$+ B_0u_{ij} + B_1u_{i+1,j} + B_2u_{i,j+1}$$
$$y_{i,j} = Cx_{ij} + Du_{ij}$$

$i, j \in Z_+$ (the set of nonnegative integers) where $x_{ij} \in R^n$ is the local semistate vector, $u_{ij} \in R^m$ is the input vector, $y_{ij} \in R^p$ is the output vector and E, A_k, B_k ($k = 0, 1, 2$) C, D are real matrices with E possibly singular. The model is called (d_1, d_2)-delay invertible if there exists a generalized 2-D model.

$$\bar{E}z_{i+1,j+1} = F_0z_{ij} + F_1z_{i+1,j} + F_2z_{i,j+1}$$
$$+ G_0y_{ij} + G_1y_{i+1,j} + G_2y_{i,j+1}$$
$$u_{ij} = Hz_{ij} + Jy_{ij}$$

with transfer matrix

$$T_L = T_L(z_1, z_2)$$
$$= H[\bar{E}z_1, z_2 - F_0 - F_1z_1 - F_2z_2]^{-1}$$
$$\times (G_0 + G_1z_1 + G_2z_2) + J$$

such that $T_L T = z_1^{-d_1} z_2^{-d_2}$, where $z_{ij} \in R^{n'}$ is the local semistate vector, $\bar{E}, F_k, G_k$ ($k = 0, 1, 2$) H, J real matrices, and d_1, d_2 are nonnegative integers.

invertible system   a system S_1 is invertible if there exists a system S_2 such that, when S_1 and S_2 are placed in series, the larger cascaded system leaves the input unchanged (x in, x out). For example, the multiplier $y(t) = k * x(t)$ is invertible with an inverse system $y(t) = x(t)/k$, while the absolute value function $y(t) = |x(t)|$ is not. *See* inverse system.

ion beam lithography   lithography performed by exposing resist with a focused beam of ions.

ion implantation   a high-energy process, usually greater than 10 keV, that injects ionized species into a semiconductor substrate. Often done for introducing dopants for device fabrication into silicon with boron, phosphorus, or arsenic ions.

ionic transition   coupling of energy levels in an ion by means of absorption or emission processes.

ionized vapor a gaseous vapor in which the atoms are ionized, usually by electron impact or by photoionization.

ionizing radiation the process in which sufficient energy is emitted or absorbed to cause the generation of electron-hole pairs.

ionosphere consists of several layers of the upper atmosphere, ranging in altitude from 100 to 300 km, containing ionized particles. The ionization is due to the effect of solar radiation, particularly ultraviolet and cosmic radiation, on gas particles. The ionosphere serves as reflecting layers for radio (sky) waves.

IPC *See* interprocess communication.

IPP Independent Power Producers, who produce electrical energy but who are not owned by utility companies.

IR *See* infrared.

IRE (1) a unit equal to 1/140 of the peak-to-peak amplitude of a video signal, which is typically 1 V. The 0 IRE point is at blanking level, with the sync tip at −40 IRE and white extending to +100 IRE.

(2) *Institute of Radio Engineers,* an organization preceding the IEEE.

iron core transformer a transformer where the magnetic core is iron or principally iron. Two classes of iron cores are iron alloy and powdered iron. Iron alloy cores are usually restricted to low-frequency applications (2 kHz or less) because of eddy current losses. Powdered iron cores are used at higher frequencies because they consist of small iron particles electrically isolated from each other with significantly greater resistivity than laminated cores, and thus lower eddy current losses.

irradiated fuel nuclear fuel which has been part of a nuclear reaction and is thus highly radioactive.

irreducible polynomial a polynomial $f(x)$ with coefficients in Gallois field $GF(q)$ that can not be factored into a product of lower-degree polynomials with coefficients in $GF(q)$.

irreversible loss a reduction in the magnetization of a permanent magnet that can only be recovered by remagnetizing the material, usually caused by temperature extremes, reversing fields, or mechanical shock.

IRVQ *See* interpolative vector quantization.

IS-54 an interim standard of the U.S.-based Telecommunications Industries Association (TIA) for a digital cellular communication system. One of a set of digital cellular systems commonly classified as second generation cellular systems. A standard originally designed to succeed the AMPS cellular standard. This interim standard has now been finalized in the standard known as IS-136. The standard is based on the use of the same channel structure as the AMPS standard (30 kHz channels), similar frequency reuse and concept of hand-off, but with the use of a digital modulation scheme, known as $\pi/4$-shifted DQPSK, to transmit a digital signal at the rate of approximately 47 kbps on a 30 kHz channel. This signal typically carries three digital voice signals encoded at the rates of 8 kbps, excluding additional overhead bits that are used for the purpose of error correction.

IS-95 an interim standard of the U.S.-based Telecommunications Industries Association (TIA) for digital cellular communication systems. One of a set of systems commonly classified as second-generation cellular systems. The standard is based on spread spectrum modulation employing CDMA. This standard makes use of the concepts of power control, soft hand-offs, pilot signals, pilot phase offsets, orthogonal CDMA, and Walsh functions.

ISDN   *See* integrated services digital network.

ISEF   *See* infinite symmetric exponential filter.

ISI   *See* intersymbol interference.

ISMA   *See* inhibit sense multiple access protocol.

ISO   (1) International Standards Organization.

(2) Independent System Operator, an entity created to operate power generation and transmission systems but which does not own these facilities.

isodata   a special elaboration of K-means used for clustering.

iso-dense print bias   the difference between the dimensions of an isolated line and a dense line (a line inside an array of equal lines and spaces) holding all other parameters constant.

isofocal bias   the difference between the desired resist feature width and the isofocal linewidth.

isofocal linewidth   the resist feature width (for a given mask width) that exhibits the maximum depth of focus.

isokeraunic map   a map which denotes the variation of lightning activity over an area by making use of observations of the incidence of thunder.

isolated   a power electronic circuit that has ohmic isolation between the input source and the load circuit.

isolated buck-boost converter   *See* flyback converter.

isolated I/O   an I/O system that is electrically isolated from where it is linked to. Isolated I/O is often found in embedded systems used for control.

isolation   (1) the separation of a part from other parts of the system so that the effects of undesired changes in the system are not seen by the separated part.

(2) a figure (usually expressed in decibels) to describe how well a transmitting device (source) and receiving device (receiver) are separated electrically; the amount of (usually undesirable) signal appearing at the receiver from the transmitting source from an undesirable signal path such as leakage in a coupler, mutual coupling, or multipath.

isolation switch   *See* bypass switch.

isolation transformer   a transformer, typically with a turns ratio of 1:1, designed to provide galvanic isolation between the input and the output.

isolator   (1) a two-port passive device based on the Faraday effect. Signals passing in one direction suffer minimal loss, however, in the reverse direction they are strongly attenuated.

(2) in optics, a device inserted in an optical fiber that prevents optical power from flowing in the reverse direction from the transmitted power. Optical fiber isolators can be made either dependent or independent of the polarization direction of the optical energy in the fiber.

isotope   a variant of an element whose nucleus contains a greater or lesser number of neutrons but the same number of protons as the base element.

isotropic antenna   a fictitious reference antenna which radiates power equally in all directions. Such an antenna cannot be realized exactly in practice.

isotropic medium   medium in which the constitutive parameters are not a function of the vector direction of the electric or magnetic fields. In such medium, the constitutive parameters are scalar quantities, and there is no magnetoelectric coupling.

ISS   *See* impedance standard substrate.

issue   the act of initiating the performance of an instruction (not its fetch). Issue policies are important designs in systems that use parallelism and execution out of program order to achieve more speed.

ITB   *See* inverse translation buffer.

iterative coordination   coordination process concerned with exchange of information between the coordinator and the local decision (control) units, which — after a number of iterations — leads to the local solutions being satisfactory from the overall (coordinator) point of view; iterative coordination is used in multilevel optimization, in particular in the direct method and the price method.

iterative decoding   decoding technique that uses past estimates to provide additional information to improve future estimates. Used in the decoding of turbo-codes.

iterative selection   a thresholding algorithm based on a simple iteration. First select a simple threshold (say, 128) and segment the image, computing the mean of the black and white pixels along the way. Next, calculate a new threshold using the midpoint between the black and white pixels as just computed. Next, repeat the entire process, using the new threshold to perform the segmentation. The iteration stops when the same threshold is used on two consecutive passes.

IWP   *See* indirect weld parameters.

J

Jacobian inverse control one of the Cartesian-based control schemes. The Jacobian inverse control is presented in the figure. In this scheme, the end-effector location in the operational space is compared with desired quantity and an error ΔX is computed. Assuming that deviation is small ΔX can be transformed into corresponding joint space deviation, Δq, through the inverse of the manipulator analytical Jacobian. The gains correspond to the joint stiffness, which is constant for each joint. *See also* Cartesian-based control.

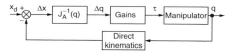

Jacobian inverse control scheme.

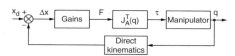

Jacobian transpose control scheme.

Jacobian of the manipulator a matrix that maps the joint velocities into end effector velocities.

Jacobian transpose from a mathematical point of view, the transpose of the Jacobian. Jacobian transpose can be used in inverse kinematics algorithm with operation space error as a feedback.

Jacobian transpose control depicted in the figure. In this case the operational space error, ΔX, is calculated first. Output of the gains block, F,

can be treated as the elastic force generated by a generalized spring. This operational space force is transformed into the joint space generalized forces, τ, through the transpose of the Jacobian. *See also* Cartesian-based control, Jacobian inverse control.

JANTX a prefix denoting that the military specification device has received extra screening and testing, such as a 100% 168-h burn-in.

JANTXV a JANTX part with an added precapsulation visual requirement.

Jensen's inequality for a function f, convex on the support $\mathcal{X}$ and a random variable $X \in \mathcal{X}$,

$$Ef(x) \leq f(EX),$$

provided the expectation exists. Named after its discoverer, Danish mathematician Johan L. W. V. Jensen (1859–1925).

jet a gas turbine peaking generator. *See* peaking unit.

JFET *See* junction field-effect transistor.

JFET oscillator device often used as the active element in any of the LC-oscillators. A special attention deserves using JFET as the active element of the tuned-gate oscillator. In this configuration, the tuned circuit is not loaded by the FET, its loading due to the transistor output impedance is exceptionally small, and the Miller effect is also minimized. This oscillator can be clamp-biased, and the oscillation amplitude may be stabilized with the FET operating completely within its square-law region.

jitter a signal sample is temporarily displaced by an unknown, usually small, interval.

JK flip-flop device that uses two inputs (J and K) to control the state of its Q and Q' outputs.

A negative edge-sensitive clock input samples J and K and changes the outputs accordingly. The simplified truth table for the JK flip flop is as follows:

J	K	Q	Q'
0	0	0	0
0	0	1	1
0	1	0	0
0	1	1	0
1	0	0	1
1	0	1	1
1	1	0	1
1	1	1	0

Johnson noise   *See* thermal noise.

joint   a lower-pair joint connects two links and it has two surfaces sliding over one another while remaining in contact. Six different lower-pair joints are possible: revolute (rotary), prismatic (sliding), cylindrical, spherical, screw, and planar. Of these, only rotary and prismatic joints are common in manipulators. Each has one degree of freedom.

joint detection   *See* multiuser detection.

Joint Electron Device Engineering Council (JEDEC)   a part of the EIA.

joint flexibility   small deformations characterized as being concentrated at the joints of the manipulator. We often refer to this situation by the term "elastic joints" in lieu of flexible joints. As an example, joint flexibility can be a dynamic, time-varying displacement that is introduced between the position of the driving actuator and that of the driven link. This is due to presence of transmission elements such as harmonic drives transmission, belts, and long shafts.

joint offset   the relative displacement between two successive links in a robotic arm.

Joint Photographic Experts Group (JPEG) (1) group that standardizes methods for still picture compression. Baseline JPEG compression uses an 8×8 *discrete cosine transform*, quantization and zigzag ordering of transform coefficients, combined runlength/value coding, and Huffman or Arithmetic lossless coding.

(2) is a still picture compression method which uses the spectral components of an image to perform compression. The JPEG baseline must be present in all JPEG modes of operation which use DCT. *See also* JPEG.

joint source-channel coding   *See* combined source-channel coding.

joint space   a position and orientation of a manipulator of n degrees of freedom is specified as a result of the forward kinematics problem. This set of variables is called as the $n \times 1$ vector. The space of all such joint vectors is referred to as joint space.

joint space control   depicted in the figure. The joint space control scheme consists of the trajectory conversion, which recalculates the desired trajectory from the operational space $(X_d, \dot{X}_d, \ddot{X}_d)$ into the joint space $(q_d, \dot{q}_d, \ddot{q}_d)$. Then, a joint space controller is designed that allows tracking of the reference inputs. Notice that only joint inputs are compared with joint positions, velocities, and accelerations. The joint space control does not influence the operational space variables which are controlled in an open-loop fashion through the manipulator mechanical structure. Joint space control is usually a single-input–single-output system; therefore, manipulator with n degrees of freedom has n such independent systems. A single-input–single-output system has PI (proportional-integral), PD (proportional-derivative), or PID (proportional-integral-derivative) structure. These schemes do not include dynamics of the manipulator. Dynamics of the manipulator can be included in, for example, computed torque control. *See also* computer torque control.

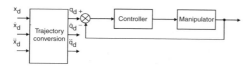

Joint space control scheme.

joint stiffness *See* stiffness of a manipulator arm.

joint time frequency analysis (JTFA) techniques the aim of which is to represent and characterize a signal in time and frequency domain simultaneously by using different kinds of transformation and kernel functions. See harmonic balance technique.

Joint Tactical Information Distribution System (JTIDS) system that uses spread spectrum techniques for secure digital communications; used for military applications.

joint transform correlator a type of optical correlator that employs two parallel paths, one for each signal, instead of an in-line cascade.

joystick an input device in the form of a control lever that transmits its movement in two dimensions to a computer. Joysticks are often used in games for control. They may also have a number of buttons whose state can be read by the computer.

JPEG *See* Joint Photographic Experts Group.

JPEG baseline *See* Joint Photographic Experts Group baseline.

JPEG DCT there are two DCT modes of JPEG, the sequential DCT mode and the progressive DCT-based mode. In the sequential mode the image components are coded individually or in blocks by a single scan in one pass. In the progressive mode several scans are taken to code parts of the quantized DCT coefficients.

JTIDS *See* Joint Tactical Information Distribution System.

JTFA *See* joint time frequency analysis.

jump instruction an instruction that causes an unconditional transfer of control to a different instruction sequence in memory.

jump linear quadratic problem optimal control problem in which a controlled process is modeled by linear system with Markov jumps and a control objective is to minimize an average quadratic criterion given by

$$J = E\left\{ \int_0^T [x'(t)Q(\xi(t))x(t) \right.$$
$$\left. + u'(t)R(\xi(t))u(t)]dt \,|\, x(0), \xi(0) \right\}$$

where $x(t), u(t), \xi(t)$ denote, respectively, process state, process control, and mode, and matrices R, Q are real valued, symmetric respectively positive definite and positive semidefinite weighting matrices of respective dimensions T is finite or infinite control horizon and E is an averaging operator. In the finite time case the optimal control law is given as

$$u(t) = p(x(t), \xi(t), t) = -P(\xi(t), t)x(t)$$

for each $\xi(t)$ taking value in finite set $\mathbf{S} = \{1, 2, \ldots, s\}$ where

$$P(\xi, t) = R(\xi)^{-1} B'(\xi) K(\xi, t) x(t)$$

where for each $\xi(t) = i$ matrices $K(i, t)$ are the unique positive semidefinite solutions of differential coupled Riccati equations. Assuming that the system is stochastically stabilizable and for each $\xi(t) = i$ the pairs $(A(i), \sqrt{Q(i)})$ are observable, then for infinite control interval $T \to \infty$ the solution of the jump linear quadratic (JLQ) problem is given by the optimal steady state control:

$$u(t) = \hat{p}(x(t), \xi(t))$$
$$= -\hat{P}(\xi(t))x(t)$$
$$= -R(\xi)^{-1}B'(\xi)\hat{K}(\xi)x(t)$$

where for each $\xi(t) = i$ matrices $\hat{K}(i)$ are the unique positive semidefinite solutions of algebraic coupled Riccati equations. This solution may be also found as the steady-state value of $K(i, t)$. *See also* linear system with Markov jumps.

jumper   a plug or wire used for setting the configuration of system. It can be used for changing the hardware configuration by forcing some line to be high or low. It is used to change software configuration (especially on embedded systems) when the status of the jumper is read by the microprocessor.

jumping jack   a voltmeter.

junction capacitance   change in charge of immobile ions in the depletion region of a diode corresponding to a change in reverse bias voltage on a diode.

junction field-effect transistor (JFET)   a type of FET in which the high input resistance at the gate is achieved by use of a reverse biased p-n junction between the gate and the drain-source channel.

junction-to-case thermal resistance   the proportionality constant at thermal equilibrium between the temperature difference of the bipolar device junction ($T_{junction}$) and a specified case surface (T_{case}) to the dissipated power in the junction (p_w), in units of $C°/W$. The specified surface is usually the most effective in reducing the temperature. It includes the thermal resistance of the chip, die attach material (solder or adhesive), packaging and mounting medium, as applicable.

$$\theta_{jc} = \left. \frac{T_{junction} - T_{case}}{P_w} \right|_{equilibrium}$$

K

K   (1) symbol for the Linville stability factor, a dimensionless quantity.

(2) abbreviation for 1024 (not for 1000).

K-factor rating   an indication of a transformer's capability to provide a specified amount of harmonic content in its load current without overheating. K-factor rated transformers typically use thinner laminations in the core to reduce the eddy current losses, larger conductors to reduce skin effect, and for three-phase transformers a larger neutral conductor to conduct zero-sequence harmonic currents.

K-means algorithm   a clustering algorithm known from the statistical literature (E. W. Forgy, 1965), that relies on the same principles as the Lloyd algorithm and the generalized Lloyd algorithm.

Given a set of patterns and K, the number of desired clusters, K-means returns the centroid of the K clusters. In a Pascal-like code, the algorithm runs as follows:

```
function K-Means(Patterns:set-of-pattern):
                Centroids;
begin
    repeat
        for h = 1 to N do
            Assign point h to class k for which the distance
            (x_h − μ_k)² is a minimum
        for k = 1 to K do
            Compute μ[k] (average of the assigned points)
        until no further change in the assignment
end
```

The algorithm requires an initialization that consists of providing K initial points, which can be chosen in many different ways. Each of the N given points is assigned to one of the K clusters according to the Euclidean minimum distance criterion. The average in the clusters is computed and the algorithm runs until no further reassignment of the points to different clusters occurs.

K-nearest neighbor algorithm   a method of classifying samples in which a sample is assigned to the class which is most represented among the k nearest neighbors; an extension of the nearest neighbor algorithm.

Kaczmarz's algorithm   the recursive least-squares algorithm has two sets of state variables, parameters vector and covariance matrix, which must be updated at each step. For large dimension, the updating of the covariance matrix dominates the computing effort. Kaczmarz's projection algorithm is one simple solution that avoids updating the matrix at the cost of slower convergence. The updating formula of the least squares algorithm using the Kaczmarz's algorithm has the form

$$\hat{\theta}(t) = \hat{\theta}(t-1) + \frac{\phi(t)}{\phi^T(t)\phi(t)}(y(t) - \phi^T(t)\hat{\theta}(t-1))$$

where $\hat{\theta}$ is parameter's estimates vector, ϕ is regressor vector. This approach is also called the normalized projection algorithm.

Kaiser window   a function defined by

$$w[n] = \begin{cases} \dfrac{I_0\left[\beta(1-[(n-\alpha)/\alpha]^2)^{\frac{1}{2}}\right]}{I_0(\beta)} & 0 \leq n \leq M \\ 0 & \text{otherwise} \end{cases}$$

where M is the length of the filter minus one, $\alpha = \frac{M}{2}$ and β is a term that depends on the desired ripple level of the filter.

Kalman decomposition of linear systems   one of several decompositions for linear systems.

Every linear, stationary finite-dimensional continuous-time dynamical system can be decomposed into four subsystems: controllable and observable, controllable and unobservable, uncontrollable and observable, uncontrollable and unobservable. This decomposition does not depend on any nonsingular transformations in the state space R^n. The transfer function matrix always describes only the controllable and observable part of the dynamical system and does not depend on any nonsingular transformations in the state space R^n. Similar statements hold true for linear stationary finite-dimensional discrete-time dynamical systems.

Kalman filter the method of recursively estimating the state vector of a linear dynamic system based on noisy output measurements. This method combines the knowledge of system parameters, statistical characteristics of initial state, process and measurement noises, and the current measurement to update the existing estimate of the state. Kalman filter is the minimum variance filter if the initial state and the noises are Gaussian distributed; otherwise, it is the linear minimum variance filter. In addition to providing the expected value of the state conditional on the measurement, this filter generates the estimation error covariance, which signifies the uncertainty associated with this estimate. This covariance can be calculated off-line before running the filter.

Kanerva memory a sparse distributed memory developed for the storage of high-dimensional pattern vectors. Memory addresses are randomly generated patterns, and a given pattern is stored at the address that is closest in terms of Hamming distance. Patterns are usually bipolar, and those stored at the same address are added together. Suitable thresholding of this sum allows a close approximation to individual patterns to be recalled, so long as the memory is not overloaded.

Karhunen–Loeve transform (KLT) an optimal image transform in an energy-packing sense.

If the images are transformed by $y = Ax$, then the corresponding quadratic term associated with the covariance matrix becomes $x'A'RAx$. Hence the image uncorrelation is reduced to finding A such that

$$A'RA = \Lambda = \mathbf{diag}(\lambda_1, \dots, \lambda_n).$$

This equation can be solved by finding in the following three steps.

(1) Solve equation $|\mathbf{R} - \lambda\mathbf{I}| = 0$ (find the $\mathbf{R}$'s eigenvalues).

(2) Determine the n solutions of $(\mathbf{R} - \lambda_i\mathbf{I})\mathbf{a_i} = 0$ (eigenvectors).

(3) Create the desired transformation $\mathbf{A}$ as follows:

$$\mathbf{A} = [\mathbf{a_1}, \dots, \mathbf{a_n}].$$

In this transform a limited number of transform coefficients are retained. These coefficients contain a larger fraction of the total energy in the image. This transform is heavily dependent on the image features and requires a covariance function estimate for performing the transform. *See also* discrete cosine transform.

Karnaugh map a mapping of a truth table into a rectangular array of cells in which the nearest neighbors of any cell differ from that cell by exactly one binary input variable.

KCL *See* Kirchoff's current law.

kcmil *See* circular mil.

KDP *See* potassium dihydrogen phosphate.

kernel (1) the set of basis functions for a transformation.

(2) the set of convolution weights for filtering.

(3) the portion of the operating system that operates in privileged mode.

Kerr coefficients describes the quadratic electro-optic effect, particularly for gases and liquids. *See* Kerr effect.

Kerr effect the quadratic electro-optic effect, particularly in the case for gases and liquids.

key (1) in a table, the value used to select the desired entry (or entries).

(2) in an access control system, a value held by a process to permit it to make access to certain objects within the system.

key point detection   a technique usually employed in specialized linear or morphological filters designed for measuring gray-level changes in several directions. A key point is an isolated image point corresponding to a peculiar physical or geometrical phenomenon in the scene from which the image arises; it can be for example a corner, a line termination, a junction, a bright or dark spot, etc. Key points are distinguished from edges by two properties: they are sparse and display strong gray-level variations in two or more directions (while edge points are grouped into lines, and have gray-level variations essentially in the direction normal to that line). *See* edge, detection, salient feature.

key punch   a device with a keyboard used for storing data in paper cards or paper tapes by punching holes. The pattern of holes punched across these cards or tapes represent the data keyed in. Now obsolete.

keyboard   an input device with a set of buttons (or keys) through which characters are input to a computer. In addition to the keys for inputing characters, a keyboard may also have function keys and special keys, such as power-on or print-screen.

keyboard controller   the device controller that processes keyboard input. Because of its importance, I/O from the keyboard controller is often handled differently from other I/O processes, with its own direct connection to the CPU.

keystone distortion   a distortion that presents an image in the shape of a trapezoidal. For projection television displays, off-axis projections of the red and blue tubes can cause keystone distortion. For direct view television displays, unequal deflection sensitivities for the two sections of the deflection yoke can result in keystone distortion. A horizontal trapezoid is the result of the vertical

yoke deflection sensitivities. Similarly, a vertical trapezoid is the result of the horizontal yoke deflection sensitivities.

KGD   *See* known good die.

Kharitonov theorem   the most popular necessary and sufficient condition of robust stability for characteristic interval polynomials. It states that robust stability of uncertain linear time-invariant system with characteristic interval polynomial $p(s, \mathbf{a})$ is stable if and only if the following four polynomials have all roots endowed with negative parts:

$$p_1(s) = a_0^+ + a_1^- s + a_2^- s^2 + a_3^+ s^3 + a_4^+ s^4 + ...$$

$$p_2(s) = a_0^+ + a_1^+ s + a_2^- s^2 + a_3^- s^3 + a_4^+ s^4 + ...$$

$$p_3(s) = a_0^- + a_1^+ s + a_2^+ s^2 + a_3^- s^3 + a_4^- s^4 + ...$$

$$p_4(s) = a_0^- + a_1^- s + a_2^+ s^2 + a_3^+ s^3 + a_4^- s^4 + ...$$

where a_i^-, a_i^+ are bounds for coefficients of the characteristic interval polynomial. The polynomials are called Kharitonov polynomials, and the required property could be checked by four Routh–Hurwitz tests. Some generalizations of this result for dependent coefficient perturbations and more general zero location regions are known.

Kilby, Jack St. Clair   (1923–) Born: Jefferson City, Missouri

Best known as the person who first suggested and then implemented the concept of integrating transistors, with capacitors and resistors, on a single silicon wafer. This concept, simple-sounding today, formed the basis for the first early integrated circuits. Kilby's employer, Texas Instruments, launched a patent suit against Fairchild Industries to establish their and Kilby's claim to the technology. The court ruled in favor of Robert Noyce and Fairchild. Despite this, Kilby is universally recognized for his pioneering work, especially in the practical implementation of IC technology.

kilovar *See* var.

kilovolt-ampere (KVA) a measure of apparent power, often in the rating of a piece of equipment or the measure of an electrical load, which is obtained by multiplying the device voltage in kilovolts by the current in amperes.

kilowatt-hour (KWH) a measure of electrical energy: 1000 watts delivered for one hour.

kinematic calibration a procedure to finding accurate estimates of Denavit–Hartenberg parameters from a series of measurements of the manipulator's end-effector locations. *See also* Denavit-Hartenberg notation.

kinematic singularity a point for which the geometric Jacobian is not invertible. In other words, those configurations at which Jacobian is rank-deficient are termed kinematic singularities. As a consequence, in singular configurations infinite solutions to the inverse kinematics problem may exist. From practical point of view small velocities in the operation space will cause large velocities in the joint space.

kinematically simple manipulator a manipulator where some or all link offset angles equal 0, $-90°$, or $90°$ and link distances and offsets equal zero. Kinematically simple manipulators are found in most of the industrial robots. For these robots, forward kinematics problems can be solved in a closed form solution and dynamics equations are the simplest. *See also* Denavit-Hartenberg notation.

kinetic energy energy of motion.

kinetic energy conservation principle any change of kinetic energy of an electron is transformed into a change in electric potential energy with which the electron is interacting, and vice versa.

kineto-static duality the static relationship combined with the differential kinematics equation defines a kineto-statics duality. For the description of differential kinematics. *See* differential kinematics. The static relationship describes the end point force and equivalent joint generalized force balance assuming that we neglect gravity and friction at the joints. This relationship is described by the transpose of the manipulator Jacobian and has the form $\tau = J^T F$, where F is a vector of exerted end point forces and τ is a vector of joint generalized forces. As a consequence of the last definition, the following properties can be defined.

(1) The range of J^T is the subspace $R(J^T)$ in R^n of the joint forces and/or torques that can balance the end-effector forces, in the given manipulator posture.

(2) The null of J^T is the subspace $N(J^T)$ in R^T of the end-effector forces and/or torques that do not require any balancing joint generalized forces, in the given manipulator posture.

Kirchhoff, Gustav Robert (1824–1887) Born: Konigsberg, Germany

Best known for discovering the laws that govern electric flow in networks, known now as Kirchhoff's laws. Kirchhoff is also famous as the inventor of spectroscopy, and his theoretical work led to the quantum theory of matter. Kirchhoff first held a teaching post in Berlin, followed by appointments in Breslau and Heidelberg. In the latter two universities, Kirchhoff had very fruitful collaborations with Robert Bunsen. Kirchhoff's studies in electricity provided a foundation for Maxwell and Lorenz's description of electromagnetic theory.

Kirchoff's current law (KCL) a fundamental law of electricity that states that the sum of the currents entering and exiting a circuit node must be equal to 0.

Kirchoff's laws laws that govern the relationships between voltages/currents in a circuit/

network. *See also* Kirchoff's voltage law, Kirchoff's current law.

Kirchoff's voltage law (KVL)   a fundamental law of electricity that states that the sum of the voltage drops and rises in a closed loop must equal 0.

Kleinman symmetry   *See* symmetries of nonlinear susceptibility.

KLT   *See* Karhunen–Loeve transform.

klydonograph   a measuring instruments for high-voltage impulses which makes use of Lichtenburg figures impressed on photographic film.

klystrode   an amplifier device for UHF-TV signals that combines aspects of a tetrode (grid modulation) with a klystron (velocity modulation of an electron beam). The result is a more efficient, less expensive device for many applications. (Klystrode is a trademark of EIMAC, a division of Varian Associates.)

knife-edge diffraction   classical diffraction model defined by the interaction of a propagating electromagnetic wave with a perfectly conducting, infinitely thin obstructing boundary. Knife-edge diffraction is often used in terrain propagation calculations as an approximation for hills and ridges.

knob-and-tube wiring   a form of residential wiring, now obsolete, in which lightly-insulated wires are supported on porcelain insulators (knobs) or porcelain bushings (tubes) through joists or studs.

knowledge engineering   the process of developing an expert system.

Knowlton's technique   a pyramidal type of hierarchical approach, in which a reversible transformation takes adjacent pairs of k-bit pixel values

and maps it into a k-bit composite value and a k-bit differentiator.

known good die (KGD)   bare silicon chips (die) tested to some known level.

Kogelnik transformation   *See* ABCD law.

Köhler illumination   a method of illuminating the mask in a projection imaging system whereby a condenser lens forms an image of the illumination source at the entrance pupil (entrance aperture) of the objective lens, and the mask is at the exit pupil of the condenser lens.

Kohonen network   a 2-dimensional array of neurons with each neuron connected to the full set of network inputs. Learning is unsupervised: the neuron whose vector of input weights is closest to an input vector adjusts its weights so as to move closer to that input vector; the neuron's neighbors in the array adjust their input weights similarly. As a result, clusters of weights in different parts of the array tend to respond to different types of input.

Kolmogorov complexity   the minimum length description of a binary string that would enable a universal computer to reconstruct the string.

Kraft inequality   a theorem from information theory that sets a restriction on instantaneous codes (codes where no codeword is a prefix of any other codeword, i.e., a code containing 0 and 01 is not an instantaneous code). The Kraft inequality states that for an instantaneous code over an alphabet with size D and codeword lengths $l_1, l_2, l_3, \ldots, l_m$ the following must be true:

$$\sum_{i=1}^{m} D^{-li} < 1$$

Kramer drive   an electric drive system in which the output of the driven frequency converter is fed to the slip rings of the wound rotor. Unlike its predecessor Leblanc system, the variable transformer is connected to the wound rotor slip rings instead of to the line.

Kramer's generalization   a sampling theory based on other than Fourier transforms and frequency.

Kramers–Kronig relations   relates the real and imaginary components of the index of refraction of a medium. *See also* Hilbert transform.

Kronecker delta function   the discrete-time unit impulse function defined as $\delta_{ij} = 1$ for $i = j$, $\delta_{ij} = 0$ for $i \neq j$. *See also* delta function.

ku-band   frequency band of approximately 11–12 GHz.

Kubo formula   a fundamental relationship, developed by Ryogo Kubo, between, e.g., the conductance and the current fluctuations. The Kubo formula is an example of the fluctuation–dissipation theorem in statistical physics, in which the average value of the current (conductance) is determined by an integral over the correlation function of the current fluctuations.

Kullback–Liebler distance   *See* relative entropy.

Kuroda's identities   four identities are used to achieve practical microwave filter implementation by using redundant transmission line sections.

kVA   *See* kilovolt-ampere.

Kvhler illumination   a method of illuminating the mask in a projection imaging system whereby a condenser lens forms an image of the illumination source at the entrance pupil (entrance aperture) of the objective lens, and the mask is at the exit pupil of the condenser lens.

KVL   *See* Kirchoff's voltage law.

KWH   *See* kilowatt-hour

L

L-band   frequency band of approximately 1–2 GHz.

L-L   *See* line to line fault.

label   a tag in a programming language (usually assembly language, also legal in C) that marks an instruction or statement as a possible target for a jump or branch.

labeling   a technique by which each pixel within a distinct segment is marked as belonging to that segment. One way to label an image involves appending to each pixel of an image the label number or index of its segment. Another way is to specify the closed contour of each segment and to use a contour filling technique to label each pixel within a contour.

ladder diagram   (1) the connection of the coils and contacts used in a control circuit shown one line after the other that looks like a ladder.
(2) a visual language for specifying the Boolean expressions, which are the core of the control law of PLC.

laddertron   a microwave vacuum tube oscillator with a slow-wave structure coupled to a single-cavity resonator.

lag   the inability of an imaging tube to respond to instantaneous changes in light. For measurement purposes, lag has two components: rise lag is the response time from dark to light, whereas decay lag is the response time from light to dark. Lag is a very short-term effect, and should not be confused with image retention, image burn, or sticking.

lag circuit   a simple passive electronic circuit designed to add a dominant pole to compensate the performance of a given system. A lag circuit is generally used to make a system more stable by reducing its high-frequency gain and/or to improve its position, velocity, or acceleration error by increasing the low frequency gain. A nondominant zero is included in the lag circuit to prevent undue destabilization of the compensated system by the additional pole.

lag network   a network where the phase angle associated with the input–output transfer function is always negative, or lagging.

lag-lead network   the phase shift versus frequency curve in a phase lag-lead network is negative, or lagging, for low frequencies and positive, or leading, for high frequencies. The phase angle associated with the input–output transfer function is always positive or leading.

Lagrange formulation   a formulation where the equations of motion are derived in a systematic way by choosing a set of generalized coordinates, forming the Lagrangian of the mechanical system (as a difference of total kinetic energy and potential energy of the system) and by solving the Lagrange equations

$$\frac{d}{dt}\frac{\partial \mathcal{L}}{\partial \dot{q}_i} - \frac{\partial \mathcal{L}}{\partial q_i} = \tau_i \quad i = 1, \dots, n$$

where $\mathcal{L}$ stands for Lagrangian, q_i is the generalized coordinate, $\dot{q}_i$ is its derivative, and τ_i is a generalized force, and n denotes number of degrees of freedom of the mechanical system. Last equations establish the relations existing between the generalized forces applied to the manipulator and the joint positions, velocities, and accelerations in so called closed form. *See also* Newton–Euler recursive algorithm.

Lagrange stable state   *See* bounded state.

Lagrangian interpolation   a classic interpolation procedure used in numerical analysis. The sampling theorem is a special case.

Laguerre polynomial   a solution to the differential equation $xy'' + (1-x)y' + ny = 0$. Laguerre polynomials $L_0(x) = 1$, $L_1(x) = 1-x$, $L_2(x) = 1 - 2x + x^2/2$, and $L_3 = 1 - 3x + 3x^2/2 - x^3/6$. Additional Laguerre polynomials may be obtained from the recursion formula $(n + 1)L_{n+1}(x) - (2n + 1 - x)L_n(x) + L_{n-1}(x) = 0$.

Laguerre–Gaussian beam   electromagnetic beam solution of the paraxial wave equation in which the field is a product of a Laguerre polynomial and a Gaussian function of distance from the beam axis.

Lamb dip   decrease in output power of a Doppler-broadened standing-wave laser oscillator as a function of length tuning when the resonant frequency is within approximately one homogeneous linewidth of gain center; results from the interaction of both the right and left travelling waves with the same atoms for line-center tuning.

lambda system   a 3-level system in which the lowest two energy states are coupled by electromagnetic fields to a common intermediate state of higher energy. This system is so named because schematic representations of it often look like the capital Greek letter lambda, Λ.

Lambertian source   a source whose directional emission pattern follows Lambert's law; a cosine variation.

Lambertian surface   a surface with perfect diffusion properties, i.e., for which the reflectance function depends only on the angle of incidence of illumination.

Lambert's cosine law   a law stating that, for a ideal matte (Lambertian) surface, the apparent brightness of the surface is proportional to the cosine of the angle of incidence, and independent of both the angle of reflection and the phase angle between the incident and reflected beams.

laminate multi-chip module (MCM-L)   a multi-chip module built using advanced PCB manufacturing techniques.

lamination   a thin sheet of metal used to build up the core of an electromagnetic device. Laminations are insulated from each other to reduce the losses associated with eddy currents.

land pattern   a combination of lands intended for the mounting, interconnection, and testing of a particular component.

Landauer formula   describes the conductance as a fundamental property of wave (electron) transmission through a structure.

Lange coupler   four coupled lines used with interconnections to provide tight coupling. A practical implementation to increase the coupling between edge-coupled lines by using several lines parallel to each other, so that the fringing fields at edges of the line contribute to coupling.

Langevin, Paul   (1872–1946) Born: Paris, France

Best known as the developer of echolocation, which is the precursor to modern sonar. Langevin was the first to describe paramagnetism and diamagnetism.

LAOS   *See* light-amplifying optical switch.

lap winding   an armature winding on a DC machine in which the two ends of each coil are connected to adjacent bars on the commutator ring. The lap winding provides "P" parallel paths through the armature winding, where P is the number of poles in the machine.

Laplace transform   the transform of a function $f(t)$ given by

$$F(s) = \int_{-\infty}^{\infty} f(t)e^{-st} dt$$

where $s = a + j\omega$ is a complex variable. The one sided or unilateral Laplace transform is given by the same equation except that the lower limit is 0 and not $-\infty$. The region of convergence of the Laplace integral is a vertical strip R in the s-plane. The inverse Laplace transform is given by

$$f(t) = 1/(2\pi j) \int_L F(s)e^{st} ds$$

where L is a vertical line in R. The Fourier Transform of $f(t)$ is given by $F(j\omega)$.

Laplacian pyramid a set of Laplacian images at multiple scales used in pyramid coding. An input image G_1 is Gaussian lowpass filtered and downsampled to form G_2. Typically G_2 is one quarter the size of G_1, i.e. it is downsampled by a factor of 2 in each direction. G_2 is upsampled and Gaussian lowpass filtered to form R_1 which is then subtracted from G_1 to give L_1. The process then repeats using G_2 as input. The sets of multiresolution images so generated are called "pyramids": $G_1 \ldots G_n$ form a Gaussian pyramid; $L_1 \ldots L_n$ form a Laplacian pyramid.

Laplace's equation   a partial differential equation mathematically described by $\nabla^2 \phi = 0$, where ∇^2 is the Laplacian and ϕ is the equation's solution.

Laplace, Pierre-Simon, Marquis de (1749–1827) Born: Beaumont-en-Auge, Normandy, France

Best known for his development of basic tools of mathematical analysis including the Laplace transform, the Laplace theorem, and the Laplace coefficients. Laplace studied in Paris with the great mathematician Jean d'Alembert. Laplace was heavily involved in politics throughout his career and held many government posts. Laplace's theoretical work was heavily in the field of celestial mechanics. He helped to establish the mathematical basis for the field and in doing so confirmed significant parts of Newton's work.

Laplacian   the second-order operator, defined in $\mathcal{R}^n$ as $\nabla^2 = \partial^2/\partial x_1^2 + \cdots + \partial^2/\partial_n^2$. The zero crossings of an image to which the Laplacian operator has been applied usually correspond to edges, as in such points a peak (trough) of the first derivative components can be found.

large cell   cell with the radius of 5–35 km (such as those found in Groupe Special Mobile systems). *See also* cell.

large disturbance   a disturbance for which the equation for dynamic operation cannot be linearized for analysis.

large-scale integration (LSI)   (1) term usually used to describe the level of integration at which entire integrated circuits can be placed on a single chip.

(2) an integrated circuit made of hundreds to thousands of transistors.

large-scale process (system)   partitioned complex process (system) composed of several subprocesses (subsystems) that are either physically interconnected or must be considered jointly due to the nature of the control objectives.

lapped orthogonal transform (LOT)   a critically sampled block transform, where the blocks overlap, typically by half a block. Equivalently the LOT is a critically sampled filter bank, where typically the filter lengths are equal to twice the number of channels or filters. The LOT was motivated by reducing the blocking effect in transform coding by using overlapping blocks. A cosine modulated filter bank is a type of LOT.

LASCR   *See* light-activated silicon controlled rectifier.

laser acronym that stands for light amplification by stimulated emission of radiation. Usually refers to an oscillator rather than an amplifier; commonly also refers to similar systems that operate at non-optical frequencies or with nonelectromagnetic wave fields.

laser amplifier usually refers to a medium that amplifies light by the process of stimulated emission; sometimes refers to amplification of some other field (nonoptical electromagnetic, phonon, exciton, neutrino, etc.) or some other process (nonlinear optics, Brillouin scattering, Raman scattering, etc.).

laser array systematic distribution of lasers intended to provide more power than a single laser.

laser beam localized electromagnetic field distribution produced by a laser.

laser efficiency output power from a laser divided by the input power (sometimes the pump power into the laser medium and sometimes the wall-plug power).

laser medium the material in a laser that emits light; it may be a gas, solid, or liquid.

laser oscillator oscillator usually producing an optical frequency output and usually based on amplification by stimulated emission in a resonant cavity.

laser pumping mechanism for obtaining a population inversion in a laser medium; the use of a laser beam to pump another laser.

laser threshold the condition under which the round-trip gain in a laser is equal to the round-trip loss.

laser transient a time-dependent laser behavior such as mode-locking, loss switching, spontaneous pulsations, relaxation oscillations.

laser transition transition in a medium that has the capability of exhibiting more stimulated emission than absorption or spontaneous emission.

last-in-first-out (LIFO) *See* first-in-last-out.

latch a small temporary holding cell for a value, the value on the input wires is buffered upon occurrence of some event, such as a clock pulse or rising edge of a separate latch signal.

latency (1) total time taken for a bit to pass through the network from origin to destination.
(2) the time between positioning a read/write head over a track of data and when the beginning of the track of data passes under the head.

lateral (1) a lateral on a primary distribution line is a short tap from the main distribution line which serves a local set of loads. Single phase laterals are common in residential districts.
(2) a three-phase or single-phase power line which supplies the distribution transformers along a street. *See* feeder.

lateral inhibition in the human visual system, the inhibitory effect between nearby cells which acts to enhance changes (temporal or spatial) in the stimulus.

lateral superlattice refers to a lithographically defined structure in which a periodic (superlattice) potential is induced onto the surface of a normal semiconductor (or metallic) system. Since the periodic potential is induced in the lateral variations of the surface, it is called a lateral superlattice.

lateral wave wave generated by a beam of bounded extent incident at an angle close to the critical angle. It manifests by producing a lateral shift of the bounded reflected wave.

lattice constant the length of the sides of the three dimensional unit cell in a crystal.

lattice structure   a filter used in linear prediction that has two outputs, the forward prediction
error $f_m[n]$ and the backward prediction error
$g_m[n]$. These two error signals are defined recursively:

$$f_m[n] = f_{m-1}[n] + K_m g_{m-1}[n-1]$$
$$g_m[n] = K_m^* f_{m-1}[n] + g_{m-1}[n-1]$$

where the K_m are the called the lattice coefficients
of the predictor.

lattice vector quantization   a structured vector quantizer where the reproduction vectors are
chosen from a highly regular geometrical structure known as a "lattice." The method is employed
mainly because of the reduction in storage capacity obtained (compared to optimal vector quantization).

lattice VQ   *See* lattice vector quantization.

law of excluded middle   a logical law stating
that for Boolean variable X, X must be either Z
or not Z.

law of the first wavefront   *See* Haas effect.

layout   specifies the position and dimension of
the different layers of materials as they would be
laid on the silicon wafer.

LC-oscillator   a type of oscillator that is practical at the frequencies above 50 kHz and up
to 500 MHz, where it is rational to use high-Q
tuned circuits for frequency selection (as a result,
sometimes they are called tuned-circuit oscillators). It is impossible to classify all circuits of
LC-oscillators, yet a wide group of LC-oscillators
are reduced to the circuit that includes one active
device (a FET characterized by transconductance
g_m and drain-source impedance r_o is shown as an
example; it can be substituted by a bipolar transistor; this essentially does not change the results),
a load resistor, and three reactances (if X_1 and

X_2 are coils, the mutual reactance X_m should be
considered as well).

The frequency of oscillation is given by the
condition

$$X_1 + X_2 + X_3 + 2X_m = 0$$

and the threshold condition for self-starting is

$$g_m \frac{r_o R_L}{r_o + R_L} \geq \frac{X_2 + X_m}{X_1 + X_m}$$

If X_1 is a tuned coil in the gate, X_2 is the drain
coil, $2X_m$ is the mutual inductance between the
two coils, the circuit is called tuned-gate oscillator. If the single tuned circuit is moved into the
drain, the circuit is called tuned-drain oscillator.
Other frequently used configurations are Colpitts
oscillator and Hartley oscillator.

LCD   *See* liquid crystal display.

LCI   *See* load-commutated inverter.

LCLV   *See* liquid crystal light valve.

LDD   *See* lightly doped drain.

lead   a conductive path, usually self-supporting;
the portion of an electrical component that connects it to outside circuitry.

lead-acid battery   a battery that consists of a
metallic lead anode and a lead oxide cathode held
in a sulfuric acid and water electrolye. Lead−acid
batteries have been used since the mid 1800s and
are still used as car batteries. *See also* nickel-
cadmium battery, nickel-metal hydride battery,
lithium battery.

lead circuit   a simple passive electronic circuit
designed to add a dominant zero to compensate
the performance of a given system. It is generally
used to make an oscillatory system more damped.
A nondominant pole must be included to make the
circuit causal. This pole also limits the high frequency gain of the lead circuit, thereby avoiding

excessive amplification and transmission of undesirable noise.

lead frame   the metallic portion of the device package that makes electrical connections from the die to other circuitry.

lead lanthanum zirconium titanate (PLZT)   a quadratic electro-optic material where the refractive index changes quadratically with applied electric field. Commonly available as a hot-pressed polycrystalline ceramic, although single-crystal film is being developed.

leader   an elongated region of ionized gas that extends from one electrode to another just before a high-voltage breakdown.

leader-follower game   *See* Stackelberg equilibrium.

leading-edge triggered   pertaining to a device that is activated by the leading edge of a signal.

leakage   the flux in a magnetic circuit that does not do any useful work.

leakage flux   the flux that does not link all the turns of a winding or in coupled circuits, flux that links one winding but not another. For example, the magnetic flux produced by the primary winding of a transformer that is not coupled to the secondary winding.

leakage reactance   the amount of inductive reactance associated with leakage flux. The leakage flux is the flux which traverses in paths farther from the designated paths such as the magnetic core in transformers and the air gap in electric machines and constitutes the non-useful flux. The electric circuit symbol of leakage reactance is X_l. It is a function of the leakage inductance and the frequency of operation. Higher values of leakage reactance affect the regulation and efficiency of the system. X_l is expressed in ohms.

leaky feeder   an antenna consisting of a cable that continuously radiates a signal from all points along the entire length of the cable. Such a cable is typically used as the radiating antenna in places such as tunnels and mines where the range of radio propagation is limited. *Compare with* distributed antenna.

leaky modes   *See* tunneling modes.

leaky wave   a wave that radiates signal power out of an imperfectly closed and shielded system. Open waveguides are able, under certain circumstances, to produce leaky waves, which can be used for radiation purposes. In this case, a leaky antenna couples power in small increments per unit length, either continuously or discretely, from a bound mode of the open waveguide.

learning   in neural networks, the collection of learning rules or laws associated with each processing element. Each learning law is responsible for adapting the input–output behavior of the processing element transfer function over a period of time in response to the input signals that influence the processing element. This adaptation is usually obtained by modification of the values of variables (weights) stored in the processing element's local memory.

Sometimes neural network adaptation and learning can take place by creating or destroying the connections between processing elements. Learning may be also achieved by replacing the transfer function of a processing element by a new one.

learning law   *See* learning rule.

learning rule   in neural networks, an equation that modifies the connection weights in response to input, current state of the processing element, and possible desired output of the processing element.

learning vector quantization (LVQ)   a supervised learning algorithm first proposed by

Kohonen that uses class information to move the Voronoi vectors slightly, so as to improve the quality of the classifier decision regions.

The training algorithm for LVQ is similar to its unsupervised counterpart with supervised error correction. After the winner is found, the weights of the winner and its neighbors will be updated according to the following rules:

If the class is correct,

$$w_i(t+1) = w_i(t) + \alpha \cdot (x_i(t) - w_i(t)),$$

otherwise

$$w_i(t+1) = w_i(t) - \alpha \cdot (x_i(t) - w_i(t)),$$

See also self-organizing system, self-organizing algorithm.

least mean square (LMS) algorithm   in some cases of parameter estimation for stochastic dynamic systems, it is not feasible to use the least squares based algorithms due to the computational effort involved in updating and storing the $P(t)$ (probability) matrix. This is especially so when the number of parameters is large. In this case, it is possible to use a variant of the stochastic gradient algorithm — the least mean square (LMS) algorithm, which has the form

$$\hat{\theta}(t+1) = \hat{\theta}(t) + \phi(t)(y(t+1) - \phi^T(t+1)\hat{\theta}(t))$$

where $\hat{\theta}$ is a parameter's estimates vector, ϕ is a regressor vector.

least recently used (LRU) algorithm   a replacement algorithm based on program locality by which the choice of an object (usually a page) to be removed is based on the longest time since last use. The policy requires bookkeeping of essential information regarding the sequence of accesses, which may be kept as LRU bits or as an LRU stack.

least squares   an approach to determining the optimal set of free parameters $\vec{w}$ of an input-output mapping $\vec{y} = \vec{F}(\vec{x}, \vec{w})$, whereby the square of the difference between the output of the function $\vec{y}$ and the desired output $\vec{d}$ is minimized.

least squares algorithm   an adaptive algorithm that adjusts the weights of a digital filter to minimize a least squares (LS) cost criterion. For equalization, this cost criterion is

$$\sum_{k=1}^{i} w^{i-k} |b_k - y_k|^2$$

where b_k is the transmitted symbol, y_k is the filter output, both at time i, and where w is an exponential weighting factor which discounts past data.

least squares filter   the optimal filter, in the least squares sense, for restoring a corrupted signal. Specifically, given observations of a signal $x[n]$, the least squares filter produces the $\hat{x}[n]$ which minimizes $E[\sum |x[n] - \hat{x}[n]|^2]$. *See* Wiener filter.

least squares solution   the set of free parameters which satisfies the least squares criterion.

least-significant bit (LSB)   in a binary word, a bit with the lowest wight associated with it.

least-square-error fit   an algorithm or set of equations resulting from fitting a polynomial or other type curve, such as logarithmic, to data pairs such that the sum of the squared errors between data points and the curve is minimized.

LED   *See* light emitting diode.

LEF   *See* lighting effectiveness factor.

left-hand circular polarization   the state of an electromagnetic wave in which the electric field vector rotates anticlockwise when viewed in the direction of propagation of the wave.

left–right models   typical hidden Markov models adopted in automatic speech recognition. The state diagram for these models is a directed acyclic graph, that is, there are no cycles apart from self-loops.

legacy system   applications that are in a maintenance phase but are not ready for retirement.

Legendre functions   a collection of functions, typically denoted as $P_\nu(x)$ and $Q_\nu(x)$, that satisfy Legendre's equation:

$$(1 - x^2)\frac{d^2 f}{dx^2} - 2x\frac{df}{dx} + \nu(\nu + 1)f = 0,$$

where f is equal to either P_ν or Q_ν; ν is the order of the function and x is its argument. Typically, Legendre functions arise in boundary value problems that are based upon a spherical coordinate system.

Leibniz, Gottfried Wilhelm   (1646–1716) Born: Leipzig, Germany

Best known for his work in mathematics. Leibniz's work was an important contribution to the development of differential calculus and logic. Leibniz also improved Pascal's early calculating machine, extending its capacity to include multiplication and division. Leibniz' later years were embittered by the charge that he had plagiarized some of Isaac Newton's work on the development of the calculus. Subsequent investigations have proved this charge to be baseless.

Leibniz's formula   a formula that is satisfied by the three types of *Lie derivatives*,

$$L_{[\mathbf{f},\mathbf{g}]}h = \langle \nabla h, [\mathbf{f}, \mathbf{g}] \rangle = L_{\mathbf{g}}L_{\mathbf{f}}h - L_{\mathbf{f}}L_{\mathbf{g}}h.$$

Lempel–Ziv coding   *See* Lempel–Ziv-Welch coding.

Lempel–Ziv compression   *See* Lempel–Ziv-Welch coding.

Lempel-Ziv-Welch (LZW) coding   a variant of the dictionary-based coding scheme invented by Ziv and Lempel in 1978 (LZ78), where strings of symbols are coded as indices into a table. The table is built up progressively from the input data, such that strings already in the table are extended by one symbol each time they appear in the data.

lengthened code   a code constructed from another code by adding message symbols to the codewords. Thus an (n, k) original code becomes, after the adding of one message symbol, an $(n+1, k+1)$ code.

lens   optical element for focusing or defocusing electromagnetic waves.

lens aberrations   any deviation of the real performance of an optical system (lens) from its ideal performance. Examples of lens aberrations include coma, spherical aberration, field curvature, astigmatism, distortion, and chromatic aberration.

lens array   a two-dimensional array of (often small) lenses fabricated on a single substrate.

lens design   the mathematical determination of the parameters of the optical elements in an optical instrument or system required to satisfy its performance goals. Rays are traced through the system to predict its performance and then changes of one or more parameters are made to improve its performance. Now performed using computer software packages.

lenslet array interconnect   free-space interconnect that uses lenslet arrays to control optical paths from sources to detectors. Each lenslet images the source array onto the detector array. It has no property of dynamic reconfiguration.

lenslike medium   beam propagation medium in which the gain and index of refraction may have linear or quadratic variations with distance away from the optical axis.

level crossing rate   *See* fading rate.

level-1 cache   in systems with two separate sets of cache memory between the CPU and standard memory, the set nearest the CPU. Level-1 cache is often provided within the same integrated circuit that contains the CPU. In operation, the CPU accesses level-1 cache memory; if level-1 cache memory does not contain the required reference, it accesses level-2 cache memory, which in turn accesses standard memory, if necessary.

level-2 cache   in systems with two separate sets of cache memory between the CPU and standard memory, the set between level-1 cache and standard memory.

level-sensitive   pertaining to a bistable device that uses the level of a positive or negative pulse to be applied to the control input, to latch, capture, or store the value indicated by the data inputs.

level-triggered   *See* level-sensitive.

Levenshtein distance   *See* edit distance.

LFSR   *See* linear-feedback shift register.

Liapunoff   *See* Lyapunov.

Liapunov   *See* Lyapunov.

Lichtenburg figure   a pattern produced on a powder-coated flat electrode or upon photographic film held between a pair of flat electrodes subjected to a high voltage impulse as from a lightning stroke. Impulse polarity, magnitude and waveform can be estimated by examination of its characteristic pattern.

lid   the fuse-holder portion of a cut-out.

Lie derivative   one of three types of special derivatives.

(1) The derivative of a vector field with respect to a vector field that is also known as the Lie bracket. If $\mathbf{f}, \mathbf{g} : \mathbb{R}^n \to \mathbb{R}^n$ are two differentiable vector fields, then their Lie bracket is

$$[\mathbf{f}, \mathbf{g}] = \frac{\partial \mathbf{f}}{\partial \mathbf{x}} \mathbf{g} - \frac{\partial \mathbf{g}}{\partial \mathbf{x}} \mathbf{f},$$

where $\frac{\partial \mathbf{f}}{\partial \mathbf{x}}$ and $\frac{\partial \mathbf{g}}{\partial \mathbf{x}}$ are the Jacobian matrices of the vector fields $\mathbf{f}$ and $\mathbf{g}$, respectively. Often the negative of the above is used.

(2) The derivative of a function h with respect to a vector field $\mathbf{f}$ defined as

$$L_{\mathbf{f}} h = \langle \nabla h, \mathbf{f} \rangle = (\nabla h)^T \mathbf{f},$$

where $\nabla h : \mathbb{R}^n \to \mathbb{R}^n$ is the gradient of h.

(3) The derivative of $dh = (\nabla h)^T$ with respect to the vector field,

$$L_{\mathbf{f}}(dh) = \left(\frac{\partial \nabla h}{\partial \mathbf{x}} \mathbf{f} \right)^T + (\nabla h)^T \frac{\partial \mathbf{f}}{\partial \mathbf{x}}.$$

See also Leibniz's formula.

lifetime broadening   a spectral line broadening mechanism that is a consequence of the finite lifetime of the excited state. Numerically, the lifetime broadened linewidth is equal to the inverse lifetime of the excited state.

LIFO   *See* first-in-last-out.

lift-off process   a lithographic process by which the pattern transfer takes place by coating a material over a patterned resist layer, then dissolving the resist to "lift off" the material that is on top of the resist.

light emitting diode (LED)   a forward-biased p-n junction that emits light through spontaneous emission by a phenomenon termed electroluminescence.

light guide   system of lenses, mirrors, graded index, or graded gain media that has the capability of overcoming diffraction and guiding an electromagnetic wave at optical frequencies.

light loss factor (LLF)   the ratio of the illumination when it reaches its lowest level at the task just before corrective action is taken, to the initial level if none of the contributing loss factors were considered.

light pen   an input device that allows the user to point directly to a position on the screen. This is an alternative to a mouse. Unlike a mouse, a light pen does not require any hand/eye coordination skills, because the users point to where they look with the pen.

light scattering   (1) spreading of the light as it passes or is reflected by an optically inhomogeneous medium.

(2) the process in which a beam of light interacts with a material system and becomes modified in its frequency, polarization, direction of propagation, or other physical property. *See also* spontaneous light scattering, stimulated light scattering, Brillouin scattering, Raman scattering.

light valve   *See* spatial light modulator.

light-activated silicon controlled rectifier   a silicon controlled rectifier in which the gate terminal is activated by an optical signal rather than an electrical signal.

light-amplifying optical switch (LAOS)   vertically integrated heterojunction phototransistor and light emitting diode that has latching thyristor-type current-voltage characteristics.

lighting effectiveness factor (LEF)   the ratio of equivalent sphere illumination to ordinary measured or calculated illumination.

lighting system   any scheme used for illuminating a scene, usually for acquisition by a digital system. Illumination is crucial to digital images, since even illumination gradients that cannot be perceived by the eye can have an influence on the results of digital processing. For inspection tasks and document digitization, a uniform, reproducable, high level of lighting is usually required. Other applications have requirements for uniformity, frequency, and intensity. Structured lighting schemes are used to collect multiple images of a scene each having different illumination (photometric stereo). Strobe lights can be used to effectively freeze motion, and are useful for many visual inspection tasks. *See* structured light.

lightly doped drain (LDD)   in an MOS transistor, an extension to the source/drain diffusion that is separated slightly from the gate region which contains lower doping than that used for the source/drain diffusions. Since the source and drain diffusions are heavily doped, the lightly doped extension tends to increase the width of the depletion region around the drain which lowers the electric field intensity, to increase the breakdown voltage of the drain region of deep submicron devices.

lightning arrestor   a voltage-dependent resistor which is connected in parallel with lightning-susceptible electrical equipment. It provides a low-resistance electrical path to ground during overvoltage conditions, thus diverting destructive lightning energy around the protected equipment.

lightning choke   one of several arrangements of conductors, usually a single or multi-turn coil, used to reduce lightning currents by increasing a power line's impedance at lightning frequencies.

lightwave communications   optical communications techniques that used guided wave optical devices and fiber optics.

lightwave technology   technology based on the use of optical signals and optical fiber for the transmission of information.

likelihood ratio   the optimum processor for reducing a set of signal-detection measurements to a single number for subsequent threshold comparison.

likelihood ratio test   a test using the likelihood ratio that can be used along with threshold information to test different information-content hypotheses. An example of a signal detection problem is the demodulation of a digital communication signal, for which the likelihood ratio test may be used to decide which of several possible transmitted symbols has resulted in a given received signal.

limit cycle   undamped but bounded oscillations in a power system caused by a disturbance.

limited-look-ahead control   predictive control policy whereby — unlike in the case of open-loop-feedback control — the decision mechanism used at each intervention instant takes into account one or more, but not all, remaining future interventions; this in particular means that such decision mechanism requires the usage of at least two scenarios of the future free input values over the considered prediction interval.

limiter   an equipment or circuit that has a function to keep output power constant. It can also be used to protect other circuits not to be overdriven.

limiting spatial resolution   for an imaging photodetector, the maximum number of black and white bar pairs of equal width and spacing that can be resolved per unit length, usually given in units of line-pairs per millimeter.

Linde–Buzo–Gray (LBG) algorithm   an algorithm for vector quantizer design, due to Y. Linde, A. Buzo, and R. M. Gray (1980). The algorithm generates a codebook for a vector quantizer using a set of training data that is representative of the source to be coded. The training procedure is based on the principles of the generalized Lloyd algorithm and the K-means clustering algorithm.

line   (1) on a bus structure, one wire of the bus, which may be used for transmitting a datum, a bit of an address, or a control signal.

(2) in a cache, a group of words from successive locations in memory stored in cache memory together with an associated tag, which contains the starting memory reference address for the group.

(3) a power-carrying conductor or group of conductors.

line broadening   nonzero spectral width of an absorbing or emitting transition; caused by many physical effects.

line code   modification of the source symbol stream in a digital communication system to control the statistics of the encoded symbol stream for purposes of avoiding the occurrence of symbol errors that may arise due to limitations of practical modulation and demodulation circuitry. Also called recording codes or modulation codes.

line conditioner   *See* power conditioner.

line detection   the location of lines or line segments in an image by computer. Often accomplished with the Hough transform.

line drop compensator   a multiply-tapped autotransformer equipped with a load-sensing relay which will adjust the line voltage to compensate for the impedance drop in the circuit between the device and the load center.

line hose   split rubber tubing which is applied over energized electric conductors as temporary insulation to protect nearby workers.

line impedance stabilization network (LISN)   a network designed to present a defined impedance at high frequency to a device under test, to filter any existing noise on the power mains, and to provide a 50-Ω impedance to the noise receiver.

line of sight (LOS) the shortest possible straight line that can be envisioned, regardless of possible obstacles in the way, between a transmitter and a receiver. If a line of sight between transmitter and receiver is not blocked, the strongest signal will be received from the line-of-sight direction.

line outage distribution factor a ratio used in contingency analysis. Given two parallel lines in a power system called x and y, assume that line y is removed from service. The line outage distribution factor of line x for the outage of line y is the ratio of the change in power flow on line x to the flow on line y before the outage.

line rate *See* horizontal rate.

line shape function shape of the spectrum of an emission or absorption line.

line spread function the response of a system to a 2-D input consisting of a single line. If the system is linear and space invariant, its output to any image with a 1-D pattern is a sum of weighted line spread functions. Such patterns are often used in vision experiments. *See* linear shift invariant system, point spread function, space invariance.

line to line fault a fault on a three phase power line in which two conductors have become connected.

line width width of the spectrum of an emission or absorption line; often full width at half maximum, but other definitions also used.

line-connected reactor *See* shunt reactor.

line-current harmonic *See* electromagnetic interference filter.

line-impact printer a printer that prints a whole line at a time (rather than a single character). An impact printer has physical contact between the printer head and the paper through a ribbon. A dot-matrix printer is an impact printer, whereas an ink-jet printer is not. A line-impact printer is both a line and impact printer.

line-reflect-match (LRM) calibration an error correction scheme (calibration) where the calibration standards used are a transmission line, a reflect load, and a matched load. Line-reflect-match (LRM) two-port calibration requires a line standard (1 picosecond line), reflect standard (open circuit is preferred, short circuit is optional), and match standard (50 ohm load). The LRM technique is similar to the TRL technique, except the reference impedance is determined by a load instead of a transmission line. It has the advantages of self-consistency, requires only two standards to contact, and has no bandwidth limitations.

line-reflect-reflect-match (LRRM) calibration an extension of line-reflect-match (LRM) calibration. The second reflect standard in LRRM is used for correcting the inductance caused by the match standard.

line-to-line run-length difference coding a coding scheme for graphics. In this approach, correlation between run-lengths in successive lines is taken into account. Differences between corresponding run-lengths of successive scan lines are transmitted.

line-to-line voltage a voltage measurement of a three phase line made between any two conductors.

linear a circuit or element in which the output spectrum is proportional through gain(s), attenuation(s) and delay(s) to the input spectrum, and in which no spectral shift, conversion or generation takes place. True linearity is seldom encountered in the real world, but often used in approximate descriptions, thereby promoting understanding and simplifying computation.

linear approximation   any technique used for the purpose of analysis and design of nonlinear systems. For example, one way of analyzing the stability of a system described by nonlinear differential equations is to linearize the equations around the equilibrium point of interest and check the location of the eigenvalues of the linear system approximation.

linear block code   a block coding scheme for which the mapping can be described by a linear transformation of the message block. The transformation matrix is referred to as the generator matrix.

linear code   a forward error control code or line code whose code words form a vector space. Equivalently, a code where the element-wise finite field addition of any two code words forms another code word.

linear constant-coefficient   equation a general Nth-order linear constant-coefficient differential equation is of the form

$$\sum_0^N a_k d^k y(t) dt^k = \sum_0^M b_k d^k x(t) dt^k$$

while an Nth-order linear constant-coefficient difference equation is of the form

$$\sum_0^N a_k y[n-k] = \sum_0^M b_k x[n-k]$$

Linear constant-coefficient equations at initial rest are linear time invariant and causal, and can be conveniently analyzed using transform techniques.

linear dynamic range of Bragg cell   regime of cell operation where the amplitude of the principal diffracted beam is approximately proportional to the acoustic signal amplitude modulating the acousto-optic interaction medium.

linear filter   a filter whose output signal is a linear function of the input (that is, input and output are related via a convolution). *See* convolution, linear system.

linear generator   *See* linear machine.

linear interpolation   linear interpolation is a procedure for approximately reconstructing a function from its samples, whereby adjacent sample points are connected by a straight line.

linear least squares estimator (LLSE)   the linear estimator $\hat{x} = Ky + c$, where matrix K and vector c are chosen to minimize the expected squared error $E[(\hat{x} - x)^T (\hat{x} - x)]$. The general LLSE solution to estimate a random vector x based on measurements y is given by

$$\hat{x}(y) = E[x] + \text{cov}(x, y) \cdot \text{cov}(y, y)^{-1} \cdot (y - E[y])$$

where "cov" represents the covariance operation. *See* least squares, covariance, expectation, minimum mean square estimation.

linear load   an electrical load with a current that is linearly proportional to the voltage supplied.

linear machine   a machine in which the moving member constitutes linear motion instead of the more conventional rotary motion. Each of the rotary machine types can be produced in linear versions. The most widely known use of linear motors are in the field of transportation, where the stator is usually the moving vehicle and the conducting rotors are the rails. In these machines, the induced currents provide levitation in addition to providing the main propulsion.

linear medium   (1) medium in which the constitutive parameters are not functions of the electric or magnetic field amplitudes.

(2) medium in which any response is directly proportional in magnitude to the magnitude of the applied field.

linear motor   *See* linear machine.

linear multistep method   this is a class of techniques for solving ordinary differential equations which is widely used in circuit simulators.

linear network   a network in which the parameters of resistance, inductance and capacitance are constant with respect to voltage or current or the rate of change of voltage or current and in which the voltage or current of sources is either independent of or proportional to other voltages or currents, or their derivatives.

linear phase system   where the phase shift produced by the filter at frequency w is a linear function of w ($\angle H(w) = dw$). If a signal $x(t)$ is passed through a unit magnitude, linear phase filter with slope d, the output signal will be $x(t + d)$, the input signal time-shifted by d seconds.

linear polarization   a polarization state of a radiated electromagnetic field in which the tip of the electric field vector remains on a line and does not rotate as a function of time for a fixed position.

linear prediction   for a stochastic process, the prediction of its samples based upon determining a linear model capable of estimating the samples with minimal quadratic error.

linear prediction based speech coding   *See* linear predictive coding.

linear predictive coding   speech coding methods where short-term redundancy of the speech signal is removed by linear prediction analysis prior to encoding. *See also* adaptive differential pulse code modulation.

linear predictor   a predictor that uses a weighted sum of K previous samples of the original signal with $\alpha_i, i = 1, \ldots, K$ as weights.

linear quadratic control   for a linear deterministic plant, the problem of determining the control structure that minimizes the performance index.

Given a plant in the form of the state-space equation

$$\dot{x}(t) = Ax(t) + Bu(t)$$

The associated performance index is the quadratic form

$$J = \frac{1}{2}x(T)S(T)x(T) + \frac{1}{2}\int_{t_0}^{T}(x^T(t)Qx(t) + u^T(t)Ru(t)dt$$

with symmetric weighting matrices $S(T) \geq 0$, $Q \geq 0$, $R > 0$. Both plant and weighting matrices can be functions of time. Linear quadratic control is the problem of determining the control $u^o(t)$ on (t_0, T) that minimizes the performance index J with $x(T)$ free and T fixed. Also called LQ control.

linear quadratic Gaussian control   for a plant, the problem of finding the control structure that minimizes the expected cost.

The linear stochastic plant is given in the following state-space form:

$$\dot{x}(t) = Ax(t) + Bu(t) + Gw(t)$$

with white noise $w(t)$ and $x(t_0)$ a random variable. The associated performance index is the quadratic form

$$J = \frac{1}{2}x^T(T)S(T)x(T) + \frac{1}{2}\int_{t_0}^{T}(x^T(t)Qx(t) + u^T(t)Ru(t)dt$$

with symmetric weighting matrices $S(T) \geq 0$, $Q \geq 0$, $R > 0$. The plant and weighting matrices can be functions of time. It is desired to determine the control u^o on (t_0, T) which minimizes the expected cost

$$j = E(J)$$

with $x(T)$ free and T fixed. This problem is called the linear quadratic Gaussian control.

linear response   the characteristic of many physical systems that some output property changes linearly in response to some applied input. Such systems obey the principle of linear superposition.

linear scalar quantization   *See* uniform scalar quantization. Also known as linear SQ.

linear scrambler   a linear one-to-one mapping of a codeword, $\mathbf{c}$, of length n onto a new codeword, $\mathbf{c}_c$, also of length n. $\mathbf{c}_c$ is determined as $\mathbf{c}_c = \mathbf{S}\,\mathbf{c}$, where $\mathbf{S}$ is the linear scrambler matrix. This matrix is full rank and, for the binary case, has all binary entries. In the binary case, modulo-2 arithmetic is performed.

linear separation   the process of determining the hyper plane that separates a given set of patterns according to their membership. Of course, such separation can only be obtained once the patterns are linearly separable.

linear shift invariant (LSI) system   a linear discrete-time system $T[]$, with $y[n] = T[x[n]]$ and $y[n - n_0] = T[x[n - n_0]]$. *See* linear system, linear time invariant system, shift invariance.

linear SQ   *See* uniform scalar quantization.

linear susceptibility   the coefficient χ relating the polarization P of a material system (assumed to exhibit linear response) to the applied electric strength E according to $P = \chi E$.

linear system   the systems in which the components exhibit linear characteristics, i.e., the principle of superposition applies. Strictly speaking, linear systems do not exist in practice; they are idealized models purely for the simplicity of theoretic analysis and design. However, the system is essentially linear when the magnitude of the signals in a control system are limited to a range in which the linear characteristics exist.

More formally, consider a system with zero initial conditions such that any two input/output signal pairs $\{f_1, y_1\}$ and $\{f_2, y_2\}$ satisfy the system equation. The system is additive if the input/output pair $\{f_1 + f_2, y_1 + y_2\}$ also satisfies the system equation. The system is homogeneous if for any real constant C, the input/output pair $\{Cf_1, Cy_1\}$ satisfies the system equation. The system is linear if it is additive and homogeneous. In other words, for any real constants C_1 and C_2, the input/output pair $\{C_1 f_1 + C_2 f_2, C_1, y_1 + C_2 y_1\}$ satisfies the system equation. If a system is not linear, then it is nonlinear. The theory of linear systems is well-developed, hence many tools exist for the analysis of system behavior. In practice, nonlinear systems are often approximated by a linear model so that the tools may be exploited.

linear systems with Markov jumps   a class of piecewise deterministic processes that follows linear dynamics between random jumps of parameters that in turn can be described by finite-state Markov processes. In the continuous-time case, linear systems with Markov jumps could be modeled by linear state equations of the form:

$$\dot{x}(t) = A(\xi(t))x(t) + B(\xi(t))u(t)$$

where $t \in [0, T]$, T being finite or infinite control horizon, $x(t) \in \mathbf{R}^n$ is the process state, $u(t) \in \mathbf{R}^m$ is the process control, A, B are real valued matrices of respective dimensions depending on the random process $\{\xi(t)\}$. This process is a continuous-time discrete-state Markov process taking values in a finite set $\mathbf{S} = \{1, 2, \ldots, s\}$ called mode with transition probability matrix $P = \{p_{ij}\}$ given by

$$p_{ij} = Pr(\xi(t + \delta t) = j | \xi(t) = i)$$
$$= q_{ij}\delta t + O(\delta t)$$

if $i \neq j$ and

$$p_{ii} = 1 + q_{ii}\delta t + O(\delta t)$$

where $\delta t > 0$, q_{ij} is the intensitivity or transition rate from i to j if $i \neq j$ and

$$q_{ii} = -q_i = -\sum_{j=1, j\neq i}^{s} q_{ij}$$

Linear systems with Markov jumps may serve as models for continuous-time processes subject to abrupt changes in parameter values because of component failures, sudden shifts in environment or subsystems connections. Generally, linear systems with jumps are hybrid in the sense that their state combines a part taking values from continuous space (process state) and a part that takes values from discrete space (mode).

linear threshold unit   a neural element that computes the weighted sum of its inputs and compares that sum to a threshold value. If the sum is greater than (or equal to) the threshold, the output of the element takes on the value $+1$. Otherwise, the output takes on the value 0 (in a binary system) or -1 (in a bipolar system).

linear time invariant (LTI) system   a linear system $T[]$, with $y(t) = T[x(t)]$ and $y(t - t_0) = T[x(t - t_0)]$. *See* linear system, linear shift invariant system (LSI) system, shift invariance.

linear   time-invariant   lumped-parameter (LTIL) system   a continuous system that can be described by an ordinary differential equation with real constant coefficients. A single loop RLC circuit is an example of a linear time-invariant lumped-parameter system (LTIL), with the following differential equation relating the input voltage $f(t)$ to the output loop current $y(t)$:

$$L\frac{d^2y}{dt^2} + R\frac{dy}{dt} + \frac{1}{C}y = \frac{df}{dt}$$

A discrete time system is a LTIL system if it can be modeled by a difference equation with real constant coefficients. For example, the following LTIL difference equation represents a discrete time approximation of a differentiator:

$$y[k] = \frac{1}{T}(f[k] - f[k-1])$$

A large number of physical systems are LTIL systems. Since the theory for modeling and analysis of such systems is well developed, LTIL systems are often used to approximately model physical processes in order to simplify the mathematical analysis.

linear-feedback shift register (LFSR)   a shift register formed by D flip-flops and exclusive OR gates, chained together, with a synchronous clock.

linear-quadratic game   one of a class of noncooperative infinite dynamic games with state equations linear with respect to state variables x and players actions u_i; $i = 1, 2, \ldots, N$ and cost functions quadratic with respect to those variables with weighting matrices semipositive definite for the state and positive definite for the players actions. The existence and uniqueness of the open-loop Nash equilibrium in such games is guaranteed under assumptions on the existence of a unique solution for respectively defined coupled Riccati equations. The equilibrium strategies appear to be linear functions of the associated state trajectories. The existence and linear form for the feedback Nash strategies could be guaranteed by the existence of positive semidefinite solution to the relevant coupled Riccati equations but it does not attribute an uniqueness feature to the solution set. For the zero-sum linear-quadratic games, the situation becomes simpler both in discrete-time and continuous-time cases. The saddle point strategies could be found by solving standard Riccati equations, and whenever both open-loop and closed-loop solutions do exist they generate the same state trajectories. For example, in the continuous-time zero-sum linear-quadratic game defined by the state equation

$$\dot{x} = Ax + B_1u_1 + B_2u_2$$

and the quadratic cost functional

$$J = \frac{1}{2}\int_0^T \left(x'Qx + u_1'u_1 + u_2'u_2\right)dt$$

the saddle point strategies are given by

$$u_i = (-1)^i B'_i K(t)x; \, i = 1, 2$$

where $K(t)$ is a unique symmetric bounded solution to the matrix differential Riccati equation

$$\dot{K} + A'K + KA + Q - K\left(B_1 B'_1 - B_2 B'_2\right) = 0$$

with $K(T) = 0$.

linearity a property of a system if that system obeys the principle of superposition. In other words, if the output $y(t)$ is a function of the input $x(t)$, i.e., $y(t) = f(x(t))$, and if $x(t) = \alpha x_1(t) + \beta x_2(t)$, then for a linear system, $y(t)$ will be $= \alpha y_1(t) + \beta y_2(t)$ where $y_1(t) = f(x_1(t))$ and $y_2(t) = f(x_2(t))$.

linearization approximation of a nonlinear evolution equation in a small neighborhood of a point by a linear equation. This is obtained by keeping only the first order terms in the Taylor series expansion of the nonlinearities about this point.

linearized machine equations state equations obtained by linearizing the nonlinear voltage and electromagnetic torque equations of induction or synchronous machines. The state variables can be either currents, or flux linkages. The linearization can be accomplished using a Taylor series expansion of the machine variables about an operating point, i.e.,

$$g(f) = g(f_0) + g'(f_0)\Delta f + \text{higher order terms}$$

where $f = f_0 + \Delta f$ (and f_0 is the value of f at a given operating point).

The small displacement characteristics are then approximated as

$$g(f) - g(f_0) = \Delta g = g'(f_0)\Delta f$$

The linearized equations are typically manipulated into a standard state model form

$$\frac{dx}{dt} = Ax + Bu$$
$$y = Cx + Du$$

and are used for eigensystem, stability, and control analysis and design.

linearizer an equipment or circuit that is used to reduce distorted components generated in nonlinear amplifiers such as traveling-wave tube amplifier (TWTA) or solid state power amplifier (SSPA). There are various kinds of linearizers, such as predistortion, feedback, and feedforward types.

linear transformation a transformation operator A which satisfies superposition,

$$A(x_1 + x_2) = Ax_1 + Ax_2$$

and homogeneity

$$A(\lambda x_1) = \lambda Ax_1$$

For a discrete linear transform A is a matrix and x_1 and x_2 are vectors. Any matrix transform is linear. *See* linear system, superposition.

lineman utility employee working on primary facilities, distribution class equipment, as opposed to customer service level facilities.

linguistic hedge *See* modifier.

linguistic variable variable for which values are not numbers, but words or sentences in a natural or artificial language. In fuzzy set theory, the linguistic values (or terms) of a linguistic variable are represented by fuzzy sets in an universe of discourse. In the example shown in the figure, the linguistic variable is *speed*, the universe of discourse is associated to the base variable s, and the fuzzy sets (through the plot of their membership

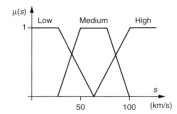

Definition of the linguistic variable speed.

functions) associated to the linguistic values *low,* *medium,* and *high.*

See also membership function, fuzzy set.

link (1) the portion of the compilation process in which separate modules are placed together and cross-module references resolved.

(2) a linkage (joint) in a manipulator arm.

link inertial parameters for a manipulator arm, consists of six parameters of the inertia tensor, three parameters of its center of mass multiplied by mass of the link (more precisely, three components of the first order moment) and mass of the link. Dynamic properties of each link are characterized by 10 inertial parameters. They appear in the dynamic equations of motion of the manipulator.

linkage flux also called magnetizing or mutual flux. In a magnetically coupled circuit such as a transformer, the linkage flux is the flux that links all the windings. For example, in a transformer the magnetic flux produced by the primary winding which is coupled to the secondary winding.

linker a computer program that takes one or more object files, assembles them into blocks that are to fit into particular regions in memory, and resolves all external (and possibly internal) references to other segments of a program and to libraries of precompiled program units.

Linville stability factor the inverse of the Rollett stability factor (K), C is a measure of potential stability in a 2-port circuit operating under small signal conditions, but stand-alone is insufficient to guarantee stability. A 2-port circuit that is matched to a positive real source and load impedance is unconditionally stable if $0 < C < 1$, $B_1 > 0$ (port 1 stability measure) and $B_2 > 0$ (port 2 stability measure). The design must provide sufficient isolation from the RF input and output ports to the bias ports to allow a reasonable interpretation of the "2-port" device criteria.

$$C = \frac{1}{K} = \frac{2 \cdot |s_{12} \cdot s_{21}|}{1 - |s_{11}|^2 - |s_{22}|^2 + |\Delta|^2}$$

where $\Delta = s_{11} \cdot s_{22} - s_{12} \cdot s_{21}$.

Lipschitz condition for a vector function $f : \Re^n \longrightarrow \Re^n$, where $\Re^n$ is the n-dimensional real Euclidean space, the condition

$$\|f(x) - f(y)\| \le b\|x - y\|$$

where $\|.\|$ is any vector norm in n and b, which is called the Lipschitz constant, is a positive real number.

Lipschitz continuous system *See* incremental gain.

liquid crystal class of organic polymeric materials made up of elongated molecules that show various degrees of order in one, two, and three dimensions.

liquid crystal display (LCD) the screen technology commonly used in notebook and smaller computers.

liquid crystal light valve (LCLV) a type of optically addressed spatial light modulator that uses twisted nematic liquid crystal material and a photosensitive layer for optical inputs.

liquid crystal on silicon (SLM) a type of electrically addressed spatial light modulator using liquid crystal material on top of a VLSI silicon circuit used for electrical signal input and signal preprocessing.

liquid laser   laser in which the active medium is a liquid, dye lasers being the most common example.

LISN   *See* line impedance stabilization network.

list decoding   decoding procedure in which the decoder, instead of producing a single estimate of the transmitted codeword, yields a list of candidate codewords, for example the L most likely.

list of capabilities   usually associated with a process, defining a set printed circuit board that can be plugged into a main board to enhance the functionality or memory of a computer.

literal   a data type consisting of alphanumeric data.

lithium battery   a battery that is similar in construction to other batteries except for a lack of any rare earth metals, which are an environmental hazard. The battery discharges by the passage of electrons from the lithiated metal oxide to the carbonaceous anode by current flowing via the external electrical circuit. See also nickel-cadmium battery, nickel-metal hydride battery, lead-acid battery.

lithium niobate ($LiNbO_3$)   a strong linear electro-optic material, also strongly piezoelectric and possessing high acousto-optic figure of merit with low acoustic attenuation.

lithography   (1) for a semiconductor manufacturing process, the process of printing images of the various circuit layers on the wafer via a photographic technique. The technique uses a radiation source, such as light, electrons, or X-rays, to generate a pattern in a radiation-sensitive material. The radiation-sensitive material is illuminated through a mask, that prevents certain portions of the material from being exposed. Exposed material is changed chemically such that it is either removed by or resistant to a solution used to develop the image.

(2) a method of producing three-dimensional relief patterns on a substrate (from the Greek *lithos*, meaning stone, and *graphia*, meaning to write).

little endian   a memory organization whereby the byte within a word with the lowest address is the least significant, and bytes with increasing address are successively more significant. Opposite of big endian. Sometimes believed (with no merit) to be either the "right" or the "wrong" memory organization, hence the name (cf. Swift's Gulliver's Travels).

For example, in a 32-bit, or four-byte word in memory, the most significant byte would be assigned address i, and the subsequent bytes would be assigned the addresses $i - 1$, $i - 2$, and $i - 3$. Thus, the least significant byte would have the lowest address of $i - 3$ in a computer implementing the little endian address assignment. *See also* big endian.

live insertion   the process of removing and/or replacing hardware components (usually at the board level) without removal of system power and without shutting down the machine.

live tank breaker   a power circuit breaker where the tank holding the interrupting chamber is not at ground potential. SF6 circuit breakers, for example, are typically live tank breakers.

livelock   a condition where attempts by two or more processes to acquire a resource run indefinitely without any process succeeding.

LLF   *See* light loss factor.

Lloyd–Max scalar quantization   a scalar quantizer designed for optimum performance (in the minimum mean squared error sense). The method, and the corresponding design algorithm, are due to S. P. Lloyd (1957) and J. Max (1960). Also referred to as PDF-optimized quantization, since the structure of the scalar quantizer is optimized to "fit" the probability density function (PDF) of the source.

Lloyd–Max SQ   *See* Lloyd–Max scalar quantization.

LMS algorithm   *See* least mean square algorithm.

LMS   *See* least mean square algorithm.

LNA   *See* low noise amplifier.

load balancing   the process of trying to distribute work evenly among multiple computational resources.

load break device   any switch, such as a circuit breaker or sectionalize capable of disconnecting a power line under load.

load buffer   a buffer that temporarily holds memory-load (i.e., memory-write) requests.

load bypass   a read (or load) request that bypasses a previously issued write (store) request. Read requests stall a processor, whereas writes do not. Therefore high-performance architectures permit load bypass. Typically implemented using write-buffers.

load center   the geographic point within a load area, used in system calculations, at which the entire load could be concentrated without affecting the performance of the power system.

load flow study   *See* power flow study.

load frequency control   the purpose of load frequency control is to maintain the power system frequency at its nominal value while maintaining the correct outputs on individual generators to satisfy the loading on the system. As the load varies, the inputs to the generator prime movers must be controlled to keep the generation in balance with the loads.

load instruction   an instruction that requests a datum from a virtual memory address, to be placed in a specified register.

load line   with a slope, also known as the permeance coefficient, determined solely by the geometry of the magnetic circuit, this line intersects with the normal demagnetization curve to indicate a magnet's operating point.

load mismatch   the load impedance does not match the device output impedance, resulting in power reflection. A perfect match occurs when the real parts of the load and device output impedance are equal and the reactive parts cancel or resonate, resulting in maximum power transfer. The magnitude of the load mismatch is usually expressed in terms VSWR, reflection coefficient or return loss.

load tap changer (LTC)   a tapped transformer winding combined with mechanically or electronically switched taps that can be changed under load conditions. The load tap changer is used to automatically regulate the output of a transformer secondary as load and source conditions vary.

load torque   the resisting torque applied at the motor shaft by the mechanical load that counterbalances the shaft torque generated by the motor and available at the shaft.

load-break device   any switch which can be opened while the circuit is loaded

load-commutated inverter (LCI)   an inverter in which the commutating voltages are supplied by the load circuit.

load-pull   the systematic variance of the magnitude and phase of the load termination of a device under test.

load-pull measurement   *See* active load-pull measurement, harmonic load-pull measurement.

load/store architecture   a system design in which the only processor operations that access memory are simple register loads and stores.

load/store unit   a computer based on the load/store architecture.

loaded Q   dimensionless ratio of the average over any period of time ($T = 1$/frequency) of the ratio of the maximum energy stored (U_{max}) to the power absorbed or dissipated ($p_{absorbed} = p_{in} - p_{out}$) in a passive component or circuit, including external loading effects, expressed as a dimensionless ratio. For most applications, the higher the Q, the better the part.

local area network   a network of computers and connection devices (such as switches and routers) that are located on a single site. The connections are direct cables (such as UTP or optical fiber) rather than telecommunication lines. The computer network in a university campus is typically a local area network.

local bus   the set of wires that connects a processor to its local memory module.

local controllability of generalized 2-D model   the generalized 2-D model

$$Ex_{i+1,j+1} = A_0 x_{ij} + A_1 x_{i+1,j} + A_2 x_{i,j+1}$$
$$+ B_0 u_{ij} + B_1 u_{i+1,j} + B_2 u_{i,j+1}$$

$i, j \in Z_+$ (the set of nonnegative integers) is called locally controllable in the rectangle $[0, N_1] \times [0, N_2]$ if for admissible boundary conditions x_{i0} for $i \in Z_+$ and x_{0j} for $j \in Z_+$, there

exists a sequence of inputs u_{ij} for $0 \leq i \leq N_1 + n_1$ and $0 \leq j \leq N_2 + n_2$ such that $x_{N_1 N_2} = 0$ where $x_{ij} \in R^n$ is the local semistate vector, $u_{ij} \in R^m$ is the input vector, E, A_k, B_k ($k = 0, 1, 2$) are real matrices with E possibly singular, and (n_1, n_2) is the index of model. *See also* local reachability of 2-D general model.

local decision unit   control agent or a part of the controller associated with a given subsystem of a partitioned system; local decision unit is usually in charge of the local decision variables and is a component of a decentralized or a hierarchical control system.

local decision variable   control inputs associated with a given subprocess (subsystem) of the considered partitioned process (system); local decision variables can be either set locally by local decision unit, or globally by a centralized controller.

local field effect   effect associated with the distinction that occurs in condensed matter between the spatially averaged electric field and the field that acts on a representative molecule of the material system. A consideration of local field effects leads to the Lorentz–Lorenz and Clausius–Mossotti relations.

local memory   memory that can be accessed by only one processor in a multiprocessor or distributed system. In many multiprocessors, each processor has its own local memory. *See also* global memory.

local minimum   a minimum of a function that is not the global minimum.

local mode oscillation   this mode of oscillation is associated with the swinging of units at a generating station with respect to the rest of the power system. The oscillations are thus localized to within a small part of the system.

local observability of 2-D Fornasini–Marchesini model

the 2-D Fornasini–Marchesini model

$$x_{i+1,j+1} = A_1 x_{i+1,j} + A_2 x_{i,j+1}$$
$$+ B_1 u_{i+1,j} + B_2 u_{i,j+1}$$
$$y_{ij} = C x_{ij}$$

$i, j \in Z_+$ (the set of nonnegative integers) is called locally observable in the rectangle $[0, N_1] \times [0, N_2]$ if there are no local states $x_{10} \neq 0$ and $x_{01} \neq 0$ such that for zero inputs u_{ij} for $0 \leq i \leq N_1$ and $0 \leq j \leq N_2$ and zero boundary conditions $x_{i0} = 0$ for $i \geq 2$ and x_{0j} for $j \geq 2$ the output is also zero $y_{ij} = 0$ for $0 \leq i \leq N_1$ and $0 \leq j \leq N_2$. The model is locally observable in $[0, N_1] \times [0, N_2]$ if and only if

$$\text{rank} \begin{bmatrix} CT_{10} \\ \vdots \\ CT_{N_1-1,0} \\ CT_{01} \\ \vdots \\ CT_{0,N_2-1} \\ CT_{11} \\ \vdots \\ CT_{N_1-1,N_1-1} \end{bmatrix} [A_1, A_2] = 2n$$

where the transition matrix T_{ij} of the model is defined by

$$T_{ij} = \begin{cases} I_n & \text{for } i = j = 0 \\ T_{ij} = A_1 T_{i-1,j} + A_2 T_{i,j-1} \\ \text{for } i, j \geq 0 \ (i + j \neq 0) \\ T_{ij} = 0 & \text{for } i < 0 \text{ or/and } j < 0 \end{cases}$$

local observability of 2-D Roesser model

the 2-D Roesser model

$$\begin{bmatrix} x_{i+1,j}^h \\ x_{i,j+1}^v \end{bmatrix} = \begin{bmatrix} A_1 & A_2 \\ A_3 & A_4 \end{bmatrix} \begin{bmatrix} x_{ij}^h \\ x_{ij}^v \end{bmatrix} + \begin{bmatrix} B_1 \\ B_2 \end{bmatrix} u_{ij}$$

$$y_{ij} = C \begin{bmatrix} x_{ij}^h \\ x_{ij}^v \end{bmatrix}$$

$i, j \in Z_+$ (the set of nonnegative integers) is called locally observable in the rectangle $[0, N_1] \times [0, N_2]$ if there is no local initial state $x_{00} \neq 0$ such that for zero inputs $u_{ij} = 0$ for $0 \leq i \leq N_1$ and $0 \leq j \leq N_2$ and zero boundary conditions $x_{0j}^h = 0$ for $j \geq 1$ and $x_{i0}^v = 0$ for $i \geq 1$, the output is also zero $y_{ij} = 0$ for $0 \leq i \leq N_1$ and $0 \leq j \leq N_2$ where $x_{ij}^h \in R^{n_1}$ and $x_{ij}^v \in R^{n_2}$ are the horizontal and vertical local state vectors, respectively, $u_{ij} \in R^m$ is the input vector, $y_{ij} \in R^p$ is the output vector, and $A_1, A_2, A_3, A_4, B_1, B_2, C$ are real matrices. The model is locally observable in $[0, N_1] \times [0, N_2]$ and only if

$$\text{rank} \begin{bmatrix} C \\ CT_{10} \\ CT_{01} \\ \vdots \\ CT_{ij} \\ \vdots \\ CT_{N_1 N_2} \end{bmatrix} = n$$

where the transition matrix is defined by

$$T_{ij} = \begin{cases} I & \text{for } i = j = 0 \\ \begin{bmatrix} A_1 & A_2 \\ 0 & 0 \end{bmatrix} & \text{for } i = 1, \ j = 0; \\ \begin{bmatrix} 0 & 0 \\ A_3 & A_4 \end{bmatrix} & \text{for } i = 0, \ j = 1 \quad \text{and} \\ T_{10} T_{i-1,j} + T_{01} T_{i,j-1} & \text{for} \\ \qquad i, j \in Z_+ \ (i + j \neq 0) \\ T_{ij} = 0 \quad \text{for } i < 0 \text{ or/and } j < 0 \end{cases}$$

local oscillator

(1) an oscillator or circuit that produces a periodic signal whose function is to be utilized in the demodulation of a received radio signal. This periodic signal is typically a sinusoid and the oscillator is typically located in a radio receiver. The tuning of the radio to a given channel, or station, typically involves the tuning of the local oscillator. The local oscillator is part of the radio frequency (RF) front end of a radio receiver and is an important component in a heterodyne receiver.

(2) the signal applied to a mixer circuit that is of a sufficient level to bias the diodes within the mixer into a nonlinear region so that the mixing process may occur.

local oxidation of silicon (LOCOS)   masked oxidation of silicon to provide electronic isolation between devices. Made possible by relatively slow oxidation of silicon nitride, which is used as a mask.

local reachability of generalized 2-D model
the generalized 2-D model

$$Ex_{i+1,j+1} = A_0 x_{ij} + A_1 x_{i+1,j}$$
$$+ A_2 x_{i,j+1} + B_0 u_{ij}$$
$$+ B_1 u_{i+1,j} + B_2 u_{i,j+1}$$

$i, j \in Z_+$ (the set of nonnegative integers) is called locally reachable in the rectangle $[0, N_1] \times [0, N_2]$ if for admissible boundary conditions $x_{i0}, i \in Z_+$ and $x_{0j}, j \in Z_+$ and every vector $x_f \in R^n$ there exists a sequence of inputs u_{ij} for $0 \le i \le N_1 + n_1$ and $0 \le j \le N_2 + n_2$ such that $x_{N_1 N_2} = x_f$, where $x_{ij} \in R^n$ is the local semistate vector, $u_{ij} \in R^m$ is the input vector, E, A_k, B_k ($k = 0, 1, 2$) are real matrices with E possibly singular. The model is locally reachable in $[0, N_1] \times [0, N_2]$ if and only if

$$\text{rank } \left[M_0, M1^1, \ldots, M^1_{\bar{N}_1}, M_1^2, \ldots, \right.$$
$$\left. M^2_{\bar{N}_2}, M_{11}, \ldots, M_{1\bar{N}_2}, M_{21}, \ldots, M_{\bar{N}_1, \bar{N}_2} \right] = n$$
$$M_0 = T_{N_1-1, N_2-1} B_0,$$
$$M_p^1 := T_{N_1-p, N_2-1} B_1 + T_{N_1-p-1, N_2-1} B_0$$
$$\text{for} \quad p = 1, \ldots, \bar{N}_1 = N_1 + n_1$$
$$M_q^2 := T_{N_1-1, N_2-q} B_2 + T_{N_1-1, N_2-q-1} B_0$$
$$\text{for} \quad q = 1, \ldots, \bar{N}_2 = N_2 + n_2$$
$$M_{pq} := T_{N_1-p-1, N_2-q-1} B_0$$
$$+ T_{N_1-p, N_2-q-1} B_1 + T_{N_1-p-1, N_2-q} B_2$$
$$\text{for} \quad \begin{cases} p = 1, \ldots, \bar{N}_1 \\ q = 1, \ldots, \bar{N}_2 \end{cases}$$

and the transition matrix T_{pq} is defined by

$$ET_{pq} = \begin{cases} A_0 T_{-1,-1} + A_1 T_{0,-1} \\ + A_2 T_{-1,0} + I \quad \text{for} \quad p = q = 0 \\ A_0 T_{p-1,q-1} + A_1 T_{p,q,-1} \\ + A_2 T_{p-1,q} \\ \text{for} \quad p \ne 0 \quad and/or \quad q \ne 0 \end{cases}$$

and

$$[Ez_1 z_2 - A_0 - A_1 z_1 - A_2 z_2]^{-1}$$
$$= \sum_{p=-n_1}^{\infty} \sum_{q=-n_2}^{\infty} T_{pq} z_1^{-(p+1)} z_2^{-(q+1)}$$

pair (n_1, n_2) of positive integers n_1, n_2 such that $T_{pq} = 0$ for $p < -n_1$ and/or $q < -n_2$ is called the index of the model.

local stability   See stable state.

local wavelength   distance between the phase fronts of a non-planewave signal inferred from measurements that are local in space; 2 pi over the magnitude of the local propagation constant, which is the gradient of the total phase. *See also* wavelength, instantaneous frequency.

locality   one of two forms of program memory relationships.
    (1) Temporal locality: if an object is being used, then there is a good chance that the object will be reused soon.
    (2) Spatial locality: when an object is being used, there is a good chance that objects in its neighborhood (with respect to the memory where these objects are stored) will be used.
    These two forms of locality facilitate the effective use of hierarchical memory. Registers exploit temporal locality. Caches exploit both temporal and spatial locality. Interleaved memories exploit spatial locality. *See also* sequential locality.

localization   refers to the "trapping" of an electron into a potential well minimum, so that the wave function ceases to be describable by a propagating wave. This localization can be "strong" localization such as when an electron is trapped by an ionized donor or other ionized potential, or "weak" localization in which it is induced by a "self" interference effect. *See also* weak localization.

lock a synchronization variable, used in shared-memory multiprocessors, that allows only one processor to hold it at any one time, thus enabling processors to guarantee that only one has access to key data structures or critical sections of code at any one time.

lock range the range of frequencies in the vicinity of the voltage controlled oscillator (VCO) free-running frequency over which the VCO will, once locked, remain synchronized with the signal frequency. Lock range is sometimes called tracking bandwidth.

lock-in amplifier a system for detecting weak, noisy periodic signals based on synchronous detection, and incorporating all the other components necessary for recording the amplitude profile of the weak incoming signal, including input AC amplifier, diode or other detectors, low-pass filter, DC amplifier, and any special filters. Such instruments are nowadays constructed with increasing amounts of digital and computerized circuitry, depending on the frequency of operation.

lock-out phenomenon exhibited during channel switching that results from a fast automatic gain control (AGC) system interacting with the horizontal automatic frequency control (AFC), thereby reducing the pull-in range of the AFC system.

lock-up-free cache *See* nonblocking cache.

locked-rotor current the current drawn by an induction motor when the shaft is not moving and rated voltage is applied. The starting current is essentially equal to the locked rotor current and may be as much as eight times the rated current of the machine.

locked-rotor torque the torque produced in an induction motor when the rotor is locked and rated AC voltage is applied to the stator.

locking *See* bus locking.

lockout the condition following fault clearing when the circuit will not attempt a reclose. Transformers, generators, and buses typically trip once and lockout immediately. Transmission lines and distribution lines will generally attempt one or more recloses, and will lockout if the fault remains following the last reclose in the sequence.

lockout relay an auxiliary relay which is operated by protective relay(s) that in turn opens the appropriate circuit breakers or other fault clearing devices. The lockout relay will remain in the trip position until manually reset, and is used in protective zones where temporary faults are unusual and the potential for equipment damage is high.

LOCOS *See* local oxidation of silicon.

log periodic antenna broadband antenna designed using physical dimensions (lengths, spacings, diameters, etc.) that vary logarithmically. The result of such designs is an antenna whose performance parameters (e.g., input impedance) is periodic with respect to the logarithm of the frequency.

log-likelihood function the likelihood function of y given x is the conditional PDF, $p(y|x)$. The log-likelihood function is the logarithm of the likelihood function, $\log(p(y|x))$.

log-normal distribution probability distribution with density

$$f(x) = \frac{1}{\sqrt{2\pi}\sigma x e^{-\frac{(\log x - \mu)^2}{2\sigma^2}}}$$

where μ and σ are the mean and standard deviation of the logarithm.

logarithmic quantization a method for non-linear scalar quantization where the input signal is transformed logarithmically and then coded using uniform quantization. The transformation is utilized to enhance performance for sources having nonuniform probability distribution, and to give robustness towards varying input signal dynamics.

logic analyser a machine that can be used to send signals to, and read output signals from, individual chips or circuit boards.

logic circuit a circuit that implements a logical function, such as AND, OR, NAND, NOR, NOT, or XOR. (DB)

logic gate a basic building block for logic systems that controls the flow of pulses. (CRC-E 1634)

logic level the high or low value of voltage variable that is assigned to be a 1 or 0 state. (CRC-E 1621)

logical operation the machine-level instruction that performs Boolean operations such as AND, OR, and COMPLEMENT.

logical register *See* virtual register.

logical shift a shift in which all bits of the register are shifted. *See also* arithmetic shift.

long code in a spread-spectrum system, a (periodic) spreading code (spreading sequence) with a period (substantially) longer than a bit duration. *See also* short code.

long duration *See* voltage variation.

long integer an integer that has double the number of bits as a standard integer on a given machine. Some modern machines define long integers and regular integers to be the same size.

long-term stability a measure of a power system's long-term response to a disturbance after all post-disturbance transient oscillations have been damped out, often associated with boiler controls, power plant and transmission system protection, and other long-period factors.

longitudinal excitation laser pumping process in which the pump power is introduced into the amplifying medium in a direction parallel to the direction of propagation of the resulting laser radiation.

longitudinal mode term (somewhat misleading) used in referring to the longitudinal structure or index of the mode of a laser oscillator.

longitudinal modelocking forcing the longitudinal modes of a laser oscillator to be equally spaced in frequency and have a fixed phase relationship; useful for obtaining very short and intense pulsations.

longitudinal phase velocity phase velocity of microwave propagation in the axial direction of a slow-wave structure of traveling wave tube.

longitudinal redundancy check an error checking character written at the end of each block of information on a track of magnetic tape. The character is calculated by counting the number of ones on a track and adding either an additional one or zero in the character so that the total number of ones in the block is even.

longitudinal section electric and magnetic modes (LSE, LSM) are alternative choices for the electromagnetic potentials. For a waveguiding structure, the most common choices are potentials directed in the propagation direction (i.e., TE and TM modes). Sometimes, especially when layered dielectric are present, it is more convenient to consider the LSE, LSM potential, perpendicular to the longitudinal section.

look-ahead carry   the concept (frequently used for adders) of breaking up a serial computation in which a carry may be propagated along the entire computation into several parts, and trying to anticipate what the carry will be, to be able to do the computation in parallel and not completely in series.

look-up table model   a model in which measured data is stored in a data base, allowing the user to access the data and interpolate or extrapolate performance based on that data. S-parameter data is an example that is encountered frequently in RF and microwave design. Sophisticated look-up tables may reduce the data by polynomial curve fitting and retaining only the coefficients. Look-up table models generally have a very low modeling valuation coefficient.

lookahead   for conditional branches, a strategy for choosing a probable outcome of the decision that must be made at a conditional branch, even though the conditions are not known yet, and initiates "speculative" execution of the instructions along the corresponding control path through the program. If the chosen outcome turns out to be incorrect, all effects of the speculative instructions must be erased (or, alternately, the effects are not stored until it has been determined that the choice was correct). Some lookahead mechanisms attempt execution of both possibilities from a conditional branch, and others will cascade lookahead choices from several conditional branches that occur in close proximity in the program flow.

lookbehind   for an instruction buffer, a means of holding recently executed instructions in a buffer within the control unit of a processor to permit fast access to instructions in a loop.

loop   (1) a set of branches forming a closed current path, provided that the omission of any branch eliminates the closed path.

(2) a programming construct in which the same code is repeated multiple times until a programmed condition is met.

loop analysis   *See* mesh analysis.

loop antenna   a two-terminal, thin wire antenna for which the terminals are close to one another and for which the wire forms a closed path.

loop feeder   a closed loop formed by a number of feeders. The resulting configuration allows a load area to be served without interruption if one feeder should fail.

loop filter   the filter function that follows the phase detector and determines the system dynamic performance.

loop gain   the combination of all DC gains in the PLL.

loop network   *See* ring network.

loop primary feeder   a feeder in a distribution system which forms a loop around the load area.

loop system   a secondary system of equations using loop currents as variables.

loop-set   *See* circuit-set.

loosely coupled multiprocessor   a system with multiple processing units in which each processor has its own memory and communication between the processors is over some type of bus.

Lorentz force   the mutually perpendicular force acting on a current-carrying element placed perpendicular to a magnetic field.

Lorentz medium   a frequency-dependent dielectric whose complex permittivity is described by an equation with a second-order Lorentz pole involving a resonant frequency and a damping constant. Some optical materials and artificial dielectrics may require several Lorentz poles to describe their behavior over the frequency band of interest.

Lorentz theorem   for isotropic and reciprocal media, the Lorentz reciprocity theorem states that, in a source-free region bounded by a surface S, the electromagnetic fields produced by two sources, denoted by subscripts a, b, satisfy the equation

$$\oint_S (\mathbf{E}_a \times \mathbf{H}_b - \mathbf{E}_b \times \mathbf{H}_a)\, d\mathbf{S} = 0$$

Lorentz, Hendrik Antoon   (1853–1928) Born: Arnhem, Holland

Best known for his work in electromagnetic theory. Along with Pieter Zeeman, he shared the 1902 Nobel Prize for Physics. His theoretical work was in the extension and refinement of Maxwell's theory. Lorentz's work was an essential cornerstone in Einstein's Special Theory of Relativity.

Lorentzian line shape   the shape of a spectral line that results from certain physical mechanisms, such as the finite lifetime of the upper level of an atomic transition. It is characterized by broad wings, that decrease as the inverse square of the frequency separation from line center.

Lorentzian lineshape function   spectrum of an emission or absorption line that is a Lorentzian function of frequency; characteristic spectrum of homogeneously broadened media.

Lorenz, Ludwig Valentin   (1829–1891) Born: Elsinore, Denmark

Lorenz was not well known. He did, however, do significant work on electromagnetic theory, on the continuous loading method for cables, and for the acceptance of the ohm as the resistance standard.

LOS   *See* line of sight.

loss   (1) decrease of intensity of an electromagnetic wave due to any of several physical mechanisms. *See also* attenuation.

(2) a term for electric power which does not register on the consumer's electric meter, e.g.,

through ohmic losses in transmission lines, iron losses in transformers, or theft.

loss coefficient   a factor used in economic dispatch calculations that relates power line losses to the power output of generating plants.

loss factor   the product of the dielectric constant and the power factor.

loss of service   the complete loss of electric power exclusive of sags, swells, and impulses.

loss tangent   *See* dissipation factor.

loss-of-field relay   a protection relay used to trip a synchronous generator when the excitation system is lost. Loss of excitation causes the generator to run as an induction generator drawing reactive power from the system. This can cause severe system voltage reductions and damage to stator due to excessive heating.

lossless coding   *See* lossless source coding.

lossless compression   compression process wherein the original data can be recovered from the coded representation perfectly, i.e. without loss. Lossless compressors either convert fixed length input symbols into variable length codewords (Huffman and arithmetic coding) or parse the input into variable length strings and output fixed-length codewords (Ziv-Lempel coding). Lossless coders may have a static or adaptive probability model of the input data. *See* entropy coding.

lossless predictive coding   a lossless coding scheme that can encode an image at a bit rate close to the entropy of the mth-order Markov source. This is done by exploiting the correlation of the neighboring pixel values.

lossless source coding   source coding methods (for digital data) where no information is "lost" in the coding, in the sense that the original can be exactly reproduced from its coded version. Such methods are used, for example, in computers to maximize storage capacity. *Compare with* lossy coding.

lossless source encoding   *See* lossless source coding.

lossy coding   *See* lossy source coding.

lossy compression   compression wherein perfect reconstruction of the original data is not possible. Aims to minimize signal degradation while achieving maximum compression. Degradation may be defined according to signal error or subjective impairment.

lossy encoding   *See* lossy source coding.

lossy source coding   refers to noninvertible coding, or quantization. In lossy source coding information is always lost, and the source data cannot be perfectly reconstructed from its coded representation. *Compare with* lossless source coding.

lossy source encoding   *See* lossy source coding.

LOT   *See* lapped orthogonal transform.

low byte   the least-significant 8 bits in a larger data word.

low level waste   nuclear waste such as gloves and towels which have comparatively low radioactivity.

low noise amplifier (LNA)   (1) an amplifier that boosts low-level radio/microwave signal received without adding substantial distortions to the signal.

(2) an amplifier in which the primary cause of noise is due to thermally excited electrons and is designed to introduce minimum internally generated noise.

low order interleaving   in memory interleaving, using the least significant address bits to select the memory module and most significant address bits to select the location within the memory module.

low output capacitance Plumbicon tube   a picture tube designed to reduce the capacitance of the target to ground, resulting in an improved signal-to-noise ratio.

low pass filter   (1) filter exhibiting frequency selective characteristic that allows low-frequency components of an input signal to pass from filter input to output unattenuated; all high-frequency components are attenuated.

(2) a filter that passes signal components whose frequencies are small and blocks (or greatly attenuates) signal components whose frequencies are large. For the ideal case, if $H(\omega)$ is the frequency response of the filter, then $H(\omega) = 0$ for $|\omega| > B$, and $H(\omega) = 1$ for $|\omega| < B$. The parameter B is the bandwidth of the filter. A filter whose impulse response is a low-pass signal.

low resistance grounded system   an electrical distribution system in which the neutral is intentionally grounded through a low resistance. Low resistance grounding will limit ground fault current to a value that significantly reduces arcing damage but still permits automatic detection and interruption of the fault current.

low side   pertains to the portion of a circuit which is connected to the lower-voltage winding of a power transformer.

low state   a logic signal level that has a lower electrical potential (voltage) than the other logic state. For example, the low state of TTL is defined as being less than or equal to 0.4 V.

low voltage holding coil   a holding coil that keeps the main-line contactor closed on low voltage conditions. Controllers that contain this feature are used in places where the motor is vital to the operation of a process, and it is necessary to maintain control of the motor under low voltage conditions.

low-level transmitter   a transmitter in which the modulation process takes place at a point where the power level is low compared to the output power.

low-level vision   the set of visual processes related to the detection of simple primitives, describing raw intensity changes and/or their relationships in an image.

low-pass equivalent (LPE) model   a method of representing bandpass signals and systems by low-pass signals and systems. This technique is extremely useful when developing discrete time models of bandpass continuous-time systems. It can substantially reduce the sampling rate required to prevent aliasing and does not result in any loss of information. This, in turn, reduces the execution time required for the simulation. This modeling technique is closely related to the quadrature representation of bandpass signals.

low-pass signal   a signal whose Fourier transform has frequency components that are small for frequencies greater than some intermediate frequency value. To define mathematically, let $X(\omega)$ be the Fourier transform of the signal then $X(\omega) = 0$ for $|\omega| > B$, for some $B > 0$. In the strict sense, if B is the smallest value for which the above holds, then B is the bandwidth of the signal.

low-power TV (LPTV)   a television service authorized by the FCC to serve specific confined areas. An LPTV station may typically radiate between 100 and 1000 W of power, covering a geographic radius of 10 to 15 miles.

low-pressure discharge   a discharge in which the pressure is less than a torr or a few torrs; low-pressure gases can be easily excited, giving spectra characteristic of their energy structure.

lower frequency band edge   the lower cutoff frequency where the amplitude is equal to the maximum attenuation loss across the band.

lower side frequency   the difference frequency that is generated during the heterodyning process or during the amplitude-modulating process. For example, if a 500 kHz carrier signal is amplitude-modulated with a 1 kHz frequency, the lower side frequency is 499 kHz.

LPE model   *See* low-pass equivalent model.

LPTV   *See* low-power TV.

LQ control   *See* linear quadratic control.

LRM calibration   *See* line-reflect-match calibration.

LRRM calibration   *See* line-reflect-reflect-match calibration.

LRU   *See* least recently used algorithm.

LRU bits   a set of bits that record the relative recency of access among pairs of elements that are managed using an LRU replacement policy. If n objects are being managed, the number of bits required is $n(n-1)/2$. Upon each access, n bits are forced into certain states; to check one of the objects to determine whether it was the least recently accessed, n bits need to be examined.

LRU replacement   *See* least recently used algorithm.

LRU stack   a stack-based data structure to perform the bookkeeping for a least recently used (LRU) management policy. An object is promoted to the top of the stack when it is referenced; the object that has fallen to the bottom of the stack is the least recently used object.

LS algorithm   *See* least squares algorithm.

LSB   *See* least-significant bit.

LSI   *See* linear shift invariant system.

LSI   *See* large-scale integration.

LTC   *See* load tap changer and tap changing under load.

LTI   *See* linear time invariant system.

LTIL system   *See* linear time-invariant lumped-parameter system.

lumen   the SI unit of illumination measurement. Also, the hollow interior of a blood vessel or airway.

luminance   (1) part of the video signal that provides the brightness information; often designated as the "Y" component in RGB systems where $Y = .3R + .59G + .11B$.

(2) formally, the amount of light being emitted or reflected by a surface of unit area in the direction of the observer and taking into account the spectral sensitivity of the human eye.

(3) the amount of light coming from a scene. candela.

luminance ratio   the ratio between the luminance of two areas in the visual field.

luminescence   emission of light caused by relaxation of an electron excited to a higher energy level. Excitation of the electron may have occurred as a result of light absorption (fluorescence or phosphorescence), a chemical reaction (chemiluminescence), or from some living organisms (bioluminescence).

luminosity   the ratio of luminous flux (total visible energy emitted) to the corresponding radiant flux (total energy emitted) usually in lumens per watt.

luminous efficiency   the measure of the display output light luminance for a given input power, usually measured in lumens per watt, which is equivalent to the nit.

lumped element   a circuit element of inductors, capacitors, and resistors; its dimension is negligible relative to the wavelength.

LVQ   *See* learning vector quantization.

Lyapunov, Aleksandr M.   (1857–1918) Born: Yaroslavl, Russia also transliterated as Liapunov or Liapunoff.

Russian mathematician noted for his contributions to the stability of dynamical systems. His most influential work is titled *The General Problem of the Stability of Motion*, published in 1892 by the Kharkov Mathematical Society. Lyapunov was a student of Chebyshev at St. Petersburg, and taught at the University of Kharkov from 1885 to 1901. He became academician in applied mathematics at the St. Petersburg Academy of Sciences. *See also* direct method.

Lyapunov equation   in statistics and linear systems theory a set of equation-classes that describe system evolution. In each case, P, Q are symmetric positive definite; A, B are arbitrary. The continuous-time algebraic Lyapunov equation:

$$0 = AP + PA^T + BQB^T$$

The discrete-time, time-dependent Lyaponov equation:

$$P(t+1) = A(t)P(t)A^T(t) + B(t)Q(t)B^T(t)$$

The discrete-time algebraic Lyanpunov equation:

$$P = APA^T + BQB^T$$

See Riccati equation.

Lyapunov function   corresponding to an equilibrium state $\mathbf{x}_{eq} \in \mathbf{R}^n$ of the system $\dot{\mathbf{x}} = \mathbf{f}(t, \mathbf{x})$ is a continuously differentiable function $V = V(t, \mathbf{x})$ such that $V(t, \mathbf{x}_{eq}) = 0$, it is positive in a neighborhood of $\mathbf{x}_{eq}$, that is, V is positive definite with respect to $\mathbf{x}_{eq}$, and the time derivative of V evaluated on the trajectories of the system $\dot{\mathbf{x}} = \mathbf{f}(t, \mathbf{x})$ is negative semidefinite with respect to $\mathbf{x}_{eq}$. For the discrete-time model, $\mathbf{x}(k + 1) = \mathbf{f}(k, \mathbf{x})$, continuous differentiablity is replaced with continuity, and the time derivative $\dot{V}$ is replaced with the first forward difference ΔV. The existence of a Lyapunov function for a given equilibrium state implies that this equilibrium is stable in the sense of Lyapunov. If the time derivative, respectively the forward difference, of V is negative definite with respect to the equilibrium, then this equilibrium is asymptotically stable in the sense of Lyapunov.

Lyapunov function candidate   for an equilibrium state of $\dot{\mathbf{x}} = \mathbf{f}(t, \mathbf{x})$, its Lyapunov function candidate is any continuously differentiable function $V = V(t, \mathbf{x})$ that is positive definite with respect to the equilibrium $\mathbf{x}_{eq}$. For an equilibrium state of the discrete-time system $\mathbf{x}(k + 1) = \mathbf{f}(k, \mathbf{x})$, a Lyapunov function candidate is any continuous function $V = V(k, \mathbf{x})$ that is positive definite with respect to the equilibrium state.

Lyapunov stability   also known as stability in the sense of Lyapunov; concerned with the behavior of solutions of a system of differential, or difference, equations in the vicinity of its equilibrium states. The concept of stability in the sense of Lyapunov can be related to that of continuous dependence of solutions upon their initial conditions. An equilibrium state is stable in the sense of Lyapunov if the system trajectory starting sufficiently close

to the equilibrium state stays near the equilibrium state for all subsequent time. Formally, the equilibrium state $\mathbf{x}_{eq} \in \mathbb{R}^n$ of the system $\dot{\mathbf{x}} = \mathbf{f}(t, \mathbf{x})$, or the system $\mathbf{x}(k + 1) = \mathbf{f}(k, \mathbf{x})$, is stable, in the sense of Lyapunov, if for any $\varepsilon > 0$ there exists a $\delta = \delta(t_0, \varepsilon)$ such that if the initial condition $\mathbf{x}(t_0)$ is within δ-neighborhood of $\mathbf{x}_{eq}$, that is,

$$\|\mathbf{x}(t_0) - \mathbf{x}_{eq}\| < \delta,$$

then the system trajectory $\mathbf{x}(t)$ satisfies

$$\|\mathbf{x}(t) - \mathbf{x}_{eq}\| < \varepsilon$$

for all $t \geq t_0$, where $\|\cdot\|$ is any Hölder norm on $\mathbb{R}^n$.

Lyapunov surface   the set of all points $x \in \mathfrak{R}^n$ (n-dimensional real Euclidean space) that satisfy $V(x) =$ positive real number, where $V(x)$ is a Lyapunov function.

Lyapunov's direct method   *See* Lyapunov's second method.

Lyapunov's first method   the method that allows one to assess the stability status of an equilibrium point of a nonlinear system based on stability investigation of the equilibrium point of the linearized version of that system. This method is also called Lyapunov's indirect method.

Lyapunov's indirect method   *See* Lyapunov's first method.

Lyapunov's second method   the method of stability assessment that relies on the use of energy-like functions without resorting to direct solution of the associated evolution equations. This is also called Lyapunov's direct method.

LZ77   refers to string-based compression schemes based on Lempel and Ziv's 1977 method. An input string of symbols that matches an identical string previously (and recently) transmitted is coded as an offset pointer to the previous occurrence and a copy length.

LZ78 refers to string-based compression schemes based on Lempel and Ziv's 1978 method. A dictionary of prefix strings is built at the encoder and decoder progressively, based on the message. The encoder searches the dictionary for the longest string matching the current input, then encodes that input as a dictionary index plus the literal symbol which follows that string. The dictionary entry concatenated with the literal is then added to the dictionary as a new string.

M

M   common notation for the number of modes in a step index fiber, given by

$$M = \frac{V^2}{2}$$

where V is the fiber parameter.

M (mega)   abbreviation for 1,048,576 (not for 1 million).

M-algorithm   reduced-complexity breadth-first tree search algorithm, in which at most M tree nodes are extended at each stage of the tree.

m-ary hypothesis testing   the assessment of the relative likelihoods of M hypotheses $H_1, H_2, \ldots, H_M$. Normally we are given prior statistics $P(H_1), \ldots, P(H_M)$ and observations y whose dependence $p(y|H_1), \ldots, p(y|H_M)$ on the hypotheses are known. The solution to the hypothesis testing problem depends upon the stipulated criterion; possible criteria include maximizing the posterior probability (MAP) or minimizing the expected "cost" of the decision (a cost C_{ij} is assigned to the selection of hypothesis j when i is true). *See* binary hypothesis testing, conditional statistic, a priori statistics, a posteriori statistics, maximum a posteriori estimation.

m-phase oscillator   *See* multi-phase oscillator.

m-sequence   maximal length sequence. A binary sequence generated by a shift register with a given number of stages (storage elements) and a set of feedback connections, such that the length of the sequence period is the maximum possible for shift registers with that number of stages over all possible sets of feedback connections. For a shift register with n stages, the maximum sequence period is equal to $2^n - 1$.

Mach band   a perceived overshoot (lighter portion) on the light side of an edge and an undershoot (darker portion) on the dark side of the edge. The Mach band is an artifact of the human visual system and not actually present in the edge. *See* brightness, simultaneous contrast.

machine code   the machine format of a compiled executable, in which individual instructions are represented in binary notation.

machine interference   the idle time experienced by any one machine in a multiple-machine system that is being serviced by an operator (or robot) and is typically measured as a percentage of the total idle time of all the machines in the systems to the operator (or robot) cycle time.

machine language   the set of legal instructions to a machine's processor, expressed in binary notation.

machine vision   *See* robot vision.

macro   *See* macroprogram.

macro cell   a cell in a cellular communication system that has a size that is significantly larger than the cell size of a typical cellular system. Macro cells are sometimes designed with the base station being located in a satellite. Macro cells typically cover large areas such as rural areas in cases where the user density is low.

macro diversity   a diversity technique, used in a cellular communication system, that is based on the transmission of multiple copies of the same signal from transmitters that have a separation that is a significant fraction of the coverage area of a cell. Macro diversity also refers to the concept in a cellular system where coverage at a particular location may be obtained from multiple transmitters, typically utilizing different frequency channels. Macro diversity is typically utilized to overcome the type of signal fading typically known as shadow fading.

macro shadowing   shadowing due to the obstacles in the propagation path of the radio wave. *See also* micro shadowing.

macroinstruction   a short code-like text, defined by the programmer, that the assembler or compiler will recognize and that will result in an inline insertion of a predefined block of code into the source code.

macroprogram   a sequence of macroinstructions.

macrotiming diagram   a graphical display showing how the waveforms vary with time but with a time scale that does not have sufficient resolution to display the delays introduced by the individual basic elements of the digital circuit.

MAD   *See* maximal area density.

made electrode   a ground electrode for a lightning rod which has been especially constructed for the purpose, as opposed to using a builiding frame or water pipe for the purpose.

magic T   *See* magic tee.

magic T junction   *See* magic tee junction.

magic tee   a combination of E-plane (series) and H-plane (shunt) tees forming a hybrid waveguide junction. Typically used to split or couple two microwave signals that are in or out of phase in the same wave guide. Also called a hybrid tee.

magic tee junction   a four-port microwave device, that couples the input signal at port 1 equally into ports 2 and 3, but not into port 4. Input signal at port 3 is coupled equally into ports 1 and 4, but

not into port 2, etc. At microwave frequency, the waveguide junction that makes up this device involves three magnetic plane arms and one electric field plane arm resembling T and hence is called magic tee junction.

maglev   *See* magnetic levitation.

magnet   any object that can sustain an external magnetic field.

magnetic actuator   any device using a magnetic field to apply a force.

magnetic bearing   a component of a machine that uses magnetic force to provide non-contact support for another component moving relative to it.

magnetic bias   a constant magnetic field on which is superimposed a variable, often sinusoidal, perturbation magnetic field in devices like magnetic bearings.

magnetic brake   any device using a magnetic field to retard motion.

magnetic charge density   a fictitious source of the electromagnetic field that quantifies the average number of discrete magnetic charges (also fictitious) per unit volume. The magnetic charge density is often introduced in problems where duality and equivalence concepts are employed.

magnetic circuit   the possible flux paths within a system consisting of a source of flux (electromagnets, permanent magnets), permeable flux carrying materials (steel, nickel) and non-flux carrying materials (aluminum, air).

magnetic clamp   a device employing a magnetic field to deliver a clamping action.

magnetic core memory   a persistent, directly addressable memory consisting of an array of ferrite toruses (cores) each of which stores a single bit. Now obsolete.

magnetic current density   a fictitious source vector in electromagnetics that quantifies the amount of magnetic charge (also fictitious) crossing some cross-sectional area per unit time. The magnetic current density is often introduced in problems where duality and equivalence concepts are employed. The direction of the magnetic current density is in the direction of magnetic charge motion. SI units are volts per square meter.

magnetic damper   any device using a magnetic field to damp motion.

magnetic dipole   an arrangement of one or more magnets to form a magnet system that produces a magnetic field with one pair of opposite poles.

magnetic disk   a persistent, random-access storage device in which data is stored on a magnetic layer on one or both surfaces of a flat disk. *See also* hard disk, floppy disk, diskette.

magnetic drive   *See* magnetic torque coupling.

magnetic drum   a persistent storage device in which data is stored on a magnetic layer on the surface of a cylinder. Now obsolete.

magnetic field   magnetic force field where lines of magnetism exist.

magnetic field integral equation (MFIE)   used in the method of moments to solve for the surface current on an object in terms of the incident magnetic field. The MFIE is valid only for closed surfaces.

magnetic field intensity   a force field that is a measure of the magnitude and direction of the force imparted upon an elemental current normalized to the elemental current's value. Depends on material characteristics. The units are amperes per meter.

magnetic flux   the integral of the component of magnetic flux density perpendicular to a surface, over the given surface.

magnetic flux density   a vector quantifying a magnetic field, so that a particle carrying unit charge experiences unit force when traveling with unit velocity in a direction perpendicular to the magnetic field characterized by unit magnetic flux density. It has the units of volt-seconds per square meter in the SI system of units.

magnetic head   *See* read/write head.

magnetic induction   the flux density within a magnetic material when driven by an external applied field or by its self demagnetizing field, which is the vector sum of the applied field and the intrinsic induction.

magnetic leakage   *See* leakage.

magnetic length   the effective distance between the north and south poles within a magnet, which varies from 0.7 (alnico) to 1.0 (NdFeB, SmCo, hard ferrite) times the physical length of the magnet.

magnetic levitation   (1) noncontact support of an object using magnetic forces. Abbreviated as maglev.

(2) a method of melting metals without contacting a surface. A cone-shaped high-frequency coil produces eddy currents in the metal which are strong enough to both suspend and melt it, generally in a neutral atmosphere.

(3) one of several techniques of suspending a driveshaft within a bearing so that no contact is made between the shaft and other surfaces. *See* magnetic bearing.

(4) one of several techniques for suspending a railroad train above its tracks so that wheels are not needed. Typically, superconducting magnets are needed, and propulsion is by a linear induction motor *cf* whose armature lies along the rails.

magnetic loss   losses in magnetic flux in a magnetic circuit, primarily due to magnetic leakage and fringing. *See also* core loss.

magnetic moment   for a current-carrying coil in an external magnetic field, the ratio of the torque sensed by the coil to the flux density of the external field. In permanent magnets, the product of the polar flux and the magnetic length; the product of the intrinsic flux density and the magnet volume.

magnetic monopole   a magnet system that produces a magnetic field of a single polarity. Although nonexistent, may be approximated by one pole of a very long magnet.

magnetic motor starter   motor starter that uses electromechanical devices such as contactors and relays.

magnetic orientation   the preferred direction of magnetization for an anisotropic magnetic material.

magnetic overload   an overload sensor in a motor controller used to shut off the motor in event of an over current condition. With a magnetic overload, the sensor uses a magnetic coil to sense the overload condition, then trips the overload contact(s). *See also* overload heater, overload relay.

magnetic permeability   tensor relationship between the magnetic field vector and the magnetic flux density vector in a medium with no hysteresis; flux density divided by the magnetic field in scalar media.

magnetic polarization vector   an auxiliary vector in electromagnetics that accounts for the presence of atomic circulating currents in a material. Macroscopically, the magnetic polarization vector is equal to the average number of magnetic dipole moments per unit volume.

magnetic quardrupole   *See* quadrupole.

magnetic recording air gap   term referring to two aspects of a magnetic recording system.

(1) The gap between the poles of a read/write head is often referred to as the "air gap." Even though filled with a solid, it is magnetically equivalent to an air gap. With a recorded wavelength of λ, and a head gap of d, the read signal varies as $\sin(\pi x)/x$ where $x = \lambda/d$. The gap must then have $d < \lambda/2$ for reliable reading.

(2) The space between the head and the recording surface is also referred to as an "air gap." The signal loss for a head at height h from the surface is approximately $55h/\lambda$ dB; if $d = \lambda/5$ the loss is 11 dB.

High recording densities therefore require heads "flying" very close to the recording surface and at a constant height – 0.1 μm or less in modern disks. Such separations are achieved by shaping the disk head so that its aerodynamics force it to fly at the correct separation.

magnetic recording code   method used to record data on a magnetic surface such as disk or tape. Codes that have been used include return-to-zero (RZ), in which two signal pulses are used for every bit — a change from negative to positive pulse (i.e., magnetization in the "negative" direction) is used for a stored 0, a change from a positive to a negative pulse (i.e., magnetization in the "positive" direction) is used for a stored 1, and a return to the demagnetized state is made between bits; non-return-to-zero (NRZ), in which a signal pulse occurs only for a change from 1 to 0 or from 0 to 1; non-return-to-zero-inverted (NRZI), in which a positive or negative pulse is used for — a sign change occurs for two consecutive 1s — and no pulse is used for a 0; double frequency (DF) (also known as frequency modulation (FM)), which is similar to NRZI but which includes an interleaved clock signal on each bit cell; phase encoding (PE), in which a positive pulse is used for a 1 and a negative pulse is used for a 0; return-to-bias (RB), in which magnetic transitions are made for 1s but not for 0s, as in NRZI, but with two transitions for each 1; and modified-return-to-bias (MRB), which is similar to RB, except that a return to the demagnetized state is made for each zero and between two consecutive 1s.

RZ is self-clocking (i.e., a clock signal is not required during readout to determine where the bits lie), but the two-signals-per-bit results in low recording density; also, a 0 cannot be distinguished from a "dropout" (i.e., absence of recorded data). NRZ is not self-clocking and there requires a clocking system for readout; it also requires some means to detect the beginning of a record, does not distinguish between a dropout and a stored bit with no signal, and if one bit is in error, then all succeeding bits will also be in error up to the next signal pulse. NRZI has similar properties to NRZ, except that an error in one bit does not affect succeeding bits. PE is self-clocking.

magnetic resonance imaging (MRI)   (1) a form of medical imaging with tomographic display that represents the density and bonding of protons (primarily in water) in the tissues of the body, based upon the ability of certain atomic nuclei in a magnetic field to absorb and reemit electromagnetic radiation at specific frequencies. Also called nuclear magnetic resonance.

(2) An imaging modality that uses a pulsed radio frequency magnetic field to selectively change the orientation of the magnetization vectors of protons within the object under study. The change in net magnetic moment as the protons relax back to their original orientation is detected and used to form an image.

magnetic saturation   the condition in a magnetic material when an increase in the magnetizing force does not result in a useful increase in the magnetic induction of the material.

magnetic separator   a device employing magnetic fields to separate magnetic materials from nonmagnetic ones.

magnetic stabilization   the act of purposely demagnetizing a magnet with reverse fields or a change in temperature so that no irreversible losses are experienced when the magnet operates under similar conditions in the field.

magnetic susceptibility   the ratio of the magnetization to the applied external field.

Tensor relationship between the magnetic field vector and the magnetization vector in a medium with no hysteresis; magnetization divided by the permeability of free space and the magnetic field in scalar media. It is an indicator of how easily a material is magnetized and has no units in the SI system of units (pure number).

magnetic suspension   *See* magnetic levitation.

magnetic tape   a polyester film sheet coated with a *ferromagnetic* powder, which is used extensively in auxiliary memory. It is produced on a reel, in a cassette, or in a cartridge transportation medium. Often used for backups.

magnetic torque coupling   any device utilizing a magnetic field to transmit torque.

magnetic vector potential   an auxiliary field used to simplify electromagnetic computations. This field satisfies a wave equation, the curl of this field is related to the magnetic field intensity vector field, and the divergence of this field is specified by some gage which is to be specified in each problem.

magnetic-tape track   each bit position across the width of a magnetic tape read/write head, running the entire length of the tape.

magnetization curve   *See* hysteresis curve.

magnetizing current   the current required to magnetize the different parts of a magnetic circuit. It is calculated as the ratio of the total magnetomotive force (F) and the number of turns (N). More or less in transformers, and AC synchronous and induction machines, the magnetizing current is the current through the magnetizing inductance. Denoted by I_m, it is calculated as the ratio of the induced EMF across the magnetizing inductance to its magnetizing reactance X_m.

magneto plasma   a plasma medium that in the presence of a static magnetic field behaves like an anisotropic dielectric medium whose dielectric function is a tensor.

magnetohydrodynamic MHD machine   a form of electric machine in which a stream of electrically conductive gas or liquid is passed through pairs of orthogonally positioned magnetic poles and electrodes. In an MHD generator, the fluid is forced by the prime mover to produce a DC across the electrodes. In the MHD motor, a current across the electrodes through the fluid forces the stream to flow.

magneto-hydrodynamic (MHD) generator   a heat-to-electricity conversion device with an intermediate kinetic energy stage. In the MHD generator, a partially conducting gas is heated by a fuel-fired source or a nuclear reactor to convert the heat energy to kinetic energy, and then passed between the poles of an electromagnet, which converts some of the kinetic energy to electrical energy. The electrical energy is collected through a pair of electrodes situated in the gas channel.

magneto-optic Bragg cell   a magnetically tunable microwave signal processing device that uses optical Bragg diffraction of light from a moving magneto-optic grating generated by the propagation of magnetostatic waves within the magnetic medium.

magneto-optic modulator   any of a class of light modulators that use magneto-optic effects, such as Faraday rotation for light modulation.

magneto-optical Kerr effect   the rotation of the plane of polarization of a linearly polarized beam of light upon reflection from the surface of a perpendicularly magnetized medium.

magneto-resistive head   *See* read/write head.

magnetoimpedance   the change in impedance of a ferromagnetic conductor experiencing a change in applied magnetic field.

magnetomotive force (MMF)   a magnetic circuit term referring to that phenomenon that pushes magnetic flux through the reluctance of the circuit path. MMF is analogous to the concept of electromotive force (voltage) in an electric circuit. For a magnetic core with a single coil of N turns, carrying current I, the MMF is NI, with units of amperes (sometimes expressed as ampere-turns).

magnetoresistance   the change in electrical resistance in a conducting element experiencing a change in applied magnetic field. This is most pronounced when the magnetic field is perpendicular to current flow.

magnetostriction   a change in the length of a ferromagnetic material as the flux changes under the influence of an applied magnetic field, or resulting from domain formation after cooling from above Curie temperature. In an AC device, the steel in the core expands and contracts twice each cycle, creating audible noise (e.g., transformer hum).

magnetostrictive smart material   one of a class of materials with self-adaptively modifiable elastic properties in response to a magnetic field applied in proportion to sensed stress–strain information.

magnetotransport   motion of electrons or holes in a conducting material in the presence of an applied magnetic field.

magnetron   any arrangement of magnets in a sputter deposition or etch system that provides the magnetic field required to trap electrons in closed loops near the cathode, thus enhancing deposition/etch rates.

magnitude (1) the absolute value of a scalar.
(2) the norm of a vector, i.e., the square root of the sum of the squares of the vector components.

magnitude response (1) the magnitude of the frequency response of a system. *See* frequency response, Manhattan distance, city-block distance.
(2) *See* frequency response.

magnitude scaling a procedure for changing the values of network elements in a filter section without affecting the voltage ratio or current ratio transfer function of the section. Resistors and inductors are multiplied by the parameter a, while capacitors are divided by a. If $1 < a < \infty$, then the magnitudes of the impedances are increased.

magnitude squared coherence (MSC) a measure of the degree of synchrony between two electrical signals at specific frequencies.

magnon a polariton in a magnetic medium.

Mahalanobis distance a distance measure used in certain decision rules.
Let $p(\mathbf{x}, c_i)$ be the probability distribution that pattern $\mathbf{x}$ belong to class c_i. A commonly used decision rule is that of looking for $\arg\max_i p(\mathbf{x}, c_i)$. Furthermore, let us assume that $p(\mathbf{x}\,|\,c_i)$ is Gaussian, that is

$$p(\mathbf{x}\,|\,c_i) = \frac{1}{\sqrt{(2\pi)^n \det R_i}}$$
$$\exp\left(-\frac{1}{2}(\mathbf{x}-\mu_i)' R_i^{-1}(\mathbf{x}-\mu_i)\right)$$

and that $\forall i = 1,\ldots,n$ $R_i = R$ (the same covariance matrices). Hence, the previously defined decision rule reduces straightforwardly to a distance evaluation where one looks for $\arg\max_i (\mathbf{x}-\mu_i)' R_i^{-1}(\mathbf{x}-\mu_i)$. This distance is commonly referred to as the Mahalanobis distance. *See also* weighted Euclidean distance.

MAI *See* multiple access interference.

mailbox an operating system abstraction containing buffers to hold messages. Messages are sent to and received from the mailbox by processes.

main beam in antenna theory, the direction in which the global maximum of the radiation pattern occurs.

main memory the highest level of memory hierarchy.

main switch a switch which controls all power to a building's wiring or other electric installation.

mains voltage European term for the voltage at the secondary of the distribution transformer.

mainframe a large centralized machine that supports hundreds of users simultaneously.

maintainability the probability that an inoperable system will be restored to an operational state within the time t.

maintenance the changes made on a system to fix errors, to support new requirements, or to make it more efficient.

major hysteresis loop for a magnetic material, the loop generated as intrinsic or magnetic induction (B_i or B) is plotted with respect to applied field (H) when the material is driven from positive saturation to negative saturation and back, showing the lag of induction with respect to applied field.

major orbit a larger helical orbit of an electron beam in a gyrotron.

majority carrier an electron in an n-type or a hole in a p-type semiconductor.

majority-logic decoding a simple, and in general suboptimal, decoding method for block and convolutional codes based on the orthogonality of the parity-check sums.

male connector   a connector presenting pins to be inserted into a corresponding female connector that presents receptacles.

MAN   *See* metropolitan-area network.

management by exception   *See* coordination by exception.

manipulability measure   is a kind of distance of the manipulator from singular configuration. Mathematically it is defined as follows: $W(q) = \sqrt{\det(J(q)J^T(q))}$, which vanishes at a singular configuration. It is clear that by maximizing this measure, redundancy is exploited to move away from singularities. Partial derivative of the manipulability measure with respect to a vector of generalized positions allows to define an arbitrary vector of joint velocities in the inverse kinematics problem for redundant manipulators.

manipulated input   a quantity influencing the controlled process from outside, available to the controller and used to meet the control objectives; attributes of the inflowing streams of material, energy, or information may serve as manipulated inputs. A manipulated input is defined by a continuous trajectory over given time interval or by a sequence of values at given time instants. Also known as control input.

manipulator workspace   a manipulator workspace defines all existing manipulator positions and orientations that can be obtained from the inverse kinematics problem. The lack of a solution means that the manipulator cannot attain the desired position and orientation because it lies outside of the manipulator's workspace.

Manley–Rowe criteria   *See* Manley–Rowe relations.

Manley–Rowe relations   relations among the intensities of optical fields interacting in a nonlinear material, which can be understood in terms of the discrete nature of the transfer of energy in terms of the emission and absorption of photons.

mantissa   the portion of a floating-point number that represents the digits. *See also* floating-point representation.

manually-controlled shunt capacitors   a bank of shunt capacitors that are controlled via SCADA signals from an operating center as opposed to local automatic control by voltage sensing.

MAP   *See* maximum a posteriori estimator.

mapping   the assignment of one location to a value from a set of possible locations. Often used in the context of memory hierarchies, when distinct addresses in a level of the hierarchy map a subset of the addresses from the level below.

MAR   *See* memory address register.

Marconi, Guglielmo   (1874–1937) Born: Bologna, Italy

Best known for work that led to the development of the commercial radio industry. Marconi's many experiments with radio waves (long wavelength electromagnetic radiation) led to many communications innovations. Marconi traveled to England to find support for his ideas, and there formed the Marconi Wireless Telegraph Company, Ltd. In 1909, Marconi shared the Nobel Prize for physics with K. F. Braun. Marconi's most famous demonstration came in 1901 when he was able to send the first Morse-coded message from the Pondhu, Cornwall, England to St. Johns, Newfoundland, Canada.

Markov chain   a particular case of a Markov process, where the samples take on values from a discrete and countable set. Markov chains are useful signal generation models for digital communication systems with intersymbol inter-ference or convolutional coding, and Markov chain theory is useful in the analysis of error pro-pogation in equalizers, in the calculation of power spectra of line codes, and in the analysis of fram-ing circuits.

Markov model   a modeling technique where the states of the model correspond to states of the system and transitions between the states in the model correspond to system processes.

Markov process   a discrete-time random pro-cess, $\{\Psi_k\}$, that satisfies $p(\psi_{k+1}|\psi_k, \psi_{k-1}, \ldots) = p(\psi_{k+1}|\psi_k)$. In other words, the future sam-ple ψ_{k+1} is independent of past samples ψ_{k-1}, $\psi_{k-2}, \ldots$ if the present sample $\Psi_k = \psi_k$ is known.

Markov random field   an extension of the def-inition of Markov processes to two dimensions. Consider any closed contour Γ, and denote by Γ_i and Γ_o the points interior and exterior to Γ, respec-tively. Then a process ψ is a Markov random field (MRF) if, conditioned on $\psi(\Gamma)$, the sets $\psi(\Gamma_i)$ and $\psi(\Gamma_o)$ are independent. That is,

$$p(\psi(\Gamma_i), \psi(\Gamma_o)|\psi(\Gamma)) = p(\psi(\Gamma_i)|\psi(\Gamma))$$
$$\cdot p(\psi(\Gamma_o)|\psi(\Gamma))$$

See Markov process, conditional statistic.

Marr–Hildreth operator   (1) edge-detection operator, also called Laplacian-of-Gaussian or Gaussian-smoothed-Laplacian, defined by

$$\nabla^2 G = -1\sqrt{2\pi}\sigma^3 \left(1 - \frac{x^2 + y^2}{\sigma^2}\right)$$
$$\times e^{-(x^2+y^2)2\sigma^2}\nabla^2 G.$$

It generates a smoothed isotropic second deriva-tive. Zero crossings of the output correspond to extrema of first derivative and thus include edge points.

(2) The complete edge scheme proposed by Marr and Hildreth, including use of the $\nabla^2 G$ op-erator at several scales (i.e., Gaussian variances), and aggregation of their outputs.

Marx generator   a high-voltage pulse genera-tor capable of charging capacitors in parallel and discharging them in series.

maser   acronym for microwave amplification by stimulated emission of radiation.

maser amplifier   usually refers to a medium that amplifies microwaves by the process of stimu-lated emission; sometimes refers to amplification of some other field (nonoptical electromagnetic, phonon, exciton, neutrino, etc.) or some other pro-cess (nonlinear optics, Brillouin scattering, Ra-man scattering, etc.).

maser oscillator   oscillator usually producing a microwave frequency output and usually based on amplification by stimulated emission in a resonant cavity.

mask   (1) in digital computing, to specify a number of values that allow some entities in a set, and disallow the others in the set, from being active or valid. For example, masking an interrupt.

(2) for semiconductor manufacturing, a de-vice used to selectively block photolithographic exposure of sensitized coating used for prevent-ing a subsequent etching process from removing material. A mask is analogous to a negative in conventional photography.

(3) a glass or quartz plate containing infor-mation (encoded as a variation in transmittance and/or phase) about the features to be printed. Also called a photomask or a reticle.

(4) in image processing, a small set of pixels, such as a 3×3 square, that is used to transform an image. Conceptually, the mask is centered above every input pixel, each pixel in the mask is mul-tiplied by the corresponding input pixel under it and the output (transformed) pixel is the sum of these products. If the mask is rotated $180°$ be-fore the arithmetic is performed, the result is a 2-D convolution and the mask represents the im-pulse response function of a linear, space-invariant

system. Also called a kernel. *See* convolution, impulse response function, kernel.

mask aligner   a tool that aligns a photomask to a resist-coated wafer and then exposes the pattern of the photomask into the resist.

mask biasing   the process of changing the size or shape of the mask feature in order for the printed feature size to more closely match the nominal or desired feature size.

mask blank   a blank mask substrate (e.g., quartz) coated with an absorber (e.g., chrome), and sometimes with resist, and used to make a mask.

mask linearity   the relationship of printed resist feature width to mask feature width for a given process.

mask programming   programming a semiconductor read-only-memory (ROM) by modifying one or more of the masks used in the semiconductor manufacturing process.

mask set   consists of the dozen or so (varies with process and company) individual masks that are required to complete a MMIC wafer fabrication from start to finish. Examples of masks or mask levels are "first level metal" (defines all the primary metal structure on the circuit), "capacitor top plate" (defines the pattern for the metal used to form the top plate of MIM capacitors), and "dielectric etch" (defines areas where dielectric (insulator) material will be removed after coating the entire wafer with it).

maskable interrupt   interrupt that can be postponed to permit a higher-priority interrupt by setting mask bits in a control register. *See also* non-maskable interrupt.

masking   a phenomenon in human vision in which two patterns P_1 and $P_1 + P_2$ cannot be discriminated even though P_2 is visible when seen alone. P_1 is said to mask P_2.

mass storage   a storage for large amounts of data.

massively parallel architecture   a computer system architecture characterized by the presence of large numbers of CPUs that can execute instructions in parallel. The largest examples can process thousands of instructions in parallel, and provide efficient pathways to pass data from one CPU to another.

massively parallel processor   a system that employs a large number, typically 1000 or more, of processors operating in parallel.

master   the system component responsible for controlling a number of others (called slaves).

master boot record   a record of the disk containing the first code and table that are loaded at the bootstrap of the computer. It is read even before the partition table sector.

master control relay (MCR)   used in programmable logic controllers to secure entire programs, or just certain rungs of a program. An MCR will override any timer condition, whether it be time-on or time-off, and place all contacts in the program to a safe position whenever conditions warrant.

master copy   in coherence protocols, the copy of the object that is guaranteed to hold the "correct" contents for the object. Coherence protocols can be designed around the tagging of master copies. The master copy can be read (rather then the copy in slower memory) to speed program execution, and it can be written, provided that all other copies are invalidated. *See also* MESI protocol.

master-oscillator-power-amplifier   (MOPA) laser system in which the output from a highly stabilized low-power laser oscillator is amplified by one or more high-power laser amplifiers.

master–slave flip-flop   a two-stage flip-flop in which the first stage buffers an input signal, and on a specific clock transition the second stage captures and outputs the state of the input.

MAT   *See* medial axis transform.

matched filter   a filter matched to a certain known signal waveform. The matched filter is the complex conjugate of the known signal waveform. If the known signal waveform falls in the matched filter, the output sample at the optimum sampling instant gives the received signal energy.

matched load   load that does not reflect any energy back into the transmission line. It could be a load equal to the characteristic impedance of the transmission line or a structure with electromagnetic absorbing properties.

matched uncertainty   *See* matching condition.

matching   when referring to circuits, the process by which a network is placed between a load and a transmission line in order to transform the load impedance to the characteristic impedance of said line and thus eliminate the presence of standing waves on the line.

matching condition   the condition that requires that the uncertainties affect the plant dynamics the same way as the control input does. For example, in the following model of a dynamical system, the matching condition is satisfied,

$$\dot{\mathbf{x}} = \mathbf{A}\mathbf{x} + \mathbf{B}(\mathbf{u} + \mathbf{d}),$$

where the vector function **d** models matched uncertainties.

matching conditions   conditions imposed on the structure of the uncertainty, which may be viewed as the assumption that uncertainty drives the state equations not stronger than a control variable. For linear systems, the matching conditions are given as the assumptions imposed on the perturbations of system matrices ΔA, ΔB and disturbances $Cv(t)$ in the following way:

$$\Delta A = BD; \quad \Delta B = BG; \quad C = BF$$

where D, G, F are unknown but bounded matrices and B is an input matrix. It enables description of the uncertain linear system by the following state equation:

$$\dot{x} = Ax + B(u + e)$$

where e is the entire system uncertainty moved into the input of the system.

matching elements   elements such as posts, screws, etc., often used in order to achieve the desired reflection coefficient of a given microwave component.

matching network   an electric circuit designed to maximize the transfer of electric power from an electrical source to an electrical load. Maximum power is transferred from a source with an output impedance that is the complex conjugate of the input impedance of the load it is driving. A matching network is connected between a source and its load. Its input impedance is the complex conjugate of the source output impedance, while its output impedance is the complex conjugate of the load impedance, thereby matching the source to the load and ensuring maximum transfer of power.

matching stub   matching technique that employs short-circuited (or open-circuited) sections of transmission lines as reactive elements.

material dispersion   wavelength dependence of the pulse velocity. It is caused by the refractive index variation with wavelength of glass.

math coprocessor   a separate chip, additional to the CPU, that offloads many of the computation-intensive tasks from the CPU.

mathematical modeling   a mathematical description of the interrelations between different quantities of a given process. In particular, a mathematical description of a relation between the input and output variables of the process. *See also* truth model and design model.

mathematical morphology   an algebraic theory of non-linear image transformations based on set-theoretical (or lattice-theoretical) operations, which is generally considered as a counterpart to the classical linear filtering approach of signal processing. This methodology in image processing arose in 1964 through the works of Matheron and Serra at Fontainebleau (France), was developed in the seventies by Sternberg at Ann Arbor, MI (Mich., USA), and became internationally known in the eighties. It has been successfully applied in various fields requiring an analysis of the structure of materials from their images; for example biomedical microscopy, stereology, mineralogy, petrography. *See* closing, dilation, erosion, morphological operator, opening.

matrix   an n by m matrix is an array of $n \times m$ numbers of height n and width m representing a linear map from an m-dimensional space into an n-dimensional space. An example of a 2×2 real matrix is

$$A = \begin{pmatrix} 2.0 & 3.1 \\ -5.4 & 6.3 \end{pmatrix}$$

See circulant matrix, Hermitian matrix, orthogonal matrix, positive definite matrix, positive semidefinite matrix, singular matrix, Toeplitz matrix.

matrix addressing   the control of a display panel consisting of individual light-producing elements by arranging the control system to address each element in a row/column configuration.

matrix element   in quantum mechanics, the expectation value of a quantum mechanical operator that is associated with a particular pair of basis states. The term matrix element derives from the fact that the expectation values associated with all possible pairs of states can be written as a matrix.

matrix of configuration of information system   in Pawlak's information system $S = (U, A)$ whose universe U has n members x_i and the set A consists of attributes $\mathbf{a}_j$, the elements of the universe are linked with each other. Connections between the elements of the universe U may be given, for example, by a function

$$\varphi : U \times U \to \{-1, 0, 1\}.$$

The pair (U, φ) is called the configuration of the information system S. The matrix of the information system S is an $n \times n$ matrix whose elements are

$$c_{ij} = \varphi \left(x_i, x_j \right).$$

matrix optics   formalism for the analysis and synthesis of optical systems in which each system element is represented by a matrix, the overall system being representable as a product of the elemental matrices.

matrixing   in a color television transmitter, the process of converting the three color signals (red, green, blue) into the color-difference signals that modulate the chrominance subcarrier. In a color TV receiver, the process of converting the color-difference signals into the red, green, and blue signals.

Mauchley, John Graham   (1907–1980) Born: Cincinnati, Ohio

Best known as one of the designers of ENIAC, an early electronic computer. It was Mauchley, in 1942, who wrote a proposal to the Army for the design of a calculating machine to calculate trajectory tables for their new artillery. Mauchley was a lecturer at the Moore School of Electrical Engineering at the time. The school was awarded the contract and he and J. Presper Eckert were the principal designers of ENIAC. They later went on to form their own company. Mauchley was the software engineer behind the development of one of the first successful commercial computers, UNIVAC I.

max operation   an operation on two or more variables where the resultant value is formed by taking the largest value, or maximum, among these variables.

Max–Lloyd scalar quantization   *See* Lloyd–Max scalar quantization.

Max–Lloyd SQ   *See* Lloyd-Max scalar quantization.

max-min composition   a frequently used method of composition of two fuzzy relations which, as the name implies, makes use of the min operation followed by the max operation.

maximal area density (MAD)   for a magnetic disk, the maximum number of bits that can be stored per square inch. Computed by multiplying the bits per inch in a disk track times the number of tracks per inch of media.

maximally flat   a circuit response in which the low-pass prototype attenuation loss (expressed in decibels) has a smooth response.

maximally flat delay (MHD) filter   a filter having a time delay that is as flat (constant) as possible versus frequency while maintaining a monotonic characteristic.

maximum a posteriori (MAP)   estimator to estimate a random x by maximizing its posterior probability; that is,

$$\hat{x}_{MAP} = \arg_x \max p(x|y),$$

where y represents an observation. x is explicitly modeled as a random quantity with known prior statistics. *See* maximum likelihood estimation, Bayesian estimation. Bayes' rule prior statistics, posterior statistics.

maximum a posteriori probability   the probability of a certain outcome of a random variable given certain observations related to the random variable, i.e., x was transmitted and y was received, then $P(x \mid y)$ is the *a posteriori* probability. The maximum *a posteriori* probability is found by considering all valid realizations of x.

For example, between two strings A and B it is defined as

$$D_{MPP} = Pr\{B|A\}$$

where $Pr(B|A)$ is the probability that A is changed into B. Sometimes called maximum posterior probability (MPP) even though, strictly speaking, it is a similarity rather than distance measure.

maximum accumulated matching   a defuzzification scheme for a classification problem in which a pattern is assigned to the class outputed by the rules that accumulates the maximum firing degree.

maximum distance separable (MDS) code   an (n, k) linear binary or nonbinary block code, with minimum distance $d_{min} = n - k + 1$. Except for the trivial repetition codes, there are no binary MDS codes. Nonbinary MDS codes such as Reed–Solomon codes do exist.

maximum effective aperture   in antenna theory, the ratio of the time-average power available at the terminals of an antenna due to power incident in the direction of the main beam that is polarized for maximum reception to the time-average power density of the incident field.

maximum entropy   a procedure that maximizes the entropy of a signal process.

maximum entropy inequality   an information theoretic inequality, upper bounding the possible value of entropy for probability distributions subject to certain moment constraints. A common example states that for any probability density function $f(x)$, subject to a power (second moment) constraint, $\int x^2 f(x) \leq \sigma^2$,

$$h(f) \leq \frac{1}{2} h(2\pi e \sigma^2).$$

See also differential entropy.

maximum entropy restoration   an iterative method of image restoration. At each iteration, an image is chosen whose Fourier transform agrees with that of the principal solution, i.e., the model of degradation, while maximizing the entropy.

maximum excitation limiter   a controller that is used to limit the maximum amount of field current, or over-excitation, at a synchronous generator. This excitation limit is set by rotor winding heating limit.

maximum input power   the maximum incident RF power that will be applied to a component or circuit without either damage or performance degradation, expressed in watts. The user normally will specify whether it is damage or degradation that is important. If performance degradation is, then how the degradation is manifested should be specified.

maximum likelihood estimation   to estimate an unknown x by maximizing the conditional probability of the observations; that is,

$$\hat{x}_{ML} = \arg_x \max p(y|x)$$

where y represents an observation. x is considered unknown; it is not a statistical quantity. *See* maximum a posteriori estimation, Bayesian estimation. Bayes' rule, prior statistics.

maximum matching   a defuzzification scheme for a classification problem in which a pattern is

signed to the class outputed by the rule that achieve the maximum firing degree.

maximum permissible exposure (MPE)   the limit adopted by a standards-setting body for exposures of unlimited duration. Generally, some allowance is made in the standard for time and/or space averaging of higher exposure conditions.

maximum posterior probability distance   *See* maximum *a posteriori* probability.

maximum rated FET gate-to-source voltage   the maximum gate-to-source voltage that the device is designed to function without damage due to breakdown, as determined by the device manufacturer. The actual breakdown voltage may be much larger than this voltage, dependent process variation and manufacturer derating criteria and margins. This voltage rating is usually in the -2.0 volt (low noise devices) to -10 volt (power devices) range.

maximum stable gain (MSG)   the maximum gain derived from a transistor, which is obtained after making a stable condition by adding some excess loss elements for the case when a simultaneous matching for input and output of the transistor causes an unstable condition.

maximum stable power gain   figure of merit specified as the maximum value of transducer gain for which stability factor K is equal to one. Equivalently, this represents the maximum value of transducer gain of a circuit or device when the device is terminated with impedance values, or by using other methods such as feedback, and so on, such that K is equal to one.

maximum transducer power gain   maximum value of transducer power gain a circuit or device exhibits; occurs when the input and output ports of the circuit are terminated with simultaneous conjugate match conditions. The transducer power gain is defined as the ratio of power delivered to a load to the power available from the source.

maximum-likelihood   the maximum of the likelihood function, $p(y|x)$ or equivalently, the log-likelihood function. Maximum-likelihood is an optimality criterion that is used for both detection and estimation.

maximum-likelihood decoding   a scheme that computes the conditional probability for all the code words given the received sequence and identifies the code word with maximum conditional probability as the transmitted word. Viterbi algorithm is the simplest way to realize maximum-likelihood decoding.

maxterm   a Boolean sum term in which each variable is represented in either true or complement form only once. For example, $x + y' + w + z'$ is a maxterm for a four variable function.

Maxwell's equations   a set of four vector equations published in 1873 that govern the generation and time evolution of electromagnetic fields of arbitrary electric source distributions. Sometimes called the Maxwell–Heaviside equations because Heaviside first wrote them in their familiar vector form.

For fictitious magnetic current density $\overline{J}$ and charge density ρ, electric field E, magnetic field H, μ is the permeability and ϵ is the dielectric constant or permittivity.

Maxwell's equations take on the following form:

$$\overline{\nabla} \cdot \overline{H} = 0$$
$$\overline{\nabla} \times \overline{H} = \mu \overline{J} \text{ (Ampere's law)}$$
$$\overline{\nabla} \times \overline{E} = -\frac{\partial \overline{H}}{\partial t} \text{ (Faraday's law of induction)}$$
$$\overline{\nabla} \times \overline{J} = -\frac{\partial \overline{\rho}}{\partial t}$$

Maxwell, James Clerk   (1831–1879) Born: Edinburgh, Scotland

Maxwell is best known as the greatest theoretical physicist of the 19th century. It was Maxwell who discovered, among other things, that light consisted of waves. He developed the fundamental equations describing electromagnetic fields in his work, *A Dynamical Theory of the Electromagnetic Field,* published in 1864. Maxwell also gave us the mathematical foundation for the kinetic theory of gases. Maxwell's life was cut short by cancer, and thus he was unable to see his greatest theoretical propositions proven by experiment.

MBE   *See* molecular beam epitaxy.

MBP   *See* morphotropic phase boundary.

McCulloch–Pitts neuron   originally a linear threshold unit that responded with a binary output at time $t + 1$ to an input applied at time t. In current usage, usually a linear threshold unit.

MCFC   *See* molten carbonate fuel cell.

MCM   a unit of area used to specify the cross-sectional area of a wire, equal to 1000 circular mils. *See* multi-chip module.

MCM-D   *See* deposited multi-chip module.

MCM-L   *See* laminate multi-chip module.

MCP   *See* motor circuit protector.

MCR   *See* master control relay.

MCT   *See* metal-oxide semiconductor controlled thyristor.

MDA   *See* monochrome display adapter.

MDR   *See* memory data register.

MDS code   *See* maximum distance separable code.

MDT   *See* mobile data terminal.

Meachem bridge   a bridge circuit where one of the arms (the reactance arm) is a series connection of an inductance, a capacitor, and a resistor, all three other arms are resistors. The Meachem bridge has a very steep phase frequency characteristics in the vicinity of the reactance arm resonance frequency. The steepness increases with higher Q-factor of the reactance arm and better tuning of the bridge balance at the resonance frequency. This combination of the bridge and reactance selectivity properties has secured the main application of the Meachem bridge as the feedback circuit of high precision crystal oscillators.

Meachem-bridge oscillator   an oscillator where the Meachem bridge is used in the amplifier feedback. This circuit is usually used in design of high-frequency stability crystal oscillators where the bridge reactance branch is substituted by a crystal. Tuning of the bridge allows to obtain an extremely steep phase frequency response of the bridge near the oscillation frequency which coincides with the crystal series-mode frequency.

mean   *See* mean value.

mean delay   the time interval between transmission of a pulse through a wideband communication channel and the instant corresponding to the centroid of its power-delay profile.

mean ergodic theorem   a mathematical theorem that gives the necessary and/or sufficient conditions for a random process to be ergodic in mean. Let $x(n)$ be a wide-sense stationary random process with autocorrelation sequence $c_x(k)$. A necessary and sufficient condition for $x(n)$ to be ergodic in the mean is:

$$\lim_{N->\infty} \sum_{k=0}^{N-1} c_x(k) = 0$$

The sufficient conditions for $x(n)$ to be ergodic in the mean are that $c_x(0) < \infty$ and $\lim_{k->\infty} c_x(k) = 0$.

mean filter   a filter that takes the mean of the various input signal components, or in the case of an image, the mean of all the pixel intensity values within the neighborhood of the current pixel.

mean free path   the length of the straight line segment travelled by a photon between two successive hits at scattering centers of an inhomogeneous medium.

mean of a random variable   *See* expected value of a random variable.

mean of a stochastic process   the expected value of a stochastic process at some point in time.

mean of max method   a method of defuzzification in which the resultant defuzzified quantity is obtained from the mean of the maximum grades of membership.

mean opinion score (MOS)   a subjective measure of the human-perceived quality of a particular telecommunication parameter, e.g., transmitted voice quality. Assessed by methodical exposure of human test subjects to stimuli corrupted with a known level of technical imperfection, and then requesting the test subjects to rate the stimuli on a subjective quality scale.

mean pyramids   the approach of forming hierarchies by averaging over blocks of pixels (typically 2×2), thus eliminating the difficulties associated with subsampling approaches.

mean squared error (MSE)   measure of the difference between a discrete time signal x_i, defined over $[1 \ldots n]$, and a degraded, restored or otherwise processed version of the signal x_i, defined as $MSE = \ln \sum_{i=1}^{n} (x_i - x_i)^2$. MSE is sometimes normalized by dividing by $\sum_{i=1}^{n} (x_i)^2$.

mean time to failure (MTTF)   given as the expected working lifetime for a given part, in a given environment, as

$$\text{MTTF} = \int_0^\infty r(t)\,dt$$

where $r(t)$ is a reliability function for the part. If the failure rate $\frac{1}{\lambda}$ is constant, then

$$\text{MTTF} = \frac{1}{\lambda}.$$

mean time to repair (MTTR)   a prediction for the amount of time taken to repair a given part or system.

mean value   the expected value of a random variable or function. The mean value of a function f is defined as

$$m_f = E(f) = \int_{-\infty}^\infty f p(f)\,df$$

where $p(f)$ is the probability density function of f. *See* expectation.

mean-q convergence   for a stochastic process, the property that the mean of the absolute difference between that process and some random variable, raised to the qth power ($q > 0$) approaches zero in time.

mean-square estimation   an estimation scheme in which the cost function is the mean-square error.

mean/residual vector quantization (MRVQ) a prediction is made of the original image based on a limited set of data, and then a residual image is formed by taking the difference between the prediction and the original image. Prediction data are encoded using a scalar quantizer and the residual image is encoded using a vector quantizer.

mean/residual VQ   *See* mean/residual vector quantization.

measurement system   the sum of all stimulus and response instrumentation, device under test, interconnect, environmental variables, and the interaction among all the elements.

mechanical degree   the spatial angle of the stator of a machine, expressed in radians or degrees. Mechanical degree also represents one revolution of the rotor ($360°$).

mechanical loss   *See* rotational loss.

mechanical power   energy per unit time associated with mechanical motion. For linear motion, it is force $\times$ speed; for rotational motion, it is torque $\times$ rotational speed. In SI units, $P = \omega T$, where P is in watts, T is in newton-meters, and ω is in rads per second.

media-access control   a sublayer of the link layer protocol whose implementation is specific to the type of physical medium over which communication takes place and which controls access to that medium.

medial axis   let X be a non-empty bounded set in a Euclidean space; assume that X is topologically closed, in other words it contains its border ∂X. For every point x in X, the Euclidean distance $d(x, X^c)$ of x to the complement X^c of X is equal to the distance $d(x, \partial X)$ of x to the border ∂X; let $B(x)$ be the closed ball of radius $d(x, \partial X)$ centered about x; it is the greatest closed ball centered about x and included in X. The medial axis transform is the operation transforming X into its medial axis or distance skeleton $S(X)$ which can be defined in several ways:

(1) $S(X)$ is the set of points x such that $B(x)$ is not included in $B(y)$ for any other y, y in X; in other words; $B(x)$ is, among all balls included in X maximal for the inclusion.

(2) $S(X)$ is the set of points x such that $B(x)$ intersects ∂X in at least two points.

(3) $S(X)$ is the set of points x at which the distance function, $f(x) = d(x, \partial X)$ is not differentiable.

These three definitions coincide up to closure, in the sense that the three skeletons that they define may be different, but have the same topological closure. When X is connected and has a connected interior, its skeleton $S(X)$ is also connected and has the same number of holes as X. For digital figures and a digital distance d, one uses only the first definition: the skeleton $S(X)$ is the set of centers of maximal "balls" (in the sense of distance d) included in X; but then the skeleton of a connected set is no longer guaranteed to be connected. *See* distance, morphological skeleton.

medial axis transform (MAT)   *See* medial axis.

median filter   a filter which takes the median of the various input signal components, or in the case of an image, the median of all the pixel intensity values within the neighborhood of the current pixel, the median being defined as the center value of the ordered signal components.

medical imaging   a multi-disciplinary field that uses imaging scanners to reveal the internal anatomic structure and physiologic processes of the body to facilitate clinical diagnoses. *See* X-ray CT, magnetic resonance imaging, ultrasound, positron emission tomography, and radiography.

medium-scale integration (MSI)   (1) an early level of integration circuit fabrication that allowed approximately between 12 and 100 gates on one chip.

(2) a single packaged IC device with 12 to 99 gate-equivalent circuits.

megacell   a cell with the radius of 20–100 km. *See also* cell.

megaflop (MFLOP)   one million floating point operations per second. Usually applied as a measurement of the speed of a computer when executing scientific problems and describes how many floating point operations were executed in the program.

meggar   a power system device for measuring high-voltage insulation or ground connections.

mel scale   the mel is the unit of pitch.

Mellin transform   a transform often arises in the study of wideband signals. The Mellin transform $F_M(s)$ of a function $f(t)$ defined on the positive real axis $0 < t < \infty$ is $F_M(s) = \int_0^\infty f(t) t^{s-1} dt$. The integral in general only exists for complex values of $s = a + jb$ for $a_1 < a < a_2$, where a_1 and a_2 depend on the function $f(t)$.

melting time   the time required for current to melt the fusible element of a fuse.

membership function   a possibility function, with values ranging from 0 to 1, that describes the degree of compatibility, or degree of truth, that an element or object belongs to a fuzzy set. A membership function value of 0 implies that the corresponding element is definitely not an element of the fuzzy set while a value of 1 implies definite membership. Values between 0 and 1 implies a fuzzy (non-crisp) degree of membership.

Rigorously, let $\mu(.)$ be a membership function defining the membership value of an element x of an element of discourse X to a fuzzy set. If A is a fuzzy set, and x is an element of A, then the membership function takes values in the interval $[a, b]$ of the real line ($\mu_A(x) : X \Rightarrow [a, b]$, $a, b \in \Re$).

Usually, fuzzy sets are modeled with a normalized membership function ($\mu_A(x) : X \Rightarrow [0, 1]$). Some examples of membership functions are shown in the figure. If A is a crisp (nonfuzzy) set, $\mu_A(x)$ is 1 if x belongs to A and 0, otherwise ($\mu_A(x) : U \Rightarrow \{0, 1\}$). *See also* crisp set, fuzzy set.

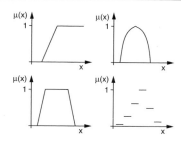

Membership function: some examples of normalized fuzzy sets.

membership grade   the degree to which an element or object belongs to a fuzzy set. It is also referred to as degree of membership.

membrane   the functional boundary of a cell. Nerve cells possess membranes that are excitable by virtue of their nonlinear electrical conductance properties.

membrane probe   used for performing high power on wafer test for large periphery MMIC power amplifiers and discrete devices with non-50-W interfaces.

memory   (1) area for storing computer instructions and data for either short-term or long-term purposes.

(2) the property of a display pixel that allows it to remain stable in an initially established state of luminance. Memory gives a display high luminance and absence of flicker.

memory access time   the time from when a read (i.e., load) request is submitted to memory to the time when the corresponding data becomes available. Usually smaller than the memory cycle time.

memory address computation   the computation required to produce an effective memory address; may include indexing and translation from a virtual to a physical address.

memory address register (MAR)   a register inside the CPU that holds the address of the memory location being accessed while the access is taking place.

memory alignment   matching data to the physical characteristics of the computer memory. Computer memory is generally addressed in bytes, while memories handle data in units of 4, 8, or 16 bytes. If the "memory width" is 64 bits, then reading or writing an 8 byte (64 bit) quantity is more efficient if data words are aligned to the 64 bit words of the physical memory. Data that is not aligned may require more memory accesses and more-or-less complex masking and shifting, all of which slow the operations.

Some computers insist that operands be properly aligned, often raising an exception or interrupt on unaligned addresses. Others allow unaligned data, but at the cost of lower performance.

memory allocation   the act of reserving memory for a particular process.

memory bandwidth   the maximum amount of data per unit time that can be transferred between a processor and memory.

memory bank   a subdivision of memory that can be accessed independently of (and often in parallel with) other memory banks.

memory bank conflict   conflict when multiple memory accesses are issued to the same memory bank, leading to additional buffer delay for such accesses that reach the memory bank while it is busy serving a previous access. *See also* interleaved memory.

memory block   contiguous unit of data that is transferred between two adjacent levels of a memory hierarchy. The size of a block will vary according to the distance from the CPU, increasing as levels get farther from the CPU, in order to make transfers efficient.

memory bounds register   register used to ensure that references to memory fall within the space assigned to the process issuing the references; typically, one register holds a lower bound, another holds the corresponding upper bound, and accesses are restricted to the addresses delimited by the two.

memory cell   a part in a semiconductor memory holding one bit (a zero or a one) of information. A memory is typically organized as a two-dimensional matrix of cells, with "word lines" running horizontally through the rows, and "bit lines" running vertically connecting all cells in that column together. *See also* bit line.

memory compaction   the shuffling of data in fragmented memory in order to obtain sufficiently large holes. *See also* memory fragmentation.

memory cycle   the sequence of states of a memory bus or a memory (sub-)system during a read or write. A memory cycle is usually uninterruptible.

memory cycle time   the time that must elapse between two successive memory operations. Usually larger than the memory access time.

memory data register (MDR)   a register inside the CPU that holds data being transferred to or from memory while the access is taking place.

memory density   the amount of storage per unit; specifically, the amount of storage per unit surface or per chip.

memory element   a bistable device or element that provides data storage for a logic 1 or a logic 0.

memory fragmentation   *See* internal fragmentation, external fragmentation.

memory hierarchy   *See* hierarchical memory.

memory interleaving   *See* interleaved memory.

memory latency   the time between the issue of a memory operation and the completion of the operation. May be less than the time for a memory cycle.

memory management unit (MMU)   a hardware device that interfaces between the central processing unit (CPU) and memory, and may perform memory protection, translation of virtual to physical addresses, and other functions. An MMU will translate virtual addresses from the processor into real addresses for the memory.

memory mapped I/O   I/O scheme in which I/O control and data "registers" and buffers are locations in main memory and are manipulated through the use of ordinary instructions. Computers with this architecture do not have specific I/O commands. Instead, devices are treated as memory locations. This simplifies the structure of the computer's instruction set.

memory mapping   (1) the extension of a processor-generated address into a longer address, in order to create a large virtual address or to extend a virtual address that is too short to address all of real memory when translated.

(2) the mapping between the logical memory space and the physical memory space, i.e., the mapping of virtual addresses to real addresses.

memory module   a physical component used in the implementation of a memory. *See also* memory bank, interleaved memory.

memory partitioning   when multiple processes share the physical main memory, the memory is partitioned between the processes. If the partitioning is static, and the amount of main memory for each process is not changed during the execution, the memory partitioning is fixed, otherwise it is said to be dynamic.

memory port   an access path to a memory unit.

memory protection   a method for controlling access to memory; e.g., a process may have no access, or read-only access, or read/write access to a given part of memory. The control is typically provided by a combination of hardware and software.

memory reference   a read of one item of data (usually a word) from memory or a write of one item of data to memory (same as memory reference).

memory reference instruction   an instruction that communicates with virtual memory, writing to it (store) or reading from it (load).

memory refresh   the process of recharging the capacitive storage cells used in dynamic RAMs. DRAMs must have every row accessed within a certain time window or the contents will be lost. This is done as a process more or less transparent to the normal functionality of the memory, and affects the timing of the DRAM.

memory select line   a control line used to determine whether a unit of memory will participate in a given memory access.

memory stride   the difference between two successive addresses presented to memory. An interleaved memory with a simple assignment of addresses performs best when reference strides are 1, as the addresses then fall in distinct banks.

memory swapping   the transfer of memory blocks from one level of the memory hierarchy to the next lower level and their replacement with blocks from the latter level. Usually used to refer to pages being moved between main memory and disk.

memory width   the number of bits stored in a word of memory. The same as the width.

memory word   the total number of bits that may be stored in each addressable memory location.

memoryless system   (1) a system whose outputs only depend on present values, not past or future values, i.e., $y[n_0]$ depends only on $x[n_0] \ \forall \ n_0$.

(2) a system whose output at any instant in time depends only on the input at that instant of time. The impulse response of a linear, time-invariant, memoryless system is the impulse itself, multiplied by a constant. *See* impulse, impulse response function, linear time invariant (LTI) system, system, system with memory.

MEMS   *See* microelectromechanical system.

MEMS for microwave   *See* microelectromechanical system for microwave.

mercurous chloride   an acousto-optic material with very slow acoustic velocity and thus very high theoretical acousto-optic figure-of-merit. The slow acoustic velocity also gives potential for high time-bandwidth deflectors.

mercury delay-line memory   *See* ultrasonic memory.

meridional ray   a ray that is contained in a plane passing through the fiber axis.

MESFET   *See* metal-electrode semiconductor field-effect transistor.

mesh analysis   a circuit analysis technique in which KVL is used to determine the mesh currents in a network. A mesh is a loop that does not contain any loops within it.

mesh networks   an interconnection network in which processors are placed in a two-dimensional grid of wires, with processors at the intersections of the wires in the x and y dimensions. "Mesh" is occasionally used to refer to similar networks of higher than two dimensionality.

meshed network a complete interconnection of all nodes in a network by means of point-to-point links. They are usually a combination of ring and star networks.

MESI protocol a cache coherence protocol for a single bus multiprocessor. Each cache line exists in one of four states, modified (M), exclusive (E), shared (S), or invalid (I).

mesopic formally, a description of luminances under which both human rod and cone cells are active. The mesopic luminance range lies between the photopic and scotopic ranges. Informally, describing twilight luminances. *See* photopic, scotopic.

message passing in object orientation, the exchange of messages between objects.

message switching a service-oriented class of communication in which messages are exchanged among terminating equipment by traversing a set of switching nodes in a store-and-forward manner. This is analogous to an ordinary postal system. The destination terminal need not be active at the same time as the originator in order that the message exchange take place.

message-passing system a multiprocessor system that uses messages passed among the processors to coordinate and synchronize the activities in the processors.

messenger cable a fully-insulated three-phase aerial cable in which three individual insulated conductors are carried on insulated looms hung upon a bare messenger wire. Such cable is used frequently in distribution work.

messenger wire a grounded wire which is used to structurally support an aerial cable.

metabolic process the method by which cells use oxygen and produce carbon dioxide and heat.

metadyne a DC machine with more than two brush sets per pair of poles. The additional brushes are located in the direct axis for the armature MMF to provide most of the excitation for higher gains.

metal 1 *See* first metal layer.

metal 2 the second layer applied in a fabrication process. In general the nth layer of the fabrication process is called metal n.

metal halide molecule formed by the reaction of metals and halogen atoms.

metal-electrode semiconductor field-effect transistor (MESFET) a specific type of FET that is the dominant active (amplifying) device in GaAs MMICs. An FET is composed of three terminals called the gate, drain, and source, and a conducting "channel." In an amplifier application, the source is connected to ground, and DC bias is applied between the drain and source causing a current to flow in the channel. The current flow is controlled and "modulated" by the AC or DC voltage applied to the gate.

metal–insulator–metal (MIM) capacitor a capacitor, which has a thin insulator layer between two metal electrodes. Generally, this capacitor is fabricated in semiconductor process, and this insulator layer provides high capacitance. Two extreme behaviors of a capacitor are that it will act as an open circuit to low frequencies or DC (zero frequency), and as a short frequency at a sufficiently high frequency (how high is determined by the capacitor value). Also called a thin film capacitor.

metal-organic chemical vapor deposition (MOCVD or OMCVD) a material growth technique that uses metal organic molecules in an atmospheric or low pressure growth chamber and a controlled chemical reaction on a heated substrate to grow a variety of II-VI, III-V, and group IV materials with atomic layer control. Used to create material structures for a variety of electronic and optical devices using quantum wells, heterostructures and superlattices (for example, TEG, TMG, and TEA).

metal-oxide semiconductor (MOS)   the basic structure in an insulated-gate field-effect transistor. In this technology, a layered capacitor, with the added property of applying a voltage of a proper polarity, causes the underlying semiconductor to be switched from n-type to p-type or vice versa. MOS technology has been responsible for the mainstream of integrated circuit technology for many years.

metal-oxide semiconductor controlled thyristor (MCT)   a voltage-controlled four-layer (pnpn) device for medium power (700 A) and medium speed (20 kHz) applications, with projections to 2000 A and 100 kHz. Unlike transistors, MCTs can only be on or off, with no intermediate operating states. The MCT has high current handling capabilities, a low forward voltage drop, and low gate drive requirements (no large negative gate current required for turn off).

metal-oxide semiconductor field-effect transistor (MOSFET)   a transistor that uses a control electrode, the gate, to capacitively modulate the conductance of a surface channel joining two end contacts, the source and the drain. The gate is separated from the semiconductor body underlying the gate by a thin gate insulator, usually silicon dioxide. The surface channel is formed at the interface between the semiconductor body and the gate insulator. Used for low power (200 A) and high speed (1 MHz) applications.

In power electronics applications, MOSFETs are typically operated as switches, in either their fully on or off states, to minimize losses. The gate is insulated from the semiconductor portion to enable faster switching. A gate-source voltage permits a current to flow between the drain and the source, where continuous gate voltage is required to be in the on-state. The primary disadvantage

is the high forward voltage drop. The "metal" in MOSFET refers to the gate electrode, which was fabricated from a metal in early MOSFETs, but is now typically fabricated from a material such as polysilicon.

metal-oxide semiconductor memory   memory in which a storage cell is constructed from metal-oxide semiconductor. Usually called MOS memory.

metal-oxide varistor   a voltage-dependent resistor frequently used to protect electronic devices from overvoltages due to lightning or switching surges. The resistance of the device drops to a few ohms if the voltage applied across it exceeds a calibrated value.

metallization   the deposited thin metallic coating layer on a microcircuit or semiconductor.

meteor burst communication   VHF radio communication using the reflective properties of the ionized trails that burning meteorite bursts leave in the atmosphere.

method of moments (MOM)   a common procedure used in order to solve integral and differential equations. The unknown function is expanded on a complete (but not necessarily orthogonal) set of functions, often called basis functions or expansion set. This expansion is truncated after a finite number of terms, N. The original functional equation is then tested, i.e., multiplied and integrated, by another set of functions, often called weighting or testing functions, hence reducing the original functional equation to a matrix equation suitable for numerical computation. The common choice of taking the same set of functions for the expansion and test is often referred to as Galerkin version of the moment method. *See also* integral equation.

method of successive projections   an iterative procedure for modifying a signal so that the modified signal has properties which match an ideal objective.

metric   (1) a measure of goodness against which items are judged.

(2) methodologies for the measurement of software features including matters as performance and cost estimation.

metrology   the process of measuring structures on the wafer, such as the width of a printed resist feature.

metropolitan-area network (MAN)   a computer communication network spanning a limited geographic area, such as city; sometimes features interconnection of LANs.

Mexican-hat function   a function that resembles the profile of a Mexican hat. According to anatomy and physiology, the lateral interaction between cells in mammalian brains has a Mexican-hat form, that is, excitatory between nearby cells and inhibitory at longer range with strength falling off with distance. According to this phenomenon, Kohonen proposes a training algorithm for self-organizing system to update not only the weights of the winner but also the weights of its neighbors in competitive learning. *See also* self-organizing algorithm.

MFD filter   *See* maximally flat delay filter.

MFIE   *See* magnetic field integral equation.

MFLOP   *See* megaflop.

MHD   *See* magneto-hydrodynamic generator and magnetohydrodynamics.

MHz   megahertz, or millions of operations per second.

micro cell   a cell in a cellular communications system having a size (or cell radius) that is significantly smaller than the cell size of a typical cellular system. Such systems have cells with radii that are at most a few hundred meters and utilize base stations placed at heights that are of the order of the height of the mobile terminal.

micro diversity   a diversity technique, used in a cellular where multiple antennas are separated by distances equal to a few wavelengths and the diversity transmission/reception is utilized to overcome the effect of multipath fading. *Compare with* macro diversity

micro shadowing   shadowing due to both the orientation with respect to transmitter and receiver. *See also* macro shadowing.

micro-stepping   a control technique — also called mini-stepping — that results in a finer positioning resolution than can be obtained with simple on/off control of the phase currents of a stepper motor. The practical implementation of micro-stepping requires the accurate and continuous control of all the phase currents of the step motor.

microbending   sharp curvatures involving local fiber axis displacements of a few micrometers and spatial wavelengths of a few millimeters. Microbending causes significant losses.

microcell   a low-power radio network that transmits its signal over a confined distance.

microchannel diode   a high-frequency planar monolithic Schottky barrier diode structure fabricated with micromachining that has very small parasitics and is used as a millimeter or submillimeter wave mixer or harmonic multiplier.

microchannel-plate spatial light modulator (MSLM)   an optically addressed spatial light modulator. When input light is incident onto a photocathode on the write-in side of the MSLM, a photoelectron image is formed, which is multiplied to about hundred thousand times by the microchannel plate. The resulting charge pattern will in turn affect the second layer, which is the $LiNbO_3$ crystal plate having electro-optic effect. The refractive index of the crystal plate is now modulated by the electric charge pattern. When output light passes through the crystal plate, the phase of output light is modulated by the varying refractive index of the crystal plate. Since the $LiNbO_3$ is birefringence, phase modulation becomes polarization modulation. Polarization modulation is visualized as intensity variation using a polarizer. The generation of electric charge and its effect to the crystal layer are nonlinear. Their combined effect has thresholding capability that can be utilized for constructing optical logic gates.

microchannel-plate spatial light modulator logic gate   an optical logic gate utilizing thresholding capability of a microchannel-plate spatial light modulator (MSLM).

microcode   a collection of low-level operations that are executed as a result of a single instruction being issued.

microcommand   an n-bit field specifying some action within the control structure of a CPU, such as a gate open or closed, function enabled or not, control path active or not, etc.

microcomputer   a computer whose CPU is a microprocessor chip, and its memory and I/O interface are LSI or VLSI chips.

microcontroller   an integrated circuit chip that is designed primarily for control systems and products. In addition to a CPU, a microcontroller typically includes memory, timing circuits, and I/O circuitry. The reason for this is to permit the realization of a controller with a minimal quantity of chips, thus achieving maximal possible minituarization. This in turn, will reduce the volume and the cost of the controller. The microcontroller is normally not used for general-purpose computation as is a microprocessor.

microelectromechanical   system   (MEMS)   micrometer-scale devices fabricated as discrete devices or in large arrays using integrated circuit fabrication techniques. Movable compact micromechanical or optomechanical structures and microactivators made using batch processing techniques.

microelectromechanical system (MEMS) for microwave   a new multidisciplinary technology field with enormous potential for various applications, including microwave. MEMS are fabricated by integrated circuit processing methods and commonly include sensors and actuators with physical dimensions of less than 1 mm on a side.

microinstruction   the set of microcommands to be executed or not, enabled or not. Each field of a microinstruction is a microcommand.

microlithography   lithography involving the printing of very small features, typically on the order of micrometers or below in size.

micromachine   a fabrication process that uses integrated circuit fabrication techniques such as diffusion, oxidation, wet and dry etching to realize mechanical and electrical structures such as resonators, membranes, cavities, waveguide structures and a variety of thermal, medical, and chemical transducers.

micromemory   *See* control memory.

microphone   a device that converts acoustical signals into electrical signals.

microprocessor   a CPU realized on an LSI or VLSI chip.

microprogram   a set of microcode associated with the execution of a program.

microprogramming   the practice of writing microcode for a set of microinstructions.

microscopy   an imaging modality that uses optical light, laser light, electrons, or another radiation source to illuminate a sample. The image is formed by gathering reflected and scattered energy.

microsensor   a sensor that is fabricated using integrated circuit and micromachining technologies.

microstrip   a transmission line formed by a printed conductor on top of a conductive-backed dielectric. It is often used in high-frequency, printed circuit board applications.

microstrip antenna   a radiating element consisting of a conducting patch formed on the surface of a dielectric slab, which in turn lies on a ground plane. Microstrip antennas are usually printed on circuit boards and fed by microstrip lines etched on the same board. Also called microstrip patch antenna.

microstrip patch antenna   *See* microstrip antenna.

microtiming diagram   a graphical display showing how the waveforms vary with time but with a time scale that has sufficient resolution to display clearly the delays introduced by the individual basic elements of the digital circuit.

microwave   term used to refer to a radio signal at a very high frequency. One broad definition gives the microwave frequency range as that from 300 MHz to 300 GHz.

microwave coplanar probe   a specially designed test probe for measuring devices from DC to microwave frequencies using a wafer probe station. The probe tip is constructed using coplanar waveguide to present a highly controlled impedance (usually 50 ohms) to a device under test.

microwave engineering   the engineering of devices in the frequency range from 1 GHz to 1000 GHz corresponding to the wavelengths from 30 cm down to 0.3 mm.

microwave transition analyzer   a device that can combine the functionality of several dedicated measurement instruments. It is a pulsed RF measurement system that operates in the time domain like a high frequency sampling oscilloscope. Using the fast Fourier transform, the information can be converted into frequency domain. It can operate as a sampling oscilloscope, pulsed network analyzer, and a spectrum analyzer.

mid-term stability   refers to system responses which are shorter than long-term *cf* but shorter than transient *cf* response, generally associated with maximum excitation limiters, load tap changers and other slow-acting devices.

Mie scattering   electromagnetic theory that describes the scattering of light by spheres.

mildly nonlinear   a circuit or element in which the output spectrum is made up of two parts, the first of which is proportional through gain(s), attenuation(s) and delay(s) to the input spectrum, and the second in which spectral shift(s), conversion(s) or generation(s) takes place in an orderly and predictable way. Most real-world circuits and elements are mildly nonlinear at some level of excitation within hyperspace, and all active devices are mildly nonlinear.

Miller capacitance   an excess amount of capacitance that appears in parallel with the input of an inverting amplifier stage.

Miller effect   the increase in the effective grid-cathode capacitance of a vacuum tube or a transistor due to the charge induced electrostatically on the grid by the anode through the grid–anode capacitance.

Miller oscillator   the name is usually applied to crystal oscillators with one active device (usually a

FET with the source AC grounded) where the crystal is connected between the gate and ground. The crystal is used like an inductor, another tunable parallel LC-circuit in the drain is used as an inductor as well, and the capacitance between the drain and gate (Miller capacitance hence the name) is used as a third reactance of LC-oscillator. The circuit becomes similar to Hartley oscillator.

Miller's rule   a semi-empirical rule, of good but not exact validity, which states that the value of the nonlinear susceptibility of order n for a given material is proportional to $n + 1$ products of the linear susceptibility of that material.

millimeter wave   an electromagnetic wave in the part of the electromagnetic spectrum that has a wavelength on the order of a millimeter. This band is centered at about 300 GHz.

MIM capacitor   *See* metal–insulator–metal capacitor.

MIMD   *See* multiple instruction multiple data architecture.

MIMO system   *See* multi-input–multi-output system. *See also* single-input–single output system.

min operation   an operation on two or more variables where the resultant value is formed by taking the smallest value, or minimum, among these variables.

min-max control   a class of control algorithms based on worst-case design methodology in which control law is chosen in such a way that it optimizes the performance under the most unfavorable possible effect of parameter variations and/or disturbances. The design procedure can be viewed as a zero-sum game with control action and uncertainty as the antagonistic players. In the case of linear models and quadratic indices, the min-max control could be found by solving zero-sum linear-

quadratic games. Mini–max operations may be performed on cost functionals, sensitivity functions, reachability sets, stability regions or chosen norms of model variables.

mini-stepping   *See* micro-stepping.

minicell   a cell with a radius of 300 m to 2 km, typically for pedestrian mobile users. *See also* cell.

minimal orientation representation   minimal orientation representation describes the rotation of the end-effector frame with respect to the base frame, e.g., Euler angles (there exists a set of 12 Euler angles). Minimal orientation representation usually has to be calculated through the computation of the elements of the rotation matrix i.e., n, o, and a vectors. *See* external space.

minimal realization   for linear stationary finite-dimensional continuous-time dynamical system, is a set of four matrices A, B, C, D that form state and output equations in the state space R^n with minimal dimension n. A minimal realization of linear stationary dynamical system is always controllable and observable. The similar statements hold true for linear stationary finite-dimensional discrete-time dynamical systems.

minimax estimate   the optimum estimate for the least favorable prior distribution.

minimum discernible signal   a signal power level equal to the noise power, usually expressed in watts or decibels. Thus measuring the system output power with no signal applied, and then increasing the input signal power until a 3 dB increase is observed results in the signal power being equal to the noise power (i.e., $S/N = 1$). Also called the minimum detectable signal.

minimum distance   in a forward error control block code, the smallest Hamming distance between any two code words. In a convolutional code, the column distance at the number of encoding intervals equal to the constraint length.

minimum energy control of generalized 2-D model   given the generalized 2-D model

$$Ex_{i+1,j+1} = A_0 x_{ij} + A_1 x_{i+1,j} + A_2 x_{i,j+1}$$
$$+ B_0 u_{ij} + B_1 u_{i+1,j} + B_2 u_{i,j+1}$$

$i, j \in Z_+$ (the set of nonnegative integers) with admissible boundary conditions $x_{i0}, i \in Z_+, x_{0j}, j \in Z_+$ and the performance index

$$I(u) := \sum_{i=0}^{\bar{N}_1} \sum_{j=0}^{\bar{N}_2} u_{ij}^T Q_{ij}$$

$$\left(\bar{N}_1 := N_1 + n_1; \ \bar{N}_2 := N_2 + n_2 \right)$$

find a sequence of inputs u_{ij} for $0 \le i \le \bar{N}_1$ and $0 \le j \le \bar{N}_2$ that transfers the model to the desired final state $x_f \in R^n$, $x_{N_1 N_2} = x_f$ and minimizes the performance index $I(u)$ where $x_{ij} \in R^n$ is the local semistate vector $u_{ij} \in R^m$ is the input vector, E, A_k, B_k ($k = 0, 1, 2$) are given real matrices, $Q \in R^{m \times m}$ symmetric and positive definite weighting matrix and (n_1, n_2) is the index of the model. *See also* local reachability of generalized 2-D model.

minimum energy control of linear systems   a design problem such that for a given initial and final condition, find a control that steers dynamical system on a given time interval from initial conditions to final conditions and has minimum energy. Minimum energy control problem has a solution for every controllable linear (continuous or discrete) dynamical system. This solution strongly depends on system parameters and the given initial and final states.

minimum excitation limiter   a controller that is used to limit the minimum amount of field current, or under-excitation, at a synchronous generator. This excitation limit is set by stability limit.

minimum free distance   for any convolutional code, it is the minimum Hamming distance between the all-zero path and all the paths that diverge from and merge with the all-zero path at a given node of the trellis diagram.

minimum mean square error (MMSE)   a common estimation criterion which seeks to minimize the mean (or expected) squared error,

$$E = E \left[e^T e \right]$$

where e represents the error.

minimum mean square estimator (MMSE)   a broad class of estimators based on minimizing the expected squared error criterion. Both the Linear least squares estimator and the Bayesian least squares estimator are special cases. *See* minimum variance unbiased estimator, linear least squares estimator, Bayesian least squares estimator, maximum a posteriori estimation, maximum likelihood estimation.

minimum noise factor   for an active circuit or device, occurs when the input terminal is terminated with an impedance which produces the minimum noise factor.

minimum phase system   a system that has all poles and zeroes inside the unit circle. It is called minimum phase because the poles and zeroes inside the unit circle cause the group delay which is the derivative of the phase of the signal to be minimized.

minimum polynomial   for a given element α of a field, and for a subfield F, the polynomial of smallest degree, with coefficients in F having α as a root. The set of roots of a minimum polynomial form a conjugacy class that is defined by the polynomial.

minimum time-to-clear   *See* clearing time.

minimum time-to-melt   *See* melting time.

minimum variance unbiased estimator an estimator $\hat{\theta}$ of a parameter θ is said to have minimum variance and to be unbiased if

$$E\hat{\theta} = \theta$$

and

$$E(\hat{\theta} - \theta)^2 \leq E(\tilde{\theta} - \theta)^2,$$

where $\tilde{\theta}$ is any other estimator of θ.

minimum-shift keying (MSK) a variant of FSK in which the separation between transmitted frequencies is $1/(2T)$, where T is the symbol duration. In addition, the initial phase associated with each bit is adjusted so that phase transitions between successive bits are continuous.

minimum-shift keying Gaussian (GMSK) a variant of MSK in which the transmitted frequency makes smooth transitions between the frequencies associated with the input bits. A GMSK signal can be generated when the input to the voltage-controlled oscillator (VCO) is a PAM signal with a Gaussian baseband pulse shape.

minisub in European usage, a miniature substation.

Minkowski distance between two real valued vectors $(x_1, x_2, \ldots, x_n)$ and $(y_1, y_2, \ldots, y_n)$ it is given as

$$D_{Minkowski} = \left(\sum_{i=1}^{n} |x_i - y_i|^{\lambda} \right)^{\frac{1}{\lambda}}$$

minor hysteresis loop a hysteresis loop generated within the major hysteresis loop when a magnetic material is not driven to full positive or negative saturation.

minority carrier a hole in an n-type or an electron in a p-type semiconductor.

minterm a Boolean product term in which each variable is represented in either true or complement form. For example, $u \cdot v' \cdot w \cdot z'$ is a minterm for a four-variable function.

MIPS millions of instructions per second, a measure of the speed of a computer.

mirror optical element that reflects and may also transmit incident light rays and beams; used to provide feedback in laser oscillators.

mirroring fault tolerance architecture for managing two or more hard disks as a unique disk by replicating the same data on all the disks in the mirroring system. The system also includes mechanism for verifying that all disks contain the same information.

MISD *See* multiple instruction single data architecture.

MISO *See* multi-input–single-output system.

miss the event when a reference is made to an address in a level of the memory hierarchy that is not mapped in that level, and the address must be accessed from a lower level of the memory hierarchy.

miss probability the probability of falsely announcing the absence of a signal.

miss rate the percentage of references to a cache that do not find the data word requested in the cache, given by $1 - h$ where h is the hit rate. Also miss ratio.

miss ratio *See* miss rate.

missile terminal guidance seeker located in the nose of a missile, a small radar with short-range capability that scans the area ahead of the missile and guides it during the terminal phase toward a target such as a tank.

443

mix-and-match lithography   a lithographic strategy whereby different types of lithographic imaging tools are used to print different layers of a given device.

mixed A/D simulator   a simulator that is capable of simulating combined analog and digital circuitry.

mixed method coordination   in case of any mixed method both direct coordination instruments — as in the direct method — and the dual coordination instruments — as in the price method — are used by the coordinator to modify local decisions until the coordination objectives are met; different combinations of direct and dual coordination instruments may be used, and so a variety of mixed methods can be conceived.

mixed mounting technology   a component mounting technology that uses both through-hole and surface-mounting technologies on the same packaging and interconnecting structure.

mixer   a nonlinear device containing either diodes or transistors, the function of which is to combine signals of two different frequencies in such a way as to produce energy at other frequencies. In a typical down converter application, a mixer has two inputs and one output. One of the inputs is the modulated carrier RF or microwave signal at a frequency f_{rf}, the other is a well controlled signal from a local oscillator or VCO at a frequency f_{lo}. The result of down conversion is a signal at the difference frequency $f_{rf} - f_{lo}$, which is also called the intermediate frequency f_{if}. A filter is usually connected to the output of the mixer to allow only the desired IF frequency signal to be passed on for further processing. For example, for an RF frequency of 10.95 GHz (=10,950 MHz) and an LO frequency of 10 GHz (=10,000 MHz), the IF frequency would be 950 MHz.

mixing amplifier   *See* harmonic amplifier.

ML   *See* maximum likelihood estimation.

MLR   *See* multilayer resist.

MMF   *See* magnetomotive force.

MMIC   *See* monolithic microwave integrated circuit.

MMSE   *See* minimum mean square estimator, minimum mean square error.

MMU   *See* memory management unit.

MMX register   a register designed to hold as many as eight separate pieces of integer data for parallel processing by a special set of MMX instructions. An MMX register can hold a single 64-bit value, two 32-bit values, four 16-bit values, or eight-byte integer values, either signed or unsigned. In implementation, each MMX register is aliased to a corresponding floating-point register. MMX technology is a recent addition to the Intel Pentium architecture.

MNOS   acronym for metal-nitride-oxide-si. A structure used in a type of nonvolatile memory device in which the oxide is sufficiently thin to permit electron/hole tunneling while the nitride and the nitrode/oxide interface are used to store charge.

mobile charge   the charge due to the free electrons and holes.

mobile data terminal (MDT)   a computer terminal installed in a service vehicle which, through some wireless link, communicates work order information between the crews and their dispatch center.

mobile ion   a charged ionic species that is mobile in a dielectric. In metal-oxide-si (MOS) devices, Na^+ is most often the source of the mobile ions.

mobile robot a wheeled mobile robot is a wheeled vehicle which is capable of an autonomous motion because it is equipped for its motion, with actuators that are driven by an onboard computer. Therefore mobile robot does not have an external human driver and is completely autonomous.

mobile station that part of a radio communications system, which is not permanently located in a given geographical location.

mobility electron mobility $\mu_n = (2LK)/(C_{ox}W)$ where

C_{ox} = capacitance per unit area of the gate-to-channel capacitor for which the oxide layer serves as a dielectric.

L = length of the channel

W = width of the channel

μ_n = the mobility of the electrons in the induced n channel

K = constant.

proton mobility $\mu_p \cong \mu_n/2$.

MOCVD *See* metal-organic chemical vapor deposition.

modal analysis the decomposition of a solution to an electromagnetic analysis problem into a linear combination (weighted sum) of elementary functions called modes. The elementary functions are usually orthogonal; typical functions used include sine and cosine functions for problems cast in rectangular coordinates, Bessel functions for cylindrical coordinates, and spherical Bessel functions for spherical coordinates.

This technique can be used in order to find the field produced by an arbitrary source inside a waveguide, be it open or closed, or to find the scattering matrix relative to the discontinuity between different waveguides. *See also* eigenfunction expansion.

modal expansion *See* modal analysis.

modal solution *See* modal analysis.

modality (1) a specific medical imaging technique, such as X-ray CT, magnetic resonance imaging, or ultrasound, that is used to acquire an image data set.

(2) a part of a computer's instruction that specifies how another part of the instruction is interpreted.

modality-specific a task that is specific to a single sense or movement pattern.

mode (1) one possible wave solution of an infinite number of time–harmonic wave solutions that exist in a waveguide or transmission line. Each mode is identified by a collection of numbers and is usually designated in electromagnetics as either transverse electric, transverse magnetic, or transverse electromagnetic.

(2) electromagnetic field distributions that match the boundary conditions imposed by a laser or other cavity.

See also propagating mode, cladding, tunneling modes, leaky modes, multimode optical fiber, single-mode fiber.

mode chart a graphical illustration of the variation of effective refractive index (or equivalently, propagation angle θ) with normalized thickness d/λ for a slab waveguide or normalized frequency V for an optical fiber.

mode field radius the radius at which the electric field in a single mode fiber falls to $1/e$ of its value at the center of the core.

mode filter a filter which takes the mode of the distribution containing the various input signal components, or in the case of an image, the mode of the distribution of all the pixel intensity values within the neighborhood of the current pixel. Complications can arise from the sparsity of the local pixel intensity distribution, and the mode should be that of the underlying rather than the actual intensity distribution.

mode matching   a particular type of modal analysis referred to the analysis of a junction between different waveguides, where the modal expansion on one side of the junction is matched to the modal expansion on the other side of the junction.

mode-locking   forcing the modes of a laser oscillator to be equally spaced in frequency and have a fixed phase relationship; sometimes also occurs spontaneously. *See also* longitudinal mode-locking, transverse mode-locking.

model based image coding   (1) image compression using stored models of known objects at both encoder and decoder, the encoder using computer vision techniques to analyze incoming images in terms of the models, and the decoder using computer graphics techniques to re-synthesize images from the models. The transmitted data are model parameters that describe at a relatively high level, the motion of the model parts. For example, if the stored model is of a human head, the transmitted data may be muscle flexions.

(2) the appreciation that all image coding relies on some model of the source, and the categorization of methods according to the type of model. In this scheme, definition (1) is termed "known-object coding" or "semantic coding."

model reference control   the control scheme in which the controlled system is made to mimic the behavior of a reference model system that possesses ideal behavioral characteristics.

model-based predictive control   *See* predictive control.

modeling   the process of creating a suitable description that emulates the performance or characteristics of the actual item being modeled, over some portion of the device hyperspace. Modeling involves all or parts of model creation, device characterization, de-embedding, parameter extraction, verification, validation, valuation and documentation. *See also* mathematical modeling, fuzzy modeling.

modem   abbreviation for modulator-demodulator. A device containing a modulator and a demodulator. The modulator converts a binary stream into a form suitable for transmission over an analog medium such as telephone wires or (in the case of a wireless modem) air. The demodulator performs the reverse operation, so two modems connected via an analog channel can be used to transfer binary data over the (analog) channel.

modem-FEC coding   error control coding (ECC), applied to a digital signal such that feed-forward error correction (FEC), can be used in the modem, thus detecting and often correcting transmission errors.

moderator   a material contained in a nuclear reactor core which slows down neutrons to thermal energies, primarily by neutron scattering.

MODFET   acronym for modulation doped FET. *See* high electron mobility transistor.

modified nodal formulation   a modification of the classical nodal formulation which allows any network to be described. The modification consists of adding extra equations and unknowns when an element not normally modeled in classical nodal analysis is encountered.

modified signed-digit computing   a computing scheme in which a number is represented by modified signed-digit. This number system offers carry free addition and subtraction. Instead of 0 and 1, numbers are represented by -1, 0, and 1 for the same radix 2. If a number is represented by 0 and 1, we may need carry in the addition. However, since the number can be represented by three possibilities -1, 0, and 1, the addition and subtraction can be directly performed without carry following a specific trinary logic truth table for this number system.

modified z-transform   a z-transform of signals and systems that contain nonzero deadtime τ in the range

$$0 \leq \tau < T$$

where T is the sampling interval. Modified z-transforms are usually derived from fundamental principles and given as separate columns in z-transform tables for standard functions.

modified-return-to-bias recording   *See* magnetic recording code.

modifier   an operation that modifies the membership of a fuzzy set. Examples of modifiers are
(1) *very* $A = \mu_{con(A)}(u) = (\mu_A(u))^2$ (concentration);
(2) *more or less* $A = \mu_{dil(A)}(u) = (\mu_A(u))^{.5}$ (dilatation).
See also fuzzy set, linguistic variable.

modular network   a network whose overall computation is carried out by subnetworks whose outputs are combined in some appropriate way. The term is most commonly applied to networks that partition the input space so that the subnetworks operate on "local" data, but is also applied to the case where a problem can be decomposed into successive tasks, each being implemented by a suitable subnetwork.

modularity   design principle that calls for design of small, self-contained unit that should lead to maintainability.

modulated filter bank   a filter bank obtained by shifting the spectrum of a prototype low pass filter in the frequency domain to cover the entire frequency band; different properties can be obtained by using different modulation schemes and imposing other conditions. Cosine-modulated filter banks are the most commonly used modulated filter banks.

modulating signal   the baseband source signal used to encode information onto a carrier wave by varying one or more of its characteristics (e.g., the amplitude, frequency, or phase of a sinusoid; or the amplitude, width, repetition rate, or position of each pulse in a periodic pulse train).

modulation   (1) variation of the amplitude or phase of an electromagnetic wave.
(2) the process of encoding an information-carrying waveform onto a carrier waveform, typically in preparation for transmission. *See* amplitude modulation, frequency modulation, phase modulation.

modulation efficiency   ratio of the baseband bit rate to the transmission bandwidth after modulation.

modulation index   for an angle-modulated signal, the modulation index is the ratio of the maximum modulation deviation to the modulating frequency, and represents the maximum phase deviation in the modulating signal.

modulation property   a property of the Fourier transform in which the Fourier transform of a modulated signal $c(t)e^{jw_ot}$ is equal to $C(w - w_o)$, where $C(w)$ is the Fourier transform of $c(t)$.

modulation transfer function (MTF)   for an imaging system the Fourier transform of the system line spread function. The MTF describes the spatial frequency resolution of the system.

modulator   device that varies the amplitude or phase of an electromagnetic wave.

modus ponens   a rule of reasoning which states that given that two propositions, A and $A \Rightarrow B$ (implication), are true, then it can be inferred that B is also true.

modus tollens   a rule of reasoning which states that if a proposition B is not true and given that

$A \Rightarrow B$, then it can be inferred that A is also not true.

Moire pattern   image caused by a combination of two effects, sampling rate and reconstruction filter shape; they occur in image signals when two conditions are met. First, when the sampling frequency is close to the Nyquist Frequency for the signal (i.e., two times the highest frequency in the image signal). Second, the cutoff frequency of the reconstruction filter is located beyond one half the sampling frequency (e.g., a first order filter). These two problems result in mirror images of frequency components around the sampling frequency causing banding in the image signal. The Moire effect is seen in practical applications due to real world problems of finite filter lengths and errors in sampling rates.

molded case circuit breaker   a low-voltage air circuit breaker that includes thermal and/or magnetic overcurrent sensing which directly trips the breaker. The molded case circuit breaker is nearly always manually closed, opened, and reset.

molecular beam   a source of molecules traveling primarily in one direction. In practice, molecular beams are usually realized by expansion of an atomic or molecular vapor into a vacuum through a small aperture. The resulting expanding cloud of molecules is usually made nearly unidirectional by a collimator that blocks or otherwise removes all molecules not propagating within a narrow range of angles.

molecular beam epitaxy (MBE)   a material growth technique that uses atomic or molecular beams in an ultra high vacuum chamber to grow a variety of II-VI, III-V, and group IV materials with atomic layer control. Individual molecules or atoms are excited from heated sources (effusion cells) and attach to the substrate in an ordered manner. High quality growth is achieved when the surface diffusion coefficient is sufficiently high that the atoms can arrange themselves coherently on the surface.

Used to create material structures for a variety of electronic and optical devices using quantum wells, heterostructures, and superlattices.

molecular transition   coupling of energy levels in an atom by means of absorption or emission processes.

molecular vapor   a material composed of molecules in the vapor phase.

Mollow gain   gain that originates when a 2-level system is driven by a strong, near resonant, electromagnetic field.

molten carbonate fuel cell (MCFC)   a fuel cell that uses various hydrocarbons as fuel, and lithium and potassium carbonate as an electrolyte. It can generate >200 kW of power, and hence can be used in power generation. *See also* alkaline fuel cell, proton exchange membrane fuel cell, phosphoric acid fuel cell, solid oxide fuel cell.

MOM   *See* method of moments.

moment   a statistic of a random variable. For example, the first moment is called the mean. In general, the nth moment is given by

$$m_n = E(X^T n) = \int_{-\infty}^{\infty} x^n X(x)dx,$$

where $f_X(x)$ is the probability density function of X. Moments are often used to aid the recognition of shapes. *See* moment feature recognition, probability density function.

moment method   *See* method of moments.

momentary interruption   a loss of voltage of less than 0.1 pu for a time period of 0.5 cycles to 3 seconds.

momentary monitoring   the duration at supply frequency from 30 cycles to 3 seconds.

momentary overvoltage an increase in voltage above the system's specified upper limit for more than a few seconds. Generally a rather loosely-defined term.

momentum relaxation time the characteristic time for loss of momentum or velocity due to scattering processes.

monaural attribute attribute of ear input signals (e.g., timbre, loudness) that require only one ear to be detected.

monitor (1) the main display device of a personal computer. Usually uses cathode ray tube (CRT) or liquid-crystal display (LCD) technology.

(2) a computer program providing basic access to the register contents and memory locations, usually for debugging purposes.

monitor display *See* monitor.

monochromatic light light that has only one frequency component.

monochrome the representation of an image, analog or digital, using only one color and black.

monochrome display adapter (MDA) a monocrome video adapter with 25 lines and supporting 80 columns, proposed by IBM in 1981.

monocular vision a vision model in which points in a scene are projected onto a single image plane.

monolayer one atomic or, in the case of materials such as GaAs, diatomic layer of atoms.

monolithic microwave integrated circuit (MMIC) integrated circuits made of gallium arsenide (GaAs), silicon, or other semiconducting materials where all of the components needed to make a circuit (resistors, inductors, capacitors, transistors, diodes, transmission lines) are formed onto a single wafer of material using a series of process steps.

Attractive features of MMICs over competing hybrid (combination of two or more technologies) circuits are that a multitude of nearly identical circuits can be processed simultaneously with no assembly (soldering) using batch processing manufacturing techniques. A disadvantage is that circuit adjustment after manufacture is difficult or impossible. As a consequence, significantly more effort is required to use accurate computer-aided-design (CAD) techniques to design MMICs that will perform as desired without adjustment. Of course, eventually assembly and packaging of MMICs is performed in order to connect them into a system such as a DBS receiver. MMICs are only cost-effective for very high volume costs, because the cost of the initial design is very high, as is the cost of wafer manufacture. These costs can only be recovered through high volume manufacture. (The word "monolith" refers to a single block of stone that does not (in general) permit individual variations).

monopole *See* magnetic monopole.

monopole antenna an antenna consisting of a straight conducting rod, wire, or other structure oriented perpendicularly to a ground plane and fed at the junction of the structure and the ground plane.

monostatic scattering the reflection of a portion of an electromagnetic wave back in the direction of the wave source. Monostatic scattering is measured by having the transmitter and receiver collocated.

Monte Carlo method a numerical technique that replaces a deterministic description of a problem with a set of random descriptions that have been chosen based on distributions that match the underlying physical description of the problem. This technique is widely used to investigate transport and terminal characteristics in small semiconductor structures.

Moore's law named for Gordon Moore, one of the founders of Fairchild Semiconductor and Intel, the observation that the number of transistors on a typical chip doubles about every 18 months.

MOPA *See* master-oscillator-power-amplifier.

MOPS acronym for millions of operations per second.

morphological duality a morphological property. There are three standard meanings of duality for morphological operators:

(1) Order-theoretic duality: To any property or concept corresponds the dual property or concept, where the relations $\subseteq, \leq$ and operations $\cup, \cap$, sup, inf, etc., are replaced by their duals $\supseteq, \geq$ and $\cap, \cup$, inf, sup, etc.

(2) Duality under complementation (or gray-level inversion): Let ψ be a morphological operator, and let N be the operation transforming an image I into its negative $N(I)$ (when I is a set, $N(I)$ is its complement, while when I is a gray-level numerical function, $N(I)$ has the gray-level inverted at each point). Then the dual of ψ is the operator ψ^* arising from applying ψ to the negative of an image; in other words ψ^* transforms I into $N(\psi(N(I)))$.

(3) Adjunction duality: The dilation and erosion by B, namely $\delta_B : X \mapsto X \oplus B$ and $\varepsilon_B : X \mapsto X \ominus B$, form an adjunction, which means that for every X, Y we have $\delta_B(X) \leq Y \iff X \leq \varepsilon_B(Y)$, the dilation of X is below Y if and only if X us below the erosion of Y. This relation constitutes a bijection between the family of dilations and that of erosions, and it can thus be considered as a duality between them.

In the latter two cases, duality inverts the ordering relation $\leq$ between operators. *See* dilation, erosion, morphological operator.

morphological filter a morphological operator ψ that is both:

Idempotent: applying it twice gives the same result as applying it only once; mathematically

speaking, given an object X, we have $\psi(\psi(X)) = \psi(X)$;

Increasing: it preserves the ordering relation $\leq$ between objects; given two objects X and Y, $X \leq Y$ implies $\psi(X) \leq \psi(Y)$. Openings and closings are morphological filters. *See* closing, morphological operator, opening.

morphological operator an operation for transforming images, which does not arise from the traditional signal processing methodology using linear filters and Fourier analysis, but which is rather based on set-theoretical operations. Such an operator for sets (binary images) combines union, intersection, translation, and sometimes complementation. For numerical functions (gray-level images), union and intersection are replaced by supremum and infimum (upper and lower envelope), while gray-level inversion takes partially the role of set complementation. However, morphological operators for gray-level images can be visualized by applying the corresponding morphological operator for binary images to the umbra of a gray-level image. *See* closing, dilation, erosion, mathematical morphology, opening, umbra.

morphological processing low-level processing technique for binarized images involving the shrinking or growing of local image regions to remove noise and reduce clutter.

morphological skeleton an archetypal stick figure that internally locates the central axis of an image.

Let B be the closed ball of unit radius centered about the origin; for any radius $r \geq 0$, let rB be the homothetic of B by factor r (thus rB is the closed ball of radius r centered about the origin). Given a non-empty bounded set X in a Euclidean space and a point p in X, p is the center of a closed ball of radius r which is included in X, but not contained in a larger ball included in X, if and only if $p \in X \ominus rB$ and . Here $\ominus$ and $\circ$ denote the erosion and opening operations respectively. The medial axis of X is the set of all points p such that there is some $r \geq 0$ for which this equation holds. The morphological skeleton is defined in the same way, except that it does not restrict itself to the isotropic Euclidean ball: take the same formula above, but consider in it the set B to be any convex, bounded, and topologically closed structuring element instead of the unit ball. For sets in digital space, the above approach is simplified as follows: For $r = 0, 1, 2, \ldots$, one takes a bounded digital structuring element B_r, with B_0 being restricted to the origin, and B_r increasing with r; the morphological skeleton is made of all pixels p such that there is some integer $r \geq 0$ for which $p \in X \ominus B_r$ and $p \notin (X \ominus B_r) \circ B_1$. See erosion, medial axis transform, opening, structuring element.

morphological system   a non-linear counterpart of linear shift invariant systems, based on morphological operators which are invariant under both spatial and gray-level shifts. See linear shift invariant system, morphological operator.

morphology   See mathematical morphology.

morphotropic phase boundary (MPB)   materials that have a MPB assume a different crystalline phase depending on the composition of the material. The MPB is sharp (a few percent in composition) and separates the phases of a material. It is approximately independent of temperature in PZT.

MOS   See mean opinion score or metal-oxide semiconductor.

MOS memory   See metal-oxide semiconductor memory.

MOSFET   See metal-oxide semiconductor field effect transistor.

MOSIS   acronym for metal-oxide semiconductor implementation service. See metal-oxide semiconductor.

motherboard   in a computer, the main printed circuit board which contains the basic circuits and to which the other components of the system are attached. See also daughter board.

mother wavelet   the wavelet from which wavelets in different scales are obtained by the translation and dilation operations. See father wavelet.

motion analysis   determination of the motion of objects from sequences of images. It plays important roles in both computer vision and image sequence processing. Examples for the latter include image sequence data compression, image segmentation, interpolation, image matching, and tracking.

motion blur   blur which is due to the motion of an object: it frequently arises when images containing moving objects are grabbed by an insufficiently rapid digitizer.

motion compensation   generation of a prediction image, an interpolated or extrapolated frame, from pixels, blocks, or regions from known frames, displaced according to estimates of the motion between the known frames and the target frame.

motion estimation   strictly, the estimation of movement within a video sequence, including camera movement and the independent motion of objects in the scene. In practice, refers to the determination of optical flow (the spatiotemporal variation of intensity) which may not correspond to real movement (when, for example, an object moves against a background of the same color). Gradient-based methods are based on the expansion of a Taylor series for displaced frame difference yielding an "optical flow constraint equation." This must be combined with other constraints to yield a variational problem whose solution requires the iterative use of spatial and temporal derivatives. Block-matching methods rely on finding minimum error matches between blocks of samples centered on the point of interest. Frequency-based techniques measure phase differences to estimate motion.

motion measurement   the measurement of velocities. Usual motion measurement techniques such as optical flow give incomplete and unstable results which must be regularized by comparing velocities over a whole moving object (the latter must first be segmented).

motion segmentation   the decomposition of a scene into different objects according to the variation of their velocities. Motion segmentation requires the measurement of velocities across the scene.

motion stereo   a specific case in motion estimation in which the 3–D scene is stationary and the only camera is in rigid motion.

motion vector   a vector displacement representing the translation of a pixel, block, or region between two frames of video, usually determined by optical flow calculation or block matching (see motion estimation). For instance, in the case of dense motion field, say, an optical flow field, a motion vector is assigned to each pixel on the image plane. In the case of block matching for image coding, a motion vector is assigned to each block. In the case of computer vision, a motion vector sometimes refers to the velocity of a point or an object in the 3–D space. Sometimes motion vectors are referred to as displacement vectors.

motional time constant   a material parameter that is inversely proportional to the product of the frequency and Q-factor for a particular mode.

motor   an electromechanical device that converts electrical energy from a DC or an AC source into mechanical energy, usually in the form of rotary motion.

motor back EMF constant   *See* EMF constant.

motor circuit   the three components of an electrical circuit are source, load, and interconnecting circuit conductors. A motor circuit is an electrical circuit designed to deliver power to a motor. It includes the overcurrent protective devices, controller, disconnect switch, circuit conductors, and the motor itself, as shown in the figure.

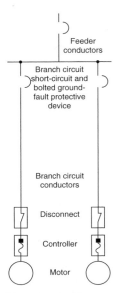

A typical electric motor circuit.

motor circuit protector (MCP)   a listed combination motor controller containing an adjustable instantaneous-trip circuit breaker and coordinated motor overload protection. MCPs can provide short-circuit and bolted ground-fault protection via the circuit breaker magnetic element, overload protection via the overload device, motor control, and disconnecting means all in one assembly.

motor control center   an enclosure with one or more sections containing motor control units that have a common power bus.

motor control circuit   a circuit containing devices such as the start/stop switches, main-line coil, main-line sealing contacts, overload contacts, timers and timer contacts, limit switches, antiplugging devices, and anything else used to control devices in the motor circuit.

motor current signature analysis   the use of the currents of an electric machine to provide diagnostic or other information on the health of the machine, coupling, or load.

motor operated switch   a switch operated by a motor that is capable of being controlled from a remote location.

motor starter   an electric controller, either manual or automatic, for accelerating a motor from rest to normal speed and for stopping the motor.

motor torque constant   *See* torque constant.

motor-generator set   a set consisting of a motor mechanically coupled to and driving one or more generators. The set used to be employed for AC-to-DC or DC-to-AC power conversion or voltage level or frequency conversion. Solid-state conversion units are replacing motor-generator sets in most applications.

mouse   an I/O device with a trackball used to produce signals which are interpreted as x/y coordinates. It is used as locator device, as are the joystick, trackball and tablet. It is the most commonly used locator for working with windowing systems for its similarity with the movements of the hand, metaphor or the moving hand.

A mouse may present 1, 2, or 3 buttons. The actions associated to the pressing of buttons can lead to draw objects, select objects, activate menu, delete objects, etc. The action of pressing is called "click." Rapidly double-clicking the mouse produces a different, program-dependent effect.

MOV   *See* metal oxide varistor.

move instruction   a computer instruction that transfers data from one location to another within a computing system. The transfer may be between CPU registers, between CPU registers and memory or I/O interface in either direction. Some systems (such as Motorola M68000) permit transfer by a move instruction between memory locations (memory-to-memory transfer).

moving average   an Nth order moving average relationship between an input x and output y takes the form

$$y[n] = \sum_{i=0}^{N-1} \alpha_i x[n-i]$$

A moving average process is any process y which can be expressed as the moving average of white noise x. *See* autoregressive.

Moving Picture Experts Group (MPEG)   group that standardizes methods for moving picture compression. MPEG-1 and MPEG-2 are very flexible generic standards for video coding, incorporating audio and system-level multiplex information. They use block-based motion-compensated prediction on 16×16 "macroblocks," and residue coding with the 8×8 discrete cosine transform. Frames types are intraframes (coded without reference to other frames), P frames (where the prediction is generated from preceding I and P frames), and B frames (where the prediction is generated bidirectionally from surrounding I and P frames).

MOVPE   *See* metal-organic chemical vapor deposition.

MOX   mixed oxide fuel, containing a mixture of U235 and PU239 oxides as the fissile material.

MPE   *See* maximum permissible exposure.

MPEG   *See* Moving Picture Experts Group.

MPP distance   *See* maximum *a posteriori* probability.

MPPC   *See* mutually pumped phase conjugator.

MQW   *See* multiple quantum well.

MRI   *See* magnetic resonance imaging.

MRVQ   *See* mean/residual vector quantization.

MSB   in a binary word, a bit with the highest weight associated with it.

MSC   *See* magnitude squared coherence.

MSE   *See* mean squared error.

MSG   *See* maximum stable gain.

MSI   *See* medium scale integration.

MSK   *See* minimum-shift keying.

MSLM   *See* microchannel-plate spatial light modulator.

MSLM logic gate   *See* microchannel-plate spatial light modulator logic gate.

MTF   *See* modulation transfer function.

MTTF   *See* mean time to failure.

MTTR   *See* mean time to repair.

MUI   multi-user interference. *See* multiple access interference.

multichannel acousto-optic device   acousto-optic device with multiple independent transducers bonded to the acousto-optic medium to introduce multiple independent acoustic signals in the device.

multilayer control   operation or structure of the control system with the (multilayer) controller decomposed into two or more control layers; usually in case of the controller with two control layers one uses the terms two-layer control and two-layer controller. Typically within the multilayer control hierarchy the frequency of interventions made by control layers decreases as one moves up the hierarchy.

multilayer resist (MLR)   a resist scheme by which the resist is made up of more than one layer, typically a thick conformal bottom layer under a thin imaging layer, possibly with a barrier layer in between.

multilayered medium   a medium composed by several different layers of dielectric materials. Such media find wide application for both microwave and optical integrated circuits.

multilevel cache   a cache consisting of two or more levels, (typically) of different speeds and capacities. *See also* hierarchical memory.

multilevel code   in this scheme each bit (level) of a signal is encoded with different error correction codes of same block length. It has inherent unequal error protection capability.

multilevel control   operation or structure of the control system, where the controller, or the given control layer, is composed of several local decision units coordinated (see coordination) by a supremal unit (coordinator unit); the number of levels can be greater than two — in this case decision units situated at an intermediate level are supremal (coordinating) units for the subordinated lower level units and at the same time are local units as perceived by the level above; the term multilevel control is equivalent to the hierarchical control, when the latter term is used in the narrow sense,

multilevel memory   *See* hierarchical memory.

multilevel optimization decision mechanism (or operation of such mechanism) whereby the decisions are made by solving a large-scale optimization problem, partitioned (decomposed) into several smaller problems; in a two-level optimization case local decisions are influenced by the coordinator — in the process of iterative coordination until overall satisfactory decisions are worked out; in a three-level or in a multilevel case the coordinating unit of the higher level coordinates the decisions of several subordinate units, which themselves can be coordinating units for the lower levels.

multiparameter sensitivity appears as an effort to introduce generalized functions that represent the influence of all circuit elements. If the circuit function depends on more than one component, then $F(s, x_1, x_2, \ldots, x_n) = F(s, \mathbf{x})$ and one may write that

$$\frac{dF(s, \mathbf{x})}{F(s, \mathbf{x})} = d \ln F(s, \mathbf{x}) = \sum_{i=1}^{n} \mathbf{S}_{x_i}^{F(s,\mathbf{x})} \frac{dx_i}{x_i}$$

One can introduce a multiparameter sensitivity row vector

$$\mathbf{S}_{\mathbf{x}}^{F(s,\mathbf{x})} = \left[\mathbf{S}_{x_1}^{F(s,\mathbf{x})} \mathbf{S}_{x_2}^{F(s,\mathbf{x})} \ldots \mathbf{S}_{x_n}^{F(s,\mathbf{x})} \right]$$

To characterize and compare the vectors of this type one introduces different vector measures that are called sensitivity indices. Sometimes these sensitivity indices are considered as multiparameter sensitivities.

multipath *See* multipath propogation.

multipath propagation the process by which a radio signal propagates from the transmitter to the receiver by way of multiple propagation paths. Depending on the frequency range employed in transmission, the physical propagation phenomena contributing to multipath propagation include reflections from terrain and man-made obstacles, diffraction and refraction. Multipath propagation leads, similar to acoustic echos, multiple replicas

of the transmitted signal to be received, each having experienced potentially different delays, attenuations, and phase shifts.

multipulse converter a three-phase converter that generates more than six pulses of DC per cycle. Multiple converters are connected so that the harmonics generated by one converter are canceled by harmonics produced by other converters, in order to reduce the line harmonics and improve the system power quality.

multispeed motor a motor that can be operated at any one of two or more definite speeds. For DC and induction motors, the speed settings are practically independent of the load, although the speed may vary with load for certain types of motors. Multispeed induction motors typically have two or more sets of windings on the stator with a different number of poles, one of which is excited at any given time.

multistage detection an iterative detection strategy where increasingly more reliable tentative decisions are made for each iteration (stage) of the detection algorithm.

multistage interconnection network an interconnection network built from small switches (often with fan-in and fan-out of 2); if 2^n modules are connected on each side of the network, $2n$ stages are required.

multistage subset decoding in this scheme decoding is done, first on the lowest partition level of the signal set and then gradually on the higher level with decoded information flow from the lower to the higher level.

multistage vector quantization a method for constrained vector quantization where several quantizers are cascaded in order to produce a successively finer approximation of the input vector. Gives loss in performance, but lower complexity than optimal (single-stage) vector quantization.

multistage VQ *See* multistage vector quantization.

multi-carrier communications a communications method where the bandwidth is subdivided into several smaller frequency bands. Either the same information is transmitted over all subbands or different information is transmitted either simultaneously or successively over these bands. By doing so, the detrimental effects of frequency selective fading can be minimized.

multi-carrier modulation a modulation technique in which the channel is divided into narrow frequency bins (subchannels), and input bits are multiplexed into substreams, each of which is transmitted over a particular subchannel. The bit rates and transmitted powers associated with the substreams can be selected to maximize the total transmitted bit rate, subject to a total average power constraint, and also to achieve a desired transmitted spectrum.

multi-input–multi-output system (MIMO) also known as multivariable (MV) systems. A system that can transform two or more input signals to two or more output signals.

multi-input–single-output system (MISO) a system which can transform two or more input signals to one output signal. *See also* single-input–single-output system and multiple-input–multiple-output system.

multi-mode code a line code where each source word is represented by a code word selected from a set of alternatives. Code words are selected according to a predefined criterion, which may depend in part on the statistics of the encoded sequence, thereby causing the same source word to be represented by a number of different code words throughout the encoded sequence.

multi-mode fiber an optical fiber with a relatively large core diameter in which more than one and usually from several hundred to several thousand modes may propagate. The optical fiber may be either a step index or a graded index fiber.

multi-mode optical fiber *See* multi-mode fiber.

multi-mode oscillation oscillation in more than a single cavity mode.

multi-phase oscillator an oscillator that provides m sinusoidal voltages shifted with respect to each other by $2\pi/m$ phase angle. It is obtained by connection of m similar isolated phase-shifting circuits in a loop. The outputs of isolating stages provide the required output voltages. These oscillators are usually used in instrumentation where light-weight auxilliary three-phase power supplies are required. Also called m-phase oscillator.

multi-resolution analysis *See* multi-scale analysis.

multi-scale analysis the analysis or transformation of a signal using analysis basis functions or analysis filters with differing time resolutions or spatial resolutions. Equivalently the basis functions have differing frequency resolutions. This is in contrast to the continuous or discrete Fourier transform whose analysis basis functions all have (roughly) the same frequency and time resolution. For discrete multi-resolution analysis a multi-level filter bank is often used: the discrete wavelet transform being a classic example. *See* multiscale.

multi-user CDMA a term that has been used to denote a CDMA system where the multiple access interference is used constructively in the receiver to enhance performance and not treated merely as interfering noise as for the conventional single-user receiver. In principle, any CDMA system is multi-user CDMA, since it supports multiple users.

multi-user detection joint detection of the data symbols for all the users in a multiple access system, as opposed to single-user detection where the data symbol for each user is detected individually.

multi-user detector   a detector in a spread spectrum multiple access system or CDMA system where the data bits from all the users are detected using a joint detection algorithm. Such a detector typically has significantly better performance (results in lower bit error probabilities) in comparison to a detector that detects the various signals individually and, in detecting a bit from a given user, treats all the other signals (from other users) as a composite source of interference.

multi-user interference (MUI)   *See* multiple access interference.

multi-user receiver   a spread spectrum or CDMA receiver that utilizes a multi-user detector.

multi-variable system (MV)   *See* multi-input–multi-output system.

multibus   a standard system bus originally developed for use in Intel's Microcomputer Development System (MDS). This standard gives a full functional, electrical, and mechanical specification for a backplane bus through which a number of circuit boards may be interconnected. A full range of devices may be involved, including computers, memory boards, I/O devices, and other peripherals.

multigrid   an efficient numerical algorithm for solving large sets of linear equations $Ax = b$, particularly for "stiff" (nearly singular) A. The algorithm defines a hierarchy of grids, with interpolation and decimation operations defined between successive grids in the hierarchy. A system of equations $A_g x_g = b_g$ is defined on each grid g: the systems on finer grids yield higher resolution solutions, however coarser grids yield much faster covergence times. Various empirical strategies have been devised which dictate the order in

which different grids are used to contribute to the final solution. *See* interpolation, decimation, multiscale.

multigrid block matching   a block matching technique that is carried out with a hierarchical structure, known as a multigrid structure. In the highest hierarchical level an image may be decomposed into large blocks each with equal size. Each block in a higher level is further equally decomposed into sub-blocks, forming the next level. The block matching is first applied to the highest level. The matching results obtained are then propagated to the next hierarchical level. That is, the matching results obtained in the highest level are utilized as initial estimates and refined in the second highest level. This process continues untill the last hierarchical level. It is noted that the term multigrid block matching is sometimes intermixed with multiresolution block matching in the literature. In the strict sense they are different. In the former, different levels in the hierarchical structure have the same resolution, while in the latter, different levels have different resolutions.

multilayer perceptron   an artificial neural network consisting of an input layer, possibly one or more hidden layers of neurons (perceptrons), and an output layer of neurons. Each layer receives input from the previous layer and the outputs of the neurons feed into the next layer.

multimedia   a generic category of computer-controlled media that combines text, sound, video, animation, and graphics into a single presentation. An example of multimedia would be an electronic encyclopedia in CD-ROM format.

multiple access channel   a multiple transmitter, single receiver communication system, in which the received signal is a (possibly non-deterministic) function of all transmitted signals. *See also* broadcast channel, interference channel, two-way channel.

multiple access interference (MAI) also multi-user interference (MUI). Interference between users in a multiple access system stemming from nonorthogonality between users, e.g., nonorthogonal code sequences in CDMA or (partially) overlapping frequency bands (FDMA) or time-slots (TDMA).

multiple dwell detector a detector typically used in spread spectrum PN sequence acquisition (i.e., PN sequence synchronization) where the detection process is carried out in stages. In the PN code acquisition process, in a spread spectrum receiver, the receiver makes a hypothesis about the correct PN code phase and then tests the hypothesis by correlating the received signal with a locally generated version with a phase equal to the hypothesized phase. If the value of the resulting correlation is less than a given threshold value, then the receiver examines a new phase. If the correlation is greater than the threshold value, then the receiver continues the examination of the same PN code phase by performing a further correlation with the same phase in a detection stage commonly referred to as the confirmation stage.

multiple instruction multiple data (MIMD) architecture a parallel processing system architecture where there is more than one processor and where each processor performs different instructions on different data values.

In an optical computer implementation, this can be done with a lenslet array, a hologram array, or a set of beamsplitters.

multiple instruction single data (MISD) architecture a parallel processing architecture where more than one processor performs different operations on a single stream of data, passing from one processor to another.

multiple quantum well (MQW) collection of alternating thin layers of semiconductors (e.g, GaAs and AlGaAs) that results in strong peaks in the absorption spectrum which can be shifted with an applied voltage.

multiple reflection algorithm a technique for the solution of nonlinear device-circuit interactions that uses a time domain representation of the nonlinear device waveforms and a frequency domain representation of the linear circuit waveforms coupled together with an ideal zero length transmission line.

multiple scattering strong interaction, with multiple hits, of the light wave with a highly inhomogenous medium. This produces large changes both in the amplitude, phase, and state of polarization of the wave,

multiple signal classification (MUSIC) a method in array processing for estimating parameters such as direction of arrival (DOA), based on estimating the noise subspace and exploiting the fact that the true signal (with the true parameters) belongs to the signal subspace (the orthogonal complement to the noise subspace).

multiple sub-Nyquist encoding (MUSE) a technique used in Japanese HDTV systems.

multiple-access time-division a method for allowing many users to communicate simultaneously with a single receiver by assigning transmissions to different time slots. The channel, which is shared among the users, is divided into successive frames, which are subdivided into time slots. Each user (or packet) is assigned a specific time slot within the current frame.

multiple-input–multiple-output system a system that transforms two or more input signals to two or more output signals. Also known as SISO system and single-variable (SV) system. *See also* system, single-input–single-output system.

multiple-operand instruction a computer instruction that contains two or more data elements that are used while executing the instruction, i.e., ADD Z, X, Y is a multiple operand instruction that specifies the values of X and Y are added together and stored into Z.

multiple-stage decision making   decision making involving future operation of the system, as in the case of closed-loop feedback control or limited-lookahead control, where possible future measurements and decision interventions are taken into account when considering the decision taken at a given time.

multiplex   (1) to use a single unit for multiple purposes, usually by time sharing or frequency sharing. *See also* multiplexer, multiplexing.

(2) the armature winding of a commutated electrical machine in which multiple, identical coil windings are placed on the rotor. In general, the number of the "plex" describes the total number of parallel windings between brush positions and, thus, also the multiplier on the number of parallel paths between brushes that would be provided by a simplex winding. For example, a duplex winding will have twice as many parallel electrical paths between brushes as a simplex winding, a triplex winding will have three times the number of paths, etc. *See also* simplex, duplex, reentrancy.

multiplexer   a combinational logic device with many input channels and usually one output, connecting one and only one input channel at a time to the output.

multiplexer channel   a computer I/O subsystem that allows multiple slow to medium speed devices to transfer data to or from the computer concurrently. *See also* byte multiplexer channel, block multiplexer channel, selector channel.

multiplexing   (1) the process of transmitting a large number of information units over a smaller number of channels or lines. For example, if we have N independent signals that we want to trans-

mit, then without using a multiplexer we need N independent channels to do so. Using a multiplexer to control the flow of these signals in only one channel reduces the number of wires, thus decreasing cost and increasing efficiency. Multiplexing is the superimposition of multiple

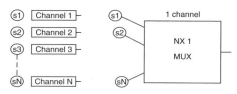

Multiplexing.

signals to make up one signal. This is done to make the transmission of the signals efficient. Signals are multiplexed at the sending end of communication systems, and demultiplexed at the receiving end, in order to obtain the original signals.

(2) of or being a communication system that can simultaneously transmit two or more messages on the same circuit or radio channel.

multiplication pattern   for arrays of similar elements, it is the product of the pattern of a single element and the array factor of the array.

multiplicative   acousto-optic   processing acousto-optic signal processing where the light is repeatedly modulated by a sequence of acousto-optic devices to result in a multiplication operation of the individual light amplitude or intensity modulations.

multiplier   an electronic system or computer software which performs the multiplication calculation.

multiplying D/A   a D/A conversion process where the output signal is the product of a digital code multiplied times an analog input reference signal. This allows the analog reference signal to be scaled by a digital code.

multipoint bus   a bus to which multiple components may be attached. The PCI bus is an example of a multipoint bus. See also point-to-point bus.

multiport   a circuit presenting multiple access ports.

multiport memory   one memory module can be accessed by two devices simultaneously. One example is a video memory, which is interfaced to a graphics co-processor as well as the video interface of the monitor.

multiprocessor   a computer system that has more than one internal processor capable of operating collectively on a computation. Normally associated with those systems where the processors can access a common main memory.

There are some additional stipulations for a "genuine multiprocessor:"

1. It must contain two or more processors of approximately comparable capabilities.

2. All processors share access to a common, shared memory.

3. All processors share access to common I/O facilities.

4. The entire system is controlled by a single operating system (OS).

See also multiple instruction multiple data architecture, operating system, shared memory architecture.

multiprogramming   a system that allows the processor to execute two or more programs concurrently.

multirate finite impulse response filter   a finite impulse response (FIR) filter in which the sampling rate is not constant.

multirate signal processing   a system in which there is at least one change of sampling rate. Typically an input signal is split into two or more sub-signals each with a sampling rate lower than the input signal.

multiresolution analysis   analysis method that decomposes a signal into components at different resolution level; the fine to coarse features are revealed in the fine to coarse resolution components.

multiresolution coding   coding schemes that involve a multiresolution structure.

multiscale   characterized by a parameterization or decomposition in scale, as opposed to (for example) time or frequency; in particular, an algorithm which computes or analyzes a function or an image at multiple resolutions. *See* wavelet, multigrid, Laplacian pyramid, quadtree, resolution.

multispectral image   an image that contains information from more than one range of frequencies. In this sense, an RGB color image is multispectral in that it contains red, green, and blue information. More usually, multispectral refers to frequencies of such number and type that it presents difficulties in interpretation and display. Adding infrared, ultraviolet, and radar information to an RGB image creates a problem, since the data cannot be simply drawn on a display. The use of many frequencies also provides a wealth of information sometimes needed to make fine distinctions between regions in an image, especially in satellite imagery. *See* data fusion, sensor fusion.

multistage depressed collector klystron   a specially designed klystron in which decreasing voltage zones cause the electron beam to be reduced in velocity before striking the collector element. The effect is to reduce the amount of heat that must be dissipated by the device, improving operating efficiency.

multithreaded   several instruction streams or threads execute simultaneously.

multitone testing   a measurement technique whereby an audio system is characterized by the simultaneous application of a combination of sine waves to a device under test.

MUSE   *See* multiple sub-Nyquist encoding.

MUSIC   *See* multiple signal classification.

mutual coupling   electromagnetic interaction between the elements in a phased array antenna in which the radiation from one element causes distortion of the current distributions or aperture field distributions of the other elements. Feed network mutual coupling occurs when the amplitude and phase of the excitation at a feed point are altered by the presence of reflected waves in the network.

mutual exclusion   a synchronization problem requiring that two or more concurrent activities do not simultaneously access a shared resource.

mutual inductance   the property that exists between two current-carrying conductors when the magnetic lines of force from one link with those from another.

mutual information   information theoretic quantity representing the amount of information given by one random variable about another. For two discrete random variables X and Y, with joint probability mass function $p(x, y)$ and marginal probability mass functions $p(x)$ and $p(y)$, respectively, the mutual information $I(X; Y)$ is given by

$$I(X; Y) = \sum_{x \in \mathcal{X}} \sum_{y \in \mathcal{Y}} p(x, y) \log \frac{p(x, y)}{p(x) p(y)}$$

where $\mathcal{X}$ and $\mathcal{Y}$ are the support sets of X and Y respectively. *See also* differential entropy, relative entropy.

mutually pumped phase conjugator (MPPC) a phase conjugator in which two phase conjugate waves are generated simultaneously when two beams are incident into the phase conjugator. With a mutually pumped phase conjugator, two incident laser beams mutually pump the photorefractive medium and produce two phase conjugate waves. There are many physical mechanisms that can yield mutually pumped phase conjugation, for example, MPPC via self-oscillation in four-wave mixing, MPPC with stimulated scattering, MPPC with ring resonator, etc.

MV   multi-variable system. *See* multi-input–multi-output system.

MVA interrupting rating   the interrupting rating of a device expressed in terms of megavolt-amperes. The conversion between fault volt-amps and fault current in three-phase systems is $VA = 1.73$ — operating RMS line-line voltage — largest phase fault current expressed in RMS symmetrical amperes. Power circuit breakers can have separate MVA and maximum current interrupting ratings, with adjustments when the circuit breaker is operated below rated voltage.

MWIR laser   a laser producing light in the midwavelength (2 – 5 micron) range, useful in chemical species identification and other applications.

N

n-channel MOSFET   a MOSFET where the source and drain are composed of heavily doped n-type semiconductor regions on a p-type surface. Electrons form drain-source current when the applied gate and substrate potentials invert the p-type surface between them.

n-well   a region of n-type semiconductor located at the surface of a p-type substrate (or larger P-well) usually created in order to contain p-channel MOSFETs.

NA   *See* numerical aperture.

Nakagami fading   a general model of fading used in the modeling of radio communication channels, introduced by Nakagami in 1960, which includes as a special case the Rayleigh model of fading.

NaN   acronym for not a number. Used in IEEE floating-point representations to designate values that are not infinity or zero or within the bounds of the representation.

NAND gate   a logic circuit that performs the operation equivalent to the AND gate followed by the inverter. The output of a NAND gate is low only if all inputs are high.

nano   prefix for metric unit that indicates division by one billion.

nanocell   cell of radius up to 10 m. *See also* cell.

nanoelectronics   a future integrated circuit technology characterized by minimum lateral feature sizes in the range 1 – 100 nanometers. Also pertains to ICs utilizing quantum devices.

nanolithography   Lithography involving the printing of ultra-small features, typically on the order of nanometers in size.

nanometer   a metric measure of distance equivalent to one billionth of a meter.

nanotechnology   any electrical, biological, chemical, mechanical or hybrid technology at nanoscale (10^{-9} meters or less). Nanotechnology promises a wide range of applications in medicine, computing, and industry.

narrow band   referring to a bandwidth of 300 hertz or less.

narrow-band fading   fading in which the communication channel amplitude can be assumed independent of fading. Occurs when the transmitted signal bandwidth is considerably less than the coherence bandwidth of the channel.

narrow band FM   frequency modulation scheme where the ratio of peak frequency deviation to the frequency of the modulating signal is smaller than 0.2.

narrow-band filter   narrow-band filters are those which pass signals undistorted in one or a set of narrow frequency bands and attenuate or totally eliminate signals in the remaining frequency bands.

narrow-band interference (NBI)   interference from a source that occupies a smaller bandwidth that the desired signal. One example is a single sinusoid (or a narrow-band communication system) interfering with a wide-band system, possibly a spread-spectrum system.

nasal   a sonorant phone that is produced with the vocal tract closed and the velum open. Nasals have a formant-like spectrum, but much information is carried out during the transition for the opening of the velum. Hence, the design of very effective automatic systems for the recognition of particular nasals (e.g., $/m/$, $/n/$) must take into account these brief transitions.

Nash equilibrium   a noncooperative equilibrium in nonzero-sum games defined by an outcome for each player that cannot be improved by altering his decision unilaterally. The roles of all players are symmetric in the sense that they make their decisions independently and none of them dominates any other. If $J_i, i = 1, 2, \ldots, N$ is a cost function of ith player in N-person game and d_i his strategy then N-tuple $d_1^*, d_2^*, \ldots, d_N^*$ constitutes a Nash equilibrium solution if

$$J_1 \left(d_1^*, d_2^*, \ldots, d_N^* \right) \leq J_1 \left(d_1, d_2^*, \ldots, d_N^* \right)$$
$$J_2 \left(d_1^*, d_2^*, \ldots, d_N^* \right) \leq J_2 \left(d_1^*, d_2, \ldots, d_N^* \right)$$
$$\vdots$$
$$J_N \left(d_1^*, d_2^*, \ldots, d_N^* \right) \leq J_N \left(d_1^*, d_2^*, \ldots, d_N \right)$$

for any admissible d_i; $i = 1, 2, \ldots, N$. The nonzero-sum game may admit more than one Nash equilibrium solution with outcomes different in each case. Since the total ordering among N-tuples of numbers does not exist, it is usually not possible to declare one of them as the most favorable. Nevertheless, some of strategies may be viewed as better than other ones if the respective outcomes are in partial order. With this notion of betterness, a Nash equilibrium strategy N-tuple is admissible if there exists no better one. In the case when the game is played, many times in the same conditions, the Nash equilibrium may be defined for average values of the cost functional and mixed strategies.

NASTRAN   a widely used computer code for mechanical and structural analysis, such as for opto-mechanical and thermal analyses.

National Electrical Code (NEC)   a standard for electrical construction, published by the National Fire Protection Association (NFPA 70-1). The National Electrical Code is often adopted by local jurisdictions and used by their electrical inspectors.

National Electrical Manufacturers Association (NEMA)   an electrical trade association that establishes standards for electrical equipment. In the case of electric motors, NEMA establishes standard frame sizes, starting torque, starting current, and other quantities for a given horsepower machine.

National Fire Protection Association (NFPA) sponsor and publisher of the National Electrical Code and other safety standards.

National Television System Committee (NTSC) a body that recommended the standard for colored television broadcasts in the U.S. in 1953. NTSC video contains 525 lines, a field rate of 59.94 fields/second, a horizontal frequency of 15,734.264 Hz, and an interlaced color subcarrier frequency of 3.579545 MHz. The NTSC format is also in use by many countries other than the U.S.

natural broadening   spectral broadening of a transition in a laser medium due to spontaneous decay; sometimes called lifetime broadening.

natural commutation   commutation of current from one switching device to another in a power electronics converter at the instant the incoming voltage has a higher potential than that of the outgoing wave, without the aid of any commutation circuitry. *See also* commutation.

natural constraint   constraint that results from the particular mechanical and geometric characteristics of the task configuration. Used to define a particular situation occurring between the manipulator end-effector and the work environment. These constraints are defined in terms of linear and angular velocities and forces and/or torques that are specified to execute the task. Natural constraints are intrinsically associated with the particular task. For example, consider a task that an end-effector moves along a rigid surface and is not free to move through that surface. In this particular situation, a natural position constraint exists. In addition, if the surface is frictionless, the end-effector is not free to apply arbitrary force tangent to the surface, and hence a natural force constraint exists.

natural frequency   (1) the frequency of any oscillating term in the response of the linear system which is due to the initial value of that system.

(2) for linear time invariant systems the total response of the system can be represented as the sum of the zero-input response plus the zero state response. The characteristic polynomial of a system (representing the zero input response),

$$Q(\lambda) = (l - l_1)(l - l_2) \dots (l - l_n) = 0$$

has n roots; these roots are called the natural frequencies. Natural frequencies are also known as the characteristic values, eigenvalues, and characteristic roots of the system.

natural laser   a laser occurring in nature.

nautical mile (nm)   1,852 meters, exactly. Approximately 1 minute of arc on the earth's surface.

NBI   *See* narrow-band interference.

NDR   *See* negative differential resistance.

near field   (1) in antennas, the electromagnetic field that is in the vicinity of an antenna where the angular field distribution is dependent on the distance from the antenna.

(2) in optics, region close to a diffracting aperture where neither the form nor the size of a transmitted beam have changed significantly from their values at the aperture.

near infrared   light in the wavelength region of the infrared electromagnetic spectrum which is adjacent to the visible spectrum. The spectral range

of the near infrared region is typically considered to be 700 – 2500 nm.

nearest neighbor algorithm   a method of classifying samples in which a sample is assigned to the class of the nearest training set pattern in feature space; a special case of the K-nearest neighbor algorithm.

near–far effect   large dissimilarities in received power from different users, thereby causing trouble in detecting the weaker users. Commonly found in DS-CDMA systems.

near–far resistant   a device, e.g., a detector or parameter estimator, that is insensitive to the near–far effect. *See also* near–far effect.

NEC   *See* National Electrical Code.

Necker cube   a classical example of ambiguous wireframe geometrical figures, it is a cube that can take either of two perspectives, as it were seen from above or from below.

NEESLA   *See* nonuniformly excited equally spaced linear array.

negation operator   *See* complement operator.

negative definite function   a scalar function $V(x, t)$ where $-V(x, t)$ is positive definite.

negative differential resistance (NDR)   a condition where the slope of the current vs. voltage characteristic is negative. Negative differential devices include quantum devices such as tunnel diodes and resonant tunneling structures. Negative differential resistance can also be obtained in a variety of circuit topologies.

negative photoresist   a photoresist whose chemical structure allows for the areas that are exposed to light to develop at a slower rate than those areas not exposed to light.

negative resistance oscillator   oscillator circuit that functions based on an active device where the device is biased under certain conditions and terminated with impedance levels such that it exhibits a negative resistance at a particular frequency or frequency range.

negative semidefinite function   a scalar function $V(x, t)$, where $-V(x, t)$ is positive semidefinite.

negative sequence   the set of balanced but reverse sequence (acb) components used in symmetrical component analysis. Normal load currents contain no negative sequence current.

negative sequence overcurrent relay   a protective relay that senses and operates on negative sequence overcurrent. Typical applications include the sensing of unbalanced faults and the protection of synchronous and induction machines from rotor overheating.

negative transition angle   the angular portion of the time based output signal that has a negative slope, expressed in degrees. This quantity could be loosely interpreted as the "trailing edge" angle.

negative-positive-zero (NPO)   an ultrastable temperature coefficient (± 30 ppm/°C from $-55°$ to 125°C) temperature-compensating capacitor.

negative-sequence impedance   the impedance offered by a circuit when negative-sequence currents alone flow through it, expressed in ohms. The impedance is complex, with its real part being the circuit resistance and imaginary part, which is a function of frequency and inductance referenced as negative-sequence reactance, also expressed in ohms.

negative-sequence reactance   inductive reactance offered by a circuit for the flow of negative-sequence currents alone. Expressed in ohms, the inductive reactance is a function of frequency and the inductance of the circuit to negative-sequence current flow. *See also* negative-sequence impedance.

neighborhood in self-organizing system   an area surrounding the winner in self-organizing competitive learning. According to so-called Mexican-hat interaction between the brain cells, Kohonen proposes that weight updating should be conducted not only for the winner but also for its neighbors. *See also* Mexican-hat function, self-organizing algorithm.

neighborhood operation   an operation, such as averaging or median filtering, that is dependent upon the locality of samples in a signal, not the signal as a whole. Also called a window operation. *See* windowing.

NEMA   *See* National Electrical Manufacturers Association.

NEMA code letter   the nameplate letter designation used to indicate the input kilowatt-amp rating of a motor under locked-rotor, or starting conditions.

NEMA size   a standard-size device, such as a motor controller. The NEMA size establishes the rating of the device. *See also* National Electrical Manufacturers Association.

NEMA type   for induction motors NEMA establishes five types of induction motors (A, B, C, D, and E) that have different torque-speed characteristics to account for various types of loads.

nematic   the type of liquid crystal in which the molecular chains align; such alignment can be controlled across the liquid crystal if it can be constrained at the boundaries.

nematic liquid crystal   one of the state of liquid crystal materials, where the elongated liquid crystal molecules are all oriented in the same direction within a layer. The molecules can be reoriented with an external electric field, allowing use in displays and spatial light modulators. *See also* liquid crystal on silicon, twisted nematic.

neocognitron   a biologically inspired hierarchical network developed primarily for the recognition of spatial images. The network has up to nine layers. The lower layers respond to simple features and the higher layers respond to more complex features. Learning can be unsupervised or supervised.

neodymium-iron-boron   a high energy magnetic material composed of the three nominal elements and other additives, characterized by a high residual induction and high coercivity. NdFeB has a high magnetic temperature coefficient, which is undesirable for high-temperature use.

neper   a natural logarithm of a ratio.

nested structure   an information structure in which each player has an access to the information acquired by all his precedents, i.e., players situated closer to the beginning of the decision process. If the difference between the information available to a player and his closer precedent involves only his actions, then the structure is ladder-nested. This structure enables decomposition of the decision process onto static games that in turn results in recursive procedure for its solution. For dynamic infinite discrete-time decision processes, the ladder-nested structure results in the classical information pattern.

nested subroutine   a subroutine called by another subroutine. The programming technique of a subroutine calling another subroutine is called nesting.

network analyzer   a test system that measures RF or microwave devices in terms of their small signal characteristics. The results are typically presented as two-port parameter matrices such as s-parameters, y-parameters, or z-parameters.

network distribution system   a distribution system in which each load is supplied by more than one path.

network function   the ratio $H(s)$ of the Laplace transform of the output function to the Laplace transform of the input function. *See also* transfer function.

network interface card (NIC)   the physical device or circuit used to interface the network with a local workstation or device.

network measurement division (NMD) connector   a special coaxial connector type developed by Hewlett Packard for test cables used in conjunction with an s-parameter test set.

network pruning   the removal from a network of interconnections and/or neural units that are identified (after training) as being either unnecessary or unimportant. After pruning, the network must be retrained. *See* artificial neural network.

network time constant   minimum of the allowable transfer delay and the time required to fill up the buffer in the node. The NTC is indicative of how long transients last inside a network node. The NTC is the interval over which network traffic must be averaged.

network weight   a scalar value representing the strength of the connection between the output of one neuron and the input of another.

neural network   a parallel distributed information processing structure consisting of processing elements, called neurons, interconnected via unidirectional signal channels called connections. Neurons can possess a local memory and can carry out localized information processing operations. Each neuron has a single output connection. The neuron output signal can be of any mathematical type desired. The information processing that goes on within each processing element can be defined arbitrarily, with the restriction that it must be completely local (it depends only on the current values of the input signals arriving at the processing element and on values stored in the processing element's local memory, e.g., weights).

neural tree a tree-structured neural network. Such networks arise in the application of certain kinds of constructive algorithm.

neuro-fuzzy control system control system involving neural networks and fuzzy systems, or fuzzy neural networks. *See also* adaptive fuzzy system.

neuron a nerve cell. Sensory neurons carry information from sensory receptors in the peripheral nervous system to the brain; motor neurons carry information from the brain to the muscles.

neutral a conductor which completes the electric circuit from the load to the source in three-phase Y-connected and single phase AC electric power systems, typically at or near the potential of the earth.

neutral axis the axis near which the direction of the velocity of armature conductors is exactly parallel to the magnetic flux lines so that the EMF induced in the conductors is zero. This axis, also referenced as the magnetic neutral axis, shifts in the direction of rotation of the machine as a generator or motor. The amount of shift depends on the armature current and hence on the load of the machine.

neutral plane *See* neutral axis.

neutral zone in permanent magnets, a plane through which all lines of flux are parallel to the direction of magnetization, i.e., the plane between north and south poles.

neutrino laser laser in which the amplified field consists of neutrinos rather than electromagnetic waves; suggested importance in the early universe.

New York City blackout a parrticularly disastrous failure of the power system in New York City in 1977.

Newton–Euler recursive algorithm used in robotics, to calculate joint generalized forces as a result of two recursions along the kinematical structure of the robot. First one, a forward recursion, calculates spatial velocities and accelerations from the base of the manipulator towards its tip. Second one, a backward recursion, calculates the spatial forces and torques along the structure in the opposite direction. Projection of the spatial force and torques at the joint axis results in the generalized force at the joint. The resulting Newton–Euler equations of motion are not in closed form as opposite to the result of using Lagrange formalism to the dynamical system of the robot itself. *See also* Lagrange formulation.

Newton–Raphson method a numerical method for finding the solution to a set of simultaneous nonlinear equations. Variations of this method are commonly used in circuit simulation programs.

Newton's method a class of numerical root-finding methods; that is, to solve $f(x) = 0$. The methods are based on iteratively approximating f using a low order polynomial and finding the roots of the polynomial. The most commonly used first-order method iterates the following equation:

$$x_{n+1} = x_n - \frac{f(x_n)}{f'(x_n)}$$

where f' represents the derivative of f.

Neyman–Pearson detector a detector that minimizes the miss probability within an upper-bound constraint on the false-alarm probability.

NFPA *See* National Fire Protection Association.

nibble four bits of information.

nibble-mode DRAM   an arrangement where a dynamic RAM can return an extra three bits, for a total of four bits (i.e., a nibble), with every row access.

The typical organization of a DRAM includes a buffer to store a row of bits inside the DRAM for column access. Additional timing signals allow repeated accesses to the buffer without a row-access time. *See also* two-dimensional memory organization.

NIC   *See* network interface card.

Nichols chart   in control systems, a plot showing magnitude contours and phase contours of the return transfer function referred to ordinates of logarithmic loop gain and abscissae of loop phase angle.

nickel-cadmium battery   a battery that is similar to the lead acid battery except that it has a longer life cycle than lead-acid batteries and can be recharged more quickly. Therefore, they are widely used in small consumer electronics. *See also* nickel-metal hydride battery, lithium battery.

nickel-metal hydride battery   a type of battery with greater power and energy capabilities than lead-acid or nickel-cadmium, but which are more expensive. Nickel-metal hydride batteries are widely used in computers and medical applications. *See also* lead-acid battery, lithium battery.

night vision   *See* scotopic vision.

NMD connector   *See* network measurement division connector.

NMI   *See* nonmaskable interrupt.

NMR   nuclear magnetic resonance. *See* magnetic resonance imaging.

NMOS   *See* n-channel MOSFET.

no fetch on write   strategy in the write-through cache policy where a line is not fetched from the main memory into the cache on a cache miss if the reference is a write reference. Also called non-allocate on write, as space is not allocated in the cache on write misses.

no load tap changer   device that provides for changing the tap position on a tapped transformer when the transformer is de-energized. Different taps provide a different turns ratio for the transformer.

no voltage holding coil   a holding coil that keeps the main-line contactor closed on zero voltage conditions. DC motor controllers that contain this feature are used in places where the motor is vital to the operation of a process. These controllers can maintain control to the motor under momentary line power loses, by using the CEMF of the coasting armature to keep power to the main-line coil/contactor. If power to the motor controller is not restored within a short period of time, the motor coasts to a speed where it can no longer keep the main-line contactor closed. At this point, the m-coil drops out to insure starting resistors are placed back in the circuit.

no-load test   measurement of input parameters of an induction motor while running at nearly synchronous speed, with zero output on the shaft. This test is used to determine the magnetizing reactance of the motor equivalent circuit. *See also* open-circuit test.

no-op   a computer instruction that performs no operation. It can be used to reserve a location in memory or to put a delay between other instruction execution.

no-write allocate   part of a write policy that stipulates that if a copy of data being updated are not found in one level of the memory hierarchy, space for a copy of the updated data will not be allocated in that level. Most frequently used in conjunction with a write-through policy.

nodal cell   cell with a radius of up to 300 m. Typically an isolated cell acting as a high capacity network node. *See also* cell.

nodal system   a secondary system of equations using nodal voltages as variables.

node   a symbol representing a physical connection between two electrical components in a circuit. *See also* graph.

node analysis   a circuit analysis technique in which KCL is used to determine the node voltages in a network.

noise   (1) any undesired disturbance, whether originating from the transmission medium or the electronics of the receiver itself, that gets superimposed onto the original transmitted signal by the time it reaches the receiver. These disturbances tend to interfere with the information content of the original signal and will usually define the minimum detectable signal level of the receiver.
(2) any undesired disturbance superimposed onto the original input signal of an electronic device; noise is generally categorized as being either external (disturbances superimposed onto the signal before it reaches the device) or internal (disturbances added to the signal by the receiving device itself). *See also* noise figure, noise power ratio. Some common examples of external noise are crosstalk and impulse noise as a result of atmospheric disturbances or manmade electrical devices. Some examples of internal noise include thermal noise, shot noise, 1/f noise, and intermodulation distortion.

noise bandwidth   an equivalent bandwidth, W_N, of a system expressed as

$$W_N = \frac{\int_0^\infty |H(\omega)|^2 \, d\omega}{|H(\omega_0)|^2},$$

where $H(\omega)$ is the transfer function of the system and ω_0 is the central frequency.

noise circles   circles of constant noise figure plotted on the Smith chart that can be used to graphically impedance match a device to achieve a desired noise figure.

noise clipping   a process by which high noise peaks are clipped or limited to eliminate most of their energy, thereby removing the worst effects of impulse noise.

noise factor   *See* noise figure.

noise figure   an indication of the contribution of that component or system to the level of noise observed at the output. The noise figure therefore gives an idea of the amount of noise generated within the component or system itself. It is usually expressed in decibels (dB) and is given by the ratio of the input signal-to-noise ratio to the output signal-to-noise ratio of the component or system. This is defined as

$$\text{Noise Figure (NF)} \triangleq \frac{(S/N)_{\text{input}}}{(S/N)_{\text{output}}}.$$

In decibels this is given by:

$$[\text{NF}]_{\textbf{dB}} \triangleq 10 \log_{10} \left[\frac{(S/N)_{\text{input}}}{(S/N)_{\text{output}}} \right]$$

noise figure meter   a system that makes accurate noise figure readings possible. It measures the amount of the noise added by a device. Noise figure is the ratio of the signal-to-noise ratio at the input of a device to the signal-to-noise ratio at the output of that device.

noise floor   the lowest input signal power level which will produce a detectable output signal in an electrical system. Noise floor is determined by the thermal noise generated in the electrical system.

noise immunity   a logic device's ability to tolerate input voltage fluctuation caused by noise without changing its output state.

noise power   the amount of power contained within a noise signal. The noise power can be found by integrating the power spectral density (PSD) of the noise signal over all possible frequencies. If the power spectral density is given by $S(\omega)$, the power, p, contained in the noise signal, is then given by

$$p \triangleq \frac{1}{2\pi} \int_{-\infty}^{\infty} S(\omega)d\omega.$$

In most cases, noise signals are bandwidth limited to between say ω_1 and ω_2 and in these cases the noise power will be given by

$$p \triangleq \frac{1}{2\pi} \int_{\omega_1}^{\omega_2} S(\omega)d\omega.$$

Generally, the power spectral density is expressed in watts per hertz (W/Hz) and the noise power is expressed in watts. *See also* noise, noise spectrum.

noise power ratio (NPR)   intermodulation distortion product generated in nonlinear transfer components such as high power amplifiers (HPAs) amplifying multiple carriers. It is defined as a ratio of an averaged power of white noise having a narrow notch, which represents multiple carriers, to an averaged power falling into a narrow bandwidth of notch.

noise rejection   the ability of a feedback control system to attenuate (reduce) the amplitude of any unwanted signal generated by the measurement of its output variable.

noise smoothing   any process by which noise is suppressed, following a comparison of potential noise points with neighboring intensity values, as for mean filtering or median filtering.

noise spectrum   indicates the frequency components in a noise signal. For ideal thermal noise (AWGN), the spectrum is flat across all frequencies and is referred to as the noise power spectral density expressed in watts per hertz (W/Hz). *See also* noise, noise power.

noise subspace   in an orthogonal decomposition of a space, the orthogonal complement to the signal subspace.

noise suppression   any process by which noise is eliminated or suppressed: a more general term than noise smoothing, since it includes such processes as noise clipping and band-pass filtering.

noise temperature   an alternate representation of noise power. It is generally accepted that noise is the result of the random motion of electrons within components in electronic systems. The greater the temperature of a component, the greater the random motion of electrons within the component, and this causes an increased instantaneous noise current to be present within the component. With an increase in noise current, the noise power also increases. The noise power generated within a component at T Kelvin, over a bandwidth of B hertz is given by

$$p = kTB$$

with $k = 1.38 \times 10^{-23}$ J/K, which is Boltzmann's constant.

From this equation, it is clear that a component at absolute zero temperature (0 K), generates no noise power. Therefore, given a particular bandwidth, B noise power can also be expressed as a temperature in Kelvin (K).

noise whitening   a process by which noise whose power spectrum is not white (i.e., not identical at all frequencies) is brought to this condition, e.g., by means of a frequency dependent filter.

noiseless source coding   *See* lossless source coding.

noiseless source coding theorem   states that any source can be losslessly encoded with a code whose average number of bits per source symbol is arbitrarily close to, but not less than, the source entropy in bits.

noisy channel vector quantization a general term for methods in vector quantizer transmission for noisy channels. *See also* channel robust vector quantization, channel optimized vector quantization, channel matched VQ, redundancy-free channel coding.

noisy channel VQ *See* noisy channel vector quantization.

noisy source coding *See* lossy source coding.

nominal voltage a number given to a system to name its classification of voltage, such as rated values.

nonallocate on write *See* no fetch on write.

nonblocking cache a cache that can handle access by the processor even though a previous cache miss is still unfinished.

noncausal system *See* causal system.

noncentrosymmetric medium *See* centrosymmetric medium.

noncoherent integration where only the magnitude of received signals is summed.

noncollinear geometry acousto-optical tunable filter an acousto-optical tunable filter (AOTF) device that operates similar to a Bragg cell where the incident light and acoustic wave are not collinear and the diffracted light can be spatially separate from the incident light.

noncollinear geometry AOTF *See* noncollinear geometry acousto-optical tunable filter.

noncooperative game one of a class of decision processes in which decision makers (players) pursue their own interests, which are at least partly conflicting with others.' A conflicting situation or collision of interests results the situation in which

a player has to make a decision and each possible one leads to a different outcome valued differently by all players. Depending on the number of decision makers involved in the process, the game may be two-person or multi-person; depending on how the outcome is viewed by the players, the process may be a zero-sum game (only two-person) or a nonzero-sum game (two- or multi-person). If the order in which the decisions are made is important, a game is dynamic, if not it is static. Dynamics of the game is usually related to the access of the decision makers to different information. Regarding the number of possible decisions available to the decision makers, a game is finite or infinite (in this case, the number of possible actions is usually a continuum). Dynamic games are usually formulated in extensive form, which for finite games leads to a finite tree structure, while for infinite ones, involves difference (in discrete-time) or differential (in continuous-time) equations describing an evolution of the underlying decision process.

non-customer-call call placed by passerby to utility, either emergency personnel or otherwise, to indicate that a scenario exists which is potentially disrupting electrical service.

nondegenerate network a network that contains neither a circuit composed only of capacitors and/or independent or dependent voltage sources nor a cutset composed only of inductors and/or independent or dependent current sources.

nondegenerate two-wave mixing a general case of two wave mixing in which the two beams are of different frequencies. In two-wave mixing, if the frequencies of the two laser beams are different, the interference fringe pattern is no longer stationary. A moving volume refractive index grating can still be induced provided that the intensity fringe pattern does not move too fast. The amplitude of the refractive index modulation decreases as the speed of the fringe pattern increases. This is related to the finite time needed for the formation of the refractive index grating in the photorefractive medium. Such a kind of two-wave mixing is referred to as nondegenerate two-wave mixing.

nondestructive readout   when data are read from a particular address in a memory device, the contents of the memory at that address remain unaltered after the read operation.

nondispersive medium   medium in which the index of refraction does not vary significantly with frequency.

nonfuzzy output   a function of the firing degrees and the fuzzy outputs of fired fuzzy rules regardless of what defuzzification method is used.

nonhomogeneous linear estimator   an estimator which is a nonhomogeneous linear function of the data.

noninteraction of photons   the nature of photons that they do not interact to each other. Unlike electrons governed by Coulomb's law, a photon cannot affect another photon. A light beam passes through another light beam without any change. If photons interacted as electrons do, the image we see from one direction would be distorted by other streams of photons, and we would never understand what we see. The noninteraction of photons provides parallelism and enables us viewing in free space. On the other hand, it makes controlling or switching light using light impossible.

noninvasive sensor   the interface device of an instrumentation system that measures a physiologic variable from an organism without interrupting the integrity of that organism. This device can be in direct contact with the surface of the organism or it can measure the physiologic quantity while remaining remote from the organism.

noninvertible system   *See* invertible system.

nonlinear dielectric property   the distinct dependence of the electric permittivity of certain dielectric materials on the intensity of an applied electric field.

nonlinear distortion   a change in signal properties due to circuit transmission characteristics not being completely linear. This may result in changes in the relative amplitude or phase of the various frequency components of the signal or may result in intermodulation.

nonlinear effect   effect that cannot be written as linear functions of the driving fields

nonlinear electro-acoustic property   nonlinear interaction between the atomic displacement and the electric field experienced in certain materials that would cause modulation effects resulting in the generation of new sideband frequencies (called Raman frequencies).

nonlinear electro-optic property   nonlinear changes in the refractive index of certain optically transparent materials with change(s) in the externally applied electric field.

nonlinear filter   *See* linear filter.

nonlinear load   an electrical load operating in the steady state that has a current that is not continuous or in which the impedance changes during each cycle of the supply voltage.

nonlinear magnetic property   nonlinear dependence of the magnetic susceptibility of certain materials on the intensity of an applied magnetic field.

nonlinear mean-square estimate   the optimum estimate under the mean-square performance criterion.

nonlinear medium   medium in which the constitutive parameters are not functions of the electric or magnetic field amplitudes.

nonlinear model a small signal model that includes internal components, essentially voltage controlled current sources and capacitors, which depend on the applied voltages to the transistors. Nonlinear models are quite useful for understanding the behavior of transistors.

nonlinear optics (1) the study of optical phenomena that occur through the use of light fields (e.g., laser beams) sufficiently intense to modify the optical properties of a material system. From a formal perspective, these phenomena are nonlinear in the sense that the material polarization depends on the applied electric field strength in a nonlinear manner.

(2) optical processes in materials capable of producing output light with wavelengths different from that of the input light.

nonlinear response the characteristic of certain physical systems that some output property changes in a manner more complex than linearly in response to some applied input.

nonlinear Schrödinger equation the fundamental equation describing the propagation of short optical pulses through a nonlinear medium, so-called because of a formal resemblance to the Schrödinger equation of quantum mechanics.

nonlinear susceptibility a quantity describing the nonlinear optical response of a material system. More precisely, the nonlinear susceptibility of order n, often designated χ^n, is defined through the relation $P^n = c\chi^n E^n$, where E is the applied electric field strength and P^n is the nth order contribution to the polarization. The coefficient c is of order unity and differs depending on the conventions used in defining the electric field strength. The nonlinear susceptibility of order n is a tensor of rank $n + 1$.

nonlinear system a system that does not obey the principle of superposition.

The superposition principle states that if

$$y(t) = f(x(t)), \quad y_1(t) = f(x_1(t))$$

and

$$y_2(t) = f(x_2(t))$$

and if

$$x(t) = \alpha x_1(t) + \beta x_2(t)$$

then

$$y(t) = \alpha y_1(t) + \beta y_2(t)$$

See also linear system.

nonlinear-optic logic gate an optical logic gate utilizing thresholding capability of nonlinear-optic materials.

nonlinear-optic material a material from the following group: GaAs, ZnS, ZnSe, CuCl, InSb, InAs, and CdS. When these transparent materials are illuminated with input light, their refractive indices depend nonlinearly on the illuminating light intensity. The relation of refractive index and illuminating intensity resembles a threshold function. The phase of output light transmitting through the nonlinear material will be modulated by the refractive index of material. The phase modulation is translated into intensity variation using an interferometer.

nonlinearity response of a medium that is not directly proportional in magnitude to the magnitude of applied fields.

nonlocal optics due to spatial dispersion, the dielectric function of a material may depend on the wave vector. In such a case, the relation between the electric field and the current density will be nonlocal. The nonlocal effect will play an important role in modeling the interaction of electromagnetic waves with metals when the frequency of the electromagnetic wave is around the plasma frequency of the metal.

nonmaskable interrupt (NMI) an external interrupt to a CPU that cannot be masked (disabled) by an instruction. *See also* maskable interrupt.

nonorthogonal projection a nonorthogonal projection is a projection of a vector **c** onto a vector **b** in a direction orthogonal to $\mathbf{g} = \mathbf{Q_g b}$, where $\mathbf{Q_g}$ is an arbitrary rotation matrix in the plane spanned by **b** and **c**. It is defined as

$$\mathbf{c_b} = \left(\frac{\mathbf{b}^T \mathbf{Q_g c}}{\mathbf{b}^T \mathbf{Q_g b}} \right) \mathbf{b}$$

nonorthogonal wavelets *See* orthogonal wavelets.

nonparabolic band refers to an energy band in which the energy dependence upon the momentum deviates from the classical quadratic behavior.

nonparametric estimation an estimation scheme in which no parametric description of the statistical model is available.

nonradiating dielectric (NRD) waveguide a rectangular dielectric waveguide bound by two infinite parallel metallic plates. The dielectric constant and the separation between the plates is chosen such that the electromagnetic wave is above cut-off and propagating in the dielectric, and is below cut-off outside the dielectric, and does not radiate out of the parallel plates.

nonreccurring engineering (NRE) costs the foundry charges the ASIC customer. These costs include engineering time, making the masks, fabricating one lot of wafers, and packaging and testing the prototype parts.

nonrecursive equation *See* recursive equation.

nonredundant number system the system where for each bit string there is one and only one corresponding numerical value.

nonremovable disk *See* removable disk.

nonrigid body motion *See* rigid body motion.

nonsaturated region *See* ohmic region.

nonseparable data *See* separable data.

nonstate variable network variable that is not a state variable.

nonuniform sampling *See* uniform sampling.

nonuniformly excited equally spaced linear array (NEESLA) an antenna array in which all the centers of the antennas are collinear and equally spaced, but each antenna can have a unique harmonic amplitude and can have a unique phasing.

non-transposition refers to a three-phase electric power transmission line whose conductors are not transposed *See* transposition.

nonunity feedback the automatic control loop configuration shown in the figure.

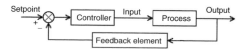

Nonunity feedback configuration.

nonvolatile *See* nonvolatile memory.

nonvolatile memory (1) memory that retains its contents even when the power supply is removed. Examples are secondary memory and read-only memory.

(2) the class of computer memory that retains its stored information when the power supply is cut off. It includes magnetic tape, magnetic disks, flash memory, and most types of ROM.

nonvolatile random-access memory (NVRAM)
SRAM or DRAM with nonvolatile storage cells.
Essentially each storage cell acts as a normal RAM
cell when power is supplied, but when power is
removed, an EEPROM cell is used to capture the
last state of the RAM cell and this state is restored
when power is returned.

nonzero-sum game   one of a class of games in
which the sum of the cost functions of the players is not constant. A number of players is not
limited to two as in zero-sum games, and in this
sense one may distinguish between two-person
and multi-person nonzero-sum games. Since the
objectives of the players are not fully antagonistic,
cooperation between two or more decision makers may lead to their mutual advantage. However,
if the cooperation between the players is not admissible because of information constraints, lack
of faith, or impossibility of negotiation, the game
is noncooperative and an equilibrium may be defined in a variety of ways. The problem of solving the nonzero-sum game differs from that of the
zero-sum game. The most natural solution is Nash
equilibrium, which is relevant to the saddle point
equilibrium in zero-sum games. Its main feature
is that there is no incentive for any unilateral deviation by any one of the players and their roles
are symmetric. If one of the players has the ability to enforce his strategy on the other ones, then
a hierarchical equilibrium called von Stackelberg
equilibrium is rational. Yet another solution for the
player is to protect himself against any irrational
behavior of the other players and adopt min-max
strategy by solving a zero-sum game, although the
original is nonzero-sum.

non-binary codes   codes in which the fundamental information units or symbols assume more
than two values. This is in contrast to binary codes,
for which the fundamental information symbols
are two-valued or binary, only. *See also* binary
code, block code, convolutional code.

non-return-to-zero recording   *See* magnetic
recording code.

non-return-to-zero inverted recording   *See*
magnetic recording code.

non-time delay fuse   *See* single-element fuse.

NOR gate   a logic circuit that performs the operation equivalent to the OR gate followed by the
inverter. The output of a NOR gate is low when
any or all inputs are high.

norator   an idealized two-terminal network element for which the voltage across it and the current
through it are determined by the network to which
it is connected.

norm   for a vector space V, a norm is a real-
valued function N defined on V, satisfying the
following requirements for every $v, w \in V$ and
scalar λ:

(1) $N(0) = 0$ and   for $v \neq 0$, $N(v) > 0$.
(2) $N(\lambda v) = |\lambda| N(v)$.
(3) $N(v + w) \leq N(v) + N(w)$.

The most usual norms are the L^1, L^2, and
L^∞ norms; for a vector x with n coordinates
$x_1, \ldots, x_n$ we have $\|x\|_1 = \sum_{k=1}^{n} |x_k|$, $\|x\|_2 =$
$(\sum_{k=1}^{n} |x_k|^2)^{1/2}$, and $\|x\|_\infty = \max_{k=1}^{n} |x_k|$; for a
function f defined on a set E, these norms become
$\|f\|_1 = \int_E |f(x)|dx$, $\|f\|_2 = (\int_E |f(x)|^2 dx)^{1/2}$,
and $\|f\|_\infty = \min\{r \geq 0 | f(x)| \leq r\}$ almost ev-
erywhere. Given a norm N, the function d defined
by $d(x, y) = N(x - y)$ is a distance function on
V. *See* chessboard distance, Euclidean distance,
Manhattan distance.

normal demagnetization curve   the second
quadrant portion of the hysteresis loop generated
when magnetic induction (B) is plotted against applied field (H), which is mathematically related to
the intrinsic curve; used to determine the performance of a magnet in a magnetic circuit.

normal dispersion   increase of index of refraction with increasing frequency; tends to occur near
the center of amplifying transitions or in the wings
of absorbing transitions.

normal distribution   *See* Gaussian distribution.

normal tree   a tree that contains all the independent voltage sources, the maximum number of capacitors, the minimum number of inductors, and none of the independent current sources.

normal vector   a vector perpendicular to the tangent plane of a surface. Also, a vector that is geometrically perpendicular to another vector.

normalization   (1) the process of reformatting a floating point number into a standard form at the completion of a floating point arithmetic operation.
(2) the process of equalizing signal energies, amplitudes, or other features prior to comparison.

normalized number   a floating-point number in which the most significant digit of the significand (mantissa) is non-zero. In the IEEE 754 floating-point standard, the normalized significands are larger than or equal to 1, and smaller than 2.

normally closed contact   contact of a contactor that is closed when the coil of the contactor is deenergized and opened when the coil is energized.

normally closed, time to close   a relay that is closed when the power to its actuator is off, but has a time delay to close when power is removed from the actuator. When power is applied, the relay immediately opens.

normally open contact   a contact that is open under normal operating conditions and closes when an action is initiated in its controller. For a contact that is part of a relay, the contact remains open when the relay is deenergized and closes when the relay is energized.

normally open, time to close   a time delay relay that is open when the power to its actuator is off. When power is applied to the actuator, the relay remains open for an adjustable time delay, after which it closes. When power is removed from the actuator, the relay opens immediately.

Northeast blackout   a 1965 failure of much of the power grid in the northeastern states of the USA.

Norton theorem   states that the voltage across an element that is connected to two terminals of a linear, bilateral network is equal to the short-circuit current between these terminals in the absence of the element, divided by the admittance of the network looking back from the terminals into the network, with all generators replaced by their internal admittances.

NOT   a Boolean operation that returns the 1's complement of the data to which it is applied.

notch   a disturbance of the normal voltage waveform of duration less than 0.5 cycles, is of a polarity that is opposite to the waveform and is hence subtracted from the normal waveform with respect to the peak value of the disturbance voltage.

Noyce, Robert N.   (1927–1990)   Born: Denmark, Iowa
Best known as one of the founders and co-chair of Intel Corporation, and a developer of the first commercially successful microchip, along with Gordon Moore. Noyce began his career with William Shockley's short-lived company. He and other engineers split with Shockley and formed Fairchild Semiconductor. Fairchild was the first of the very successful "Silicon Valley" high-tech firms. Noyce's patent on the microchip was challenged by Jack Kilby and Texas Instruments. A court ruled in favor of Noyce, although it is commonly accepted that the original ideas were Kilby's.

NPO   *See* negative–positive zero.

NPR   *See* noise power ratio.

NRC   the Nuclear Regulatory Commission, a US regulatory body which oversees civilian power reactors.

NRD waveguide   *See* nonradiating dielectric waveguide.

NRE   *See* nonrecurring engineering.

NTSC   *See* National Television System Committee.

NuBus   an open bus specification developed at MIT and used by several companies. It is a general-purpose backplane bus, designed for interconnecting processors, memory, and I/O devices.

nuclear fuel   fissile material, including natural uranium, enriched uranium, and and some plutonium prepared for use in a nuclear reactor.

nuclear fuel management   the process of managing the degree of enrichment, the timing of insertion into the core, and the placement and possible relocation of fuel within the core during the lifetime of the fuel.

nuclear magnetic resonance   the phenomenon in which the resonant frequency of nuclear spin is proportional to the frequency of an applied magnetic field. *See* magnetic resonance imaging.

nuclear power plant   a thermal electric power plant in which the heat for steam turbines is produced by nuclear fission.

nuclear reaction   a reaction which causes changes in the nucleus of an atom, thus changing elements to another element or isotope, usually with the release of energy.

nuclear reactor   (1) an apparatus designed to facilitate, contain, and control a nuclear chain reaction.

(2) any heat-producing array of fissile radioactive materials constructed so as to produce a controlled chain-reaction.

null   a point on the radiation pattern that corresponds to zero or minimum values.

null controllability   a dynamical system that is controllable to zero end state. In many cases, null controllability is equivalent to controllability. This is always true for linear finite-dimensional continuous-time dynamical system. However, in discrete case, controllability may be the stronger notion than null controllability. In this case, the two concepts are equivalent if and only if $\text{rank}(A) = n$. For dynamical systems with delays, these two notions are essentially different. For infinite-dimensional systems the relations between null controllability and controllability depend on the properties of the semigroup $S(t)$ generated by the operator A.

null space   the set of vectors that are orthogonal to every vector of a set forming vector space. For an (n, k) code, the dimension of the null space is $n - k$.

null space of the Jacobian   the subspace $N(J)$ in R^n of joint generalized velocities that do not produce any end-effector velocity, in the given manipulator posture. Nonzero subspace for a redundant manipulator allows to handle redundant degrees of freedom. The existence of nonzero subspace allows for redundant manipulator to generate internal motions of the manipulator structure that do not change the end-effector position and orientation. As a consequence allows manipulator reconfiguration into more dexterous postures for execution of a given task.

null steering   a technique used to form an antenna's radiation pattern in such a way that there is no radiation in the direction of an interfering signal. This technique has been used with great success, for example, where CDMA systems have been overlaid with fixed microwave communication systems.

null-to-null bandwidth the width of the main lobe of a signal or system transfer function in the frequency domain.

nullator an idealized two-terminal network element that conducts no current and yet maintains zero volts across itself. This element is often used to represent a virtual connection.

number system the representation of numbers as a sequence of digits with an interpretation rule which assigns a value to each sequence. The conventional number systems are fixed-radix and positional systems, where the digit in position i has a weight of r^i, where r is the radix.

numerical aperture (NA) a parameter describing the light-gathering capacity of an optical fiber. It is defined as the sine of half the maximum angle of light acceptance into the fiber

$$NA = \sqrt{n_{co} - n_{cl}}$$

where n_{co} is the refractive index of the core, and n_{cl} is the refractive index of the cladding.

Also used as a measure of the maximum angle of the cone of light entering or emerging from an optical fiber.

numerical methods methods useful for obtaining quantitative solutions of electromagnetic and microwave problems that have been expressed mathematically, including the study of the errors and bounds on errors in obtaining such solutions.

NUREG a contraction for Nuclear Regulations, which are published by the NRC.

NVRAM *See* nonvolatile random-access memory.

nybble *See* nibble.

Nyquist A/D converter A/D converter that samples analog signals that have a maximum frequency that is less than the Nyquist frequency. The Nyquist frequency is defined as one-half of the sampling frequency. If a signal has frequencies above the Nyquist frequency, a distortion called aliasing occurs. To prevent aliasing, an antialiasing filter with a flat passband and very sharp rolloff is required.

Nyquist band for pulse (or quadrature) amplitude modulation, the frequency band which is the support of the Nyquist pulse shape. For a symbol rate of $1/T$, the Nyquist band is the interval $-1/(2T)$ to $1/(2T)$.

Nyquist criterion in analog-to-digital conversion, the stipulation that the bandwith of the sampling frequency must be greater than twice the bandwidth of the frequency of the signal being sampled.

Nyquist frequency *See* Nyquist rate.

Nyquist I criterion the Nyquist I criterion for zero inter-symbol interference states that if $h(t)$ is the signaling pulse and the symbol period is T, then we must have $h(kT) = 1$ for $k = 0$ and 0 for $k \neq 0$, where k is an integer.

Nyquist noise *See* thermal noise.

Nyquist plot a parametric frequency response plot with the real part of the transfer function on the abscissa and the imaginary part of the transfer function on the ordinate.

Nyquist pulse for pulse (or quadrature) amplitude modulation, the minimum bandwidth pulse which satisfies the Nyquist I criterion. Given a symbol rate $1/T$, the Nyquist pulse is $T^{-1} \sin(\pi t/T)/(\pi t/T)$, which has two-sided bandwidth $1/T$.

Nyquist rate the lowest rate at which recovery of an original signal from its sampled signal is possible. In digital transmission systems, the analog signal is transformed to a digital signal using an A/D converter. If the highest frequency of the analog signal is f_H, the Nyquist frequency is the frequency of the samples f_S for proper recovery of the signal at the receiving end.

$$f_{\textbf{Nyquist}} = f_S \text{ should be } > 2f_H$$

Nyquist sampling theorem fundamental signal processing theorem that states that in order to unambiguously preserve the information in a continuous-time signal when sampled, the sampling frequency, $f_s = 1/T$, must be at least twice the highest frequency present in the signal. *See* aliasing.

O

object   in object-orientation, an instance of a class definition.

object based coding   video compression based on the extraction, recognition and parameterization of objects in the scene. Unknown-object coding finds geometrical structures representing 2-D or 3-D surfaces and codes these areas and their movement efficiently. Known-object coding relies on the detection of objects for which the system has a high-level structural model. *See* model based coding.

object code   a file comprising an intermediate description of a program segment.

object detection   the detection of objects within an image. This type of process is cognate to pattern recognition. Typically, it is used to locate products ready for inspection or to locate faults during inspection. An object usually denotes a larger characteristic in an image, such as an alignment of points, a square, a disk, or a convex portion of a region, etc.; its detection involves non-local image transforms, such as the *Hough transform* or the medial axis transform. *See* edge detection, Hough transform, key point detection, medial axis transform. Feature detection.

object measurement   the measurement of objects, with the aim of recognition or inspection to determine whether products are within acceptable tolerances.

object orientation   measurement of the orientation of objects, either as part of the recognition process or as part of an inspection or image measurement process.

object recognition   the process of locating objects and determining what types of objects they are, either directly or indirectly through the location of sub-features followed by suitable inference procedures. Typically, inference is carried out by application of Hough transforms or association graphs.

object type   the type of an object determines the set of allowable operations that can be performed on the object. This information can be encoded in a "tag" associated with the object, can be found along an access path reaching to the object, or can be determined by the compiler that inserts "correct" instructions to manipulate the object in a manner consistent with its type.

object-oriented analysis   a method of analysis that examines requirements from the perspective of the classes and objects found in the problem domain.

object-oriented design   a design methodology viewing a system as a collection of objects with messages passed from object to object.

object-oriented methodology   an application development methodology that uses a top-down approach based on a decomposition of a system in a collection of objects communicating via messages.

object-oriented programming   object-oriented programming or task-level programming allows the user to command desired subgoals of the task directly, rather that to specify the details of every action the robot is to take. In another words, task-level programming describes the assembly task as a sequence of positional goals of the objects rather than the motion of the robot needed to achieve these goals, and hence no explicit robot motion is specified. Task-oriented programming is the most advanced programming system for robots. Not many industrial robots are equipped with such a system.

objective function when optimizing a structure toward certain result, i.e., during optimization routines, the objective function is a measure of the performance which should be maximized or minimized (to be extremized). *See also* optimization of microwave networks.

objective lens a well-corrected lens of high numerical aperture, similar to a microscope objective, used to focus the beam of light onto the surface of the storage medium. The objective also collects and recollimates the light reflected from the medium.

observability (1) the property of a system that ensures the ability to determine the initial state vector by observing system outputs for a finite time interval. For linear systems, an algebraic criterion that involves system and output matrices can be used to test this property. *See also* observability conditions.

(2) the ability to determine the signal value at any node in a circuit by applying appropriate input states to the circuit and observing its outputs.

observability conditions a linear finite-dimensional continuous and stationary dynamical system is observable if and only if

$$\text{rank } [C^T | A^T C^T | (A^T)^2 C^T$$
$$| \dots | (A^T)^{n-1} C^T] = n$$

It should be pointed out that observability of stationary dynamical systems does not depend on the length of time interval.

observer an algorithm to estimate the state variables from the input and output variables of the system.

OCC *See* one-cycle control.

occlusion the hiding of one object by another. In images of scenes, occluded objects are invisible and irrelevant unless their presence can be inferred, e.g., by motion analysis: partially visible objects are often said to be occluded, though strictly speaking they are partially occluded. Occlusion and disocclusion of objects in the 3-D space add difficulties to motion analysis and structure estimation related problems. For instance, when occlusion takes place, some motion information will be lost, while as disocclusion takes place, some areas that are originally occluded will become newly visible. Often no previous information will be available regarding these areas, thus making analysis more difficult.

occupancy of energy levels number of atoms (or molecules, holes, electrons, etc.) per unit volume occupying an energy state.

OCR *See* optical character recognition.

octal number system a number system consists of eight digits from 0 to 7; it is also referred to as base 8 system.

octave a frequency ratio of two.

octave filter bank a filter bank with octave spaced subbands. That is any given subband has twice the bandwidth of the adjacent lower frequency subband, except for the subband adjacent to the lowpass subband. The discrete wavelet transform is an octave-band filter bank.

octtree the 3-dimensional generalization of the *quadtree*, it gives a representation of a volumetric image as a tree, where each parent node has 8 children nodes and each leaf node has a label corresponding to the color of the corresponding region. It is built recursively as follows. The root node corresponds to the whole image; if all voxels are of the same color, then this node receives that color's label, and is a leaf; otherwise, we subdivide the image into 8 sub-images whose dimensions are half of those of the whole image, and each smaller image corresponds to a node which is child of the root. Each of the 8 children nodes is the root of a sub-tree which is the octtree of the corresponding sub-image; thus: if the sub-image has all its voxels of the same color, this node is a leaf to which the color's label is attached, otherwise we subdivide this sub-image into 8 smaller sub-images, to each of which corresponds a child node of the actual node. This recursive subdivision ends when one reaches a sub-image of a single color, giving a leaf node with the corresponding label. *See* quadtree.

odd function   a real-valued function $x(t)$ in which $x(-t) = -x(-t)$ for all values of t. *Compare with* even function.

odd order response   a circuit gain or insertion loss versus frequency response in which there are an odd number of peaks in the ripple pattern, due to an odd number of paired elements in the circuit. Odd order circuits exhibit a peak for each element pair and a peak equal to L_{Amin} (minimum attenuation loss across the band) at DC (low pass) or at w_o (band pass).

odd-mode characteristic impedance   characteristic impedance of a circuit due to an odd-mode current or voltage excitation. Often applied in the context of a transmission line coupler where the odd-mode excitation consists of applying equal amplitude but opposite polarity voltages or currents on two conductors. The resulting impedance under this excitation is defined as the odd-mode characteristic impedance.

ODP   *See* open drip-proof.

ODP code   *See* optimum distance profile code.

OEIC   *See* optoelectronic integrated circuit.

Oersted, Hans Christian   (1777–1851) Born: Rudkobing, Langeland, Denmark

Oersted is best known as the discoverer of electromagnetism. Oersted was a strong teacher and did much to bring Danish science up to world-class standards. Oersted predicted the magnetic effect of electric current in 1813, but was unable to prove it until 1820. The publication of his results spurred the work of Faraday and Ampere. Oerstad went on to make other contributions in other sciences. He did not, however, return to his study of electricity.

OFD code   *See* optimum free distance code.

OFDM   *See* orthogonal frequency division multiplex.

off-axis illumination   illumination that has no on-axis component, i.e., that has no light which is normally incident on the mask. Examples of off-axis illumination include annular and quadrupole illumination.

off-line error detection   techniques of detecting faults by device testing (e.g., with the use of BIST) such that the device is allowed to perform useful work while under test.

off-line testing   testing process carried out while the tested circuit is not in use.

offset   a sustained derivation or error due to an inherent characteristic of positioning controller action. The difference exists at any time between the set point and the value of the controlled variable.

offset short   a short circuit impedance standard with a phase offset from the reference plane used in the process of calibrating vector network analyzers.

offset voltage   for an ideal differential pair or op-amp, zero output corresponds to zero differential input. In reality, some nonzero input called the "offset voltage" is required for the output to be zero. This means that in the presence of an offset voltage error, a zero input will produce a nonzero output. Offset voltage is caused primarily by mismatch in the two transistors of a differential pair. *See also* common centroid.

Ohm's Law   a fundamental law which states that the voltage across a resistance is directly proportional to the current flowing through it. The constant of proportionality is known as the resistance.

This concept can be generalized to include the relationship between the voltage and current in all situations, including alternating voltages and currents. In this case, all the quantities are measured as complex numbers, known as phasors, that are functions of frequency. This broadens the basic definition of resistance, which is a real number measured in ohms, to that of impedance, which is a complex number with magnitude measured in ohms and phase angle in degrees. The real part of the complex number representing impedance is the resistance while the imaginary part is the reactance. Ohm's Law is a central concept to most electrical engineering theories.

Ohm, Georg Simon   (1789–1854) Born: Erlangen, Germany

Best known for his discovery of what we now call Ohm's Law. Ohm held a variety of teaching posts at secondary schools as well as universities. In 1827 he published his greatest work, *Die Galvanische Kette*. Along with Andre Ampere, Ohm was the first to publish rigorously mathematical and theoretical work on electricity. Ohm's famous law states that current in a resistor is proportional to the applied voltage and inversely proportional to the resistance. Ohm's work was initially scorned because it lacked the experimental evidence. Worldwide acclaim changed Ohm's fortunes several years later. He is honored by having his name used as the unit of resistance, the ohm, and the unit of conductivity, the mho.

ohmic contact   a heavily doped and/or low barrier height metal to semiconductor interface or contact that has a very low resistance relative to the remainder of the device, such that the device performance is not significantly degraded. At lower doping levels, the ohmic contact is described by Ohm's Law, while at higher doping levels, tunneling dominates.

ohmic loss   a term used to describe the power dissipated due to the finite conductivity of the metallic structure of an antenna, waveguide, transmission line, etc.

ohmic medium   a medium in which conductivity is independent of the applied field.

ohmic region   the voltage-controlled resistance region of operation of a transistor. Also referred to as the triode region, nonsaturated region, and pinch-off region. This region is in effect up to the point that the channel of the transistor is completely depleted of charge carriers.

oil circuit breaker   a power circuit breaker that uses oil as an insulating and arc-clearing medium.

oil-filled transformer   a transformer in which the magnetic core and the windings are submerged in an insulating oil. In addition to serving as an insulator, the oil provides a heat exchange medium to cool the transformer.

oil-paper insulation   an insulation scheme used in transformers and cables in which conductors are insulated with heavy paper impregnated with a dielectric oil.

OLAP   *See* optical linear algebraic processor.

OLE   *See* optical logical etalon.

OMCVD   *See* metal-organic chemical vapor deposition.

omnidirectional antenna   an antenna that radiates power equally in all directions in at least one plane through its radiation center, but whose radiation pattern may vary for other such planes.

on-line error detection   a real-time detection capability that is performed concurrently with useful work (e.g., parity checkers, comparators in duplicated systems).

on-line memory   memory that is attached to a computer system.

on-line optimization interval   time interval over which optimization of the decisions required to be made by the considered decision mechanism at a particular time instant during the control system operation is performed; setting of the on-line optimization interval is one of the essential decisions during design of a controller with the model-based predictive control mechanism.

on-line testing   concurrent testing to detect errors while circuit is in operation.

on–off keying (OOK)   a binary form of amplitude modulation in which one of the states of the modulated wave is the absence of energy in the keying interval (the "off" state). The "on" state is represented by the presence of energy in the modulated wave.

on-wafer measurements   electrical measurements made using a wafer probe station by directly contacting the device under test using special test probes.

one decibel desensitization point   a reduction in a device output signal power by one decibel due to one or more additional signals that compress the device output.

one's complement   (1) a representation of integer numbers in which a data word is organized such that negative numbers all contain a binary "one" in the leftmost bit while positive numbers contain a "zero" in the leftmost bit, and in which

the negative numbers are the bit-by-bit inverse of their positive equivalent.

(2) the operation of inverting a data word so that all ones become zeros and vice versa.

one-cycle control (OCC)   the switched signal is sensed to control the pulse timing such that the error between the control reference and the average value of the switched signal is zero in each cycle. In a constant frequency implementation, the switch is turned on by a constant frequency clock. The output of the switch is integrated and compared to a reference; when the reference is reached, the switch is turned off and the integrator is reset. One-cycle control achieves zero error within each cycle under steady-state or transient conditions. In addition, this method effectively rejects input perturbations, corrects switching errors, and provides high linearity.

one-dimensional coding scheme   a scheme in which a run of consecutive pixels of the same gray scale is combined together and represented by a single code word for transmission.

Scan lines are coded as "white" or "black," which alternate along the line. The scans are assumed to begin with white and are padded with white if this is not the case. Generally, run-length coding is a one-dimensional coding scheme.

one-dimensional correlator   a correlator where both input signals are one dimensional, such as temporal signals.

one-dimensional space integrating correlator   a correlator where a single lagged product of two input signals is performed at a given instant and integrated spatially, such as with a lens onto a single photodetector. A series of such operations are performed for each lag value to give a temporal sequence that is the correlation function.

one-gun color display   a color CRT in which a single electron gun produces one electron beam that is controlled to produce the proper excitation of each of the three color phosphors.

485

one-level memory   an arrangement of different (in terms of speed, capacity, and medium) types of memory such that the programmer has a view of a single flat memory space. *See also* hierarchical memory, virtual memory.

one-level storage   *See* one-level memory.

one-line diagram   an abbreviated schematic representation of a power system in which three-phase transmission lines are shown as single lines between principal circuit components and from which circuit parameters are often omitted.

one-out-of-N coding   a method of training neural networks in which the input vector and/or the output vector have only one nonzero element (usually equal to unity) for each training example.

one-port device   an electrical network in which only one external terminal is available for analysis. Antennas can be modeled as one-port devices.

one-port oscillator   a device in a class of circuits in which the amplifying device and its non-linearity are lumped into a single, reactance-free, controlled source. The oscillations in this approach are most successfully found using the harmonic balance method (especially when the nonlinearity can be approximated by a polynomial) or by the phase plane methods (for second-order systems).

one-shot multivibrator   a circuit that is obtained from a closed-loop regenerative bistable system including two similar amplifiers connected by coupling circuits. In one-shot multivibrator, one of the circuits is purely resistive, another, which is frequently called "toggling" circuit, includes a reactance (capacitor or coil). The circuit is normally in its stable state; by an external pulse it is transferred in its quasi-stable state, and the toggling circuit helps to preserve the quasi-stable condition for the "timing-out" time. Then the circuit returns to its normal state.

one-step-ahead control   the control method to drive the value of the d-step-ahead output, where d is the inherent time delay of the system, to its desired level in one step.

ontogenic network   a network that adapts its topology during training through the addition or deletion, as appropriate, of connections and neurons, until the problem of interest is satisfactorily accommodated. Learning can be either supervised or unsupervised.

OOK   *See* on–off keying.

op-amp   *See* operational amplifier.

OPC   *See* optical proximity correction.

opcode   a part of an assembly language instruction that represents an operation to be performed by the processor. Opcode was formed from the contraction of "operational" and "code."

open circuit impedance   the impedance into an N-port device when the remaining ports are terminated in open circuits.

open drip-proof (ODP)   pertaining to a ventilated machine whose openings are constructed to prevent drops of liquid or solid particles falling on the machines at an angle less than $15°$ from the vertical from entering the machine either directly or by rolling along a horizontal or inwardly inclined surface of the machine.

open kinematic chain   a chain that consists of one sequence of links connecting two ends of the chain.

open loop gain   *See* open-loop gain.

open system architecture   a layered architectural design that allows subsystems and/or components to be readily replaced or modified; it is achieved by adherence to standardized interfaces between layers.

open systems interconnection (OST) model   a framework for organizing networking technology developed by the International Standards Organization (ISO). Generally called the OSI model.

open waveguide   a type of waveguide whose cross section is not bounded by perfect electric conductor, i.e., a waveguide for which the boundary value problem is on an infinite domain. A few examples are optical fiber, microstrips, coplanar waveguides, dielectric waveguides.

open-circuit test   a transformer test conducted by applying nominal voltage on the low voltage side while keeping the high voltage side open. By measuring the power in, current, and voltage, the magnetizing reactance of the transformer equivalent circuit can be determined.

open-circuit quarter-wave transmission line   transmission line of ninety degrees electrical length where one end of the line is terminated in an open-circuit impedance. The properties of the transmission line result in the non-open circuited end to exhibit a short-circuit impedance.

open-delta transformer   a connection similar to a delta–delta connection, except that one single-phase transformer is removed. It is used to deliver three-phase power using only two single-phase transformers. The normal capacity of the open-delta transformer is reduced to 57.7% of its delta rating.

opening   for structuring element B, the composition of the erosion by B followed by the dilation by B; it transforms X into $X \circ B = (X \ominus B) \oplus B$. The opening by B is what one calls an algebraic opening; this means that: (*a*) it is a morphological filter; (*b*) it is anti-extensive, in other words it can only decrease an object. *See* dilation, erosion, morphological filter, structuring element.

open-loop control system   a control system in which the system outputs are controlled by the system inputs only, and no account is taken of the actual system output.

open-loop gain   the gain of an operational amplifier with no feedback applied (with the negative feedback loop "open").

open-loop-feedback control   predictive control policy with repetitively used decision mechanism, that at a given time instant, considers current state of the controlled process — or the current estimate of this state — and computes the values of the control inputs for the next intervention instant or interval by performing an open-loop optimization of the process operation over specified prediction (optimization) interval. The open-loop optimization can be defined as an open-loop stochastic optimization problem or as a deterministic optimization problem using a single forecast of the free inputs over specified prediction interval; other forms of open-loop optimization problems can also be defined. Model-based-predictive control, widely used as an industrial standard, is usually realized in the form of the open-loop-feedback control.

operand   specification of a storage location that provides data to or receives data from the operation.

operand address   the location of an element of data that will be processed by the computer.

operand address register   the internal CPU register that points to the memory location that contains the data element that will be processed by the computer.

operating point   *See* set point.

operating system   a set of programs that manages the operations of a computer. It oversees the interaction between the hardware and the software and provides a set of services to system users.

operating temperature   the ambient case temperature to which a device or circuit is exposed during operation under all supply bias and RF signal conditions, at which it must meet all specified requirements unless otherwise called out. The operating temperature range is the minimum to maximum operating temperatures of the device.

operation   specification of one or a set of computations on the specified source operands placing the results in the specified destination operands.

operational amplifier   a high-gain DC-coupled amplifier with a differential input and single-ended output. In nearly all amplifier applications, the op-amp is used with negative feedback ("closed-loop"), so that the closed-loop gain of the amplifier depends primarily on the feedback network components, and not on the op-amp itself. It is widely used as a basic building block in electronic designs. Abbreviated as "op-amp."

operational control   control or decision making activity that requires, or in some instances may require, a human operator (dispatcher) to approve — or modify — automatically computed decisions before they are actually implemented; such a situation is typical in production management, in control of supply and environmental systems, at upper layers of process control systems.

operational impedance   a representation in which the impedance of a system is expressed as a function of the Heaviside operator $p = d/dt$ or the Laplace operator $s = j\omega$. In the modeling of synchronous machines, the Park's transformed stator flux linkages per second are often expressed in terms of impedances $X_q(p)$, and $X_d(p)$, termed the quadrature- and direct-axis operational impedances, respectively. Using these, the dynamics of the rotor windings are represented within the operational impedances, and therefore the rotor of a synchronous machine can be considered as either a distributed or lumped parameter system.

operational space   *See* external space.

operational space control   *See* Cartesian-based control.

operator   *See* Canny operator, Laplacian operator, Marr-Hildreth operator, morphological operator, Sobel operator.

opposition effect   *See* weak localization of light.

optical adder   an optical device capable of performing the function of arithmetic addition using binary signals. It can be constructed using a series of cascaded full adders where carry has to ripple through each full adder from least significant bit to most significant bit. It can also be constructed using cascaded layers. Each layer consists of a series of half adders. Carry also has to ripple through the cascaded layers. It is also known as digital adder.

optical addition   adding operation using light. Two incoherent light beams have intensities A and B, respectively. When two beams are combined, i.e., illuminating the same area, the resultant intensity is $A + B$.

optical amplifier   amplifier of electromagnetic waves at optical frequencies, usually by the process of simulated emission or nonlinear optics. *See* laser amplifier.

optical beam   beam of electromagnetic power at optical frequencies.

optical bistability   the property of certain nonlinear optical system to possess two possible output states for a given input state. In one typical example, the bistable optical device has the form of a third-order nonlinear optical material placed inside of an optical cavity, and the device can display two possible different transmitted intensities for certain values of the input intensity.

optical bus   an optical channel used for transmitting a signal from a source to one or more detectors. A bus allows only the same interchange of information to take place at different detectors. A source is connected to many detectors.

optical cavity   resonant structure for maintaining oscillation modes at optical frequencies.

optical character recognition (OCR)   a process in which optically scanned characters are recognized automatically by machine. It is widely utilized in document storage, processing and management.

optical circulator   a four-ports optical device that can be used to monitor or sample incident light (input port) as well as reflected light (output port) with the two other unidirectional coupling ports.

optical communications   communication of information at optical carrier frequencies. There are two general categories:

(1) free-space, such as with lasers and optical modulators, and

(2) guided-wave, such as over optical fiber and using guided lightwave devices.

optical computer   (1) a general purpose digital computer that uses photons as an electronic computer uses electrons. The technology is still immature. The nonlinear operations that must be performed by a processor in computer are difficult to implement optically.

(2) an analog optical processor capable of performing computations such as correlation, image subtraction, edge enhancement, and matrix-vector multiplication that are usually performed by an electronic digital computer. Such a computer is usually very specific and inflexible, but well adapted to its tasks.

optical computing   the use of optics to aid in any type of mathematical computation. Categories of optical computing include:

(1) analog processing such as using optical Fourier transforms for spectral analysis and correlation,

(2) optical switching and interconnection using nonlinear, e.g., bistable, optical devices, and

(3) use of optical devices in digital computers.

optical demultiplexer   a device that directs an input optical signal to an output port depending on its wavelength.

optical disk   a disk on which data are stored optically. Data are written by altering the reflectivity of the surface, and read by measuring the surface reflection of a light source. Storage is organized in the same way as on a magnetic disk, but higher storage density can be achieved.

optical disk track   the region on a compact disk in which pits or other features are located to store digital data. CD-ROM pit sizes are approximately 0.5 microns in width and from 0.8 to 3.6 microns in length, depending on the number of ones and zeros represented by the length. The track-to-track spacing is 1.6 microns.

optical energy   energy of an electromagnetic wave oscillating at optical frequencies. *See also* energy, electromagnetic.

optical expert system   an expert system that utilizes optical devices for performing logic operations. An expert system mostly requires logic gates to perform inferences. In an optical expert system, not only discrete optical logic gates are used, but the parallelism and capability of performing matrix-vector multiplication of optical processor are exploited as well. Sequential reasoning in an electronic expert system is replaced by parallel reasoning in an optical expert system, so its speed can be increased substantially. As with the optical computer, it is still immature.

optical fiber   a single optical transmission fiber usually comprised of a cylindrical core

(5–100 mm diameter) in which the light is guided of higher index of refraction surrounded concentrically by a cladding (125–250 mm diameter) with a lower index of refraction. More properly defined as an optical waveguide. Some optical fibers may have multiple concentric cores and/or claddings. Optical fibers made be of all glass, all plastic, or a combination of glass core and plastic cladding construction. Optical glass fibers may be silica- or fluoride-based glass.

optical fiber signal distortion   a change in the temporal shape of an optical signal transmitted through an optical fiber caused by a combination of wavelength effects (dispersion) and multimode and polarization effects. The wavelength effects include material dispersion, profile dispersion, and waveguide dispersion. The multi-mode effects cause distortion by the differential time delays between the various modes propagating in a multimode fiber. The polarization effect causes distortion by the differential time delay between the two polarizations of a single mode.

optical flow   (1) a method, termed after the biological, quasi-continuous, flow of information across the retina and used to study motion, in which every point in the field of view is assigned a 2-D velocity vector representing the point motion across the visual field.

(2) the 2-D field of apparent velocities of pixels on an image plane; i.e., the raw motion information arising from the displacement of points in the visual (optical) field. Let each image point p have an intensity I and a velocity $v = (v_x, v_y)$, which are both functions of p and time t, where the velocity represents the image plane projection of the point's 3-dimensional velocity. Under the assumption that the point's intensity does not change along its trajectory, it follows that $\frac{\partial I}{\partial x} v_x + \frac{\partial I}{\partial y} v_y + \frac{\partial I}{\partial z} v_z + \frac{\partial I}{\partial t} = \nabla I \cdot v + \frac{\partial I}{\partial t} = 0$. *See also* aperture problem.

optical flux   *See* optical flow.

optical Fourier transform   the implementation of the Fourier transform using the transform properties of a lens, spatial light modulators, such as acousto-optic modulators, to generate the optical input information, and photodetector arrays to detect the optical field representing the Fourier transform. *See also* Fourier transform, two-dimensional Fourier transform.

optical full adder   an optical device forming part of an adder and able to receive three inputs, augend, addend, and carry from the previous stage, and deliver two outputs, sum and carry. Several logic gates are required to provide the sum and the carry.

optical gain   increase in the amplitude of an optical signal with propagation distance in an optical amplifier.

optical guided-wave device   an optical device that transmits or modifies light while it is confined in a thin-film optical waveguide.

optical half adder   an optical device forming part of an adder and able to receive two inputs, augend and addend, and deliver two outputs, sum and carry. The sum is the XOR function of two inputs, and the carry is the AND function of two inputs. An optical half adder consists of an optical XOR gate and an optical AND gate. It is also known as one-digit adder.

optical inference engine   *See* optical expert system.

optical integrated circuit   guided wave optics on a single substrate. If also incorporating active and electrical devices, also known as optoelectronic integrated circuit (OEIC).

optical interconnect   an optical communication system in an electronic computer that consists of three primary parts:
    (1) sources,

(2) optical paths with switches or spatial light modulators, and

(3) detectors.

An optical signal from a source of an array of sources is transmitted to a detector of an array of detectors in optical interconnects. The transmitted optical signal from the source is converted from an electronic signal. The detector in turn reverts the optical signal into an electronic signal. The merits of optical interconnects include large bandwidth, high speed of light propagation (the velocity of electric signals propagating in a wire depends on the capacitance per unit length), no interferences, high interconnection density and parallelism, and dynamic reconfiguration.

optical invariant a parameter that remains constant thoughout an optical system. A consequence of the optical invariant is that the product of numerical aperture and magnification in the system is constant.

optical isolator a unidirectional optical device that only permits the transmission of light in the forward direction. Any reflected light from the output port is blocked by the device from returning to the input port with very high extinction ratio.

optical Kerr effect *See* Kerr effect.

optical linear algebraic processor (OLAP) an optical processor that performs specific matrix algebraic operations as fundamental building blocks for optical computation and signal processing.

optical lithography lithography method that uses light to print a pattern in a photosensitive material. Also called photolithography.

optical logic logic operations, usually binary, that are performed optically.

optical logic gate an optical device for performing Boolean logic operations. The basic idea is that since a computer is built by Boolean logic gates, if

we can make optical logic gates, then we can eventually build a complete optical computer. Since light does not affect light, it cannot directly perform nonlinear operations represented by sixteen Boolean logic gates. A thresholding device is required if intensities of two input beams are simply added in the logic gate. For example, to perform AND operation, 00-0, 01-0, 10-0, 11-1, the threshold is set at 2. Only when both beams have high intensities, the output is 1. Otherwise, the output is 0. To perform OR operation, 00-0, 01-1, 10-1, 11-1, the threshold is set at 1. The gate can be constructed by directing two input light beams onto the same detector. After passing through an electronic thresholding device, the resulting electric signal from the detector shows the logic output. To implement optical logic gate without using electronic circuit, optical thresholding devices such as MSLM or nonlinear optic materials can be used.

optical logical etalon (OLE) pulsed nonlinear Fabry–Perot etalon that requires two wavelengths ($\lambda_1 =$ signal, $\lambda_2 =$ clock).

optical maser early equivalent name for laser.

optical matrix–matrix multiplication an operation performing matrix–matrix multiplication using optical devices. For two-by-two matrices A and B, the matrix–matrix multiplication produces two-by-two matrix C, whose elements are

$$lc_{11} = a_{11}b_{11} + a_{12}b_{21}$$
$$c_{12} = a_{11}b_{12} + a_{12}b_{22}$$
$$c_{21} = a_{21}b_{11} + a_{22}b_{21}$$
$$c_{22} = a_{21}b_{12} + a_{22}b_{22}$$

An optical implementation is as follows. Matrix elements b_{12}, b_{22}, b_{11}, and b_{21} are represented by four vertical lines on a spatial light modulator. Matrix A is represented by a two-by-two source array. Two images of matrix A are generated by an optical means, for example, two lenslets. An image of A covers two lines of b_{12} and b_{22}, another image of A covers lines b_{11} and b_{21}. Using two cylindrical lenses, light passing through the spatial light modulator is focused into four points at the four corners of a square representing c_{11}, c_{12}, c_{21}, and c_{22}. Various optical arrangements for implementing matrix–matrix multiplication have been proposed. Optical matrix–matrix multiplication is needed for solving algebraic problems, which could be faster than an electronic computer because of parallel processing.

optical matrix-vector multiplication   an operation performing matrix–vector multiplication using optical devices. For matrix W and vector X, the matrix–vector multiplication produces vector Y, whose elements are

$$y_1 = w_{11}x_1 + w_{12}x_2 + w_{13}x_3$$
$$y_2 = w_{21}x_1 + w_{22}x_2 + w_{23}x_3$$
$$y_3 = w_{31}x_1 + w_{32}x_2 + w_{33}x_3$$

A simple coherent optical processor to implement this operation is as follows. Vector elements x_1, x_2, and x_3 are represented by the transmittance of three vertical lines on a first spatial light modulator. A collimated coherent light beam passes through the first spatial light modulator. The modulated light then passes through a second spatial light modulator displaying three-by-three squares with transmittances of w_{11} to w_{33}. Two spatial light modulators are aligned such that the vertical line x_1 covers three squares of w_{11}, w_{21}, w_{31}, line x_2 covers w_{12}, w_{22}, w_{32}, and line x_3 covers w_{13}, w_{23}, w_{33}. The light passing the second spatial light modulator is focused by a cylindrical lens to integrate light in horizontal direction. In other words, $w_{11}x_1$, $w_{12}x_2$, and $w_{12}x_3$ are summed up at a point on the focal line of cylindrical lens. A detector placed at this point will produce y_1. Two other detectors will provide y_2 and y_3 in a similar way. Various optical arrangements are possible to implement matrix–vector multiplication, including correlators. Optical matrix–vector multiplication is important for crossbar switch and neural networks. It can also be used for finding eigenvalues and eigenvectors, solving linear equations, and computing the discrete Fourier transform.

optical modulator   device or system that modulates the amplitude or phase of an optical signal.

optical multiplication   multiplying operation using light. A light beam has intensity A. When the beam passes through a transparency with transmittance B, the intensity of transmitting light beam is AB.

optical neural network   optical processor implementing neural network models and algorithms. A neural network is an information processing system that mimics the structure of the human brain. Two features of neural networks are recognition capability and learning capability. In recognizing process, the neural net is formulated as

$$z_j = f\left(\sum_{ji} x_i + \theta_j\right)$$

where z_j is the output of the jth neuron, w_{ji} is the interconnection weight between the ith input neuron and the jth neuron, x_i is the input coming from the ith input neuron, θ_j is the bias in the jth neuron, and f is a nonlinear transfer function. Notice that z_j and x_i are binary. Nonlinear transfer function f could simply be a threshold function. The formation of interconnection weight matrix is called learning process. It appears that in recognizing process, the neural network has to perform a matrix–vector multiplication and a nonlinear operation. The matrix–vector multiplication can be easily performed using an optical system. However, the nonlinear operation can not be performed by optical means. It may be performed using electronic circuits. For two-dimensional neural processing, tensor–matrix multiplication is needed, which can also be realized by optical means.

optical parametric oscillator   a nonlinear optical device that can produce a frequency-tunable output when pumped by a fixed-frequency laser beam. The device consists of a second-order nonlinear optical crystal placed inside of an optical resonator as well as additional components for precise control of the output characteristics. When pumped by a laser beam at the pump frequency ω_p, it produces two output frequencies, one at the signal frequency ω_s and one at the so-called idler frequency ω_i, where $\omega_p = \omega_s + \omega_i$.

optical path   optical elements in the path of the laser beam in an optical drive. The path begins at the laser itself and contains a collimating lens, beam shaping optics, beam splitters, polarization-sensitive elements, photodetectors, and an objective lens.

optical path length   the distance an incident photon travels within a material before it emerges and impinges on a detector. The path may be directly through the material, in which case the pathlength will equal the thickness of the light absorbing material, or may involve scattering due to heterogeneous structures, causing the pathlength to exceed the material's thickness (effective pathlength).

optical phase conjugation   an optical process in which a time reversed replica of the incident wave is generated. The time-reversed replica is identical to the incident wave, except the direction of propagation. Optical phase conjugation can be achieved by using a deformable mirror whose mirror surface matches exactly the wavefront of the incident wave. For example, consider the reflection of a spherical wave from a spherical mirror with an identical radius of curvature. The reflected wave is a time-reversed replica of the incident wave. Optical phase conjugation can also be obtained by using optical four-wave mixing, stimulated Brillouin scattering (SBS) in nonlinear media. In optical phase conjugation via four-wave mixing, the nonlinear medium is pumped by a pair of counter-propagating beams. When a signal beam is incident into the medium, a phase conjugate beam is generated which is a time-reversed replica of the incident beam.

optical potential transformer (OPT)   a potential transformer that uses a voltage-sensitive

optical device, typically in conjunction with optical fibers, to avoid the need for the heavy insulation required of electromagnetic or capacitive potential transformers.

optical processing   *See* optical signal processing.

optical proximity correction (OPC)   a method of selectively changing the shapes of patterns on the mask in order to more exactly obtain the desired printed patterns on the wafer.

optical proximity effect   proximity effect that occurs during optical lithography.

optical pumping   excitation of an atom or molecule resulting from absorption of optical frequency electromagnetic radiation; the electromagnetically assisted accumulation of population into or out of one or more states of a quantum mechanical system. In practice, this generally involves selective absorption of the electromagnetic field to populate an excited state, followed by a less selective decay into more than one ground state. For example, a system having ground state spin sublevels can be optically pumped by circularly polarized light into a single ground state spin sublevel. In a multilevel system, more selective transfer of population from one state to another can be achieved by adiabatic passage.

optical rectification   the second-order nonlinear optical process in which a material develops a static electric field in response to and proportional to the square of the strength of an applied optical field.

optical repeater   optoelectric device that receives a signal and amplifies it and retransmits it. In digital systems, the signal is regenerated.

optical representation of binary numbers   the representation of binary numbers 0 and 1 using light. Since an optical detector is sensitive to light intensity, it is a very logical choice to represent binary numbers 0 and 1 with dark and bright states or low and high intensity, respectively. However, some difficulty would occur when 1 has to be the output from 0, because no energy can be generated for 1 from 0 without pumping light. Binary numbers 0 and 1 can be represented by a coded pattern instead of the intensity of a single spot. They can also be represented by two orthogonal polarizations, although the final states should be converted to intensity, which is the only parameter that can be detected by a detector.

optical resonator   electromagnetic cavity designed to have low loss at optical frequencies.

optical signal processing   the use of light with optical devices to process signals and images, exploiting the high bandwidths and inherent parallelism offered by optical systems.

optical soliton   a pulse that propagates without change in shape through a dispersive nonlinear optical medium as a consequence of an exact balance between dispersive and nonlinear effects. The propagation of such a pulse is described by the nonlinear Schrödinger equation.

optical switching   a process in which one optical beam is controlled by another optical wave or by an electro-optical signal.

optical tensor–matrix multiplication   an operation performing tensor–matrix multiplication using optical devices. For tensor W and matrix X, the tensor–matrix multiplication produces matrix Y, whose elements are

$$y_{11} = w_{1111}x_{11} + w_{1112}x_{12} + w_{1211}x_{21}$$
$$+ w_{1212}x_{22}$$
$$y_{12} = w_{1121}x_{11} + w_{1122}x_{12}$$
$$+ w_{1221}x_{21} + w_{1222}x_{22}$$
$$y_{21} = w_{2111}x_{11} + w_{2112}x_{12} + w_{2211}x_{21}$$
$$+ w_{2212}x_{22}$$
$$y_{22} = w_{2121}x_{11} + w_{2122}x_{12}$$
$$+ w_{2221}x_{12} + w_{2222}x_{22}$$

An incoherent optical processor can implement this operation as follows. The two-by-two 0 matrix

$$\begin{pmatrix} x_{11} & x_{12} \\ x_{21} & x_{22} \end{pmatrix}$$

represented by four LEDs is imaged by a lenslet onto a part of the tensor that is a two-by-two matrix

$$\begin{pmatrix} w_{1111} & w_{1112} \\ w_{1211} & w_{1212} \end{pmatrix}$$

represented by a spatial light modulator. Light passing through the spatial light modulator is integrated by another lenslet to provide y_{11}. Thus, two sets of two-by-two lenslet arrays with four LEDs and a spatial light modulator displaying all tensor elements as four matrices are sufficient to perform this tensor–matrix multiplication. Various optical arrangements can implement tensor–matrix multiplication. Optical tensor–matrix multiplication is important for two-dimensional neural networks.

optical time domain reflectometry (OTDR)   device used to locate faults or determine attenuation in a length of optical fiber. The technique relies on launching a pulse of light into the fiber and measuring the backward scattered power with time, which is then related to distance along the fiber.

optical time-domain reflectometer optical fiber   an instrument capable of launching optical pulses into an optical fiber transmission link or network, detecting the backscattered optical pulses, and displaying these backscattered pulses as a function of distance. Used to measure optical fiber length, attenuation, connector and splice losses, and to detect fault in optical fiber networks.

optical transfer   function the normalized Fourier transform of the incoherent point spread function, i.e.,

$$H(\omega_x, \omega_y)$$

$$= \int_{-\infty}^{\infty} \int_{-\infty}^{\infty} h(x, y) \exp[-i(\omega_x x + \omega_y y)] dx dy$$

$$\times \int_{-\infty}^{\infty} \int_{-\infty}^{\infty} h(x, y) dx dy$$

where $h(x, y)$ is a point spread function which represents the image intensity response to a point source of light. *See* modulation transfer function, impulse response function, point spread function.

optically inhomogeneous medium a medium whose refractive index randomly varies either in space, in time, or in both. It produces scattering of the light transmitted or reflected at it.

optimal decision the best decision, from the point of view of given objectives and available information, that could be taken by the considered decision (control) unit; the term optimal decision is also used in a broader sense — to denote the best decision that can be worked out by the considered decision mechanism, although this decision mechanism may itself be suboptimal. An example is model-based optimization as a decision mechanism at the upper layer of a two-layer controller for the steady-state process — the results of this optimization will be referred to as the optimal decisions.

optimal sensitivity minimal, in the sense of the H_∞ norm, value of the sensitivity function found using H infinity design (H_∞) techniques. For single-input–single-output systems, the H_∞ norm is simply the peak magnitude of the gain of the sensitivity function. In the multi-input–multi-output case, it is expressed by the maximal singular value of the sensitivity matrix function. To meet the design objectives, the optimal sensitivity is found for the sensitivity function multiplied by left and right weighting functions. It enables one to specify minimum bandwidth frequency, allowable tracking error at selected frequencies, the shape of the sensitivity function over selected frequency range, and maximum allowed peak magnitude.

optimization determining the values of the set of free parameters that minimizes or maximizes an objective function. The minimization or maximization may be subject to additional constraints. *See* gradient descent, graph search, genetic algorithm, minimax estimate, relaxation labeling, simulated annealing.

optimization of microwave networks the procedure used in the design of a microwave system in order to maximize or minimize some performance index. May entail the selection of a component, of a particular structure, or a technique. *See also* objective function.

optimizing control layer operation or structure of the control layer of a multilayer controller, usually situated above the direct control layer, where the objective of the decision mechanism is to minimize (or maximize) a given performance function associated with the control system operation; typical example of optimizing control is the set point control of an industrial process concerned with optimal operation of this process in the steady state conditions. *See also* steady-state control.

optimum combining a technique for weighted addition of the output of several communication channels where the weights are chosen to maximize the ratio between the desired signal and the sum of the interference and noise in combined output. Used in adaptive antenna systems. *See also* angle diversity, antenna diversity.

optimum coupling choice of output coupling mirror reflectivity that yields the largest output power from a laser oscillator.

optimum distance profile (ODP) code a convolutional code with superior distance profile given the code rate, memory, and alphabet size. Optimum distance profile codes are suitable if, e.g., sequential decoding is used.

optimum free distance (OFD) code a convolutional code with the largest possible free distance given the code rate, memory, and alphabet size.

optode a fiber optic sensor used to determine the concentration of a particular chemical species present in the sensor's environment by utilizing spectroscopic changes in a sensing element placed at the end of the optical fiber.

optoelectronic integrated circuit (OEIC) opto-electronic device combining optical and electrical devices on a common substrate. This includes combinations of semiconductor lasers, modulators, photodetectors, and electrical processing circuitry.

optoelectronics the interaction of light and electrons in which information in an electrical signal is transferred to an optical beam or vice versa, e.g., as occurs in optical fiber communications components.

optrode *See* optode.

opus sweep an amplifier large signal stability test in which a signal within the passband is held constant in both frequency and power such that compression is achieved. A second signal operated at a much smaller level is then injected and swept across the entire frequency band, observing the gain of this small signal through the compressed amplifier. Sharp gain peaks in this small signal response are indications of large signal stability problems. The test is dependent on bias conditions, large signal frequency, large signal compression, and amplifier source and load impedance.

OR the Boolean operator that implements the disjunction of two predicates. The truth table for $\vee \equiv X$ OR Y is

X	Y	$X \vee Y$
F	F	F
F	T	T
T	F	T
T	T	T

n-ary ORs can be obtained as disjunction of binary ORs.

OR gate a logic circuit that performs the OR operation. The output of the gate is high if one or all of its inputs are high.

oral cavity the human cavity where speech is produced and where the articulations of the speech organs makes it possible to produce different sounds.

Orange book *See* IEEE Color books.

ordinary refractive index the refractive index that is invariant with light direction, affecting one particular optical polarization component.

ordinary wave polarization component of a light wave that is affected by the ordinary refractive index.

organic LED *See* organic light emitting diode.

organic light emitting diode a group of recently developed organic material that emits light in response to electrical input. Although lower in efficiency, they have greater manufacturing flexibility than semiconductor LED.

orientation for an orthonormal coordinate frame attached to an object, the direction of the three axes of the frame relative to the base orthonormal coordinate frame.

orthogonal for two signals or functions, the condition that their inner product is zero. For example, two real continuous functions $s_1(t)$ and

$s_2(t)$ are orthogonal if

$$\int s_1(t)s_2(t)\,dt = 0$$

orthogonal code division multiple access scheme a code division multiple access scheme (CDMA), i.e., a spread spectrum system with multiple users, where the PN sequences utilized by the different users are orthogonal sequences. In such a system, the multiple access interference is zero. The orthogonal sequences can be obtained from the rows of a Hadamard matrix, for example.

orthogonal CDMA *See* orthogonal code division multiple access scheme.

orthogonal filter bank a biorthogonal filter bank whose polyphase transfer functions of the analysis and synthesis filters satisfy $F(z) = H^T(z^{-1})$.

orthogonal frequency division multiplex (OFDM) a frequency division multiplex scheme where the subcarriers for each of the frequency divided bands (subbands) are orthononal to each other. This allows for spectral overlap between successive subbands.

orthogonal hopping sequences a set of frequency hopping sequences (hopping patterns) for which no two sequences have the same value for a given position in the sequence. The set of sequences can be represented as the rows of a matrix. In such a case for a given column of the matrix, the elements are distinct.

orthogonal matrix a matrix A whose inverse is A^T; that is, $A^T A = I$, where I is the identity matrix. More generally, A is a unitary matrix if its inverse is A^H (that is, the complex-conjugate transpose); consequently a real unitary matrix is an orthogonal matrix. Although not recommended, the phrase "orthogonal matrix" is sometimes used to refer to a unitary matrix.

orthogonal projection for a vector $\mathbf{c}$ onto a vector $\mathbf{b}$, the orthogonal projection is

$$\mathbf{c_b} = \left(\frac{\mathbf{b}^T \mathbf{c}}{\mathbf{b}^T \mathbf{b}} \right) \mathbf{b}$$

orthogonal set of functions for real signal set $f_m(t)$ over $[t_1, t_2]$ such that

$$\int_{t_1}^{t_2} f_m(\tau) f_n(\tau) d\tau = \begin{cases} 0 & m \neq n \\ 1 & m = n \end{cases}$$

Orthogonal functions are necessary for transforms such as the FFT, or the DCT.

orthogonal transform a transform whose basis functions are orthogonal. The transform matrix of a discrete orthogonal transform is an orthogonal matrix. Sometimes orthogonal transform is used to refer to a unitary transform. Orthogonal real transforms exhibit the property of energy conservation.

orthogonal wavelet wavelet functions that form orthogonal basis by translation and dilation of a mother wavelet.

orthographic projection a form of projection in which the rays forming an image are modeled as moving along parallel paths on their way to the image plane: usually the paths are taken to be orthogonal to the image plane. Orthographic projection suppresses information on depth in the scene. A limiting case of perspective projection.

orthonormal functions an orthogonal signal set $f_m(t)$ on $[t_1, t_2]$ such that

$$\int_{t_1}^{t_2} |f_m(t)|\,dt = 1$$

for all m.

orthonormal transform *See* orthogonal transform.

oscillation condition   (1) condition that must be satisfied for an active circuit or device to exhibit a stable frequency oscillation. For a stable oscillation to be achieved, a condition is placed on the impedance presented to the active device, the impedance exhibited by the active device, and the derivative of these impedances with respect to both frequency and voltage.

(2) requirement that the electromagnetic field in a laser oscillator be self-sustaining including all loss and gain elements; also subconditions on self-consistency of amplitude, phase, polarization, and frequency.

oscillation frequency   frequency produced by a laser oscillator, including effects of the optical cavity, active medium, and any other elements in the cavity.

oscillation threshold   condition under which unsaturated gain is equal to loss.

oscillator   (1) a circuit that generates a repetitive series of pulses at a certain frequency.

(2) an amplifier that has an output for zero input, i.e., it is providing an output signal even though there is no signal applied at the input.

oscillatory transient   a rapid change in frequency (not power frequency) during steady state that is bipolar for voltage or current.

oscillograph   a continuous recording of the waveforms of an electric power line, formerly made with a cathode-ray tube but currently with a digital signal recorder, kept updated to record abnormalities during switching operations and fault conditions.

OSI model   *See* open systems interconnection model.

OTDR   *See* optical time domain reflectometry.

OTDR optical fiber   *See* optical time-domain reflectometry, optical fiber.

OTP   one-time programmable. *See* programmable read-only memory.

out-of-order issue   the situation in which instructions are sent to be executed not necessarily in the order that they appear in the program. An instruction is issued as soon as any data dependencies with other instructions are resolved.

out-of-step   an abnormal condition when generators in a power system cannot operate in synchronism.

out-of-step relay   a protective relay that senses that a synchronous generator has pulled out of step, and is operating at a frequency different than the system frequency.

outage   (1) the percentage of time or area for which a communication system does not provide acceptable quality.

(2) loss of power from all or part of a power system.

outage inferencing   the act of identifying the probable location of an outage based on information received from customer trouble calls and power monitoring units.

outlier   a statistically unlikely event in which an observation is very far (by several standard deviations) removed from the mean. Also refers to points which are far removed from fitted lines and curves. In experimental circumstances, "outlier" frequently refers to a corrupt or invalid datum.

output backoff   *See* backoff.

output block   when the processor writes a large amount of data to its output, it writes one block, a certain amount of data, at a time. That amount is referred to as an output block.

output buffer   when the processor writes to its output device, it must make sure that device will be able to accept the data. One common technique is that the processor writes to certain memory addresses, the output buffer, where the output device can access it.

output dependency the situation when two sequential instructions in a program write to the same location. To obtain the desired result, the second instruction must write to the location after the first instruction. Also known as write-after-write hazard.

output device a device that presents the results sent to the user, and typical output devices are the screen and the printer. (According to some interpretations, it also includes the stable storage where the processor saves data.)

output filter a lowpass filter used to attenuate the switching ripple at the output of switching circuits to a tolerable level.

output impedance the ratio of the drop in voltage to the current drawn is known as the output impedance of the electric source and is measured in units of ohms.

output neuron layer a neuron (layer of neurons) that produces the network output (outputs). In feedforward networks, the set of weights connected directly to the output neurons is often also referred to as the output layer.

output power (1) the difference in the power available under perfectly matched conditions and the reflected power taking the output return loss into account, expressed in watts.

(2) in lasers, the useful output from a laser oscillator.

output return loss negative ratio of the reflected power to the incident power at the output port of a device, referenced to the load or system impedance, expressed in decibels. The negative sign results from the term "loss." Thus an output

return loss of 20 dB results when 1/100th of the incident power is reflected.

output routine low-level software that handles communication with output devices. Handles the formatting of data as well as eventual protocols and timings with the output devices.

output swing for a semiconductor device, the difference between the output high voltage and the output low voltage.

output vector a vector formed by the output variables of a network.

outstar configuration consists of neurons driving a set of outputs through synaptic weights. An outstar neuron produces a desired excitation pattern to other neurons whenever it fires. *See also* outstar training, instar configuration.

outstar rule a learning rule that incorporates both Hebbian learning and weight decay. A weight is strengthened if its input signal and the activation of the neuron receiving the signal are both strong. If not, its value decays. A similar rule exists for instars, but has found less application.

outstar training neuron training where the weights are updated according to

$$w_i(t+1) = w_i(t) + v\,(y_i - w_i(t))$$

where v is the training rate starting from about 1 and gradually reduced to 0 during the training. *See also* instar configuration.

over-compounded a compound DC generator in which the terminal voltage increases as load current increases. Extra turns are added in the series winding to generate the additional voltage after compensating for the armature voltage drop and the armature resistance drop.

overcurrent (1) current in a circuit that exceeds a preset limit.

(2) motor current magnitude in the normal circuit path exceeding the full-load current.

overcurrent protection   (1) the act of protecting electrical and electronic devices or circuits from a dangerous amount of input current.

(2) the effect of a device operative on excessive current.

overcurrent relay   a protective relay that operates when fed a current larger than its minimum pick-up value.

overdamped   more damping than a critically damped system. For a characteristic equation of the form: $s^2 + 2\zeta\omega_n s + \omega_n^2$; the system is overdamped if $\zeta < 1.0$ the roots of the characteristic equation are complex conjugate pairs.

overexcited   the condition where the field winding current is greater than a rated value that produces the rated MMF in the armature. With a synchronous or a DC machine, the excitation current is a direct current in the field windings.

If a machine is overexcited, the excess MMF must be counterbalanced in the armature. In the case of a DC motor, the overexcitation is counterbalanced by the increase of armature current, which is translated by the increase of both torque and speed. In the case of synchronous motor, a leading component of the armature current is present and the machine operates at a leading power factor.

overflow   a data condition in arithmetic operations of signed numbers where the magnitude of a result exceeds the number of bits assigned to represent the magnitude. The result changes the sign bit, thus, making the result incorrect.

overlap angle   *See* commutation angle.

overlapped execution   processing several instructions during the same clock pulses.

overlay   (1) in wireless communications, refers to a modulated signal (communication scheme) that occupies an RF channel that is already occupied by other signals. The classical example is that of a spread spectrum signal that is (spectrally) placed on top of a set of narrow band channels that carry narrow band signals. The wideband spread spectrum signal has a low power spectral density; hence the overlay scheme can coexist with the narrow band signals occupying the frequency band.

(2) a vector describing the positional accuracy with which a new lithographic pattern has been printed on top of an existing pattern on the wafer, measured at any point on the wafer. Also called registration.

overload   a situation that results in electrical equipment carrying more than its rated current. Placing too much electrical load on a generator or too much mechanical load on a motor would cause an overload.

overload heater   a term used to describe the thermal sensors that detect motor overload currents. Usually located on the motor starter, the heaters cause the overload relay to operate.

overload protection   a protective device which opens the circuit to a piece of electrical equipment or power line in the event of current exceeding the upper design limit.

overload relay   a device designed to detect and interrupt motor overload conditions. Motor overload relays may be actuated by thermal (temperature), magnetic (current), or electronic (voltage and current) sensors.

overmoded   the condition of a waveguide at a frequency where two or more modes are above cutoff (propagating).

overmodulation   technique used in pulse-width-modulated (PWM) switching schemes to obtain a higher output voltage. Overmodulation occurs when the control (or modulation) signal magnitude exceeds the magnitude of the triangle carrier signal that it is being compared to and the number of pulses begin to drop out. This does introduce greater output voltage distortion.

oversampling   sampling a continuous-time signal at more than the Nyquist frequency.

oversampling converter   A/D converter that samples frequencies at a rate much higher than the Nyquist frequency. Typical oversampling rates are 32 and 64 times the sampling rate that would be required with the Nyquist converters.

overshoot   the amount by which an output value momentarily exceeds the ideal output value for an underdamped system.

overstress failure   failure mechanisms due to a single occurrence of a stress event when the intrinsic strength of an element of the product is exceeded.

overvoltage   a voltage having a value larger than the nominal voltage for a duration greater than 1 minute.

overvoltage relay   a protective relay that operates on overvoltage.

over-excitation limiter   *See* maximum excitation limiter.

oxidation   for a semiconductor manufacturing process, the process of growing silicon dioxide on a silicon wafer subjected to elevated temperature in an oxygen-containing environment.

oxide charge   the charge in the oxide, which may be grown in, or may be introduced by charge injection from, external sources or by ionizing radiation. Its presence in the oxide of a MOS device causes a shift in the flatband voltage and the threshold voltage.

oxide trap   a defect, impurity, or disordered bond in the oxide that can trap an electron or a hole.

P

P commonly used symbol for power in watts or milliwatts.

P$_{DC}$ common symbol for DC power in watts.

P$_{input}$ common symbol for power input to a device in watts.

P$_{load}$ common symbol for power delivered to the load.

P$_{ref}$ common symbol for power reference level in watts or milliwatts.

p-channel MOSFET a MOSFET where the source and drain are composed of heavily doped p-type semiconductor regions in a n-type surface. Holes form drain-source current when the applied gate and substrate potentials invert the n-type surface between them.

P-I-N photodiode a photodiode (detector) in which a layer of intrinsic (undoped) material is added between the p-n junction. This has the effect of increasing the amount of incident optical power absorbed in the device and hence the efficiency in converting optical power into electrical current.

p-n junction (1) a junction between regions of the same bulk material that differ in the concentration of dopants, n-type on one side and p-type on the other. The diode is based on a single p-n junction.

(2) metallurgical interface of two regions in a semiconductor where one region contains impurity elements that create equivalent positive charge carriers (p-type) and the other semiconductor region contains impurities that create negative charge carriers (n-type).

P-well a region of p-type semiconductor located at the surface of a n-type substrate (or larger N-well) usually created in order to contain n-channel MOSFETs.

P1dB acronym for 1 dB compression power. This gives a measure of the maximum signal power level that can be processed without causing significant signal distortion or saturation effects. Technically, this refers to the power level at the input or the output of a component or system at which the saturation of active devices like transistors causes the gain to be compressed by 1 dB from the linear gain.

PAC learning a supervised learning framework in which training examples x are randomly and independently drawn from a fixed, but unknown, probability distribution on the set of all examples. Each example is labeled with the value $f(x)$ of the target function to be learned. A PAC (probably, approximately correct) learning algorithm is one which, on the basis of a finite number of examples, is able, with high probability, to learn a close approximation to the target function.

package in MMIC technology, die or chips have to ultimately be packaged to be useful. An example of a package is the T07 "can." The MMIC chip is connected within the can with bond wires connecting from pads on the chip to lead pins on the package. The package protects the chip from the environment and allows easy connection of the chip with other components needed to assemble an entire system, such as a DBS TV receiver.

packed decimal a data format for the efficient storage and manipulation of real numbers, similar to BCD, with digits stored in decimal form, two per byte.

packet   a unit of data which is sent over a network. A packet comprises a payload containing some data, and either a header or a trailer containing control information.

packet reservation multiple access (PRMA)   a user transmitting a packet in a particular slot will have the corresponding slot reserved in the next frame. The base station will broadcast the state of the slots (reserved and free for contention) at the beginning of each frame.

packet switching   means of switching data among the ports (inputs and outputs) of a switch such that the data is transferred in units of variable size.

packet-switched bus   *See* split-transaction.

pad   (1) a device (network) that impedance matches and/or attenuates. Typically used to refer to a coax attenuator.

(2) a concrete foundation, usually prefabricated and used to support power transformers in underground residential distribution work.

pad-mount transformer   a heavily-enclosed distribution transformer mounted at grade level upon a concrete slab or pad.

Pade approximation   a pole-zero approximation of deadtime based on the expansion

$$e^{-s\tau} = \frac{1 - \theta + \frac{1}{2!}\theta^2 - \frac{1}{3!}\theta^3 + \ldots + \frac{1}{n!}\theta^n + \ldots}{1 + \theta + \frac{1}{2!}\theta^2 - \frac{1}{3!}\theta^3 + \ldots + \frac{1}{n!}\theta^n + \ldots}$$

where $\theta = s\tau/2$. When the infinite series in both numerator and denominator are truncated to m terms, the approximation contains m zeros and m poles and is known as the mth order Pade approximation to deadtime. This formula allows pole-zero methods like root locus or pole placement which cannot normally deal with system containing deadtime, to be applied to such systems.

PAFC   *See* phosphoric acid fuel cell.

page   *See* virtual memory.

page fault   event that occurs when the processor requests a page that is currently not in main memory. When the processor tries to access an instruction or data element that is on a page that is not currently in main memory, a page fault occurs. The system must retrieve the page from secondary storage before execution can continue.

page frame   a contiguous block of memory locations used to hold a page. *See also* virtual memory.

page miss penalty   when a page miss occurs, the processor will manage the load of the requested page as well as the potential replacement of another page. The time involved, which is entirely devoted to the page miss, is referred to as the page miss penalty.

page offset   the page offset is the index of a byte or a word within a page, and is calculated as the physical as well as virtual address modulus of the page size.

page printing   a printing technique where the information to be printed on a page is electronically composed and stored before shipping to the printer. The printer then prints the full page nonstop. Printing speed is usually given in units of pages per minute (ppm).

page replacement   at a page miss, when a page will be loaded into the main memory, the main memory might have no space left for that page. To provide space for that new page, the processor will have to choose a page to replace.

page table   a mechanism for the translation of addresses from logical to physical in a processor equipped with virtual memory capability. Each row of the page table contains a reference to a logical block of addresses and a reference to a corresponding block of physical storage. Every memory reference is translated within the CPU before storage is accessed. The page table may itself be stored in standard memory or may be stored within a special type of memory known as associative memory. A page table stored in associative memory is known as a translation lookaside buffer.

page-fault-frequency replacement   a replacement algorithm for pages in main memory. This is the reciprocal of the time between successive page faults. Replacement is according to whether the page-fault frequency is above or below some threshold.

page-mode DRAM   a technique that uses a buffer like a static RAM; by changing the column address, random bits can be accessed in the buffer until the next row access or refresh time occurs.

This organization is typically used in DRAM for column access. Additional timing signals allow repeated accesses to the buffer without a row-access time. *See also* two-dimensional memory organization.

page-printing printer   a human-readable output device used for producing documents in a written form. The printer stores a whole page in memory before printing it (e.g., a laser printer).

paged segmentation   the combination of paging and segmentation in which segments are divided into equal sized pages. Allows individual pages of a segment rather than the whole segment to be transferred into and out of the main memory.

paged-segment   a segment partitioned into an integral number of pages.

paging   the process of transferring pages between main memory and secondary memory.

paging channel   a channel in a wireless communication system used to send paging messages.

Paging messages are typically used to set up telephone calls and also to send control messages to terminals that are involved in a communication session.

Pagourek–Witsenhausen paradox   the best known result concerning comparison of performance sensitivity between open-loop and closed-loop nominally optimal systems. The result provides the somewhat deceptive answer to a question of what is the deviation of the performance index for a closed-loop and open-loop implementations of nominally optimal control systems due to parameter deviations from its nominal value (for which the optimal control has been calculated). The answer considered either obvious or paradoxical can be expressed as follows: the infinitesimal sensitivity of the performance index expressed by its first variation caused by a variation of the parameter vector is the same whether an open-loop or a closed-loop implementation of the nominally optimal control is used. The result primary found for linear-quadratic problem has been generalized for nonlinear time-varying sufficiently smooth optimal control problems with free terminal state.

PAL   *See* phase alternate line or programmable array logic.

PAM   *See* pulse amplitude modulation.

PAM system   *See* prism | air | metal system.

panelboard   an assembly of one or more panel units containing power buses, automatic overcurrent protective devices, that is placed in a cabinet or cutout box located in or flush on a wall. The assembly can only be accessed from the front and may contain switches for operation of light, heat, or power circuits. *See* switchboard.

pantograph   an apparatus for applying sliding contacts to the power lines above an electric railroad locomotive.

paper tape   strips of paper capable of storing or recording information, most often in the form of punched holes representing the values. Now obsolete.

Papoulis' generalization   a sampling theory applicable to many cases wherein signal samples are obtained either nonuniformly and/or indirectly.

parabolic index profile   quadratic transverse variation of the index of refraction; leads to analytic solutions of the paraxial equations for rays and beams.

parabolic reflector   a reflecting surface defined by a paraboloid of revolution or section of a paraboloid of revolution.

paraelectric   the nonpolar phase into which the ferroelectric transforms above T_c, frequently called the paraelectric phase.

parallel adder   a logic circuit that adds two binary numbers by adding pairs of digits starting with the least significant digits. Any carry generated is added with the next pair of digits. The term "parallel" is misleading, since all the digits of each numbers are not added simultaneously.

parallel architecture   a computer system architecture made up of multiple CPUs. When the number of parallel processors is small, the system is known as a multiprocessing system; when the number of CPUs is large, the system is known as a massively parallel system.

parallel bus   a data communication path between parts of the system that has one line for each bit of data being transmitted.

parallel computing   computing performed on computers that have more then one CPU operating simultaneously.

parallel computing system   a system whose parts are simultaneously running on different processors.

parallel data transfer   the data transfer proceeds simultaneously over a number of paths, or a bus with a width of multiple bits, so that multiple bits are transferred every cycle. A technique to increase the bandwidth over that of serial data transfer.

parallel feed   *See* corporate feed.

parallel I/O   *See* parallel input/output.

parallel I/O interface   I/O interface consisting of multiple lines to allow for the simultaneous transfer of several bits. Commonly used for high-speed devices, e.g., disk, tape, etc. *See also* serial I/O interface.

parallel input/output   generic class of input/output (I/O) operations that use multiple lines to connect the controller and the peripheral. Multiple bits are transferred simultaneously at any time over the data bus.

parallel interference cancellation   a multiple access interference cancellation strategy for CDMA based on multistage detection. In the first stage, tentative decisions are made for each user in parallel. For each user, an estimate of the resulting multiple access interference is made and substracted from the received signal. In the succeeding stage, a more reliable tentative decision can then be made for each user. This proceeds successively for a predetermined number of stages.

parallel manipulator   manipulator that consists of a base platform, one moving platform and various legs. Each leg is a kinematic chain of the serial type, whose end links are the two platforms. Parallel manipulators contain unactuated joints, which makes their analysis more complex than those of serial type. A paradigm of parallel manipulators is the flight simulator consisting of six legs actuated by hydraulic pistons.

parallel paths the number of separate paths through the armature winding that exist between the brushes of a DC machine. In a DC machine's armature, the conductors and coils are placed in their slots and connected to the commutator using either the lap winding method or the wave winding method. The number of conductors that are connected in parallel depend on the number of poles the machine has, and whether the winding connections are lap or wave. For the lap wound armature, the number of parallel paths is found by multiplying the number of poles by the number of revolutions it takes to fill all the slots of the armature. The number of revolutions it takes to fill the slots is known as the machine's "plex" value. In a simplex wound armature, the "plex" value is 1, duplex has a "plex" value of 2, triplex has a "plex" value of 3, and so on. For the wave wound armature, the number of parallel paths is two times the "plex" value. This same concept can also be applied to AC machinery.

parallel plate waveguide a type of waveguide formed by two parallel metallic plates separated by a certain distance. It supports propagation of several types of modes: the TEM mode and the TE and TM mode families.

parallel port a data port in which a collection of data bits are transmitted simultaneously.

parallel processing processing carried out by a number of processing elements working in parallel, thereby speeding up the rate at which operations on large data sets such as images can be achieved. Often used in the design of real-time systems.

parallel transmission the transmission of multiple bits in parallel.

parallel-to-serial conversion a process whereby data, whose bits are simultaneously transferred in parallel, is translated to data whose bits are serially transferred one at a time. During the translation process, some timing information may be included (such as start and stop bits) or is implicitly assumed.

parallel-transfer disk a disk in which it is possible to simultaneously read from or write to multiple disk surfaces. Advantageous in providing high data transfer rates.

parallelism the possibility to simultaneously execute different parts of a system on different processors.

Although parallelism can be found in many electronic computing systems, parallelism is the inherent property of an optical system. For example, a lens, the simplest optical system, forms the whole image at once and not point-by-point or part-by-part. If the image consists of one million points, the lens processes one million data in parallel.

parallelogram mechanism a manipulator that has a kinematic chain of the serial type and part of its kinematic chain forms a closed kinematic chain.

paramagnetic materials with permeability slightly greater than unity. Sodium, potassium, and oxygen are examples.

parameter coding also called vocoding. In parameter coding, the signal is analyzed with respect to a model of the vocal mechanism and the parameters of the model are transmitted.

parameter estimation the procedure of estimation of model parameters based on the model's response to certain test inputs.

parameter matrix an $N \times N$ matrix of complex linear parameters which describe the behavioral relationships between the N ports of a circuit or network. The most commonly utilized parameter matrices are conductance (G), impedance (Z), admittance (Y), (H), scattering (S), chain ($ABCD$) and scattering chain (Φ), all of which can be readily transformed from one to another through simple algebraic manipulations. Any set of these parameters are referenced to given port source and/or load impedance, at fixed input frequencies and fixed power levels.

parameter space   a domain formed by all possible values of the given parameters. *See* Parseval's equation, Parseval's theorem.

parametric amplification   a nonlinear optical process in which a signal wave of frequency ω_s and a higher-frequency pump wave of frequency ω_p propagate through a second-order nonlinear optical material, leading to the amplification of the signal wave and the generation of an idler wave of frequency $\omega_p - \omega_s$. *See also* optical parametric oscillator.

parametric coding   refers to the class of signal compression methods that are based on a criterion where parameters, or features, of the signal are extracted and coded. Contrasts waveform coding techniques, since the reproduced waveform can be (analytically) quite different from the input, but will still be a subjectively (in terms of vision or hearing) good, even indistinguishable, replica of the input signal.

parametric fixed form control rule   control rule given in a predeclared form, with a number of parameters to be tuned; tuning of those parameters can be performed either off-line, prior to control rule implementation, or on-line. A classical example of a parametric fixed form control rule is PID industrial controller; another example is a neural network based control rule with a number — usually large — of parameters of the network to be tuned.

parametric oscillator   *See* optical parametric oscillator.

parasitic capacitance   the generally undesirable and not-designed-for capacitance between two conductors in proximity of one another.

parasitic inductance   the generally undesirable and not-designed-for inductance associated with a conductor, or path of current on a conductor.

parasitic reactance   the generally undesirable and not-designed-for reactance associated with one or more conductors in a circuit.

parasitic resistance   the generally undesirable and not-designed-for resistance associated with a conductor, or path of current on a conductor.

paraxial approximation   neglect of the second derivative of the nearly plane-wave amplitude in the direction of propagation; makes possible analytic solutions for diffracting beams.

paraxial optics   formalism for optics in which the paraxial approximation is employed.

paraxial ray   ray propagating so nearly parallel to the z axis that its length can be considered equal to the z propagation distance.

paraxial ray equation   set of second-order differential equations in the propagation distance for the trajectory of a light ray propagating almost parallel to a fixed axis.

paraxial wave   solution of the scalar wave equation in the paraxial approximation.

PARCOR coefficients   *See* partial-correlation parameters.

parity   property of a binary sequence that determines if the number of 1's in the sequence is either odd or even.

parity bit   an extra bit included in a binary sequence to make the total number of 1's (including itself) either odd or even. For instance, for the following binary sequence 101, one would insert a parity bit $P(odd) = 1$ to make the total number of 1's odd; a parity bit $P(even) = 0$ would be inserted to make this number even. *See also* error detecting code.

parity check matrix   a matrix whose rows are orthogonal to the rows in the generator matrix of a linear forward error control block code. A nonzero result of element-wise finite field multiplication of the demodulated word by this matrix indicates the presence of symbol errors in the demodulated word.

   Is generated from the parity check polynomial of any linear (n, k) code and has dimension of $(n - k \times n)$. It is used by the decoder for error detection by checking the parity bits.

parity detection circuit   a parity check logic incorporated within the processor to facilitate the detection of internal parity errors (reading data from caches, internal buffers, external data, and address parity errors).

parity-check code   a binary linear block code.

Park's transformation   a change of variables represented by a linear matrix multiplication used in the analysis of electric machines. *See* rotor reference frame.

parking   on a bus, a priority scheme that allows a bus master to gain control of the bus without arbitration.

parse tree   the tree that is used for parsing strings of a given language.

Parseval's theorem   a relationship that states that the integral of the square of the magnitude of a periodic function is the sum of the square of the magnitutde of each harmonic component.

   Rigorously, suppose that two continuous time signals $f_1(t)$ and $f_2(t)$ have corresponding Fourier transforms $F_1(\omega)$ and $F_2(\omega)$, and that $\overline{F_2(\omega)}$ is the complex conjugate of $F_2(\omega)$. Then Parseval's theorem states that

$$\int_{-\infty}^{\infty} f_1(t) f_2(t) dt = \frac{1}{2\pi} \int_{-\infty}^{\infty} F_1(\omega) \overline{F_2(\omega)} d\omega.$$

If $f_1(t) = f_2(t)$, then the left-hand side of the above equation provides an expression of the energy of a signal, which can be related to its Fourier transform as follows:

$$\int_{-\infty}^{\infty} f(t)^2 dt = \frac{1}{2\pi} \int_{-\infty}^{\infty} |F(\omega)|^2 d\omega.$$

parsing   the process of detecting whether a given string belongs to a given language, typically represented by grammars.

partial coherence   the ratio of the sine of the maximum half-angle of illumination striking the mask to the numerical aperture of the objective lens. Also called the degree of coherence, coherence factor, or the pupil filling function, this term is usually given the symbol s.

partial element equivalent method   an integral equation technique in which the electromagnetic problem is reduced to a lumped circuit problem by defining some regions in space associated with a node in the lumped circuit. This method takes electric field interactions in the original problem into account by finding (through the integral equations) either a capacitor to ground at infinity and a summation of current controlled current sources or a capacitor connected in series to a summation of voltage controlled voltage sources. The magnetic field interactions are taken into account by finding an inductance in series with a summation of current controlled voltage sources which is placed between nodes.

partial fraction expansion   the method of partial fraction expansion consists of taking a function that is the ratio of polynomials and expanding it as a linear combination of simpler terms of the same type. This tool is useful in inverting Fourier, Laplace, or z-transforms and in analyzing linear time invariant systems described by linear constant-coefficient differential or difference equations.

partially coherent illumination   a type of illumination resulting from a finite size source of light that illuminates the mask with light from a limited, nonzero range of directions.

partial restoration   refers to the situation where not all customers who are part of a larger outage can have power restored. Utilities typically make efforts to restore power to all customers, but situational factors may dictate that the restoration process proceed in stages.

partial-correlation parameters   the parameters that are obtained when solving the autocorrelation equations for the problem of linear prediction.

partially decorrelating noise whitening   the process of transforming the matched-filtered sufficient statistic of a CDMA signal into a corresponding sufficient statistic which is affected by uncorrelated (white) noise and is partially decorrelated, i.e., the resulting sufficient statistic for a K-user system is described by

$$\mathbf{x} = \mathbf{Fd} + \mathbf{n},$$

where $\mathbf{x}$ is a length K column vector of sufficient statistics, $\mathbf{F}$ is a $K \times K$ lower left triangular matrix, $\mathbf{d}$ is a length K column vector of data symbols, and $\mathbf{n}$ is a length K column vector of AWGN noise samples.

participation factor   the ratio of the the change in the power output of a generator versus the total change in power demand. The participation factor of each generator in a power system is found in the solution of the economic dispatch problem.

partition noise   electrical noise generated within a vacuum tube when the electron stream hits obstacles and is divided.

partition table   a table that enables the grouping of states into equivalent sets.

partitioned process   controlled process that is considered as consisting of several sub-processes that can be interconnected; process partitioning is an essential step in control problem decompo-

sition — for example, before decentralized or hierarchical control can be introduced. The term "partitioned system" is used when one is referring to the partitioned control system, but is also often used to denote just the partitioned controlled process.

passband   (1) the range of frequencies, or frequency band, for which a filter passes the frequency components of an input signal.
(2) the frequency difference between the higher and lower band edges, expressed in radians/second. The band edges are usually defined as the highest and lowest frequencies within a contiguous band of interest at which the loss equals the maximum attenuation loss across the band. *Compare with* stop band.

passband edge   the frequency at which the signal becomes significantly attenuated; typically the frequency at which the signal is attenuated at 3 dB from the maximum response.

passband ripple   difference between the maximum attenuation loss and the minimum attenuation loss across the band, expressed in decibels. This parameter is also known as the loss attenuation in ratio form. The band edges are usually defined as the highest and lowest frequencies within a contiguous band of interest at which the loss equals the maximum attenuation loss across the band.

passivation   the process in which an insulating dielectric layer is formed over the surface of the die. Passivation is normally achieved by thermal oxidation of the silicon and a thin layer of silicon dioxide is obtained in this manner. Other passivation dielectric coatings may also be applied, such as silicon glass.

passive backplane   in printed circuit boards, a circuit board in which other boards are plugged, which contains no active circuit elements to control signal quality.

passive filter a filter circuit that uses only passive components, i.e., resistors, inductors, and capacitors. These circuits are useful at higher frequencies and as prototypes for ladder filters that are active.

passive magnetic bearing a magnetic bearing that does not require input energy for stable support during operation. Generally implemented using permanent magnets.

passive network an electronic circuit made up of passive elements. Passive elements are capacitors, resistors and inductors, and have no gain characteristics.

passive optical network a network where the optical fiber plays the role of a broadband, passive interconnect, and the functions of switching and control are made using electronics.

passive redundancy a circuit redundancy technique which assures fault masking by error correcting codes or N-modular redundancy with voting.

passivity naturally associated with power dissipation. It can be defined for linear as well as nonlinear systems. A formal definition of passivity requires a representation of systems by an operator mapping signals to signals. The signal space under consideration is assumed to be extended L_2 space with a scalar product defined by

$$\langle x | y \rangle = \int_0^\infty x(s) y(s) ds$$

A system with input u and output y is passive if

$$\langle x | y \rangle \geq 0$$

The system is input strictly passive if there exists $\varepsilon > 0$ such that

$$\langle x | y \rangle \geq \varepsilon \| u \|^2$$

and output strictly passive if there exists $\varepsilon > 0$ such that

$$\langle x | y \rangle \geq \varepsilon \| y \|^2$$

path the space curve that the manipulator end-effector moves along from the initial location (position and orientation) to the final location is called the path. Notice that the path is a pure geometric description of motion. Path can be specified in the joint space or in the operational space.

path-delay testing any one of several possible techniques to verify that signal transitions created by one clock event will travel through a particular logic/path in a subcircuit, IC component, or system and will reach their final steady-state values before a subsequent clock event.

path-set the set of all edges in a path.

patrolling in overhead power lines, action taken when the location of a fault is not known. The crew will typically follow overhead spans until the location of the outage is found.

pattern a plot of the distribution of radiated power. Typically, the pattern consists of a main lobe (major lobe), in which most of the radiated power is confined, and a number of progressively weaker sidelobes (minor lobes).

pattern classification assignment of a pattern (typically a vector of measurements or a feature derived from measurements) to a class.

patterning the processes of lithography (producing a pattern that covers portions of the substrate with resist) followed by etching (selective removal of material not covered by resist) or otherwise transferring the pattern into the substrate.

pattern (template) matching the detection, in an image (signal), of a subimage (time window) in which pixel (sample) values are structured according to a predefined schema.

511

pattern recognition   the feature extraction, clustering, and classification processes associated with assigning meaning to measurements. There are statistical, syntactic, and structural methods for pattern recognition, and neural network methods are often considered as a subset of pattern recognition.

pattern sensitive   integrated circuits in which an error may result from a certain data pattern being encountered.

pattern synthesis   the process of designing an antenna such that its radiation characteristics meet desired specifications.

pause instruction   an assembly language instruction whose execution causes a momentary pause in program execution.

Pawlak's information system   a system model denoted S can be viewed as a pair $S = (U, A)$, where $U = \{x_1, \ldots, x_n\}$ is a nonempty finite set of objects. The elements of U may be interpreted, for example, as concepts, events, goals, political parties, individuals, states, etc. The set U is called the universe. The elements of the set A, denoted $\mathbf{a}_j, j = 1, \ldots, m$, are called the attributes. The attributes are vector-valued functions. For example,

$$\mathbf{a}_j : U \to \{-1, 0, 1\}.$$

An example of a simple information system is shown in the following table.

	$\mathbf{a}_1$	$\mathbf{a}_2$	$\mathbf{a}_3$
x_1	-1	0	0
x_2	0	0	1
x_3	0	1	1
x_4	1	1	1

The first component of $\mathbf{a}_1$ being -1 may mean that x_1 is opposed to the issue $\mathbf{a}_1$, while x_4 support $\mathbf{a}_1$, etc. Such data can be collected from newspapers, surveys, or experts. Also known as rough set theory.

payload   the portion of a packet that is neither the header nor the trailer and is either user- or protocol-specific data.

PC   *See* program counter, personal computer.

PC-relative addressing   an addressing mechanism for machine instructions in which the address of the target location is given by the contents of the program counter and an offset held as a constant in the instruction added together. Allows the target location to be specified as a number of locations from the current (or next) instruction. Generally only used for branch instructions.

PCA   *See* principal component analysis.

PCB   *See* printed circuit board process, control block, or polychlorinated biphenyls.

PCBA   acronym for printed circuit board assembly.

PCM   *See* pulse-code modulation, phase coded modulation.

PCN   *See* personal communications network.

PCS   *See* personal communications service.

PDC   *See* personal digital cellular.

PDF   *See* probability density function.

PDF-optimized scalar quantization   *See* Lloyd–Max scalar quantization.

PDMA   *See* polarization division multiple access.

PDP   *See* piecewise deterministic process.

PDU   *See* protocol data unit.

PE   *See* processing element.

peak accumulation mode   an operating feature of digital storage oscilloscope in which the maximum and minimum excursions of the waveform are displayed for a given point in time.

peak detector   an electronic circuit which outputs a DC voltage which indicates the peak amplitude of an alternating waveform.

peak let-through current   the maximum value of the available short-circuit current that is let through a current-limiting fuse. *See also* current-limiting fuse.

peak output power   the instantaneous peak power delivered to a load, expressed in watts. It is simply the maximum power at any instant in time taken during the total time period being considered. This term is often used when a device or application is subjected to time-varying signals such as pulse modulation.

peak power   the maximum value of the square of the absolute value of a signal over the duration of the signal; i.e., for the signal $x(t)$, $\max_t |x(t)|^2$. The maximum instantaneous power of a signal.

peak signal to noise ratio (PSNR)   objective distortion measure of the difference between a discrete time signal x_i, defined over $[1 \ldots n]$, and a degraded, restored or otherwise processed version of the signal x_i, defined as PSNR $= 10 \log_{10} \frac{x_{max}^2 n}{n \sum_{i=1}^{n} (x_i - x_j)}$. BPSNR differs from the standard signal to noise ratio by using the square of the signal's maximum value rather than its variance in the numerator. In image coding this value is often assumed to be 255, for 8-bit images.

peaking generator   (1) a utility generating unit, typically driven by a gas-fired turbine, available to rapidly come on line when the system demand reaches its highest levels.

(2) a generator used by a plant to reduce the peak demand drawn in a given period of time (typically during a one-month interval).

peaking unit   a generator used only to supply peak periods of electric power demand.

Pearl Street Station   the first investor-owned electric utility plant, started by Edison in 1882 in New York City. It provided low-voltage, DC electric service to 85 customers with an electric load of 400 lamps.

PEB   *See* post-exposure bake.

PEC   *See* perfect electric conductor.

pel   a picture element that has been encoded as black or white, with no gray scale in between.

pellicle   a thin, transparent membrane placed above and/or below a photomask to protect the photomask from particulate contamination. Particles on the pellicle are significantly out of focus, and thus have a much reduced chance of impacting image quality.

PEMFC   *See* proton exchange membrane fuel cell.

penalty function   *See* cost function.

penstock   a water tube that feeds the turbine. It is used when the slope is too steep for using an open canal.

pentode   vacuum tube with five active electrodes: cathode, control grid, screen grid, suppressor grid, plate.

per-unit system   a dimensionless system for expressing each quantity in terms of a fractional part of a "base" value, often the nominal or rated value of the system. Typical electrical calculations require four base quantities (voltage, current, impedance, and apparent power), any two of which may be chosen arbitrarily. The per-unit system greatly simplifies calculations in electrical systems containing transformers with non-unity turns ratios, making the voltage differences transparent.

percent impedance   the per-unit impedance expressed as a percentage on a certain MVA and voltage base.

percent system   a variation of the per-unit system in which the ratios expressing system quantities are expressed as a percentage of the base quantity.

perceptron   one of the earliest neural algorithms demonstrating recognition and learning ability. Since the output of neuron is binary, the neuron can classify an input into two classes, A and B. The perceptron model containing a single neuron is expressed as follows: $z = 1$, if $\sum w_i x_i > T$ and $z = 0$, if $\sum w_i x_i < T$ where z is the output, x_i is an input, w_i is an element of the interconnection weight matrix, and T is a generalized threshold. If the interconnection weight matrix is known, an input vector (x_i) can be classified into A or B according to the result of z. On the other hand, the interconnection weight matrix can be formed if a set of input vectors (x_i) and their desired outputs z are known. The process of forming the interconnection weight matrix from known input–output pairs is called learning. The perceptron learning is as follows. First, the weight w_i and the threshold T are set to small random values. Then the output z is calculated using a set of known input. The calculated z is compared with the desired output t. The change of weight is then calculated as follows: $\delta w_i = \eta(t-z)x_i$ where η is the gain term that controls the learning rate, which is between 0 and 1. The learning rule determines that the corrected weight is

$$w_i = w_i(\text{old}) + \delta w_i$$

The process is repeated until $(t - z) = 0$. The whole learning process is completed after all input–output pairs have been tested with the network. If the combination of inputs are linearly separable, after a finite number of steps, the iteration is completed with the correct interconnection weight matrix. Various optical systems have been proposed to implement the perceptron, including correlators.

The original perceptron was a feedforward network of linear threshold units with two layers of weights, only one layer of which (the output layer) was trainable, the other layer having fixed values.

perceptron convergence procedure   a supervised learning technique developed for the original perceptron. If the output y_i of unit i in the output layer is in error, its input weights, w_{ij}, are adjusted according to $Dw_{ij} = h(t_i - y_i)x_j$, where t_i is the target output for unit i and h is a positive constant (often taken to be unity); x_j is the output of unit j in the previous layer which is multiplied by weight w_{ij} and then fed to unit i in the output layer.

perceptual coding   involves the coding of the contextual information of the image features by observing the minimum perceptual levels of the human observer.

perfect code   a t-error correcting forward error control block code in which the number of nonzero syndromes exactly equals the number of error patterns of t or fewer errors. Hamming codes and Golay codes are the only linear nontrivial perfect codes.

perfect electric conductor (PEC)   a conductor that has infinite conductivity or zero resistivity.

perfect reconstruction   condition in which the output of a filter bank is a delayed version of the corresponding input. In other words, there are no aliasing and (phase and magnitude) distortions for the output of this filter bank.

perfect shuffle   interconnects that connect sources 1, 2, 3, 4, 5, 6, 7, 8 to detectors 1, 5, 2, 6,

3, 7, 4, 8, respectively. The operation divides 1, 2, 3, 4, 5, 6, 7, 8 into two equal parts 1, 2, 3, 4 and 5, 6, 7, 8 and then interleaves them. The size of the array must be 2^n. The array returns to its original order after n operations. When the option to exchange pairs of neighboring elements is added to a perfect shuffle network, any arbitrary permutation of the elements is achievable.

performability   the probability that a system is performing at or above some level of performance, L, at the instant of time t.

periodic convolution   a type of convolution that involves two periodic sequences. Its calculation is slightly different from that of discrete linear convolution in that it only takes summations of the products within one period instead of taking summations for all possible products. Circular convolution is an operation applied to two finite-length sequences. In order to make its result equal to that of the linear convolution, the two sequences have to be zero-padded appropriately. These two zero-padded sequences can then serve as periods to formulate two periodic sequences. At this time, the result of the periodic convolution is equal to circular convolution.

periodic coordination   coordination process occurring when a sequence of control decisions is required over a given time interval and these decisions are directly made by the local units but the operation of these units is periodically adjusted by the coordinator; an example is operation of an industrial process controller with several independent regulators, which at specified time instants are provided with new parameters or additive compensation signals so as to achieve better results from the overall point of view, for example, to achieve better transient responses to the changes of the free inputs.

periodic signal   a continuous time signal $f(t)$ with period $T \geq 0$ such that

$$f(t) = f(t + T) \text{ for all } -\infty < t < \infty.$$

The smallest such T is its *fundamental* period. For example, the signal

$$f(t) = A \sin(\omega t + \phi)$$

is periodic with fundamental period $2\pi/\omega$.

Let T be a positive integer. A discrete time signal $f[k]$ is periodic with period T if

$$f[k] = f[k + nT] \text{ for all integers } k.$$

The smallest such T is its fundamental period. For example, $f[k] = \cos[\frac{\pi}{2}k]$ is periodic with period 4. On the other hand, $f[k] = \cos(k)$ is not periodic, as a positive *integer* T does not exist to satisfy the definition.

If a signal is not periodic, then it is *aperiodic*.

periodic structure   structure consisting of successive identically similar sections, similarly oriented, the electrical properties of each section not being uniform throughout. Note that the periodicity is in space and not in time. The analysis of infinite periodic structure is significantly simplified by the Floquet's theorem. *See also* Floquet's theorem.

periodic waveform   phrase used to describe a waveform that repeats itself in a uniform, periodic manner. Mathematically, for the case of a continuous-time waveform, this characteristic is often expressed as $x(t) = x(t \pm kT)$, which implies that the waveform described by the function $x(t)$ takes on the same value for any increment of the kT, where k is any integer and the characteristic value $T > 0$, describes the fundamental period of $x(t)$. For the case of the discrete-time waveform, we write $x(n) = x(t \pm kN)$, which implies that the waveform $x(n)$ takes on the same value for any increment of sample number kN, where k is any integer and the characteristic value integer $N > 0$ describes the fundamental period of $x(n)$.

peripheral   an ancillary device to a computer that generally provides input/output capabilities.

peripheral adapter   a device used to connect a peripheral device to the main computer;

515

sometimes called an I/O card, I/O controller, or peripheral controller.

peripheral control unit   a device used to connect a peripheral device to the main computer; sometimes called an I/O card, I/O controller, or peripheral controller.

peripheral controller   *See* peripheral control unit.

peripheral device   a physical mechanism attached to a computer that can be used to store output from the computer, provide input to the computer, or do both. *See also* I/O device.

peripheral processor   a computer that controls I/O communications and data transfers to peripheral devices. It is capable of executing programs much like a main computer. *See also* I/O channel.

peripheral transfer   a data exchange between a peripheral device and the main computer.

peripheral unit   a physical mechanism attached to a computer that can be used to store output from the computer, provide input to the computer, or do both. *See also* I/O device.

permalloy   a family of ferromagnetic alloys consisting of iron, nickel, and molybdenum that saturate at moderate flux density levels and have a low coercive force.

permanent fault   a fault that remains in existence indefinitely if no corrective actions are taken.

permanent magnet (PM)   a magnet that produces an external magnetic field by virtue of the alignment of domains inside the material and retains its magnetism after being subjected to demagnetizing fields.

permanent magnet AC motor   a generic term used to describe both permanent magnet synchronous motors and brushless DC motors.

permanent magnet brushless DC machine   a machine that is similar in structure to a permanent magnet synchronous machine, containing armature windings on the stator and permanent magnets on the rotor. The permanent magnet brushless DC machine, however, is characterized by a trapezoidal flux density distribution in the airgap instead of the sinusoidal distribution of the synchronous machine. In operation, a DC voltage is applied sequentially to the stator coils to create a rotating field that pulls the rotor with it. To correctly operate, the brushless DC machine requires sensors to determine the rotor position so that the proper stator phases may be excited.

permanent magnet DC machine   a DC machine in which the field excitation in the stator is provided by permanent magnets instead of electromagnets.

permanent magnet DC motor   *See* permanent magnet DC machine.

permanent magnet machine   a machine that uses permanent magnets to establish the field. In DC machines, the permanent magnets are placed on the stator, while on AC synchronous machines they are placed on the rotor.

permanent magnet stepper motor   a stepper motor that has a permanent magnet assembly on the rotor. *See also* stepper motor.

permanent magnet synchronous machine   a polyphase AC motor with rotor mounted permanent magnets and sinusoidal distribution of stator phase windings. The field windings in the rotor are replaced by permanent magnets to provide the field excitation in these machines.

permanent split-capacitor (PSC) motor   a induction motor that operates from a single-phase supply. The motor contains two phase windings in quadrature; however, one of them has a capacitor in series with it to create a phase shift between the winding currents. Both windings and the capacitor operate continuously so the machine acts like a two-phase machine when running at its operating speed, producing less vibration and noise than a single-phase motor. Since the capacitor runs continuously, it is sized smaller than the capacitor used in a capacitor-start induction motor (CSIM). Thus, the PSC motor produces a lower starting torque than the CSIM.

permeability   tensor relationship between the magnetic field vector and the magnetic flux density vector in a medium with no hysteresis; flux density divided by the magnetic field in scalar media. Permeability indicates the ease with which a magnetic material can be magnetized. An electromagnet with a higher permeable core material will produce a stronger magnetic field than one with a lower permeable core material. Permeability is analogous to conductance, when describing electron flow through a material. *See also* reluctance.

permeameter   making use of Hall effect gaussmeters, search coils, and flux meters, the permeameter, or hysteresigraph, records the major hysteresis loop of a material, from which its basic material properties can be determined: residual induction, coercivity, energy product, saturation flux density, and recoil permeability.

permeance   the magnetic analog for conductance, indicating the ease with which magnetic flux will follow a certain path, which can be approximated by calculations based purely on magnetic circuit geometry.

permeance coefficient   the slope of the load line for a magnetic circuit, determined solely by physical geometry of the magnet and permeable materials around it; the ratio of magnetic induction (B) and applied field (H) at the operating point.

permission   *See* access right.

permittivity   *See* electric permittivity.

persistent current   a current circulating in a closed structure without applied potential. Exam-ples are the supercurrent in a superconducting magnet and the current in a closed mesoscopic ring in a magnetic field.

persistent spectral hole burning   spectral hole burning with a long lifetime, usually on the scale of seconds or longer.

personal computer (PC)   a desktop or laptop computer.

personal communications network (PCN)   a telecommunications network designed to provide services to a person rather than a geographic location. The network may comprise a range of different technologies from end to end and contains within it intelligence to enable the communication to be directed to the appropriate terminal or device carried with the person.

personal communications services (PCS)   a mobile telephone service with an essential urban and suburban coverage characterized by low cost pocket terminals, communications at a price comparable to a cable telephone, and distribution of the services and products to the general public. This definition is independent of the technology used.

PCS is the proposed next generation of wireless network services, providing voice communication services similar to today's cellular services, only with smaller cells, lower power, and cheaper rates.

personal digital cellular (PDC)   one type of digital cellular phone system. PDC is 800 MHz/1.5 GHz band, FDD, TDMA system, and it handles 1/4p shift QPSK modulated signal with 32 kbit/s. This cellular system was developed and operated in Japan.

personal handy phone system (PHS)   a digital microcell system designed in Japan that operates in the 1.9 GHz band. PHS provides cordless telephone or telepoint (q.v.) services similar to other digital cordless technologies, such as CT2 (q.v.). The spectrum allocation for PHS is 1895–1918 MHz. There are 77 radio channels, each 300 kHz wide and divided into 2 × 4 time slots. This provides a total of 308 duplex traffic channels, with duplexing via TDD. The carrier bit rate is 384 kb/s, and PHS terminals have a peak transmit power of 80 mW.

perspective distortion   a type of object distortion that results from projecting 3-D shapes onto 2-D image planes by convergence of rays towards a center of projection: perspective projection. This type of distortion is also called foreshortening.

perspective inversion   a property of perspective projection in which planar (typically silhouetted) objects which are not perpendicular to the axis of projection will appear, in the absence of additional information, to have either of two possible orientations in space.

perspective projection   projection of a 3-D object on a plane, termed "projection plane," as if it were imaged by an ideal pinhole camera located in a point termed "center of perspective."

perspective transformation   a matrix transformation which represents the perspective projection of objects in 3-D scenes into 2-D images, or the projection of one 2-D image into another 2-D image. Perspective transformations are conveniently carried out using 4 × 4 matrices representing homogeneous co-ordinate transformations.

perturbations of controllable systems   the set of controllable linear stationary continuous-time finite-dimensional dynamical systems is dense and open in the set of all linear stationary continuous-time finite-dimensional dynamical systems with the same state and input dimensions.

PET   *See* positron emission tomography.

Petersen coil   another term for a ground fault neutralizer.

petticoat   another name for shed, a feature of an insulator.

PFN   *See* pulse forming network.

PGA   *See* field-programmable gate array.

phantom   an artificial target, sometimes designed to mimic the size, shape, and attenuation characteristics of actual tissue, that is used to test and calibrate imaging hardware and software.

phase   (1) a notion used extensively in interpreting complex quantities such as Fourier series, Fourier transforms, etc. Given a complex number $c = x + iy = r \cos \phi + i r \sin \phi$, then r represents the magnitude of c and ϕ the phase.

(2) a horizontal translation parameter of the signal. Given a sinusoidal signal $s(t) = A \sin(2\pi f t + \phi)$, then f represents the frequency (in Hz) of s and ϕ the phase.

phase alternate line (PAL)   a color television system that inverts the (R-Y) color signal on alternate lines. The color burst is located on the back porch of the composite video signal and the phase alternates every line for identification of the (R-Y) difference signal phase. The color demodulator uses the shift in the color burst phase and the (R-Y) phase reversals to eliminate any hue changes caused by phase errors. The PAL color demodulation method eliminates the tint control but does cause desaturated colors.

The vertical scan parameters for the PAL television system are based on the European power line frequency of 50 Hz. The different PAL standards in existence in different countries allow for a slightly wider video bandwidth. The primary PAL system parameters (with the allowed bandwidth variance) are

Vertical Fields Frequency	50 Hz
Number of Interlaced Fields	2
Vertical Frame Frequency	25 Hz
Number of Horizontal Lines/ Frame	625
Horizontal Scan Frequency	15,625 Hz
Video Bandwidth	5 to 6 MHz
Color Subcarrier Frequency	4.433618 MHz
Sound Modulation	FM
RF Channel Bandwidth	7 to 8 MHz

[* Alternative definitions for the 3 television systems that you may enjoy]

[NTSC - Never Twice the Same Color, SE-CAM - System Essentially Contrary to the American Method, PAL - Perfect At Last]

phase angle   the angle in the complex plane of a complex value $x = a + jb$ calculated as $\angle x = \arctan \frac{b}{a}$.

phase angle meter   meter used to measure the phase angle difference between two AC quantities. In power systems, typical meters use perpendicular moving coils to measure the phase angle between an AC current and an AC voltage. More accurate devices typically measure the time interval between zero crossings of the two input signals.

phase breaking   the process by which the quantum mechanical phase, which is related to energy, of a particle is destroyed. The most common example is the inelastic scattering due to electron–electron and electron–phonon interactions.

phase comparator   often referred to as a phase detector; a three-port device that produces an analog output proportional to the phase difference between its two inputs. Since both inputs are periodic, the relative output voltage (or current) as a function of input phase difference (i.e., the transfer function) is also periodic; the shape of the transfer function (sometimes called the "output

characteristic" of the phase detector) depends upon the particular technique used to accomplish the phase detection. These include sinusoidal, triangular, and sawtooth shape factors. Analog/digital implementation, required linearity, and range of input phase difference are primary factors in determining a suitable output characteristic for a specific phase comparator application.

phase comparison relay   a phase comparison relay is a protective relay used on transmission lines which operates by comparing phase angles of signals generated at opposite ends of the line. They employ a dedicated communications channel to make the comparison. The signals compared are typically corresponding phase currents or sequence currents.

phase conjugate mirror   *See* phase conjugator.

phase conjugation   a technique for providing dispersion compensation that relies on spectral inversion of the optical signal using a nonlinear process known as four-wave-mixing.

phase conjugator   an optical set-up or system that can generate the time-reversed replica of the incident wave. Phase conjugators play an important role in many optical systems that require the transmission of optical waves through scattering media such as atmosphere, optical fibers. Photorefractive crystals such as $BaTiO_3$, SBN, BSO are by far the most efficient media for the generation of phase conjugate waves.

phase constant   a constant, which is generally complex, that is important in the study of electromagnetic waves. The phase constant is equal to the frequency of excitation of the wave times 2 times pi times the square root of the product of the permeability and the permittivity of the medium that the wave is traveling in. Also called propagation constant or wavenumber.

phase control   a method for controlling the amount of power delivered to a load by varying the delay angle. This controls the delay between the instant when the voltage across the power semiconductor goes positive and the actual start in conduction of the device.

phase delay   the difference in the absolute angles between a point on a wavefront at the device output and the corresponding point on the incident input wavefront, expressed in seconds or degrees. The delay can exceed 360 degrees.

phase detector gain   the ratio of the DC output voltage of the phase detector to the input phase difference. This is usually expressed in units of volts per radian.

phase deviation   the peak difference between the instantaneous angle of the modulated wave and the angle of the carrier. In the case of a sinusoidal modulating function, the value of the phase deviation, expressed in radians, is equal to the modulation index.

phase discriminator   a device for detecting the phase deviation of a phase modulated signal.

phase distortion   problem that occurs when the phase shift in the output signal of an amplifier is not proportional to the frequency.

phase error   the difference in phase between two sinusoidal waveforms having the same frequency.

phase grating   an optical grating characterized by a spatially periodic variation in the refractive index of a medium.

phase interrupting collision   a collision that interrupts the phase of the wavefunctions of lasing atoms and consequently broaden their emission spectra.

phase matching   the condition that the phase of the nonlinear polarization bears a spatially fixed relation to that of the optical field generated by the polarization. This condition is a requirement for high efficiency in nonlinear optical generation processes.

phase modulation   a type of angle modulation whereby information is encoded onto a carrier wave by modifying its phase angle as a function of time in proportion to the intelligence signal amplitude.

phase modulator   a device that alters the phase of a signal.

phase noise   frequency variation in a carrier signal that appears as energy at frequencies other than the carrier frequency.

phase only binary filter   transmission or reflection phase plate in which neighboring regions differ in phase shift by pi radians.

phase parameter   complex parameter representing corrections to the gain and phase of a Gaussian beam.

phase plane   a two-dimensional state space. *See also* state space.

phase plate   transparent medium that introduces different phase shifts to different transverse regions of an optical wave for the purpose of introducing or reducing phase or amplitude structure on the wave; often having only two phase shift values differing by π.

phase portrait   many different trajectories of a second-order dynamical system plotted in the phase plane.

phase response   the way in which a system alters the phase of an input sinusoid.

phase ripple   the variation in phase response across the operating bandwidth of an optical or electrical device.

phase sensitive measurement measurement in which the phase of an AC signal is tracked in a feedback loop in order to improve detection sensitivity. The most common examples are phase-locked loops in control systems and lock-in amplifier measurements in electronics. In the latter, a small AC signal is added to the bias voltage supply, and this signal is then detected in any measured quantity with an amplifier whose phase can be varied to "lock onto" that of the initial AC signal. This effectively mixes the two AC signals, and their difference (at DC) is used to characterize the measurement. The effective bandwidth is determined by the bandwidth at the DC level.

phase sequence describes the rotational orientation of the voltage phasors in a 3-phase electrical power system. A positive phase sequence, designated by the nomenclature ABC, indicates a 3-phase connection in which the B phase voltage lags the A phase voltage by 120 degrees, and the C phase voltage leads the A phase voltage by 120 degrees. A negative phase sequence, designated by ACB, reverses this relationship so that the B phase leads the A phase, and the C phase lags the A phase. *See also* phase sequence indicator.

phase sequence indicator device used to detect the phase sequence of a 3-phase electrical power system. *See also* phase sequence.

phase shift a time displacement of a waveform with respect to another waveform of the same frequency.

phase shifter a device that changes the phase angle between two buses in a power system. Conventional phase shifters are special autotransformers in with each phase voltage is connected in series with a variable component of voltage from another phase. By adjusting the variable component, the phase angle can be changed. Newer phase shifters are built with power electronic devices. Phase shifters are often used in antenna arrays.

phase spectrum the phase angle of the Fourier transform $\angle F(\omega)$, $-\infty < \omega < \infty$ of a signal $f(t)$. For example, the phase spectrum of a rectangular pulse of unit width is given in the figure. *See also* Fourier transform.

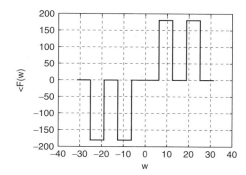

Phase spectrum.

phase velocity the velocity, at a given frequency, and for a given mode, of an equiphase surface in the direction of propagation. In other words, it is the velocity at which an observer travelling in the propagation direction should move in order to see the phase unchanged. It can be greater than the speed of light, since no transportation of energy is actually involved.

phase-coded modulation (PCM) a form of angle modulation. The modulated wave, $s(t)$, is given by: $s(t) = A_c \cos[2\pi f_c t + k_p m(t)]$, where m(t) is the message signal f_c is the carrier frequency, and k_p is the phase sensitivity of the modulator.

phase-matching conditions in waveguides formed by two or more different media, the phase-matching condition states the phase velocity along the propagation direction in the different media should be the same.

phase-controlled converter converter in which the power devices are turned off at the natural crossing of zero voltage in AC to DC conversion applications.

phase-controlled oscillator   a voltage controlled oscillator whose output frequency is determined by some type of phase difference detecting circuitry. *See also* phase error.

phase-locked loop (PLL)   (1) a circuit for synchronizing a variable local oscillator with the phase of a feedback portion of that oscillator so as to operate at a constant phase angle relative to the reference signal source. May be categorized as 1st, 2nd, 3rd, and so on, order loops according to the number of integrators in the so-called loop filter. The number of integrators is one less than the order of the loop. This loop is used for demodulators, frequency synthesizers, and timing recovery circuits.

(2) a closed-loop feedback control system where the feedback signal is a phase variation of the signal frequency rather than a voltage or a current.

phase-only filter   an optical mask imposes a pattern of phase variations over an image passing through it; used frequently in image correlators.

phase-sensitive detection   synchronous detection. *See also* frequency response.

phase-shift keying (PSK)   an information encoding method that uses changes in the phase of the signal carrier.

phase-shifting mask   a mask that contains a spatial variation not only in intensity transmittance but phase transmittance as well.

phase-to-phase fault   a fault with two transmission lines being short circuited.

phased array   *See* phased array antenna.

phased array antenna   an antenna composed of an aperture of individual radiating elements. Beam scanning is implemented by changing the phases of the signals at the antenna elements with the weights remaining fixed as the beam is steered.

phasor   a complex number representing the amplitude and phase of a sinusoidal function.

phoneme   the smallest units in phonemics. Phonemes are produced by different manners of articulation (e.g., plosives, fricatives, vowels, liquids, nasals). The automatic speech recognition on large lexicons is often based on the recognition of units like phonemes in order to break the complexity of the global recognition process, from speech frames to words and sentences.

phonemics   the study of sound units in the framework of descriptive linguistics. Basically, unlike phonetics, the sounds are studied by taking into account the language and not only observable features in the signal.

phones   the smallest units in phonetics, where the emphasis is placed on observable, measurable characteristics of the speech signal.

phonetic knowledge   knowledge of the acoustic structure of phones. It is of fundamental importance for the design of effective automatic speech recognition systems.

phonetics   the study of the acoustic sounds where the emphasis is placed on observable, measurable characteristics of the speech signal.

phonon   a quantized packet of energy associated with material lattice vibrations that have been excited by an incident photon.

phonon maser   *See* acoustic laser.

phosphorescence   emission of light from a long-lived electronically excited state. Phosphorescent emission is a quantum mechanically forbidden transition between electronic levels of different spin states.

phosphoric acid fuel cell (PAFC)   a fuel cell that uses hydrogen as fuel, and phosphoric acid as an electrolyte. It can generate 5 to 250kW of power, and hence can be used in automotive applications. *See also* akaline fuel cell, proton exchange membrane fuel cell, molten carbonate fuel cell, solid oxide fuel cell.

photochemistry   chemical reactions that occur as a result of electronically exciting at least one of the reactant molecules with light.

photodetector   device capable of producing or modifying an electrical signal in proportion to the amount of light falling on the active area of the device.

photoelastic effect   mechanical strain in a solid causes a change in the index of refraction that can affect the phase of a light wave traveling in the strain medium.

photogalvanic effect   *See* photovoltaic effect.

photolithography   *See* optical lithography.

photoluminescence   the process by which light is emitted from solids, atoms, gases, after excitation by an additional light source. The input light excites electrons to higher energy states, and as they relax they emit light (through electron-hole recombination) whose frequency is characteristic of the statistical properties of the carriers.

photomask   *See* mask.

photon   a minimum energy quantum of light energy proportional to the frequency of the radiation.

photon echo   (1) an optical field emitted by a macroscopic polarization that has been generated by reversing the dephasing process in a material having an inhomogeneously broadened spectrum.

(2) complex output pulses that are generated when two intense input pulses interact with the same semiclassically described laser medium.

photon lifetime   *See* cavity lifetime.

photon noise   fundamental noise due to the quantum nature of light; a statistical variation in optical intensity due to measurements of discrete number of photons.

photopic   formally, a description of luminances under which human cone cells are active. Informally, describing daylight luminances.

photorefractive beam fanning   a photorefractive phenomenon in which a beam of coherent light is scattered into a fanned pattern by a photorefractive crystal (e.g., barium titanate, lithium niobate). When a laser beam passes through a photorefractive crystal, significant scattering often occurs. The scattered light appears to be asymmetrical with respect to the beam except for propagation along the c-axis. For laser beams of moderate power, the scattered light appears to develop slowly in time and eventually reaches a steady-state scattering pattern. This is known as photorefractive beam fanning. The beam fanning originates from an initial scattering due to crystal imperfections. The initially scattered light is amplified due to the physical overlap and the energy coupling between the incident beam and the scattered beam. Beam fanning often occurs in highly efficient photorefractive media, even if the material is near-perfect. Photorefractive beam fanning plays an important role in the initiation of many phase conjugators and resonators, even through the fanning itself can be a source of noise in many experimental measurements.

photorefractive crystal   crystalline solids that exhibit photorefractive effect. The photorefractive effect is observed in many electro-optic crystals, including $BaTiO_3$, $KNbO_3$, $LiNbO_3$, $Sr_{1-x}Ba_x xNb_2O_6$ (SBN), $Ba_{2-x}Sr_xK_{1-y}Na_y$

Nb_5O_{15} (KNSBN), $Bi_{12}SiO_{20}$ (BSO), $Bi_{12}GeO_{20}$ (BGO), GaAs, InP, CdTe, etc. These crystals are referred to as photorefractive crystals. They are by far the most efficient medium for the generation of phase conjugation and real time holography using relatively low intensity levels (e.g., 1 W/cm^2). In addition to the efficient holographic response, beam coupling known as two-wave mixing also occurs naturally in photorefractive crystals.

photorefractive effect   in the broad sense, any optical process in which the refractive index of a material system is modified by an optical field. More precisely, a particular nonlinear optical process in which loosely bound charges are redistributed within a crystalline material in response to an incident light field, thereby producing a static electric field, and modifying the refractive index of the material by the linear electro-optic effect. Changes in refractive index upon exposure to a light pattern in certain materials such as BSO, lithium niobate, and barium titanate. A manifestation of the electro-optic effect on a microscopic scale due to spatial charge transport.

photorefractive grating   a refractive index grating often produced by using the illumination of two coherent laser beams, having sufficient photon energy, in photorefractive crystals.

photoresist   a photosensitive material that forms a three-dimensional relief image by exposure to light and allows the transfer of the image into the underlying substrate (for example, by resisting an etch).

photoresist contrast   a measure of the resolving power of a photoresist, the photoresist contrast is defined in one of two ways. The measured contrast is the slope of the standard H-D curve as the thickness of resist approaches zero. The theoretical contrast is the maximum slope of a plot of log-development rate versus log-exposure energy (the theoretical H-D curve). The photoresist contrast is usually given the symbol g.

photovoltaic effect   a photoelectric phenomenon in certain photorefractive crystals, for example, $LiNbO_3$, $BaTiO_3$, $LiTaO_3$, in which the illumination of light leads to the generation of electric current along certain direction in the crystals. This leads to the accumulation of charges on the surfaces of the crystals, causing an open circuit voltage.

photovoltaics   conversion of insolation into DC electricity by means of solid state p-n junction diodes.

PHPS   *See* personal handy phone system.

PHS   *See* personal handy phone system.

physical address   *See* real address.

physical medium   the communication channel over which signals are transmitted. Broadcast media, in which all stations receive each transmission, are primarily used in local-area networks. Common media are optical fibers, coaxial cable, twisted copper-wire pairs, and airwaves.

physical model   a mathematical model based on device physics. Physical models generally have a mid-range to high modeling valuation coefficient.

physical optics   a high-frequency technique to approximate electromagnetic scattering from an object by representing the smooth, electrically large parts of the object by equivalent currents at the object's surface.

physical page number   the page-frame address (in main memory) of a page from virtual memory.

physical placement of logic (PPL)   a design entry method between full custom and standard cell. It begins at the block diagram level where the detailed block specification and the corresponding layout are done simultaneously. PPL is targeted at the occasional designer.

physical sensor   an interface device at the input of an instrumentation system that quantitatively measures a physical quantity such as pressure or temperature.

physical theory of diffraction   a correction to physical optics that includes diffracted fields due to edges and corners.

physical vapor deposition   a process in which a conductive or insulating film is deposited on a wafer surface without the assistance of chemical reaction. Examples are vacuum evaporation and sputtering.

pickup current   the specified value that, if exceeded, causes the relay to act on its contact and cause a circuit breaker action. It is the threshold current for system protection, and a magnitude above this is considered a fault or abnormal condition.

pickup voltage   *See* pickup current.

picocell   cell with radius of a few meters. *See also* cell.

picture–carrier   the carrier frequency of 1.25 MHz above the lower frequency limit of a standard NTSC television signal (luminance carrier in color TV).

PID control   *See* proportional-integral-derivative control.

picture description language   a language in which parts of scenes are labeled and their relative positions are described in a special symbolic form. Typical relationships between objects are "inside," "adjacent to," "underneath," and so on. Such a language can be parsed and interpreted symbolically to build up a meaningful understanding of the picture.

piecewise deterministic process (PDP)   a class of Markov stochastic processes that follow deterministic dynamics between random jumps. A generator of piecewise deterministic process (PDP) is an integro-differential operator. Roughly speaking PDPs may be characterized by three functions:

(1) deterministic dynamics between jumps driven by a differential equation,
(2) jump intensity,
(3) jump magnitude probabilistic distribution.
PDPs can be used to describe some queueing, storage, and renewal processes. Moreover, they may also be applied in modeling control processes in the presence of failures, abrupt changes of working conditions, jumping deterioration of quality of working plants, as well as inventory and maintenance systems. The important class of PDPs is produced by linear systems with Markov jumps.

Pierce gun   an electron gun with which the focusing electrode is tilted 67.5° from the axis of the electron beam and the accelerator anode is also tilted forward to produce a rectilinear flow of an electron beam.

Pierce oscillator   usually, an FET crystal oscillator where the crystal is connected between gate and drain. The crystal is used like an inductor, the source of the FET is AC grounded, and the gate-source capacitance of the device and drain-source capacitance are used as parts of the LC-oscillator similar to Colpitts oscillator.

piezoelectric   pertaining to a material that possesses a noncentrosymetric crystal structure that will generate charge on the application of a mechanical stress. As in the case of a pyroelectric materials, this can be detected as either a potential difference or as a charge flowing in an external circuit.

piezoelectric transducer   device that converts electric signals to ultrasonic waves, and vice versa, by means of the piezoelectric effect in solids.

pig   slang for line hose *See* line hose.

pig tail a type of hot stick that can slide over an overhead conductor.

piggyback board *See* daughter board.

pigtail short electrical conductor used to connect the brushes of an electrical motor or generator to the external electrical connections of the machine.

pilot carrier a means of providing a carrier at the receiver, which matches the received carrier in phase and frequency. In this method, which is employed in suppressed carrier modulation systems, a carrier of very small amplitude is inserted into the modulated signal prior to transmission, extracted and amplified in the receiver, and then employed as a matching carrier for coherent detection.

pilot sequence a spreading code sequence used in a CDMA cellular network such as IS-95 to facilitate the PN code synchronization and coherent demodulation of the received signal. In the most typical case, the pilot sequence is utilized in the forward channel, i.e., transmission by the base station.

pin the electronic connection that allows connection between an integrated circuit or circuit board and some socket into which it is plugged.

PIN diode a diode with a large intrinsic region sandwiched between p- and n-doped semiconducting regions.

pin insulator an electric insulator which is concentric with a hollow, threaded hole so that it can be screwed onto a steel pin mounted on a utility pole or crossarm.

pinch effect the collapse of a hollow conductor due to the magnetic effects of very large currents. Sometimes observed in cable shields which have been struck by lightning.

pinch resistance the resistance of a fully depleted channel of a junction field effect transistor (JFET).

pinch-off region *See* ohmic region.

pinch-off voltage the gate-to-source voltage at which the channel current is reduced to a very small predetermined level specified in milliamperes per millimeter. The effect is caused by carrier depletion of the channel area.

pincushion distortion the geometric distortion present in an image in which both horizontal and vertical lines appear to collapse toward the display center. For a CRT image system, pincushion distortion is a result of the interaction of the long face plate radius of curvature and the short radius from the deflection center to the face plate. The interaction causes the top, bottom, left, and right raster center points to be closer to the center of the CRT face plate deflection than the top left, top right, bottom left, and bottom right raster corners. Consequently, a square grid has the appearance of a pin cushion.

pipe cable a paper-insulated high-voltage electric power transmission cable laid within a rigid steel pipe containing pressurized insulating oil.

pipe flush systems that fetch streams of data or instructions sometimes have to interrupt the stream when an unusual event occurs. When this happens, the pipeline containing the stream of instructions or data must be emptied before execution can continue; this is called flushing the pipe.

pipeline chaining a design approach used in computers whereby the output stream of one arithmetic pipeline is fed directly into another arithmetic pipeline. Used in vector computers to improve their performance.

pipeline interlock a hardware mechanism to prevent instructions from proceeding through a pipeline when a data dependency or other conflict exists.

pipeline processor   a processor that executes more than one instruction at a time, in pipelined fashion.

pipelined bus   *See* split transaction.

pipelined cache   a cache memory with a latency of several clock cycles that supports one new access every cycle. A new access can be started even before finishing a previous one. The access to the cache is divided into several stages whose operation can be overlapped. For instance, the cache can be pipelined to speed up write accesses: tags and data are stored in independently addressable modules so that the next tag comparison can be overlaped with the current write access. Read accesses are performed in a single cycle (tag and data read at the same time).

pipelining   a technique to increase throughput. A long task is divided into components, and each component is distributed to one processor. A new task can begin even though the former tasks have not been completed. In the pipelined operation, different components of different tasks are executed at the same time by different processors. Pipelining leads to an increase in the system latency, i.e., the time elapsed between the starting of a task and the completion of the task.

pitch   commonly used by physicists and musicians, defined with reference to the frequency. Given two signals with frequencies f_1 and f_2, the difference in pitch is defined by $1200 \log_2 \frac{f_2}{f_1}$. *See also* coil pitch.

pitch angle   an angle between a tangent to a helix and another tangent to a cylinder that contains the helix and is perpendicular to the cylinder axis at a common tangential point on the helix.

pitch factor   in an electric machine, the ratio of the fractional pitch in electrical degrees to the full pitch, also in electrical degrees.

pivoting   when applying Gaussian elimination to solve a set of simultaneous linear equations, the natural solution order is sometimes varied. The process of varying the natural solution order is termed pivoting. Pivoting is used to avoid fill-in and to maintain the accuracy of a solution.

pixel   contraction of "picture element:" each sample of a digital image, i.e., a square or rectangular area of size $\Delta x \times \Delta y$ of constant intensity, located at position $(k\Delta x, l\Delta y)$ of the image plane. Also called pel.

pixel adjacency   the property of pixels being next to each other. The adjacency of pixels is ambiguous and is defined several ways. Pixels with four-adjacency or four-connected pixels share an edge. Pixels with eight-adjacency or eight-connected pixels share an edge or a corner. *See* chain code, connectivity, pixel.

pixel density   a parameter that specifies how closely the pixel elements are spaced on a given display, usually a color display.

PLA   *See* programmable logic array.

placement   a placement routine determines an optimal position on the chip for a set of cells in a way that the total occupied area and the total estimated length of connections are minimized.

planar array   in addition to placing elements along a line to form a linear array, one can position them on a plane to form a planar array. For example, a circular array is a special form of planar array, where the elements are placed along a circle that is usually located on a horizontal plane. Planar arrays provide additional variables that can be used to control and shape the array's beam pattern. The main beam of the array can be steered towards any point in its half space.

planar doped barrier   a material growth technique that allows control of doping on a single atomic layer, rather than the more conventional doping over a thicker region. This technique is used for bandstructure control in heterojunction transistors and is also used in the fabrication of planar doped barrier detector diodes.

planar magnetron   a permanent magnet arrangement consisting of a steel base on which an outer ring of one polarity surrounds an inner ring or island of the opposite polarity, creating a magnetic flux "tunnel" for trapping electrons in the sputtering process.

planar mirror   an optically flat mirror, the flatness generally specified as a fractional number of optical wavelengths.

planar–optic interconnect   an interconnect in which optical signals travel along zigzag path in a transparent substrate with a source and a detector array, and other optical elements such as mirrors, gratings, and holograms. All optical elements on the substrate can be fabricated by lithographic techniques. The merits are compact packaging, easy alignments, large heat removal capacity, and suitability for integration with opto-electronic devices.

plane earth loss   the propagation loss experienced when isotropic transmit and receive antennas are operated in the vicinity of a flat reflecting plane. Frequently used as an approximation for propagation over Earth's surface.

plane wave   (1) an electromagnetic wave in which each wavefront (surface of constant phase) forms a plane of infinite extent and propagates in a direction perpendicular to the wavefront. A uniform plane wave has the same amplitude over an entire wavefront; a nonuniform plane wave has varying amplitude.

(2) a wave having as equiphase surface a plane. It is commonly represented by the functional dependence $\exp^{j(\omega t - \mathbf{k}\cdot\mathbf{r})}$ where ω is the angular frequency, t is the time, $\mathbf{r}$ is the position vector, and $\mathbf{k}$ is the wave vector. Sometimes monochromaticity is also implied.

plane-parallel resonator   a laser resonator in which the mirrors are flat and parallel to each other.

planned outage   an interruption in service to consumers which has been caused by a sequence of events which were pre-planned by the utility, for example to perform maintenance or construction.

plasma   the fourth state of matter comprised of positive ions and negative electrons of equal and sufficiently high density to nearly cancel out any applied electric field. Not to be confused with blood plasma.

plasma frequency   a critical frequency in a plasma (e.g., ionosphere) below which radio waves cannot propagate through the plasma. Usually, in the ionosphere for an $f > f_p$ the wave gets refracted towards Earth.

plasmon   a polariton in a plasma medium.

plastic   a term for a flexible roll-on polymer insulation layer.

plate   the positive electrode or anode in a vacuum tube.

platinotron   a strap feed starting radial M-type backward wave oscillator.

platter   *See* disk platter.

plausibility function   a function that gives a measure of plausibility or belief.

PLC   *See* programmable logic controller power line conditioner.

PLD   *See* programmable logic device.

PLL   *See* phase-locked loop.

plosives   sounds produced when the vocal tract, completely shut off at a certain point, is subsequently reopened so as a small explosion occurs. The "silence" that precedes the explosion can either be voiced or voiceless. In the first case we have voiced plosives ($/b/$, $/d/$, $/g/$), while in the second case we have voiceless plosives ($/p/$, $/t/$, $/k/$).

plossl   an eyepiece consisting of four lens elements, a pair of cemented doublets facing each other in a symmetrical configuration.

Plotkin's upper bound   for any (n, k) linear block code, the minimum distance is bounded asymptotically as, for $n \rightarrow \infty$

$$\left(1 - \frac{k}{n}\right) \geq 2\frac{d}{n}$$

plug fuse   a small fuse with a threaded base designed to be installed in a mated screw-type receptacle. Plug fuses are rated 125 V and are typically applied on branch circuits in ratings of 15 A, 20 A, and 30 A. Type S plug fuses are designed to prevent overfusing.

plug-in board   a printed-circuit board that can be plugged into another board or module to provide some functionality.

plugging   a procedure to bring a three-phase motor to an abrupt stop by reversing the direction of the rotating magnetic field in the airgap. The reversal is accomplished by reversing two of the phase connections to the motor.

Plumbicon tube   a type of camera tube first developed in the 1960s by Philips.

PLZT   *See* lead lanthanum zirconium titanate.

PM   *See* permanent magnet.

PMA   *See* post-metal anneal.

PMN   generic name for electrostrictive materials of the lead (Pb) magnesium niobate family.

PMOS   acronym for P-channel MOSFET.

PMU   *See* power monitoring unit.

PN   *See* pseudo-noise.

PN code   *See* pseudo-noise code.

PN code tracking   *See* pseudo-noise code tracking.

PN sequence   pseudo-noise sequence. *See* spreading sequence.

Pockels coefficients   denotes the strength of the linear effect in electro-optic materials. Named after the pioneer researcher F. Pockels.

Pockels effect   change in refractive index in noncentrosymmetric materials upon application of an electric field. Same as the linear electro-optic effect. Named after the pioneer researcher F. Pockels. *See also* electro-optic effect.

Pockels readout optical modulator (PROM)   an optically addressed spatial light modulator device using the electro-optic effect in BSO.

Poincaré sphere   a conceptual aid in which the state of polarization is represented by a point on a sphere. For example, points along the equator correspond to linear polarization (horizontal, vertical, or slant), the poles correspond to circular polarization (right-hand for upper pole and left-hand for lower pole), and all other points correspond to elliptical polarization.

529

point feature   a small feature which can be regarded as centered at a point, so that inter-feature distances can be measured, as part of a process of inspection, or prior to a process of inference of the presence of an object from its features. Commonly used point features include corners and small holes, or fiducial marks (e.g., on printed circuit boards). *See* salient feature, key point detection.

point operation   an image processing operation in which individual pixels are mapped to new values irrespective of the values of any neighboring pixels.

point process   a type of image processing in which the enhancement at any pixel depends only on the gray level at that pixel. This is in contrast to an operation whose value depends on the location of the pixel or on the values of neighboring pixels. *See* gray level, neighborhood operation.

point source   a light source so small that its size and shape cannot be determined from the characteristics of the light emanating from it. The light emitted has a spherical wave front and is spatially coherent.

point spread function (PSF)   multidimensional impulse response. Output of a multidimensional system when a point function (δ−function) is input.

point-to-point bus   a bus that connects and provides communication capability between two, and only two, components. It should be noted that buses support communications within the CPU, as well as external to the CPU. Most internal buses are point-to-point buses. In contrast, see also multipoint bus.

point-to-point motion   in the point-to-point motion, the manipulator has to move from an initial to a final joint configuration in a given time t_f.

pointer   in programming languages, a variable that contains the address of another variable.

poison   any material or process which absorbs neutrons and thus dampens a nuclear fission reaction, e.g., control rods.

Poisson counting process   *See* Poisson process.

Poisson distribution   a probability distribution widely used in system modeling. A non-negative integer-valued random variable X is Poisson-distributed if $-aa^k/k!$, where a is a positive parameter sometimes called the intensity of the distribution. For example, the number of 'purely' random events occurring over a time interval of t often follows a Poisson distribution with parameter a proportional to t.

Poisson process   a random point process denoting the occurrence of a sequence of events at discrete points in time. The time difference between the different events is a random variable. For a given time interval of length T, the number of events (points in the process) is a random variable with Poisson distribution given by the following probability law $P[N = k] = \frac{(\lambda T)^k e^{-\lambda T}}{k!}$, where N is the number of events that occur in the interval of length T, and λ is the expected number of occurrences of events per unit time. The time interval between any two events is a random variable with exponential distribution given by $F(\tau) = 1 - e^{-\lambda \tau}$, $\tau > 0$.

Poisson's equation   a partial differential equation that expresses the relation between the scalar potential V and the charge density at any point ρ. Mathematically described by $\nabla^2 V = \frac{\rho}{\epsilon}$, where ∇^2 is the Laplacian, ρ is the forcing function, V is the equation's solution, and ϵ is the dielectric constant of the medium.

polariton   term originally used to designate the polarization field particles analogous to photons. It is currently used to denote the coupling of the electromagnetic field with the polar excitations in solids.

polarity the notation used in the assignment of voltages. In DC generators, the polarity of the armature voltage can be reversed by either reversing its field current or by rotating the generator in reverse direction.

polarization (1) the shape traced out by the tip of the electric field vector as a function of time and the sense in which it is shaped.

(2) a description of the form of the temporal variation of the electric field vector of a light field. In general, the polarization state can be described by the ellipse that the tip of the electric field vector traces each optical cycle. Commonly encountered limiting forms are linear polarization and circular polarization.

(3) the response of material systems, an applied light field by developing a time varying dipole moment. The response is described quantitatively in terms of the dipole moment per unit volume, which is known as the polarization vector.

(4) description of the direction and motion of the electric field vector of a wave. Plane waves may be linearly or elliptically (including circularly) polarized.

polarization controller a device that alters only the polarization state of the incident light.

polarization distortion distortion in the temporal shape of an optical signal transmitted through an optical fiber caused by the differential time delay between the two polarizations of each propagating mode. Usually of significance only in a single mode fiber.

polarization diversity the use of at least two antennas with different polarization characteristics in a depolarizing medium so as to produce communication channels with substantially uncorrelated fading characteristics.

polarization division multiple access (PDMA) a multiple access technique where user channels are separated in the polarization domain of the transmitted signal (horizontal–vertical for linear polarization and left–right for circular polarization).

polarization ellipse the most general form of polarization specification in which the polarization of an electromagnetic wave is completely specified by the orientation and axial ratio of the polarization ellipse and by the sense of rotation (right-hand or left-hand).

polarization logic gate an optical logic gate using polarization modulating devices. Polarization modulating devices include MOSLM (magneto-optic spatial light modulator), LCTV (liquid crystal television), LCLV (liquid crystal light valve), MSLM (microchannel-plate spatial light modulator). These devices are very useful in performing XOR and XNOR operations. The basic operation is as follows. Two devices represent inputs A and B, respectively. When a beam of light passes through the first device, its polarization is rotated according to input A. The beam then passes through the second device resulting that its polarization is either rotated further away or rotated back according to input B. To perform other logic operations, an additional thresholding device is necessary whether it is based on optical addition or multiplication.

polarization loss also called polarization mismatch. It occurs when the polarization of the incident wave is different than the polarization of the receiving antenna.

polarization maintaining fiber a "single mode" fiber that supports two (linearly polarized) orthogonal modes.

polarization mismatch *See* polarization loss.

polarization vector that part of the flux density attributable to orientation of bound charges inside a dielectric by an applied electric field. It is also the electric dipole moment per unit volume inside the dielectric and is equal to the product of the electric susceptibility and the applied electric field intensity.

polarization-sensitive device   device that exhibits behavior dependent on the polarization of the incident electromagnetic wave. A polarizing filter exhibits transmission as a function of the polarization of the light incident on it.

polarized capacitor   an electrolytic capacitor in which the dielectric film is formed on only one metal electrode. The impedance to the flow of current is then greater in one direction than in the other. Reversed polarity can damage the part if excessive current flow occurs.

pole   (1) the values of a complex function which cause the value of the function to equal infinity (positive or negative). The poles are all natural frequencies of vibration, or resonances of the circuit described by the equation, but occur at infinite (finite if loss is present) attenuation. They are properties of the function itself, and are not influenced by any other elements in the system (immune to load pulling).

(2) one end of a magnet or electromagnet in electrical machines, created by the flux of the machine.

(3) the root s of the denominator $D(s) = 0$, for irreducible rational function $X(s) = N(s)D(s)$ at which points $X(s)$ become unbounded. Rational functions important to signal processing applications include Laplace transforms; the pole locations of Laplace transforms provide valuable information on system stability and other behaviors.

pole line   any power line which is carried overhead on utility poles.

pole of 2-D transfer matrix   a pair of complex numbers (p_1, p_2)

$$T(z_1, z_2) = \frac{N(z_1, z_2)}{d(z_1, z_2)}$$

$$N(z_1, z_2) \in R^{p \times m} [z_1, z_2]$$

that are the root of the 2-D polynomial $d(z_1, z_2)$, i.e., $d(p_1, p_2) = 0$, where $R^{p \times m} [z_1, z_2]$ is the set of $p \times m$ polynomial matrices in z_1 and z_2 with real coefficients.

pole pitch   the angular distance (normally in electrical degrees) between the axes of two poles in an electrical machine.

pole top pin   a steel pin onto which a pin insulator is screwed.

pole-coefficient sensitivity   when a coefficient d_k of the polynomial $D(s) = d_0 + d_1 s + d_2 s^2 + \ldots$ is a variable parameter then the roots of the polynomial are functions of this parameter. If these roots are simple, one can use the pole-coefficient sensitivity

$$\mathbf{S}_{d_k}(p_i) = \frac{\partial p_i}{\partial d_k / d_k} = -d_k \frac{p_i^k}{D'(p_i)}$$

The pole-coefficient sensitivity is frequently used in evaluation of active circuit stability.

pole-top transformer   generally a distribution transformer which is mounted atop a utility pole near the customer.

pole-zero plot   a graphical representation of a Laplace transform is known as a pole-zero plot. Except for a scale factor, the numerator and denominator polynomials in a rational Laplace transform $X(s) = N(s)D(s)$ can be specified by their roots. Thus, marking the location of the roots of $N(s)$ and $D(s)$ in the s-plane provides a powerful way of visualizing a Laplace transform. Pole-zero plots are used extensively in signal processing design and analysis, e.g., to determine the stability of a system function.

polled interrupt   a mechanism in which the CPU identifies an interrupting device by polling each device. *See also* vectored interrupt.

polling   sequencing through a group of peripheral devices and checking the status of each. This is typically done to determine which device(s) are ready to transfer data.

polychlorinated biphenyls chemical compounds added to insulating oils to improve stability and insulating capability.

polygon detection the detection of polygon shapes, often from corner signals or from straight edges present in an image. Polygon detection is important when locating machined parts in images, e.g., prior to robot assembly tasks. *See* rectangle detection and square detection.

polygonalization a method of representing a contour by a set of connected straight line segments; for closed curves, these segments form a polygon.

polymer membrane type of structure used to mechanically modulate light.

polynomial warp a type of commonly used image processing operation designed to modify an image geometrically. This type of operation is useful when an image is subject to some unknown physical spatial distortion, and then measured over a rectangular array. The objective is then to perform a spatial correction warp to produce a corrected image array.

polyphase filter bank a filter bank where the filters are implemented in terms of their polyphase representation. The decimators can be moved ahead of the polyphase filters in the analysis section, thereby reducing computation.

polyphase representation a sequence representation. A sequence $H(z)$ can be represented by the summation of M sequences $E_k(z)$ or $R_k(z)$ as $H(z) = \sum_{k=0}^{M-1} z^{-k} E_k(z^M)$ Type I polyphase, or $H(z) = \sum_{k=0}^{M-1} z^{-(M-1-k)} R_k(z^M)$ Type II polyphase.

polyphase system electrical system that has more than one phase, which are separated by angles of $360°/n$, where n is the number of phases. For example, three phase systems are polyphase systems where the three phases are separated by 120 electrical degrees from each other. A six-phase system is a polyphase system where each successive phase is separated by 60 electrical degrees from the other.

polysalicide a variation in the formation of polysilicide where the metal is deposited after the polysilicon and reacted with the silicon during a subsequent thermal annealing cycle.

polysilicon a polycrystalline or amorphous form of silicon deposited on the surface of a wafer during integrated circuit fabrication. In modern technologies, it most often forms the MOSFET gate, the bipolar emitter, the plates of a capacitor or a resistor.

pop instruction an instruction that retrieves contents from the top of the stack and places the contents in a specified register.

popcorning a plastic package crack or delamination that is caused by the phase change and expansion of internally condensed moisture in the package during reflow soldering, which results in stress the plastic package cannot withstand.

population dynamics a variety of models used to describe evolution, growth, kinetics, and dynamics of diverse populations. Population dynamics models may be stochastic or deterministic, discrete or continuous, differential or integral, and cover a number of mathematical tools from ordinary differential and difference equations, through partial differential, integro-differential, functional and integral equations to particle systems, cellular automata, neural networks and genetic algorithms. The type of the model depends on the type of population, the objective of modeling, available data, knowledge of phenomena, etc. The most often modeled and analyzed populations include human and animal populations for demographic and epidemiologic purposes, cell populations including cancer, blood, bone marrow, eukaryotic cells, virus, bacteria, fungi, genomes, and biomolecules. Control problems in population dynamics may be formulated in terms of optimal treatment protocols (for example cell-cycle-specific control), vaccination, harvesting strategies, modulation of growth and so on. The simplest models are found by clustering distributed in reality systems into lumped compartments. It leads to compartmental models of population dynamics. Linear models could be obtain under hypotheses of Malthusian (exponential) growth of the population. If, however, such a model is used to describe the evolution of the population under control (for example treatment by drugs, vaccination, harvesting), the model is no longer linear, and the simplest class of control models that could be used is given by bilinear control systems. Since real populations never grow unboundedly, more realistic models are given by nonlinear models with saturation effects. The simplest nonlinear differential models represent logistic, Pearl–Verlhurst, Michaelis–Menton, and Gompertz dynamics.

population inversion   usually the density of atoms or molecules in the higher state of a laser transition minus the density in the lower state.

population of energy level   number or density of atoms or molecules in an energy state.

population trapping   optical pumping into a noninteracting state or states of a quantum mechanical system.

porous silicon   noncrystalline silicon grown so as to produce porous, sponge-like structures that seem to enhance optical emission.

port   (1) a terminal pair.

(2) a place of connection between one electronic device and another.

(3) a point in a computer system where external devices can be connected.

port protection device   device in line with modem that intercepts computer communication attempts and requires further authentication.

portability   the possibility to move software to another hardware/operating system environment without changes.

portable system   a small, grid independent electric power unit ranging from a few watts to one kilowatt, which serves mainly a purpose of convenience rather than being primarily a result of environmental or energy-saving considerations.

position error   the final steady difference between a step function setpoint and the process output in a unity feedback control system. Thus it is the asymptotic error in position that arises in a closed loop system that is commanded to maintain a constant position. *See also* acceleration error and velocity error.

position error constant   a gain K_p from which the position error e_p is readily determined. It is a concept that is useful in the design of unity feedback control systems, since it transforms a constraint on the final error to a constraint on the gain of the open loop system. The relevant equations are

$$e_p = \frac{1}{1 + K_p} \quad \text{and} \quad K_p = s \overset{\lim}{\to} \infty \, q(s)$$

where $q(s)$ is the transfer function model of the open loop system, including the controller and the process in cascade. *See also* acceleration error constant and velocity error constant.

position sensor   a device used to detect the position of the rotor with respect to the stator. The most commonly used position sensors for electric motors are Hall effect devices, encoders, and resolvers with resolver-to-digital converters.

position servo   a servo where mechanical shaft position is the controlled parameter. *See also* servo.

positioner   a mechanical device used to move an antenna or target to a desired position for measurement purposes. Positioners can be single or multi-axis and are usually controlled by computers and automated measurement equipment.

positive 2-D Roesser model   the 2-D Roesser model

$$\begin{bmatrix} x^h_{i+1,j} \\ x^v_{i,j+1} \end{bmatrix} = \begin{bmatrix} A_1 & A_2 \\ A_3 & A_4 \end{bmatrix} \begin{bmatrix} x^h_{ij} \\ x^v_{ij} \end{bmatrix} + \begin{bmatrix} B_1 \\ B_2 \end{bmatrix} u_{ij}$$

$$y_{ij} = C \begin{bmatrix} x^h_{ij} \\ x^v_{ij} \end{bmatrix} + Du_{ij}$$

$i, j \in Z_+$ (the set of nonnegative integers) is called positive if for all boundary conditions $x^h_{0j} \in R^{n_1}_+, j \in Z_+$, and $x^v_{i0} \in R^{n_2}_+, i \in Z_+$ and all inputs sequences $u_{ij} \in R^m_+, i, j \in Z_+$ we have $x^h_{ij} \in R^{n_1}_+, x^v_{ij} \in R^{n_2}_+$ for $i, j \in Z_+$ and $y_{ij} \in R^p_+$ for $i, j \in R^p_+$ where R^n_+ is the set of n-dimensional vectors with nonnegative components and x^h_{ij} and x^v_{ij} are horizontal and vertical state vectors, respectively. The 2-D Roesser model is positive if and only if all entries of the matrices $A_1, A_2, A_3, A_4, B_1, B_2, C$, and D are nonnegative.

positive definite function   a scalar function $V(x)$ with continuous partial derivatives with respect to the components of the vector x is said to be positive definite if
(1) $V(0) = 0$.
(2) $V(x) > 0$ whenever $x \neq 0$.

positive definite matrix   a symmetric matrix A such that $x^T A x > 0$ for any vector x not identically zero. The eigenvalues of a positive definite matrix are all strictly greater than zero.

positive photoresist   a photoresist whose chemical structure allows for the areas that are exposed to light to develop at a faster rate than those areas not exposed to light.

positive real (PR) function   a rational function $H(s)$ of the complex variable $s = \sigma + j\omega$ is said to be positive real (PR) if it satisfies the following properties:
(1) $H(s)$ is a real number whenever s is a real number, and
(2) $Re[H(s)] \geq 0$ whenever $Re[s] > 0$, where $Re[\cdot]$ represents the real part of $[\cdot]$.

positive semidefinite   a scalar function $V(x, t)$ with continuous partial derivatives with respect to all of its arguments is said to be positive semidefinite if
(1) $V(0, t) = 0$.
(2) $V(x, t) \geq 0$ whenever $x \neq 0$.

positive semi-definite matrix   a symmetric matrix A such that $x^T A x >= 0$ for any vector x. The eigenvalues of a positive semi-definite matrix are all greater than or equal to zero.

positive sequence   the set of balanced normal (abc) sequence components used in symmetrical components. Balanced load currents, for example, are strictly positive sequence.

positive transition angle   the angular portion of the time-based output signal (in degrees) that has a positive slope. This quantity could be loosely interpreted as the "leading edge" angle.

positive-sequence reactance   the inductive reactance offered by a circuit to the flow of positive-sequence currents alone. The positive-sequence reactance is a function of the operating frequency of the circuit and the inductance of the circuit to positive-sequence currents.

positivity   a system $H : X \to X$ where

$$\langle x, Hx \rangle \to \geq 0 \qquad \forall x \in \mathcal{X}$$

See also inner product space, and passivity.

positron emission tomography (PET)   (1) a form of tomographic medical imaging based upon the density of positron-emitting radionuclides in an object.

(2) an imaging modality that uses injected positron-emitting isotopes as markers for physiological activity. The isotopes emit pairs of gamma photons which are detected using a gamma camera and coincidence detector.

possibility theory a theory evolved from the concepts of fuzzy sets and approximate reasoning which provides a mathematical framework for possibility analysis.

POST *See* power on self-test.

postbake *See* hard bake.

post-apply bake *See* prebake.

post insulator an electrical insulator which is supported by its firmly-bolted base, either in an upright position or cantilevered out horizontally from a utility tower.

post-compensator a compensator positioned on the process output signal. For MIMO systems, a given compensator will have a different effect depending on whether it is positioned before or after the process. Hence the importance of the prefix "post." *See also* compensation, compensator, and pre-compensator.

post-exposure bake (PEB) the process of heating the wafer immediately after exposure in order to stimulate diffusion of the products of exposure and reduce the effects of standing waves. For a chemically amplified resist, this bake also causes a catalyzed chemical reaction.

post-metal anneal (PMA) a process used in semiconductor processing after metallization to improve device properties and circuit performance. Typically it is performed at a modest temperature (300–450 C°) in a forming gas (e.g., 5% hydrogen with 95% nitrogen) ambient.

posterior statistics the empirical statistics of a random quantity (scalar, vector, process etc.), based on the a priori statistics supplemented with experimental or measured observations. *See* prior statistics, Bayes' rule.

postfix *See* postfix notation.

postfix notation a notational or programming scheme in which both operands of a two-operand operation are written before the operator is specified. Example: $ab+$ is the postfix representation of a sum; this could be implemented in a programming model based upon an evaluation stack by the operation sequence PUSH a, PUSH b, ADD. Postfix notation is used in programming zero-address computers.

postincrementation an assembly language addressing mode in which the address is incremented after accessing the memory value. Used to access elements of arrays in memory.

pot head a fork-shaped transition between a three-phase buried cable and an overhead threee-phase electric power line.

potassium dihydrogen phosphate (KDP) a strong linear electro-optic material, belonging to the same crystal class as ammonium dihydrogen phosphate (ADP). Its chemical form is KH_2PO_4.

potential an auxiliary scalar or vector field that mathematically simplifies the solution process associated with vector boundary value problems. *See also* Hertzian potential.

potential coil a long, finely wound, straight coil, similar in operation to a Chattuck coil, that is used with a fluxmeter to measure magnetic potential difference between a point in a magnetic field and a flux-free point in space.

potential source rectifier exciter a source of energy for the field winding of a synchronous machine obtained from a rectified stationary AC potential source. The AC potential can be obtained from the machine phase voltages, or from an auxiliary source. The components of the exciter are the potential source transformer and the rectifiers (including possible gate-circuitry).

potential transformer (PT) a device which measures the instantaneous voltage of an electric power line and transmits a signal proportional to this voltage to the system's instruments. *See* voltage transformer.

potentially unstable active circuit or device, where particular impedance termination placed on its input and/or output ports causes it to become unstable and oscillates.

potentiometric sensor a chemical sensor that measures the concentration of a substance by determining the electrical potential between a specially prepared surface and a solution containing substance being measured.

Potier reactance the leakage reactance obtained in a particular manner from a test on a synchronous machine at full load with a power factor of zero lagging. The test requires little power but supplies the excitation for short circuit and for normal rated voltage both at full-load current at zero power factor. The Potier reactance is determined by a graphical manner from the open circuit characteristic and the short circuit point for full-load current.

power (1) a measurable quantity that is the time rate of increase or decrease in energy. Units are in watts.

(2) ratio of energy transferred from one form to another (i.e., heat, radio waves, light, etc.) to the time required for the transfer, expressed in watts.

power added efficiency dimensionless ratio of RF power delivered from a device to the load minus the input incident RF power versus the total DC power dissipated in the device.

power amplifier amplifier operating at either audio or radio frequency range delivering power to a termination such as speaker, antenna, or resistive load.

power angle the angular displacement of the rotor from the stator rotating magnetic field while the machine is on load. The power angle is also the angle between the terminal voltage V_t of a synchronous machine and the generated voltage E_g or E_m, respectively, for a generator or motor. This angle denoted by δ is also referenced as power angle or torque angle or the load angle in a synchronous machine. It signifies the limits of the machine to remain in synchronism. *See also* torque angle.

power angle curve a curve shown the relationship between the active power output of a generator and its power angle.

power broadening increase in linewidth of a homogeneously broadened transition due to stimulated emission.

power conditioner a device designed to suppress some or all electrical disturbances including overvoltages, undervoltages, voltage spikes, harmonics, and electromagnetic interference (EMI). Example power conditioners are active filters for the reduction of harmonics, metal-oxide varistors (MOVs) and isolation transformers for the protection against voltage spikes, and EMI filters.

power control (1) the use of a mechanism to adjust a transmitter's power, usually in order to improve the overall performance of a communications network.

(2) in CDMA, a technique to increase the radio capacity. This is due to the fact that a CDMA system is interference-limited and that all users in a cell operate at the same frequency. The power control scheme used at the forward link of each cell can reduce the interference to the other adjacent cells. The less the interference generated in a cell, the higher the capacity.

(3) in a TDMA system, used to reduce co-channel interference or interference in adjacent cells. By having all cells (which operate at different frequencies with a cell reuse factor of 7 for a TDMA system) at approximately the same power, interference is reduced.

power density generally refers to the average power density, which is a measure of the power per unit area of a propagating EM wave. Mathematically, it is defined as the time average of the Poynting vector.

power disturbance a variation of the nominal value of the voltage or current.

power divider passive electronic circuit consisting of one input and two or more outputs. A signal applied to the input is divided into equiphase output signals, generally of equal amplitude.

power factor in an AC system, the ratio of the (active component) real power P to the apparent power S; it is given by the cosine of the angle subtended by S on the real, P axis. *See also* apparent power, real power, reactive power.

power factor correction the addition of reactive load to bring the combined power factor nearer unity. Since most industrial loads are inductive, capacitors are often employed as passive devices for power factor correction.

power fault arc an arc through soil extending from a power lines's lightning ground to a buried, grounded structure. These may form when lightning strikes an energized overhead electric power line.

power flow studies solutions of transmission line active and reactive power flow and bus voltages giving system load.

power flow study the circuit solution of an electric power system which yields the voltage of each bus and thus the power flows throughout the system.

power follow a fault condition, especially through a lightning arrester, in which power line current flows along a path through air or other insulation broken down by a high voltage impulse such as a lightning stroke to a conductor. *See* power fault arc.

power follow transformer a rugged, high-current power transformer used in tests of lightning arresters to test the arrester's power follow arc suppression capability.

power flux density a vector that gives both the magnitude and direction of an electromagnetic field's power flow. The units are watts per square meter.

power fuse a protective device that consists of a fusible element and an arc quenching medium. An overload or fault current in the fuse melts the fusible element, which creates an arc. The quenching medium then interrupts the current at a current zero, and prevents the arc from restriking.

power gain dimensionless ratio of RF power delivered to a load versus the input incident RF power, normally expressed in decibels. Thus a power gain of 100 is expressed as 20 dB.

power line conditioner (PLC) equipment installed on the customer's side of the meter to eliminate momentary over- and under- voltages to critical loads.

power monitoring unit (PMU) a device that is installed in at a consumer's site that will indicate to the utility whether or not an outage condition exists. The units commercially available at the present time typically have the facility to call power off and power on status to a central facility via a telephone link.

power noise factor for a photodetector, the ratio of signal-to-noise power ratio at the input (SNPR$_{\text{in}}$) to the signal-to-noise power ratio at its output (SNPR$_{\text{out}}$). If the power noise factor is denoted by k, then

$$k = \frac{\text{SNPR}_{\text{in}}}{\text{SNPR}_{\text{out}}}$$

power on self-test (POST) a series of diagnostic tests performed by a machine (such as the personal computer) when it powers on.

power optical power carried by an optical frequency beam.

power output useful output from a laser oscillator.

power quality (1) the concept of maintaining appropriate voltage and current waveforms and frequency in transmission, distribution, and generation systems, and usually taken to mean undistorted and balanced waveforms.

(2) a measure of an electric supply to meet the needs of a given electrical equipment application. As delivered by the utility, power quality is the faithfulness of the line voltage to maintain a sinusoidal waveform at rated voltage and frequency.

power signal suppose that $f(t)$ is a continuous time signal and let

$$P_T = \frac{1}{2\pi} \int_{-T}^{T} f(t)^2 dt$$

represent the power dissipated by $f(t)$ during the interval $[-T, T]$. The signal $f(t)$ is a power signal if $\lim_{T \to \infty} P_T$ exists and satisfies

$$0 < P \lim_{T \to \infty} P_T < \infty.$$

Furthermore, $\lim_{T \to \infty} P_T$ is the average power of the signal. Notice that an energy signal is not a power signal since necessarily $\lim_{T \to \infty} P_T = 0$.

Suppose that $f[k]$ is a discrete time signal and let

$$P_T = \frac{1}{2T} \sum_{k=-T}^{T} f(tk)^2.$$

The signal $f[k]$ is a power signal if $\lim_{T \to \infty} P_T$ exists and satisfies

$$0 < P \lim_{T \to \infty} P_T < \infty.$$

Furthermore, $\lim_{T \to \infty} P_T$ is the average power of the signal. All discrete time periodic signals are power signals. *See also* energy signal, periodic signal.

power spectral analysis computation of the energy in the frequency distribution of an electrical signal.

power spectral density (1) a function associated with a random process that specifies the distribution of the signal power over the different frequency components of the spectrum of the random process. The integral of the power spectral density function is the average power of the signal. The Wiener–Khinchine theorem states that the power spectral density $S_X(\omega)$ of a random process $X(t)$ is equal to the Fourier transform of the autocorrelation function $R_X(\tau)$ of the process; i.e., for the process $X(t)$ $S_X(\omega) = \int_{-\infty}^{\infty} R_X(\tau)e^{-j\omega\tau}d\tau$, where $j^2 = -1$.

(2) the power spectral density function of a signal is its power spectrum.

power spectrum for a wide sense stationary random process is the Fourier transform of the autocorrelation function. The power spectrum is a measure of the average power of a random process as a function of frequency. The expected value of the periodogram tends to the power spectrum as the periodogram window length tends to infinity.

power splitter   passive device that accepts an input signal and delivers multiple outputs with equal power.

power supply   an electronic module that converts power from some power source to a form which is needed by the equipment to which power is being supplied.

power supply unit (PSU)   *See* power supply.

power system damping   a torque action which works to reduce power system oscillations as time progresses.

power system oscillation   variations in the machine rotor angle caused by small disturbances in the system. The oscillations are usually at low frequency.

power system security   an operational index that determines the capability of a system to withstand equipment failures and external disturbances. It could be measured by the reserve capacity available or the number of operating constraint violations under a given prevailing condition or under a contingent probability of disturbances.

power system stabilizer   a control device that provides an additional input signal to the AVR to damp power system oscillations.

power transfer function   any function of input to output power in ratio form, expressed as a dimensionless ratio.

power transformer   a transformer that is used to transmit power from one voltage level to another. Power transformers can be of either single phase or three phase design, and include either two or three windings.

Poynting vector   a vector, proportional to the cross product between the electric and magnetic field intensity vectors, indicating the density and flow of electromagnetic power.

PPL   *See* physical placement of logic.

PR function   *See* positive real function.

practical intrafield predictor   technique used to improve the image quality. In practical intrafield predictors, several issues are considered. For example, in two-dimensional images, prediction may not lead to significant improvements in entropy of the prediction error; however, there is a decrease in the peak prediction error. By choosing proper coefficients, it is possible to get improved prediction and a decay of the effects of the transmission bit errors in the reconstructed picture.

practical stability   *See* uniform ultimate boundedness.

practical stabilization   deterministic approach to robust systems design based on the use of Lyapunov functions. Design feedback control law ensures desirable behavior of the systems for all admissible variations of bounded uncertainty in the following sense.

There exists a neighborhood of the origin $\mathbf{B}(d)$ such that

(1) For any initial condition there exists a solution of the state equation for the closed-loop system in the given time interval.

(2) All solutions for a given set of initial conditions can be extended over infinite time interval.

(3) All solutions are uniformly bounded for the given set of initial conditions (*See* uniform boundedness).

(4) All solutions for the given set of initial conditions are uniformly ultimately bounded with respect to a ball $\mathbf{B}(d_0)$ with $d_0 > d$ (*See* ultimate boundedness).

(5) All solutions from the ball $\mathbf{B}(d_0)$ are uniformly stable.

If the given set of initial conditions could be extended to the whole state space, then the closed-loop system is globally practically stable for the designed control law.

PRAM model   a multiprocessor model in which all the processors can access a shared memory for reading or writing with uniform cost.

Prandtl number   a nondimensional characteristic of fluids, relating the rate of momentum diffusion to heat diffusion.

praseodymium doped fiber amplifier   optical amplifier based on the rare-earth element praseodymium used to amplify signals in the 1310 nm telecommunications window. The region of useful gain is 1300–1350 nm.

prebake   the process of heating the wafer after application of the resist in order to drive off the solvents in the resist. Also called softbake and post-apply bake.

pre-compensator   a compensator positioned on the process input signal. For MIMO systems, a given compensator will have a different effect depending on whether it is positioned before or after the process. Hence the importance of the prefix "pre-." *See also* compensation, compensator, and post-compensator.

pre-emphasis   a technique of processing baseband signals through a network that provides a frequency-dependent amplitude transfer characteristic.

preamplifier   an amplifier connected to a low level signal source to present suitable input and output impedances and to provide gain so that the signal may be further processed without appreciable degradation in the signal-to-noise ratio. A preamplifier may include provisions for equalization and frequency discrimination.

precise interrupt   an implementation of the interrupt mechanism such that the processor can restart after the interrupt at exactly where it was interrupted. All instructions that have started prior to the interrupt should appear to have completed before the interrupt takes place and all instructions after the interrupt should not appear to start until after the interrupt routine has finished. *Compare with* imprecise interrupt.

predecrementation   an addressing mode using an index or address register in which the contents of the address are reduced "decremented" by the size of the operand before the access is attempted.

predicate   a logical expression that is to be evaluated, often to determine whether an important condition is satisfied. For example, "IF speed is high THEN reduce input supply."

prediction   (1) an estimation procedure in which a future value of the state (see the definition) is estimated based on the data available up to the present time.

(2) in branching, the act of guessing the likely outcome of a conditional branch decision. Prediction is an important technique for speeding execution in overlapped processor designs. Increasing the depth of the prediction (the number of branch predictions that can be unresolved at any time) increases both the complexity and speed.

predictive coding   compression of a signal by coding differences between samples and predictions from previously coded values. For example, in still image coding, a predictive encoder may predict a pixel by taking the average of the pixel's left neighbor and its above neighbor. With raster-order coding these values are already available in the decoder, which can form the same prediction. The difference or prediction error values may be quantized. Their probability density function is approximately Laplacian, and further compression can therefore be achieved with entropy coding. Also called differential pulse code modulation.

predictive control   control policy (scheme), realized at a given control layer, involving repetitive usage of a decision mechanism based upon considering, at each intervention instant, the future operation of the controlled process (or the control system as a whole) over specified period of time (prediction interval). Usually, predictive control involves the use of optimization-based decision tools and of the free input forecasting; predictive control is the term describing a variety of possible control schemes, in particular open-loop-feedback control and limited-look-ahead-control.

predictive pyramid a limited amount of data is used to form a prediction image, and then the difference between the predicted image and the original image is used to form a residual image. This can then further be iterated to form a pyramid of residuals called the Laplacian pyramid.

predictive scalar quantization *See* differential pulse-code modulation.

predictive SQ *See* differential pulse-code modulation.

predictive vector quantization the generalization of scalar predictive coding to vector coding. *See* differential pulse-code modulation.

predictive VQ *See* differential pulse-code modulation.

predistorter predistortion type linearizer used to compensate the distortion component generating at high power amplifier (HPA). Since the predistorter adds the distortion component to the signal at 180 degrees out of phase, the distortion components, generated in predistorter and HPA, are canceled out at the output of the HPA.

prefetch *See* fetch policy.

prefetch queue in the CPU, a queue of instructions that has been prefetched prior to being needed by the CPU.

prefetching in the CPU, the act of fetching instructions prior to being needed by the CPU. *See also* fetch policy.

preformat information such as sector address, synchronization marks, servo marks, etc., embossed permanently on the optical disk substrate.

preincrementation an assembly language addressing mode in which the address is incremented prior to accessing the memory value. Used to access elements of arrays in memory.

preliminary breakdown an electrical discharge in the cloud that initiates a cloud-to-ground flash.

preprocessing a series of image enhancements and transformations performed to ease the subsequent image analysis process through, e.g., noise removal or feature extraction/enhancement.

pressure broadening spectral broadening of a transition in a laser medium due to elastic or inelastic collisions.

pressure vessel a steel tank which encloses the core of a nuclear reactor and is generally filled with coolant under pressure.

pressurized water reactor (PWR) a nuclear reactor which uses liquid water under pressure for a primary coolant and moderator.

preventive congestion control protocols that control the system and user traffic parameters so that the probability of a congestion is minimized. *See also* congestion.

price method coordination coordination by the price method amounts to iterating the coordination variables (dual coordination instruments) defined as the vector of prices by which the local interaction inputs to the subprocesses as well as their outputs are multiplied and added to the local performance criteria; these local criteria are then minimized with respect to the local decisions and the results are passed to the coordinator. The coordinator iterates the values of prices until the interaction equations and, if needed, any other coupling constraints are satisfied.

primal sketch   a hierarchical image representation that makes explicit the amount and disposition of the intensity changes in the image. At the lowest levels of the representation, primitives just describe raw intensity changes and their local geometrical structure; at the higher levels they represent groupings and alignments of lower-level ones.

primary   (1) the source-side winding.
(2) refers to the portion of a nuclear power plant containing the reactor and its equipment.

primary coolant   the medium used to remove energy, in the form of heat, from a nuclear reactor core, e.g., water, helium, or liquid metal.

primary feeder   *See* feeder

primary system of equations   a system of algebraic and differential equations obtained by applying the Kirchhoff's current and voltage laws and the element I-V relations.

primary voltage   in power distribution the voltage at the primary winding of the distribution transformer.

primary winding   the transformer winding connected to the energy source.

prime mover   the system that provides the mechanical power input for a mechanical-to-electrical energy conversion system (generator), e.g., the diesel engine of an engine–generator set.

prime-number interleaving   *See* interleaved memory.

primitive polynomial   a polynomial $p(x)$ of degree m that gives a complete table of 2^m distinct symbols containing 0 and 1. The reciprocal of the primitive polynomial is also primitive. *See also* irreducible polynomial.

primitive-based coding   any scheme to detect edges, lines, and other local features of images, then use them to code the image. For example, edges may be used to segment the image into regions which are then independently coded as simple surfaces, while the boundaries are compressed with a chain code.

Princeton architecture   a computer architecture in which the same memory holds both data and instructions. This is contrasted with the Harvard architecture, in which the program and data are held in separate memories.

principal axis   the optical axis of a lens or camera, usually normal to the image plane.

principal component   notionally, the direction of greatest variability of a random vector or among a set of sample vectors. More specifically, the principal component is the direction of the eigenvector associated with the largest eigenvalue of the covariance matrix of the random vector (or the sample covariance of a sample set). More generally, the n principal components of a distribution are the eigenvectors corresponding to the n largest eigenvalues. Principal components are frequently used for data clustering, pattern analysis, and compression.

principal component analysis (PCA)   a technique applied to n-dimensional data vectors with the aim of finding a set of m-orthogonal vectors (typically $m \ll n$) that account as much as possible for the data's variance. PCA can be carried out in an unsupervised fashion by implementing a normalized version of Hebb's rule on m-linear neural units.

principal point   the point at which the optical axis of the lens in a camera meets the image plane: also, the corresponding point in the image.

principle of locality   *See* locality. *See also* sequential locality.

principle of superposition   in a linear electrical network, the voltage or current in any element resulting from several sources acting together is the sum of the voltages or currents from each source acting alone.

printed circuit board (PCB)   a substrate made from insulating material that has one or more sandwiched metallic conductor layers applied that are etched to form interconnecting traces useful for interconnecting components.

printer   an output device for printing results on paper.

prior statistics   the statistics of a random quantity (scalar, vector, process etc.) before any experimental or measured knowledge of the quantity is incorporated. *See* posterior statistics.

prioritization coding   a coding scheme whereby the position of the symbol in the data steam indicates its weight.

priority encoder   an encoder with the additional property that if several inputs are asserted simultaneously, the output number indicates the numerically highest input that is asserted.

prism | air | metal (PAM) system   the two-interface model of an ATR (attenuated total reflection) system comprised of prism | air | metal. Commonly known as PAM system.

prismatic joint   a joint characterized by a translation that is the relative displacement between two successive links. This translation is sometimes called the joint offset.

private key cryptography   also known as secret key cryptography. In such a cryptographic system, the secret encryption key is only known to the transmitter and the receiver for whom the message is intended. The secret key is used both for the encryption of the plaintext and for the decryption of the ciphertext. *See also* public key cryptography.

privileged instruction   an instruction that can be executed only when the CPU is in privileged mode.

privileged mode   a mode of execution of machine instructions in the CPU in which certain special instructions can be executed or data accessed that would otherwise be prohibited. *See also* user mode.

PRMA   *See* packet reservation multiple access.

probabilistic metric space   a generalization of the notion of metric spaces onto the uncertain systems by replacing a metric on a given set S by a distance distribution function F, and a triangle inequality by a generalized inequality defined by triangle function τ. A distance distribution functions between two elements $p, q \in S$ is defined as a real function F_{pq} whose value $F_{pq}(x)$ for any real number x is interpreted as the probability, the membership function, or the grade of membership (depending on the type of the uncertainty model) that the distance between p and q is less than x. The simplest distance distribution function is given by the unit step (Heaviside) function **1** as follows:

$$F_{pq}(x) = \mathbf{1}(x - d(p, q))$$

where d is a standard metric. Then a probabilistic metric space reduces to the standard metric space. More precisely, a probabilistic metric space (PMS) is defined as a triple (S, F, τ) endowed with the following properties:

$$F_{pq}(x) = \mathbf{1}(x) \iff p = q$$
$$F_{pq} = F_{qp}$$
$$F_{pr} \geq \tau\left(F_{pq}, F_{qr}\right)$$

for all $p, q, r \in S$. The types of the distribution function and the triangle function are related to the model of uncertainty and the way of composition of the standard operations in the model.

probabilistic neural network   a term applied loosely to networks that exhibit some form of probabilistic behavior but also applied specifically to a type of network developed for pattern classification based upon statistical techniques for the estimation of probability densities.

probability density function (PDF)   (1) a function describing the relative probability of outcomes of an experiment. For experiments with discrete outcomes, the PDF is analogous to a relative frequency histogram. For experiments with continuous outcomes, the PDF is analogous to a relative frequency histogram where the category bin widths are reduced to zero. The total area underneath a PDF must always be unity.

(2) the derivative of the cumulative distribution function (when the derivative exists). More formally, for a random x and any probabilistic event A, the probability density function $p_x(x)$ satisfies

$$\Pr(x \in A) = \int_A \mathrm{d}p_x(x).$$

See also cumulative distribution function.

procedure   a self-contained code sequence designed to be re-executed from different places in a main program or another procedure. *See also* call instruction, return instruction.

procedure call   in program execution, the execution of a machine-language routine, after which execution of the program continues at the location following the location of the procedure call.

process   the context, consisting of allocated memory, open files, network connections, etc., in which an operating system places a running program.

process control block (PCB)   an area of memory containing information about the context of an executing program. Although the PCB is primarily a software mechanism used by the operating system for the control of system resources, some computers use a fixed set of process control blocks as a mechanism to hold the context of an interrupted process.

process environment   part of the control scene that is outside of the controlled process; within this environment are formed the uncontrolled, free inputs to the process. Specified quantities related to the environment may be observed and used by the controller, for example, when performing free input forecasting.

process interaction   a stream of energy, material, or information exchanged between the sub-processes of a large-scale process. Relevant attributes of those streams are, respectively, interaction inputs or interaction outputs. Interactions are described by the interaction equations, which relate interaction inputs to a given subprocess to interaction outputs from other subprocesses.

process monitor   in semiconductor manufacturing, wafers processed at a particular process step to permit the quality or some other attribute of the step to be monitored. For example, in a photolithographic etching process, a process monitor would be a wafer exposed and etched that permits the physical dimensions of the resulting etched layer to be monitored and measured.

process state   the set of information required to resume execution of a process without interfering with the results of the computation.

process swap   the act of changing the execution point from one process to another.

process window   a window made by plotting contours corresponding to various specification limits as a function of exposure and focus. One simple process window, called the critical dimension (CD) process window, is a contour plot of the high and low CD specifications as a function of focus and exposure. Other typical process windows include sidewall angle and resist loss. Often, several process windows are plotted together to determine the overlap of the windows.

process-oriented analysis   a method of analyzing application transformation processing as the defining characteristic of applications.

processing amplifier   high-performance amplifier that regenerates as well as amplifies the signal being processed.

processing element (PE)   a processing module, comprising at least a control section, registers, and arithmetic logic, in a multiprocessor system. A processing element may be capable of operating as a stand-alone processor.

processing ensemble   a collection of processors under control of a single control unit.

processing gain   See spreading gain.

processor element   See processing element.

processor farm   a collection or ensemble of processing elements to which parallel processing tasks are assigned and distributed for concurrent execution. In this model, tasks are distributed, "farmed out," by one "farmer" processor to several "worker" processors, and results are sent back to the farmer. This arrangement is suitable for applications that can be partitioned into many separate, independent tasks. The tasks are large and the communications overhead is small.

processor status word   a register in the CPU that stores a collection of bits that, taken as a group, indicate the status of the machine at a given period of time.

product code   two-dimensional burst and random error correcting code in the form of a matrix with each row and column are code words of two different linear codes.

product of sums (PS)   the AND combination of terms, which are OR combinations of Boolean variables.

program counter (PC)   a CPU register that contains the address in memory of the next instruction to be fetched and executed.

program status word (PSW)   a combination program counter and status-flag register provided in IBM mainframe computers.

programmable array logic (PAL)   a programmable logic array with no OR array, but with a fixed set of OR gates into which are fed sets of product terms.

programmable gate array (PGA)   See field-programmable gate array.

programmable logic array (PLA)   a programmable logic device that consists of an AND array forming logical products of the input literals and an OR array that sums these products to form a set of output functions.

programmable logic controller (PLC)   a microprocessor based system comprised of a set of modules for acquiring signals from environment and other for producing effects on the environment. These effects are typically used for controlling electromechanical machines by means of actuators (motors, heating element, etc.). The rules for specifying the control law are Boolean expressions. These sequentially and cyclically are executed. More complex PLCs allow the description of the control law by means of ladder diagram.

programmable logic device (PLD) (1) an integrated circuit able to implement combinational and/or sequential digital functions defined by the designer and programmed into this circuit.

(2) A one-time customizable logic device. *See* PAL, PLA. *See also* FPGA.

programmable radio system radios based on digital waveform synthesis and digital signal processing to allow simultaneous multiband, multiwaveform performance.

programmable read-only memory (PROM) a semiconductor memory device that has a primary function of storing data in a nonvolatile fashion that can be programmed to contain predetermined data by means other than photomasking. PROMs may be one-time programmable (OTP) or may be either UV or electrically eraseable, depending on the particular semiconductor process technology used for manufacturing.

programmed I/O transferring data to or from a peripheral device by running a program that executes individual computer instruction or commands to control the transfer. An alternative is to transfer data using DMA.

progressive coding ordering of coded values such that the original signal can be recovered progressively. For example, in transform coding of a picture, transmission of the zero sequency coefficients for all blocks first, rather than transmission of all coefficients for the first block first. This allows the receiver to generate an approximate reconstruction early in the reception of data.

progressive scan method of scanning a display on a horizontal line-by-line basis in strict sequence during one vertical sweep of the scanning beam/element. Each scan provides one complete video frame, and interlaced fields do not occur in either the production or display of the picture element sequence. Also known as sequential scan.

progressive transmission partial information of an image is transmitted. At each stage, an approximation of the original image is reconstructed at the receiver. The quality of the reconstructed image improves as more information is received.

projection (1) in signal and image processing, the conversion of an n-dimensional signal into an $n - 1$ dimensional version through some integration in the continuous case, or some summation in the discrete case. For instance, a 2-D image can be viewed as a (perspective) projection of 3-D scene via a camera. Another example exists in computed tomography. There a projection is a line integral along a straight ray.

(2) a set of parallel line integrals across the image oriented at a particular angle.

projection of a fuzzy set for fuzzy set Q in a Cartesian product X^n, $X^n = X_1 \times \cdots \times X_n$, the, fuzzy set subspace R in X^i of X^n, $(i < n)$. The projection is usually denoted $R = \text{proj}(Q; X^i)$, in X^m with membership function obtained as the supremum of the membership function for the dimensions to be eliminated. *See also* cylindrical extension of a fuzzy set, fuzzy set, membership function.

projection printing a lithographic method whereby the image of a mask is projected onto a resist-coated wafer.

projection receiver multiple-access receiver technique in which users are estimated by first projecting the received signal onto the null space of the unwanted or interfering users.

projection television system a TV system in which a small but very bright image is optically projected onto a large viewing screen.

projective invariant   a measure that is independent of the distance and direction from which a particular class of object is viewed, under perspective projection. The cross-ratio is an important type of projective invariant, which is constant for four collinear points viewed under perspective projection from anywhere in space. Projective invariants are important in helping with egomotion (e.g., automatic guidance of a vehicle) and for initiating the process of object recognition in 3-D scenes.

PROM   *See* programmable read-only memory, Pockels readout optical modulator.

prompt critical   a nuclear reaction which can maintain criticality without the contribution of delayed neutrons, i.e., the sort of reaction which takes place in nuclear weapons.

propagating mode   pattern of an electromagnetic field with most of the energy contained in the core of the fiber that is sustained for long distances in an optical fiber (a few kilometers to 10's of kilometers) with only the amplitude of the electromagnetic field gradually decreasing while the shape of the mode remains constant.

propagating wave   a wave that exhibits a spatial shift as time advances. Mathematically, a propagating wave is described by $f(t - r/v_p)$, where f is some function, t is the temporal variable, r is the spatial variable, and v_p is the phase velocity of the wave.

propagation   the motion of electromagnetic waves through a medium or free space.

propagation constant   a complex constant whose real part is the wave attenuation constant (nepers attenuation per unit length) and whose imaginary part is the wave phase constant (radian phase shift per unit length). *See also* phase constant.

propagation delay   (1) the delay between transmission and reception of a signal. Caused by the finite velocity of electromagnetic propagation.

(2) the delay time between the application of an input signal to a chain of circuit elements and the appearance of the resulting output signal.

(3) the time it takes for a transistor switch to respond to an input signal, symbolized t_{pd}. It is calculated between the 50% rise point to the 50% fall point or vice-versa (see graph of typical inverter gate).

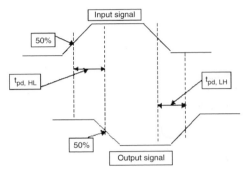

Propagation delay for a typical inverter gate.

The time from when the input logic level to a device is changed until the resultant output change is produced by that device.

propagation delay time   *See* propagation delay.

propagation factor   the ratio of the electric field intensity in a medium to its value if the propagation took place in free space.

propagation mode   (1) in radio communication, refers to the manner in which radio signals propagate from the transmitting antenna to the receiving antenna. Radio waves can be reflected from the ionized layers within the ionosphere. Scatter modes of propagation also exist, and in these modes, propagation from the transmitting to receiving antennas is largely the result of scattering of the radio waves by dust and other particles in the troposphere. The ionized trails of meteors can also act as scatterers.

(2) in waveguides, used at microwave frequencies, propagation mode refers to the arrangement of electrical and magnetic fields within the waveguide. Modes can be classified as being transverse electric (TE) or transverse magnetic (TM), depending on whether the electric field or magnetic field is transverse to the direction of wave propagation within the waveguide. The mode that exists for the lowest frequency that can be transmitted by the waveguide is known as the dominant mode of the waveguide.

propagation path the route along which a radio wave propagates from the transmitting antenna to the receiving antenna. If this is a straight line, the path is a line of sight path. Otherwise, reflection, diffraction, and other phenomena may change the direction of propagation so that the path can be envisioned as made up of several straight-line segments. *See also* line of sight, multipath propagation.

proper 2-D transfer matrix the 2-D transfer matrix

$$T(z_1, z_2) = \frac{N(z_1, z_2)}{d(z_1, z_2)}$$

$$N(z_1, z_2) \in R^{p \times m}[z_1, z_2]$$

$$d_{z_1, z_2} = \sum_{i=0}^{n_1} \sum_{j=0}^{n_2} d_{ij} z_1^i z_2^j$$

is called proper if $d_{n_1, n_2} \neq 0$ ($d(z_1, z_2)$ is acceptable) and

$$\deg_{z_1} n_{ij}(z_1, z_2) \leq n_1$$
$$\deg_{z_2} n_{ij}(z_1, z_2) \leq n_2$$
$$\text{for } i = 1, \ldots, p; \quad j = 1, \ldots, m;$$

where $n_{ij}(z_1, z_2)$ are the entries of $N(z_1, z_2)$ and $\deg_{z_1}$ ($\deg_{z_2}$) denotes the degree with respect to z_1 (z_2).

proper mode a mode as obtained from a boundary value problem defined over a finite interval and relative to a second-order differential equation with real coefficients. *See also* improper mode.

proper subgraph a subgraph that does not contain all of the edges of the given graph.

proportional control control scheme whereby the actuator drive signal is proportional to the difference between the input/desired output and measured actual output.

proportional-integral-derivative (PID) control a control scheme whereby the signal that drives the actuator equals the weighted sum of
(1) the difference,
(2) time integral of the difference, and
(3) time derivative of the difference between the input and the measured actual output.

It is a widely used control scheme in industry that can be tuned to give satisfactory performance based on knowledge of dominant system time constant.

protection control access to information in a computer's memory, consistent with a particular policy or mechanism. Ring numbering was introduced in the Multics system as one basis for limiting access and protecting information. The term "security" is used when the constraints and policies are very restrictive.

protection fault an error condition detected by the address mapper when the type of request is not permitted by the object's access code.

protective relay a device that monitors the condition of the electric power system and determines the presence of faults or other system anomalies. The protective relay monitors current flow, voltage level, or other parameter. When it operates due to a fault or other event, it initiates a trip signal intended to open the appropriate circuit breaker(s) or other protective devices.

protective relaying   a unit for discriminating normal operating condition from faulted conditions in power systems.

protocol   *See* bus protocol.

protocol data unit (PDU)   the unit of exchange of protocol information between entities. Typically, a PDU is analogous to a structure in C or a record in Pascal; the protocol in executed by processing a sequence of PDUs.

protocol spoofing   a technique used by VSAT networks to reduce the network delay. The satellite network emulates the host computer front-end processor at the VSAT locations and emulates the multiple cluster controllers at the hub location.

proton exchange membrane fuel cell (PEMFC) a fuel cell that uses a plastic membrane as its electrolyte. Also called solid polymer fuel cell (SPFC). *See* alkaline fuel cell, phosphoric acid fuel cell, molten carbonate fuel cell, solid oxide fuel cell.

prototype filter   a low-pass filter which is modulated to form a modulated filter bank.

prototyping   building an engineering model of all or part of a system to prove that the concept works.

proximity effect   a variation in the size of a printed feature as a function of the sizes and positions of nearby features.

proximity printing   a lithographic method whereby a photomask is placed in close proximity (but not in contact) with a photoresist coated wafer and the pattern is transferred by exposing light through the photomask into the photoresist.

pruning self-generating neural network   a methodology for reducing the size of self-generating neural network (SGNN).

During SGNN training, the network may grow very quickly, and some parts of the network may become useful in neither training nor classification, and the weights of these parts of the neurons of the network never change after some training stage. We call these parts of the network dead subnets. It is obvious that the dead subnets of the network should be pruned away to reduce the network size and improve the network performance. One way to achieve this is to check the weights of each neuron in the network to see whether they have been changed since last training epoch (or during the last few epochs). If they are unchanged, the neuron may be evidently dead and should be removed from the network. If a neuron is removed from an SGNN, all of its offspring should also be removed. *See also* self-generating neural network, training algorithm of self-generating neural network.

PS   *See* product of sums.

PSC   *See* permanent split-capacitor motor.

PSC motor   *See* permanent split-capacitor motor.

pseudo code   a technique for specifying the logic of a program in an English-like language. Pseudo code does not have to follow any syntax rules and can be read by anyone who understands programming logic.

pseudo color   the display in color of gray-level pixels in order to make certain gray-level pixels or gray level patterns more visible. Typical pseudocoloring schemes are:

(1) displaying all pixels in a range of gray levels in one color;

(2) displaying each gray level at a fixed interval in a different color to produce colored contour lines;

(3) displaying the entire gray-level range as a rainbow. Pseudocoloring is usually carried out by look-up tables.

See false color, look-up table (LUT).

pseudo-exhaustive testing a testing technique that relies on various forms of circuit segmentation and application of exhaustive test patterns to these segments.

pseudo-noise (PN) describes a class of pseudo-randomly generated processes that exhibit the statistical properties of noise.

pseudo-noise code phase the starting symbol of a pseudo-noise (PN) code with respect to some time reference in the system. In the IS-95 CDMA system, the same PN code with different PN code phases (typically multiples of 64 code symbols, or chips) is assigned as the pilot sequence for different base stations. *See also* spreading code.

pseudo-noise code tracking the act of accurately generating a local pseudo-noise (PN) code signal whose phase (timing relative to some reference time) accurately tracks the phase of an incoming PN code signal.

pseudo-noise sequence *See* spreading sequence.

pseudo-operation in assembly language, an operation code that is an instruction to the assembler rather than a machine-language instruction.

pseudo-random describes a process that has the statistical properties of being random, but is deterministic and can be recreated given the requisite properties of its generator.

pseudo-QMF a cosine-modulated filter bank with the prototype filter being even-length symmetric spectral factor of an M-th band filter. The aliasing of the adjacent bands is cancelled while the aliasing of non-adjacent bands is very small.

pseudo-random pattern generator device that generates a binary sequence of patterns where the bits appear to be random in the local sense

(1 and 0 are equally likely), but they are repeatable (hence only pseudo-random).

pseudo-random skewing *See* interleaved memory.

pseudo-random testing (1) a testing technique (often used in BISTs) that bases on pseudo-randomly generated test patterns. Test length is adapted to the required level of fault coverage.

(2) a technique that uses a linear feedback shift register (LFSR) or similar structure to generate binary test patterns with statistical distribution of values (0 and 1), across the bits; these patterns are generated without considering the implementation structure of the circuit to which they will be applied.

PSF *See* point spread function.

PSK *See* phase shift keying.

PSNR *See* peak signal-to-noise ratio.

PSU power supply unit. *See* power supply.

PSW *See* program status word.

PT potential transformer. *See* voltage transformer.

PU-239 a man-made, fissile element suitable for use in nuclear weapons as well as power reactors.

psychovisual redundancy the tendency of certain kinds of information to be relatively unimportant to the human visual system. This information can be eliminated without significantly degrading image quality, and doing so is the basis for some types of image compression. *See* image compression, joint photographics expert group (JPEG), quantization matrix.

public key cryptography   two different keys are used for the encryption of the plaintext and decryption of the ciphertext. Whenever a transmitter intends sending a receiver a sensitive message that requires encryption, the transmitter will encrypt the message using a key that the receiver makes available publically to anyone wanting to send them encrypted messages. On receiving the encrypted message, the receiver applies a secret key and recovers the original plaintext information sent by the transmitter. In contrast to secret key cryptography, public key cryptography does not suffer from the problem of having to ensure the secrecy of a key. *See also* private key cryptography.

public key cryptosystem   system that uses a pair of keys, one public and one private, to simplify the key distribution problem.

pull-in time   the time required by an automatic frequency control (AFC) system to lock or stabilize after a frequency change.

pull-in torque   the amount of torque needed to change a synchronous motor's operation from induction to synchronous when self-started.

pull-out torque   the maximum value of torque that an AC motor can deliver. An induction motor operating at the pull-out torque will operate at maximum slip, and loading it beyond the pull-out torque will cause the motor to stall. Synchronous motors remain at synchronous speed up to the pull-out torque. Exceeding the pull-out torque for a synchronous machine will lead to pole slipping and destruction of the machine.

pull-up torque   the minimum torque generated by an AC motor as the rotor accelerates from rest to the speed of breakdown torque. For an induction motor, this value usually is less than the locked-rotor torque, and thus establishes the maximum load that can be started.

pulse amplitude modulation (PAM)   a process in which an analog signal is sampled, the sampled values are then used to modulate the amplitudes of pulses. PAM is one type of analog pulse modulation, which is different from pulse code modulation.

pulse   a sudden change of an electrical value of short duration with a quick return to the original value. A pulse injects a short sharp burst of energy into a system and is usually quantified by its area rather than its amplitude or its duration. In the limit as the amplitude tends to infinity and the duration to zero, it approaches the Dirac delta function whose Laplace transform is unity.

pulse broadening   *See* pulse distortion.

pulse compression   technique of correlating identically coded signals, such as chirp signals, to produce a sharply peaked correlation function, the peak width being much shorter than that of the original coded pulses by a ratio equal to the time-bandwidth of the coded signals.

pulse distortion   a spreading or lengthening of the temporal shape of an optical signal transmitted through an optical fiber caused by a combination of wavelength effects (dispersion) and multimode and polarization effects. Also called pulse broadening. *See also* signal distortion.

pulse Doppler   a coherent radar, usually having high pulse repetition rate and duty cycle and capable of measuring the Doppler frequency from a moving target. Has good clutter suppression and thus can see a moving target in spite of background reflections.

pulse forming network (PFN)   (1) a transmission line with different impedance along its length.

(2) a lumped element circuit consisting of inductors and capacitors designed to deliver a square pulse.

pulse number of a converter   number of ripples in DC voltage per cycle of AC voltage. A three-phase, two-way bridge is a six-pulse converter.

pulse propagation   time and space dependence of a travelling electromagnetic pulse.

pulse shape   refers to the time domain characteristics of a pulse that is typically used to transmit data symbols over a channel. In most cases the pulse is shaped using a transmitter filter so that it satisfies the Nyquist I criterion for zero intersymbol interference.

pulse spreading   increase with time or distance of the length of a propagating electromagnetic pulse.

pulse width   width of the temporal region over which a pulse has significant amplitude; sometimes represented more specifically as, for example, full width at half maximum.

pulse width modulated inverter   an inverter that uses pulse width modulation to create AC voltages from a DC supply. *See also* pulse width modulation.

pulse-code modulation (PCM)   (1) a digital encoding technique whereby an analog signal is converted into a sequence of digital words in a periodic pulse train. The PCM process starts by bandlimiting the analog input waveform to one half the sampling rate (*See* aliasing). The analog signal is periodically sampled, using flat-top PAM (*See* pulse amplitude modulation). The height of each pulse is then quantized (estimated at an integer value) using an analog-to-digital converter, with the number of discrete quantizing levels determined by the resolution of the ADC used. The integer number associated with each sample is converted into binary format, then serially transmitted (MSB first). Note that for an 8-bit resolution ADC, each amplitude sample from the original analog waveform requires a

sequence of 8 data bits in the output pulse train to represent it; as such, the output data clock rate must be at least eight times faster than that of the clock used to sample the analog input waveform. *See also* analog-to-digital conversion and quantization.

(2) in image processing, a system that transmits quantized amplitudes of each pixel. Each pixel is assigned a unique binary word of finite length. Coding of image intensity samples in this manner is called pulse code modulation or PCM. In the case of the DPCM, the difference resulting from the subtraction of the prediction and the actual value of the pixel is quantized into a set of L discrete amplitude levels. *See also* differential pulse-code modulation encoding.

(3) in color signal encoding, PCM digitization of R, G, and B using full bandwidth for each component is common practice, as the PCM signals can be used for filtering, storage, compression, etc. But PCM encoding of color signals may not always use all the R, G, and B signals. However, if all the components R, G, B have the same bandwidth then fewer bits are used to quantize the red and blue compared with the green signal.

pulse-echo ultrasound   using a probe containing a transducer to generate a short ultrasound pulse and receive echoes of that pulse, associated with specular reflection from interfaces between tissues or scattering from inhomogeneities within the tissue, to form a display of tissue backscatter properties.

pulse-width modulated inverter   a self-commutated inverter where the amplitude and frequency of the output are controlled by PWM. The switching frequency is generally much higher than the output frequency to ensure a smooth output waveform. In the following example, PWM is performed by comparing a sine reference with a triangle carrier. The output before and after the lowpass filter is shown in the figure.

553

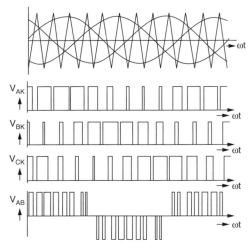

Waveforms of a PWM inverter.

pulse-width modulated switch   an active switch driven by a pulse-width modulated (PWM) pulse train.

pulse-width modulation (PWM)   a control technique used in variable speed DC, AC, or other electrical variable speed drives to control the harmonic content of the applied voltage or current. Typically, the pulse width is modulated in three ways, trailing-edge modulation, leading-edge modulation, and double-edge modulation. Most popular is sinusoidal PWM for AC drives.

PWM is most frequently used in switching converter technology as the drive signal for active switching elements.

pulsed laser   laser designed to produce its output in the form of isolated or periodic pulses that may be either long or short compared to the length of the laser cavity.

pulsed network   a network that conveys information between neurons by means of pulse streams. The information may be coded using conventional pulse modulation techniques or by using random bit streams, in which case the information is coded into the statistics of the stream.

pulsed response   the response of a system, or circuit to a pulsed input signal.

pumping   process for achieving population inversion and gain in a laser medium; process for providing energy to an arbitrary nonlinear optical system; process in which input energy is used to obtain a population inversion in a laser medium.

punched card   a method, now obsolete, used to represent data and programs as cardboard cards where the values were represented by punched holes at appropriate places.

punctured code   a code constructed from another code by deleting one or more coordinates from each codeword. Thus an (n, k) original code becomes an $(n - 1, k)$ code after the puncturing of one coordinate.

punctured convolutional code   a code where certain symbols of a rate $1/n$ convolutional code are periodically punctured or deleted to obtain a code of higher rate. Because of its simplicity, it is used in many cases, particularly in variable rate and high-speed applications. *Compare with* punctured code.

puncturing   periodic deletion of code symbols from the sequence generated by a forward error control convolutional encoder, for purposes of constructing a higher rate code. Also, deletion of parity bits in a forward error control block code.

pure ALOHA   type of multiple access protocol. Any user is allowed to transmit at any time. The possible collisions result in destroyed packets. Destroyed packets are retransmitted at a later time. *See also* ALOHA.

pure longitudinal wave   ultrasonic plane wave in which the particle motion is parallel to the wave vector and for which energy flow is parallel to the wave vector.

pure procedure a procedure that does not modify itself during its own execution. The instructions of a pure procedure can be stored in a read-only portion of the memory and can be accessed by many processes simultaneously.

pure shear wave ultrasonic plane wave in which the particle motion is parallel to the wave vector and for which energy flow is perpendicular to the wave vector.

purity in a color display, the production of red image content (only) by the red picture signal, blue image content (only) by the blue picture signal, and green image content (only) by the green picture signal. Picture impurity may be due to a color phosphor being excited by an inappropriate color electron beam.

Purkinje region the region of light intensity where, as illumination is reduced slowly enough to allow the eye to adapt, sensitivity to long wave light (red–orange) decreases more rapidly than sensitivity to short wave light (blue–violet).

pursuit behavior the human operator's outputs depend on system errors, as in compensatory behavior, but may also be direct functions of system inputs and outputs. The human response pathways make the man–machine system a combined open-loop, closed-loop system.

pursuit display in the simplest case, a display that shows input command, system output, and the system error as separable entities.

push brace a rigid brace, typically another utility pole, which is angled against a utiltity pole to serve the purpose of a guy where a guy cannot be placed.

push instruction an instruction that stores the contents of specified register/s on the stack.

push–pull amplifier an amplifier made up of two identical class B amplifiers operated as a balanced pair (180 degrees out of phase with each other) into a common load, resulting in each amplifier operating over alternating half cycles of the input signal, and having the output signal combined such that full cycles are dissipated across the load. Odd order harmonics are present at the load, while even order harmonics are suppressed, resulting in an output signal that is an odd order multiple of the input frequency (either a fundamental frequency amplifier or an odd order frequency multiplier).

push–push amplifier an amplifier made up of two identical class B amplifiers operated as a balanced pair (180 degrees out of phase with each other) into a common load, resulting in each amplifier operating over alternating half cycles of the input signal, and having the output signal combined such that rectified half cycles are dissipated across the load. Even order harmonics are present at the load, while odd order harmonics are suppressed, resulting in an output signal that is an even order multiple of the input frequency (even order frequency multiplier).

pushdown stack a data structure containing a list of elements that are restricted to insertions and deletions at one end of the list only. Insertion is called a push operation and deletion is called a pull operation.

PV bus *See* voltage-controlled bus.

PWM *See* pulse-width modulated.

PWM inverter *See* pulse-width modulated inverter.

PWM switch *See* pulse-width modulated switch.

PWR *See* pressurized water reactor.

pyramid coding any compression scheme which repeatedly divides an image into two subbands, one, a lowpass representation that is subsampled and used as input for the next level, and the other an error (difference) image. A small lowpass image plus a pyramid of difference images of increasing size is generated, allowing the lowpass information to be coded accurately with few codewords, and the highpass information to be coded with coarse quantization. *See* Laplacian Pyramid.

pyramidal horn   a horn antenna where the aperture is formed by flaring the walls in both the E-plane and the H-plane. The flare angles for the E-plane and H-plane can be adjusted independently.

pyroelectric   a polar dielectric material in which the internal dipole moment is temperature dependent. This leads to a change in the charge balance at the surface of the material which can be detected as either a potential difference or as a charge flowing in an external circuit.

pyrolytic grid   a grid structure made of pyrolytic (oriented) graphite that is laser-cut to the proper geometry for use in a power vacuum tube.

pyrosensitive smart material   material that self-adaptively (smartly) manages the electromagnetic surface characteristics of active surfaces constituted by pyrosensitive inclusions, in response to an external temperature-inducing stimulus applied per the feedback information on electromagnetic characteristics.

PZT   generic name for piezoelectric materials of the lead (Pb) zirconate titanate family.

Q

q unit of electric charge. $q = 1.602 \times 10^{-19}$ coulombs.

Q *See* quality factor.

q axis *See* quadrature axis.

Q_L common symbol loaded quality factor. Q_L is dimensionless.

Q_U common signal for unloaded quality factor. Q_U is dimensionless. *See also* quality factor.

Q-factor a figure of merit that represents the ratio of stored to dissipated energy per cycle.

Q-switching rapid increase in the quality of a laser cavity from below threshold to above threshold for the purpose of obtaining a large output optical pulse; also called loss-switching.

QAM *See* quadrature amplitude modulation.

QOS *See* quality of service.

quad tree a tree data structure used to represent an image, where each node is a square subset of the image domain. The root represents the whole image domain, which is subsequently subdivided in square subsets until all pixels in each node have the same value, i.e., until all nodes become "pure."

quadratic detector a detector that makes use of the second-order statistical structure (e.g., the spectral characteristics) of measurements. The optimum structure for detecting a zero-mean

Gaussian signal in the presence of additive Gaussian noise is of this form.

quadratic gain profile gain that varies quadratically with distance away from the axis of an optical medium.

quadratic index profile index of refraction that varies quadratically with distance away from the axis of an optical medium.

quadratic measure of sensitivity when the multiparameter sensitivity row vector is known and the components x_i have tolerance constants ϵ_i (these are considered to be positive numbers), i.e.,

$$x_{i0}(1 - \epsilon_i) \le x_i \le x_{i0}(1 + \epsilon_i)$$

then the quadratic measure of sensitivity is defined as

$$M_Q = \int_{\omega_1}^{\omega_2} \left[\sum_{i=1}^{n} \left(\mathbf{ReS}_{x_i}^{F(j\omega,\mathbf{x})} \right)^2 \epsilon_i^2 \right] d\omega$$

quadratic media propagation media in which the gain and/or index of refraction vary quadratically with distance away from the axis.

quadratic phase coupling a measure of the degree to which specific frequencies interact to produce a third frequency.

quadratic stabilizability a property of a dynamical (possibly uncertain) system that could be stabilized using state feedback designed via a quadratic Lyapunov function. If the system is linear, then it is usually required that it could be quadratically stabilizable via linear feedback. It means that there exists such a linear in state x feedback control law and a positive definite matrix P that the resulting closed-loop system has the property that the derivative of the Lyapunov function $V(x) = x'Px$ is negative definite for all possible uncertainties. The suitable matrix P may be found, for example, by solving a respective game type Riccati equation or parametrically scaled linear matrix inequalities.

quadrature a condition where there is a 90 degree separation between items. That is, they are at right angles to one another.

quadrature amplitude modulation (QAM) a modulation technique in which the incoming symbols are split into two substreams, which are modulated in quadrature. In the conventional implementation, the "in-phase" symbols are modulated by $\cos 2\pi f_c t$, and the "quadrature" symbols are modulated by $\sin 2\pi f_c t$, so that the transmitted phases associated with the two substreams differ by $\pi/2$.

quadrature axis (q axis) an axis placed 90 degrees ahead of the direct axis of a synchronous machine. *See* direct axis.

quadrature axis magnetizing armature reactance a reactance that represents all the inductive effects of the q-axis stator current of a synchronous machine, except for that due to the stator winding leakage reactance. In Park's q-axis equivalent circuit of the synchronous machine, this reactance is the only element through which both the stator and rotor currents flow. Its value may be determined by subtracting the stator winding leakage reactance from the steady-state value of the q-axis operational impedance or from the geometric and material data of the machine, using expressions described in.

quadrature axis magnetizing reactance *See* quadrature axis magnetizing armature reactance.

quadrature axis subtransient open circuit time constant a constant that characterizes the initial decay of transients in the q-axis variables of the synchronous machine with the stator windings open-circuited. The interval characterized is that immediately following a disturbance during which the effects of all amortisseur windings are considered. A detailed (derived) closed-form expression for the subtransient open-circuit time constant of a machine with two q-axis amortisseur windings is obtained by taking the reciprocal of the smallest root of the denominator of the q-axis operational impedance. An approximate (standard) value is often used, in which it is assumed one amortisseur winding resistance is very small relative to the other and the detailed expression simplified.

quadrature axis subtransient reactance the high-frequency asymptote of the q-axis operational impedance of a synchronous machine. The reactance characterizes the equivalent reactance of the q-axis windings of the machine during the initial time following a system disturbance. In models in which the rotor windings are represented as lumped parameter circuits, the q-axis subtransient reactance is expressed in closed form as the sum of the stator winding leakage reactance, and the parallel combination of the q-axis magnetizing reactance and the q-axis rotor amortisseur leakage reactances.

quadrature axis synchronous reactance the sum of the stator winding leakage reactance and the q-axis magnetizing (armature) reactance of a synchronous machine. This reactance represents the balanced steady-state value of the q-axis operational impedance of the synchronous machine.

quadrature axis transient reactance a value that characterizes the equivalent reactance of the q-axis windings of the synchronous machine between the initial time following a system disturbance (subtransient interval) and the steady state. This reactance cannot be directly mathematically related to the q-axis operational impedance. However, in models in which the q-axis contains two amortisseur windings and the rotor windings are represented as lumped parameter circuits, the q-axis transient reactance is expressed in closed form as the sum of the stator winding leakage reactance, and the parallel combination of the q-axis magnetizing reactance and the primary q-axis amortisseur winding leakage reactances.

quadrature detector   mixer in a FM receiver where the output voltage is a function of the original modulation of two carrier inputs that are 90 degrees apart in phase.

quadrature distortion   interference in signal transmission caused by phase error in the reference carrier.

quadrature FM demodulator   FM demodulation of two carrier inputs that are 90 degrees apart in phase.

quadrature hybrid   a directional coupler that accepts an input signal and delivers two equal power outputs that are 90 degrees out of phase.

quadrature modulation   a modulation scheme that involves the modulation of two sinusoidal carriers with a 90-degree phase difference by two independent message signals. The two carriers are typically represented as $A \cos(\omega_c t)$ (the in-phase carrier) and $A \sin(\omega_c t)$ (the quadrature carrier), where ω_c is the carrier frequency. The modulated signal is written as $s(t) = A m_c(t) \cos(\omega_c t + \phi) + A m_s(t) \sin(\omega_c t + \phi)$, where $m_c(t)$ and $m_s(t)$ are the in-phase and quadrature modulating signals and ϕ is a random phase.

quadrature multiplexing   a type of multiplexing technique, in which two message signals $m_1(t)$ and $m_2(t)$ are transmitted on the same carrier frequency with two quadrature carriers $A_c \cos 2\pi f_c t$ and $A_c \sin 2\pi f_c t$. That is, the quadrature multiplexed signal, also known as quadrature amplitude

modulated signal, is

$$v(t) = A_c m_1(t) \cos 2\pi f_c t + A_c m_2(t) \sin 2\pi f_c t$$

It is noted that the term quadrature indicates a 90° phase difference between the two carriers. For the demodulation process, refer to quadrature demodulation. The term quadrature multiplexing is sometimes also referred to as quadrature carrier multiplexing.

quadrature phased signals   two independent signals that have identical frequencies and have a fixed 90-degree phase difference.

quadrature signal   in quadrature modulation, the signal component that multiplies $\sin 2\pi f_c t$, where f_c is the carrier frequency.

quadrature spreader   a device that takes an information signal and two independent PN sequences (two spreading codes) to produce two baseband direct sequence spread spectrum signals. These signals constitute the modulation signals for a quadrature modulation scheme.

quadrature-axis subtransient short-circuit time constant   a constant that characterizes the initial decay of transients in the q-axis variables of the synchronous machine with the stator windings short-circuited. The interval characterized is that immediately following a disturbance, during which the effects of all amortisseur windings are considered. A detailed (derived) closed-form expression for the subtransient short-circuit time constant of a machine with two q-axis amortisseur windings is obtained by taking the reciprocal of the largest root of the numerator of the q-axis operational impedance. An approximate (standard) value is often used, in which it is assumed one amortisseur winding resistance is small relative to the other and the detailed expression simplified.

quadrature-axis transient open-circuit time constant   a constant that charaterizes the decay of transients in the q-axis variables of the synchronous machine with the stator windings open-circuited. The interval characterized is that following the subtransient interval, but prior to steady-state, during which the fastest q-axis rotor circuit dynamics have subsided. A detailed (derived) closed-form expression for the transient short-circuit time constant of a machine with two q-axis amortisseur windings is obtained by taking the reciprocal of the smallest root of the denominator of the q-axis operational impedance. An approximate (standard) value is often used, in which it is assumed one amortisseur winding resistance is infinite and the detailed expression simplified.

quadrature-axis transient short-circuit time constant   a constant that characterizes the decay of transients in the q-axis variables of the synchronous machine with the stator windings short-circuited. The interval characterized is that following the subtransient interval, but prior to steady-state, during which the transients of the fastest dynamic q-axis rotor circuits have subsided. A detailed (derived) closed-form expression for the transient short-circuit time constant of a machine with two q-axis amortisseur windings is obtained by taking the reciprocal of the smallest root of the numerator of the q-axis operational impedance. An approximate (standard) value is often used, in which it is assumed one amortisseur winding resistance is infinite and the detailed expression simplified.

quadrupole   a magnet system that produces a pair of dipoles with like poles opposite each other.

quadrupole illumination   a type of off-axis illumination where four circles of light are used as the source. These four circles are spaced evenly around the optical axis.

quadword   a data unit formed from four words.

qualification   all activities that ensure that the nominal design and manufacturing specifications will meet or exceed the reliability goals.

qualitative robustness   an estimation scheme in which we optimize performance for the least favorable statistical environment among a specified statistical class.

quality factor (Q)   measure of the persistence of damped oscillations in a resonator; 2π times the energy stored in an oscillating system divided by the energy lost in one cycle of the oscillations. Q is a measure of the precision of the frequency selectivity of an electrical resonator or filter. Q is dimensionless.

quality of service (QOS)   a means of specifying the level of service provided by a network or required by a user of the network. Typical QOS parameters include bandwidth, delay, jitter, and error rate.

quantization   (1) the process of converting amplitude values that can take on many different values (infinitely many for analog signals) into a finite (or more coarse) representation. Quantization is necessary in digital processing of inherently analog signals, such as voice or image. *See also* lossy source coding, analog-to-digital conversion.

(2) aspects of particle properties related to their wavelike nature as described by quantum mechanics.

(3) dividing a signal value into equally sized portions, known as quanta.

quantization matrix   a matrix of numbers in the JPEG compression scheme that specifies the amount by which each discrete cosine transform coefficient should be reduced. The numbers are based on the human contrast sensitivity to sinusoids and are such that coefficients corresponding to frequencies which people are not very sensitive to are reduced or eliminated. *See* contrast sensitivity, discrete cosine transform, joint photographics expert group , human visual system , psychovisual redundancy.

quantization of transform coefficients   technique that exploits the statistical redundancy in transform coefficients and also the subjective redundancy in pictures using transform coding. Unlike DPCM, quantization of transform coefficients affects every pixel in the transformed block. There is a spreading of the quantization error, thus making the design of transform coefficients quantizers quite different from that of DPCM.

quantizer   a device that performs quantization (sense (1)). A quantizer takes an analog input signal and produces a discrete valued output signal. The number of possible levels in the output signal typically determines the error incurred in the quantization process. The output of the quantizer is typically encoded into a binary number.

quantizer characteristics   the performance of a scalar quantizer Q is typically characterized by its mean-squared quantization error

$$E[(Q(x) - x)^2] = \int (Q(x) - x)^2 f_X(x)dx,$$

where $f_X(x)$ represents the probability distribution of the input X. In the special case where X is uniformly distributed, and where Q is a uniform quantizer (that is, the interval between successive quantization levels is Δ, and $Q(x)$ is the rounding of x to the nearest level), then the mean- squared quantization error is $\frac{\Delta^2}{12}$.

quantum coherence   a condition that exists when a quantum mechanical system is in a linear combination of two or more basis states. The polarization produced when a 2-level atom interacts with a resonant electromagnetic field is an example of a quantum coherence.

quantum dot   a particle of semiconductor material having all dimensions comparable to an average free carrier Bloch wavelength.

quantum effect   macro effect on carrier movement created by the confinement of the carriers within very small dimensions along the direction of current flow (length). When the length dimension starts to approach an appreciable portion of the mean-free carrier path length, then odd things start to happen. This effect is somewhat analogous to the transition between low RF and microwave theory.

quantum efficiency   the ratio of the average number of free electrons produced in the photodetector per photon. Since this is a quantized process, some photons produce no free electrons, while other photons produce one or more free electrons when their energy is absorbed in the photodetector. Quantum efficiency is usually denoted Q_e.

quantum electronics   the field of study of the interactions of electromagnetic fields with matter, with emphasis placed on interactions that cannot be described classically.

quantum hall effect   quantization of the Hall resistivity observed in two-dimensional high-mobility semiconductor systems, caused by novel many-body effects.

quantum interference   the enhancement or suppression of a transition in a quantum mechanical system that arises when there are multiple paths from the initial to the final state.

quantum limit   the minimum number of photons that on average must be received in an optical communication system for a specified error probability (usually 10^{-9}). Using on–off keying and an ideal photon counting receiver, the quantum limit is approximately 10 photons/bit.

quantum numbers   a group of integrating constants in a solution of Schrödinger's equation.

quantum well infrared photodetector mid-to long-wavelength photodetector exploiting optically induced transitions between confined quantum well states. Amenable to focal plane array manufacture due to compatibility with GaAs epitaxial growth and processing techniques.

quantum well laser a laser that uses quantum wells as the gain medium.

quantum wire a piece of semiconductor having two dimensions comparable to an average free carrier Bloch wavelength, the other dimension being much larger.

quarter-wave symmetry a periodic function that displays half-wave symmetry and is even symmetric about the 1/4 and 3/4 period points (the negative and positive lobes of the function).

quarter-wave transformer a piece of transmission line with an electrical length of 1/4 of the operating frequency wave length (or odd multiple) used for impedance matching.

quasi-ballistic motion in a metal or semiconductor in which the scattering is infrequent, and for which the phase coherence is mostly maintained during the time in which the carrier is in the active volume.

quasi-Fermi levels energy levels that specify the carrier concentration inside a semiconductor under nonequilibrium conditions.

quasi-linear nearly linear, easily approximated as linear with insignificant errors. Many circuits, elements and devices operated under small signal conditions are considered quasi-linear for many parameters or applications. A low noise amplifier under small signal operation still exhibits signal distortion (nonlinearity), but the gain characteris-

tics can be determined from S-parameters, which are linear, with little or no error.

quasi-optics a low loss method of transmission and manipulation of millimeter and sub-millimeter waves.

quasi-TEM a mode of a structure comprising multiple dielectrics that behaves like the TEM mode of an equivalent structure comprising a single material with an appropriate effective dielectric constant. In certain types of transmission lines, for example, microstrip lines, the propagating waves are not pure TEM waves. Therefore, the general static approximations cannot be applied; however, in most practical cases one can make an approximation by assuming that the fields are quasi-TEM. In this case, good approximation for the phase velocity, propagation constant, and impedance is obtained.

quasiquantum device electron device in which the principle of the device function is more easily explainable by quantum mechanics than by Newtonian classical dynamics.

queue a data structure maintaining a first-in-first-out discipline of insertion and removal. Queues are useful in many situations, particularly in process and event scheduling. *See also* FIFO memory.

quick gun a tool for installing crimped wire connectors.

quincunx five points, four at the corners of a square and one in the center.

quincunx lattice a horizontally and vertically repeated pattern of quincunxes, which is identical to a square lattice oriented at 45 degrees.

quincunx sampling the downsampling of a 2-D or 3-D signal on a quincunx lattice by removing every even sample on every odd line and every odd sample on every even line. In the frequency domain, the repeat spectrum centers also form a quincunx lattice. Quincunx sampling structures have been used for TV image sampling on the basis that they limit resolution on diagonal frequencies but not on horizontal and vertical frequencies.

R

R_S commonly used symbol for source impedance.

R_T commonly used symbol for transformation ratio.

R-ALOHA   *See* reservation ALOHA.

R_L Typical symbol for load resistance.

Rabi frequency   the characteristic coupling strength between a near-resonant electromagnetic field and two states of a quantum mechanical system. For example, the Rabi frequency of an electric dipole allowed transition is equal to μE/hbar, where μ is the electric dipole moment and E is the maximum electric field amplitude. In a strongly driven 2-level system, the Rabi frequency is equal to the rate at which population oscillates between the ground and excited states.

race condition   a situation where multiple processes access and manipulate shared data with the outcome dependent on the relative timing of these processes.

raceway   a channel within a building which holds bare or insulated conductors.

radar   an instrument that transmits electromagnetic waves and receives properties of the reflected electromagnetic wave from the target, which can be used to determine the nature and distance to the target. Radar is an acronym that stands for radio detection and ranging.

radar cross section (RCS)   a measure of the reflective strength of a radar target; usually represented by the symbol s, measured in square meters, and defined as 4π times the ratio of the power per unit solid angle scattered in a specific direction of the power unit area in a plane wave incident on the scatterer from a specified direction.

RADHAZ   radiation hazards to personnel as defined in ANSI/C95.1-1991 IEEE Standard Safety Levels with Respect to Human Exposure to Radio Frequency Electromagnetic Fields, 3 kHz to 300 GHz.

radial basis function network   a fully connected feedforward network with a single hidden layer of neurons each of which computes a nonlinear decreasing function of the distance between its received input and a "center point." This function is generally bell-shaped and has a different center point for each neuron. The center points and the widths of the bell shapes are learned from training data. The input weights usually have fixed values and may be prescribed on the basis of prior knowledge. The outputs have linear characteristics, and their weights are computed during training.

radial intensity histogram   a histogram of average intensities for a round object in circular bands centered at the center of the object, with radial distance as the running index. Such histograms are easily constructed, and, suitably normalized, form the basis for scrutinizing round objects for defects, and for measuring radius and radial distances of cylindrically symmetrical features.

radial system   a network of straight wires or other conductors radiating from the base of a vertical monopole antenna that simulates the presence of a highly conducting ground plane beneath the antenna. Typically, radial wires are approximately a quarter wavelength long and are arranged to have equiangular spacing between them. The radial wire ends at the base of the monopole are electrically bonded together and to one conductor of the feedline.

radiated emission   an electromagnetic field propagated through space.

radiating aperture   a basic element of an antenna that in itself is capable of effectively radiating or receiving radio waves. Typical example of a radiating apertures are a slot, and horn antennas or a truncated metallic waveguide.

radiation   the phenomenon by which sources generate energy, which propagates away from them in the form of waves.

radiation boundary condition (RBC)   a boundary condition that is imposed to truncate the computational domain of a differential equation method so that it satisfies the Sommerfield radiation condition.

radiation condition   the condition that specifies the field behavior at infinity; in fact, for an unbounded domain it is necessary to specify the field behavior on the surface at infinity. By assuming that all sources are contained in a finite region, only *outgoing waves* must be present at large distances from the sources; hence the field behavior at large distances from the sources must meet the physical requirement that energy travel away from the source region. This requirement is the Sommerfield "*radiation condition*" and constitutes a boundary condition on the surface at infinity. Let us denote with A any field component transverse to the radial distance r and with k the free-space wavenumber. The transverse field of a spherically diverging wave in a homogeneous isotropic medium decays as $1/r$ at large distances r from the source region; locally, the spherical wave behaves like a plane wave travelling in the outward r direction. As such, each field component transverse to r must behave like $\exp(-jkr)/r$; this requirement may be phrased mathematically as

$$\lim_{r \to \infty} r \left(\frac{\partial A}{\partial r} + jkA \right) = 0$$

radiation efficiency   in antenna theory, the ratio of radiated power to the amount of input power to the antenna. Has a value between 0 and 1, inclusive.

radiation intensity   in antenna theory, a far-field quantity that is a function of angle that gives the level of radiation in a specific direction. Radiation intensity is the radial component of the time average Poynting vector with all the terms associated with the distance from the antenna normalized out. The units are watts per square radian.

radiation pattern   a plot of the far-field radiation as a function of the angle phi or theta while the other angle is held constant. A radiation pattern can be either polar or rectangular, and can be either logarithmic or linear.

radiative broadening   a spectral line broadening mechanism that arises due to spontaneous decay of the excited state.

radiative heat transfer   the process by which long-wave electromagnetic radiation transports heat from a surface to its surroundings.

radio astronomy   the study of celestial objects based on the investigation of their radio frequency electromagnetic waves spectrum.

radio frequency (RF)   a general term used to refer to radio signals in the general frequency range from thousands of cycles per second (kHz) to millions of cycles per second (MHz). It is also often used generically and interchangeably with the term microwave to distinguish the high frequency AC portion of a circuit or signal from the DC bias signal or the downconverted intermediate frequency (IF) signal.

radio frequency choke   an inductance (a coil of wire) intended to present a very high impedance at radio frequencies. Such frequencies span the range of kilohertz to hundreds of megahertz.

radio frequency integrated circuit (RFIC) integrated circuit designed to operate at radio frequencies as amplifiers, mixers, oscillators, detectors or combinations of above. Typically, RFICs are configured for specific application to operate as a complete RF system.

radio frequency interference (RFI) electromagnetic phenomenon that either directly or indirectly contributes to degradation in the performance of a receiver or other RF system, synonymous with electromagnetic interference. *See also* electromagnetic inteference.

radio horizon the maximum range, from transmitter to receiver on Earth's surface, of direct (line-of-sight) radio waves. This is greater than the optical horizon, because the radio waves follow a curved path as a result of the continuous refraction it undergoes in the atmosphere.

radio local loop *See* wireless local loop.

radio waves electromagnetic radiation suitable for radio transmission in the range of frequencies from about 10 kHz to about 300 Mhz.

radiography an imaging modality that uses an X-ray source and collimator to create a projection image. The image intensity is proportional to the transmitted X-ray intensity.

radiology the monitoring and control of radioactivity, especially in regard to human exposure, in a nuclear power plant.

radiometer a passive receiver that detects energy from a transmitting source or reradiated energy from a target. Typically, radiometers are used in remote sensing applications.

radius of curvature one of the parameters characterizing the reflecting surface of a spherical mirror.

radix the base number in a number system. Decimal (radix 10) and binary (radix 2) are two example number systems.

radix complement in a system that uses binary (base 2) data negative numbers, can be represented as the two's complement of the positive number. This is also called a true complement.

Radon transform for a function $f(x, y)$, $r(d, \phi)$, is the line integral along a line inclined at angle ϕ from the y axis and at a distance d from the origin.

radwaste a contraction for "radioactive waste," usually referring to mildly radioactive sludge removed from the coolant in a nuclear reactor.

RAID *See* redundant array of inexpensive disks.

RAKE receiver a receiver type in which the received signal from each (or a few dominating) resolvable propagation ray is individually demodulated and subsequently combined into one decision variable according to some criterion. RAKE receivers are commonly used in wideband transmission systems such as spread-spectrum systems where the large bandwidth allows several rays to be resolved, thus creating a diversity gain (frequency diversity).

RAM *See* random access memory.

RAM neuron a random access memory with n inputs and a single output. The inputs define $2n$ addresses and presentation of a particular input vector allows the contents of the 1-bit register at that address to be read, or to be written into. Training consists of writing 1s or 0s into the 1-bit registers, as required for the various input vectors in the training set.

RAMA *See* resource auction multiple access.

ram's head   the top of a transmission line tower

Raman echo   a type of photon echo in which the stimulated emission is assisted by a Raman transition.

Raman laser   laser in which the amplification mechanism is considered to be Raman scattering.

Raman scatter   frequency-shifted light scatter, utilized as powerful analytical chemistry technique.

Raman scattering   scattering of light by means of its interactions with the vibrational response of a molecular system. The scattered light is typically shifted to lower frequencies (*See* Stokes law); the frequency shift is equal to the vibrational frequency of the molecule, typically $10^1 3$ to $10^1 4$ Hz. In spontaneous Raman scattering (*See* spontaneous light scattering), the scattered light is emitted in nearly all directions. In stimulated Raman scattering (*See* stimulated light scattering), the scattered light is emitted in the form of an intense beam. The emitted beam tends to be intense because it experiences large amplification by an amount given by e^{gIL}, where g is the gain factor of stimulated Raman scattering, I is the intensity of the incident laser beam in units of power per unit area, and L is the length of the interaction region.

Raman, Sir Chandrasekhara Venkata   (1888–1970) Born: Trichinopoly, India

Author of 500 articles and independent investigator of light scatter and acoustics. Winner of a Nobel Prize for the discovery of Raman scattering.

Raman–Nath diffraction   in acousto-optics, the regime where many diffraction orders exist due to the thinness of the acoustic grating relative to the acoustic wavelength in the direction of light.

Raman–Nath diffraction regime   regime where the acoustic beam width is sufficiently narrow to produce many diffracted beams with significant power.

Raman–Nath mode acousto-optic spectrum analyzer   similar to the acousto-optic (AO) Bragg-mode spectrum analyzer, but uses illumination at normal incidence to the face of the AO cell that results in multiple diffracted orders, with the two first order beams being used at the Fourier plane to obtain the input RF signal spectrum.

Raman–Nath scattering   the scattering of light from a periodically varying refractive index variation in a thin medium, as contrasted with Bragg scattering, for instance in the operation of an acousto-optic modulator.

ramp   a linear function of grey level, usually connecting two contrasting regions. Named after its appearance in one dimension, it is often used as one model of an edge.

Ramsey fringe   the spectral feature generated when a quantum mechanical transition is excited by two identical-frequency, time-separated electromagnetic pulses. Ramsey fringes are used in cesium atomic beam clocks.

random access   (1) term describing a type of memory in which the access time to any cell is uniform.

(2) a method for allowing multiple users to access a shared channel in which transmissions are not coordinated (or perhaps are partially coordinated) in time or frequency.

random access device   *See* random access.

random access memory (RAM)   direct-access read/write storage in which each addressable unit has a unique hardwired addressing mechanism. The time to access a randomly selected location is constant and not dependent on its position or on any previous accesses. The RAM has a set of k address lines ($m = 2^k$), n bidirectional data lines,

33

and a set of additional lines to control the direction of the access (read or write), operation and timing of the device.

RAM is commonly used for the main memory of a computer and is said to be static if power has to be constantly maintained in order to store data and dynamic if periodic absences of power do not cause a loss of data. RAM is usually volatile. *See also* static random-access memory, dynamic random-access memory, non-volatile random-access memory.

random behavior   response without a spectral or amplitude pattern or relationship to the excitation. The excitation may be internal, and may be thermal in nature.

random coding   coding technique in which codewords are chosen at random according to some distribution on the codeword symbols. Commonly a tool used in the development of information theoretic expressions.

random logic   a digital system constructed with logic gates and flip-flops and other basic logic components interconnected in an non-specific manner. *See also* microprogramming.

random process   a mathematical procedure for generating random numbers to a specific rule called a process, x which is defined on continuous $x(t)$, $t \in R^n$, or discrete $x(k)$, $k \in Z^n$ space/time. The value of the process at each point in space or time is a random vector. *See* random variable, random vector. *See also* correlation, covariance, autocorrelation, autocovariance.

random replacement algorithm   in a cache or a paging system, an algorithm that chooses the line or page in a random manner. A pseudo-random number generators may be used to make the selection, or other approximate method. The algorithm is not very commonly used, though it was used in the translation buffers of the VAX11/780 and the Intel i860 RISC processor.

random signal   a signal $X(t)$ that is either noise $N(t)$, an interfering signal $s(t)$, or a sum of these:

$$X(t) = s_1(t) + \cdots + s_m(t) + N_1(t) + \cdots + N_n(t)$$

random testing   the process of testing using a set of pseudo-randomly generated patterns.

random variable   a continuous or discrete valued variable that maps the set of all outcomes of an experiment into the real line (or complex plane). Because the outcomes of an experiment are inherently random, the final value of the variable cannot be predetermined.

random vector   a vector (typically a column vector) of random variables. *See* random variable, random process.

randomized decision rule   a hypothesis decision / classification rule which is not deterministic (that is, the measurement or observation does not uniquely determine the decision). Although typically not useful given continuous observations, a randomized rule can be necessary given discrete observations. *See* Neyman-Pearson lemma, receiver operating characteristic.

range filter   an edge detection filter that finds edges by taking the difference between the maximum and minimum values in a local region of the image. The range filter also accepts a weight mask the size of the local image region that controls pixel values before they enter the minimum and maximum calculations. The weight mask allows edges in certain directions to be searched for.

range image   an image in which the intensity of point $\mathbf{x}$ is a function of the distance between $\mathbf{x}$ and the corresponding point in the scene (object) projected on the image plane.

range of Jacobian   denoted J, it is defined as a subspace $R(J)$ in R^r of the end-effector-velocities that can be generated by the joint velocities, in the given manipulator posture. *See* full rank Jacobian.

rank filter   an image transform used in mathematical morphology. Assume that to every pixel p one associates a window $W(p)$ containing it. Let k be an integer > 1 which is less than or equal to the size of each window $W(p)$. The rank filter with rank k and windows $W(p)$ transforms an image I into a filtered image I' whose gray-level $I'(p)$ at pixel p is defined as the k-th least value among all initial gray-levels $I(q)$ for q in the window $W(p)$. In a dual version, the k-th greatest value is selected. When each $W(p)$ is the translate by p of a structuring element W of size n, three particular cases are noteworthy:

(1) $k = 1$, the rank filter is the erosion by W.
(2) $k = n$, the rank filter is the dilation by $\widetilde{W}$, the symmetric of W.
(3) n is odd and $k = (n + 1)/2$ the rank filter is a median filter.

See dilation, erosion, median filter, structuring element.

rare gas   one of the rare gases specified in the periodic table, He, Ne, Ar, Kr, Xe.

rare gas halides   excimer molecule formed by a reaction of a rare gas atom and a halogen atom: e.g., XeF, ArF, KrF.

rare gas molecule   excimer molecule formed by a reaction of an excited and a neutral rare gas atom, usually with a third atom present. For example, Ar_2, Kr_2.

rare gas oxides   an excimer molecule formed by the reaction of a rare gas atom and an oxygen atom.

rare-earth magnet   a magnet that has any of the rare-earth elements in its composition. Typically stronger than other magnet materials, these include neodymium iron boron and samarium cobalt.

rare-earth permanent magnet   magnet made of compounds of iron, nickel, and cobalt with one or more of the rare-earth elements such as samarium. These materials combine the high residual flux density of the alnico-type materials with greater coercivity than ferrites.

raster   a predetermined pattern of scanning lines used to provide uniform coverage of the area used for displaying a television picture.

raster coordinates   coordinates in a display system that specifically identify a physical location on the display surface.

raster graphics   a computer graphics system that scans and displays an image periodically in a raster, or left-to-right, top-to-bottom fashion.

raster image   *See* bitmapped image.

raster width   (1) physical distance between raster lines on a display surface and between distinguishable points in the same raster line; the two distances are frequently different.

(2) the physical distance between raster lines on a display surface and between distinguishable points in the same raster line. The two distances are frequently different.

rate distortion function   the minimum rate at which a source is represented by one of a set of discrete points.

rate distortion theory   a theory aimed at quantifying the optimum performance of source coding systems. Using information theory, for several source models and distortion measures, rate distortion theory provides the optimum distortion function and the optimum rate function. The distortion function is optimum in that the distortion for a given rate is the theoretical minimum value of distortion for encoding the source at the given or lower rate. The rate function is optimum in that the rate at a given distortion is the minimum possible rate for coding the source at the given or lower distortion.

rate equation approximation assumption, in a semiclassical model for the interaction of light with atoms, that all fields and populations change negligibly within the coherence time of the wave functions, loses all information about the phase of the fields and wave functions.

rate equation model model for the interaction of light with atoms in which the atoms are represented only by their populations or population densities and the electromagnetic field is represented only by its intensity, power, energy, or photon density.

rate equations coupled ordinary nonlinear differential equations governing the interaction of an electromagnetic field (represented by an intensity, energy density, or photon density) with an atomic or molecular laser medium (represented by populations of the energy states); phase information relating to the fields and wavefunctions is absent from these equations.

rate split multiple access coding technique for the multiple-access channel in which each user splits their information stream into two or more streams, which are independently encoded. These encoded streams are multiplexed according to some rule prior to transmission. Used to show that time-sharing is not required to achieve certain points in the capacity region of multiple access channels.

rate-adaptive digital subscriber line (RDSL) a digital subscriber line (DSL) in which the rates in each direction are adjusted according to the quality of the channel. In general, longer loops are associated with lower rates.

rate-compatible punctured convolutional (RCPC) code one of a family of punctured convolutional codes derived from one low-rate convolutional parent code by successively increasing the number of punctured symbols, given that the previously punctured symbols should still be punctured (rate-compatibility). These codes have applications in, for example, variable error protection systems and in hybrid automatic repeat request schemes using additional transmitted redundancy to be able to correctly decode a packet. Also called RCPC code.

rate-distortion theory Claude Shannon's theory for source coding with respect to a fidelity criterion, developed during the late 1940s and the 1950s. Can be viewed as a generalization of Shannon's earlier theory (late 1940s) for channel coding and information transmission. The theory applies to the important methods for vector quantization and predicts the theoretically achievable optimum performance.

rated voltage the voltage at which a power line or electrical equipment is designed to operate.

ratio detector a circuit for recovering (demodulating) baseband information (usually audio) from a frequency modulated (FM) wave.

rational function a function that is the ratio of two polynomials. Rational functions often arise in the solution of differential equations by Laplace transforms. *See* Laplace transform.

ray one of a family of lines (rays) used to represent the propagation of an electromagnetic wave; most useful in real media when the wave amplitude varies slowly compared to the wavelength (see geometrical optics).

ray equation set of second-order differential equations governing the trajectory of a light ray propagating along an arbitrary path.

ray optics approximate representation of electromagnetic wave propagation in terms of light rays, most useful in real media when the wave amplitude varies slowly compared to the wavelength.

ray tracing (1) a high-frequency electromagnetic analysis technique in which the propagation path is modeled by flux lines or "rays." The ray density is proportional to the power density, and frequently, bundles of these rays are called ray tubes.

(2) a rendering technique in which the paths of light rays reaching the viewpoint are computed to obtain realistic images. Given a 3-D description of a scene as a collection of surfaces characterized by different optical properties, rays are traced backward from the viewpoint through the image plane until they hit one of the surfaces or go off to infinity.

ray transfer matrix real two-by-two matrix governing the transformation of the ray displacement and slope with respect to a fixed axis.

Rayleigh criterion a method of distinguishing between rough and smooth surfaces in order to determine whether specular reflection will occur. A surface is considered smooth if the phase difference between waves reflected from the surface is less than ninety degrees.

Rayleigh distribution the probability distribution of the magnitude of a complex quantity whose real and imaginary parts are independent Gaussian random variables with zero mean. Frequently used to approximate multipath fading statistics in non-line-of-sight mobile radio systems.

Rayleigh length distance over which the spot size of a Gaussian beam increases from its value at the beam waist to a value 2.5 larger; a measure of the waist size of a Gaussian beam, π times waist spot size squared divided by wavelength; half of the confocal parameter.

Rayleigh noise the envelope of a zero mean, wide-sense stationary, narrowband Gaussian noise

process. The probability density function of a sinusoid in narrowband noise is a generalized Rayleigh distribution, $p(z) = z\sigma e^{-z^2 2\sigma^2}$, $z \geq 0$. Also known as a Rician distribution.

Rayleigh scattering (1) theory for the interaction between light and a medium composed of particles whose size is much smaller than the wavelength. According to it, the scattering cross section is proportional to the fourth power of the wavelength of the scattered light. This explains both the red and blue colors of the sky.

(2) an intrinsic effect of glass that contributes to attenuation of the guided optical wave. The effect is due to random localized variations in the molecular structure of the glass which acts as scattering centers.

Rayleigh–Ritz procedure a procedure for solving functional equations. *See also* moment method.

Rayleigh-wing scattering of light the scattering of light with no change in central frequency, and with moderate (of the order of $10^1 1$ Hz) broadening of spectrum of the light. Rayleigh-wing scattering occurs when light scatters from anisotropic molecules.

RBC *See* radiation boundary condition.

RC time constant the time needed for signal traveling from an end to the other end of a wire is constant when the wire and the whole chip is scaled down. As the length of a wire shrinks by a factor of k and the cross-sectional area of the wire is reduced by a factor of k^2, the capacitance of the wire decreases by a factor of k while the resistance increases by a factor of k. The RC time remains constant, and thus the input charging time remains the same, independent of scaling. Consequently, the scaling down of the chip cannot increase the speed of the chip if wire is used. Optical interconnects can speed up the chip.

RCPC code *See* rate-compatible punctured convolutional code.

RCS *See* radar cross section.

RCT *See* reverse-conducting thyristor.

RDSL *See* rate-adaptive digital subscriber line.

re-entrancy term describing the number of times that a multiplex armature winding of a commutated machine closes upon itself via the commutator ring. For example, duplex windings can be either singly or doubly re-entrant. In a doubly re-entrant duplex winding, the ends of the two winding circuits close only on themselves and not on each other, creating two distinct circuits through the commutator and two distinct circuit closures. Conversely, in a singly re-entrant duplex winding, the two windings are connected in series through the commutator ring creating only a single circuit closure.

reachability a term that indicates that a dynamical system can be steered from zero initial state to any final state in a given time interval. For many dynamical systems, reachability is equivalent to controllability. This is always true for linear finite-dimensional continuous-time dynamical systems. However, in discrete case, controllability may be the stronger notion than reachability. In this case, the two concepts are equivalent if and only if rank $A = n$. For dynamical systems with delays, these two notions are essentially different. For infinite-dimensional systems, the relations between reachability and controllability depend on the properties of the semigroup $S(t)$ generated by the operator A.

reactance grounded an electrical system in which the neutral is intentionally grounded through a reactance. Frequently used in the neutral of generators and transformers to limit the magnitude of line to ground fault currents.

reactance modulator modulator normally using phase or frequency modulation where the reactance of the circuit is dependent on changes in the input modulating voltage.

reaction a functional in electromagnetics that relates a set of fields and sources to one another. Reaction concepts are often used in the discussion of field reciprocity.

reaction range sum of end-to-end round-trip delay and processing time.

reactive compensation process of counteracting the reactive component of a device by means of capacitors and inductors. Both series and shunt compensation are prevalent.

reactive congestion control in packet networks, a congestion control system whose actions are based on actual congestion occurrence.

reactive ion etching the process of etching materials by the use of chemically reactive ions or atoms. Typically, the reactive ions or atoms are generated in a RF plasma environment or in a microwave discharge.

reactive load a load that is purely capacitive or inductive.

reactive matching impedance transformation achieved by employing a matching network constructed of only reactive elements.

reactive near field the region close to an antenna where the reactive components of the electromagnetic fields from charges on the antenna structure are very large compared to the radiating fields. Considered negligible at distances greater than a wavelength from the source (decay as the square or cube of distance). Reactive field is important at antenna edges and for electrically small antennas.

reactive power (1) electrical energy per unit time that is alternately stored, then released. For example, reactive power is associated with a capacitor charging and discharging as it operates on an AC system. Symbolized by Q, with units of volt-amperes reactive (VAR), it is the imaginary part of the complex power.

(2) the power consumed by the reactive part of the load impedance, calculated by multiplying the line current by the voltage across the reactive portion of the load. The units are vars (volt-ampere reactive) or kilovars.

reactor a container where the nuclear reaction takes place. The reactor converts the nuclear energy to heat.

reactor containment *See* containment building.

reactor core an array of nuclear fuel rods that are arranged so as to encourage a chain reaction and thus heat water to supply a power for the steam turbine in a nuclear power plant.

reactor refueling the process of shutting down a nuclear reactor for maintenance and fuel replacement, typically every 12 to 24 months.

read ahead on a magnetic disk, reading more data than is nominally required, in the hope that the extra data will also be useful.

read instruction an assembly language instruction that reads data from memory or the input/output system.

read phase the first portion of a transaction during which the executing process obtains information that will determine the outcome of the transaction. Any transaction can be structured so that all of the input information is obtained at the outset, all the computation is then performed, and finally all results are stored (pending functionality checks based on the locking protocols in use).

read-after-write hazard *See* true dependency.

read-modify-write cycle a type of memory device access that allows the contents at a single address to be read, modified, and written back without other accesses taking place between the read and the write.

read-mostly memory memory primarily designed for read operations, but whose contents also can be changed through procedures more complex and typically slower than the read operations. EPROM, EEPROM, and flash memory are examples.

read-only (ROM) memory semiconductor memory unit that performs only the read operation; it does not have the write capability. The contents of each memory location is fixed during the hardware production of the device and cannot be altered. A ROM has a set of k input address lines (that determine the number of addressable positions 2^k) and a set of n output data lines (that determine the width in bits of the information stored in each position). An integrated circuit ROM may also have one or more enable lines for interconnecting several circuits and make a ROM with larger capacity. Plain ROM does not allow erasure, but programmable ROM (PROM) does. Static ROM does not require a clock for proper operation, whereas dynamic ROM does. *See also* random access memory, programmable read-only memory.

read/modify/write an uninterruptible memory transaction in which information is obtained, modified, and replaced, under the assurance that no other process could have accessed that information during the transaction. This type of transaction is important for efficient implementations of locking protocols.

read/write head conducting coil that forms an electromagnet, used to record on and later retrieve data from a magnetic circular platter constructed of metal, plastic, or glass coated with a magnetizable material. During the read or write operation, the head is stationary while the platter rotates beneath it. The write mechanism is based on the magnetic field produced by electricity flowing through the coil. The read mechanism is based on the electric current in the coil produced by a magnetic field moving relative to it.

Less common are magnetoresistive heads, which employ noninductive methods for reading. A system that uses such a head requires an additional (conventional) head for the writing. *See also* disk head, magnetic recording code.

real address   the actual address that refers to a location of main memory, as opposed to a virtual address that must first be translated. Also called a physical address. *See also* memory mapping, virtual memory.

real power   consider an AC source connected at a pair of terminals to an otherwise isolated network. The real power, equal to the average power, is the power dissipated by the source in the network.

real-time   refers to systems whose correctness depends not only on outputs, but the timeliness of those outputs. Failure to meet one or more of the deadlines can result in system failure. *See also* soft real-time system, firm real-time system, hard real-time system.

real-time clock   a hardware counter that records the passage of time.

real-time computing   support for environments in which response time to an event must occur within a predetermined amount of time. Real-time systems may be categorized into hard, firm, and soft real-time.

realization   for a linear continuous or discrete stationary finite-dimensional dynamical system, a set of four constant matrices A, B, C, D of the state and output equations. The matrices may be calculated using certain algorithms. The realization is said to be minimal if the dimension n of the square matrix A is minimal.

realization problem for 2-D Fornasini–Marchesini model   a problem in control. The transfer matrix $T(z_1, z_2)$ of the 2-D Fornasini–Marchesini model

$$x_{i+1,j+1} = A_1 x_{i+1,j} + A_2 x_{i,j+1}$$
$$+ B_1 u_{i+1,j} + B_2 u_{i,j+1}$$
$$y_{ij} = C x_{ij} + D u_{ij} \qquad (1)$$

$i, j \in Z_+$ (the set of nonnegative integers) is given by

$$T(z_1, z_2) = C [I z_1 z_2 - A_1 z_1 - A_2 z_2]^{-1}$$
$$\times (B_1 z_1 + B_2 z_2) + D \qquad (2)$$

where $x_{ij} \in R^n$ is the local state vector, $u_{ij} \in R^m$ is the input vector, $y_{ij} \in R^p$ is the output vector, and A_k, B_k ($k = 1, 2$) are real matrices. Matrices

$$A_1, A_2, B_1, B_2, C, D \qquad (3)$$

are called a realization of a given transfer matrix $T(z_1, z_2)$ if they satisfy (2). A realization (3) is called minimal if the matrices A_1 and A_2 have minimal dimension among all realizations of $T(z_1, z_2)$. The (minimal) realization problem can be stated as follows. Given a proper transfer matrix $T(z_1, z_2) \in R^{p \times m}(z_1, z_2)$, find matrices (3) (with minimal dimension of A_1 and A_2) that satisfy (2).

realization problem for 2-D Roesser model   the transfer matrix $T(z_1, z_2)$ of the 2-D Roesser model

$$\begin{bmatrix} x^h_{i+1,j} \\ x^v_{i,j+1} \end{bmatrix} = \begin{bmatrix} A_1 & A_2 \\ A_3 & A_4 \end{bmatrix} \begin{bmatrix} x^h_{ij} \\ x^v_{ij} \end{bmatrix} + \begin{bmatrix} B_1 \\ B_2 \end{bmatrix} u_{ij} \quad (1)$$

$$y_{ij} = C \begin{bmatrix} x^h_{ij} \\ x^v_{ij} \end{bmatrix} + D u_{ij}$$

$i, j \in Z_+$ (the set of nonnegative integers) is given by

$$T(z_1, z_2) = C \begin{bmatrix} I_{n_1}z_1 - A_1 & -A_2 \\ -A_3 & I_{n_2}z_2 - A_4 \end{bmatrix}^{-1}$$

$$\times \begin{bmatrix} B_1 \\ B_2 \end{bmatrix} + D \qquad (2)$$

where $x_{ij}^h \in R^{n_1}$ and $x_{ij}^v \in R^{n_2}$ are the horizontal and vertical state vectors, $u_{ij} \in R^m$ is the input vector, and $y_{ij} \in R^p$ is the output vector. Matrices

$$A = \begin{bmatrix} A_1 & A_2 \\ A_3 & A_4 \end{bmatrix}, \quad B = \begin{bmatrix} B_1 \\ B_2 \end{bmatrix}, \quad C, \ D \quad (3)$$

are called a realization of a given transfer matrix $T(z_1, z_2)$ if they satisfy (2). A realization (3) is called minimal if the matrix A has minimal dimension amongst all realizations of $T(z_1, z_2)$. The (minimal) realization problem can be stated as follows. Given a proper transfer matrix $T(z_1, z_2) \in R^{p \times m}(z_1, z_2)$, find matrices (3) (with minimal dimension of A) that satisfy (2).

received signal strength indicator (RSSI) ratio of signal power level for a single frequency or a band of frequencies to an established reference; the reference is typically 1 mW, and the resultant value is expressed in decibels. RSSI is often used in mobile communications to make assessments such as to which base station a call should be connected or which radio channel should be used for communication.

receiver noise   thermal (Boltzmann-type) noise in a receiver, a function of its physical temperature above absolute zero and the noise bandwidth of the receiver's electronic devices. Receiver noise causes finite receiver sensitivity. *See also* thermal noise.

receiver operating characteristics curve   plot of the probability of detection (likelihood of detecting the object when the object is present) versus the probability of false alarm (likelihood of detecting the object when the object is not present) for a particular processing system.

receiver sensitivity   the minimum radio signal power at the input to a receiver that results in signal reception of some stated quality.

reciprocity   (1) a consequence of Maxwell's equations, stipulating the phenomenon that the reaction of the sources of each of two different source distributions with the fields generated by the other are equal, provided the media involved have certain permeability and permittivity properties (reciprocal media). Referring to reciprocal circuits, reciprocity states that the positions of an ideal voltage source (zero internal impedance) and an ideal ammeter (infinite internal impedance) can be interchanged without affecting their readings.

(2) in antenna theory, the principal that the receive and transmit patterns of an antenna are the same.

reciprocity in scattering   law according to which the source and detector points can be exchanged, providing the source amplitude and phase are preserved.

reciprocity theorem   in a network consisting of linear, passive impedances, the ratio of the voltage introduced into any branch to the current in any other branch is equal in magnitude and phase to the ratio that results if the positions of the voltage and current are interchanged.

recloser   a self-contained device placed on distribution lines that senses line currents and opens on overcurrent. Reclosing is employed to reenergize the protected line segment in the case of temporary faults. Reclosers have the capability for fast tripping for fuse saving, and slow tripping to allow sectionalizing fuse operation for faults on laterals. The recloser will retrip on permanent faults and go on to lockout. Reclosers are suitable for pole mounting on overhead lines.

reclosing relay an auxiliary relay that initiates circuit breaker closing in a set sequence following fault clearing. Reclosing relays are typically employed on overhead lines where a high proportion of the faults are temporary.

recoil permeability the average slope of the minor hysteresis loop, which is roughly the slope of the major hysteresis loop at zero applied field (H), and is most often used to determine the effect of applying and removing a demagnetizing field to and from a magnetic material.

recombination the process in which an electron neutralizes a hole. Sometimes this process causes light emission (i.e., through radiative recombination), and sometimes it doesn't (i.e., through nonradiative recombination).

recombination X-ray laser an X-ray laser made by gain from an inverted population where the upper level is inverted due to recombining ions and electrons.

reconstruction (1) the process of forming a 3-D image from a set of 2-D projection images. Also applies to the formation of a 2-D image from 1-D projections. *See* image reconstruction, tomography and computed tomography.

(2) from marker in a binary image, this is the operation extracting all connected components having a non-empty intersection with a marker. This operation can be generalized to gray-level images by a morphological operator applying such a reconstruction on the gray-level slices of the image.

record unit of data, corresponding to a block, sector, etc., on a magnetic disk, magnetic tape, or other similar I/O medium.

recording code a line code optimized for recording systems. *See also* line code.

recording density number of bits stored per linear inch on a disk track. In general, the same number of bits are stored on each track, so that the density increases as one moves from the outermost to the innermost track.

recovery action that restores the state of a process to an earlier configuration after it has been determined that the system has entered a state which does not correspond to functional behavior. For overall functional behavior, the states of all processes should be restored in a manner consistent with each other, and with the conditions within communication links or message channels.

rectangle detection the detection of rectangle shapes, often by searching for corner signals, or from straight edges present in an image. Rectangle detection is important when locating machined parts in images, e.g. prior to robot assembly tasks. *See* polygon detection and square detection.

rectangular cavity a section of rectangular waveguide closed on both ends by conducting plates.

rectangular window (1) in finite impulse response (FIR) filter design, the rectangular window constituting the most straightforward window function used usually as a reference in studying other window functions. It is defined as 1 within an even interval centered at the origin and 0 elsewhere.

(2) in image processing, a rectangular area centered at a pixel under consideration. This area is known as a window, or a mask, or a template. A square window of size 3×3 is used most often. *See* neighborhood operation.

rectifier a circuit that changes an AC voltage to DC. Switching elements or diodes are used to create the DC voltage. Diode rectifiers and thyristor rectifiers are the two most commonly used rectifiers.

recurrent coding old name for convolutional coding.

recurrent network   a neural network that contains at least one feedback loop.

recursion   the process whereby a program calls itself. *See also* recursive procedure.

recursive equation   a difference equation that is of the form

$$y(k) = \sum_{i=0}^{i=m} a_i x(k - i) - \sum_{i=1}^{i=n} b_i y(k - i)$$

where a_i and b_i are some proper real constants. When all the $b_i = 0$, it is called a nonrecursive equation.

recursive filter   a digital filter that is recursively implemented. That is, the present output sample is a linear combination of the present and past input samples as well as the previously determined outputs. Traditionally, the term recursive filter is closely related to infinite impulse filter. In a nonrecursive filter the present output sample is only a linear combination of the present and past input samples.

recursive function   *See* recursive procedure.

recursive method   method that estimates local displacements iteratively based on previous estimates. Iterations are performed at all levels, as in every pixel, each block of pixels, along scanning line, from line to line, or from frame to frame.

recursive procedure   a procedure that can be called by itself or by another program that it has called; effectively, a single process can have several executions of the same program alive at the same time. Recursion provides one means of defining functions.

The recursive definition of the factorial function is the classic example:
if n = 0
    factorial(n) = 1
else
    factorial(n) = n * factorial (n − 1)

recursive self-generating neural network (RSGNN)   a recursive version of self-generating neural network (SGNN) that can discover recursive relations in the training data. It can be used in applications such as natural language learning/understanding, continuous spoken language understanding, and DNA clustering/classification, etc. *See also* self-generating neural network.

Red book   *See* IEEE color books.

red head   another name for a hot tap.

reduced characteristic table   a tabular representation used to illustrate the operation of various bistable devices.

reduced instruction set computer (RISC) processor   relatively simple control unit design with a reduced menu of instructions (selected to be simple), data and instructions formats, addressing modes, and with a uniform streamlined handling of pipelines.

One of the particular features of a RISC processor is the restriction that all memory accesses should be by load and store instructions only (the so called load/store architecture). All operations in a RISC are register-to-register, meaning that both the sources and destinations of all operations are CPU registers. All this tends to significantly reduce CPU to memory data traffic, thus improving performance. In addition, RISCs usually have the following properties: most instructions execute within a single cycle, all instructions have the same standard size (32 bits), the control unit is hardwired (to increase speed of operations), and there is a CPU register file of considerable size (32 registers in most systems, with the exception of SPARC with 136 and AMD 29000 with 192 registers).

Historically, the earliest computers explicitly designed by these rules were designs by Seymour Cray at CDC in the 1960s. The earliest development of the RISC philosophy of design was given by John Cocke in the late 1970s at IBM. However, the term RISC was first coined by Patterson *et al.* at the University of California at Berkeley to describe a computer with an instruction set designed for maximum execution speed on a particular class of computer programs. Patterson and his team of researchers developed the first single-chip RISC processor.

Compare with complex instruction set computer.

reduced-order model   a mathematical representation of a system that is obtained by neglecting portions of a more explicit (detailed) model. In large-scale power system analysis, this term is typically used to indicate a model derived by neglecting the electric transients in the stator voltage equations of all machines and in the voltage equations of power system components connected to the stators of the machines.

reduced-voltage motor starter   a device designed to safely connect an electric motor to the power source while limiting the magnitude of its starting current. Various electromechanical configurations may be used: primary resistor, delta-wye, part-winding (requires special motor or dual voltage windings). Power electronic devices may also be utilized to gradually increase the applied voltage to system levels. The complete starter must also include fault and overload protection.

redundancy   (1) the use of parallel or series components in a system to reduce the possibility of failure. Similarly, referring to an increase in the number of components which can interchangeably perform the same function in a system. Sometimes it is referred to as hardware redundancy in the literature to differentiate from so called analytical redundancy in the field of FDI (fault detection and isolation/identification). Redundancy can increase the system reliability.

(2) in robotics, the number n degrees of mobility of the mechanical structure, the number m of operational space variables, and the number r of the operational space variables necessary to specify a given task. Consider the differential kinematics mapping $v = J(q)\dot{q}$ in which v is $(r \times 1)$ vector of end-effector velocity of concern for the specific tasks and J is $r \times n$ Jacobian matrix. If $r < n$, the manipulator is kinematically redundant and has $(n-r)$ redundant degrees of mobility. Manipulator can be redundant with respect to a task and nonredundant with respect to another. *See also* redundant manipulator.

redundancy encoding   any digital encoding scheme which takes advantage of redundancy in the digital signal. For example, in run-length encoding, a gray scale digital image is represented by the gray level of a pixel and the number of times adjacent pixels with that gray level appear. So an image containing large regions of a single gray level can be represented with a great reduction in digital information.

redundancy statistics model   refers to statistical similarities such as correlation and predictability of data. Statistical redundancy can be removed without destroying any information.

redundancy-free channel coding   refers to methods for channel robust source coding where no "explicit" error protection is introduced. Instead, knowledge of the source and source code structure is utilized to counteract transmission errors (for example, by means of an efficient index assignment).

redundant array of inexpensive disks (RAID) standardized scheme for multiple-disk data base systems viewed by the operating system as a single logical drive. Data is distributed across the physical drives allowing simultaneous access to data from multiple drives, thereby reducing the gap between processor speeds and relatively slow electromechanical disks. Redundant disk capacity can also be used to store additional information to guarantee data recoverability in case of disk failure (such as parity or data duplication). The RAID scheme consists of six levels (0 through 5),

RAID$_0$ being the only one that does not include redundancy.

redundant manipulator   the manipulator is called redundant if more degrees of mobility are available than degrees of freedom required for the execution of a given task. *See also* differential kinematics, redundancy.

redundant number system   the system in which the numerical value could be represented by more than one bit string.

Reed switch   a magnetomechanical device composed of two thin slats of ferromagnetic material within a hermetically sealed capsule that attract each other when an external magnetic field (from an electromagnet or permanent magnet) induces opposite poles at the overlapping ends of both slats.

Reed–Solomon code   an extension of BAH codes to nonbinary alphabets developed by Driving Reed and Gustave Solomon independently of the work by Bose, Chaudhuri, and Hocquenghem. Arguably, the most widely used of any forward error control code.

reentrancy   the characteristic of a block of software code that, if present, allows the code in the block to be executed by more than one process at a time.

reentrant   a program that uses concurrently exactly the same executable code in memory for more than one invocation of the program (each with its own data), rather than separate copies of a program for each invocation. The read and write operations must be timed so that the correct results are always available and the results produced by an invocation are not overwritten by another one.

reference black level   picture signal level corresponding to a specified maximum limit for black peaks.

reference frame   *See* base frame.

reference impedance   impedance to which scattering parameters are referenced.

reference matrix   a triangular array of bits used to implement the least recently used algorithm in caches. When the ith line is referenced, all the bits in the ith row are set to a 1 and then all the bits in the ith column are set to a 0. Having 0s in the jth row and 1s in the jth column identifies the jth line as the least recently used line.

reference monitor   a functional module that checks each attempt to access memory to determine whether it violates the system's security policy, intercepting it if a violation is imminent. The memory management unit can provide this service, provided that the access control information contained there is known to be consistent with the security policy.

reference node   one node in a network that is selected to be a common point, and all other node voltages are measured with respect to that point.

reference point   *See* set point.

reference white   in a color matching process, a white with known characteristics used as a reference. According to the trichromatic theory, it is possible to match an arbitrary color by applying appropriate amounts of three primary colors.

reference white level   picture signal level corresponding to a specified maximum limit for white peaks.

reflectance   a property of the object and is independent of the illumination on that object. It is often important, especially in industrial inspection, to be able to determine the reflectance under varying illumination. *See* illumination, industrial inspection.

reflected power   power in the reflected part of an electromagnetic wave.

reflected wave   the result that ensues when a high-speed electromagnetic wave reaches the end of a transmission line, when the line is not terminated with an impedance matching the surge impedance of the line. When a surge reaches an open circuited line terminal, the reflected voltage wave equals the incident voltage wave, resulting in a doubling of the level of the voltage surge at that point.

reflection   in electromagnetic wave propagation, the change in direction of propagation of a plane wave due to the wave being incident on the surface of a material. Typically, the effect is greater in the case of a material that has a high electrical conductivity.

reflection coefficient   (1) the ratio of the reflected field to the incident field at a material interface.

(2) another way of expressing the impedance. The reflection coefficient is defined as how much signal energy would be reflected at a given frequency. Like impedance, the reflection coefficient will vary with frequency if inductors or capacitors are in the circuit. The reflection coefficient is always defined with respect to a reference or characteristic impedance $(= (Z - Z_0)/(Z + Z_0))$. For example, the characteristic impedance of one typical TV transmission line is 75 ohms, whereas another type of TV transmission line has a characteristic impedance of 300 ohms. Hooking up a 75-ohm transmission line to a 300-ohm transmission line will result in a reflection coefficient of value $(300 - 75)/(300 + 75) = 0.6$, which means that 60% of the energy received from the antenna.

reflection grating   a diffraction grating that operates in reflection, i.e., the diffracted light is obtained by reflecting off the grating.

reflective notching   an unwanted notching or feature size change in a photoresist pattern caused by the reflection of light off nearby topographic patterns on the wafer.

reflectivity   a property that describes the reflected energy as a function of the incident energy of an EM wave and a material body. The property may be quantified in terms of the magnitude of the reflection coefficient or the ratio of the incident to the reflected field.

reflectometer   instrument that measures reflected power.

reflector antenna   an antenna comprised of one or more large reflecting surfaces used to focus intercepted power around a small region where a primary feed antenna is then used to input the power to the receiver electronics. Also used to transmit power in narrow angular sectors.

reflex klystron   a high-power microwave tube oscillator.

refraction   the process undertaken by an electromagnetic wave wherein the wave changes direction of propagation as it is incident on the edge of a material. The wave undergoes a "bending" action, sometimes referred to as knife edge refraction, and the "bending" angle is less than 90 degrees. Refraction may also occur as a wave propagates through a media such as the atmosphere.

refractive index   a parameter of a medium equal to the ratio of the velocity of propagation in free space to the velocity of propagation in the medium. It is numerically equal to the square root of the product of the relative permittivity and relative permeability of the medium. *See also* index of refraction.

refractivity   the refractive index minus 1.

refractory period   a period of time after the initiation of an action during which further excitation is impossible (absolute refractory period) or requires a greater stimulus (relative refractory period).

refresh   refers to the requirement that dynamic RAM chips must have their contents periodically refreshed or restored. Without a periodic refresh, the chip loses its contents. Typical refresh times are in the 5–10 millisecond range. *See also* memory refresh.

refresh cycle   (1) a periodically repeated procedure that reads and then writes back the contents of a dynamic memory device. Without this procedure, the contents of dynamic memories will eventually vanish.

(2) the period of time taken to "refresh" a portion of a dynamic RAM chip's memory. *See also* refresh.

refresh period   the time between the beginnings of two consecutive refresh cycles for dynamic random access memory devices.

regeneration   the process of returning energy back into a system during a portion of the machine's operating cycle.

regeneration loop   a water purification system used to maintain proper conditions of the cooling liquid for a power vacuum tube.

regenerative braking   (1) a method for extracting kinetic energy from the load, converting it back to electricity, and returning it to the supply. Used widely in electric train drives and electric vehicles.

(2) in electric vehicles, recapturing electric energy during braking. In regenerative braking, kinetic energy of the vehicle is processed by the electric machine and returned to the energy source. In this scheme, the induction machine works as a generator when it is operated with a negative slip, that is, the synchronous speed is less than the motor speed. Negative slip makes the electromagnetic torque negative during regeneration,

inducing voltages and currents. Electromagnetic torque acts on the rotor to oppose the rotor rotation, thereby decelerating the vehicle. *See* electric vehicle.

region growing   the grouping of pixels or small regions in an image into larger regions. Region growing is one approach to image segmentation. *See* dilation, erosion, image segmentation, mathematical morphology.

region of absolute convergence   the set of complex numbers s for which the magnitude of the Laplace transform integral is finite. The region can be expressed as

$$\sigma_+ < Re(s) < \sigma_-$$

where σ_+ and σ_- denote real parameters that are related to the causal and anticausal components, respectively, of the signal whose Laplace transform is being sought. $Re(s)$ represents the real part of s.

region of asymptotic stability   *See* region of attraction.

region of attraction   the region around an equilibrium state of a system of differential or difference equations such that the trajectories originating at the points in the region converge to the equilibrium state. Trajectories starting outside the region of attraction of the given equilibrium state may "run away" from that equilibrium state.

region of convergence (ROC)   (1) the set of complex numbers $s = a + jb$ for which the magnitude of the Laplace transform integral is finite. The region can be expressed as: $\sigma_- < Real(s) < \sigma_+$, $where\ \sigma_-$ and σ_+ denote real parameters that are related to the causal and anticausal components, respectively, for the signal whose Laplace transform is being sought.

(2) the set of complex numbers $z = e^{st}$ for which the magnitude of the z-transform sum is finite. The region can be expressed as:

$R_- < |z| < R_+$, where R_- and R_+ denote real parameters that are related to the causal and anticausal components, respectively, for the signal whose z-transform is being sought.

(3) an area on a display device where the image displayed meets an accepted criteria for raster coordinate deviation. *See* region of absolute convergence.

region of interest (ROI)   a restricted set of image pixels upon which image processing operations are performed. Such a set of pixels might be those representing an object that is to be analyzed or inspected.

region of support   the region of variable or variables where the function has non-zero value.

register   a circuit formed from identical flip-flops or latches and capable of storing several bits of data.

register alias table   *See* virtual register.

register direct addressing   an instruction addressing method in which the memory address of the data to be accessed or stored is found in a general purpose register.

register file   a collection of CPU registers addressable by number.

register indirect addressing   an instruction addressing method in which the register field contains a pointer to a memory location that contains the memory address of the data to be accessed or stored.

register renaming   dynamically allocating a location in a special register file for an instance of a destination register appearing in an instruction prior to its execution. Used to remove antidependencies and output dependencies. *See also* reorder buffer.

register transfer notation   a mathematical notation to show the movement of data from one register to another register by using a backward arrow. Notation used to describe elementary operations that take place during the execution of a machine instruction.

register window   in the SPARC architecture, a set or window of registers selected out of a larger group.

registration   (1) *See* overlay.

(2) the process of aligning multiple images obtained from different modalities, at different time-points, or with different image acquisition parameters. *See* fusion.

regression   the methods which use backward prediction error as input to produce an estimation of a desired signal. Quantitatively, the regression of y on X, denoted by $r(y)$, is defined as the first conditional moment, i.e.,

$$r(y) = E(X|y)$$

regular controllability   a dynamical system is said to be regularly controllable in time interval $[t_0, t_1]$ if every dynamical system of the form

$$x'(t) = Ax(t)x(t) + b_j u_j(t)$$
$$j = 1, 2, \ldots, m$$

is controllable where b_j is the jth column of the matrix B and $u_j(t)$ is the jth scalar admissible control.

regular cue   any regular recurring point/element of a signal that can be used to signal the start of a new signal sequence; e.g., the leading edge of a 60-Hz square wave is a regular cue.

regular form   a particular form of the state space description of a dynamical system. This form is obtained by a suitable transformation of the system state. The regular form is useful in control design.

regularization   a procedure to add a constraint term in the optimization process that has a stabilizing effect on the solution.

regulation   the change in voltage from no-load to full-load expressed as a percentage of full-load voltage.

regulator   a controller designed to maintain the state of the controlled variable at a constant value, despite fluctuations of the load.

reinforcement learning   learning on the basis of a signal that tells the learning system whether its actions in response to an input (or series of inputs) are good or bad. The signal is usually a scalar, indicating how good or bad the actions are, but may be binary.

rejection criteria   criteria such as poor surface texture, existence of scratch marks, out-of-tolerance distance measures, which constitute reasonable grounds for rejecting a product from a product line.

related color   color perceived to belong to an area of object seen in relation to other colors. *See* unrelated color.

relational model   a logical data structure based on a set of tables having common keys that allows the relationship between data items to be defined without considering the physical database organization.

relative addressing   an addressing mechanism for machine instructions in which the address of the target location is given by the contents of a specific register and an offset held as a constant in the instruction added together. *See also* PC-relative addressing, index register, base address.

relative-address coding   in facsimile coding, represents the transition between levels on a particular scan line relative to transitions on the preceding scan line. A relative-address coding system has a pass mode codeword for indicating where a pair of transitions on the previous line does not have corresponding transitions on the current line, and a runlength coding mode applied when there is no nearby suitable transition on the previous line. CCITT Group IV facsimile uses a form of relative-address coding.

relative controllability   a dynamical system with delays in which for a given time interval $[0, t_1]$ if for any initial condition $(x(0), x_0)$ and any final vector $x_1 \in R^n$ there exists an admissible control $u(t), t \in [0, t_1]$, such that the corresponding trajectory satisfies the condition

$$x(t_1, x(0), x_0, u) = x_1$$

relative entropy   information theoretic quantity representing the "distance" between two probability distribution functions. Also known as the Kullback–Liebler distance. For two probability mass functions $p(x)$ and $q(x)$, the relative entropy $D(p||q)$ is given by

$$D(p||q) = \sum p(x) \log \frac{p(x)}{q(x)}.$$

This quantity is only a pseudo-distance, as $D(p||q) \neq D(q||p)$.

relative intensity noise   noise resulting from undesirable fluctuations of the optical power detected in an optical communication system.

relative permeability   the complex permeability of a material divided by the permeability of free space: $\mu_r = \mu/\mu_0$.

relative permittivity   the complex permittivity of a material divided by the permittivity of free space: $\epsilon_r = \epsilon/\epsilon_0$.

relative refractive index difference   the ratio $(n_1^2 - n_2^2)/2n_1^2 \approx (n_1 - n_2)/n_1$ where $n_1 > n_2$ and n_1 and n_2 are refractive indices.

relative sensitivity   denoted $\mathbf{S}_x^y(y, x)$ and defined as follows:

$$\mathbf{S}_x^y(y, x) = \frac{\partial y}{\partial x}\frac{x}{y} = \frac{\partial y/y}{\partial x/x} = \frac{\partial \ln y}{\partial \ln x}$$

It is usually used to establish the approximate relationship between the relative changes $\delta y = \Delta y/y$ and $\delta x = \Delta x/x$. Here Δy and Δx are absolute changes. If these relative changes are small, one writes that

$$\delta y \approx \mathbf{S}_x^y \delta x$$

This relationship assumes that $\mathbf{S}_x^y$ is different from zero. If $\mathbf{S}_x^y = 0$, the relative changes δy and δx may be independent. The properties of relative sensitivity established by differentiation only are tabulated and may be found in many textbooks and handbooks. *See also* absolute sensitivity, sensitivity, sensitivity measure, semi-relative sensitivity.

relaxation   (1) a general computational technique where computations are iterated until certain parameter measurements converge to a set of values.

(2) the response of a linear time invariant system can be represented as the sum of the zero-input response (system response to a zero input function) plus the zero-state response (system response to an input function when the system is in the zero state). Relaxation is the process of putting a system into its zero-state, i.e. all initial conditions are zero and there are no internal energy stores. A system is considered relaxed if it is in the zero state. *See* relaxation labelling, optimization.

relaxation labelling   an iterative mathematical procedure in which a system of values is processed, e.g. by mutual adjustment of adjacent or associated values, until a stable state is attained. Especially useful for achieving consistent optimal

estimates of pixel intensities or deduced orientation values for points on the surface of an object. *See* relaxation, optimization.

relaxation oscillations   the damped output oscillations that occur in some laser oscillators when they are perturbed from steady state.

relaxation time   the time in which the initial distribution of charge will diminish to 1/e of its original value.

relay   a device that opens or closes a contact when energized. Relays are most commonly used in power systems, where their function is to detect defective lines or apparatus or other abnormal or dangerous occurrences and to initiate appropriate control action. When the voltage or current in a relay exceeds the specified "pickup" value, the relay contact changes its position and causes an action in the circuit breaker. A decision is made based on the information from the measuring instruments and relayed to the trip coil of the breaker, hence the name "relay." Other relays are used as switches to turn on or off equipment.

relay channel   a multiterminal channel in which the receiver observes the transmitted signal through two channels: one direct to the transmitter, the other via an intermediate transmitter/receiver pair.

reliability   the probability that a component or system will function without failure over a specified time period, under stated conditions.

reliability criteria   a set of operating conditions that the system operator adheres to in order to guarantee secure operation.

relocatability   the capability for a program to be loaded into any part of memory that is convenient and still execute correctly.

relocation register   register used to facilitate the placement in varying locations of data and instructions. Actual addresses are calculated by adding program-given addresses to the contents of one or more relocation registers.

reluctance the resistance to magnetic flux in a magnetic circuit; analogous to resistance in an electrical circuit.

reluctance motor a motor constructed on the principle of varying reluctance of the air gap as a function of the rotor position with respect to the stator coil axis. The torque in these motors arises from the tendency of the rotor to align itself in the minimum reluctance position along the length of the air gap.

reluctance torque the type of torque a reluctance machine's operation is based upon. A reluctance torque is produced in a magnetic material in the presence of an external magnetic field, which makes it line up with the external magnetic field. An induced field due to fringing flux develops a torque that eventually twists the magnetic material around to align itself with the external field.

remanent coercivity the magnetic field required to produce zero remanent magnetization in a material after the material was saturated in the opposite direction.

remanent magnetism *See* residual magnetism.

remanent polarization the residual or remanent polarization of a material after an applied field is reduced to zero. If the material was saturated, the remanent value is usually referred to as the polarization, although even at smaller fields a (smaller) polarization remains.

remanence (1) in a ferromagnetic material, the value of the magnetic flux density when the magnetic field intensity is zero.
(2) the magnetic induction (B) of a magnet after the magnetizing field is removed and an air gap (hence self-demagnetizing field) is introduced

to the magnetic circuit. Also called retentivity or residual induction.

remote sensing the use of radar or radiometry to gather data about a distant object. Usually, the term refers to the use of microwaves or millimeter waves to map features or characteristics of planetary surfaces, especially the Earth's. Applications include military, meteorological, botanical, and environmental investigations.

remote terminal unit (RTU) hardware that gathers system-wide real-time data from various locations within substations and generating plants for telemetry to the energy management system.

removable disk disk that can be removed from disk drive and replaced, in contrast with a nonremovable disk, which is permanently mounted. *See also* exchangeable disk.

rename register *See* virtual register.

rendering preparation of the representation of an image to include illumination, shading depth curing, coloring, texture, and reflection. Common techniques include Phong and Gouraud shading; more complex rendering models such as raytracing and radiosity emphasize realistic physics models for calculating light interactions and texture interactions with objects.

renewable fuse a fuse consisting of a reusable cartridge and a fusible element that can be replaced.

reorder buffer a set of storage locations provided for register renaming for holding results of instructions. These results may be generated not in program order. At some stage, the results will be returned to the true destination registers.

repeatability the ability of a sensor to reproduce output readings for the same value of measurand, when applied consecutively and under the same conditions.

repeater   electromagnetic device that receives a signal and amplifies it and retransmits it. In digital systems, the signal is regenerated.

repetition coding   the simplest form of error control coding. The information symbol to be transmitted is merely repeated an uneven number of times. A decision regarding the true value of the symbol transmitted is then simply made by deciding which symbol occurred the greatest number of times.

representation singularity of ϕ a set of orientations for which the determinant of the transformation matrix in the analytical Jacobian vanishes. *See also* analytical Jacobian.

representative level   one of the discrete output values of a quantizer used to represent all input values in a range about the representative level. *See* decision level.

reprocessing   the recycling of reactor fuel by separation into fissile, non-fissile, radioactive and non-radioactive components such that wastes can be isolated and fissile material re-used to make more reactor fuel.

repulsion–induction motor   a single-phase motor designed to start as a repulsion motor, then run as an induction motor. The rotor has a DC-type winding with brushes shorted together, in addition to the normal squirrel cage winding. Although it is an expensive design, it provides excellent starting torque with low starting current (similar to a universal motor) and relatively constant speed under load.

requirements analysis   a phase of software development life cycle in which the business requirements for a software product are defined and documented.

reservation ALOHA (R-ALOHA)   a class of multiple access control protocol in which the user transmits the first packet in random access fashion. If successful, the user will have a fixed part of the channel capacity allocated.

reservation station   storage locations placed in front of functional units and provided to hold instructions and associated operands when they become available. Used in a superscalar processor.

reset scrambling   a technique of randomizing a source bit sequence by adding the sequence to a pseudo-random bit stream using element-wise modulo-2 arithmetic. The source sequence is recovered at the decoder through addition of the demodulated bit stream with the same pseudo-random sequence. The pseudo-random sequences are usually generated with linear feedback shift registers which must be aligned with a reset or framing signal in order to recover the source sequence accurately.

reset time   for a line recloser or circuit breaker, a time begins when the device successfully recloses following a temporary fault. After the reset time elapses, the fault cycle is considered over, and any subsequent fault will be treated as the first fault in a new cycle.

residential loop   two radial sources coming in to an open point used for switching.

residual current circuit breaker   European term for ground fault interrupter.

residual error   the degree of misfit between an individual data point and some model of the data. Also called a "residual."

residual inductance   *See* remanence.

residual magnetism   a form of permanent magnetism, referring to the flux remaining in a ferromagnetic material after the MMF that created the flux is removed. For example, if a bar of steel is surrounded by a coil and current is applied to the coil, the steel bar will create a magnetic field due to the rotation of the domains in the steel. After the current is removed, some of the domains will remain aligned, causing magnetic flux.

residual overcurrent relay   an overcurrent relay that is connected to sense residual current. Residual current is the sum of the three phase currents flowing in a current transformer secondary circuit, and is proportional to the zero sequence current flowing in the primary circuit at that point.

residual pyramid   *See* predictive pyramid.

residual vector quantization   *See* interpolative vector quantization.

resistivity   (1) the product of the resistance of a given material sample times the ratio of its cross-sectional area to its length.

(2) an electrical material property described by a tensor constant indicating the impedance of free electron flow in the material. Resistivity relates the electric field strength to the conduction current, and can be expressed as the inverse of the conductivity.

resistance   ratio of the potential of an electrical current applied to a given conductor to the current intensity value.

resistance ground   a grounding scheme in which the neutral of Y-connected machines is connected to ground through a resistance such that ground-fault currents are limited.

resistance grounded   *See* low-resistance grounded, high-resistance grounded.

resistive mixer   a device used to convert microwave frequencies to intermediate frequencies. Depending on the frequency of microwaves to be converted, the intermediate frequencies can be UHF, VHF, or HF.

resolution   (1) the act of deriving from a sound, scene, or other form of intelligence, a series of discrete elements from which the original may subsequently be reconstructed. The degree to which nearly equal values of a quantity can be discriminated.

(2) the fineness of detail in a measurement. For continuous systems, the minimum increment that can be discerned.

(3) the ability to distinguish between two units of measurement.

(4) the number of pixels per linear unit (or per dimension) in a digital image.

(5) the smallest feature of a given type that can be printed with acceptable quality and control.

resolvable component   a component of a signal that exists in a group of components (a sum of signals) such that the amplitude of the component can be approximately determined by correlation of the overall signal with a component signal of unit amplitude. Typically, the different signal components are delayed versions of the same signal received at a receiver in a channel with multipath propagation. The different delay components are then resolvable if the relative delay between any two components is greater than the reciprocal of the transmitted signal bandwidth.

resolver   an electric machine that is used to provide information about the position of a motor-driven system. The shaft of the resolver is connected to the main motor either directly or through gears. The rotor contains a single winding, while the stator typically contains two windings in quadrature. One stator winding and the rotor winding are excited, while the remaining stator winding is shorted. When the resolver is rotated by the main motor, voltages are produced in the stator windings that can be used to determine the position. For multi-turn systems, two resolvers (coarse and fine) may be required.

resonance   in an RLC circuit, the resonance is the state at which the reactance of the inductor,

X_L, and the reactance of the capacitor, X_C, are equal.

$$X_L = 2\pi f_L$$
$$X_C = 1/(2\pi f_C)$$

resonance fluorescence the modified fluorescence produced when a quantum mechanical system is strongly driven by one or more near-resonant electromagnetic fields.

resonance Raman system Raman systems that have a near-resonant intermediate state. Resonance Raman systems are sometimes referred to as lambda systems and can exhibit coherent population trapping.

resonant in any circuit or system under excitation, the frequency at which a pair of reactive components cancels (pole or zero) resulting in a natural mode of vibration.

resonant antenna linear antennas that exhibit current and voltage standing wave patterns formed by reflections from the open end of the wire.

resonant cavity cavity with reflecting surfaces or mirrors that can support low-loss oscillations. By closing a metallic waveguide by two metallic surfaces perpendicular to its axis, a cylindrical cavity is formed. Resonant modes in this cavity are designated by adding a third subscript so as to indicate the number of half-waves along the axis of the cavity. When the cavity is a rectangular parallelepiped, the axis of the cylinder from which the cavity is assumed to be made should be designated, since there are three possible cylinders out of which the parallelepiped may be made. More generally, each closed cavity may sustain a discrete infinity of resonant field distributions.

resonant frequency (1) a frequency at which the input impedence of an device is nonreactive, since the capacitive and inductive stored energy cancel each other.

(2) an oscillation frequency of the modes of a resonator.

resonant link inverter an inverter that uses a resonant circuit to convert a constant DC voltage to a pulsating DC voltage. The switching elements in the inverter are then turned off during the times that the input voltage is zero, a technique referred to as soft-switching. Resonant switching techniques reduce the switching losses and allow high switching frequency operation to reduce the size of magnetic components in the inverter unit.

resonant tunneling refers to the process of resonant enhancement of electron tunneling by intermediate energy states. In the simplest case, it occurs when incoming electrons coincide in energy with the states created in the well.

resonator (1) circuit element or combination of elements, which may be either lumped or distributed, that exhibit a resonance(s) at one or more frequencies. Generally, a resonant condition coincides with the frequency where the impedance of the circuit element(s) is only resistive.

(2) cavity with reflecting surfaces designed to support low-loss oscillation modes.

See also bidirectional resonator, concentric resonator, confocal resonator, high-loss resonator, plane-parallel resonator, ring resonator, standing-wave resonator, unidirectional resonator, unstable resonator.

resonator stability perturbation stability of an axial light ray in a resonator; boundedness of ray trajectories; corresponds to confinement of the resonator modes; not the same as mode stability (unstable resonators have stable modes).

resource auction multiple access (RAMA) a multiple access protocol that stipulates part of the frame for contention. Unlike PRMA and D-TDMA, the contention is not performed by an ALOHA-type of protocol but by an auction — a tree-sorting type of algorithm.

resource conflict the situation when a component such as a register or functional unit is required by more than one instruction simultaneously. Particularly applicable to pipelines.

response formula for general 2-D model the solution to the general 2-D model

$$x_{i+1,j+1} = A_0 x_{ij} + A_1 x_{i+1,j} + A_2 x_{i,j+1}$$

$$+ B_0 u_{ij} + B_1 u_{i+1,j} + B_2 u_{i,j+1} \quad \text{(1a)}$$

$$y_{ij} = C x_{ij} + D u_{ij} \quad \text{(1b)}$$

$i, j \in Z_+$ (the set of nonnegative integers) with boundary conditions x_{i0}, $i \in Z_+$ and x_{0j}, $j \in Z_+$ is given by

$$x_{ij} = \sum_{p=1}^{i} \sum_{q=1}^{j} (T_{i-p-1,j-q-1} B_0$$

$$+ T_{i-p,j-q-1} B_1 + T_{i-p-1,j-q} B_2) u_{pq}$$

$$+ \sum_{p=1}^{i} (T_{i-p,j-1}[A_1 \quad B_1]$$

$$+ T_{i-p-1,j-1}[A_0 \quad B_0]) \begin{bmatrix} x_{p0} \\ u_{p0} \end{bmatrix} \quad \text{(2)}$$

$$+ \sum_{q=1}^{j} (T_{i-1,j-q}[A_2 \quad B_2]$$

$$+ T_{i-1,j-q-1}[A_0 \quad B_0]) \begin{bmatrix} x_{0q} \\ u_{0q} \end{bmatrix}$$

$$+ T_{i-1,j-1}[A_0 \quad B_0] \begin{bmatrix} x_{00} \\ u_{00} \end{bmatrix}$$

where $x_{ij} \in R^n$ is the local state vector, $u_{ij} \in R^m$ is the input vector, A_k, B_k $(k = 0, 1, 2)$ are given real matrices, and the transition matrix T_{pq}

is defined by

$$T_{pq} = \begin{cases} I \text{ for } p = q = 0 \\ A_0 T_{p-1,q-1} + A_1 T_{p,q-1} \\ + A_2 T_{p-1,q} \text{ for } p, q \geq 0 \, (p + q \neq 0) \\ 0 \text{ for } p < 0 \text{ or/and } q < 0 \end{cases}$$

Substitution of (2) into (1b) yields the response formula.

restart the act of restarting a hardware or software process.

restoration the act of restoring electric service to a consumer's facility. A distinction is made between restoration and repair as repair implies that whatever was the cause of the outage has been corrected, whereas restoration implies that the power was restored but it may have been through some means other than a repair, for example through switching.

restore instruction an assembly language instruction that restores the machine state of a suspended process to the active state.

retentivity *See* remanence.

reticle *See* mask.

retiming the technique of moving the delays around the system. Retiming does not alter the latency of the system.

retire unit in modern CPU implementations, the module used to assure that instructions are completed in program order, even though they may have been executed out of order.

retrace blanking blanking of a display during vertical retrace to prevent the retrace line from showing on the display.

retrace switch electronic method used to blank a display during retrace blanking.

retrace time   amount of time that a blanked vertical retrace takes for a display device. Note that this time is less than the time the display is blanked.

retrograde channel   for a MOSFET semiconductor device, the channel region whose doping is less at the surface of the channel than it is at some depth below the surface.

retry   a repetition of an operation, disturbed by a transient fault, to obtain good result.

return address   the address of an instruction following a Call instruction, where the program returns after the execution of the Call subroutine.

return difference matrix at the input   the transfer function matrix relating the difference between the injected and the returned signal in a feedback loop that has been opened at the input. This transfer function matrix characterizes the effect of closing the loop and is the mathematical entity that is central to numerous control engineering theories. For a nonunity feedback configuration with both pre- and postcompensation, it is given by

$$1 + FGK(s)$$

See also return difference matrix at the output.

return difference matrix at the output   the transfer function matrix relating the difference between the injected and the returned signal in a feedback loop that has been opened at the output. For a nonunity feedback configuration with both pre- and postcompensation, it is given by

$$1 + GKF(s)$$

See also return difference matrix at the input.

return instruction   an instruction, when executed, gets the address from the top of the stack and returns the program execution to that address.

return loss   usually expressed in decibels, it is the magnitude of the reflection coefficient; the return loss is a measure of the power reflected due to impedance mismatch of an antenna or other device.

return stroke   in lightning, the upward propagating high-current, bright, potential discontinuity following the leader that discharges to the ground some or all of the charge previously deposited along the channel by the leader.

return-to-bias recording   *See* magnetic recording code.

return-to-zero recording   *See* magnetic recording code.

reusability   the possibility to use or easily adapt the hardware or software developed for a system to build other systems.

reuse   programming modules are reused when they are copied from one application program and used in another. Reusability is a property of module design that permits reuse.

reuse ratio   the ratio of the physical distance between the centers of radio communication cells and the nominal cell radius. This term is usually used with reference to the reuse of particular radio channels (e.g., cochannel reuse ratio).

reverberation   inhomogeneities, such as dust, sea organisms, schools of fish, and sea mounds on the bottom of the sea, form mass density discontinuities in the ocean medium. When an acoustic wave strikes these inhomogeneities, some of the acoustic energy is reflected and reradiated. The sum total of all such reradiations is called reverberation. Reverberation is present only in active sonar, and in the case where the object echoes and is completely masked by reverberation, the sonar system is said to be "reverberation-limited."

reverse breakdown   the diode operating region in which significant current flows from cathode to anode, due to an applied voltage exceeding the breakdown voltage.

reverse breakdown region   the region of the I-V curve(s) of a device in which the device is operating in avalanche or zener breakdown.

reverse breakdown voltage   *See* reverse breakdown region.

reverse conducting thyristor (RCT)   a variety of asymmetric silicon controlled rectifier in which the diode is integrated into the thyristor structure.

reverse engineering   the reverse analysis of an old application to conform to a new methodology.

reverse generation-recombination current part of the reverse current in a diode caused by the generation of hole-electron pairs in the depletion region. This current is voltage-dependent because the depletion region width is voltage-dependent.

reverse isolation   device or circuit loss in a path that is the reverse of the path normally desired, expressed as the power ratio in decibels of the RF power delivered to a load at the input port versus the RF power incident at the output port. The negative sign results from the term "loss."

reverse leakage current   a nondestructive current flowing through a capacitor subjected to a voltage of polarity opposite to that normally specified.

reverse link   *See* uplink.

reverse saturation current   part of the reverse current in a diode caused by diffusion of minority carriers from the neutral regions to the depletion region. This current is almost independent of the reverse voltage.

reverse translation buffer   *See* inverse translation buffer.

reverse voltage   the voltage across the device when the anode is negative with respect to the cathode.

reversible   a motor capable of running in either direction, although it may be necessary to rewire the connections to the motor to change the direction of rotation. *See also* reversing.

reversible loss   a decrease in magnetic induction (B) of a permanent magnet when subjected to thermal or magnetic demagnetization that is fully recovered (without remagnetization) when the detrimental conditions are removed.

reversible temperature coefficient   for permanent magnets, a quantity that indicates the reversible change in magnetic induction with temperature.

reversing   a motor that can be run in either direction through the use of suitable switches or contactors.

reversing motor starter   a motor controller capable of accelerating a motor from rest to normal speed in either direction of rotation. Some reversing motor starters can go directly from forward to reverse (or vice versa), while others must be stopped before a reversal of direction can take place. Both electromechanical and electronic reversing starters are available.

revolute joint   a joint characterized by a rotation angle that is the relative displacement between two successive links.

revolving field   the magnetic field created by flow of a set of balanced three-phase currents through three symmetrically displaced windings. The created field revolves in the air-gap of the machine at an angular velocity corresponding to the synchronous speed of the machine. The revolving field theory is the basis of functioning of synchronous and induction machines.

Reynolds number   a nondimensional parameter used to determine the transition to turbulence in a fluid flowing in pipes or past surfaces.

RF   *See* radio frequency.

RF amplifier   amplifier capable of providing gain at radio frequencies.

RF choke   a large-valued inductor that exhibits a large reactance at the operating frequency that effectively blocks the RF signal.

RF input power   the difference between incident power and reflected power at the input into a device or circuit, expressed in watts. The specific point in the design at which RF input power will be measured is important, since it will affect gain and efficiency calculations.

RF output power   the difference in the power available under perfectly matched conditions and the reflected power taking the output return loss into account, expressed in watts. This is the RF power delivered to the load.

RF quadrature   part of a radio frequency receiver that contains circuits that can be tuned/varied to pass a desired signal carrier wave while rejecting other signals or carriers.

RF quadrature demodulation   radio frequency demodulation of a signal that was produced by two signals separated in phase by 90 degrees.

RF quadrature modulation   radio frequency modulation of a signal that is produced by two signals separated in phase by 90 degrees.

RF tuner   the part of a radio frequency receiver that contains circuits that can be tuned/varied to pass a desired signal carrier wave while rejecting other signals or carriers.

RFI   *See* radio frequency interference.

RFIC   *See* radio frequency integrated circuit.

RGB   the most widely used image representation, where color is represented by the combination of the three primary colors of the additive light spectrum (tristimulus value). The RGB space is represented in Cartesian coordinates as a unit cube, where the origin represents black and the point $(1, 1, 1)$ represents white. *See* color spaces.

rheobase   the minimum current necessary to cause nerve excitation — applicable to a long duration current (e.g., several milliseconds).

rib waveguide   a type of dielectric waveguide formed by several planar layers of dielectric media; the upper layer, instead of being planar, presents a ridge (rib) where the field is mostly confined. Used in integrated optics.

Riccati equation   a class of equations which arises frequently in statistics and linear systems theory, describing the evolution of the statistics of measured systems. As with the Lyapunov equation, several types exist. In each case, P, Q, R are symmetric positive definite; A, B, C are arbitrary.
Continuous-time, time-varying Riccati Equation:

$$\dot{P} = A(t)P(t) + P(t)A^T(t) + B(t)Q(t)B^T(t) - P(t)C^T R^{-1}(t)C(t)P(t)$$

Continuous-time algebraic Riccati Equation:

$$0 = AP + PA^T + BQB^T - PC^T R^{-1}CP$$

Discrete-time algebraic Riccati Equation:

$$P = APA^T + BQB^T - APC^T (CPC^T + R)^{-1} CPA^T$$

See Lyapunov equation, Kalman filter.

Rice distribution   the probability distribution of the magnitude of a complex quantity whose real and imaginary parts are independent Gaussian random variables with nonzero mean. Frequently used to approximate multipath fading statistics in line-of-sight mobile radio systems.

Rice factor   for a Rice-distributed signal, the parameter giving the ratio of the power of the static (direct) signal component to the power of the remaining signal components. The remaining signal components (which follow a Rayleigh distribution) are often referred to as the diffuse signal. The Rice factor is a parameter of the Rician probability density function. *See also* Rician fading.

Rician distribution   *See* Rayleigh noise.

ridge detection   *See* edge detection.

right hand circular polarization   the state of an electromagnetic wave in which the electric field vector rotates clockwise when viewed in the direction of propagation of the wave.

rigid body motion   motion of bodies which are assumed not to change their shape at all — i.e., deformation is absent or is neglected. In contrast, non-rigid body motion takes deformation into consideration.

rigid link   *See* flexible link.

ring bus   a power transmission scheme in which a region is supplied by a continuous, closed loop of power transmission lines.

ring coupler   a type of planar 180 degree hybrid that can easily be constructed in planar (micro strip or stripline) form.

ringing   (1) the phenomena in discrete-time (sampled data) systems in which a system containing a single pole on the negative real axis in the z-plane can oscillate with a period of twice the sampling interval. Mathematically, this discrete effect can be related back to the discontinuity that exists along the negative real axis in the z-plane.

(2) in image processing, the occurrence of ripples near edges in an image processed by a lowpass filter with a steep transition band.

ring network   a network topology where all nodes are connected in a loop. The topology is resilient to breaks in the loop as traffic can be rerouted in either direction. Also known as loop network.

ring numbering   for access control, protection scheme in which every memory object is assigned a set of ring numbers and every executing process is assigned a number. The legality of an access attempt is determined by numeric comparisons between the execution ring number and the ring numbers of the object to be accessed. A typical design assigns more privilege to lower numbers, where the system programs reside. If there are three access modes (execute, read, write), three numbers (e, r, w) are assigned to each object. Let p denote the ring number of the executing process. Then execute is permitted if $p < e$, read if $p < r$, and write if $p < w$. This scheme is simple to explain, but does not support general access control policies, since they cannot be mapped to a linear sequence of integer values.

ring resonator   resonator in which for much of the mode volume the electromagnetic waves are described in terms of travelling rather than standing waves. Since the ring has no open ends, radiative losses are very small. In such a resonator, the mode volume of the electromagnetic waves are described in terms of travelling rather than standing waves.

ripple   the AC (time-varying) portion of the output signal from a rectifier circuit.

ripple current   the total amount of alternating and direct current that may be applied to an electrolytic capacitor under stated conditions.

ripple-carry adder   a basic *n*-bit adder that is characterized by the need for carries to propagate from lower- to higher-order stages.

RISC   *See* reduced instruction set computer processor.

rise time   the time required for a digital signal to make the transition from a "low" value to a "high" value.

rise time degradation   a measure of the slowing down of the pulse as it passes through an I&P element. It includes both the increase in rise time of the pulse, as well as loss in amplitude.

rising edge   in a clock or data signal, that portion of the signal that denotes the change from the "low" state to the "high" state.

RLL   *See* run-length limited code.

RMS   *See* root-mean-squared error.

RMS delay spread   *See* root-mean-squared delay spread.

RMS Doppler spread   *See* root-mean-squared Doppler spread.

RMS gain ripple   *See* root-mean-squared gain ripple.

RMS phase ripple   *See* root-mean-squared phase ripple.

RMS power   *See* root-mean-squared power.

RMW memory cycle   *See* read/modify/write.

robot   a term originated from the Chech word *robota,* meaning work.
A definition used by the Robot Institute of America gives a more precise description of industrial robots: "A robot is a reprogrammable multifunctional manipulator designed to move materials, parts, tools, or specialized devices, through variable programmed motions for the performance of a variety of tasks."
The British Robot Association defines a robot as "A reprogrammable device designed to both manipulate and transport parts, tools or specialized manufacturing implements through variable programmed motions for the performance of specific manufacturing tasks."

robot programming language   a computer programming language that has special features which apply to the problems of programming manipulators. Robot programming is substantially different from traditional programming. One can identify several considerations that are typical to any robot programming method: The objects to be manipulated by a robot are three-dimensional objects; therefore, a special type of data is needed to operate an object. Robots operate in a spatially complex environment. The description and representation of three-dimensional objects in a computer are imprecise. Also, sensory information has to be monitored, manipulated, and properly utilized. Robot programming languages can be divided into three categories:

(1) Specialized manipulator languages built by developing a completely new language. An example is the VAL language developed by Unimation, Inc.
(2) Robot library for an existing computer language. It is a popular computer language augmented by a library of robot-specific subroutines. An example is PASRO (Pascal for Robots) language.
(3) Robot library for a new general-purpose language. These robot programming languages have been developed by first creating a new general purpose language, and then supplying a library of predefined robot-specific subroutines. An example is AML language developed by IBM.

robot vision   a process of extracting, characterizing, and interpreting information from images of a three-dimensional world. This process is also called machine or computer vision.

robot-oriented programming   using a structured programming language that incorporates high-level statements and has the characteristics of an interpreted language, in order to obtain an interactive environment allowing the programmer to check the execution of each source program statement before preceding to the next one. Robot-oriented programming incorporates the teaching-by-doing method, but allows an interaction of the environment with physical reality. *See also* teaching-by-showing programming.

robust control   control of a dynamical system so that the desired performance is maintained despite the presence of uncertainties and modeling inaccuracies.

robust controller design   a class of design procedures leading to control systems that are robust in the sense of required performance. Robust design is a feedback process involving robustness analysis. A specific technique used in robust controller design depends on the type of model describing a system and its uncertainty, control objective, and a set of admissible controllers. The first requirement is to ensure robust stability; this could be followed by guaranteed cost, disturbance rejection, robust poles localization, target sets or tubes reachability, or other demands.

Ackermann's three basic rules of robust controller design are as follows:

(1) Require robustness of control system only for physically motivated parameter values and not with respect to arbitrarily assumed uncertainties of the model.

(2) When you close a loop with actuator constraints, leave a slow system slow and leave a fast system fast.

(3) Be pessimistic in analysis; then, you can afford to be optimistic in design.

See also worst-case design, min-max control, practical stabilization, guaranteed cost control, H infinity design.

robust estimation   an estimation scheme in which we optimize performance for the least favorable statistical environment among a specified statistical class.

robust fuzzy controller   a fuzzy controller with robustness enhancement or robust controller with fuzzy logic concepts.

robust fuzzy filter   a fuzzy filter with robustness enhancement or robust filter with fuzzy logic concepts.

robust stability   a property of the family of models for the system with uncertainty which ensures that it remains stable for all possible operating conditions for uncertain variables ranging over their sets. For time-invariant linear systems with uncertainty, robust stability means that the family of characteristic polynomials generated by uncertain parameters defined over the operating sets has all roots endowed with negative real parts. *See* Kharitonov theorem. For nonlinear systems, robust stability could be checked by some techniques based on direct Lyapunov method or Popov criterion. In some cases, the requirement of robust asymptotic stability may be weakened by more realistic practical stability or ultimate boundedness demand. *See also* practical stabilization.

robust statistics   the study of methods by which robust measures may be extracted from statistical or numerical data, thereby excluding measurements which are unlikely to be reliable and weighting other measurements appropriately, thereby increasing the accuracy of finally assessed values. Of specific interest is the systematic elimination of outliers from the input data. *See* robustness, median filter.

robustness   (1) a control system quality of keeping its properties in the admissible range in spite of disturbances and other environmental perturbations as well as uncertainties in the system model. The most frequent requirements deal with robust stability and robust performance expressed, for example, in terms of guaranteed cost. Robust systems have the property of being insensitive to changes in the model parameters as well as external disturbances. Usually the system robustness is reached via robust controller design, although systems could be robust in some sense (for example, robustly stable) without use of any special design techniques. It is, for example, well known that for the majority of real-world plants, standard PID controller suitably tuned ensures sufficient robustness. Nevertheless, in many situations, robustness can be guaranteed only by sophisticated design techniques such as H infinity design, min-max control, practical stabilization, guaranteed cost control, and others.

(2) the property of a process which results in its being able to suppress the effects of noisy or unreliable data, thereby arriving at reliable measures or interpretations, and degrading gracefully as more and more unreliable data is included. With image data, robust procedures are those that are able to detect objects without becoming confused by partial occlusions, noise, clutter, object breakages, and other distortions.

ROC *See* receiver operating characteristics curve, region of convergence.

ROI *See* region of interest.

rollback *See* backward error recovery.

Rollett stability factor the inverse of the Linville stability factor (C), K is a measure of potential stability in a 2-port circuit operating under small signal conditions, but stand-alone is insufficient to guarantee stability. A 2-port circuit that is matched to a positive real source and load impedance is unconditionally stable if $K > 1$, $B_1 > 0$ (port 1 stability measure), and $B_2 > 0$ (port 2 stability measure). The design must provide sufficient isolation from the RF input and output ports to the bias ports to allow a reasonable

interpretation of the "2-port" device criteria.

$$K = \frac{1}{C} = \frac{1 - |s_{11}|^2 - |s_{22}|^2 + |\Delta|^2}{2 \cdot |s_{12} \cdot s_{21}|}$$

where $\Delta = s_{11} \cdot s_{22} - s_{12} \cdot s_{21}$.

rolling ball a method of determining the lightning protection to nearby structures afforded by a tall, well-grounded structure like a steel tower. A shpere with a radius of 45 meters is imagined rolled against the tower. Any structure which can fit within the space defined by the sphere's point of contact with the earth, the base of the tower, and the sphere's point of contact with the tower is considered to be protected by the tower against lightning strikes. *See* cone of protection.

rolloff transition from pass-band to stop-band.

ROM *See* read-only memory.

Romex cable a heavily insulated, non-armored cable used in residential wiring.

root locus the trajectory of the roots of an algebraic equation with constant coefficient when a parameter varies.

root sensitivity the dependence of the poles and zeros of a lumped parameter circuit function on the circuit elements.

Let a lumped parameter circuit function be represented as

$$F(s) = \frac{a_m \prod_{i=1}^m (s - z_i)}{d_n \prod_{i=1}^n (s - p_i)}$$

where z_i are zeros and p_i are poles. If $F(s)$ is also a function of the circuit element x, the location of these poles and zeros will depend on this element. This dependence is described by the semirelative root sensitivities

$$\mathbf{S}_x(z_i) = x \frac{\partial z_i}{\partial x} \qquad \mathbf{S}_x(p_i) = x \frac{\partial p_i}{\partial x}$$

The calculation of these sensitivities is simplified when zeros and poles are simple (not multiple). For example, for a simple pole p_i, the denominator of the circuit function satisfies the relationship

$$D(p_i) = D_1(p_i) + x D_2(p_i) = 0$$

and one can find that

$$\mathbf{S}_x(p_i) = x \frac{\partial p_i}{\partial x} = -x \frac{D_2(p_i)}{D'(p_i)}$$

where $D'(p_i)$ is the derivative of the polynomial $D(p)$ calculated at the point p_i. For multiple poles, the calculation of the root sensitivity is more involved. The relationship between the function sensitivity and poles and zeros sensitivities can also be established.

root-mean-squared delay spread   a measure of the width of a delay power spectrum. Computed for the delay power spectrum in a similar fashion as the standard deviation is computed for a probability density function. Usually known as RMS delay spread.

root-mean-squared Doppler spread   a measure of the width of a Doppler power spectrum. Computed for the Doppler power spectrum like the standard deviation is computed for a probability density function. Also known as RMS Doppler spread.

root-mean-squared (RMS) error   the square root of the mean squared error.

root-mean-squared gain ripple   the difference of the root-mean-squared (RMS) values of the power gain peaks and gain dips relative to the RMS power across a specified band.

root-mean-squared phase ripple   the difference of the root-mean-squared (RMS) values of the phase peaks and phase dips relative to a best fit linear phase response across a specified band.

root-mean-squared power   the average power, expressed in watts, delivered to the load over a complete period of time (T), as if the current and voltage were constant over that time period.

$$P_{RMS} = \frac{1}{T} \int_0^T p \cdot dt.$$

Also known as RMS power or average power.

rotate   a logical operation on a data element that shifts each bit one position to the left or right. The bit at the end of the location is transferred to the opposite end of the element.

rotating excitation system   an excitation system derived from rotating AC or DC machines. The output of the system is still DC and connected to the rotor.

rotating wave approximation   assumption in a semiclassical model for the interaction of light with atoms that all populations, field amplitudes, and polarization amplitudes change negligibly within one optical cycle.

rotating-rectifier exciter   an AC generator, with rotating armature and stationary field, whose output is rectified by a solid-state device located on the same shaft to supply excitation to a larger electrical machine, also connected to the same shaft.

rotational latency   the time it takes for the desired sector to rotate under the head position before it can be read or written.

rotational loss   one of several losses in a rotating electric machine that are primarily due to the rotation of the armature and include the friction and windage losses. Also called mechanical loss. They can be determined by running the machine as a motor at its rated speed at no load, assuming the armature resistance is negligible.

rotational position sensing   a mechanism used in magnetic disks, whereby the disk interrupts the I/O controller when the desired sector is under the read/write head. Used to recognize the different sectors in a track and synchronize the different bits in a sector.

rotational transition   transition between rotational states of a molecule.

Rotman lens   a constrained cylindrical-lens antenna over a wide instantaneous bandwidth as derived by W. Rotman and P. Franchi (1980). The antenna configuration consists of a stripline lens and line feed. The lens layers can use a microstrip printed-circuit construction. Microstrip lines interconnect radiating elements.

rotor   the rotating part of an electrical machine including the shaft, such as the rotating armature of a DC machine or the field of a synchronous machine.

rotor power developed   the amount of power developed by the rotor. In DC machines, the developed power, frequently denoted by P_d, is calculated as the product of the induced EMF E_a and the armature current I_a. In induction machines, the rotor power developed is obtained by subtracting the rotor copper losses from the air gap power.

rotor power input   represents the total power delivered from the armature coil of an induction motor across the air gap to the rotor via the air gap magnetic flux. It is represented, on a per-phase basis, by

$$RPI = I_s^2(R_{2eq}/s)$$
$$= I_s^2 R_{2eq} + I_s^2(R_{2eq}(1-s)/s)$$

where I_s is the armature coil current, R_{2eq} is the equivalent per-phase rotor resistance reflected to the armature, and s is the rotor slip speed. The first term on the right-hand side represents the ohmic heating power dissipated in the rotor windings, and the second term represents the conversion of electrical to mechanical power by the rotor. *See also* rotor power loss.

rotor power loss   represents the portion of the power transferred across the air gap to the rotor of an induction motor that is lost either through ohmic heating of the rotor windings or due to friction and windage losses in the rotor. The mechanical power available at the motor shaft is the difference between rotor power input and rotor power losses. *See also* rotor power input.

rotor reference frame   a two-dimensional space that rotates at the electrical angular velocity of a specified machine rotor. In electric machines/power system analysis, an orthogonal coordinate axis is established in this space upon which fictitious windings are placed. A linear transformation is derived in which the physical variables of the system (voltage, current, flux) are referred to variables of the fictitious windings. *See also* Park's transformation, transformation equations, arbitrary reference frame, synchronous reference frame, stationary reference frame.

rotor speed   quantification of the rotational operation of the moving part of a rotating electrical machine. The rotor speed is measured either in SI units in radians per second (rad/s) or in practical units in revolutions per minute (rev/min).

rough sets   *See* Pawlak's information system.

rough surface   surface whose corrugation is random and appreciable compared with the light wavelength so as to produce light scattering. It is commonly characterized by the root mean square and the correlation length of the random height profile.

round-robin arbitration   a technique for choosing which of several devices connected to a bus will get control of the bus. After a device has had control of the bus, it is not given control again until all other devices on the bus have been given the opportunity to get control in a predetermined order. The opportunity to get control of the bus circulates in a predetermined order among all the devices.

rounding   an operation that modifies a floating-point representation considered infinitely precise in order to fit the required final format. Common rounding modes include round to nearest, round toward zero, and round toward positive or negative infinity.

router   a node, connected to multiple networks, that forwards packets from one network to another. It is much more complex than bridges that work between networks having compatible protocols. Also called a gateway.

routing   given a collection of cells placed on a chip, the routing routine connects the terminals of these cells for a specific design requirement.

row decoder   logic used in a direct-access memory (ROM or RAM) to select one of a number of rows from a given row address. *See also* two-dimensional memory organization.

row-access strobe   *See* two-dimensional memory organization.

RS flip-flop   a single-bit storage element, usually formed by connecting two NOR or NAND gates in series. RS stands for reset–set. For state variable Q and next state variable Q', the simplified truth table is given as

R	S	Q	Q'
0	0	0	0
0	0	1	1
0	1	0	0
0	1	1	0
1	0	0	1
1	0	1	1
1	1	0	X
1	1	1	X

the symbol "X" is used to denote an unknown state for the flip-flop. *See also* JK flip-flop.

RS-170A   technical standard developed by the Electronics Industry Association that describes in detail the relationship between vertical, horizontal, and subcarrier components within a video signal. The standard permits synchronization of two or more video signals.

RS-422   technical standard developed by the Electronics Industry Association that defines the exact physical, electrical, and functional characteristics for a 40-pin connector that links a computer to communication equipment.

RSGNN   *See* recursive self-generating neural network.

RSSI   *See* received signal strength indicator.

RTB   reverse translation buffer. *See* inverse translation buffer.

RTU   *See* remote terminal unit.

rubbers   personal protective wear for line workers, including insulating gloves, sleeves, and rubber boots.

ruby   amplifying medium employed in the first man-made optical frequency laser.

ruby laser   first man-made optical frequency laser.

run winding   the main winding of a single-phase induction motor.

run-length coding   the assignment of a codeword to each possible run of 0s (white pel sequence) or run of 1s (black pel sequence) in a scan of the subject copy.

run-length encoding   *See* run-length limited code.

run-length limited (RLL) code   a line code that restricts the minimum and/or maximum number of consecutive like-valued symbols that can appear in the encoded symbol sequence.

running digital sum   the difference between cumulative totals of the number of logic 1s and number of logic 0s in a binary sequence. It is a common measure in the performance of a line code.

running integral   for the function $x(t)$, the running integral, $y(t)$ is $y(t) = \int_{-\infty}^{t} x(\tau)d\tau$. An example of a running integral is the unit step $u(t) = \int_{-\infty}^{t} \delta(\tau)d\tau$, where $\delta(t)$ is the unit impulse function.

S

S-100 a 100-pin bus formerly used by computer hobbyists and experimenters.

S-matrix *See* scattering matrix.

S-parameters *See* scattering parameters.

S-plane the domain of the Laplace transform $F(s)$ of a complex-valued function $f(t)$. Since s is a complex number, the domain of $F(s)$ is the complex plane. The line $s = j\omega$, where ω is a real number, corresponds to where $F(s)$ is the Fourier transform of $f(t)$.

S/H *See* sample-and-hold amplifier.

saddle a U-shaped piece of wire which is crimped to a main conductor so that a hot tap *cf.* can be readily attached at a later time.

saddle-point equilibrium an equilibrium in zero-sum game constituted by a pair of security strategies of the players. The necessary and sufficient condition for existence of the saddle point is equality of the security levels for both players. If J_i denotes a cost function of the zero-sum game and $d_i; i = 1, 2$ a strategy of the ith player, then a pair d_1^*, d_2^* is in saddle-point equilibrium iff

$$J(d_1^*, d_2) \leq J(d_1^*, d_2^*) \leq J(d_1, d_2^*)$$

for all admissible (d_1, d_2). Depending of the type of game, the meaning of the cost functional and strategies and the specific conditions for the existence of the saddle point may vary. If the equilibrium in pure strategies does not exist and the game is played many times in the same conditions, then the saddle point may be defined for average values of the cost function and for mixed strategies defined as probability distributions on the spaces of pure strategies.

Safe Medical Devices Act (SMDA) a public law that imposes reporting requirements on "device-user facilities" including hospitals, ambulatory surgical facilities, nursing homes, and outpatient clinics. They are required to report information that "reasonably suggests" the probability that a medical device has caused or contributed to the death, serious injury, or serious illness of a patient at that facility.

Safeguard a program administered by the International Atomic Energy Agency comprising procedures and inspections which assure that fissile materials from power reactors are not diverted to nuclear weapons use.

safety the probability that a system will either perform its functions correctly or will discontinue its functions in a well-defined, safe manner.

safety-critical system a system that is intended to handle rare unexpected, dangerous events.

sag a decline ranging from 0.1 to 0.9 pu in RMS voltage or current at the supply frequency for a time period of 0.5 cycles to 1 minute.

Sagittal projection a projection of a 3-dimensional object onto a 2-dimensional plane which intersects the object in a front to back direction dividing the objects into right and left halves. Typically with reference to an animal or human body.

Sagnac interferometer a common path interferometer devised by Georges Sagnac to measure the ether wind, then adapted as an optical gyroscope, and most recently utilized for hyperspectral imaging. A Sagnac interferometer is composed of two coils of optical fiber arranged so that light from a single source travels clockwise in one and counterclockwise in the other.

Sagnac logic gate   an all-optical gate based on a Sagnac interferometer.

Sagnac, Georges   (1869–1928)
    French scientist and professor at the University of Lille and the University of Paris, active in the investigation of radiation produced by X-rays, and the optics of interference. The inventor of the common path Sagnac interferometer.

sail switch   a device used in control systems that detects the flow of air, or other gas, and causes a relay to open or close as a result of the motion of the sail.

salient feature   a characteristic often local feature on an object which can be detected and used as part of the process of inferring the presence of an object from its features. Typical salient features include point features such as corners and small holes, or fiducial marks (e.g., on printed circuit boards), but may in addition include large-scale straightforwardly detected features such as large circular holes which can also aid the inference process.

salient-pole drive   *See* synchronous drive.

salient-pole rotor machine   AC motor/ generator design in which the rotor is constructed of outward-projecting pole pieces mounted on a shaft-mounted central spider assembly. Spider assemblies are typically spoked. Pole pieces are built up from laminated sheets, which are bolted together between a pole shoe on the outer end and dovetail fixture on the inner end. The dovetails are keyed into slots on the spider to mount the pole pieces to the rotor. Rotor windings are generally constructed from preformed, insulated coils that are fit over the pole pieces during assembly. Salient rotors are typically low-speed designs with short axial length and large diameter.

salt and pepper noise   *See* impulse noise.

SAM   *See* standard additive model.

samarium cobalt   a brittle, high-energy magnetic material that is best known for its performance at high temperatures, which comes in two compositions, $SmCo_5$ and a higher energy Sm_2Co_{17}.

sample   a single measurement that is taken to be representative of the measured property over a wider area, frequency range, or time period. When recording digital sound, a sample is a voltage measurement that reflects the intensity of the acoustic signal at a particular moment, and has a time period associated with it; that is, the sample represents the signal until the next measurement is made. In a digital image, a sample is a single measurement of light intensity at a particular point in the scene, and that measurement is used to represent the actual but unmeasured intensity at nearby points.

sample complexity   the number of training examples required for a learning system to attain a specified learning goal.

sample space   the set of all possible samples of a signal, given the particular parameters of the sampling scheme.

sample-and-hold   *See* sample-and-hold amplifier.

sample-and-hold amplifier (SHA)   a unity gain amplifier with a mode control switch where the input of the amplifier is connected to a time-varying signal. A trigger pulse at the mode control switch causes it to read the input at the instance of the trigger and maintain that value until the next trigger pulse. Usually the signal is electrical, but other forms, such as optical and mechanical forms, are possible. Also denoted S/H.

sample-and-hold circuit   a device with a mode control switch that causes reading the input at the instance of the trigger and maintaining that value until the next trigger pulse.

sampling   the act of turning a time-continuous signal into a signal that is time-discrete or time-discontinuous. See the following diagrams, where T_S is the sampling period.

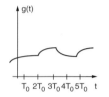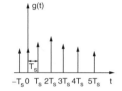

The sampling process.

In order to maintain frequency components of interest in the time-discontinuous signal, the Nyquist sampling criterion needs to be satisfied. This criterion states that the rate of sampling of the time-continuous signal has to be at least twice as great as the frequency of the signal component with the highest frequency which is of interest in the time-continuous signal. Sampling is very common in digital recording and digital communication systems.

sampling frequency   in analog-to-digital conversion, the rate or frequency at which an analog signal is sampled and converted into a digital signal.

sampling function   a mathematical function used when sampling a signal. In particular, a sampling function $S(t)$ can be multiplied by the continuous function to be sampled, $F(t)$, to obtain the sampled version of F. S is most often a collection of equally spaced impulses.

sampling period   the period for which the sampled variable is being held constant.

sampling process   *See* sampling.

sampling rate   *See* sampling frequency.

sampling theorem   states that samples of a bandlimited signal, if taken close enough together, exactly specify the continuous time signal from which the samples were taken.

SAR   *See* synthetic aperture radar or specific absorption rate.

satellite cell   cell with the radius larger than 500 km where the cell is controlled by a satellite. *See also* cell.

satellite imagery   the acquisition of pictures of the earth from space. Satellite imagery can be used to enhance maps, collect resource inventories (eg., forestry, water, land use), assess environmental impact, appraise damage following a disaster, and collect information on the activities of humans. Satellite imagery tends to be multi-spectral, including a wide range of optical frequencies and, more recently, infrared and radar. *See* remote sensing.

satistical pattern recognition   methods for carrying out the recognition of patterns on the basis of statistical analysis. These methods are typically based on the learning of unknown pattern probability distributions from examples.

saturable absorber   the nonlinear optical phenomenon in which the absorption coefficient of a material decreases as the intensity of the light used to measure the absorption increases.

saturable absorption   the effect of there being less absorption in a material for larger values of the incident illumination.

saturated gain   value of the gain in a saturable amplifier for a particular value of intensity.

saturated logic   logic gates whose output is fully on or fully off, determined principally by the external circuit.

605

saturating control a controller producing a bounded control signal. Finite limits on the magnitude of the control signals that are provided by the actuators are due to the fact that the actuators are physical devices and as such are subject to physical constraints. Thus, the actuator saturates, that is, it has "limited authority."

saturation (1) the failure of the output to increase as fast as the input.

For example, often the current regulator used in variable-speed drives is unable to track the commanded current because of insufficient voltage difference between the motor back EMF and the supply.

(2) In an amplifier, saturation results in a reduction of gain in an amplifier or loss in an absorber due the intensity of the signal being amplified or absorbed.

(3) In ferromagnetic circuits, the magnetic flux initially increases linearly with the applied magnetomotive force (MMF), but eventually most of the domains in the ferromagnetic material become aligned, and the rate of increase in flux decreases as the MMF continues to increase. See figure below. *See also* saturation flux density.

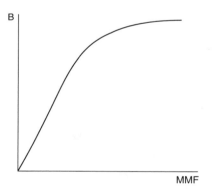

Saturation effect in a ferromagnetic circuit.

(4) with respect to color, the amount or purity of the color seen. A pure color is said to be fully saturated, and the saturation decreases as white is added to the mix. The color "pink," for example, is a less saturated version of "red."

saturation angle the angular portion of the time-based output signal (in degrees) over which the device is saturated. It is always less than or equal to the conduction angle, since the device must conduct before it can saturate.

saturation flux density the maximum value of intrinsic induction (B_i) beyond which an increase in magnetizing field yields no further improvement, indicating that all magnetic moments in the material have been aligned.

saturation intensity the intensity of a beam of light above which saturation effects become appreciable. *See also* saturable absorption.

saturation magnetization the magnetic moment per unit volume of a material when the magnetization in the sample is aligned (saturated) by a large magnetic field.

saturation parameter reciprocal of the value of intensity for which the gain of an amplifying medium or the loss of an absorbing medium is reduced to one half of its unsaturated value.

saturation polarization the value to which the externally measured electrical dipole moment of a ferroelectric body tends when subjected to an external electrical field greater than the coercive field.

save instruction an assembly language instruction that saves information about the currently executing process.

SAW *See* surface acoustic wave.

SAW device *See* surface acoustic wave device.

saw-tooth coupler transmission line coupler consisting of two parallel transmission lines placed in close proximity to one another. The adjacent edges of the two transmission lines are shaped in a notch or saw-tooth pattern to equalize the phase velocity of the even and odd mode voltage components.

SCA   *See* subsidiary communication authorization and station control error.

SCADA   acronym for supervisory control and data acquisition. A system which measures critical power system parameters (e.g., voltage, power flow, circuit breaker status, and generator outputs) at remote points in an electric power system and transmits the data to a central control site where these conditions may be monitored.

scalable video   coding compression of video such that transmission at different data rates, or reception by decoders with differing performance, is possible merely by discarding or ignoring some of the compressed bitstream, i.e., without recoding the data. The compressed data are prioritized such that low-fidelity reconstruction is possible from the high-priority data alone; addition of lower-priority data improves the fidelity.

scalar network analyzer   a test instrument designed to measure and process only the magnitude of transmitted and reflected waves. Used to measure such microwave characteristics as insertion loss, gain, return loss, SWR, and power.

scalar processor   a CPU that dispatches at most one instruction at a time.

scalar quantization (SQ)   quantization of a scalar entity (a number; as opposed to vector quantization), obtained, e.g., from sampling a speech signal at a particular time-instant. Each input value to the quantizer is assigned a reproduction value, chosen from a finite set of possible reproductions. A device performing scalar quantization is called a (scalar) quantizer.

scalar wave   wave that can be described by a single scalar function of space and time.

scalar wave equation   in optics, a simplification of the Maxwell–Heaviside equations that governs a single scalar function representing an electromagnetic wave; sometimes a complex equation if the waves are harmonic in time.

scale   (1) a property of an image relating the size of a pixel in the image to the size of the corresponding sampled area in the scene. A large-scale image shows object features in more detail than a small-scale image. (resolution).

(2) to change the size (i.e., enlarge or shrink) of an image or object while maintaining the overall proportions.

(3) one of two parameters of a wavelet, the other being translation. The scale specifies the duration of the wavelet.

scaled processor architecture (SPARC)   name for a proprietary class of CPUs.

scaling function   the solution to the multiscale equation; it can be obtained by iterating a low pass filter in the two-channel filter bank an infinite number of times.

scan design   a technique whereby storage elements (i.e., flip-flops) in an IC are connected in series to form a shift-register structure that can be entered into a test mode to load/unload data values to/from the individual flip-flops.

scan line   in a digital image, a contiguous set of intensity samples reflecting one row or column of the image. A class of image processing algorithms, called scan line or scan conversion methods, looks at the image one or two scan lines at a time in order to achieve the goal.

scan tape   *See* helical scan tape.

scan-based testing   a mechanism for accessing all the data in a hardware module by treating it as one long shift register and then shifting the data out of the module one bit at a time. A device with this capability can also be set to any desired state by shifting in the desired state. The method can also be applied to software objects.

scan-test path   a technique that enhances circuit observability and controllability by using a register with shift (in test mode) and parallel load (during normal operation) capabilities.

scanner   (1) a device used for scanning written documents or printed pictures by tracing light along a series of many closely spaced parallel lines.

(2) any device that deflects a light beam through a range of angles, using mechanisms such as diffraction from electro-optic or acousto-optic gratings or mechanical deflectors.

(3) a type of projection printing tool whereby the mask and the wafer are scanned past the small field of the optical system that is projecting the image of the mask onto the wafer.

scanning   process for converting attributes of a display at raster coordinate locations, such as color and intensity, into a fixed set of numerical attributes for manipulation, transmission, or storage of the display.

scanning tunneling microscope   extremely sensitive method for measuring atomic position at a surface by monitoring the electron current due to tunneling between a moveable metal tip and the surface semiconductor.

scara manipulator   a robot with three parallel revolute joints allowing it to move and orient in a plane (q_1, q_2, q_3) with a fourth prismatic joint (q_4) for moving the end-effector normal to the plane. Usually scara manipulators can move very fast and they are used to assemble parts.

scattering   (1) the process by which a radio wave experiences multiple reflections from a surface or a volume of the propagation medium.

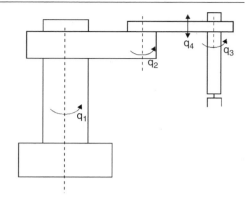

A scara manipulator.

(2) the area of electromagnetics that deals with finding the total fields in a region that contains an incident field and one or more objects (scatters) in the region. The total field is the sum of the incident field when no objects are present, plus the additional field produced when the objects are present.

scattering cross section   total energy scattered in all directions, normalized to the wavenumber squared. It has dimensions of area.

scattering function   the scattering function is a function of two variables, delay and Doppler shift, characterizing the spread of average received signal energy over time and frequency. In a nondispersive channel (no multipath and no Doppler shift), the scattering function is simply an impulse at zero delay and zero Doppler shift.

scattering holography   recovery of the phase of the scattered light by means of its interference with a reference beam, both emanating from the same coherent source, such as the same laser beam.

scattering matrix   a $n \times n$ square matrix **S** of complex numbers used to relate an array of incident waves, **a**, and one of reflected waves, **b**, for an n-port network representing a waveguide component or discontinuity. Generally, each matrix element (called a scattering parameter) is the amplitude of a reflected or transmitted wave to that of an incident wave. The subscripts of a typical coefficient S_{ij} refer to the output and input ports related by the coefficient. These coefficients are generally frequency-dependent and are relative at a specified set of input and output reference planes. The scattering matrix is defined with respect to a specific set of port terminations (normalizations).

Physical interpretations can be given to the scattering coefficients; for example, S_{ij}^2 is the fraction of available power at port i due to a source at port j. The incident and reflected wave amplitudes at port i, a_i, and b_i, respectively, are obtained from the voltage and current, V_i, I_i, at the same port as

$$a_i = \frac{1}{2} |Re Z_i|^{-1/2} (V_i + Z_i I_i)$$

$$b_i = \frac{1}{2} |Re Z_i|^{-1/2} (V_i - Z_i^* I_i)$$

where Z_i is the reference impedance and the asterisk denotes the complex conjugate. It is also known as S-matrix.

scattering parameters   parameters that characterize a microwave network with an arbitrary number of ports by relating the voltage waves incident on the ports to those reflected from the ports. Also known as S-parameters. The scattering parameters are often represented in terms of a scattering matrix. *See also* scattering matrix.

scattering resonance   sharp increase (or decay) of the scattered energy as a function of either frequency or observation direction, as for the glory or the rainbow.

scatterometer   a device to measure the angular distribution of scattered intensity. Its main component contains a photodetector mounted over a goniometer.

scene analysis   the process of analyzing a 3-D scene from 2-D images. Typically, this process will involve object and feature detection, inference

of the presence of objects from their features, projective invariance properties, analysis of the play of light on surfaces, including approaches such as texture analysis, and many other types of procedure.

scheduler   a part of the operating system for a computer that decides the order in which programs will run.

scheduling   in an operating system, scheduling of CPU time among competing processes.

schematic   a diagram that shows how an electronic device is constructed.

schematic capture   a design entry method wherein the designer draws the schematic of the desired circuit using a library of standard cells. The program outputs a netlist of the schematic.

schematic diagram   a circuit diagram, divorced of biasing subcircuits, that depicts only the dynamic signal flow paths of an electronic circuit.

Schmoo plot   an X-Y plot giving the pass/fail region for a specific test while varying the parameters in the X and Y coordinates.

Schottky barrier diode   a two-terminal junction barrier device formed by a junction of a semiconductor and a metal. These diodes are widely used in integrated circuit applications and in very high frequency mixer and multipliers. Also called hot-carrier diode.

Schottky contact   a metal-to-semiconductor contact where, in order to align the Fermi levels on both sides of the junction, the energy band forms a barrier in the majority carrier path.

Schottky noise   *See* shot noise.

Schrödinger wave equation (SWE)   an important relationship in quantum mechanics that describes the energy states of electrons. In its time-independent form it is expressed as

$$\frac{(h/2\pi)^2}{2m} = \nabla^2\phi_n + (E_n + qV)\phi_n = 0$$

where ϕ_n is the wave function corresponding to the subband n whose minimum energy is E_n, V is the potential of the region, m is the particle mass, and h and q are Planck's constant and the electronic charge, respectively.

Schur matrix a square matrix with real elements and whose all eigenvalues have absolute values less than one.

scientific visualization the use of computer graphics techniques to represent complex physical phenomena and multidimensional data in order to aid in its understanding and interpretation.

scintillation the variation of electromagnetic signal strength with time due to random changes in time of refractive index of the atmosphere. Apparent at optical frequencies as the twinkling of stars.

scoreboard term originally used for a centralized control unit in the CDC 6600 processor which enabled out-of-order issue of instructions. The scoreboard unit held various information to detect dependencies. Now sometimes used for the simpler mechanism of having a single valid bit associated with each operand register.

scotopic formally, a description of luminances under which human rod cells are active. Informally, describing dim or nighttime luminances.

scotopic vision vision in the eye determined by the number of and condition of the rods in the eye. Also called night vision.

SCP *See* service control point.

SCR *See* silicon controlled rectifier.

scram the emergency shutdown of a nuclear reactor by the rapid insertion of all control rods. The term is attributed to Enrico Fermi.

scrambling randomization of a symbol sequence using reversible processes that do not introduce redundancy into the bit stream. *See also* reset scrambling, self-synchronizing scrambling.

scrubber a means of removing sulfur dioxide from coal-burning power plant exhaust gas by forcing it through a chemical solution.

SCSI *See* small computer systems interface.

SDH *See* synchronous digital hierarchy.

SDMA *See* spatial division multiple access.

SDS *See* structured distribution systems.

seal-in relay an auxiliary relay that remains energized through one of its own contacts, which bypasses the initiating circuit until deenergized by some other device.

sealing current the current necessary to complete the movement of the armature of a magnetic circuit closing device from the position at which the contacts first touch each other.

sealing voltage the voltage necessary to complete the movement of the armature of a magnetic circuit closing device from the position at which the contacts first touch each other.

search coil a solenoid that is wound with an air core or around a magnet or permeable component of a magnetic circuit to measure the change of flux within the coil; used with a fluxmeter.

SECAM *See* sequential color and memory.

second harmonic component the signal component of a periodic signal whose frequency is twice the fundamental frequency.

second-harmonic generation the process in which a laser beam of frequency ω interacts with a material system to produce a beam at frequency 2ω by means of the second-order susceptibility. Under carefully controlled circumstances, more than 50% of the incident intensity can be converted to the second harmonic. *See also* harmonic generation.

second order discrete time system a discrete system for which the difference between the input and output signals is of second order.

second order system a continuous time system described by a second order differential equation.

second-order susceptibility a quantity, often designated χ^2, describing the second-order nonlinear optical response of a material system. It is defined through the relation $P^2 = C\chi^2 E^2$, where E is the applied electric field strength and P^2 is the 2nd-order contribution to the material polarization. The coefficient C is of order unity and differs depending on the conventions used in defining the electric field strength. The second order susceptibility is a tensor of rank 3, and describes nonlinear optical processes such as second-harmonic generation, sum- and difference-frequency generation, and optical rectification. *See also* nonlinear susceptibility.

secondary (1) the load-side winding.

(2) refers to the portion of a nuclear power plant containing non-radioactive components such as turbines and generators.

secondary cache a buffer element between slow-speed peripheral devices, such as disks, and a high-speed computer.

secondary distribution system a distribution system in which a significant the subdivision of power to customers is done on the secondary, or low-voltage side of the distribution transformer, as opposed to the practice of assigning a small distribution transformer to each customer or small group of customers.

secondary memory generic term used to refer to any memory device that provides backup storage besides the main memory. Secondary memory is lower-level, larger capacity, and usually a set of disks.

Only data and programs currently used by the processor reside in main memory. All other information (not needed at a specific time) is stored in secondary memory and is transferred to main memory on a demand basis. It is the highest (big but slow) level in the memory hierarchy of modern computer systems.

secondary resistor a resistor connected to the rotor of a wound-rotor induction machine to permit variation of the effective rotor resistance. By varying the resistance, machine characteristics may be optimized for starting or varying load conditions.

secondary selective service a redundant electric service in a critical load is supplied by two distribution transformers, each of which is served by a separate, independent distribution primary circuit.

secondary service refers to areas serviced by skywaves and not subject to objectionable interference.

secondary side that side of the packaging and interconnecting structure farthest from layer number one. (Also called the solder side in through-hole component mounting technology.)

secondary storage *See* secondary memory.

secondary system of equations a system of algebraic and differential equations obtained from the primary system of equations by transformation of network variables.

secondary voltage   in power distribution work the voltage at the secondary of the distribution transformer.

secondary voltage control   an automatic voltage control scheme that is similar in function to the automatic voltage regulator, but its purpose is to control a bus voltage which need not have a synchronous generator connected at the bus.

secondary winding   the transformer winding to which the loads are connected. *See also* primary winding.

sectionalizer   a switch placed in distribution lines and programmed to open during a line dead time. The sectionalizer will sense the presence of current surges due to faults, and is programmed to open after a set number of faults occur during a short period of time. When the fault is cleared by the protecting recloser or circuit breaker, the sectionalizer will open, allowing the recloser or breaker to successfully reenergize the portion of the line upstream from the sectionalizer.

sectionalizing fuse   a sectionalizing fuse is a fuse employed on the primary distribution system to isolate laterals from the main feeder in the event of a fault on that lateral.

sectionalizing switch   a switch on primary distribution systems used to isolate laterals and segments of main feeder lines. On radial distribution systems, sectionalizing switches are placed to allow rerouting of power to minimize extended outages following a line segment failure.

sector   *See* disk sector.

sector mapping   a cache organization in which the cache is divided into sectors where each sector is composed of a number of consecutive lines. A complete sector is not transferred into the cache from the memory; only the line requested. A valid bit is associated with each line to differentiate between lines of the sector that have been transferred and lines from a previous sector. Originally appeared in the IBM System/360 Model 85.

sectorization   the action of modifying an omni-directional antenna in a cellular system so that it is replaced by a number of directional antennas each having a radiation pattern approximately covering a sector of a circle. Common examples in cellular systems are those of sectorization with three sectors per cell (120 degree sectors) and sectorization with six sectors per cell (60 degree sectors).

security   the ability of the power systems to sustain and survive planned and unplanned events without violating operational constraints.

security analysis   *See* contingency analysis.

SEED   *See* self-electro-optic device.

seek time   the time that it takes to position the read/write device over a desired track of information.

segment   a region in computer memory defined by a segment base, stored in a segment base register and, usually, a segment limit, stored in a segment limit register. *See also* virtual memory.

segment mapping table   a memory table within a computer that is used to translate logical segment addresses into physical memory addresses.

segment register   a register that stores the base, or starting memory address, of a memory segment.

segment table   a table that is used to store information (e.g., location, size, access permissions, status, etc.) on a segment of virtual memory.

segmentation   (1) an approach to virtual memory when the mapped objects are variable-size memory regions rather than fixed-size pages.

(2) the partitioning of an image in mutually exclusive elements in which visual features are

homogeneous. Region-based segmentation relies on the analysis of uniformity of grey level or color; contour-based segmentation relies on the analysis of intensity discontinuities.

segmentation-based coding   a coding scheme that is based on segmentation. *See* segmentation.

segmented architecture   in computer architecture, a scheme whereby the computer's memory is divided up into discontinuous segments.

selective controllability   a dynamical system is said to be selectively controllable with respect to the ith state variable by jth control variable in $[t_0, t_1]$ if all the dynamical systems with scalar admissible controls $u_j(t)$, $j = 1, 2, \ldots, m$ have the ith state variable $x_i(t)$, $i = 1, 2, \ldots, n$ controllable in $[t_0, t_1]$.

selective fuse coordination   *See* fuse coordination.

selectivity   the ability of a receiver to receive only its desired band of frequencies and reject those on either side. This not only improves the signal properties but also the noise characteristics of the receiver.

selector channel   I/O channel that handles only one I/O transaction at a time. Normally used for high-speed devices such as disks and tapes. *See also* multiplexor channel.

self organizing map (SOM)   a neural network which transforms a multi-dimensional input to a 1- or 2-D topologically ordered discrete map. Each neuron has a weight vector $\vec{w}_j$ and for a given input $\vec{x}$, the output is the index of the neuron with minimum Euclidean distance. During training, the weights are updated as

$$\vec{w}_j(n+1) = \begin{cases} \vec{w}_j(n) + n[\vec{x} - \vec{w}_j(n)], & C_j \in N(C_j, n) \\ \vec{w}_j(n) & C_j \notin N(C_j, n) \end{cases}$$

where α is a learning parameter in the range $0 < \alpha < 1$, and $N(C_i, t)$ is the set of classes which are

in the neighborhood of the winning class C_i at time t. Initially, the neighborhood may be quite large during training, e.g., half the number of classes or more. As the training progresses, the size of the neighborhood shrinks until, eventually, it only includes the one class.

self-arbitrating bus   a communication bus that is capable of resolving conflicting requests for access to the bus.

self-bias   a technique employed whereby a transistor only needs a single bias supply voltage between the drain terminal and ground. This is commonly accomplished by placing a parallel combination of a resistor and capacitor between the source terminal and ground.

self-biased amplifier   an amplifier that utilizes a voltage-controlled current source as the active device (such as a MESFET), and in which a series resistive feedback element in the DC current path creates the DC voltage required to control the quiescent bias point, thereby resulting in the need for a single bias supply.

self-checking   pertaining to a circuit, with respect to a set of faults, if and only if it is fault-secure and self-testing.

self-commutated   *See* natural commutation.

self-demagnetizing field   a field inside of a permanent magnet that is opposed to its own magnetization, which is due to internal coupling of its poles following the introduction of an air gap in the magnetic circuit.

self-electro-optic device (SEED)   a bistable device that is a PIN photodetector and also an optical modulator; the intrinsic region is generally constructed as a quantum-well stack. Detection of light alters the electrical bias on the PIN, which in turn alters the transmission through the device; the optical transmission change exhibits hysteresis and a two-state transmission character.

613

self-focusing   focusing of an electromagnetic beam in a nonlinear medium by the gain or index profile resulting from the action of the beam on the medium.

self-generating neural network (SGNN)   networks of self-organizing networks, each node network of which is an incomplete self-organizing network. For this kind of network of neural networks, not only the weights of the neurons but also the structure of the network of neural networks are learned from the training examples.

SGNN can be as complex as acyclic directed graph, but the most frequently used SGNN takes a tree structure and is called a self-generating neural tree (SGNT), which is very similar to self-organizing tree but with much higher ratio of neuron utilization. Since many fewer neurons participate in the competition during the training and classification, the speed of SGNT is much faster. SGNN has found applications in diagnosis of communication networks, image/video coding, large-scale Internet information services, and speech recognition. *See also* self-generating neural tree, self-organizing neural tree.

self-generating neural tree (SGNT)   a simplified version of self-generating neural network with a tree structure. SGNT is normally much faster in training and classification, but with less descriptive power compared with the corresponding SGNN because of its simple topological structure. However, if the number of network nodes is the same, SGNT has the same descriptive power, higher ratio of neuron utilization, higher speed, and may end up with higher accuracy, since large-scale networks can be generated and trained quickly. *See also* self-generating neural network, self-organizing neural tree.

self-modifying code   a program using a machine instruction that changes the stored binary pattern of (usually) another machine instruction in order to create a different instruction which will be executed subsequently. Definitely not a recommended practice and not supported on all processors.

self-organizing algorithm   a training algorithm for a self-organizing system consisting of the following main steps:

(1) Calculate the similarities of the training vector to all the neurons in the system and compare them to find the neuron closest to the training vector, i.e., the winner.

(2) Update the weights of the winner and its neighborhood according to

$$w_i(t+1) = w_i(t) + \alpha(x_i(t) - w_i(t)),$$

where $w_i(t)$ is the ith weight of the neuron at time t, $x_i(t)$ is the ith component of the training vector at t, and α is a training rate. The neighborhood of the winner starts from a bigger area and reduces gradually during the training period.

self-organizing neural tree (SONT)   a tree-like network of self-organizing neural networks, each node of which is a Kohonen network. Each of the neurons in the higher level networks has its child network in the lower level of the tree hierarchy. The training method is similar to that of Kohonen's method, but is conducted hierarchically. From the top (root) of the tree down, the winner of the current self-organizing network is found as the closest neuron to the training example. The weights of the winner and its neighbors are updated, and then the child network of the winner will be selected as the current network for further examination until a leaf node network is encountered. This kind of network of neural networks can be useful for complex hierarchical clustering/classification. However, the utilization of the neurons may become poor as the network size grows if the uniform tree structure is adopted. The utilization may be improved if carefully designed structure is used but how to obtain an optimum structure remains an issue. Self-generating neural network (SGNN) may be a solution to this problem. *See also* self-organizing system, self-generating neural network.

self-organizing system   a class of unsupervised learning systems that can discover for itself patterns, features, regularities, correlations, or categories contained in the training data, and organize itself so that the output reflects these discovery.

self-phase modulation   the nonlinear optical process in which a pulse of light traveling through a material characterized by an intensity-dependent refractive index undergoes spectral broadens as a result of the time-varying phase shift imparted on the beam.

self-pumped phase conjugator (SPPC)   a phase conjugator that does not require the external pumping beams. Optical four-wave mixing is a convenient method for the generation of phase conjugate waves. However, the process requires a pair of counterpropagating pump beams. With a self-pumped phase conjugator, an incident beam will pump the photorefractive medium and generate its own phase conjugate wave. There are many physical mechanisms that can yield self-pumped phase conjugation, for example, SPPC with resonators, SPPC with semi- resonators, SPPC with stimulated backward scattering, etc.

self-synchronizing scrambling   a technique that attempts to randomize a source bit stream by dividing it by a scrambling polynomial using arithmetic from the ring of polynomials over $GF(2)$. Descrambling is performed with only bit-level synchronization through continuous multiplication of the demodulated sequence by the same scrambling polynomial. The division and multiplication procedures can be implemented with simple shift registers, enabling this technique to be used in very high bit rate systems.

self-test   a test that a module, either hardware or software, runs upon itself.

self-test and repair   a fault-tolerant technique based on functional unit active redundancy, spare switching, and reconfiguration.

self-testing   pertaining to a circuit, for a set of faults, if and only if for any fault in this set there exists a valid input code such that the circuit output is noncorrect (does not belong to the valid output codes, i.e., can be detected with a code checker).

Selfoc lens   a type of gradient-index lens, using a refractive-index profile across the cross section of the element, typically a cylinder; the profile is generally produced by implantation. *See also* gradient-index optics.

selsyn   *See* synchro.

semaphore   a synchronization primitive consisting of an identifier and a counter, on which two operations can be performed: lock, to decrease a counter, and unlock, to increase a counter.

semiconductor device   a transistor, resistor, capacitor, or integrated circuit made from a semiconductor material.

semi-classical model   model for the interaction of light with atoms or molecules in which the atomic or molecular wave functions are described by Schrödinger's equation, while the electromagnetic fields are described by the Maxwell–Heaviside equations.

semi-guarded machine   a machine in which some of the ventilating openings, usually in the top half, are guarded as in the case of a guarded machine to prevent accidental contact with hazardous parts, but the others are left open.

semi-insulator   *See* semiconductor.

semi-magnetic semiconductor   semiconductor alloy or superlattice, usually from the II-VI columns of the periodic table, in which there is a concentration of magnetically active atoms such as manganese that introduce new magneto-optical and magneto-transport properties.

semi-relative sensitivity   one of two sensitivity measures used in circuit and system theory. These are

$$\mathbf{S}_x(y, x) = x\frac{\partial y}{\partial x} = \frac{\partial y}{\partial x/x} = \frac{\partial y}{\partial \ln x}$$

which is frequently denoted by $\mathbf{S}_x$, and

$$\mathbf{S}^y(y, x) = \frac{1}{y}\frac{\partial y}{\partial x} = \frac{\partial y/y}{\partial x} = \frac{\partial \ln y}{\partial x}$$

which is also denoted by $\mathbf{S}^y$. Both these quantities are used in a way similar to using relative sensitivity. *See also* absolute sensitivity, relative sensitivity, sensitivity, sensitivity measure.

semi-rigid cable   a coaxial cable with a solid metal outer-conductor. Typically used where the cable is bent to fit the application only once.

semiconductor   a material in which electrons in the outermost shell are able to migrate from atom to atom when a modest amount of energy is applied. Such a material is partially conducting (can support electrical current flow), but also has properties of an insulator. The amount of current conduction that can be supported can be varied by "doping" the material with appropriate materials, which results in the increased presence of free electrons for current flow. Common examples are silicon and GaAs. Also called semi-insulator.

semiconductor device   a transistor, resistor, capacitor, or integrated circuit made from a semiconductor material.

semiconductor diode laser   laser in which the amplification takes place in an electrically pumped semiconducting medium.

semiconductor laser   *See* semiconductor diode laser.

semiconductor laser amplifier   semiconductor laser in which feedback from the end facets or other reflecting surfaces is unimportant.

semiconductor laser oscillator   a laser oscillator in which the amplification takes place in a semiconducting medium.

semiorthogonal wavelets   wavelets whose basis functions in the subspaces are not orthogonal but the wavelet and scaling subspaces spanned by these basis functions are orthogonal to each other.

sense amplifier   in a memory system, circuitry to detect and amplify the signals from selected storage cells.

sensitivity   a property of a system indicating the combined effect of component tolerances on overall system behavior, the effect of parameter variations on signal perturbations and the effect of model uncertainties on system performance and stability. For example, in radio technology, sensitivity is the minimum input signal required by the receiver to produce a discernible output.

The sensitivity of a control system could be measured by a variety of sensitivity functions in time, frequency or performance domains. A sensitivity analysis of the system may be used in the synthesis stage to minimize the sensitivity and thus aim for insensitive or robust design. *See also* robustness, robust controller design, objective function.

sensitivity bound   a lower or upper limit on the sensitivity index for a system.

Such a bound exists, for example, for filters whose passive elements are limited to resistors, capacitors, and ideal transformers, and whose active elements are limited to gyrators (characterized by two gyration resistances considered different in sensitivity calculations), and active voltage and current controlled voltage and current sources.

sensitivity function   a measure of sensitivity of signals or performance functions due to parameter variations or external signals (disturbances, controls). For small changes in parameters, sensitivity functions are partial derivatives of signals or performance functions with respect to the parameters and could be found from linearized models of the system model under study.

For linear systems, the sensitivity functions in semilogarithmic form could be obtained from the original system model in some special nodes of its block scheme called sensitivity points. Moreover, for linear time-invariant feedback systems, sensitivity may be defined in frequency domain both for small and large parameter deviations. One way to do this is to compare the output errors due to plant parameter variations in the open-loop and closed-loop nominally equivalent systems. The resulting sensitivity function or for multi-input–multi-output systems, sensitivity matrix $S(j\omega)$ is given by a return difference function or its matrix generalization, i.e.,

$$S(j\omega) = (I + K(j\omega))^{-1}$$

where $K(j\omega)$ is an open-loop frequency transfer function (respectively matrix) and I is a unit matrix of appropriate dimension. In the single-input–single-output systems and for the small variations, S is equal to the classical differential logarithmic sensitivity function defined by Bode. *See also* sensitivity index, sensitivity invariant.

sensitivity index a sensitivity measure. For the multiparameter sensitivity row vector one can determine the following two characteristics:

(1) the worst-case sensitivity index

$$\nu(F) = \sum_{i=1}^{n} \left| \mathrm{Re}\mathbf{S}_{xi}^{F(j\omega,\mathbf{x})} \right|$$

(2) the quadratic sensitivity index

$$\rho(F) = \left[\sum_{i=1}^{n} \left(\mathrm{Re}\mathbf{S}_{xi}^{F(j\omega,\mathbf{x})} \right)^2 \right]^{1/2}$$

These two are used most frequently in sensitivity comparisons, yet other indices, for example, for the case of random circuit parameters, can be derived as well. *See also* sensitivity function, sensitivity invariant.

sensitivity invariant feature of a circuit or system that is insensitive.

Let a circuit function be $F(s, \mathbf{x})$ where $\mathbf{x} = [x_1, x_2, \ldots, x_n]^t$ is a vector of circuit parameters. A sensitivity invariant exists for this circuit if there is a relationship

$$\sum_{i=1}^{n} \mathbf{S}_{x_i}^{F(s,\mathbf{x})} = k$$

Here $\mathbf{S}_{x_i}^{F(s,\mathbf{x})}$ is a component of a multiparameter sensitivity row vector and k is a constant. In most cases, this constant can have one of three possible values, namely, 1, 0, and -1. The sensitivity invariants are useful to check the sensitivity calculations. *See also* sensitivity, sensitivity function, sensitivity bound.

sensitivity measure a number used to characterize and compare circuits the functions of which depends of more than one component. They are used when the components of multiparameter sensitivity row vector and the circuits component tolerances are known. The most frequently used are worst-case measure and quadratic measure of sensitivity. *See also* sensitivity, sensitivity index.

sensitivity reduction reducing sensitivity is very desirable in active filter realization. Some general suggestions, at the stage of approximation, ensuring that filter realizations will have low sensitivities to component variations include the following:

(1) Increasing the order of approximation and introducing a redundancy.
(2) Using predistorted (tapered) specifications in the vicinity of the passband edge.
(3) Using transfer functions with a limited value of the maximum Q of the transfer function poles. The realization itself gives the circuits lower sensitivities if the filter is realized as the doubly terminated lossless structure or its active equivalent.

sensor a transducer or other device whose input is a physical phenomenon and whose output is a quantitative measurement of that physical phenomenon. Physical phenomena that are typically measured by a sensor include temperature or pressure to an internal, measurable value such as voltage or current.

sensor alignment   alignment of sensors so as to correct the time delay differences arising from spatial differences.

sensorless control   a control method in which mechanical sensors are replaced by an indirect estimation of the required variable.

separability   the separable property for the signal or system such that the signal or system representation can be expressed by the product of component terms, each depending on fewer independent variables.

separable data   a 2-D signal that can be written as a product of two 1–D signals.

separable filter   a filter which can be applied in two or more operations without any change in overall function, thereby gaining some computational advantage. In particular, a 2-D mean filter can be reimplemented identically as two orthogonal 1-D mean filters, and is therefore separable. However, a 2-D median filter is nonseparable, as its action is not in general identical to that of two orthogonal 1-D median filters. *See* separable transform.

separable image transform   a 2-D separable transform used to transform images.

separable kernel   for a 2-D transform a kernel that can be written as the product of two 1-D kernels. For higher dimension transforms a separable kernel can be written as the product of several 1-D kernels. *See* separable transform.

separable transform   a 2-D transform that can be performed as a series of two 1-D transforms. In this case the transform has a separable kernel. The 2-D continuous and discrete Fourier transforms

are separable transforms. In higher dimensions a separable transform is one that can be performed as a series of 1-D transforms. *See* separable filter.

separately excited DC machine   a DC machine where the field winding is supplied by a separate DC source. Separately excited generators are often used in feedback control systems when control of armature voltage over a wide range is required.

sequence (012) quantities   symmetrical components computed from phase (abc) quantities. Can be voltages, currents, and/or impedances.

sequential-access storage   storage, such as magnetic tape, in which access to a given location must be preceded by access to all locations before the one sought. *See also* random-access device.

sequencer   a programmable logic array (PLA) that has a set of flip-flops for storage of outputs that can be fed back into the PLA as inputs, enabling the implementation of a finite state machine.

sequency   in a transform, the number of zero-crossings of a particular basis function. By extension used to refer to the transform coefficient that corresponds to a particular basis function. For example, in the discrete cosine transform the zero sequency coefficient is the one for which the basis function is flat (and therefore has no zero crossings), often called the DC coefficient. Sequency is roughly analogous to frequency; higher sequency basis functions correspond to higher frequency components of signal energy.

sequential access   data stored on devices such as magnetic tape must be accessed by first moving the media to a particular location and then reading or writing the information. Information cannot be accessed directly; a sequential search must be done first.

sequential color and memory (SECAM)   a color television system that transmits the two

SECAM

Vertical Fields Frequency	50 Hz
Number of Interlaced Fields	2
Vertical Frame Frequency	25 Hz
Number of Horizontal Lines/Frame	625
Horizontal Scan Frequency	15,625 Hz
Video Bandwidth	6 MHz
Color Subcarrier Frequency	4.433618 MHz
Sound Modulation	FM
	(AM in France)
RF Channel Bandwidth	8 MHz

(B - Y) and (R - Y) color difference signals on alternate horizontal lines as a constant amplitude FM subcarrier. A one-line memory in the receiver allows reconstruction of the color signals on all lines. The SECAM system requires no color controls. The vertical scan parameters for the SECAM television system are based on the European power line frequency of 50 Hz. The SECAM system parameters are shown in the table.

sequential consistency   the situation when any arbitrary interleaving of the execution of instructions from different programs does not change the overall effect of the programs.

sequential decoding   a suboptimum decoding method for trellis codes. The decoder finds a path from the start state to the end state using a sparse search through the trellis. Two basic approaches exist: depth-first algorithms and breadth-first algorithms.

sequential detection   a detection algorithm for tree or trellis structured problems based on depth-first tree/trellis search. *See also* depth-first search.

sequential fault   a fault that causes a combinational circuit to behave like a sequential one.

sequential locality   part of the principle of (spatial) locality, that refers to the situation when locations being referenced are next to each other in memory. *See also* principle of locality.

sequential logic   a digital logic in which the present state output signals of a circuit depend on all or part of the present state output signals fed back as input signals as well as any external input signals if they should exist.

sequential scan   *See* progressive scan.

serial bus   a data communication path between parts of the system that has a single line to transmit all data elements.

serial I/O interface   I/O interface consisting of a single line over which data is transferred one bit at a time. Commonly used for low-speed devices, e.g., printer, keyboard, etc. *See also* parallel I/O interface.

serial operation   data bits on a single line are transferred sequentially under the control of a single signal.

serial port   a communications interface that supports bit by bit data transmission.

serial printing   printing is done one character at a time. The print head must move across the entire page to print a line of characters. The printer may pause or stop between characters. Printing speed is usually given in units of characters per second (cps).

619

serial transmission   a process of data transfer whereby one bit at a time is transmitted over a single line.

series equalizer   in a single-loop feedback system, a series equalizer is placed in the single loop, generally at a point along the forward path from input to output where the equalizer itself consumes only a small amount of energy.

series feed   a method of feeding a phased array antenna in which the element feedpoints are located at even or uneven spacings along a single transmission line. Unless phase shifters are placed in the line between elements, the phase shift between elements changes with frequency.

series feedback   with reference to a three-terminal device or grounded amplifier, the application of an electrical element in series with the device or amplifier to ground, thereby causing some of the output signal to be fed back in series with the input signal.

series field   a field winding of a DC machine consisting of a few turns of thick wire, connected in series with the armature and carrying full armature current.

series resonant converter   a power converter that uses a series resonant tank circuit. It has high efficiency from full load to part load, and transformer saturation is avoided due to the series blocking capacitor. The major problem with the series resonant converter is that it requires a very wide change in switching frequency to regulate the load voltage, and the output filter capacitor must carry high ripple current.

series-connected DC machine   a direct current machine in which the field winding is connected in series with the armature winding.

serif   a small ancillary pattern attached to the original pattern on a mask (usually at the corners) in order to improve the printing fidelity of the pattern.

service control point (SCP)   an on-line, real-time, fault-tolerant, transaction-processing database that provides call-handling information in response to network queries.

service drop   the wire which extends from the street to the customer's electric meter.

service entrance   the point at which the electric power service drop enters a building.

service management system (SMS)   an operations support system that administers customer records for the service control point.

service primitive   the name of a procedure that provides a service; similar to the name of a subroutine or procedure in a scientific subroutine library.

servo   *See* servomechanism.

servo drive   an automatic control system in which position, speed, or torque are the control variables.

servomechanism   a closed-loop control system consisting of a motor driven by a signal that is a function of the difference between commanded position and/or rate and measured actual position and/or rate to achieve the conformance. Usually a servomechanism contains power amplification and at least one integrating element in the forward circuit.

session   an instance of one or more protocols that provides the logical endpoints through which data can be transferred.

set associative cache   a cache in which line or block from main memory can only be placed in a restricted set of places in the cache. A set is a group of two or more blocks in the cache. A block is first mapped onto a set (direct mapping defined by some bits of the address), and then the block can be placed anywhere within the set (fully associative within a set). *See also* direct mapped, fully associative cache.

set partitioning   rules for mapping coded sequences to points in the signal constellation that always result in a larger Euclidean distance for a trellis-coded modulation system than for an encoded system, given appropriate construction of the trellis. Used in coded modulation for optimizing the squared Euclidean distance.

set point   (1) a specified constant value of the controlled variable of a dynamical process that a controller is required to maintain. The controller must generate a control signal that drives the controlled variable to the set point and keeps it there, once it is reached. The set point is often referred to as reference point or operating point. In aircraft flight control, the set point is also called the trim condition.

(2) the intersection of the load line and the normal B-H curve, indicating the flux density and energy a permanent magnet is delivering to a given magnetic circuit geometry.

set-associative cache   a cache that is divided into a number of sets, each set consisting of groups of lines and each line having its own stored tag (the most significant bits of the address). A set is accessed first using the index (the least significant bits of the address). Then all the tags in the set are compared with that of the required line to find whether the line is in the cache and to access the line. *See also* cache, direct mapped, and associativity.

set-membership uncertainty   a model of uncertainty in which all uncertain quantities are unknown except that they belong to given sets in appropriate vector spaces. The sets are bounded and usually compact and convex. The estimation problem in this case becomes one of characterizing the set of states consistent both with the observations received and the constraints on the uncertain variables. Control objectives are usually formulated in terms of worse-case design tasks, target sets reachability, guaranteed cost control, robust stability or practical stabilization of the uncertain systems.

For linear systems with energy-type ellipsoidal constraints imposed on the uncertain variables representing initial conditions, additive disturbances and observation noises solution of the state estimation problem is given by the estimator similar to the Kalman filter, and a control problem in the form of min-max optimization of a given quadratic criterion leads to the linear-quadratic game. In the case of instantaneous ellipsoidal constraints, the exact solution of estimation and control problems is difficult to obtain, nevertheless, by bounding recursively the sets of possible state approximating ellipsoids leading to suboptimal filtering and control laws similar to the optimal ones for the energy-type constraints.

Generally, efficient results might be found only for bounding sets parameterized by a little number of parameters. Except for the ellipsoids, such property is endowed only by polyhedral sets bounding uncertain variables. In this case, efficient results could be reached by the use of linear programming algorithms.

settling time   (1) the time required for a signal to change from one value to another.

(2) refers to the time that it takes stable transients of a dynamic system to decay to a negligible amplitude. This can be quantified to the time it takes an exponential transient mode to decay to a band that is $\pm 37\%$, $\pm 5\%$, or $\pm 2\%$ of its initial value. *See also* time constant.

setup   a video term relating to the specified base of an active picture signal. In NTSC, the active picture signal is placed 7.5 IRE units above blanking (0 IRE). Setup is the separation in level between the *video blanking* and *reference black* levels.

setup time   *See* hold time.

SF6   *See* sulfur hexaflouride gas.

SF6 circuit breaker   a power circuit breaker where sulfur hexaflouride (SF_6) gas is used as an insulating and arc clearing agent.

SGNN   *See* self-generating neural network.

SGNT   *See* self-generating neural tree.

SGVQ   *See* shape-gain vector quantization.

SHA   *See* sample-and-hold amplifier.

shaded pole   a magnetic pole-face in which part of the pole is encircled by a shorted conductor (usually copper). The flux through the encircled portion will be out of phase with the flux through the other portion. Shaded pole motors use the phase shift to produce a quasi-rotating magnetic field which develops a weak torque, suitable primarily for small fans. In AC relays, shaded poles are used to prevent chatter (the attempted opening and subsequent closing each time the flux passes through zero).

shaded-pole motor   a single-phase induction type motor that uses shaded poles on the stator to create a weak quasi-rotating magnetic field. Shaded-pole motors are only built in small fractional horsepower sizes and produce a very low starting torque that is suitable only for fan-type loads. *See also* shaded pole.

shadow casting logic gate   an optical logic gate originally using shadow casting technique. The principle of shadow casting logic gate can be explained as follows. First, NOT A and NOT B are generated from inputs A and B. Second, four products of AB, A (NOT B), (NOT A) B, and (NOT A) (NOT B) are produced by passing a light beam through two transparencies that could be spatial light modulators representing A and B, A and NOT B, NOT A and B, and NOT A and NOT B. Third, the four products are added optically. The sixteen combination of four products are the sixteen Boolean logic operations.

shadowing   (1) excess propagation loss resulting from the blocking effect of obstacles such as buildings, trees etc.

(2) the statistical variation of propagation loss in a mobile system between locations the same distance from a base station, usually described by a lognormal distribution.

shaft torque   the component of the motor generated electromagnetic torque available at the shaft of the motor after overcoming the operational losses of the motor during the electromechanical energy conversion process.

Shannon, Claude   (1916–1989) Born: Petoskey, Michigan

Considered to be the founding father of modern electronic communications theory. His contributions include the application of Boolean algebra to analyze and optimize switching circuits and, in his classic paper "The Mathematical Theory of Communication," established the field of information theory by developing the relationship between the information content of a message and its representation for transmission through electronic media.

Shannon information   the information content of an event x with a probability of occurrence of $p(x)$ defined as $I(x) = -\log p(x)$. The units of $I(x)$ depends on the base of the algorithm — "bits" for base 2, "nats" for the natural algorithm. *See* entropy.

Shannon's law   fundamental relationship of information theory, which states that the lower bound on the average code-word length for coding a discrete memoryless source is given by the source entropy. *See* entropy.

Shannon's sampling theorem   mathematical theorem which states that when an analog signal is sampled, there is no loss of information and the analog signal can be reconstructed by low-pass filtering, if and only if the largest (absolute value) frequency present in that signal does not exceed the Nyquist frequency, this being half the sampling frequency.

Shannon's source coding theorem a major result of Claude Shannon's information theory. For lossy source coding, it gives a bound to the optimal source coding performance at a particular rate ("rate" corresponds to "resolution"). The theorem also says that the bound can be met by using vector quantization of (infinitely) high dimension. For lossless source coding, the theorem states that data can be represented (without loss of information) at a rate arbitrarily close to (but not lower than) the entropy of the data. *See also* rate-distortion theory.

shape analysis the analysis of shapes of objects in binary images, with a view to object or feature recognition. Typically, shape analysis is carried out by measurement of skeleton topology or by boundary tracking procedures including analysis of centroidal profiles.

shape from ... the recovery of the 3-D shape of an object based on some feature (e.g., shading) of its (2-D) image.

shape measure a measure such as circularity measure (compactness measure), aspect ratio, or number of skeleton nodes, which may be used to help characterize shapes as a preliminary to, or as a quick procedure for, object recognition.

shape-gain vector quantization (SGVQ) a method for vector quantization where the magnitude (the gain) and the direction (the shape) of the source vector are coded separately. Such an approach gives advantages for sources where the magnitude of the input vector varies in time.

shape-memory effect mechanism by which a plastically deformed object in the low-temperature martensitic condition regains its original shape when the external stress is removed and heat is applied.

shape-memory smart materials include three categories, namely shape-memory alloys (SMA), shape-memory hybrid composites (SMHC), and shape-memory polymers (SMP).

shaping a traffic policing process that controls the traffic generation process at the source to force a required traffic profile.

shared memory characteristic of a multiprocessor system: all processors in the system share the access to main memory. In a physically shared–memory system, any processor has access to any memory location through the interconnection network.

shared memory architecture a computer system having more than one processor in which each processor can access a common main memory.

sharpening the enhancement of detail in an image. Processes that sharpen an image also tend to strengthen the noise in it. *See* edge enhancement, gradient, image enhancement, Laplacian operator, noise, Sobel operator.

SHDTV *See* super high definition television.

shed a circular roof-like feature of an electrical insulator which presents a long electrical leakage path while permitting the drainage of rainwater.

sheet resistance lateral resistance of an area of thin film in the shape of a square.

shell-type transformer a power transformer in which the magnetic circuit surrounds and normally encloses a greater portion of the electrical winding.

shield wire (1) a ground wire placed above an electric transmission line to shield the conductors from lightning strokes.

(2) a ground wire buried directly above a buried communications cable for lightning protection.

shielding effectiveness a measure of the reduction or attenuation in the electromagnetic field strength at a point in space caused by the insertion of a shield between the source and that

623

point. Typically expressed in decibels: $SE = 20 \log_{10}(E_1/E_2)$ dB, where $E_1 =$ field strength measured without shield and $E_2 =$ field strength measured after shielding is applied.

shift instruction   a program instruction in which data in a register or memory location is shifted one or more bits to the left or right. Data shifted off the end of the register or memory location is either shifted into a flag register, used to set a condition flag, or dropped, depending on implementation of the instruction. *See also* rotate instruction.

shift invariance   a characteristic of a property that in some domain it is invariant to displacements within that domain. Particularly important sorts of shift invariance are space invariance and time invariance. The impulse response of a system is independent of the spatial (or temporal) location of the impulse.

shift register   a register whose contents can be shifted to the left or right.

Shockley, William   (1910–1989) Born: London, England

Shockley is best known as one of the developers of the transistor. In 1956 Shockley, John Bardeen, and Walter Brattain received the Nobel Prize for their work. Schockley led the group at Bell Labs responsible for the semiconductor research that led to the development of the "point-contact transistor." In later life, Shockley became known for his public pronouncements on various political and genetic issues.

Shoeffler sensitivity   when the multiparameter sensitivity row vector is known, then the measure of this vector determined as

$$\rho^2(F) = \sum_{i=1}^{n} \left| \mathbf{S}_{x_i}^{F(j\omega, \mathbf{x})} \right|^2$$

is called Shoeffler multivariable sensitivity.

short circuit   a condition on the power system where energized conductors come in contact (or generate an arc by coming in close proximity) with each other or with ground, allowing (typically large) fault currents to flow.

short circuit admittance   the admittance into an N-port device when the remaining ports are terminated in short circuits. For port 1 of a 2-port device, it is the input admittance into port 1 when port 2 is shorted.

short circuit gain–bandwidth product   a measure of the frequency response capability of an electronic circuit. When applied to bipolar circuits, it is nominally the signal frequency at which the magnitude of the current gain degrades to one.

short code   in a spread-spectrum system, periodic spreading code (spreading sequence) with a period equal to a bit duration. *See also* long code.

short duration   *See* voltage variation.

short-circuit protection   the beneficial effect provided by an overcurrent device when it acts to interrupt short-circuit current.

short-circuit test   a transformer test conducted by placing a few percent of rated voltage on the voltage side while the low voltage winding is shorted. By measuring the voltage, current, and input power, it is possible to calculate the equivalent winding impedance for the transformer equivalent circuit.

shortened code   a code constructed from another code by deleting one or more message symbols in each message. Thus an (n, k) original code becomes an $(n - 1, k - 1)$ code after the deleting of one message symbol.

shot noise   noise voltage developed at internal device boundaries, such as solid-state junctions, where charges cross from one type of material into another. Also known as Schottky noise.

shotgun   a specialized hot stick that is used to install a hot tap.

Shubnikov–de Haas oscillation   quantum oscillations in resistance as a function of applied magnetic field.

shunt   (1) a device having appreciable impedance connected in parallel across other devices or apparatus and diverting some of the current from it. Appreciable voltage exists across the shunted device or apparatus, and an appreciable current may exist in it.

(2) an inductive element connected across a power line or bus. Those connected to buses are known as bus-connected reactors, while those connected across a power line are called line-connected reactors.

shunt capacitor   a capacitor or group of capacitors which are placed across an electric power line to provide a voltage increase or to improve the power factor of the circuit. A switchable shunt may be disconnected from the circuit when conditions warrant, while a fixed shunt is permanently connected to the power line.

shunt DC machine   a DC machine with the field winding connected in shunt with the armature. In shunt generators, residual magnetism must be present in the machine iron in order to initiate the generation process. These machines are also known as self-excited, since they supply their own excitation.

shunt field   a field winding of a DC machine consisting of many turns of fine wire, connected in parallel with the armature circuit. It may be connected to the same source as the armature or a separate source.

shunt peaking   use of a peaking coil in a parallel tuned circuit branch connecting the output load of one amplifier stage to the input load of the following stage, in order to compensate for high frequency loss due to the distributed capacitance of the two stages.

shunt reactor   a reactor intended for connection in shunt to an electric system to draw inductive current.

Si   periodic table symbol for silicon. *See* silicon.

Si/SiGe/SiGeC   silicon-based alloy system providing band offsets that enable heterostructures that can be utilized for heterojunction transistor design and quantum confinement.

sideband   the signal produced when a carrier signal is modulated. They may be one single sideband, one set of upper and lower sidebands, or a series of sidebands whose number is dependent on the modulation index of the modulation system being used.

sidelobe   a lobe in an antenna radiation pattern apart from the main lobe and any grating lobes. Sidelobes have peak amplitudes less than that of the main lobe.

side lobe level   the ratio of a local maximum in a radiation pattern to the global maximum (main beam) of the radiation pattern.

sidelobe level (SLL)   the peak amplitude of a sidelobe relative to the peak amplitude of the main lobe. The SLL is usually expressed as the number of decibels below the main lobe peak.

Siemens, Ernst Werner von   (1816–1892) Born: Lenthe, Hanover, Germany

Best known for the German and British companies that bear his name. Siemens was a strong believer in basic research, as well as an avid inventor. His early inventions included an improved gutta-percha wrapped telegraph cable that allowed his companies to secure a number of lucrative cable contracts. His discovery of the dynamo principle, and his use of this in heavy-current applications, allowed his companies to become pioneers in devices to generate electricity and rail applications. Siemens' belief in basic research made him a champion of standards and research institutions that he helped to establish.

sifting property   the ability of the impulse function to select a particular value of another function. In discrete time $x(n) = \sum_{-\infty}^{\infty} x(n)\delta(n - i)$. In continuous time $x(t) - \int_{-\infty}^{\infty} x(s)\delta(t - s)ds$.

SiGe   in order to increase the speed of Si semiconductor devices without compromising on Si's ease of device processing, SiGe heterostructures can be used.

sigma-delta A/D conversion   an oversampling A/D conversion process where the analog signal is sampled at rates much higher (typically 64 times) than the sampling rates that would be required with a Nyquist coverter. Sigma-delta modulators integrate the analog signal before performing the delta modulation. The integral of the analog signal is encoded rather than the change in the analog signal, as is the case for traditional delta modulation. A digital sample rate reduction filter (also called a digital decimation filter) is used to provide an output sampling rate at twice the Nyquist frequency of the signal. The overall result of oversampling and digital sample rate reduction is greater resolution and less distortion compared to a Nyquist coverter process.

sigma-delta modulation   a method for scalar quantization, similar in principle to delta modulation but somewhat more sophisticated. Employed in, e.g., compact-disk players.

sigmoidal characteristic   a widely used type of activation function, especially in networks trained using schemes like backpropagation that are based upon gradient descent. The most common functions used are the arctan, tanh, and logistic functions, with appropriate variations for binary and bipolar variables.

sigmoid function   a compressive function that maps inputs less than -1 to approximately zero, inputs greater than 1 to approximately 1 and maps values from -1 to 1 into the range 0 to 1. A common sigmoid function is $\frac{1}{1+e^{-\frac{x}{T}}}$. Sigmoid functions are often used as activatoin functions in neural nets. *See* activation function.

sign flag   a bit in the condition code register that indicates whether the numeric result of the execution of an instruction is positive or negative (1 for negative, 0 for positive).

sign-magnitude representation   a number representation that uses the most significant bit of a register for the sign and the remaining bits for the magnitude of a binary number.

signal   a real or complex function, $f(t)$ of a time variable t. If the domain is the real line $t \in \mathcal{R}$, or an interval of the real line, then the signal is a *continuous time* signal. For example, $f(t)$ could represent the magnitude of a voltage applied to a circuit, as shown in the figure.

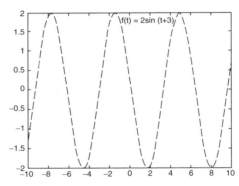

Continuous time signal.

If the domain is a discrete set $\{k, k = 0, 1, 2, \ldots\}$, then the signal is a *discrete time* signal. Discrete time signals commonly arise in engineering applications by sampling a continuous

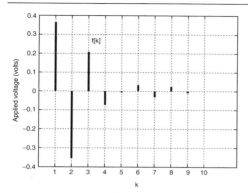

Discrete time signal.

time signal, in control and filtering applications. For example, $f[k]$ could represent the above voltage sampled every 0.5 seconds, as shown in the figure.

signal averaging an averaging process which is used to enhance signals and suppress noise, thereby improving the signal to noise ratio. *See* averaging.

signal conditioning a process which is used to improve the appearance or effectiveness of a signal, either by suppressing noise or by transforming the signal into a more suitable form. This latter category includes image enhancement. Signal conditioning is often appropriate in preparing signals for recognition.

signal decimation *See* decimation.

signal detection detecting the presence of a signal in noise.

signal constellation in digital communications, the set of transmitted symbols represented as points in Euclidean space. For example, the signal constellation for (uncoded) quadrature amplitude modulation is a set of points in the complex plane.

signal flow graph graphical representation of the relationships between a set of independent input variables that are linearly related to a set of dependent output variables.

signal level the value of a signal at a particular place and time.

signal processing a generic term which refers to any technique that manipulates the signal, including but not limited to signal averaging, signal conditioning and signal recognition. When applied to images, it is normally referred to as image processing, the term signal processing usually being reserved for 1-D signals.

signal recognition the recognition of signals by appropriate analysis, often with the help of filters such as matched filters or frequency domain filters.

signal recovery the process of extraction of signals from a background of noise or clutter, often in situations where the signal-to-noise ratio is so low that matched filters, synchronous detectors, or lock-in amplifiers have to be used. *See* synchronous detection.

signal reference subsystem this subsystem provides the reference points for all signal grounding to control static charges, noise, and interference. It may consist of any one or a combination of the lower frequency network, higher frequency network, or hybrid signal reference network.

signal restoration the restoring of data that has been corrupted by instrumentation dynamics and noise.

signal subspace in an orthogonal decomposition of a space, the part to which the desired signal belongs. *See also* noise subspace.

signal transfer point (STP) a packet switch found in the common-channel signaling network; it is used to route signaling messages between network access nodes such as switches and SCPs.

signal variance   *See* variance.

signal-to-interference ratio (SIR)   the ratio of the average power of the signal component to the average power of the interference component in a case where an information-bearing signal of interest has been corrupted by interfering signals.

signal-to-noise plus interference ratio (SNIR) the ratio of total signal power to the sum of total noise power and total interference power at a receiver. The SNIR is a more complete indicator of received signal quality than either SIR or SNR, where the relative contribution of receiver noise and external sources of interference are either unknown or widely varying. It is a unitless quantity. *See also* signal-to-noise ratio, signal-to-interference ratio.

signal-to-noise ratio (SNR)   the ratio of the average power of the information signal component to the average power of the noise component in a signal consisting of the sum of an information signal component and a corrupting noise component. It is a unitless quantity.

signaling   procedures used to control (set up and clear down) calls and connections within a telecommunication network.

signaling system 7 (SS7)   a communications protocol used in common-channel signaling networks.

signature   a characteristic easily computed feature or function by which a particular object or signal may be at least tentatively identified. An example is the centroidal profile for an object having a well defined boundary.

signature analysis   (1) a test where the responses of a device over time are compared to a characteristic value called a signature, which is then compared to a known good one.

(2) a analysis of the signature to extract the desired (signal) information.

signed-digit representation   a fixed-radix number system in which each digit has a sign (positive or negative). In a binary signed-digit representation, each digit can assume one of the values -1, 0 and 1.

significand   the mantissa portion of a floating-point number in the IEEE 754 floating-point standard. It consists of an implicit or explicit leading integer bit and a fraction.

signum function   the function

$$\text{sgn}(t) = \begin{cases} -1, & t < 0 \\ 0, & t = 0 \\ +1, & t > 0 \end{cases}.$$

Used in modeling numerous types of system functions.

SIL   *See* surge impedance loading.

silicon   most common element in the earth's crust and a type IV (from periodic table of elements) semiconductor material. The bipolar carriers, both holes (p dopants) or electrons (n dopants) are roughly in proportion to each other, resulting in nearly equal currents in the same direction. They move at about half the speed of comparable GaAs unipolar carriers. The thermal resistance is also about half that of GaAs.

Silicon has an indirect band gap of 1.11 eV, density-of-states masses of 1.1 times the free-electron mass for the conduction band and 0.56 for the valence band.

silicon compiler   a set of software programs intended to start with design equations and output the corresponding GDS2 data. Silicon compilers are currently used to translate a standard cell design from one set of design rules to another or to create a new set of standard cells.

silicon controlled rectifier (SCR)   a current controlled four-layer (pnpn) device for high power (3000 A) and low speed (500 Hz) applications. SCRs can only be on or off, with no intermediate operating states like transistors. The SCR acts as a switch that is turned on by a short current pulse to the gate, provided that the device is in its forward blocking state. Once latched on, the gate current can be removed and the device will remain on until the anode current goes negative, or the current through the SCR becomes less than its designated holding current. A disadvantage is that a commutation circuit is often needed for forced turn-off (forced commutation).

silicon dioxide (SiO$_2$)   an amorphous compound of silicon and oxygen with a resistivity of 10^{14} to 10^{17} ohm-cm and a bandgap of 8.1 eV. Essential as a dielectric and insulator for silicon devices.

silicon nitride (SiN$_4$)   for a semiconductor manufacturing process, a compound formed of silicon and nitrogen that is deposited and etched back to provide a masking layer appropriate for withstanding subsequent high temperature processing such as oxidation.

SIMD   *See* single-instruction stream, multiple-data stream.

similarity between symbol strings   *See* distance between symbol strings.

similarity measure   the reciprocal concept of distance measure. *See* distance measure.

similarity of 2-D system matrices   two system matrices

$$ S_i = \begin{bmatrix} P_i & -Q_i \\ C_i & D_i \end{bmatrix} \quad i = 1, 2, $$

of the same dimensions (See input–output equivalence of 2-D system matrices) where there exists a nonsingular matrix M such that

$$ \begin{bmatrix} P_1 & -Q_1 \\ C_1 & D_1 \end{bmatrix} = \begin{bmatrix} M^{-1} & 0 \\ 0 & I \end{bmatrix} \begin{bmatrix} P_2 & -Q_2 \\ C_2 & D_2 \end{bmatrix} \begin{bmatrix} M & 0 \\ 0 & I \end{bmatrix} $$

The transfer matrix $T_i(z_1, z_2) = C_i P_i^{-1} Q_i + D_i$ ($i = 1, 2$) and the characteristic polynomial det P_i are invariant under the similarity.

SIMM   *See* single in-line memory module.

simple medium   a medium that is linear, isotropic, homogeneous, and time-invariant.

simplex   term used to describe a method of winding the armature of a commutated electric machine in which consecutive coils are placed in adjacent coil slots around the periphery of the rotor. In a lap winding, this produces two parallel electrical path between brushes for each pole pair. In a wave winding, a simplex arrangement produces two parallel electrical paths between brushes regardless of the number of poles. *See also* duplex, multiplex.

simulated annealing   an optimization technique that seeks to avoid local minima by allowing the search trajectory to follow paths that not only decrease the objective function but also sometimes increase it. The probability that an increase in the objective function is allowed by the technique is governed by a quantity that is analogous to temperature. The scheme commences with a high temperature, under which the probability of allowing increases in the objective function is high, and the temperature is gradually reduced to zero, and from then on no further increases in the function are allowed. *See also* annealing.

simulation   *See* computer simulation.

simulation model   *See* truth model or computer simulation.

simulator   a program used to predict the behavior of a circuit. Simulators may be transistor level, gate level, behavioral level, analog, digital, unit delay, timing, or various combinations.

simulcast systems   systems that simultaneously broadcast over two or more different frequency channels or modes of broadcast signals.

simultaneous contrast   the phenomenon in which the brightness (perceived luminance) of a region on a dark background is greater than the brightness of an identical region on a light background. Illustrates that brightness (perceived luminance) is different from lightness (actual luminance). *See* brightness, brightness constancy, human visual system, Mach band.

SiN_4 *See* silicon nitride.

sinc function   this function is defined by the relation: $\text{sinc}(t) = \sin c(t) = \frac{\sin \pi t}{\pi t}$. A scaled sinc function is the impulse response of an ideal low-pass filter.

sine-squared pulse   pulse string made from a standard sinewave with an added DC component equal to one-half of the peak-to-peak value of the sine wave. The pulse string is, therefore, always positive in value.

sine transform   usually refers to the discrete sine transform. It also refers to a continuous time transform similar to the Fourier transform.

sinewave brushless DC   a permanent magnet brushless motor with sinusoidally distributed stator phase windings. More commonly known as permanent magnet synchronous machine.

single dwell detector   a detector in a communications receiver based on a decision on a transmitted symbol being made after a single correlation of the received signal with a reference signal. *Compare with* multiple dwell detector.

single electron transistor   solid state device that performs electronic functions using a single transistor's electron.

single in-line memory module (SIMM)   a miniature circuit board that contains memory chips and can be plugged into a suitable slot in a computer motherboard in order to expand the physical memory.

single in-line packaging (SIP)   a method of packaging memory and logic devices on small PCBAs with a single row of pins for connection.

single layer perceptron   an artificial neural network consisting of a single layer of neurons (perceptrons) with an input layer. *See* multilayer perceptron, perceptron.

single line to ground fault   a fault on a three phase power line in which one conductor has become connected to ground.

single machine infinite bus system   a model of a power system consisting of a single generator working into an infinite bus which represents the remainder of the system.

single mode   single frequency resonance of a cavity that is usually associated with a unique field distribution.

single phase to ground fault   *See* single line to ground fault.

single photon emission computed tomography (SPECT)   a form of tomographic medical imaging based upon the density of gamma ray-emitting radionuclides in the body.

single precision   floating point numbers that are stored with fewer rather than more bits. Often refers to numbers stored in 32 bits rather than 64 bits.

single scattering   weak interaction of the light wave with the medium. This occurs when this is weakly inhomogeneous. This process yields low changes in the phase and amplitude of light, and no variation in its state of polarization.

single sideband modulation (SSB)   a method of amplitude modulation in which only one of the sidebands (upper or lower) is transmitted. This method can potentially double the capacity of a single channel.

single variable system   *See* single-input–single-output system.

single-address computer   a computer based on single-address instructions.

single-address instruction   a CPU instruction defining an operation and exactly one address of an operand or another instruction.

single-chip microprocessor   a microprocessor that has additional circuitry in it that allows it to be used without additional support chips.

single-electron tunneling   the name given to very small capacitors with thin insulators so that tunneling can occur through this insulator. When the capacitor is small, it is possible that the energy change for the tunneling of one electron is larger than the thermal energy, so that fluctuations cannot support the tunneling. In this case, an external source must provide the energy needed for the tunneling process, which occurs usually (in these very small capacitors of order $< 10–18$ F) by the transfer of a single charge from one plate to the other.

single-element fuse   a fuse that is constructed with a single fusible element. It does not meet the standard definition of time-delay.

single-ended amplifier   an amplifier that has only one signal path and only one set of input and output ports.

single-input–single-output (SISO) system   a system that transforms one input signal to one output signal. Also known as single variable (SV) system. *See also* system, multiple-input–multiple-output system.

single-instruction stream, multiple-data stream (SIMD)   a parallel computer architecture in which a collection of data is processed simultaneously under one instruction. Example in optics is imaging by a lens.

single-instruction stream, single data stream (SISD)   a processor architecture performing one instruction at a time on a single set of data. Same as uniprocessor.

single-layer network   a feedforward network consisting of input units connected directly to the output units. Thus, the network has a single layer of weights and no hidden units.

single-mode fiber   an optical fiber with a relatively small diameter in which only one mode may propagate. However, this mode may have two orthogonal states of polarization that propagate unless a polarization maintaining optical fiber is used.

single-mode optical fiber   relatively thin fiber that has low loss for one mode and much higher losses for all other modes. *See also* Flynn's taxonomy.

single-phase inverter   an inverter with a single-phase AC voltage output. Half-bridge and full-bridge configurations are commonly used.

single-phase rectifier   a rectifier with a single-phase AC voltage input. *See also* half-wave rectifier and full-wave rectifier.

single-phasing   a condition that occurs when a three-phase motor has an open circuit occur in one of the three lines. The motor continues to operate with one line to line voltage as a single-phase motor, with an increase in noise, vibration, and current. Proper overload protection should detect the higher current and shut down the motor after some time delay.

single-pole reclosing   the practice of clearing a fault which appears on one phase of a three-phase electric power line by disconnecting and reclosing only that phase as opposed to opening and reclosing all three phase conductors.

single-pole double-throw (SPDT)   a switch that has a common port and two output ports. Among these two ports, only one selected port can be connected to the common port.

single-pole single-throw (SPST)   a switch that has a pair of input–output ports. By changing its status, the switch works as short or open circuit.

single-sided assembly   a packaging and interconnecting structure with components mounted only on the primary side.

single-stage decision making   decision making involving future operation of the system, as in the case of open-loop feedback control, where no future measurements and decision interventions are assumed when considering the decision taken at a given time.

single-step   to operate a processor in such a way that only a single instruction or machine memory access cycle is performed at a time, enabling the user to examine the status of processor registers and the flags. A common debugging method for small machines.

single-tuned circuit   a circuit which is tuned by varying only one of it's components, e.g., an IF transformer in which only the secondary coil (rather than both primary and secondary) is tuned.

single-valued   a function of a single variable, $x(t)$, which has one and only one value $y_0 = x(t_0)$ for any t_0. The square root is an example of a function that is not single-valued.

singular 2-D Attasi-type model   a 2-D model described by the equations

$$Ex_{i+1,j+1} = -A_1A_2x_{ij} + A_1x_{i+1,j}$$
$$+A_2x_{i,j+1} + Bu_{ij}$$
$$y_{ij} = Cx_{ij} + Du_{ij}$$

$i, j \in Z_+$ (the set of nonnegative integers) is called singular 2-D Attasi-type model, where $x_{ij} \in R^n$ is the local semistate vector, $u_{ij} \in R^m$ is the input vector, $y_{ij} \in R^p$ is the output vector, and E, A_1, A_2, B, C, D are real matrices with E singular (det $E = 0$ if E is square or rectangular).

singular 2-D Fornasini–Marchesini-type model   a 2-D model described by the equations

$$Ex_{i+1,j+1} = A_0x_{ij} + A_1x_{i+1,j}$$
$$+A_2x_{i,j+1} + Bu_{ij} \quad (1a)$$
$$y_{ij} = Cx_{ij} + Du_{ij} \quad\quad (1b)$$

$i, j \in Z_+$ (the set of nonnegative integers) is called the first singular 2-D Fornasini–Marchesini-type model, where $x_{ij} \in R^n$ is the local semistate vector, $u_{ij} \in R^m$ is the input vector, $y_{ij} \in R^p$ is the output vector, E, A_k $(k = 0, 1, 2)$, B, C, D are real matrices with E singular (det $E = 0$ if E is square or rectangular). An 2-D model described by the equations

$$Ex_{i+1,j+1} = A_1x_{i+1,j} + A_2x_{i,j+1}$$
$$+B_1u_{i+1,j} + B_2u_{i,j+1}$$

and (1b) is called the second singular 2-D Fornasini–Marchesini-type model, where x_{ij}, u_{ij}, and y_{ij} are defined in the same way as for (1), E, A_1, A_2, B_1, B_2 are real matrices with E singular.

singular 2-D general model   a 2-D model described by

$$Ex_{i+1,j+1} = A_0x_{ij} + A_1x_{i+1,j} + A_2x_{i,j+1}$$
$$+B_0u_{ij} + B_1u_{i+1,j} + B_2u_{i,j+1}$$
$$y_{ij} = Cx_{ij} + Du_{ij}$$

$i, j \in Z_+$ (the set of nonnegative integers) is called singular 2-D general model, where $x_{ij} \in R^n$ is the local semistate vector, $u_{ij} \in R^m$ is the input vector, $y_{ij} \in R^p$ is the output vector, E, A_k, B_k $(k = 0, 1, 2)$, C, D are real matrices with E singular (det $E = 0$ if E is square or rectangular). In particular case for $B_1 = B_2 = 0$ we obtain the first singular 2-D Fornasini–Marchesini-type model and for $A_0 = 0$ and $B_0 = 0$ we obtain the second singular 2-D Fornasini–Marchesini-type model.

singular 2-D Roesser-type model   a 2-D model described by

$$E \begin{bmatrix} x^h_{i+1,j} \\ x^v_{i,j+1} \end{bmatrix} = \begin{bmatrix} A_1 & A_2 \\ A_3 & A_4 \end{bmatrix} \begin{bmatrix} x^h_{ij} \\ x^v_{ij} \end{bmatrix} + \begin{bmatrix} B_1 \\ B_2 \end{bmatrix} u_{ij}$$

$$y_{ij} = C \begin{bmatrix} x^h_{ij} \\ x^v_{ij} \end{bmatrix} + D u_{ij}$$

$i, j \in Z_+$ (the set of nonnegative integers) is called singular 2-D Roesser type model, where $x^h_{ij} \in R^{n_1}$ and $x^v_{ij} \in R^{n_2}$ are horizontal and vertical semistate vector, respectively, $u_{ij} \in R^m$ is the input vector, $y_{i,j} \in R^p$ is the output vector, E, A_1, A_2, A_3, A_4, B_1, B_2, C, D are real matrices with E singular (det $E = 0$ if E is square or rectangular). In particular case for $E = I$ we have 2-D Roesser model.

singular matrix   a square matrix A is singular if its rows (or columns) are not linearly independent. Singular matrixes cannot be inverted, have zero determinants, and have linearly dependent columns and rows.

singular perturbation   for a dynamical system, a state-space model in which the derivatives of some of the states are multiplied by a small positive parameter ε,

$$\left. \begin{aligned} \dot{\mathbf{x}} &= \mathbf{f}(t, \mathbf{x}, \mathbf{z}, \varepsilon) \\ \varepsilon \dot{\mathbf{z}} &= \mathbf{g}(t, \mathbf{x}, \mathbf{z}, \varepsilon) \end{aligned} \right\}$$

The parameter ε represents the parasitic elements such as small masses, inertias, capacitances, inductances, etc. The system properties undergo a discontinuous change when the perturbation parameter ε is set to zero since the second differential equations becomes an algebraic, or a transcendental equation

$$\mathbf{0} = \mathbf{g}(t, \mathbf{x}, \mathbf{z}, 0).$$

The singular perturbation methods are used in the analyses of high-gain nonlinear control systems as well as *variable structure sliding mode* control systems.

singular value decomposition (SVD)   useful decomposition method for matrix inverse and pseudoinverse problems, including the least-squared solution of overdetermined systems. SVD represents the matrix A in the form $A = U \Lambda^{\frac{1}{2}} V$, where Λ is a diagonal matrix whose entries are the singular values of A, and U and V are the row and column eigenvector systems of A. Any matrix can be represented in this way. In image processing, SVD has been applied to coding, to image filtering, and to the approximation of non-separable 2-D point spread functions by two orthogonal 1-D impulse responses.

singularity   a location in the workspace of the manipulator at which the robot loses one or more DOF in Cartesian space, i.e., there is some direction (or directions) in Cartesian space along which it is impossible to move the robot end effector no matter which robot joints are moved.

singularity function   a function-like operation that is not a proper function in the analytic sense. This is because it has a point at which the function or its derivative is infinite or is undefined. In particular, the Dirac delta function is defined as:

$$\delta(t) = \begin{cases} \infty, & m = 0 \\ 0, & \text{otherwise} \end{cases}$$

Other examples are the step function and ramp function.

633

sintered magnet   magnet made from powdered materials that are pressed together and then heated in an oven to produce desired shapes and magnetic properties.

sinusoid   a periodic signal $x(t) = \cos(\omega t + \theta)$ where $\omega = 2\pi f$ with frequency in hertz.

sinusoidal amplitude modulation   amplitude modulation where the carrier signal is a sinusoid. *See* amplitude modulation and carrier Signal.

sinusoidal coding   parametric speech coding method based on a speech model where the signal is composed of sinusoidal components having time-varying amplitudes, frequencies and phases. Sinusoidal coding is mostly used in low bit rate speech coding.

sinusoidal signal   a signal of the form x(t) = A $x(t) = A\sin(2\pi f t + \theta)$A where A is the amplitude and θ is the phase angle.

sinusoidal steady-state response   response of a circuit a sine wave input as $t \rightarrow \infty$. The output then has two components, a magnitude and a phase, that are the magnitude and the phase of the transfer function itself for $s = j\omega$.

sinusoidal–Gaussian beam   electromagnetic beam in which the transverse field distribution is describable in terms of sinusoidal and Gaussian functions.

SiO$_2$ *See* silicon dioxide.

SIP   *See* single in-line packaging.

SIR   *See* signal-to-interference ratio.

SISD   *See* single-instruction, stream, single data architecture.

SISO   *See* single-input–single-output system.

SISO system   *See* single-input–single-output system.

SIT   *See* static induction transistor.

site diversity   the combination of received signals at widely separated locations having substantially different propagation paths to the transmitter. The resultant signal has reduced fading depth and therefore higher quality communication is possible. Often used in Earth–satellite link to overcome the effects of scintillation and rain fading. *See also* space diversity.

SITH   *See* static induction thyristor.

six connected   *See* voxel adjacency.

size distribution   the size distribution of a family of objects is a function measuring the number or volume of all objects in any size range. In mathematical morphology, this notion is developed by analogy with a family of sieves: each sieve retains objects larger than a given size, and lets smaller objects go through; when two sieves are put in succession, this is equivalent to using the finest sieve. This idea is formalized by taking a one-parameter family of morphological operators γ_r $(r > 0)$ such that for all objects X, Y and $r, s > 0$, one has

(1) $\gamma_r(X) \leq X$, that is, γ_r is anti-extensive;

(2) X $\leq$ Y implies that $\gamma_r(X) \leq \gamma_r(Y)$, that is, γ_r is increasing;

(3) $\gamma_r(\gamma_s(X)) = \gamma_{\max(r,s)}(X)$. In particular, each γ_r is an algebraic opening, and for $s > r$ we have $\gamma_s(X) \leq \gamma_r(X)$.

See mathematical morphology, morphological operator, opening.

skeleton   (1) the set of arcs, enclosing a region, resulting from the successive application of a thinning operator on the region.

(2) a shape representation consisting of a connected set of pixel locations of unit width running along the centers of the object limbs. This is a natural representation, where (a) limbs are of

approximately uniform width and (b) the width is unimportant for shape analysis, so that all the shape topology is embedded in the skeleton. The prime application relates to interpretation of hand-drawn characters and script.

skeletonization   a procedure, usually thinning, that produces an image skeleton.

skew   (1) an arrangement of slots or conductors in squirrel cage rotors so that they are not parallel to the rotor axis.

(2) in computer buses, a condition where values on certain bus lines have slightly different transmission times than values on other lines of the same bus. *See also* tape skew.

skewed addressing   *See* interleaved memory.

skewed symmetry   the nonperpendicular appearance of a symmetry-axes system for an object, when the plane of the object is not perpendicular to the line of sight from the viewpoint.

skewing   (1) the bending of a curve away from it's original shape.

(2) In a differential amplifier, the offset between two signals.

skin depth   for a lossy material, the distance at which electromagnetic fields experience one neper of attenuation. For a good conductor, the skin depth is given by

$$\frac{1}{\sqrt{\pi \mu f \sigma}}$$

where f is the frequency, μ is the permeability, and σ is the conductivity.

skin effect   the tendency of an alternating current to concentrate in the areas of lowest impedance.

skinny minnie   a telescoping fiberglass pole with interchangeable tools mounted at its end. It

can be extended sufficiently to allow a line worker to service cut-outs and similar pole-top equipment from the ground.

skip instruction   an assembly language instruction that skips over the next instruction without executing it.

sky wave   a wave that propagates into the ionosphere. It undergoes several reflections and refractions before it returns back to Earth.

slab waveguide   a dielectric waveguide useful for theoretical studies and for approximating other types of waveguide such as the rib waveguide. *See* rib waveguide.

slant angle   also called "dip angle"; the angle by which a plane slants or dips away from the frontal plane of the observer.

sleeve   (1) rubber cover for a line worker's arms.

(2) a type of wire connector.

slew rate   the rate of variation of an AC voltage in terms of volts per second.

In an op-amp, if the signal at the op-amp output attempts to exceed this limit, the op-amp cannot follow and distortion ("slew rate limiting") will result.

slice   *See* wafer.

SLG   *See* single line to ground fault.

slicer   a device that estimates a transmitted symbol given an input that is corrupted by (residual) channel impairments. For example, a binary slicer outputs 0 or 1, depending on the current input.

sliding correlation   a principle of operation of a correlation receiver in channel measurement, where pseudo-random sequences are utilized. The transmitted signal consists of a carrier modulated (typically employing phase shift keying) by a pseudorandom sequence. The received signal is correlated (multiplied) by a similar reference signal, in which the pseudorandom sequence has a clock rate slightly lower than in the transmitted signal. The difference in clock frequencies causes the relative phase (chip position) of the pseudorandom sequences to slide by each other. The output of a sliding correlator is a time-scaled version of the auto-correlation function of the pseudorandom sequence.

The time-scaling factor depends on the difference of the clock frequencies, and typically cannot be lower than 1000 without significant distortion in the resulting correlation function. This sets an upper bound to the rate of producing complete autocorrelation functions, making sliding correlation not ideally suited to measurement of channels with fast time variance. A sliding correlator must be implemented using analogue signal processing. *See also* stepping correlation.

sliding mode   the motion of a dynamical system's trajectory while confined to a sliding surface.

sliding mode domain   a sliding mode domain D where for each $\varepsilon > 0$ there exists a $\delta > 0$ such that any trajectory starting in the n-dimensional δ-neighborhood of Δ may leave the n-dimensional ε-neighborhood of Δ only through the n-dimensional ε-neighborhood of the boundary of Δ.

sliding mode observer   *See* sliding mode state estimator.

sliding mode state estimator   state estimators of uncertain dynamical plants in which the error between the state estimate and the actual state exhibits sliding mode behavior on a sliding surface in the error space.

sliding surface   a surface in the state space specified by a designer of a variable structure sliding mode controller. The role of a sliding mode con-

troller is to drive the system's trajectories to the surface and to maintain them on the surface for all subsequent time. Alternative terms for sliding surface are switching surface, discontinuity manifold, equilibrium manifold.

sliding termination   a precision air-dielectric coaxial transmission line that consists of a moveable, tapered termination used as an impedance standard for calibrating vector network analyzers and in other precision microwave measurement applications.

sliding window   in an ARQ protocol, the (sliding) window represents the sequence numbers of transmitted packets whose acknowledgments have not been received. After an acknowledgement has been received for the packet whose sequence number is at the tail of the window, its sequence number is dropped from the window and a new packet whose sequence number is at the head of the window is transmitted, causing the window to slide one sequence number.

sliding-mode control   a bang-bang control technique that confines the state space trajectory to the vicinity of a sliding line. Assuming a second-order system, the sliding line is defined as $ax_1 + bx_2 = 0$, where x_1 and x_2 are the state variables and a and b are constant coefficients determined by the desired control law. The sliding line exists if the trajectories of the subcircuits on either side of the line are directed toward the sliding line. The sliding line is stable if the motions along the sliding line are toward an operation point. The ideal overall trajectory is independent of the trajectories of the subcircuits.

slip   in an induction motor, slip is defined as the ratio of the slip speed to the synchronous speed. The slip speed is the difference between the synchronous speed and the speed of the rotor. *See also* synchronous speed.

slip frequency   the frequency of the rotor induced currents in an induction machine. Denoted by f_{sl}, the slip frequency is given by slip × stator frequency (f_s) and is the prime frequency used in slip frequency control of induction machines.

slip power recovery control   a method of controlling the speed of a wound rotor induction machine by recovering the slip frequency power from the rotor to an AC power source or mechanical shaft through the converter connected to the rotor windings of the motor. Slip power recovery control reduces the losses that occur with rotor resistance control.

slip-ring contact   a rotating, brush-contacted ring electrode connected to one end of a coil in an AC generator.

SLL   *See* sidelobe level.

SLM   *See* spatial light modulator or liquid crystal on silicon.

slope detector   a circuit consisting of an LC tuned circuit, a detector diode, and a filter circuit that has an IF set to be on the most linear portion of the response curve. The circuit converts FM to AM by having the frequency changes from the FM signal cause the signals to move up and down the response curve which results in amplitude variations.

slope transform   transform that plays for morphological operators a role which is to some extent analogous to that of the Fourier transform for linear shift invariant systems. *See* dilation, erosion, Fourier transform, linear shift invariant system, morphological operator.

slot   a space between the teeth used to place windings in electrical machines.

slot pitch   the angular distance (normally in electrical degrees) between the axes of two slots.

slotless motor   permanent magnet brushless DC motor in which stator teeth are removed and the resulting space is partially filled with copper. The slotless construction permits an increase in rotor diameter within the same frame size, or alternatively an increase in electric loading without a corresponding increase in current density.

slotted ALOHA   a multiple access protocol. In slotted ALOHA, time is divided into frames. Any user is allowed to transmit during any frame. The possible collisions result in retransmission at a later time. *See also* pure ALOHA.

slotted line   coax or waveguide with a longitudinal slot that accommodates a voltage probe that can measure the voltage anywhere along the slot. Typically used to measure standing wave ratio (SWR).

slow start   a congestion control algorithm that rapidly determines the bandwidth available to a transmitter by doubling the number of packets sent each round trip until losses are detected. This algorithm is "slow" when compared to the alternative of sending packets at the maximum rate achievable by the transmitter.

slow wave   a wave whose phase velocity is slower than the velocity of light. For example, for suitably chosen helixes the wave can be considered to travel on the wire at the velocity of light, but the phase velocity is less than the velocity of light by the factor that the pitch is less than the circumference. Slow wave may also be present on structures like coplanar waveguides.

slow-wave structure   a short microwave transmission line in a traveling wave tube in which the longitudinal phase velocity of traveling microwave is slowed down to almost equal speed of electrons in the interacting electron beam of the tube.

slowly varying envelope approximation   neglect of the second time and/or space derivatives in the wave equation governing nearly monochromatic/nearly plane-wave electromagnetic fields.

slowness surface   a plot of the reciprocal of the phase velocity as a function of direction in an anisotropic crystal.

SMA connector   a subminiature coaxial connector with both male and female versions capable of an upper frequency limitation of about 26 GHz.

small computer systems interface (SCSI)   a high-speed parallel computer bus used to interface peripheral devices such as disk drives.

small disturbance   a disturbance for which the equation for dynamic operation can be linearized for analysis.

small disturbance stability   power system stability under small disturbances, which can be studied by using linearized power system models.

small gain theorem   a sufficient condition for the robust stability of the closed-loop system. It requires the open-loop operator of the system to have a norm less than one. For linear systems, the small gain theorem guarantees well posedness while in the nonlinear case it should be assumed. The theorem may be highly conservative for structured uncertainties. In some cases the conservatism could be decreased by the use of structured norms.

small scale integration (SSI)   an early level of integration circuit fabrication that allowed approximately between 1 and 12 gates on one chip.

small signal amplifier   amplifier designed for amplifying very low level signals. Typically, small signal amplifiers have an AC signal magnitude that is 1/10 the DC value and operate under class A amplifier biasing conditions.

small-signal stability   *See* dynamic stability.

smart antenna   a set of antennas used in an intelligent way in one receiver to improve the performance of a communication link. *See also* beamforming, spatial diversity, spatial division multiple access.

smart card   credit-card-sized device containing a microcomputer, used for security-intensive functions such as debit transactions.

smart material   one of a class of materials and/or composite media having inherent intelligence together with self-adaptive capabilities to external stimuli applied in proportion to a sensed material response. Also called intelligent material.

smart pixel   an element in an array of light detectors that contains electronic signal processing circuitry in addition to the light detector; a spatial light modulator in which each pixel is controlled by a local electronic circuit. Smart pixels are fabricated with VLSI technology. Each light modulating pixel is connected to its own tiny electronic circuit adjacent to the pixel. The circuit may consist of detector, switching or logic circuit, memory, and source or additional shutter. It is an advanced, optically addressed spatial light modulator and still immature.

smart sensor   sensor with inherent intelligence via built-in electronics.

smart structural material   material in which the mechanical (elastic) properties can be modified adaptively through the application of external stimuli.

SMB connector   a subminiature coaxial connector with snap-on mating and typical upper frequency limit of 10 GHz.

SMDA   *See* Safe Medical Devices Act.

SMIB   *See* single machine infinite bus system

Smith chart   a graphical polar plot of the voltage reflection coefficient superimposed on a plot of two families of circles, with one family representing the real part of a complex impedance (admittance) and the other family representing the imaginary part of the complex impedance (admittance). A Smith chart can be used for many tasks including visualizing transmission line equations as well as matching loads to transmission lines, finding lengths and placements of stub tuners (both single and double stub tuners), finding the voltage standing wave ratio on a transmission line, and finding the reflection coefficient at any point on the transmission line.

Smith predictor   a control scheme designed to deal with processes that contain significant deadtime in their responses. Its block diagram configuration is where model one is an exact replica

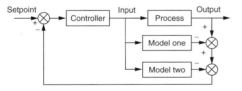

Block diagram configuration for the Smith predictor.

of the process under control, while model two is the same model with the deadtime term removed. The controller is designed to control model two, thereby allowing faster closed loop responses than would otherwise have been possible. *See also* internal model control.

SMM   *See* sub-millimeter.

smoothing   (1) an estimation procedure in which a past value of the state vector (see the definition) is estimated based on the data available up to the present time.

(2) the removal from an image (signal) of high-frequency components obtained, e.g., through a convolutional averaging or Gaussian filter, usually performed to remove additive speckle noise.

(3) any process by which noise is suppressed, following a comparison of potential noise points with neighboring intensity values, as for mean filtering and median filtering. Also, a process in which the signal is smoothed, e.g., by a low-pass filter, to suppress complexity and save on storage requirements.

SMS   *See* service management system.

SMT   *See* surface mount technology.

snake   *See* active contour.

Snell's Law   the law that gives the angles of reflection and refraction of a plane electromagnetic wave when the wave is incident on a boundary between two media.

Consider a plane wave impinging on a surface between two media with different dielectric constants: part of the incident power will be transmitted to the other region, while part will be reflected. Let us denote the angle of reflection by θ_1^r, the angle of transmission by θ_2^t, and the angle of incidence by θ_1^i. We also denote the electric permittivity and the magnetic permeability by ϵ, μ, respectively; the subscripts 1, 2 refer to the particular medium under consideration. The angle of reflection is equal to the angle of incidence, i.e.,

$$\theta_1^r = \theta_1^i$$

and the Snell's Law states that

$$\sqrt{\mu_1 \epsilon_1} \sin \theta_1^i = \sqrt{\mu_2 \epsilon_2} \sin \theta_2^t$$

SNIR *See* signal-to-noise-plus interference ratio.

snoop in hardware systems, a process of examining values as they are transmitted in order to possibly expedite some later activity.

snooping bus a multiprocessor bus that is continuously monitored by the cache controllers to maintain cache coherence.

snow noise noise composed of small, white marks randomly scattered throughout an image. Television pictures exhibit snow noise when the reception is poor.

SNR *See* signal-to-noise ratio.

snubber an auxiliary circuit or circuit elements used to control the rate of rise or fall of the current flowing into a power electronic device or the rate of fall of the voltage across the device during turn-off. Snubbers are used to limit dv/dt and di/dt and eliminate ringing in a switching circuit during switching transients. Both dissipative and nondissipative snubbers are used. *See also* soft-switching.

snubber circuit *See* snubber.

Sobel operator a common digital approximation of the gradient ∇f, often used in edge detection. It is specified by the pair of convolution mask

$$\frac{\partial f}{\partial x} = \begin{pmatrix} -1 & 0 & 1 \\ -2 & 0 & 2 \\ -1 & 0 & 1 \end{pmatrix}$$

$$\frac{\partial f}{\partial y} = \begin{pmatrix} -1 & 2 & -1 \\ 0 & 0 & 0 \\ 1 & 2 & 1 \end{pmatrix}$$

responses, g_x, g_y, are components of the local vector gradient, from which edge magnitude and direction can be calculated straightforwardly.

sodium-cooled reactor a nuclear reactor in which the heat is removed from the core by means of circulating liquid metallic sodium.

SOFC *See* solid oxide fuel cell.

soft computing is an association of computing methodologies centering on fuzzy logic, artificial neural networks, and evolutionary computing. Each of these methodologies provide us with complementary and synergistic reasoning and searching methods to solve complex, real-word problems. *See also* fuzzy logic, neural networks.

soft contaminant a contaminant which to first order is not absorbed by X-rays, and which therefore tends to remain undetected in X-ray images: typical soft contaminants are rubber, wood and many types of plastic (though much depends on the particular atomic composition of the material).

soft decision demodulation that outputs an estimate of the received symbol value along with an indication of the reliability of that value. It is usually implemented by quantizing the received signal to more levels than there are symbol values.

soft facimile for low-capacity channels, images are transmitted and displayed in a progressive (stage by stage) manner. A crude representation is first transmitted and then details are added at each stage. This is referred to as soft facimile. *See also* progressive transmission.

soft fault *See* transient fault.

soft hand-off a hand-off scheme in a CDMA cellular system such as that specified by the IS-95 standard where signal transmission occurs through multiple base stations during the hand-off process. The multiple signal components received from the different base stations are combined using some type of diversity combining. *See also* hand-off, IS-95.

soft iron   a term used to describe iron that has a low coercivity. Note that soft refers to the magnetic properties of the material, not necessarily the physical properties.

soft magnetic material   a magnetic material that does not retain its magnetization when the magnetizing field is removed; a material with low coercivity and high permeability.

soft real-time   *See* soft real-time system.

soft real-time system   a real-time system in which failure to meet deadlines results in performance degradation but not necessarily failure. *Compare with* firm real-time, hard real-time.

soft switching   the control of converter switching in order to utilize device and component parasitics and resonance conditions to enable zero current switching (ZCS) or zero voltage switching (ZVS), thereby reducing switching losses, stress and EMI. Typically this is performed with additional resonant components and switches that are activated only during the switching transients.

Soft switching also allows higher switching frequencies in order to reduce the converter size and weight, thus increasing the power density.

softbake   *See* prebake.

soft-decision decoding   decoding of encoded information given an unquantized (or finely quantized) estimate of the individually coded message symbols (for example, the output directly from the channel). *Compare with* hard-decision decoding.

soft-starter   a motor starter that provides a ramp-up of voltage supplied to the motor at starting with the objective of reducing the starting current and torque.

soft X-ray lithography   *See* EUV lithography.

softer hand-off   similar to the concept of soft hand-off except that it involves transmission/reception through multiple antenna sectors of the same cell as opposed to multiple base stations. In contrast to soft hand-off, softer hand-off need not involve the mobile telephone system switch in the hand-off process.

software design   a phase of software development life cycle that maps what the system is supposed to do into how the system will do it in a particular hardware/software configuration.

software development life cycle   a way to divide the work that takes place in the development of an application.

software engineering   systematic development, operation, maintenance, and retirement of software.

software evolution   the process that adapts the software to changes in the environment where it is used.

software interrupt   a machine instruction that initiates an interrupt function. Software interrupts are often used for system calls, because they can be executed from anywhere in memory and the processor provides the necessary return address handling. Also known as a supervisor call instruction (SVC) (IBM mainframes) or INT instruction (Intel X86).

software reengineering   the reverse analysis of an old application to conform to a new methodology.

soil electrode   an electrical connection to the soil, often in the form of a metal stake driven into the earth.

solar–thermal–electric conversion   collection of solar energy in thermal form using flat-plate or concentrating collectors and its conversion to electrical form.

solder duration   *See* solder profile.

solder profile   the time versus temperature profile required to properly solder a connection without damaging the component to either side of the connection, and without leaving any defects in the solder joint. The solder profile will include preheating, ramp-up, time at solder reflow temperature, ramp-down, and any relevant limits in time and temperature. The profile will be specific to a given system (components, materials, plating, fluxes or gases, solder type, surface preparation, etc.).

solder temperature   *See* solder profile.

sole   a nonemitting cathode.

solenoid   a wound cylindrical and magnetic material assembly used typically for producing linear motions.

solid laser   laser in which the amplifying medium is a solid.

solid oxide fuel cell (SOFC)   a fuel cell that uses various hydrocarbons as fuel, and solid oxide electrolytes. It can generate >200 kW of power, and hence can be used in power generation. *See also* akaline fuel cell, proton exchange membrane fuel cell, phosphoric acid fuel cell, molten carbonate fuel cell.

solid polymer fuel cell (SPFC)   *See* proton exchange membrane fuel cell.

solid state disk (SSD)   very large-capacity, but slow, semiconductor memory that may be used as a logical disk, extended main memory, or as a logical cache between main memory and conventional disk. SSD is typically constructed from DRAM and equipped with a battery to make it nonvolatile. First used in IBM 3090 and Cray-XMP computer systems.

solid state laser   laser in which the amplifying medium is a solid, sometimes considered to exclude semiconductor lasers.

solid state power amplifier (SSPA)   a high-power, multistage amplifier using semiconductor devices.

solid state relay   a protective relay that employs analog electronics, logic electronics, and magnetics to implement the operating logic.

solid state ultraviolet laser   an ultraviolet laser made from a solid state or semiconductor material, e.g., Nd:LaF3, Ce:LaF.

solid state UV laser   *See* solid state ultraviolet laser.

solidly grounded   an electrical distribution system in which one of the normal current-carrying conductors, often the neutral, is intentionally connected to ground with no impedance other than that of the conductor comprising the connection.

solidly grounded system   a grounding scheme in which the neutral wire of a power system is connected to ground at frequent intervals so as to minimize the impedance between neutral and ground.

soliton   an optical pulse that preserves its shape while propagating by balancing fiber dispersion and nonlinearity.

soliton transmission system   often termed the fifth generation of fiber optic communication systems. *See also* soliton.

solution domain   electromagnetic fields can be represented as a function of time, or a *time-domain* description, or as a function of frequency using a (usually) Fourier transform, which produces a frequency-domain description.

SOM   *See* self-organizing map.

sonar   acronym for "sound navigation and ranging" adopted in the 1940s, involves the use of sound to explore the ocean and underwater objects.

sonar equation   *See* figure of merit.

SONET   *See* synchronous optical network.

sonorant   the class of phonemes with a formant-like spectrum. For example, vowels and nasals exhibit a spectrum that is based on formants.

SONT   *See* self-organizing neural tree.

SOP   *See* sum of products.

sound carrier   in a TV signal, the FM carrier that transmits the audio part of the program.

sound velocity profile (SVP)   description of the speed of sound in water as a function of water depth.

source   (1) refers to the signal generator/device that generates the RF, microwave, or micromilliwatt frequencies.
(2) the terminal of a FET from which electrons flow (electrons in the FET channel flow from the source, and current flow is always in the negative direction of electron movement, since electrons are negative). It is usually considered to be the metal contact at the surface of the die.

source code   (1) software code written in a form or language meant to be understood by programmers. Must be translated to object code in order to run on a computer.
(2) a set of codewords used to represent messages, such that redundancy is removed, in order to require less storage space or transmission time.

source coding   the process of mapping signals onto a finite set of representative signal vectors referred to as codewords.

source compression   *See* source coding.

source encoder   a device that substantially reduces the data rate of linearly digitized audio signals by taking advantage of the psychoacoustic properties of human hearing, eliminating redundant and subjectively irrelevant information from the output signal. Transform source encoders work entirely within the frequency domain, while time-domain source encoders work primarily in the time domain. Source decoders reverse the process, using various masking techniques to simulate the properties of the original linear data.

source follower   *See* common drain amplifier.

source operand   in ALU operations, one of the input values.

source routing   method of routing packets in which the entire route through the network is prepended to the packet. From any node in the route, the next entry in the source root determines the node to which the packet should next be forwarded.

source-coupled pair   *See* differential pair.

space division multiple access (SDMA)   multiple access technique where the users' channels are separated into spot-beams with highly directional or adaptive antennas allowing reuse of time and frequency resources between users. Used in conjunction with FDMA/TDMA/CDMA.

space invariance   a system which a shift (translation) in the spatial coordinates for all inputs yields the same output except for an identical shift. Thus in two dimensions, for an input $I(x, y)$, output $O(x, y)$ and system S, if $O(x - x_0, y - y_0) = S\{I(x - x_0, y - y_0)\}$ for all $(x(x_0, y_0))$ and all inputs $I(x, y)$, the system S is space invariant.

spacer cable   another name for messenger cable.

SPARC   *See* scaled processor architecture.

spark gap   a pressurized high-current switch using a principle of electric field disruption to start the electron flow.

sparse equation   when a set of linear simultaneous equations has very few nonzeros in any row, the system is said to be sparse. Normally for a system to be considered sparse, less than 10% of the possible entries should be nonzero. For large integrated circuits, less than 1% of the possible entries are nonzero.

sparse matrix   a rectangular array of numbers, most of whose elements are zero or null.

sparse vector   in computer instruction processing, a matrix in which most elements have such small values that are treated as zeros. Special representation schemes can be used to save memory space, with a cost of increased execution time to access single elements of the matrix.

spatial coherence   *See* coherence.

spatial dispersion   occurs in a medium when its dielectric function depends on the wave number.

spatial domain   the representation of a signal, usually an image, as a function of spatial coordinates. *See* frequency domain.

spatial filtering   a technique used to either filter out interfering signals in a communication system, or as a multiple access technique that enables two or more subscribers, controlled by the same base station to use the same time, frequency and code resources on the grounds of their physical location or spatial separation. *See also* beamforming.

spatial frequency   the variables of the 2-D Fourier transform of a function of spatial coordinates are referred to as horizontal and vertical spatial frequencies. The spatial frequency of a 2-D sinusoidal signal in a given dimension is the number of cycles per unit distance in that dimension.

spatial hole burning   spatially localized reduction in the gain of a laser amplifier due to saturation by an intense signal; transverse spatial hole burning due to the transverse beam profile is distinguished from longitudinal spatial hole-burning due to the standing wave nature of the fields and possibly also to high gain per pass; spatial relaxation (or cross-relaxation) reduces spatial hole burning.

spatial light modulator (SLM)   a device that alters both the spatial and temporal character of a light beam. A three-port device with input, readout, and control or modulation ports. Modulation signals can be applied either electrically, i.e., an electrically addressed spatial light modulator, or optically, i.e., optically addressed spatial light modulator. Also called a light valve.

spatial light modulator in optical computing   a device for modulating amplitude or phase of light passing through it. If the input signal is a light beam, it is called optically addressed spatial light modulator. If the input signal is electronic, it is called electrically addressed spatial light modulator. Light modulation is usually based on the electro-optic effect. In optical computing, an optically addressed spatial light modulator is a device used as a medium for controlling or switching light using light. Since a light beam cannot directly affect another light beam, a spatial light modulator is required. The modulation process can be seen, first, as the modulating light affecting the spatial light modulator; then the affected spatial light modulator modulates another light beam.

spatial locality   *See* locality.

spatial power combining   the power generated from many devices can be combined coherently into space. These techniques are used in order to alleviate circuit losses at high microwave frequencies.

spatial redundancy   the redundancy between samples of an image or random process that is a function of spatial coordinates. Images typically exhibit a high degree of spatial redundancy which can be exploited to obtain a high compression ratio.

spatial resolution   (1) the ability to resolve two closely spaced points or a periodic pattern. Rayleigh proposed the criterion that two stars could be resolved when the maximum in the image pattern from one star coincides with the first minimum in the other. Units of spatial resolution are lines or line pairs per millimeter.

(2) a measure of the ability of a system to resolve spatial details in a signal. For a discrete image, spatial resolution generally refers to the number of pixels per unit length, giving possibly different horizontal and vertical spatial resolutions. *See* frequency resolution.

spawn   to create a new process within a multitasking computing system.

SPDT   *See* single-pole double-throw.

speaker identification   a task that consists of identifying which speaker (of a closed set) pronounced a given portion of speech signal. The basic assumption is that no speaker different from the defined closed set is considered. The emphasis is on the discrimination of the given speakers, whereas no strong rejection constraints are commonly required.

speaker verification   a task that, unlike speaker identification, the speaker set for this problem is open. As a consequence, one has to verify the given speaker against any potential impostor

that is not known in advance. Basically, one cannot rely on the knowledge of the probability distribution of the "negative examples," since there is no restriction on who is supposed to use the verification system.

SPEC   *See* System Performance and Evaluation Cooperative.

SPEC benchmarks   suites of test programs created by the System Performance and Evaluation Cooperative.

special-purpose digital signal processor   digital signal processor with special feature for handling a specific signal processing application, such as FFT.

specific absorption rate (SAR)   the deposition of energy over time into a body. The units are generally watts per kilogram of body mass. This is the attribute on which findings by various researchers can be compared and on which the exposure standards base their guidelines.

specific inductive capacity   *See* dielectric constant.

specification   a statement of the design or development requirements to be satisfied by a system or product.

speckle   granular image noise due to fluctuations in the number of photons arriving at an image sensor. Speckle often occurs in night-vision equipment and X-ray images. Also called quantum mottle. *See* quantum mottle speckle.

speckle pattern   grainy appearance of the intensity of scattered light due to random interference. The grain size is inverse of the illuminated area of the scattering medium in wavelength units.

SPECT   *See* single photon emission computed tomography.

spectral completeness   characteristic of a linear dynamical system whose eigenfunctions connected with eigenvalues form a basis in the state space. Spectral completeness depends on the matrix A_1. System is spectrally complete if and only if

$$\operatorname{rank} A_1 = n$$

spectral controllabillity   a linear dynamical system characteristic where every subsystem connected with an eigenvalue is controllable.

spectral density function   the Fourier transform of the covariance for a wide-sense stationary process.

spectral domain   the transform domain obtained by taking a Fourier transform in order to solve a boundary value problem. This technique is particularly convenient for the analysis of microstrip circuits and antennas.

spectral hole burning   a technique used to render an absorbing material transparent at select frequencies by bleaching a portion of the (inhomogeneously broadened) absorption spectrum.

spectral linewidth   *See* linewidth.

spectral quantum efficiency   for a photodetector, the ratio of the average number of free electrons produced per monochormatic input photon of wavelength λ. The relationship between spectral quantum efficiency ($Q_e(\lambda)$), wavelength (λ) is

$$S(\lambda) = (124) \left(Q_e(\lambda) \right) / \lambda$$

where $S(\lambda)$, $(Q_e(\lambda))$, and λ are given in units of milliamperes per watt, percentage, and nanometers, respectively.

spectral representation   *See* spectral domain.

spectral width   *See* linewidth.

spectrometer   optical instrument that disperses broadband light into its component wavelengths, allowing the measurement of light intensity at each individual wavelength. Spectrometers may use prisms or gratings for wavelength dispersion and any of a variety of light detectors including photomultiplier tubes or charge-coupled devices.

spectroscopy   the measurement of the intensities of wavelength dispersed light to identify a chemical component or measure its concentration.

spectrum   (1) a range of electromagnetic energy ordered in accordance with its relative periodicity.

(2) the magnitude of the Fourier transform of a (deterministic) signal. The word spectrum is also used to refer to the power spectrum of a random process.

spectrum analyzer   a test system that measures RF or microwave devices in terms of signal frequency and signal power.

spectrum reuse   reusing frequencies over and over again in a confined area, resulting in more efficient utilization and higher radio network capacity.

specular   *See* specular reflection.

specular intensity   the energy reflected from a rough surface in the specular direction. Sometimes called coherent component of the scattered intensity.

specular reflection   (1) the process by which a radio wave reflects from an electrically "even" surface experiencing changes only in amplitude, phase, and polarization, comparable to light reflecting from a mirror.

(2) the part of an electromagnetic wave that is reflected in the direction specified by Snell's law of reflection.

specular scattering   *See* specular reflection.

specular transmittance   the effect on a signal passing through a diffusely transmitting surface such as that the signal scattered in all directions.

speculative execution   a CPU instruction execution technique in which instructions are executed without regard to data dependencies. *See also* lookahead.

speech activity factor   the fraction of time for which a speech signal is nonzero-valued, over a long period of time. Zero-valued speech time segments occur as a result of pauses in the speech process. The speech activity factor is an important concept in the theory of statistical multiplexing of voice signals in a telephone switch. It is also an important concept in the IS-95 CDMA cellular system.

speech analysis   process of extracting time-varying parameters from the speech signal that represent a model for speech production.

speech coding   source coding of a speech signal. That is, the process of representing a speech signal in digital form using as low rate (in terms of, e.g., bits per second) as possible.

speech compression   the encoding of a speech signal into a digital signal such that the resulting bit rate is small and the original speech signal may be reproduced with as little distortion as possible. The transformation of a coded speech signal into another coded speech signal of lower bit rate in such a way that there is insignificant loss in speech quality of the decoded and play-back signal.

speech enhancement   improvement of perceptual aspects of speech signals.

speech preprocessing   the first step in all problems of speech processing. In preprocessing, the objective is to condition the signal so as to come up with more compressed and informative representations. Within portions of about 10 ms (frames), in practice the speech signal turns out to be quasi-stationary. For each frame, all relevant speech preprocessing approaches return a vector of parameters that make it possible to reconstruct the signal. Speech preprocessing is mainly carried out by using frequential approaches (e.g., the short-time Fourier transform) or linear prediction.

speech recognition   the process of recognizing speech portions carrying out linguistic information. The recognition can involve phonemes, single and connected words. Because of the crucial role of time, most successful approaches to automatic speech recognition are currently based on HMM (hidden Markov models) that incorporate very naturally the time dimension.

speech recognizer   system for performing speech recognition.

speech synthesis   the process of turning information into synthesized speech. When the synthesis involves restrictive linguistic domains (e.g., announcements in railway stations), the process often consists simply of playing back speech recorded in EPROM memories using proper coding (e.g., ADPCM). However, if one makes no restrictions on the information to synthesize, only artificial speech production is possible, which is commonly based on systems that predict phonetic units from linguistic information.

speech synthesizer   system for performing speech synthesis.

speech understanding   the process of understanding the meaning of a given portion of speech containing one or more sentences. Unlike speech recognition, the problem is not that of translating spoken to written units (e.g., words), but to extract the meaning. Models of speech understanding cannot simply be based on the recognition of linguistic units, but must also take into account the domain semantics.

speed droop   a linear characteristic that is provided to governors of two or more units operating in parallel for stable load division in case of load increase.

speed of light   (1) a scalar constant in vacuum roughly equal to 3×10^8 meters per second.

(2) the phase velocity representing the rate of advance of the phase front of a monochromatic light wave.

speed range   the minimum and maximum speeds at which a motor must operate under constant or variable torque load conditions. A 4:1 speed range for a motor with a top speed of 1800 rpm means that the motor must be able to operate as low as 450 rpm and still remain within regulation specifications. The controllable speed range of a motor is limited by the ability to deliver required torque below base speed without additional cooling.

speed regulation   the variation of the output speed of an electromechanical device as the load on the shaft is increased from zero to some specified fraction of the full load or rated load. Usually expressed as a percentage of the no-load speed. A large speed regulation is most often considered as a bad regulation from a control point of view.

speed sensor   a device used to detect the speed of the rotor of an electric machine. Optical (strobe) and electromagnetic tachometers are commonly used.

speed servo   a servo where the speed is the controlled parameter. *See also* servo.

speed-power product   an overall performance measurement that is used to compare the various logic families and subfamilies.

speedup factor   the ratio of execution time for a problem on a single processor using the best sequential algorithm to the execution time on a multiprocessor using a parallel algorithm under consideration. Provides a performance measure for the parallel algorithm and the multiprocessor.

spent fuel   irradiated fuel whose fissile component has been reduced such that it is no longer useful as reactor fuel.

SPFC   acronym for solid polymer fuel cell. *See* proton exchange membrane fuel cell.

sphere gap   a spark gap whose electrodes are metal spheres. A sphere gap with carefully-calibrated electrode spacing is used as a measuring instrument for voltages in the kilovolt to megavolt range.

spherical mirror   a mirror in which the reflecting surface is spherical.

spherical wave   an electromagnetic wave in which each wavefront (surface of constant phase) forms a sphere and propagates in toward or away from the center of the sphere. A uniform spherical wave has the same amplitude over an entire wavefront; a nonuniform spherical wave has varying amplitude.

spherical wrist   a wrist where all of its revolute axes intersect at a single point. Such a wrist is typically thought of as mounted on a three-degree-of-mobility arm of a six-degree-of-mobility manipulator. For manipulators with a spherical wrist it is possible to solve the inverse kinematics from the arm separately from the inverse kinematics for the spherical wrist. This is equivalent to the inverse kinematics problem subdivided into two subproblems, since the solution for the position is decoupled from that for the orientation.

SPICE   a computer simulation program developed by the University of California, Berkeley, in 1975. Versions are available from several companies. The program is particularly advantageous for electronic circuit analysis, since DC, AC, transient, noise, and statistical analysis is possible.

spike suppressor any of several devices e.g., metal-oxide varistors that clamp short-duration power line overvoltages to an acceptable level.

spillover phenomenon that occurs when radiation from a feed extends past the reflector edges and is not intercepted by the reflector.

spin echo an oscillating electromagnetic field emitted by a macroscopic orientation of atomic or nuclear spins, generated by reversing the dephasing process in an inhomogeneously broadened material.

spin coating the process of coating a thin layer of resist onto a substrate by pouring a liquid resist onto the substrate and then spinning the substrate to achieve a thin uniform coat.

spin lock a mutual exclusion mechanism where a process spins in an infinite loop waiting for the value of a variable to indicate a resource availability.

spindle *See* disk spindle.

spiral computed tomography (CT) an imaging modality that uses a rotating X-ray source and detector revolving around a continuously moving gantry. As viewed from the gantry, the X-ray source appears to travel in a spiral. A continuous set of projection images is gathered around the spiral and is interpolated to obtain traditional transverse cross-section images. Also known as helical CT.

spiral CT *See* spiral computed tomography.

spiral inductor an integrated circuit implementation of a common electrical element that stores magnetic energy. Two extreme behaviors of an inductor are that it will act as a short circuit to low frequency or DC energy, and as an open circuit to energy at a sufficiently high frequency (how high is determined by the inductor value). In an MMIC, a spiral inductor is realized by a rectan-

gular or circular spiral layout of a narrow strip of metal. The value of the inductance increases as the number of turns and total length of the spiral is increased. Large spiral inductors are very commonly used as "bias chokes" to isolate the DC input connection from the RF circuit. Since a large valued inductor essentially looks like an open circuit to high frequency RF/microwave energy, negligible RF/microwave energy will leak through and interact with the DC bias circuitry.

splice a permanent connection between two fibers made by melting or fusing the two fibers together in an electric arc or gas flame. Or they may be held together in a variety of mechanical devices that align the two fiber cores. In fusion splicing, connections can be achieved with losses < 0.1 dB.

spline a continuous function, interpolating a set of data points p_i, that is composed of segments, having p_i and p_{i+k} as extremes, that are linked together in such a way that the continuity constraint is satisfied.

spline wavelet wavelet that is in the form of a spline.

split and merge procedure often used in image or signal segmentation. The procedure involves splitting, iteratively applied if needed, the inhomogeneous regions of an image or sections of a discrete signal and followed by merging similar regions or sections is a split and merge

split transaction a bus transaction (e.g., memory read or write) in which a request and the corresponding response are sent in two different bus transactions.

SPM laser *See* synchronously pumped-mode-locked laser.

spontaneous decay process by which an atom or molecule in the absence of outside influence undergoes a transition from one energy state to another lower state.

spontaneous emission radiation resulting when an atom or molecule in the absence of outside influence undergoes a transition from one energy state to another lower state. *Contrast with* stimulated emission.

spontaneous lifetime coefficient representing the time after which a population of isolated atoms in an excited state may be expected to fall to one over e of its initial value, transition lifetime.

spontaneous light scattering scattering of light from thermally produced refractive index variations, e.g., spontaneous Brillouin scattering and spontaneous Raman scattering.

spontaneous polarization the internal electrical dipole moment of ferroelectric crystal.

spontaneous pulsations periodic or chaotic pulsations in the output of a laser oscillator when there is no modulation of the laser excitation or cavity loss.

spool (1) acronym for "simultaneous peripheral operation on-line." Area managed by a process (called a spooler) where data from slow I/O operations are stored in order to allow their temporal overlapping with other operations.

(2) a cylindrical ceramic insulator, typically used for secondary conductors in distribution work.

spooler *See* spool.

spooling sending printer output to a secondary storage device, such as a disk, rather than directly to the printer. This is done because disk devices can accept data at a much higher rate than printers.

spot size the $1/e$ amplitude radius of the electromagnetic field of a Gaussian beam, squeezed light.

SPPC *See* self-pumped phase conjugator.

SPR *See* strictly positive real.

SPR function *See* strictly positive real function.

spread-spectrum a modulation procedure in which the spectrum of an information bearing signal is spread by some techniques such as multiplication by a pseudo-noise sequence. The result is a signal with much wider bandwidth that has better protection again interference.

spread-spectrum multiple-access a multiple-access system in which each sender transmits their data using a frequency bandwidth significantly greater than the information bandwidth of the signal.

spreading code a sequence used for spreading the spectrum of the information signal in a spread-spectrum system, commonly done either by direct multiplication of the faster-varying spreading signal with the data sequence (direct-sequence, DS), or by hopping the carrier-frequency (frequency-hopping, FH). Also known as spreading sequence. *See also* short codes, long codes.

spreading gain in a spread-spectrum system, the number of dimensions used for transmitting the signal divided by the number of dimensions actually needed if spreading was not used. This is approximately equal to the ratio between the bandwidth after spreading and the bandwidth before spreading. In a BPSK system, it is equal to the number of chips per bit in a direct sequence system. Also called processing gain.

spreading sequence *See* spreading code.

SPST *See* single-pole single-throw.

spur a conductor which branches off of a main line.

spur-free dynamic range of Bragg cell regime of Bragg cell $f_1 + f_2$ multifrequency drive condition given by the ratio of the diffracted light intensity at the true frequency spatial/spectral locations f_1 or f_2 to the intensity of the intermodulation products at $2f_1 - f_2$ and $2f_2 - f_1$.

spurious   undesired, nonharmonically related, nonrandom signals or spectral content generated internal to a nonlinear circuit. Generally, spurious signals are created by internal mixing of multiple input signals, by internally generated oscillations, and by combinations thereof.

spurious interrupt   unwanted, random interrupt.

SQ   *See* scalar quantization.

square detection   a special case of rectangle detection.

square pixel   *See* pixel.

square-law detector   the square-law behavior for detector diodes is the usual operating condition, but can only be obtained over a restricted range of input powers. If the input power is too large, small-signal conditions will not apply, and the output will become saturated and approach a linear, and then a constant, i versus p characteristic. At very low signal levels, the input will be lost in the noise floor of the device.

square-wave   a waveform of square shape which is usually periodic with known periodicity. Often used as a test signal.

square-wave brushless DC motor   a permanent magnet brushless DC motor with concentrated stator phase windings. The concentrated windings create a square wave flux distribution across the air gap and a trapezoidal shaped back-EMF.

square-wave inverter   a self-commutated inverter with a square-wave output. The frequency is set by the switching frequency while the ampli-tude may be controlled by adjusting the input DC voltage.

squelch   to automatically reduce the gain of the audio amplifier of a receiver in order to suppress background noise when no input signal is being received. The circuit performing this function is called the squelch circuit, and it acts as a controllable receiver input switch to allow reception of strong signals and block the weak and noisy signals.

squeeze-on   a large crimped connector which requires a special press for installation.

squirrel-cage induction motor   an induction motor in which the secondary circuit (on the rotor) consists of bars, short-circuited by end rings. This forms a squirrel cage conductor structure, which is disposed in slots in the rotor core. *See also* cage-rotor induction motor.

SRAM   *See* static random access memory.

SS7   *See* signaling system 7.

SSB   *See* single sideband modulation.

SSD   *See* solid state disk.

SSI   *See* small scale integration.

SSPA   *See* solid state power amplifier.

stability   (1) the condition of a dynamic or closed-loop control system in which the output or controlled variable always corresponds, at least approximately, to the input or command within a limited range. In most devices, this is a measure of the inherent ability of the circuit to avoid internally generated oscillations.

In oscillators, stability denotes the ability of the circuit to maintain a stable internally generated amplitude and frequency. The circuit components, bias, loading, drive and environmental conditions, and possible variations therein, must be accounted for. *See also* Linville stability factor and Rollet stability factor.

(2) in electronic drives, the ability of a drive to operate a motor at constant speed (under varying load), without hunting (alternately speeding up and slowing down). It is related to both the characteristics of the load being driven and electrical time constants in the drive regulator circuits.

stability circles   circles plotted on the Smith chart that graphically indicate the regions of instability for an RF device.

stability criteria   boundaries on regions of stable and unstable behavior in laser parameter space.

stability factors   two factors, K and $B1$, that specify the necessary and sufficient conditions for a linear circuit or device to be conditionally or unconditionally stable when the input and output ports are terminated in arbitrary impedances. For unconditional stability, factors K must be greater than unity, and $B1$ must simultaneously be greater than 0.

stability limit   the maximum power flow possible through a point in a power system if the system is to remain stable. *See* transient stability, steady-state stability.

stability of 2-D Fornasini–Marchesini model
the second 2-D Fornasini–Marchesini model

$$x_{i+1,j+1} = A_1 x_{i+1,j} + A_2 x_{i,j+1}$$
$$+ B_1 u_{i+1,j} + B_2 u_{i,j+1}$$

$i, j \in Z_+$ (the set of nonnegative integers) is called asymptotically stable if for zero inputs $u_{ij} = 0$ $i, j \in Z_+$ and bounded $\|X_0\|$ we have $\lim_{i \to \infty} \|X_i\| = 0$, where $x_{ij} \in R^n$ is the local state vector, $u_{ij} \in R^m$ is the input vector,

$$X_k := \{x_{ij} : i + j = k; i, j \in Z_+\}$$
$$\|X_i\| = \sup_{k \in Z+} \|x(i - k, k)\|$$

and $\|x\|$ denotes the Euclidean norm of x. The model is asymptotically stable if and only if

$$\det \left[I_n - A_1 z_1^{-1} - A_2 z_2^{-1} \right] \neq 0$$

for $\|z_1^{-1}\| \leq 1$, $\|z_2^{-1}\| \leq 1$.

stability of 2-D Roesser model   the 2-D Roesser model

$$\begin{bmatrix} x_{i+1,j}^h \\ x_{i,j+1}^v \end{bmatrix} = \begin{bmatrix} A_1 & A_2 \\ A_3 & A_4 \end{bmatrix} \begin{bmatrix} x_{ij}^h \\ x_{ij}^v \end{bmatrix} + \begin{bmatrix} B_1 \\ B_2 \end{bmatrix} u_{ij}$$

$i, j \in Z_+$ (the set of nonnegative integers)

$$y_{ij} = C \begin{bmatrix} x_{ij}^h \\ x_{ij}^v \end{bmatrix} + D u_{ij}$$

is called asymptotically stable if for zero inputs $u_{ij} = 0$, $i, j \in Z_+$ and bounded $X^h := \sup_{j \in Z_+} \|x_{0j}^h\|$ and $X^v := \sup_{j \in Z_+} \|x_{i0}^v\|$ we have bounded $\sup_{j \in Z_+} \|x_{ij}\|$ and $\lim_{i,j \to \infty} \|x_{ij}\| = 0$, where $x_{ij}^h \in R^{n_1}$ and $x_{ij}^v \in R^{n_2}$ are the horizontal and vertical state vectors, $u_{ij} \in R^m$ is the input vector, $y_{ij} \in R^p$ is the output vector, $A_1, A_2, A_3, A_4, B_1, B_2, C, D$ are real matrices, and $\|x\|$ denotes the Euclidean norm of x. The model is asymptotically stable if and only if

$$\det \begin{bmatrix} I_{n_1} - A_1 z_1^{-1} & -A_2 z_1^{-1} \\ -A_3 z_2^{-1} & I_{n_2} - A_4 z_2^{-1} \end{bmatrix} \neq 0$$

for

$$\left\| z_1^{-1} \right\| \leq 1, \quad \left\| z_2^{-1} \right\| \leq 1$$

stability study   the determination of conditions which will cause a power system to become unstable so that these conditions can be avoided or corrected.

stabilizability   the property of a system concerning the existence of a stabilizing state feedback or output feedback control. For linear systems, it is characterized as the controllability (see the definition) of all the unstable modes.

stabilization of linear 2-D systems   the 2-D Roesser model

$$\begin{bmatrix} x^h_{i+1,j} \\ x^v_{i,j+1} \end{bmatrix} = \begin{bmatrix} A_1 & A_2 \\ A_3 & A_4 \end{bmatrix} \begin{bmatrix} x^h_{ij} \\ x^v_{ij} \end{bmatrix} + \begin{bmatrix} B_1 \\ B_2 \end{bmatrix} u_{ij}$$

$i, j \in Z_+$ (the set of nonnegative integers)

$$y_{ij} = [C_1 \quad C_2] \begin{bmatrix} x^h_{ij} \\ x^v_{ij} \end{bmatrix} + Du_{ij}$$

is called stabilizable by state feedback

$$u_{ij} = [K_1 \quad K_2] \begin{bmatrix} x^h_{ij} \\ x^v_{ij} \end{bmatrix}$$

if there exists $K_1 \in R^{m \times n_1}$ and $K_2 \in R^{m \times n_2}$ such that the closed-loop system is asymptotically stable, i.e.,

$$\det \begin{bmatrix} I_{n_1} - (A_1 + B_1 K_1) z_1^{-1} \\ - (A_3 + B_2 K_1) z_2^{-1} \\ \\ - (A_2 + B_1 K_2) z_1^{-1} \\ I_{n_2} - (A_4 + B_2 K_2) z_2^{-1} \end{bmatrix} \neq 0$$

$$\text{for } \left\| z_1^{-1} \right\| \leq 1, \left\| z_2^{-1} \right\| \leq 1$$

where $x^h_{ij} \in R^{n_1}$ and $x^v_{ij} \in R^{n_2}$ are the horizontal and vertical state vectors, respectively, $u_{ij} \in R^m$ is the input vector and $y_{ij} \in R^p$ is the output vector, $A_1, A_2, A_3, A_4, B_1, B_2, C_1, C_2, D$ are real matrices. Similarly, the model is called stabilizable by output feedback $u_{ij} = F y_{ij}$ if there exists $F \in R^{m \times p}$ such that the closed-loop system is asymptotically stable, i.e.,

$$\det \begin{bmatrix} I_{n_1} - (A_1 + B_1 F C_1) z_1^{-1} \\ - (A_3 + B_2 F C_1) z_2^{-1} \\ \\ - (A_2 + B_1 F C_2) z_1^{-1} \\ I_{n_2} - (A_4 + B_2 F C_2) z_2^{-1} \end{bmatrix} \neq 0$$

$$\text{for } \left\| z_1^{-1} \right\| \leq 1, \left\| z_2^{-1} \right\| \leq 1$$

stabilized beam current   the amount of beam current required to stabilize the target when a given amount of light is incident on the target. The beam current is normally set at two times picture white.

stable   a system characteristic in which the transients all decay to zero in finite time is said to be stable. If any transient term grows with time, then the system is unstable. If the transient persists, then the system is marginally stable. (An oscillator is a common example of marginal stability.)

Much of control engineering theory deals with the problem of classifying closed-loop systems into those that are stable and those that are unstable, with marginally stable systems defining the boundary between the two.

stable equilibrium   an equilibrium point (see the definition) such that all solutions that start "sufficiently close," stay "close" in time. If the point is not stable, it is called unstable.

stable state   (1) the equilibrium state of a dynamic system described by a first-order vector differential equation is said to be stable if given $\epsilon > 0$ there exists a $\delta = \delta(\epsilon, t_0)$, such that

$$\| x(t_0) - x_e \| < \delta \Rightarrow \| x(t) - x_e \| < \epsilon \quad \forall t \geq t_0$$

(2) in storage elements, being in a condition that is highly unlikely to undergo a spontaneous transition to another state.

stable system   a system is stable if the output of the system is bounded for all bounded inputs. *See* bounded-input bounded-output stability.

stack   a hardware or software data structure in which items are stored in a last-in-first-out manner, similar to a cafeteria plate dispenser.

stack algorithm   a sequential decoding algorithm for the decoding of convolutional codes, proposed by Zigangirov in 1966.

stack architecture   *See* zero-address computer.

stack machine   *See* zero-address computer.

stack pointer   a register in a processor that holds the address of the top of the stack memory location. The address varies as information is stored on or retrieved from the stack; it always points to the top of the stack.

stack program concept   a class of CPU or data structure in which items are stored in a last-in-first-out manner, similar to a cafeteria plate dispenser.

stacked microstrip antenna   a microstrip patch configuration where two or more patches are stacked on top of each other separated by one or more dielectric layers. Typically, the lower patch is fed directly and the upper patch is electromagnetically coupled to the lower patch. This arrangement results in improved bandwidth compared to that of a single layer microstrip patch antenna.

Stackelberg equilibrium   a hierarchical equilibrium solution in non-zero-sum games in which one of the players has the ability to force his strategy on the other players. The player who holds the powerful position is called the leader, while the other players who react to the leader's strategy are called the followers. In the case of multiperson games, there exists a variety of possible multilevel decision making structures with many leaders and followers. Thus, the definition of the Stackelberg equilibrium is uniquely and clearly set only for two-person decision problems, but it could be adopted for any given hierarchical structure. If J_1, J_2 denote cost functions of leader and the follower, respectively, and d_1, d_2 their admissible strategies, then the set $R(d_1)$ defined as: $\{d$ (admissible for the follower): $J_2(d_1, d) \leq J_2(d_1, d_2)$ for each admissible $d_2\}$ is called the optimal response or rational reaction set of the follower. Then a strategy d_1^* is a Stackelberg strategy for the leader if

$$J_1^* = \max_{d_2 \in R(d_1^*)} J_1\left(d_1^*, d_2\right)$$
$$\leq \max_{d_2 \in R(d_1)} J_1\left(d_1, d_2\right)$$

for all admissible d_1. J_1^* is the Stackelberg cost of the leader and any $d_2^* \in R(d_1^*)$ is an optimal strategy for the follower that is in equilibrium with d_1^*. The pair (d_1^*, d_2^*) is a Stackelberg solution and corresponding values of the cost functions give the Stackelberg equilibrium outcome. The Stackelberg outcome of the leader may be lower than his Stackelberg cost. If the rational reaction set of the follower is a singleton, then they are equal and they are not worse than the outcome that could be achieved by the leader in the Nash equilibrium if it exists. *See also* Nash equilibrium.

Stackelberg game   *See* Stackleberg equilibrium.

stacking factor   a design factor for the core of an electromagnetic device that accounts for the effects of the insulating material on the surface of laminations. The stacking factor gives the percentage of cross-sectional area of the core that is actually ferromagnetic material. Usually expressed as the ratio of the thickness of the laminations without the coating to the thickness with the coating.

stall   a pause in processing instructions in a pipeline, usually caused by a data dependency or resource conflict. Instructions in the pipeline before the condition causing the stall are prevented from proceeding through the pipeline.

standard additive model (SAM)   a fuzzy system that stores IF-THEN rules that approximate a function $F : X \rightarrow Y$. In a simple SAM, the rules may have the form "IF $x = A_j$ THEN $y = B_j$," where $x \in X$, $y \in Y$, and A_j, B_j are fuzzy sets. The SAM then computes the output $F(x)$ given the input x using a centroidal defuzzifier. An example of a centroidal defuzzifier is

$$F(x) = \frac{\sum a_j(x)c_j}{\sum a_j(x)},$$

where a_j is a membership function of the fuzzy set A_j and c_j is the centroid of the fuzzy set B_j. The term SAM was coined by B. Kosko. *See also* fuzzy system.

standard array decoding   during decoding of a forward error correction code, the process of associating an error pattern with each syndrome by way of a look-up table.

standard cell   an element of a standard cell library designed using rules from the targeted wafer fabricator. Standard cells are usually designed to be of constant height and variable width with interconnection points located along the bottom and possibly the top of the cell. This is done to facilitate use of an auto-place-and-route program.

standing wave   (1) the phenomenon where waves propagating in opposite directions interfere and result in diminished, or eliminated, energy transfer.

(2) class of laser resonators (often having only two mirrors) in which the right and left waves are largely overlapping.

standing wave effect   caused by standing waves of light intensity in the resist, this is horizontal, periodic ridges formed along the sides of a resist profile.

standing wave pattern   a pattern of the envelope of the wave resulting from interference of two same-frequency waves travelling in opposite directions.

standing wave ratio (SWR)   the ratio of the magnitudes of the incident to reflected signal levels for a traveling wave. The SWR has a value between one and infinity inclusive.

standing-wave laser   a class of lasers (often having only two mirrors) in which the right and left waves are largely overlapping.

standing-wave resonator   superposition of equal amplitude right and left travelling waves. One of a class of laser resonators (often having only two mirrors) in which the right and left waves are largely overlapping. Also called a Fabry–Perot resonator.

standstill frequency response test   a test in which the rotor of a machine is held fixed, and the appropriate windings are energized over a frequency range large enough to determine machine parameters.

star connection   *See* Y connection.

star network   a network topology where a central node broadcasts radially to all subscribers. The central node is a vulnerable element on which the whole network depends.

star–star transformer   *See* wye–wye transformer.

Stark broadening   inhomogeneous spectral broadening of a transition in a laser medium due to Stark shifts that vary among the laser atoms or molecules in the medium.

start bit   the first bit (low) transmitted in an asynchronous serial transmission to indicate the beginning of the transmission.

starting torque   the torque at zero speed obtained at the very beginning of the starting process of an electrical machine. The condition to obtain the rotation of the rotor is that the starting torque has to be greater than the load torque at zero speed.

starvation   a condition when a process is indefinitely denied access to a resource while other processes are granted such access.

state   a set of data, the values of which at any time t, together with the input to the system at the time, determine uniquely the value of any network variable at the time t.

state automaton   *See* finite state machine.

state diagram   (1) a form of diagram showing the conditions (states) that can exist in a logic system and what signals are required to go from one state to another state.

(2) a simple diagram representing the input–output relationship and all possible states of a convolutional encoder together with the possible transitions from one state to another. Distance properties and error rate performance can be derived from the state diagram.

state equations   equations formed by the state equation and the output equation.

state feedback   the scheme whereby the control signal is generated by feeding back the state variables through the control gains.

state machine   a software or hardware structure that can be in one of a finite collection of states. Used to control a process by stepping from state to state as a function of its inputs. *See also* finite state machine.

state plane   *See* phase plane.

state space conditional codec   an approach where the number of codes is much less than with conditional coding. The previous $N - 1$ pixels are used to determine the state s_j. Then the jth variable word-length is used to code the value.

state space model   a set of differential and algebraic equations defining the dynamic behavior of systems. Its generic form for linear continuous-time systems is given by

$$\frac{d}{dt}x(t) = Ax(t) + Bu(t)$$
$$y(t) = Cx(t) + Du(t)$$

where $u(t)$ is the system input signal, $x(t)$ is its internal or state space variable, and $y(t)$ is its output. Matrices A, B, C, D of real constants define the model. The internal variable is often a vector of internal variables, while in the general multivariable case all the input and output signals are also vectors of signals. Although not identically equivalent, the state space model can be related to the transfer function (or transfer function matrix)

model by

$$G(s) = C[sI - A]^{-1}B + D$$

assuming that the initial conditions on all internal variables are zero. The Laplace variable is denoted by s. Similar equations, based on difference and algebraic equations, define state space model for linear discrete-time (digital) dynamic systems.

$$x_{t+1} = Ax_t + Bu_t$$
$$y_t = Cx_t + Du_t$$

See also transfer function.

state space variable   the internal variable (or state) in a state space model description of a dynamic process. These internal variables effectively define the status or energy locked up in the system at any given instant in time and hence influence its behavior for future time.

state transition diagram   a component of the essential model; it describes event-response behaviors.

state variable   one of a set of variables that completely determine the system's status in the following sense: if all the state variables are known at some time t_0, then the values of the state variables at any time $t_1 > t_0$ can be determined uniquely provided the system input is known for. The vector whose components are state variables is called the state vector. The state space is the vector space whose elements are state vectors.

state vector   a vector formed by the state variables.

state-space averaging   a method of obtaining a state-model representation of a circuit containing switching elements by averaging the state models of all the switched topologies.

state-space averaging model   a small-signal dynamic modeling method for PWM switching circuits. The circuit is viewed as two linear sub-circuits, one with the switch on and one with the switch off. A duty-ratio weighted average of the state-space equations for the two subcircuits is then linearized and used to obtain the small-signal transfer function for the switching circuit.

static excitation system   an excitation system derived from solid state devices such as thyristors that convert the AC terminal voltage to DC before application to the rotor.

static induction thyristor (SITH)   a self-controlled power device with high switching frequency. The structure is similar to the static induction transistor (SIT) (hence, not really a thyristor), but has an additional p-layer added to the anode side. It is a normally-on device with the n-region saturated with a minority carrier. The device does not have reverse blocking capability.

static induction transistor (SIT)   a high-power, high-frequency device that is essentially a solid-state version of the triode vacuum tube. It is a normally-on device, and a negative gate voltage holds it off. The current ratings of the SIT can be up to 300 A, 1200 V, and the switching frequency can be as high as 100 kHz.

static prediction   a method of branch prediction that relies of the compiler selecting one of the two alternative instructions for after the branch instruction (either the next instruction or that at the target location specified in the branch instruction). A bit is provided in the branch instruction, which is set to a 0 for one alternative and 1 for the other alternative. The processor then follows this advice when it executes the branch instruction.

static random access memory (SRAM)   random access memory that, unlike dynamic RAM, retains its data without the need to be constantly refreshed.

static system   a system whose output does not depend upon past or future input is a *static,* or

memoryless system. For example, consider a voltage $v_{in}(t)$ applied to an amplifier with gain K that yields the output

$$v_{out}(t) = K v_{in}(t).$$

The output voltage at a particular instant in time depends only on the input applied at that same instant, thus the amplifier is a static system. If a system is not static, then it is a system *with memory,* or a *dynamic* system.

static var compensator   a device for fast reactive compensation, either inductive or capacitive, brought about by thyristor-based control of an effective shunt susceptance. It is typically used to regulate voltage at a bus on the high voltage transmission system.

static VAR regulator   also called a static VAR compensator. A nonrotating electrical device designed to adjust the reactive power flow of an AC power system. It typically consists of a reactive load (either inductive or capacitive) and a series electronic switch (thyristor) that controls the reactive power.

static-column DRAM   DRAM that is organized in the same manner as a page-mode DRAM but in which it is not necessary to toggle the column access strobe on every change in column address.

station battery   a battery used to provide operating energy for the protective relay operations and to initiate circuit breaker operations in a generating station. The battery is necessary, as the equipment must work reliably during severe voltage sags and outages on the AC system.

station control error   in economic dispatch studies, the difference between the desired generation of all plants in a control area and the actual generation of those plants.

station insulator   refers to a large-sized insulator used in substations.

stationarity interval the interval of either time (temporal stationarity interval) or space (spatial stationarity interval) over which the conditions required for a WSSUS approximation is valid. That is, the stationarity interval is the period of time or spatial separation over which the scattering function of the channel, and consequently also the delay and Doppler power spectra, stays fixed. This requires that the significant scatterers should remain the same.

stationary a dynamic system described by a first-order vector differential equation that does not depend explicitly on time. In other words, such a system is governed by an equation of the form

$$\dot{x}(t) = f(x(t), u(t))$$

stationary process a stochastic process $x(t)$, for which the joint probability distribution of $x(t_1)$ and $x(t_2)$ depends only on $|t_1 - t_2|$.

stationary reference frame a two-dimensional space that is fixed (nonrotational). In electric machines/power system analysis, an orthogonal coordinate axis is established in this space upon which fictitious windings are placed. A linear transformation is derived in which the physical variables of the system (voltage, current, flux) are referred to variables of the fictitious windings.

statistical multiplexing multiplexing of a number of variable bit rate (VBR) sources. A result of statistical multiplexing is that for a sufficiently large number of VBR sources, the aggregate bit rate is less than the sum of the peak bit rates of the individual sources.

statistical quality control methods of quality improvement based on statistical techniques. The main idea is to use statistical methods for identification of unusual variations of the controlled process and to pinpoint the causes of such variations. By collecting data at every stage of the production process and statistical analysis of those data (*See* control charts), the process is maintained in a state of statistical control. The main difference between the statistical quality control and quality inspection is that the latter enables only quality control while the former leads to quality improvement. This in turn results in increased productivity.

statistical sensitivity a statistic derived from statistical interpretation of the variation of the circuit function, $F(s, x)$ around the average value $\underline{x}$.

Let $F(s, x)$ be a circuit function that depends on a random parameter x. Then the statistical sensitivity can be approximated as

$$F(s, x) \approx F(s, \underline{x}) + (x - \underline{x}) \left. \frac{\partial F(s, x)}{\partial x} \right|_{x=\underline{x}}$$

The values $\underline{x}$, $F(s, \underline{x})$, and $\partial F(s, x)/\partial x$ calculated at $x = \underline{x}$ are considered constants (the last one is usually denoted as $\partial F(s, \underline{x})/\partial x$). If instead of x and $F(s, x)$ are used, their relative deviations from the average values, namely $\delta x = (x - \underline{x})/\underline{x}$ and $\partial F(s) = (F - \underline{F})/\underline{F}$, one obtains that

$$\delta F(s) \approx \left[\frac{\underline{x}}{F(s, \underline{x})} \frac{\partial F(s, \underline{x})}{\partial x} \right] \delta x = \mathbf{S}_x^{F(s,\underline{x})} \delta x$$

Hence, in the first-order approximation the random variables $\delta F(s)$ and δx are proportional. Then this equation is interpreted statistically. For example, on the $j\omega$ axis the averages of $\delta F(j\omega)$ and δx are related by

$$\mu_{\delta F} \approx \mathbf{S}_x^{F(j\omega, \underline{x})} \mu_{\delta x}$$

The relationships for other statistical parameters can be obtained as well.

statistical spectral compression a common approach to compression used in picture coding. In this approach, the statistical redundancy of the image is exploited and the compression is also obtained by coding the spectral components of the transformed image.

stator   the portion of a motor that includes and supports the stationary active parts. The stator includes the stationary portions of the magnetic circuit and the associated windings and leads.

status callback   a request made by a consumer for the utility to give them a phone call which indicates the change of status of their service request. An example of this would be calling the consumer once a crew has arrived on the scene of an outage or has located the root cause of an outage.

status register   a register in a processor that holds the status of flags; individual bits in the register represent flag status.

steady-state control   operation and mechanisms of the control system in which the main objective is to keep the controlled process in the condition where the state variables relevant to the controlled process performance are constant — i.e., to keep the process in a required operating point. Steady-state control structure may be composed of several control layers, including direct regulatory layer, optimization layer and, eventually, other layers; steady-state control is widely used in chemical and power industries.

steady-state error   the difference between the desired reference signal and the actual signal in steady state, i.e., when time approaches infinity.

steady-state gain   the gain that a system applies to DC (constant) input signals.

steady-state response   in network analysis, a condition that the response reaches a constant value with respect to the independent variable. In control system studies, it is more usual to define steady state as the fixed response at infinity with respect to the fixed input under the stable circumstances.

steady-state stability   a power system is steady-state stable if it reaches another steady-state oper-

ating point after a small disturbance. *See* dynamic stability.

steepest descent algorithm   *See* gradient descent.

steering vector   in an antenna array, the complex weights associated with each antenna element to form a specific radiation pattern are called the steering vector, since these weights steer the radiation pattern in a specified direction.

Steinmetz constant   a constant n that relates the area of the hysteresis loop of a magnetic material to the maximum flux density in the material.

$$A_h = K_h B_{\max}^n$$

where A_h is the area of the hysteresis loop, K_h is a constant of the material, $B_{\max}$ is the maximum flux density, and n is the Steinmetz constant.

step and repeat camera   *See* stepper.

step edge   an idealized edge across which the luminance profile takes the form of a step function: i.e., a line separating two regions having different average gray-levels. *See* edge.

step index fiber   a type of optical fiber where there is an abrupt transition from the core to cladding region, each region having a different refractive index; optical fiber in which the a homogeneous core region has a higher index of refraction than a homogeneous cladding region, in contrast to a graded index fiber. This configuration is more typical of single mode fibers, than multimode fibers, which suffer from modal dispersion effects.

step index optical fiber   *See* step index fiber.

step response   the output of a linear time-invariant system when the inputs are varied as a step signal.

659

step size   when solving for the transient behavior of an electrical circuit, the associated differential equations are solved at specific points in time. The difference between two adjacent solution time points is known as the step size.

step-and-scan   a type of projection printing tool combining both the scanning motion of a scanner and the stepping motion of a stepper.

step voltage   in power system safety studies, the voltage measured across two points on the ground which are separated by a distance equal to an average person's step while walking over the area in question.

step-down converter   *See* buck converter.

step-up converter   *See* boost converter.

stepped leader   in lightning, a discharge following the preliminary breakdown that propagates from a cloud toward the ground in a series of intermittent luminous steps with an average speed of 10^5 to 10^6 m/s. Negatively charged leaders clearly step, while positively charged leaders are more pulsating than stepped.

stepper   a type of projection printing tool that exposes a small portion of a wafer at one time, and then steps the wafer to a new location to repeat the exposure. Also called a step-and-repeat camera.

stepping correlation   a principle of operation of a correlation receiver in channel measurement, where pseudo-random sequences are utilized. The transmitted signal consists of a carrier modulated (typically employing phase-shift keying) by a pseudo-random sequence. The received signal is correlated (multiplied) by an exact replica of the transmitted signal by stepping the chip position of the reference signal with respect to the received signal through all or part of the chip positions. The output of a stepping correlator is a time-scaled version of the autocorrelation function of the pseudo-random sequence, or a part of it. The time-scaling factor depends on the rate with which all the chip positions of interest can be stepped. The rate of producing autocorrelation functions can be made much higher than in sliding correlation. *See also* sliding correlation.

stereo imaging   *See* binocular imaging.

stereo vision   *See* binocular vision.

stereospecific   directional covalent bonding between two atoms.

sticky bit   the least significant guard digit in floating-point representations. It is an indicator bit obtained through a logical "OR" operation of the discarded bits, indicating whether at least one of the discarded bits was equal to 1.

stiction   in variable-speed drives, the initial static friction that must be overcome when the load is at rest.

stiff system   when an electrical circuit has widely separated time constants, the circuit is said to be stiff. The system of equations associated with the circuit is known as a stiff system, and special numerical methods must be used to maintain stability and accuracy when simulating a stiff system.

stiffness   as applied to a tie-line between generators, a low-impedance connection which forces the two generators to run in synchronization regardless of load variations on one or the other.

stiffness control   in stiffness control a generalized joint force and/or torque is generated in response to small position error as to a constant task space stiffness matrix. *See also* stiffness matrix.

stiffness matrix   the stiffness matrix of the arm endpoint is the inverse of the compliance matrix. *See also* compliance matrix.

stiffness of a manipulator arm   an attribute of a robot arm.

Assume that a force is applied to the end-effector of a manipulator arm. The end-effector will deflect by an amount that depends on the stiffness of the arm and the force applied. In other words, the stiffness of the arm's end-effector determines the strength of the manipulator arm. Usually, the actuator itself has a limited stiffness determined by its feedback control system, which generates the drive torque based on the discrepancy between the reference position and the actual measured position. We model the stiffness by a spring contact that relates the small deformation at the joint to the force or torque transmitted through the joint itself. It is called the joint stiffness. *See also* stiffness matrix.

still image   stationary image or single frame as opposed to moving image or video. Includes photographic images, natural images, medical images, remote sensing images. Usually implies multilevel (grayscale or color) rather than bilevel.

still image coding   compression of a still image. A coder consists of the four steps: data representation (typically by transform, decomposition into subbands or prediction), quantization (in which data is approximated or discarded according to some measure of its importance), clustering of nulls (in which runs or blocks of zero values are coded compactly), entropy coding (in which the statistical properties of the data are exploited in lossless compression).

stimulated emission   enhanced emission of electromagnetic radiation due to the presence of radiation at the same frequency; also called induced emission.

stimulated light scattering   scattering of light from refractive index variations that are produced or amplified by the interaction of laser light with the material system, e.g., stimulated Brillouin scattering and stimulated Raman scattering.

stirrup   *See* saddle.

stochastic ARMA (ARMAX) model   a generalized ARMA model in which the uncertain environmental effects are included as an independent noise input.

stochastic independence   independence of two random variables or two random processes.

stochastic neuron   an artificial neuron whose activation determines the probability with which its output will enter one of its two possible states. The most commonly used expression for the probability that the neuron output y takes on the value $+1$ is $Pr\{y = +1\} = 1/(1 + e^{-2\text{net}/T})$, where net represents the activation of the neuron and T is a quantity analogous to temperature that controls the uncertainty in the neuron output. When T is infinite, a positive activation leads to an output of $+1$ with probability 0.5, and when T is zero, a positive activation leads to an output of $+1$ with probability 1.0.

stochastic process   a collection of vector random variables defined on a common probability space and indexed by either the integers (discrete stochastic process) or the real numbers (continuous stochastic process). A stochastic process $x = x(t)$ is a vector function of both time t and the sample path.

stochastic sampling   a type of sampling that varies the time intervals between samples. Stochastic sampling allows for a signal to be sampled at a lower apparent sampling frequency achieving equal results to a signal sampled at a much higher sampling frequency. The apparent benefits of stochastic sampling are counterbalanced by the fact that the sampling interval, since it is changing, must be recorded in addition to the signal samples, in order to reconstruct the signal correctly.

stochastic signal processing   the branch of signal processing which models and manipulates signals as stochastic processes rather than as deterministic or unknown functions. *See* random process.

Stokes Law of light scattering   the statement that the scattering of light is typically accompanied by a shift to lower (not higher) frequencies.

Stokes scattering   *See* Stokes Law of light scattering and Raman scattering.

Stokes theorem   let $\mathbf{A}(\mathbf{r})$ be any vector function of position, continuous together with its first derivative throughout an arbitrary surface S bounded by a contour C, assumed to be resolvable into a finite number of regular arcs. Stokes theorem states that

$$\oint_C \mathbf{A}(\mathbf{r}) \cdot d\ell = \int_S [\nabla \times \mathbf{A}(\mathbf{r})] \cdot \mathbf{n} \, dS$$

where $d\ell$ is an element of length along C and $\mathbf{n}$ is a unit vector normal to the positive side of the element area dS. This relationship may also be considered as an equation defining the curl.

stop band   the band of frequencies in a filter or application at which substantial attenuation or suppression is required relative to a passband. Stop band filtering is utilized to eliminate known high-level signals, which will disrupt system operation. *Compare with* passband.

stop bit   the last bit (high) transmitted in an asynchronous serial transmission to indicate the end of a character. In some serial transmissions, one and a half to two bits are used as stop bits.

stopband edge   the frequency at which the attenuation of a signal diminishes; typically the frequency at which the signal is attenuated at 3 dB from the maximum response.

storage temperature   the maximum non-operating long-term temperature that a device or assembly will be exposed to or stored at without experiencing permanent degradation or damage.

store   (1) the act of placing a value into storage.
   (2) the place where data and instructions are stored.

store instruction   a machine instruction that copies the contents of a register into a memory location. *Compare with* load instruction.

stored program computer   a computer system controlled by machine instructions stored in a memory; the instructions are executed one after the other unless otherwise directed.

STP   *See* signal transfer point.

straight edge detection   the location of straight edges in an image by computer. Often accomplished with the Hough transform.

strain   semiconductors, strained either by external forces or due to lattice mismatched epitaxial growth, have modified band structures, especially the band gap and effective masses.

strain insulator   an insulator which forms an insulated tensile link between two conductors in overhead line work.

strained layer superlattices   epitaxially grown lattice mismatched alternating layers, usually designed to optimize a desirable property such as band gap, effective mass, quantum confinement, etc.

strained-layer laser diode   surface emitting laser diode.

stranded cost   a facility like a nuclear power plant which cannot be charged to ratepayers after electric utility de-regulation takes place.

strap   a conducting ring that ties tips of poles of magnetron or magnetron-like devices in a specified fashion for microwave potential and phase equalization.

strap fed device   strapped magnetron-like device that operates by microwaves fed through the strapping such as amplitron amplifiers and platinotron oscillators.

stray light analysis   a computation to determine the intensity of unwanted light at various locations in an optical system, combining factors such as diffraction, surface scatter, spurious reflections, and optical design.

streak camera   a camera that performs one-dimensional imaging while also measuring the temporal evolution of the image.

stream   the sequence of data or instructions that flows into the CPU during program execution.

stream cipher   an encryption system or cipher in which the information symbols comprising the plaintext are transformed into ciphertext individually. An important property of a stream cipher is that like-valued plaintext symbols are not necessarily transformed into the same ciphertext. A stream cipher normally acts in an additive sense and in the case of bits being encrypted, the information bits X_n are added modulo-2 to the bits, Z_n, generated by the so-called running-key generator. The ciphertext Y_n is therefore given by $Y_n = X_n \oplus Z_n, n = 1, 2, \ldots, N$, where the plaintext consists of N bits and $\oplus$ denotes modulo-2 addition. Generally, the running key bits and the encryption key bits are not the same. The encryption key merely specifies the mechanism used to generate the running-key bits. Such a mechanism could be a number of linear feedback shift registers whose outputs are combined to form the running-key bits. *See also* block cipher, encryption.

stream line   *See* direction line.

streamer   a precursor of the high-voltage electrical breakdown of a gas which consists of a linked series of local electron avalanches forming a finger-like structure extending from one electrode toward another. Before a lightning strike, streamers extend from points on the earth up towards the thundercloud.

strength duration curve   a curve expressing the functional relationship between the threshold of excitation of a nerve fiber and the duration of a unidirectional square-wave electrical stimulus.

strict consistency   the situation when a processor reads a shared variable and obtains the value produced by the most recent write to the shared variable irrespective of the processor that did the write operation.

strict equivalence of 2-D system matrices   two 2-D system matrices

$$S_i = \begin{bmatrix} P_i & -Q_i \\ C_i & D_i \end{bmatrix} \quad i = 1, 2$$

(*See also* input–output equivalence of 2-D system matrices) of the same dimensions are called strictly equivalent if

$$\begin{bmatrix} M & 0 \\ K & I \end{bmatrix} \begin{bmatrix} P_1 & -Q_1 \\ C_1 & D_1 \end{bmatrix} = \begin{bmatrix} P_2 & -Q_2 \\ C_2 & D_2 \end{bmatrix} \begin{bmatrix} N & L \\ 0 & I \end{bmatrix}$$

holds for M, N unimodular matrices and K, L polynomial matrices. The transfer matrix $T_i(z_1, z_2) = C_i P_i^{-1} Q_i + D_i$ and the degree of det P_i are invariant under the strict equivalence, i.e., $T_1 = T_2$ and deg det $P_1 =$ deg det P_2.

strict passivity   a system $H : X_e \rightarrow X_e$ where there exists a $\delta > 0$ such that

$$\langle x, Hx \rangle_T \rightarrow \; \geq \delta \langle x, x \rangle_T \rightarrow \quad \forall x \in X_e$$

See also extended spaces, inner product space, and passivity.

strictly Hurwitz polynomial   *See* Hurwitz polynomials.

strictly positive real (SPR)   a rational transfer function $G(s)$ with real coefficients such that

$$\text{Re}\, G(s) \geq 0 \quad \text{for} \quad \text{Re}\, s \geq 0$$

A transfer function G is strictly positive real if $G(s - \varepsilon)$ is positive real for some real $\varepsilon > 0$.

A rational transfer function $G(s)$ with real coefficients is strictly positive real if and only if the following conditions hold.

(1) The function has no poles in the right half-place.

(2) The function has no poles or zeros on the imaginary axis.

(3) The real part of G is nonnegative along the $i\omega$ axis.

strictly positive real function   *See* strictly positive real.

strictly proper transfer matrix   a 2-D transfer matrix

$$T(z_1, z_2) = \frac{N(z_1, z_2)}{d(z_1, z_2)}$$

$$N(z_1, z_2) \in R^{p \times m}[z_1, z_2]$$

$$d(z_1, z_2) = \sum_{i=0}^{n_1} \sum_{j=0}^{n_2} d_{ij} z_1^i z_2^j$$

such that $d_{n_1, n_2} \neq 0$ ($d(z_1, z_2)$ is acceptable) and $\deg_{z_1} n_{ij}(z_1, z_2) < n_1, \deg_{z_2} n_{ij}(z_1, z_2) < n_2$ for $i = 1, \ldots, p;\ j = 1, \ldots, m$, where $n_{ij}(z_1, z_2)$ are the entries of $N(z_1, z_2)$ and $\deg_{z_1} (\deg_{z_2})$ denotes the degree with respect to z_1 (z_2).

strictness   attribute of a function whereby one can compute the value error whenever one or more of their arguments have the value error.

stride   the spacing (measured in memory address space) between the addresses of consecutive elements of a vector that are accessed during the execution of a program loop. If the stride is one, all elements are accessed in order; if it is two, every other element is skipped. *See also* memory stride.

stripline   a transmission line formed by a printed conductor sandwiched between two conductive-backed dielectrics.

strong inversion   the range of gate biases corresponding to the "on" condition of the MOSFET. At a fixed gate bias in this region, for low drain-to-source biases, the MOSFET behaves as a simple gate-controlled resistor. At larger drain biases, the channel resistance can increase with drain bias, even to the point that the current saturates or becomes independent of drain bias.

strong localization of light   confinement of light inside a highly inhomogeneous medium due to very strong scattering.

strong SPR function   *See* strictly positive real function.

strong strictly positive real function   *See* strictly positive real function.

structural controllability   a dynamical system where for a structured pair of state space matrices (A, B) there exists an admissible pair (A^-, B^-) that is controllable.

structural pattern recognition   methods for carrying out the recognition of pattern on the basis of a structured representation. For instance, in many interesting problems, the patterns can effectively be given linguistic descriptions based on grammars.

structure estimation   determination of the structure of objects, i.e., the 3-D coordinates of surface points of objects, from sequences of images. It is a task sometimes closely related to motion estimation.

structured cell   an element of a standard cell library designed using rules from the targeted wafer fabricator. Structured cells are integral multiples of a unit cell with interconnection points on all four sides of the cell. Structured cells normally interconnect simply by being placed next to another structured cell. Unwanted connections are broken as opposed to desired connections being made.

structured distribution systems (SDS)   a topology that advocates cabling saturation of a desired environment to accommodate all potential personnel movements and reconstructions within that office.

structured light   patterns of light projected onto objects which are to be viewed by cameras and interpreted by computer. For example, a grid of parallel straight lines of light projected on to a curved object will appear from a separate viewpoint to be curved and will provide information on the 3-D shape of the object.

structured matrix   a matrix whose entries are either zeros or independent free parameters.

structured noise   noise that is not random but which is typically periodic, or contains elements of some unwanted signal. This category of noise includes clutter, crosstalk, easily recognized spikes, and so on.

structured uncertainty   low-order parameter perturbations or unmodeled variations represented by a family of models with uncertain parameters ranging within a prespecified set. In the case of linear systems with models in frequency domain, an uncertain system with structured uncertainties is represented by a family of rational matrices given the highest order and a prespecified set for each uncertain parameter. In the state space counterpart, an uncertain system with structured uncertainties is represented by a family of matrices (in the state equations of known dimensions depending on uncertain parameters from the prespecified set).

structuring element   an image or shape which is used in a morphological operator as a probe interacting with the image to be analyzed, leading thus to a transformation of that image. It can be either a set of points (a colorless shape), or a gray-level image (a shape with a gray-level profile on it). In contrast with the natural image to be processed, the structuring element is chosen by the user and generally has a small support. *See* morphological operator.

stub   a short section of transmission line, usually short-circuited or open-circuited at one end, designed to present a specific impedance at the other end. Stubs are typically employed as impedance matching elements.

stub tuner   matching network, either double-stub or triple-stub, used to match all load admittances.

stuck-at fault   a fault model represented by a signal stuck at a fixed logic value (0 or 1).

stuck-open   in logic circuits, refers to a fault wherein the value of a signal is "stuck" at the open-circuit value.

subband analysis   decomposition of a signal into a set of subbands by using a filter bank, followed by an appropriate subsampling. *See* subband synthesis.

sub-band coding   (1) a method for source coding where the input signal is divided into frequency sub-bands, through the use of, e.g., a filter bank. The sub-bands are then quantized separately. Such methods utilize the fact that most real-world signals contain low amounts of information in some frequency regions and much information in others. Hence, enhanced compression can be obtained by focusing (only) on "important" frequency regions.

(2) image coding scheme in which the image is first filtered to create a set of images containing

a limited range of frequencies. These images are down sampled and encoded using one or more coders. The reverse is carried out at the receiver to reconstruct the original image.

sub-band pyramid   sub-band coding using quadrature mirror filters (QMF) provides a natural hierarchical structure and is called sub-band pyramid. This is quite similar to the Laplacian pyramid.

subband signal   the outputs of subband analysis are referred to as subband signals.

subband synthesis   a process in which a signal is generated from the subband signals through upsampling and filtering. *See* subband analysis.

sub-block   a part of a cache line that can be transferred to or from the cache and memory in one transaction. This is applicable in the cases where the complete line cannot be transferred in one transaction. Each sub-block requires a valid bit.

sub-millimeter (SMM)   the portion of the electromagnetic spectrum corresponding to wavelengths less than a millimeter, but longer than those of the long-wave infrared (> 20–30 μm).

subband coding   *See* sub-band coding.

subchannel I/O   the portion of a channel subsystem that consists of a control unit module, the connections between the channel subsystem and the control unit module, and the connections between the control unit module and the devices under its control. In earlier versions of the IBM channel architecture, the subchannel was known as an I/O channel.

subcircuit   a simulation approach that allows an efficient description of repetitive circuitry.

subjective contour   illusory contours perceived by the visual system even in the presence of no real

intensity change. A typical example is Kanisza triangle.

submersible transformer   a transformer, used in underground distribution work, which is capable of of operation while submerged in water.

subroutine   a group of instructions written to perform a task, independent of a main program; can be accessed by a program or another subroutine to perform the task.

subroutine call and return (IE)   the subroutine call is a specialized JUMP or BRANCH instruction that provides a means to return to the instruction following the call instruction after the subroutine has been completed. A RETURN instruction is usually provided for this purpose.

subsampling pyramid   a spatial domain hierarchy is generated by repeatedly subsampling the original image data. The reconstruction at any level simply uses the subsampled points from all previous levels in conjunction with the new points from the current level.

subsidiary communication authorization (SCA) services for paging, data transmission, specialized foreign language programs, radio readings services, utility load management and background musing using multiplexed subcarriers from 53–99 kHz in connection with broadcast FM.

subspace based   algorithm based on splitting the whole space into two orthogonal complements, the signal and noise subspaces, and exploiting properties of the desired signal in these two subspaces. *See also* MUSIC, ESPRIT, signal subspace, noise subspace.

substation   a junction point in the electric network. The incoming and outgoing lines are connected to a busbar through circuit breakers.

substation battery   a battery used to provide operating energy for the protective relay operations and to initiate circuit breaker operations in a generating substation. The battery is necessary, as the equipment must work reliably during severe voltage sags and outages on the AC system.

substrate   a dielectric or semiconductor slab over which active devices, planar transmission lines, and circuit components are fabricated. This can be a PCB, a ceramic, or a silicon or other semiconductor wafer that has electronic components interconnected to perform a circuit function. *See also* wafer.

subsynchronous resonance   an electric power system condition where the electric network exchanges energy with a turbine generator at one or more of the natural frequencies of the combined system below the synchronous frequency of the system.

subthreshold   the range of gate biases corresponding to the "off" condition of the MOSFET. In this regime, the MOSFET is not perfectly "off" but conducts a leakage current that must be controlled to avoid circuit errors and power consumption.

subtracter   a circuit that subtracts two values.

subtractive polarity   polarity designation of a transformer in which terminals of the same polarity on the low- and high-voltage coils are physically opposite each other on the transformer casing. With subtractive polarity, a short between two adjacent terminals results in the difference of the two coil voltages appearing between the remaining terminals. Subtractive polarity is generally used on transformers larger than 500 kVA and higher than 34.5 kV. Smaller units use additive polarity. See the diagram below. *See also* additive polarity.

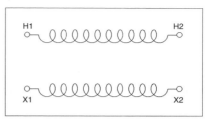

Transformer with subtractive polarity.

subtransient current   the fault current that flows during the subtransient period when the generator and motor apparent impedances are their respective subtransient impedances.

subtransient impedance   the series impedance that a generator or motor exhibits during the subtransient period, typically the first few cycles of a fault. Subtransient impedances are generally used in calculating fault currents for determining instantaneous relay settings.

subtransient open-circuit time constant   *See* quadrature-axis open-circuit time constant and direct-axis subtransient open-circuit time constant.

subtransmission   the circuits which connect bulk power substations to distribution substations.

subway transformer   another name for a submersible transformer.

sub-critical   the state of a fission chain reaction which is not self-sustaining because an insufficient number of neutrons are produced at each fission.

successive approximation   an A/D conversion process that systematically evaluates the analog signal in *n* steps that produce an *n*-bit code. The analog signal is successively compared to determine the digital code, beginning with the determination of the most significant bit of the code.

successive cancellation multiple-access receiver technique in which users are estimated one by one, first subtracting previously estimated data from the received signal.

sudden pressure relay a protective relay that senses the internal pressure in a transformer tank, and operates on sudden changes in this pressure. These sudden pressure changes reliably indicate a fault inside the tank.

sufficient statistic for a parameterized family of probability distributions $f(x; \theta)$, depending upon some parameter θ, it is a common problem to estimate θ from observation of a sample, X drawn according to an unknown member of this family. A statistic $S(X)$ is called a sufficient statistic of X for θ if it retains all the information available in X for the estimation of θ.

Sugeno fuzzy rule a special fuzzy rule in the form if x is A and y is B then $z = f(x, y)$ where "if x is A and y is B" is the *antecedent, A* and *B* are fuzzy sets, and the *consequent* is the crisp (nonfuzzy) function $z = f(x, y)$. *See also* fuzzy if-then rule, fuzzy inference system.

sulfur hexaflouride a heavy, highly-electronegative gas used as a high-voltage, self-healing insulation.

sum of products (SOP) a standard form for writing a Boolean equation that contains product terms (input variables or signal names either complemented or uncomplemented ANDed together) that are logically summed (ORed together).

sum-frequency generation the process in which two light beams of frequencies ω_1 and ω_2 interact with a nonlinear optical material to produce a beam at frequency $\omega_3 = \omega_1 + \omega_2$ by means of the second-order susceptibility.

super high definition television (SHDTV) television at resolutions of 2000×2000 pixels and higher.

supercomputer at any given time, the most powerful class of computer available.

superconducting magnetic bearing a magnetic bearing utilizing levitation between a magnet and a superconductor.

superconductivity a state of matter whereby the correlation of conduction electrons allows a static current to pass without resistance and a static magnetic flux to be excluded from the bulk of the materials.

superconductor a material that loses all electrical resistance below a certain temperature. Superconductors prevent externally applied magnetic fields from penetrating their interior. They are considered perfect diamagnetic materials. Once the externally applied field exceeds a critical value, the materials revert back to a nonsuperconducting status.

super-critical the state of a nuclear fission reaction which more neutrons are produced than are necessary to compensate for neutron absorbtion and leakage.

superdirectivity a condition of a phased array in which the excitation of the array elements is adjusted to obtain a directivity greater than that achievable with uniform excitation. Such antennas are often impractical, because high excitation currents are usually required, leading to ohmic losses that more than offset the additional directivity. Superdirective antennas also typically have high reactive fields and thus exhibit very narrow bandwidth.

superfluorescence usually refers to the enhanced spontaneous emission that occurs due to self-organization into a coherent state by a system of atoms or molecules.

superheater a heat exchanger that increases the steam temperature to about 1000 degrees F. It is heated by the flue gases.

superheterodyne an architecture used in virtually all modern-day receivers. In the early days of radio, tuned stages of amplification were cascaded in order to secure a sufficiently high level of signal for detection (demodulation).

superheterodyne receiver most receivers employ the superheterodyne receiving technique, which consists of either down-converting or up-converting the input signal to some convenient frequency band, called the intermediate frequency band, and then extracting the information (or modulation) by using an appropriate detector. This basic receiver structure is used for the reception of all types of bandpass signals, such as television, FM $<$ AM $<$ satellite, and radar signals.

superinterleaving *See* interleaved memory.

superlattice a stack of ultrathin layers of material. Layer thicknesses are sufficiently thin to produce quantum-confined effects, typically 100–1000 angstroms; generally, there are two different layer compositions, and the superlattice is built with layer composition in an alternating scheme.

supernode a cluster of nodes, interconnected with voltage sources, such that the voltage between any two nodes in the group is known.

superparamagnetism a form of magnetism in which the spins in small particles are exchange coupled by may be collectively switched by thermal energy.

superpipelined processor a processor where more than one instruction is fetched during a cycle in a staggered manner. That is, in an n-issue superpipelined processor, an instruction is fetched every $1/n$ of a cycle. For example, in the MIPS R4000, which is two-issue superpipeline, a new instruction is fetched every half cycle. Thus, in effect, the instruction pipeline runs at a frequency double than the system (in the R4000 the pipeline frequency is 100 MHz, while the external frequency is 50 MHz). Superpipeline processors usually have a relatively deep pipeline, of about 7 stages or more (8 stages on the R4000).

superpipelining a pipeline design technique in which the pipeline units are also pipelined internally so that multiple instructions are in various stages of processing within the units. The clock rate is increased accordingly.

superpolish methods for producing a surface of low RMS roughness, typically 10 angstroms RMS or less; methods include special mechanical, chemical, and ion polishing techniques.

superposition coding multiple-access channel coding technique in which each user encodes independently, such that at the receiver, the transmitted signals may be estimated using successive cancellation. *See also* successive cancellation.

superradiance usually refers to the strongly enhanced spontaneous emission that is emitted by a coherently prepared system of atoms or molecules.

superscalar processor a processor where more than one instruction is fetched, decoded, and executed simultaneously. If n instructions are fetched and processed simultaneously, it is called an n-issue superscalar processor. For example, the Pentium is a two-issue, and the DEC 61164 is a four-issue superscalar processor. This feature was implemented both on CISC (Pentium) and RISC (61164) processors.

superposition for a system $T[]$, the property that $T[a_1 x_1(t) + a_2 x_2(t)] = a_1 t[x_1(t)] + a_2 t[x_2(t)]$.

superposition integral for a linear shift-invariant system characterized by an impulse response, $h(t)$, the output, $y(t)$, for a given input, $x(t)$, is calculated as $y(t) = \int_{-\infty}^{\infty} x(s)h(t-s)ds$. Also called the convolution integral.

super-resolution the process of combining data from multiple, similar images of the same object to form a single image with increased spatial resolution.

supervised learning (1) a procedure in which a network is trained by comparing its output, in response to each training data item, with a target value (label) for that item. Network weights are adjusted so as to reduce the differences between outputs and targets until these differences reach acceptable values.

(2) a training technique in statistical pattern recognition or artificial neural networks in which the training set includes a predefined desired output.

supervised learning for self-generating neural network there are two ways for supervised learning in SGNN. The first is the same as that of supervised learning for a self-organizing system. The second is to make use of information gains of the attributes to the classification. That is, use the inner product of the training vector and the information gain vector corresponding to its attributes to train the network. Experiments show that this way of supervised learning for SGNN can significantly improve both the performance of the network and the training speed. *See also* self-generating neural networks, information gain, learning vector quantization.

supervised learning for self-organizing system *See* learning vector quantization.

supervised neural network neural network that requires input–output pairs to form the interconnection weight matrix of a network. The Hopfield model, perceptron, and backpropagation algorithm are supervised neural networks.

supervisor call instruction (SVC) *See* software interrupt.

supervisor instructions processor instructions that can be executed when the processor is running in supervisor mode. The separation of supervisor instructions is required to isolate the system's control information from tampering by user programs.

supervisor mode one of two CPU modes, the other being user mode. Sometimes called privileged mode, this mode allows access to privileged system resources such as special instructions, data, and registers.

supervisor state one of two CPU states, the other being user state. When the CPU is in supervisor state, it can execute privileged instructions.

support of a fuzzy set the crisp set of all points x in X with membership positive ($\mu_A(x) > 0$), where A is a fuzzy set in the universe of discourse X. *See also* fuzzy set, membership function.

supporting plane a planar structure that is an external support for a packaging and interconnecting structure, used to alter the structure's coefficient of thermal expansion.

supremal decision unit control agent or a part of the controller of the partitioned system, which perceives the objectives and the operation of this system as a whole and is concerned with following these overall objectives; in case of a large-scale system with hierarchical multilevel (two-level) controller, the coordinator unit is often regarded as the supremal decision unit.

supremum operator an operation that gives the least upper bound function. For example, if S is the supremum of a set A, then S is the upper bound of A, and no value less than S is an upper bound of A.

surface acoustic wave (SAW) a surface acoustic wave (also known as a Rayleigh wave) is composed of a coupled compressional and shear wave. On a piezoelectric substrate, there is also an electrostatic wave that allows electroacoustic coupling. The wave is confined at or near the surface and decays rapidly away from the surface.

surface acoustic wave (SAW) device   in this device, electrical signals are converted to acoustic signals, processed, and then converted back to electrical energy. Due to their small propagation velocity, acoustic waves have small wavelengths; thereby, one can construct miniature high performance components such as filters using SAW devices.

surface impedance   the impedance exhibited by the surface of a conductor/dielectric due to the variation in its conductivity with frequency.

surface mount technology (SMT)   the electrical connection of components to the surface of a conductive pattern without component lead holes.

surface mounting   *See* surface mount technology.

surface plasmon   a surface polariton in a plasma medium.

surface polariton   a polariton that propagates as a wave along the interface between two media.

surface rendering   *See* rendering.

surface roughness   the RMS value of the peaks and valleys in the profile of a solid surface. High frequency currents flow near the skin of conductors due to a skin effect phenomenon. Therefore, high frequency currents follow the contours on the surface of conductors. For this reason, surface roughness should be minimized so the patch of current flow is as short as possible.

surface scattering   scattering at the rough boundary between two media of different refractive index.

surface texture   *See* texture, texture analysis, texture modeling.

surface wave   a wave that propagates with dissipation in one direction and exponentially decays (without propagating) in the other directions. Most of the field is contained within or near the interface.

Surface waves are supported, for example, by a dielectrically coated conductor or by a corrugated conductor.

surface-emitting laser logic (SELL)   a device that integrates a phototransistor with a low-threshold vertical-cavity surface-emitting laser.

surface-mounted package   in both electrical and mechanical devices, a mounting technique between chip and substrate using solder joints between pads on the two surfaces. The advantage is that higher circuit densities can be achieved on the board.

surge   a short-duration (microsecond to millisecond) increase in power line voltage. Also called a spike or an impulse.

surge arrestor   a device that limits overvoltages by conducting large currents in response to an overvoltage. Surge arrestors are typically connected line to ground in transmission and primary distribution systems. They can be employed in a variety of connections in secondary distribution, and can be necessary in communications, sensing, and control circuits.

surge impedance   the ratio of voltage to current on that line for a high speed wave propagating down the line. The surge impedance of a line is a constant which depends on the line geometry and conductor characteristics. On power transmission lines, these waves are typically generated by lightning strokes, circuit breaker switching, etc. Also called characteristic impedance.

surge impedance loading (SIL)   of a transmission line, the characteristic impedance with resistance set to zero (resistance is assumed small compared to reactance). The power that flows in a lossless transmission line terminated in a resistive load equal to the line's surge impedance is denoted as the surge impedance loading of the line.

surge response voltage   the voltage that appears at the output terminals of surge protection equipment and is seen by loads connected to that device both during and after a surge condition.

surge suppressor   *See* transient voltage suppressor.

surge tank   an empty vessel located at the top of the penstock. It is used to store water surge when the turbine valve is suddenly closed.

susceptibility   the part of the permittivity or permeability that is attributable to the electromagnetic behavior of the medium. In a linear, isotropic medium, the electric susceptibility is numerically equal to the relative permittivity minus one, and the magnetic susceptibility is equal to the relative permeability minus one. *See also* electrical susceptibility.

sustained interruption   all interruptions that are not momentary. Generally used when referring to long duration voltage interruptions of greater than 1 minute.

SV system   single variable system. *See also* single-input–single-output system.

SVC   supervisor call instruction. *See also* software interrupt.

SVD   *See* singular value decomposition.

SVP   *See* sound velocity profile.

Swan, Joseph   (1828–1914) Born: Sunderland, England

Best known as the inventor of the incandescent lamp. During his life he acquired seventy patents in many areas. He was a devoted experimentalist with interests in photography, the development of miner's lamps, batteries, electroplating, and artificial silk. Swan teamed with Edison in 1883 to form the Edison and Swan United Electric Light Company after Edison's suit for patent infringement was dismissed. J. W. Starr and W. E. Staite were the early pioneers who inspired Swan to research that led to his knighthood in 1904.

swap   in assembly language, an instruction that swaps two values one for the other.

SWE   *See* Schrödinger wave equation.

sweep generator   a frequency source that can be setup to sweep from a start frequency to a stop frequency in a specified time interval.

swell   a voltage or current RMS value at supply frequency that increases for a time period from 0.5 cycles to 1 minute.

swing curve   a sinusoidal variation of a parameter, such as the critical dimension or the dose-to-clear, as a function of resist thickness caused by thin film interference effects.

swing bus   in power-flow studies, a bus in the power system which is assigned unknown real and reactive power so as to compensate for losses in the system.

swing equation   a nonlinear differential equation utilized in determining the dynamics of synchronous machines. *See also* electromechanical equation.

switch   (1) a device that allows current flow when closed and provides isolation when open. The switch provides similar functions to the circuit breaker, but cannot interrupt fault currents. Some switches are capable of making and breaking load currents, while others are only able to break charging current. Switches can be either manually controlled or motor controlled. *See also* single-pole single-throw switch, single-pole double-throw switch, transmit/receive switch, all optical switch.

(2) a device comprising a number of input and output ports and circuitry to switch packets from one input port to one or more output ports based on the addressing information contained in the packet header.

switchable shunt   *See* shunt capacitor.

switchboard   literally, a large panel or board upon switches were mounted in early electrical systems.

switched combining   a method of diversity combining in which the receiver is switched between alternative communication channels to find the channel that yields the best signal quality. *See also* angle diversity, antenna diversity.

switched reluctance machine   a doubly salient, singly excited electrical machine that contains a different number of poles on the stator and rotor. Since there are a different number of poles on the rotor and stator, only one stator phase can be aligned at a time with the rotor.

When operated as a motor, the stator phases are sequentially switched on and off to pull the rotor into alignment with them. This requires knowledge of the rotor position to properly excite the stator phases. The switched reluctance machine can also operate as a generator. In this case the stator windings are charged with a current as the rotor comes into alignment. When the current reaches a determined level, the windings are reconnected to send current out of the machine. As the rotor is driven, the inductance drops, causing an increase in current.

This type of machine requires an external capacitor bank, switches and diodes in each phase, and a sophisticated control system to operate.

switched-mode power supply   *See* switching power supply.

switching amplifier   a type of amplifier that utilizes switching between the cutoff and saturated states to minimize the time in the lossy transition states, thus achieving high efficiencies. All class D, E, and S amplifiers fit into this general group. Parameters such as device characteristics, quiescent bias point, RF load line, significant harmonic and/or mixing frequencies, and amplitude and waveform of the applied signal(s) should be included with the class definition, thus defining the major contributors to the physical actions taking place in one of these amplifiers.

switching flow graph model   a large-signal dynamic modeling method for PWM switching circuits. The circuit is viewed as two linear subcircuits, one with the switch on and one with the switch off. Flow graphs are obtained for the subcircuits and then combined using switching branches to form a switching flow graph. The switching flow graph provides a graphical representation of the dynamic switching circuit from which the large-signal, small-signal, and steady-state behaviors can be extracted.

switching frequency   the frequency at which converter switches are switched. In sine-triangle PWM switching applications, the switching frequency is defined by the triangle wave frequency, i.e., the carrier frequency.

switching node   a computer or computing equipment that provides access to networking services.

switching order   procedure which includes the sequence of switching operations to shift load.

switching power supply   a power supply, with one or more outputs, based on switching converters. The output(s) may be regulated via a control technique.

switching surface   *See* sliding surface.

switching surge   a momentary overvoltage in a power system which results from energy stored in the magnetic filed of a long power line being injected into the system at the instant that the line is switched out of service.

switching time   the time required for an entity to change from one state to another.

SWR   *See* standing wave ratio.

symbol error rate   a fundamental performance measure for digital communication systems. The symbol error rate is estimated as the number of errors divided by the total number of demodulated symbols. When the communication system is ergodic, this is equivalent to the probability of making a demodulation error on any symbol.

symbol synchronization   a technique to determine delay offset or rate of symbol arrival from the received signal. Can be based on either closed or open loop methods.

symbolic substitution   *See* symbolic substitution logic gate.

symbolic substitution logic gate   an optical logic gate using a specific algorithm developed for optical computing called symbolic substitution. In symbolic substitution, one or more binary input data are together represented by an input pattern. In its original method, four identical patterns are duplicated from the input pattern. The four patterns are shifted to different directions. The shifted patterns are added. The added pattern is thresholded. The previous procedure is then repeated. The thresholded pattern is split into four identical patterns. The four identical patterns are shifted to different directions and then combined as output patterns. All these steps are equivalent to first, recognizing input pattern, and then substituting it with output pattern. In the improved method,

input pattern can be substituted with output pattern using two correlators, the first correlator for recognition and the second for generating output pattern, or using a holographic associative memory. Output pattern can be any of sixteen Boolean logic operations or their combinations.

symmetric half plane field   the class of image models which can be implemented recursively pixel by pixel. That is, if the pixels in an image are ordered lexicographically (either by rows or by columns), then a symmetric half plane model is one in which a pixel p is a function of only those pixels preceding p in the ordering. *See* Markov random field.

symmetric multiprocessor   a multiprocessor system where all the processors, memories, and I/O devices are equally accessible without master–slave relationship.

symmetric resonator   a standing-wave resonator with identical right and left mirrors; usually refers to the mirror curvatures and not the mirror transmissions.

symmetrical component   the method by which unbalanced three-phase power system operation (particularly unbalanced fault performance) can be efficiently analyzed. Symmetrical components convert unbalanced line currents and voltages to three sets of balanced sequence components: positive sequence, negative sequence, and zero sequence.

The transformed phasor variables f_{+-0s} are obtained by applying the appropriate Fortescue transformation to any multiphase set of phasor variables. Denoted, f_{+s}, f_{-s}, f_{0s}, these are the positive sequence, negative sequence, and zero sequence components, respectively. The variables are so named because any unbalanced set of currents can be expressed (in phasor form) in terms of

(1) a balanced set of currents with magnitude i_{+s} that has a phase sequence which produces counterclockwise rotor rotation in a machine (positive sequence).

(2) a balanced set of phasor currents with magnitude i_{-s} that has a phase sequence which produces clockwise rotor rotation in a machine (negative sequence), and

(3) a set of three equal phasors with magnitude f_0 (zero sequence) which does not produce rotor rotation in machine.

symmetrical fault another term for a three-phase fault, a fault in which all three conductors of a three-phase power line are short-circuited together. System faults are symmetrical and can be analyzed by using single phase circuit.

symmetrical fault current the total current flowing to a fault less the DC offset current. In many cases, fault current calculations are expressed in terms of symmetrical amps.

symmetries of nonlinear susceptibility the elements of the nonlinear susceptibility tensor are not all independent. For instance, any rotational symmetries of the material medium will be reflected in tensor properties of the susceptibility. In addition, there are symmetries that depend on the frequency dependence of the susceptibility. For instance, intrinsic permutation symmetry states that the susceptibility is unchanged under simultaneous interchange of two input frequencies and two input tensor indices. Likewise, full permutation symmetry states that the susceptibility is unchanged under simultaneous interchange of two frequencies and two tensor indices, input and output considered interchangeably. Kleinman symmetry states that the susceptibility is unchanged under interchange any two tensor indices, input and output considered interchangeably.

sync distribution a system that allows for the distribution of sync pulses to multiple devices.

sync generator (1) signal generator that is designed to produce a specified signal waveform in order to synchronize a specific electronic device or system.

(2) an electronic unit used to generate the sync (synchronizing) information used in a video signal. Sync generators typically provide signals, such as horizontal sync, vertical sync, composite sync, and blanking. The sync generator signals are used in a video facility to keep all video signals properly aligned with each other.

sync interval the time period between neighboring sync pulses.

sync separator electronic circuitry used to separate the horizontal and vertical sync information that are contained in a composite video or composite sync signal.

sync tip the sync level that represents the peaks of the sync signal.

sync-locking a condition in which a circuit will continue to follow the sync pulse even with variations in amplitude and phase.

synchro also called a selsyn (for self-synchronous). An AC servo machine used in pairs primarily for remote sensing and shaft positioning applications. Its construction is essentially that of a wound-rotor induction machine with either a single-phase or 3-phase rotor winding. Various stator and rotor interconnections are possible, depending on desired function and required torque.

synchro-check relay a device used to monitor the frequency and phase angle of the voltages across an open circuit breaker. Synchro-check relays are commonly used to prevent breaker closing or reclosing on excessive voltage or frequency difference.

synchronization (1) a situation when two or more processes coordinate their activities based upon a condition.

(2) the process of determining (usually channel) parameters from a received signal, for

example carrier frequency offset, carrier phase, or symbol timing.

synchronized CDMA   a CDMA system where all the users are time-synchronized, i.e., the signals associated with all users arrive at the receiver with identical time delays.

synchronizing coefficient   electrical torque component in phase with the rotor angle.

synchronizing relay   a relay that monitors the voltage across an open circuit breaker to determine the frequency and phase relationship of the voltage sources on either side of the breaker. Synchronizing relays are used on generator breakers to bring the generator to the system frequency and to match the phase angle between the generator and system prior to closing the breaker.

synchronous   an operation or operations that are controlled or synchronized by a clocking signal.

synchronous bus   a bus in which bus transactions are controlled by a common clock signal and a fixed number of clock periods is allocated for specific bus transactions. *Compare with* asynchronous bus.

synchronous circuit   a sequential logic circuit that is synchronized with a system clock.

synchronous condenser   an unloaded, overexcited synchronous motor that is used to generate reactive power.

synchronous demodulation   a form of a phase sensitive angle demodulation in which local oscillator is synchronized or locked in frequency and phase to the incoming carrier signal.

synchronous detection   demodulation scheme using a balanced modulator to translate the center frequency of an IF signal down to DC (i.e., zero Hz). A local oscillator (LO) tuned to the IF

center frequency is injected into one of the input ports of the balanced detector, while the AM or SSB signal containing the information is applied to the other. When used in this manner, the LO is often referred to as a beat frequency oscillator (BFO). Low-pass filtering the output results in retrieval of the intelligence signal, superimposed upon a DC voltage (or current, dependent upon the actual device). The DC value may either be discarded via high-pass filtering, or used as a received signal strength indicator for use in automatic gain control circuits.

synchronous digital hierarchy (SDH)   an international interface specification for high-speed optical fiber transmission networks that allows different manufacturers' equipment to be interconnected with full maintenance and signal transparency. Specifies the optical parameters and the basic rates and formats of the signal. Emphasizes protection from faults and fast restoration of service after service interrupts.

synchronous drive   a magnetic drive characterized by synchronous transmission of torque, typically using a salient pole structure. There is no slip between the driver and the follower.

synchronous machine   an AC electrical machine that is capable of delivering torque only at one specific speed (n_s), which is determined by the frequency of the AC system (f) and the number of poles (P) in the machine. The relationship between synchronous speed and the other variables is

$$n_s = 120 f / P$$

synchronous motor   an AC motor in which the average speed of normal operation is exactly proportional to the frequency to which it is connected. A synchronous motor generally has rotating field poles that are excited by DC.

synchronous operation   an operation that is synchronized to a clocking signal.

synchronous optical network (SONET) a U.S. interface specification for high-speed optical fiber transmission networks that allows different manufacturers' equipment to be interconnected with full maintenance and signal transparency. SONET emphasizes protection from faults and fast restoration of service after service interrupts.

synchronous reactance the inductive reactance of the armature windings in synchronous machines under steady-state conditions. Designated by the symbol X_s, expressed in ohms per phase, the synchronous reactance is a function of the stator inductance and the frequency of the stator currents.

synchronous reference frame a two-dimensional space that rotates at an angular velocity corresponding to the fundamental frequency of the physical stator variables (voltage, current, flux) of a system. In electric machines/power system analysis, an orthogonal coordinate axis is established in this space upon which fictitious windings are placed. A linear transformation is derived in which the physical variables of the system (voltage, current, flux) are referred to variables of the fictitious windings. *See also* transformation equations, arbitrary reference frame, rotor reference frame, stationary reference frame.

synchronous reluctance machine a type of synchronous machine that has no rotor winding. The rotor consists of salient poles, which causes the reluctance to vary as a function of position around the airgap. When operated as a motor, a rotating magnetic field is established by the stator windings that causes a reluctance torque on the rotor as the path of lowest permeability stays aligned with the peak of the stator flux wave.

synchronous reluctance motor a synchronous motor that depends on a reluctance variation on the rotor for the mechanism of torque production. The rotor shape is designed to provide a high dif-

ference in the reluctances between the d and q axes.

synchronous speed speed of the rotating magnetic flux produced by three-phase currents in stationary coils in three-phase AC machines. The synchronous speed is calculated by a knowledge of the number of poles of the machine and the frequency of the stator currents as

$$N_s = 120 f_s / P$$

synchronous transfer mode a method of multiplexing messages onto a channel in which each period of time (also called a frame) on the channel is divided into a number of slots; one slot is allocated to each source for the transmission of messages, and a slot's worth of data from each source is sent every period.

synchronous updating all units in a neural network have the values of their outputs updated simultaneously.

synchronously pumped-modelocked (SPM) laser a laser in which periodic pump pulsations arrive at the amplifying medium of a laser oscillator in synchrony with the circulating mode-locked pulses, a standard technique for obtaining sub-picosecond pulsations.

synchroscope a device used to determine the phasor angle between two 3-phase systems. It is normally used to indicate when two systems are in phase so that they can be connected in parallel.

syndrome bit pattern used for error-detection and correction that is formed by multiplying the received vector by the parity-check matrix. Any two n-tuples that have the same syndrome can differ at most by a code word.

synonym in a virtual addressed cache, when a real address has more than one virtual address, the name given to the virtual addresses.

syntax   the part of a formal definition of a language that specifies legal combinations of symbols that make up statements in the language.

synthesis filter   a bank of filters that recombines the components decomposed by analysis filters from different frequency bands.

synthesizer   a software program that creates GDS2 data from a hardware description language specification such as VHDL or Verilog.

synthetic aperture radar (SAR)   a technique for overcoming the need for large antennas on side-looking airborne radar (SLAR) systems. The effect of a large antenna is synthesized by using Doppler shifts to classify the return signals, generating a very small effective beamwidth. The process is quite similar to that of holography, since the amplitude and frequency of the signals is recorded over time.

synthetic diamond   diamond grown artificially, usually as a film, for industrial purposes such as hardness, thermal conductivity, or optical properties.

system   a physical process or device that transforms an input signal to an output signal. For example, the following figure describes a system consisting of a RC circuit with an input of voltage, and an output of measured voltage across the capacitor:

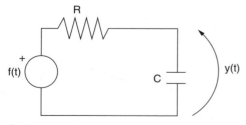

System example.

Often the behavior of systems is governed by a physical law, which when applied yields a math-ematical description of the behavior. For example, application of Kirchoff's voltage law and Ohm's Law leads to the following differential equation description of the above electric circuit input/output behavior:

$$\frac{d}{dt} y(t) + \frac{1}{RC} y(t) = \frac{1}{RC} f(t)$$

system bus   in digital systems, the main bus over which information flows.

system identification   a field of control engineering dealing with the derivation of mathematical models for the dynamics of processes, often by a detailed study of its input and output signals. It includes the design of experiments for enhancing the accuracy of the models.

system implementation   a phase of software development life cycle during which a software product is integrated into its operational environment.

system interaction   a stream of energy, material, or information exchanged between the sub-systems of a large-scale system. Relevant attributes of those streams are, respectively, interaction inputs or interaction outputs. Interactions are described by the interaction equations, which relate interaction inputs to a given subsystem to interaction outputs from other subsystems.

system noise factor   a value, in decibels, representing the ratio of the signal-to-noise ratio (S/N) appearing at the input of a system to that appearing at the output.

$$\text{system NF}_{dB} = 10 \log_{10} \left(\frac{\text{S/N}_{\text{input}}}{\text{S/N}_{\text{output}}} \right)$$

where S/N is the ratio (not in decibels) of the signal power to noise power at a given temperature. This value indicates the amount of signal-to-noise degradation from input to output of a system of components. If the S/N and power gain of each individual component in the system is known, then Friis' formula can be used to predict the overall system noise factor:

$$
\begin{aligned}
\text{system NF}_{dB} = &\ 10\log_{10} \text{NR}_1 \\
&+ (\text{NR}_2 - 1)\,/\,P_{G1} \\
&+ (\text{NR}_3 - 1)\,/\,(P_{G1} \times P_{G2}) \\
&+ (\text{NR}_4 - 1)\,/\,[P_{G1} \times P_{G2} \times P_{G3}] + \cdots \\
&+ (\text{NR}_n - 1)\,/\,\left[P_{G1} \times P_{G2} \ldots P_{G(n-1)}\right]
\end{aligned}
$$

where all values for S/N and power gain are in ratio (non-decibel) format, and the noise ratio, NR, of each stage is defined as NR = $(\text{S/N}_\text{input})/(\text{S/N}_\text{output})$. Also referred to as system noise figure.

System Performance and Evaluation Cooperative (SPEC)   a cooperative formed by four companies, Apollo, Hewlett-Packard, MIPS, and Sun Microsystems, to evaluate smaller computers.

systematic code   a code for which the information sequence itself is a part of the coded sequence. For block codes, it is common to assume that the information sequence is the first (or last) part of the codeword.

system transfer function   the result of sending a known test signal (often an impulse function or sine wave) through a system that defines what a system will do when presented with an input signal. Test signals often must be varied in frequency since system transfer functions are often frequency dependent (e.g., a stereo amplifier or speakers).

system with memory   a system whose output at time t depends on the input at other times (and possibly including) that instant t. If the output of the system at time t depends only on the input to the system at time t the system is said to be memoryless.

systems engineering   an approach to the overall life cycle evolution of a product or system. Generally, the systems engineering process comprises a number of phases. There are three essential phases in any systems engineering life cycle: formulation of requirements and specifications, design and development of the system or product, and deployment of the system. Each of these three basic phases may be further expanded into a larger number. For example, deployment generally comprises operational test and evaluation, maintenance over an extended operational life of the system, and modification and retrofit (or replacement) to meet new and evolving user needs.

systolic   flow of data in a rhythmic fashion from a memory through many processors, returning to the memory just as blood flows from and to the heart.

T

T   common symbol for temperature, usually expressed in degrees Kelvin.

T-bracket   a metal frame which holds a lightning arrester and a cut-out to the top of a utility pole.

T-connection   term often used to describe, with some ambiguity, two distinct transformer connections — one to simply convert voltage levels in a 3-phase power system, and the other to convert between 3-phase and a 2-phase voltages. Both connections use only two single-phase transformers, one called the main and the other the teaser, arranged in a T configuration. Details of each configuration are described below. These connections are also often referred to as Scott connections, since they were first proposed by Charles F. Scott.

Conversion of 3-phase voltage levels: In this configuration, the main transformer in the T-connection is a center-tapped unit that is connected between two lines of the three-phase system. The teaser transformer is connected between the center-tap of the main transformer and the third line of the 3-phase power system. Additionally, the coils of the teaser transformer have 86.6% of the turns in the corresponding coils in the main transformer. The result is a balanced three-phase voltage on the secondary. In most applications, the main and teaser transformers are actually identical, full voltage units with center taps and two 86.6% taps, one with respect to each terminal. This allows main and teaser units to be interchanged, plus it provides for true, 3-phase, 4-wire system with a neutral connection. This is illustrated in the diagram.

Conversion of 3-phase to 2-phase voltage: The main and teaser transformers in this connection are arranged as they are in the Scott connection. However, while the voltage connections to the primary of the teaser coil are made at the 86.6% tap, the secondary voltage is taken from the full coil. This produces two equal secondary voltages with a 90 degree phase difference, as illustrated in the diagram. *See also* teaser transformer.

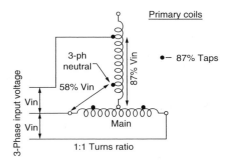

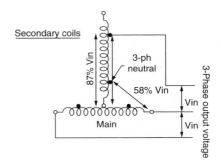

Scott transformer connection.

t-conorm   *See* triangular conorm.

t-norm   *See* triangular norm.

T/R switch   *See* transmit/receive switch.

TAB   *See* tape automated bonding.

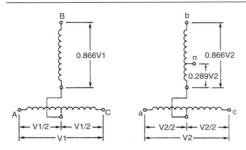

T transformer connection.

tableau formulation   a method for formulating the equations governing the behavior of electrical networks. The tableau method simply groups the KVL, KCL, and branch relationships into one huge set of equations.

taboo channels   channels that the FCC does not currently assign in order to avoid interference from adjacent channels.

tachometer   a instrument used to measure the speed of a rotating device. Several types of tachometers are available. Friction devices are placed against the shaft of the device being measured. Others used magnetic variation or reflected light pulses to determine the speed. Tachometer generators are mounted on the shaft of the device being measured and provide a voltage proportional to the speed. Tach generators are often used in servo systems.

tachometer-generator   a small generator that is connected to the shaft of a rotating machine and produces an output voltage directly proportional to the rpm of the machine. Typically used for closed-loop speed control.

tag   (1) that part of a memory address held in a direct mapped or set associative cache next to the corresponding line, generally the most significant bits of the address.

(2) a field attached to an object to denote the type of information stored in the object. The tag can flag control objects to prevent misuse. Tags can be used to identify the type of each object and thereby to simplify the instruction set, since, for example, only one ADD instruction would be necessary if each numeric object were tagged with its type (integer vs. real, for example).

(3) a temporary sign which is affixed to a network device to identify particular instructions. An example of this might be placing a tag which indicates "Do Not Close" on a circuit breaker which has been opened to permit downstream work.

tagged image file format (TIFF)   a popular image-file format that is very flexible. TIFF can hold compressed or uncompressed images, or different types of pixels, and is usable on different operating systems. *See* file format, image compression, Lempel-Ziv-Welch coding.

tail biting   a frame-by-frame transmission scheme for which the data is convolutionally encoded so that the encoder begins and ends in the same state, which, however, is unknown to the decoder. The advantage of this scheme is that no tail (overhead) is added to the data to force the encoder into a (by the decoder) known state. *Compare with* fractional rate loss.

Takagi–Sugeno–Kang fuzzy model   a fuzzy model that was studied by Takagi, Sugeno, and Kang. It is called the TSK or just TS fuzzy model, and it can be viewed as a special case of SAM. *See also* fuzzy system.

take-off   *See* spur.

tangential sensitivity   amount of input signal power to a two-port network to produce the output signal-to-noise power ratio unity.

Tanimoto similarity   for vectors x and y, defined as

$$S_T = \frac{(x, y)}{\| x \|^2 + \| y \|^2 - (x, y)}$$

where (x, y) is the inner product of x and y and $\| x \|$ is the Euclidean norm of x.

tank the container for the coils and core of a transformer, which is usually oil-filled for insulation and cooling.

transverse-excitation-atmospheric (TEA) laser high pressure (sometimes atmospheric pressure or higher) gas laser excited by a discharge in which the current flow is transverse to the direction of propagation of the laser beam; useful for very high power pulsed applications.

tap a connection (actually one of several) to a coil, allowing the number of turns in the coil to be varied.

tap changer a device to change the tap setting on a transformer coil, allowing voltage control. *See also* tap, tap changing under load.

tap changing under load (TCUL) a type of transformer in which the output voltage can be adjusted while the load is connected to the transformer. The voltage is adjusted by changing the turns ratio of the primary and secondary coils. That, in turn, is accomplished by bringing out connections (taps) at several points on the coil. Changing from one tap to another either adds or subtracts turns from the coil and raises or lowers the voltage, respectively.

tape automated bonding (TAB) a manufacturing technique in which leads are punched into a metal tape, chips are attached to the inside ends of the leads, and then the chip and lead frame are mounted on the MCM or PCB.

tape skew misalignment of magnetic tape during readout, leading to a difference between bits positions as written and as recognized for reading. Generally not a serious problem for low recording density or low tape speed; otherwise, a correction is required, e.g., by the use of "deskewing buffers."

tape-wound core a ferromagnetic core constructed by winding ribbon-like steel instead of stacking thicker, punched lamination. Usually used for higher frequency devices, or where it is desired to reduce the eddy current losses.

tapered mirror mirror in which the reflection profile varies across the mirror surface; useful for discriminating against high-order transverse modes in a resonator. *See also* variable reflectivity mirror.

tapped delay line a realization of a digital filter which represents a unit time delay by unit spatial difference in a transmission line.

TAR *See* top antireflective coating.

target architecture the architecture of a system that is being emulated on a different (host) architecture.

task-level programming *See* object-oriented programming.

task-oriented space *See* external space.

TCSC *See* thryristor-controlled series compensator.

TCUL *See* tap changing under load.

TDD *See* total demand distortion or time division duplexing.

TDM *See* time division multiplexing.

TDMA *See* time division multiple access.

TE mode *See* transverse electric mode.

TE polarizaion *See* transverse electric polarization.

TE wave *See* transverse electric wave.

TEA laser *See* transverse-excitation-atmospheric laser.

teaching-by-doing *See* teaching-by-showing programming.

teaching-by-showing programming a programming technique in which the operator guides the manipulator manually or by means of a teach pendant along the desired motion path. During this movement, the data read by joint position sensors (all robots are equipped with joint position sensors) are stored. During the execution of the motion (playing back), these data are utilized by the joint drive servos. Typical applications of this kind of programming are spot welding, spray painting, and simple palletizing. Teaching-by-doing does not require special programming skill and can be done by a plant technician. Each industrial robot is equipped with these capabilities. Also called teaching-by-doing.

team decision decision taken independently by several decision makers being in charge of a given process (or a decision problem) and forming a team, i.e., contributing to a commonly shared goal.

teaser coil *See* teaser transformer.

teaser transformer one of two single-phase transformers used to make up a Scott or T-connection transformer. The teaser transformer is connected between one line of the three-phase voltage system and the center-tap of the main transformer in the T-connection. Also called teaser coil. *See also* T-connection.

tee-structured VQ A method for structured vector quantization, where the input signal is successively classified and coded in a manner described by a mathematical structure known as a "tree."

TEGFET *See* high electron mobility transistor.

telepoint a generic term used to describe public-access, cordless, telephone systems.

television (1) literally, seeing at a distance.
(2) representation and transmission of moving images electrically (video).
(3) receiver for displaying such pictures.

Tellegen's theorem states the orthogonality of $\mathbf{v}$, $\mathbf{i}'$ and of $\mathbf{v}'$, $\mathbf{i}$ for two different networks, N and N', with identical topologies, i.e., with the same directed graph and with $\mathbf{v}$, $\mathbf{i}$ the branch quantities of network N and $\mathbf{v}'$, $\mathbf{i}'$ the branch quantities of network N'.

TEM *See* transverse electromagnetic wave.

TEM wave *See* transverse electromagnetic wave.

TEM00 mode term sometimes used to describe the fundamental Gaussian beam mode, though this mode has small longitudinal components of the electric and magnetic field vectors.

temperate plasma a preferred term used to describe the "cold plasma" to convey the following limits on the thermal velocity of the electrons. The electron thermal velocity is much less than the phase velocity of the wave in the medium but much greater than the induced velocity increments produced by the electromagnetic fields.

temperature coefficient of resistance the change in electrical resistance of a resistor per unit change in temperature.

template a pattern, often in the form of a mask, which can be used to locate objects and features in an image. For large objects which might appear in many orientations, this procedure is very computation intensive, and it is normal to use small templates to search just for features, and then infer the presence of the objects. Template matching is commonly performed for tasks such as edge detection and corner detection.

template mask   a mask which forms a pixellated template of an object or feature, and which may then be used for template matching. *See* template.

template matching   a technique in which a model and an optimization method is used to deform a template to a study in order to find the best match for the purpose of detection or recognition.

temporal alignment   the process of aligning two sequences by using dynamic programming. *See also* DTW algorithm.

temporal averaging   averaging a signal in the time domain. For discrete signals, temporal averaging by a finite impulse response filter is a way to smooth out the signal.

temporal frequency   a frequency that represents the change of an image with time; temporal frequency components can result from motion between completed images or from the methods used to construct a complete image. A monochrome interlaced television frame requires two (2) fields or 30 Hz temporal frequency in constructing a complete monochrome television frame. Similarly, the NTSC color subcarrier frequency is interlaced with the horizontal line frequency and creates a 15 Hz temporal frequency component to the color television frame.

temporal locality   *See* locality.

temporal resolution   the ability to resolve two closely spaced targets in the time domain. *See* resolution.

temporary fault   a fault that will not re-occur if the equipment is deenergized and then reenergized. An example of a temporary fault is when a lightning stroke causes an uninsulated overhead line to arc over an insulator, with no equipment damage.

temporary interruption   a loss of voltage of less than 0.1 pu for a duration of 3 seconds to 1 minute.

terahertz (THz)   a frequency unit, 10^{12} hertz.

terminal bushing   *See* bushing.

termination   a circuit element or device placed at the end of a transmission line that reflects and/or absorbs signal energy.

terminator   (1) a device connected to the physical end of a signal line that prevents the unwanted reflection of the signal back to its source.

(2) a data item in a stream that marks the end of some portion or all of the data.

ternary logic   digital logic with three valid voltage levels.

tertiary winding   a third winding on a transformer. A tertiary winding may be used to obtain a second voltage level from the transformer. For example, in a substation it may be necessary to have low voltage power for the substation equipment in addition to the distribution voltage. Another application of a tertiary winding is in a wye-wye three-phase transformer. Here the tertiary is connected in delta, to provide a path for the triple harmonic components of the exciting current and prevent distortion of the phase voltages.

tesla   a unit of magnetic flux density equal to one weber per square meter, i.e., one volt-second per square meter. Denoted by T. The unit is named in honor of Nikola Tesla, an early pioneer in the electric industry, who is most commonly credited with building the first practical induction motor.

Tesla, Nikola   (1856–1943) Born: Smiljan, Croatia

Best known as the electrical pioneer who championed the use of alternating current. When Tesla first came to the United States he worked for Edison. He soon split with Edison, because Tesla approached invention from a theoretical standpoint, whereas Edison was a "trial and error" type experimentalist. Together with his financial backer, George Westinghouse, they battled with Edison, who championed the use of direct current for electrifying the world. Tesla is also known for his many inventions including the Tesla coil and the AC induction motor. It was Westinghouse who made a fortune from Tesla's inventions. Tesla was known for his eccentricities and died a recluse in New York City.

tesselation   in the Euclidean plane, a subdivision of that plane into polygonal cells which cover the whole of it, and such that two neighboring cells have disjoint interiors (in other words they have in common either a vertex, or a side with its two end-vertices). When the cells are isometric regular polygons, one of the following three cases occurs:

(1) each cell is a regular hexagon and has 6 neighboring cells each having a side in common with it.

(2) each cell is a square and has 8 neighboring cells, of which 4 have a side in common with it, and 4 have a vertex in common with it.

(3) each cell is an equilateral triangle and has 12 neighboring cells, of which 3 have a side in common with it, 3 have a vertex in common with it in such a way that that the 2 neighboring cells are symmetric with respect to the common vertex, and 6 have a vertex in common with it but without symmetry with respect to the common vertex.

The tesselation of the plane into regular cells is a mathematical model of the subdivision of an image into pixels, and the corresponding digital space is made of the centers of all the cells. In both the hexagonal and square tesselations, there is a vector basis such that the cell centers coincide with points with integer coordinates. Modern technology accords with Cartesian tradition in favoring the square tesselation, but the hexagonal tesselation has a simpler topology (with fewer neighboring cells, and all of the same type), which simplifies certain types of algorithm—such as thinning algorithms. *See* pixel adjacency.

test access port   a finite state machine used to control the boundary scan interface.

test fixture   a device or software module that is attached to another device or module so that tests can be run on the unit in question.

test function   *See* moment method.

test pattern   input vector such that the faulty output is different from the fault-free output.

test point   (1) a physical contact for a hardware device that can be monitored with an external test device.

(2) a data element within a software module that is accessible to an external test module.

test register   a register used in the processor to ease testing of some functional blocks (e.g., cache memory) by simplifying accesses to their internal states.

test response compaction   the process of reducing the test response to a signature. Common compaction techniques use signal transition counting, accumulated addition, CRC codes, etc.

test set   specialized sets of instruments used to verify the operation of relays, fault indicators, or other instrumentation.

test vectors   a test scheme that consists of pairs of input and output. Each input vector is a unique set of 1s and 0s applied to the chip inputs and the corresponding output vector is the set of 1s and 0s produced at each of the chip's output.

test-and-set instruction   an atomic instruction that tests a Boolean location and if FALSE, resets it to TRUE. *See also* atomic instruction.

testability   the measure of the ease with which a circuit can be tested. It is defined by the circuit controllability and observability features.

testing   a phase of software development life cycle during which the application is exercised for the purpose to find errors.

testing function one of a set of functions used in the method of moments to multiply both sides of the integral equation (in which the current has been expanded in a set of basis functions) to form a matrix equation that can be solved for the unknown current coefficients.

textural edgedness a measure of the mean edge contrast at every position in an image, where the average is taken over a significant region so as to smooth out small scale variations, thereby providing an indication of the type of texture present.

textural energy a measure of the amount of statistical, periodic or structural variation at a location in a texture, "energy" being a suitable square-law unit corresponding to the variance imposed on the mean intensity at that location in the texture.

texture quantitative measure of the variation of the intensity of a surface that can be described in terms of properties such as regularity, directionality, smoothness/coarseness, etc.

texture analysis the process of analyzing textures which appear at various positions in images. The term also includes the process of demarcating the boundaries between different textural regions, and leads on to the interpretation of visual scenes.

texture modeling the process of modeling a texture with a view to (a) later recognition or (b) generating a similar visual pattern in a graphics or virtual reality display. Textures are usually partly random in nature, and texture models usually involve statistical measures of the intensity variations.

THD *See* total harmonic distortion.

thermal control *See* thermal management.

thermal expansion mismatch the absolute difference in thermal expansion of two components.

thermal fin an extension of the surface that is in contact with a heat transfer fluid, usually in the form of a cylinder or rectangular prism protruding from the base surface.

thermal light light generated by spontaneous emission, such as when a group of excited atoms or molecules drops to a lower energy state in a random and independent manner emitting photons in the process; contrasted with laser light.

thermal management the process or processes by which the temperature of a specified component or system is maintained at the desired level. Also called thermal control.

thermal noise a noise process that affects communication channels and electrical circuits which is due to the random motion of electrons in materials and more specifically resistors. In such a circuit, the resistor produces a level of noise that is proportional to the resistance of the component, the ambient temperature, and the bandwidth of the circuit. Also known as Johnson, Nyquist, or white noise.

thermal neutrons neutrons which move at the same velocity as the random thermal motions of the atoms of the ambient medium.

thermal reactor a reactor which maintains a critical reaction with thermal neutrons.

thermal resistance a thermal characteristic of a heat flow path, establishing the temperature drop required to transport heat across the specified segment or surface; analogous to electrical resistance.

thermionics direct conversion of thermal energy into electrical energy by using the Edison effect (thermionic emission).

thermit welding a welding process that produces coalescence of metals by heating them with superheated liquid metal from a chemical reaction between a metal oxide and aluminum with or without the application of pressure.

thermomagnetic process the process of recording and erasure in magneto-optical media, involving local heating of the medium by a focused laser beam, followed by the formation or annihilation of a reverse-magnetized domain. The successful completion of the process usually requires an external magnetic field to assist the reversal of the magnetization.

thermomagnetic recording recoding method used with magneto-optical disks. It involves first using a focused laser beam to heat the disk surface and then forming or annihilating magnetized domains.

thermometer coding a method of coding real numbers in which the range of interest is divided into nonoverlapping intervals. To code a given real number, say x, the interval in which x lies is assigned the value $+1$, as are all intervals containing numbers less than x. All other intervals are assigned the value 0 (in the binary case) or -1 (in the bipolar case).

thick lens lens inside of which internal ray displacements and beam profile evolution are too large to be neglected.

thin film capacitor *See* metal-insulator-metal capacitor.

thin lens lens inside of which internal ray displacements and beam profile evolution are so small that they may be neglected.

thinning image operator that clears, somehow symmetrically, all the interior border pixels of a region without disconnecting the region. Successively applying a thinning operator results in a set of arcs termed "skeleton."

third rail a method of transmitting power to an electric locomotive. An insulated steel rail is laid along the railbed just outside the traction rails. This third rail is maintained at (typically) 600 volts DC by the railroad power supply, and contact is made to the locomotive by a shoe which slides atop the rail. Ground return is through the traction rails.

third-harmonic generation the process in which a laser beam of frequency ω interacts with a nonlinear optical system to produce a beam at frequency 3ω. *See also* harmonic generation.

third-order intercept (TOI) point this gives a measure of the power level where significant undesired nonlinear distortion of a communication signal will occur. It is related to the maximum signal that can be processed without causing significant problems to the accurate reproduction of the desired information (e.g., TV signal). Technically, the TOI is the hypothetical power in decibel-meter at which the power of the "third-order intermodulation" nonlinear distortion product between two signals input to a component would be equal to the linear extrapolation of the fundamental power.

third-order susceptibility a quantity, often designated χ^3, describing the third-order nonlinear optical response of a material system. It is defined through the relation $P^3 = C\chi^3 E^3$, where E is the applied electric field strength and P^3 is the 3rd order contribution to the material polarization. The coefficient C is of order unity and differs depending on the conventions used in defining the electric field strength. The third-order susceptibility is a tensor of rank 4 and describes nonlinear optical processes including third harmonic generation, four-wave mixing, and the intensity dependent refractive index. *See also* nonlinear susceptibility.

Thomson, William (Lord Kelvin) (1824–1907) Born: Belfast, Ireland

Best known as a physicist who championed the absolute temperature system that now bears his name (Kelvin). Thomson did significant work to expand Faraday's ideas. It was Thomson's work that Maxwell would extend into his seminal publications on electromagnetics. Thomson received a knighthood for his theoretical suggestions for the use of low-voltage signals in the trans-Atlantic telegraph cable. Thomson was proved correct when on a third attempt, a cable was laid and worked. An earlier high-voltage cable had failed.

thoriated   pertains to a metal to which the element thorium has been added.

thrashing   in a paging system, the effect of excessive and continual page transfers that can occur because the memory is overcommitted and programs cannot obtain sufficient main memory.

thread   in software processes, a thread of control is a section of code executed independently of other threads of control within a certain program.

three-antenna gain measurement method
technique based on Friis transmission formula in which the gain of each of three different antennas is calculated from a measurement of three insertion loss values (corresponding to all three combinations of antenna pairs) and the calculated propagation loss between the antenna pairs.

three-gun color display   a color-TV picture tube having a separate gun for each primary color (red, green, and blue).

three-lamp synchronizing   a method used to connect a three-phase power system in parallel to another one. In order to connect two systems, they must have the same voltage magnitude, frequency, and phase-shift. To determine that is the case, an open switch is connected between the phases of the two systems and a lamp is connected across the open switch pole in each phase. If the criteria previously listed are met, the lamps will all be dark. If there is a difference in voltage, the lamps will glow. If there is a difference in frequency, the

lamps will alternately glow and go dark in unison. Finally, if the two sides have different phase rotations, the lamps will blink sequentially as only one phase can be aligned at a time. In order to synchronize the two systems, it is necessary to close the contactor when the phase-shift is minimum, which means that the three lights are dark.

three-level laser   laser in which the most important transitions involve only three energy states; usually refers to a laser in which the lower level of the laser transition is separated from the ground state by much less than the thermal energy kT. *Contrast with* four-level laser.

three phase fault   a fault on a three phase power line in which all three conductors have become connected to each other and possibly the ground as well.

Three-Mile Island   typically refers to a cooling failure at a nuclear power plant on the Susquehanna River in central Pennsylvania, USA in 1979.

three-phase inverter   an inverter with a three-phase AC voltage output.

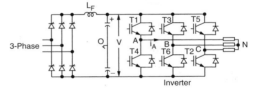

Three-phase inverter.

three-phase rectifier   a rectifier with a three-phase AC voltage input.

three-point starter   a manual DC motor starter in which a handle is pulled to start the motor. The motion of the handle causes a contact to move across a variable resistance in the armature circuit to limit the starting current. When the handle is moved to its fullest extent, the resistance is out of the armature circuit and an electromagnet holds the handle in place. With a three-point starter, the electromagnet is in series with the shunt field and loss of the field will shut down the motor. The disadvantage is that if it is desired to weaken the field for speed control, the motor starter may drop out. Also, the three-point starter cannot be used on a series DC motor.

three-point tracking   a tracking error reduction technique in which the preselector and local oscillator have trimmed capacitors added in parallel to the primary tuning capacitor and the local oscillator has an additional padder capacitor in series with the tuning coil.

three-state circuit   *See* tri-state circuit.

three-terminal line   a branched transmission line with three ends, each of which is terminated with circuit breakers such that all three breakers must trip to clear a fault anywhere in the line.

three-terminal device   an electronic device that has three contacts, such as a transistor.

three-winding transformer   a transformer with three windings, typically primary, secondary, and tertiary. Common three phase, three winding transformers employ wye connected primary and secondary windings and a delta connected tertiary winding. In some cases, an autotransformer is used to form the primary–secondary combination.

three-wire method   in the three-wire method, the remote resistor (which plays the role of active gauge) is connected to the bridge circuit by three wires. One end of this resistor is connected by one wire to the bridge power supply node; at another end of the resistor there are two wires, the first of which is connected to the detector and the second is to the adjacent resistor of the bridge. Other arrangements are possible as well. The goal

is to reduce the errors introduced by the long wires while measuring the remote resistor value.

threshold   (1) the limiting value of some variable of interest.

(2) the condition under which the unsaturated round-trip gain in a laser is equal to the loss.

(3) that point at which the indication exceeds the background or ambient.

threshold coding   a coding scheme in transform coding in which transform coefficients are coded only if they are larger than a selected threshold.

threshold current   the current in an electrically pumped laser that is necessary for the unsaturated round-trip gain to be equal to the loss.

threshold decoding   a special form of majority-logic decoding of block or convolutional codes.

thresholding   (1) the transformation of an image I with depth > 1 into a binary image I_b in which $I_b(\mathbf{x}) = 1$ if $I(\mathbf{x})$ is greater than a given threshold (possibly dependent on $\mathbf{x}$), $I_b(\mathbf{x}) = 0$ otherwise.

(2) any technique involving decision making based on certain deliberately selected value(s), known as threshold(s). For an example, refer to threshold coding. These techniques are also often utilized in image segmentation: specifically, thresholding groups of pixels into black and white based on a numerical value; pixel levels below the value (threshold) become black, those above become white.

threshold inversion   population inversion that provides an unsaturated round-trip gain in a laser equal to the loss.

threshold sample selection   in transform coders, quantization could be nonadaptive from block to block, where a percentage p of coefficients transmitted could vary according to spatial activity. To avoid large reconstruction errors as in coefficient reduction used by zonal selection, a threshold is chosen. Only the coefficients whose values are above the threshold are quantified and encoded.

threshold voltage when applied between the gate and source of a MOSFET, the voltage, VT, that results in an inversion of the charge carrier concentration at the silicon surface.

through via a via that connects the primary side and secondary side of a packaging and interconnecting structure.

through-reflect-line calibration a network analyzer two-port probe calibration technique. A TRL two-port calibration requires a through standard, reflect (open circuit is preferred, short is optional), and one or more lines (transmission lines). It has the advantages of self-consistency and it requires electrically simple standards.

throughput the amount of flow per unit time. Generally refers to information flow.

thumper slang for a time-domain reflectometer used to detect and locate defects in buried electric power cables.

thyratron (1) gas-filled triode in which the voltage on the grid can trigger ionization of the gas in the tube. Once the gas is ionized, current flows from cathode to anode until the potential across the two falls below a certain level. In Linac (a linear accelerator), the thyratrons are used as high-voltage relays in the chopper power supplies and in the RF modulators to trigger the ignitrons.

(2) an electronic tube containing low pressure gas or metal vapor in which one or more electrodes start current flow to the anode but exercise no further control over its flow.

thyristor a controllable four-layer (pn-pn) power semiconductor switching device that can only be on or off, with no intermediate operating states like transistors. *See also* silicon controlled rectifier.

thyristor rectifier a rectifier where the switches are thyristors. Thyristors are turned on by a gate trigger signal, and turned off by natural commutation. The output voltage is controllable by adjusting the firing angle α of the trigger signal. The direction of the power flow is reversible when an inductive load is used. When the average power flow is from DC to AC ($\alpha > \pi/2$), the rectifier is said to be operating in the line-commutated inverter mode.

thyristor-controlled phase angle regulator a phase shifting device used in transmission systems. The phase angle change is brought about by thyristor-based control.

thyristor-controlled series compensator a capacitor bank installed in series with an electric power transmission line in which each capacitor is placed in parallel with a thyristor device. Each capacitor may thus be switched in or out of the line for some variable portion of the AC cycle so as to maintain the line's maximum power-carrying ability under varying load conditions.

Ti sapphire laser laser in which the active medium is sapphire with titanium substituted for some of the aluminum atoms, important for its large bandwidth in the visible spectrum.

tiepoint a point in an input image whose corresponding point in a tranformed image is known. Tiepoints are often used to specify transformations in which the locations of transformed pixels change, as in geometric transformations and morphing. *See* geometric transformation.

tie switch a disconnect switch used on feeders and laterals to reconfigure distribution circuits to allow for line maintenance.

tier a group of cellular network cells of similar distance from some specific central cell. The tier number is the number of cell radii distant that the tier is from the central cell.

TIFF *See* tagged image file format.

tightly coupled multiprocessors a system with multiple processors in which communication between the processors takes place by sharing data in memory that is accessible to all processors in the system.

tilt angle the angle by which a surface slants away from the viewer's frontal plane.

time constant mathematically, the time required for the exponential component of a transient response (input as the step function) to decay to 37% ($1/e$) of its initial value, or rise to 63% ($1 - 1/e$) of its final value, where e is the mathematical constant 2.718281828 . . . .

In electronic circuits, the time constant is often related directly to the circuit RC value (i.e., the product of the resistance in ohms and the capacitance in farads) or to its L/R value (i.e., the ratio of the inductance in henrys to its resistance in ohms). *See also* settling time.

In a control system transfer function factor, T is the time constant and is equal to where f is the corner frequency in the bode plot. A closed-loop control system commonly has more than one time constant.

time correlation function a function characterizing the similarity of a received signal with respect to a shift in time. *See also* correlation.

time delay a time-current response characteristic, established by national standards, which means that a time-delay fuse is designed to carry five times rated current for 10 seconds before opening. *See also* envelope delay.

time diversity a way to try to obtain uncorrelated received signals to improve the performance of the system by transmitting the signals in different time instants. Interleaving is one way to implement time diversity.

time division duplexing (TDD) a technique for achieving duplex (i.e., two-way) communication. One direction of transmission is conducted within specific segments of time, and the reverse direction of transmission is conducted within different segments of time.

time division multiple access (TDMA) a technique for sharing a given communication resource amongst a number of users. The available communication resource is divided into a number of distinct time segments, each of which can then be used for transmission by individual users. Sometimes used in cellular and personal communications applications.

time division multiplexing (TDM) refers to the multiplexing of signals by taking rounds of samples from the signals. Each round consists of one sample from each signal, taken as a snapshot in time. *See also* time division multiple access.

time domain the specification of a signal as a function of time; time as the independent variable.

time domain analysis a type of simulation that allows the user to predict the circuit response over a specified time range. The result of the simulation is a graph of amplitude against time.

time domain storage an optical data storage technique in which time-dependent information is stored as a Fourier transform in an inhomogeneously broadened spectral hole burning material. This is usually accomplished with photon echoes or spin echoes. The maximum storage density is given by the ratio of inhomogeneous to homogeneous widths of the absorption spectrum.

time frequency analysis any signal analysis method that examines the frequency properties of a signal as they vary over time.

time hopping a type of spread spectrum wherein the transmission of the signal occurs as bursts of pulses in time. Each burst has a random starting time and may have a fixed or random duration.

time invariance   a special case of shift invariance, applying in the time domain. In particular, the impulse response of a system is independent of the time at which the impulse occurs. *See* space invariance.

time invariant channel   a communication channel for which the impulse response and transfer function are independent of time. Strictly time-invariant channels do not exist in practice, but many communication channels can be regarded as time invariant for observation intervals of practical interest.

time invariant system   a system with zero initial conditions for which any input $f(t)$ applied at a time $t - \tau$ will simply result in a time shift of the output $y(t)$ to $y(t - \tau)$. For example, a RC circuit with relationship between loop current $f(t)$ and output voltage $y(t)$ described by

$$y(t) = Rf(t) + \frac{1}{C} \int_{-\infty}^{t} f(\tau)d\tau$$

is time invariant.

A system that is not time invariant is *time varying*, or *time variant*. For example, a system described by

$$y[k] = kf[k]$$

is time varying.

time overcurrent (TOC) relay   an overcurrent relay that has intentional, selectable, time delay. The time delay is chosen so that the relay will operate more slowly than downstream relays or fuses, and more quickly than upstream relays or transformer fuses. Relay and fuse curves are generally displayed on time-current curves.

time response   the system response in the time domain when a reference input signal is applied

to a system. The time response of a control system is usually divided into two parts: the transient response and the steady-state response.

time shift   for a signal $x(t)$ a displacement in time t_0. The time shift is given by $x(t - t_0)$.

time slot   in time-division multiple access (TDMA), a time segment during which a designated user transmits, or control information is transmitted. In time-division multiplexing (TDM), each time slot carries bits associated with a particular call, or control information.

time stability   the degree to which the initial value of resistance is maintained to a stated degree of certainty under stated conditions of use over a stated period of time. Time stability is usually expressed as a percent or parts per million change in resistance per 1000 hours of continuous use.

time variant channel   a communication channel for which the impulse response and transfer function are functions of time. All practical channels are time-variant, providing the observation interval can be arbitrarily long. *See also* fading channel.

time variant system   system in which the parameters vary with time. In practice, most physical systems contain time varying elements.

time varying system   (1) *See* time invariant system.

(2) a system which not exhibiting time invariance. In particular, one in which the impulse response varies as a function of the time at which the impulse occurs.

time–bandwidth product   (1) in an acousto-optic deflector, the product of the acoustic-wave propagation time across the optical beam and the electrical bandwidth for optical diffraction; equivalent to the number of independent resolvable spots for the acousto-optic deflector. For coded signals such as chirp signals, the signal duration times the bandwidth of the signal; equivalent to the pulse compression ratio obtained in autocorrelation. *See also* pulse compression.

(2) the product of a signal's duration and bandwidth approximates the number of samples required to characterize the signal.

time-current characteristic curve (1) a relay time-current curve is a curve showing the time versus current characteristic of a time overcurrent (TOC) relay.

(2) a fuse time-current curve shows the melting and clearing times of a given fuse or family of fuses.

(3) a coordination time-current curve shows the relationship of the operating and clearing characteristics of coordinating devices (TOC relays and fuses) on a power system.

time-current curve *See* time-current characteristic curve.

time-delay fuse *See* dual-element fuse.

time-delay neural network a multilayer feedforward network in which the output is trained on a sequence of inputs of the form $x(t), x(t - D), \ldots, x(t - mD)$, where $x(.)$ is, in general, a vector. By specifying the required output at sufficient times t, the network can be trained (using backpropagation) to recognize sequences and predict time series.

time-delay relay relay that responds with an intentional time delay.

(1) in control circuits, time-delay relays are used to cause a time delay in the state of the relay when power is applied or removed to the relay actuator;

(2) in power system protective relays, the response time usually depends on the magnitude of the measured value. If the measured value is a large multiple of the pickup value, then the relay operates or trips after a short time delay. For smaller multiples of pickup, the relay trips after a longer time delay.

time-dependent dielectric breakdown breakdown of a dielectric is marked by a sudden increase in current when an electric field is applied. The breakdown does not occur immediately upon application of the electric field, but at a period of time later that depends exponentially upon the magnitude of the field.

time-invariant system the system in which the parameters are stationary with respect to time during the operation of the system.

time-of-arrival the time instant of the arrival of the first signal component to the radio receiver. *See also* propagation delay.

time-to-close contact a contact in which the desired time to close the contactor could be set by the user.

time-to-open contact a contact in which the desired time to open the contactor could be set by the user.

timeout the concept of allowing only a certain specified time interval for a certain event. If the event has not occurred during the interval, a timeout has said to have occurred.

timer a circuit that records a time interval.

timing the temporal relationship between signals.

timing diagram a diagram showing a group of signal values as a function of time. Used to express temporal relationships among a series of related signals.

timing error an error in a system due to faulty time relationships between its constituents.

tin whisker   a hairlike single crystal growth formed on the metallization surface.

tint   the intensity of color. The name for a non-dominant color.

TLB   *See* translation lookaside buffer.

TLM   *See* transmission line measurement.

TM mode   *See* transverse magnetic mode.

TM wave   *See* transverse magnetic wave.

TMI   refers to an accident at the Three Mile Island nuclear plant in 1979.

TOC relay   *See* time overcurrent relay.

Toeplitz matrix a   matrix with the property that it is symmetric and the i, j^{th} element is a function of $(i - j)$. The Toeplitz nature of autocorrelation matrices of wide-sense stationary discrete time random processes is exploited extensively in minimum mean square error prediction/estimation algorithms.

toggle   change of state from logic 0 to logic 1, or from logic 1 to logic 0, in a bistable device.

TOI point   *See* third-order intercept point.

token   device that generates or assists in generation of one-time security code/passwords.

token bus   a method of sharing a bus-type communications medium that uses a token to schedule access to the medium. When a particular station has completed its use of the token, it broadcasts the token on the bus, and the station to which it is addressed takes control of the medium. Also called token ring.

token ring   *See* token bus.

Tokomak   an experimental power reactor that uses fusion, in which the hot plasma is contained and compressed with a magnetic field.

tolerance   (1) the total amount by which a quantity is allowed to vary; thus the tolerance is the algebraic difference between the maximum and minimum limits.

(2) In the design of microwave component, it is important to perform a tolerance analysis in order to ascertain if a given component will also satisfy specifications when taking manufacturing tolerances into account.

(3) the amount of error allowable in an approximation.

Tomlinson precoding   a transmitter precoding method that compensates for intersymbol interference introduced by a dispersive channel. The purpose is to move the feedback filter $F(z)$ of a decision-feedback equalizer to the transmitter. The purpose of the mod$2M$ operation is to satisfy the transmitted power constraint.

tomography   (1) the process of reconstructing an image from projection data.

(2) the process of forming a cross-sectional view of an object by irradiating it from many directions and deducing from the transmitted energies the interior structure of the object. This latter process is known as reconstructing an image from its projections. Tomography can provide a very detailed map of the inside of an object and has revolutionized medical diagnostics. Also known as Computed tomography (CT). *See* projection, fan beam reconstruction, Radon transform.

tone control   a resistance-capacitance network used to alter the frequency response of an amplifier by accentuating or attenuating the bass or treble portion of the audio-frequency spectrum. See figure.

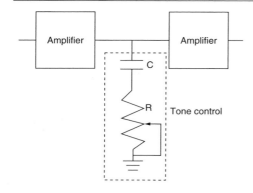

Tone control

tool space   space of a 6×1 vector representing the positions and orientations of the tool or end effector of the robot.

top-down development   an application development methodology that begins at a high level of abstraction and works through successively more detailed levels.

top antireflective coating (TAR)   a thin film coated on top of the photoresist used to reduce reflections from the air–resist interface and thus reduce swing curves.

top hat transform   a transform used in mathematical morphology. Let A be a structuring element centered about the origin, and for every pixel p, let A_p be its translate by p. The top hat transform measures the extent by which in a given gray-level image I the gray-levels of pixels in A_p are higher than those in the portion surrounding A_p. One way is to take the arithmetical difference $I - (I \circ A)$ between I and the opening $I \circ A$ of I by A; in a dual form, one takes the arithmetic difference $(I \bullet A) - I$ between the closing $I \bullet A$ of I by A and I. Another method considers a second structuring element B which forms a ring surrounding A, and at each pixel one computes the arithmetic difference between a "representitive" gray-level in A_p (either minimum, maximum, median, or average) and a "representitive" gray- level in B_p

(either minimum, maximum, median, or average). *See* closing, opening, structuring element.

top surface imaging   a resist imaging method whereby the chemical changes of exposure take place only in a very thin layer at the top of the resist.

topological map   an organization of nodes in which the similarity of any two nodes is a function of their distance from each other on the map. For example, with a 2-dimensional grid, Euclidean distance can be used as the distance between two nodes. Used in the self-organizing maps.

topology preserving skeleton   result of an an operation transforming a digital figure into a one-pixel wide skeleton having the same "topology," in other words whose connected components and holes correspond in a one-to-one way with those of the figure. This is generally achieved by homotopic thinning. The medial axis transform or distance skeleton preserves the topology of a figure (binary image) in the Euclidean case, but not in the digital case; the same defect arises with the morphological skeleton. *See* thinning/thickening.

tornadotron   a gyrotron with a simple helical beam.

toroidal deflection yoke   magnetic deflection yoke wound on a toroid core containing a split winding; one-half winding placed on opposite sides of the toroid. The yoke winding interconnection creates an opposing flux from each winding within the toroid and an aiding flux within the toroid inner diameter that causes the deflection of the electron beam. The toroid deflection yoke has a low inductance for compatibility with semiconductor deflection systems and permits precise control of the winding placement to obtain a uniform magnetic deflection field. A horizontal toroidal deflection yoke, in combination with the in-line electron gun, permits the construction of an inherent self-converging color TV display system.

torque the product of a force acting at a distance. The output of an electric motor.

torque angle the displacement angle between the rotor and rotating magnetic flux of the stator due to increases in shaft load in a synchronous machine. *See also* power angle.

torque pulsation oscillating torque produced by the interaction between the air gap flux, consisting mainly of the fundamental component, and the fluxes produced by harmonics in the rotor. Torque pulsations can stimulate complex mechanical vibrations that can flex and damage rotor and turbine elements.

torque ripple in variable speed motor drives, refers to the torque not being smooth as the rotor moves from one position to another. Torque ripple may be produced from space harmonics within the machine or time harmonics generated by the supply.

torque servo a servo where the output torque is the controlled variable and the operating speed depends on the load torque. *See also* servo.

torus a donut-shaped magnetic core used in electric transformers.

total demand distortion (TDD) an index that quantifies the amount of distortion in the voltage or current waveform with respect to the maximum 60-Hz demand load current I_{ML} over a 15–30 minute demand period.

$$\%TDDi = 100 \times \sqrt{\sum_{h \neq 1} \left(\frac{I_h}{I_{ML}} \right)^2}$$

total efficiency dimensionless ratio of the total RF power delivered to a load versus the total DC and RF incident power into the amplifier.

total harmonic distortion (THD) an index that quantifies the amount of distortion in the voltage or current waveform with respect to the fundamental component.

$$\%THDi = 100 \times \sqrt{\sum_{h \neq 1} \left(\frac{I_h}{I_1} \right)^2}$$

total harmonic distortion disturbance level an electromagnetic disturbance level due to all emissions from equipment in a system. This is expressed as a ratio of the RMS value of the harmonic content to the RMS of the fundamental and is calculated as a percentage of the fundamental component.

total internal reflection when light is incident on a boundary between two media from the one having the higher refractive index, n_1, then the angle of refraction is larger than the angle of incidence. At an angle of incidence times arc $\sin(n_2/n_1)$ the light is totally reflected and remains in the denser medium.

total reflection the phenomenon where a wave impinging on a certain medium interface is totally reflected without being damped by and without penetrating the boundary medium.

totem-pole output the standard output of transistor–transistor logic (TTL) gates consisting of two bipolar junction transistors in series between the source voltage and common. In normal operation, either one is driven into saturation while the other is at cutoff; alternating these conditions changes the output logic level.

touch input a means for selecting a location on the surface of the display unit using a variety of technologies that can respond to the placing of a finger or other pointing device on the surface. These are essentially data panels placed either on the display surface or between the user and the display surface.

touch screen a specialized type of video display where a control circuit is actuated when areas of the display are touched by ones finger or similar object. The types of touch screen technology include capacitive overlay, force vector, guided acoustic wave, resistive overlay, scanning infrared, strain gage and surface acoustic wave.

touch voltage   used in power system safety studies, the voltage between any two conductive surfaces which can be touched simultaneously by a person.

tower   a structure for elevating electric transmission lines, distinguished from a pole *cf* by its greater height and structural complexity.

TPBVP in optimal control   two-point boundary value problem resulting from necessary conditions for control optimality given by the Pontryagin maximum principle. For continuous-time control systems described by the state equation in the form

$$\dot{x} = f(x, u, t); x(t_0) = x_0$$

with performance index defined as

$$J = \int_{t_0}^{t_f} L(x, u, t)dt + q(x(t_f), t_f)$$

where x, u are respectively state and control vectors, $[t_0, t_f]$ is an optimization horizon and f, L, q are sufficiently smooth functions of appropriate dimensions, the TPBVP is given by the state equation and costate equation defined as follows:

$$\dot{p} = -\frac{\partial H(x, u, p, t)'}{\partial x}; p(t_f) = \frac{\partial q(x(t_f), t_f)'}{\partial x}$$

where $H(x, u, p, t) = L(x, u, t) + p'(t)f(x, u, t)$ is a Hamiltonian, p is a costate vector, u should be found by minimizing the Hamiltonian with x found from the state equation driven by the optimal u.

trace length   the physical distance between electronic components connected by a circuit path.

trace loading   the electronic load on a circuit path.

tracing   in software engineering, the process of capturing the stream of instructions, referred to as the trace, for later analysis.

track   a narrow annulus or ring-like region on a disk surface, scanned by the read/write head during one revolution of the spindle; the data bits of magnetic and optical disks are stored sequentially along these tracks. The disk is covered either with concentric rings of densely packed circular tracks or with one continuous, fine-pitched spiral track. *See also* magnetic disk track, optical disk track, magnetic tape track.

track buffer   a memory buffer embedded in the disk drive. It can hold the contents of the current disk track.

trackball   the earliest version of an input device using a roller ball, differing from the mouse in that the ball is contained in a unit that can remain in a fixed position while the ball is rotated. It is sometimes referred to as an upside-down mouse, but the reverse is more appropriate, as the trackball came first.

tracking   conduction along the surface of an insulator and especially the establishment of a carbonized conduction path along the surface of a polymer insulator.

tracking bandwidth   *See* lock range.

traffic channel   a channel in a communication network that is used to carry the main information or service, which is typically voice, data, video, etc. *Compare with* control channel.

traffic decomposition   the decomposition approximates the steady-state behavior of the traffic by "decomposing" it into long- and short-term behavior.

trailing edge   when a pulse waveform switches from high to low, it is called a trailing edge.

trailing-edge triggered   a device that is activated by the trailing edge of a signal.

trainability   the property of an algorithm or process by which it can be trained on sample data and thus rendered adaptable to different situations. *See* training procedure, supervised learning, unsupervised learning.

training algorithm of self-generating neural network   given as follows:

1. Initially, the network contains nothing. When the first training example comes in, the system creates a neuron for it.

2. When a new training example comes in, the neurons in the hierarchy are examined for similarity to the training example, hierarchically from the root(s). Each time, the last winner and all the neurons in its child networks are examined to find a new winner. If the new winner is the old winner, the examination stops and the old winner is the final winner; otherwise, the new winner and its child network are examined until a leaf node is found to be the winner. During this process, the weights of all the winners and their neighbors should be updated according to the same rule as that of self-organizing network. After the final winner is found, the network structure should be updated according to the following rules:

3. If the winner is a terminal node (i.e., it has no child network(s)), generate two new neurons, and copy the training vector and the original winner into them, respectively. Make them two neurons in a new child network of the updated winner.

If the winner is a nonterminal node (i.e., it has a child network), generate only one neuron, copy the training vector into it, and put it in the child network of the updated winner.

4. Repeat the above process until all the training examples are exhausted. One training epoch has been completed at this point, and more than one epoch may be required for the network of neural networks to reach an equilibrium. *See also* self-generating neural network.

training procedure   the method of calculating the set of free parameters of a function given a set of training data.

training sequence   a sequence of transmitted symbols known at the receiver, which is used to train an adaptive equalizer or echo canceller.

training set   set of data used as the basis for determining the best set of free parameters. Typically used in iterative training algorithms.

trajectory   a path on which a time law is specified for instance in terms of velocities and/or accelerations at each point.

trajectory planning   a trajectory planning algorithm is the path description, the path constraints imposed by manipulator dynamics as inputs and position, velocity, and accelerations of the joint (end-effector) trajectories as the outputs. *See also* path.

transceiver   a device that can serve as both transmitter and receiver.

transco   contraction of "transmission company," a firm which owns electric power transmission lines but does not engage in power generation or distribution.

transconductance   a quantity used to specify a field-effect transistor, defined as the differential change in drain current upon a differential change in gate voltage.

transducer   a device that converts a physical quantity into an electrical signal. Typically, transducers are electromechanical energy conversion devices used for measurement or control. Transducers generally operate under linear input-output conditions and with relatively small signals. Examples include microphones, pickups, and loudspeakers.

transducer gain ratio of the power delivered to the load to the power available from the source.

transducer power gain *See* transducer gain.

transfer characteristic curve one of a set of device curves derived from the expression for the output current as a function of the input voltage.

transfer function (1) a mathematical model that defines the relationship between the output and the input of a linear system. It is usually expressed as the ratio of two co-prime polynomials in the Laplace variable s and can be interpreted as a shorthand notation for a differential equation in which the initial conditions are assumed to be zero. A typical transfer function model has the form

$$g(s) = \frac{b_0 + b_1 s + b_2 s^2 + b_3 s^3 + \cdots + b_m s^m}{a_0 + a_1 s + a_2 s^2 + a_3 s^3 + \cdots + a_m s^m}$$

without deadtime, or

$$g(s) = \frac{b_0 + b_1 s + b_2 s^2 + b_3 s^3 + \cdots + b_m s^m}{a_0 + a_1 s + a_2 s^2 + a_3 s^3 + \cdots + a_m s^m} e^{-s\tau}$$

with deadtime. The numerator and denominator polynomials are co-prime and the coefficients are real constants. For causality, the order of the numerator polynomial must not exceed that of the denominator ($m \leq n$). For larger systems with multiple inputs and outputs, the concept is generalized to a transfer function matrix in which each (i, j) element is a simple transfer function that defines how output (i) is affected by input (j). Although not uniquely defined, it is possible to transform any transfer function or transfer function matrix to a state space model having the same input–output behavior. *See also* state space model. Equivalent pulse transfer function models exist for

digital (discrete-time) systems:

$$g(z) = \frac{b_0 + b_1 z + b_2 z^2 + b_3 z^3 + \cdots + b_m z^m}{a_0 + a_1 z + a_2 z^2 + a_3 z^3 + \cdots + a_m z^m}$$

where the polynomials are based on the z-transform. Though not as common, these can also be collected into matrices for MIMO systems to form pulse transfer function matrices. *See also* z-transform and modified z-transform.

(2) the Fourier transform of the output signal divided by the Fourier transform of the input signal. Alternatively, the Fourier transform of the impulse response function.

transfer matrix of linear stationary continuous time dynamical system, $K(s)$ represents controllable and observable part of dynamical system and is defined by the equality

$$K(s) = C(sI - A)^{-1} B + D$$

for systems with m inputs and q outputs $K(s)$ is $q \times m$ dimensional matrix.

transfer rate a measure of the number of bits that can be transferred between devices in a unit of time.

transfer time in a hierarchical memory system, the time required to move a block between two levels.

transform the process of converting data from one form into another. Often used to signify a system that rotates the coordinate axes. Examples of transforms include the Fourier transform and the discrete Fourier transform. A discrete linear transform can be described as a product of the input vector with a transform matrix. *See* transform kernel.

transform coding a method for source coding similar to subband coding. The input signal is transformed into an alternative representation, using an invertible transform (e.g., the Fourier transform), and the quantization is then performed in the transform domain. The method utilizes the fact that enhanced compression can then be obtained by focusing (only) on "important" transform parameters.

transform kernel   a function that is multiplied with an input function: the result of which is integrated or summed to form a transformed output. For example in the definition of the continuous Fourier transform the kernel is $e^{-j\omega t}$ and in the definition of the discrete Fourier transform the kernel is $e^{-j2\pi nm/N}$.

transform vector quantization   the generalization of scalar transform coding to vector coding. *See also* transform coding.

transform VQ   *See* transform vector quantization.

transform-based heirarchical coding   use of heirarchically compacted image energies to allow recognizable reconstruction with a relatively small amount of data.

transformation ratio   dimensionless ratio of the real parts of the load and source impedance.

transformed circuit   an original circuit with the currents, voltages, sources, and passive elements replaced by transformed equivalents.

transformer   a device that has two or more coils wound on an iron core. Transformers provide an efficient means of changing voltage and current levels, and make the bulk power transmission system practical. The transformer primary is the winding that accepts power, and the transformer secondary is the winding that delivers power. The primary to secondary voltages are related by the turns ratio of the coils. The corresponding currents are related inversely by the same ratio.

transformer differential relay   a differential relay specifically designed to protect transformers.

In particular, transformer differential relays must deal with current transformer turns ratio error and transformer inrush and excitation current.

transformer fuse   a fuse employed to isolate a transformer from the power system in the event of a transformer fault or heavy overload.

transformer vault   a fireproof enclosure in which power transformers containing oil must be mounted if used underground or indoors.

transient   (1) the behavior exhibited by a linear system that is operating in steady state in moving from one steady state to another. For stable systems, the transient will decay while for unstable system it will not, and thus the latter never reach another steady-state operation. *See also* settling time and time constant.

(2) any signal or condition that exists only for a short time.

(3) an electrical disturbance, usually on a power line.

(4) refers to momentary overvoltages or voltage reductions in an electric power system due to lightning, line switching, motor starting, and other temporary phenomena.

transient current   the fault current that flows during the transient period when the machine apparent impedance is the transient impedance.

transient fault   a fault that can appear (e.g., caused by electrical noise) and disappear within some short period of time.

transient impedance   the series impedance that a generator or motor exhibits following the subtransient period but prior to the steady-state situation.

transient open-circuit time constant   *See* quadrature-axis open-circuit time constant and direct-axis transient open-circuit time constant.

transient operation   a power system operating under abnormal conditions because of a disturbance.

transient reactance   the reactance offered for the transient currents in synchronous machines. Referred to by the symbol X'_s, the transient reactance is a function of the stator frequency and the transient inductance. X'_s is comparatively smaller in comparison to the steady-state inductive reactance of the machine.

transient short-circuit time constant   *See* quadrature-axis short-circuit time constant and direct-axis transient short-circuit time constant.

transient stability   the ability of a power system to remain stable following a system disturbance.

transient suppressor   a device connected to a piece of sensitive electrical equipment to reduce the amplitude of transient voltage excursions, thus protecting the equipment.

transistor–transistor logic (TTL)   a transistor technology in which the output of a logic gate is amplified in going from 0 to 1 as well as from 1 to 0.

transit time   the average time in seconds required for an electron to move between two specified surfaces.

transition band   the portion of the frequency spectrum where a filter changes from a stop filter to a pass filter or vice versa. The steepness of the transition band is often a measure of the quality of a filter, where the pass and stop bands are critical. A longer filter length generally implies that the filter can have a steeper transition band.

transition lifetime   coefficient representing the time after which a population of atoms in an excited state may be expected to fall to $1/e$ of its initial value due to stimulated and spontaneous processes as well as inelastic collisions. *See also* spontaneous lifetime, transition rate.

transition matrix of 2-D Fornasini–Marchesini model   for

$$x_{i+1,j+1} = A_1 x_{i+1,j} + A_2 x_{i,j+1} + B_1 u_{i+1,j} + B_2 u_{i,j+1}$$

$i, j \in Z_+$ (the set of nonnegative integers) is defined as follows:

$$T_{ij} = \begin{cases} I_n \text{ for } i = j = 0 \\ A_1 T_{i,j-1} + A_2 T_{i-1,j} \\ \quad \text{for } i, j \in Z_+ \ (i + j \neq 0) \\ 0 \text{ for } i < 0 \quad \text{or/and} \quad j < 0 \end{cases}$$

where $x_{ij} \in R^n$ is the local state vector, $u_{ij} \in R^m$ is the input vector, A_k, B_k $(k = 1, 2)$ are real matrices.

transition matrix of 2-D general model   for

$$x_{i+1,j+1} = A_0 x_{ij} + A_1 x_{i+1,j} + A_2 x_{i,j+1} + B_0 u_{ij} + B_1 u_{i+1,j} + B_2 u_{i,j+1}$$

$i, j \in Z_+$ (the set of nonnegative integers) is defined as follows:

$$T_{ij} = \begin{cases} I \text{ for } i = j = 0 \\ A_0 T_{i-1,j-1} + A_1 T_{i,j-1} + A_2 T_{i-1,j} \\ \quad \text{for } i, j \in Z_+ \ (i + j \neq 0) \\ 0 \text{ for } i < 0 \quad \text{or/and} \quad j < 0 \end{cases}$$

The transition matrix of the first 2-D Fornasini–Marchesini model is defined in the same way.

transition matrix of 2-D Roesser model   denoted T_{ij},

$$\begin{bmatrix} x^h_{i+1,j} \\ x^v_{i,j+1} \end{bmatrix} = \begin{bmatrix} A_1 & A_2 \\ A_3 & A_4 \end{bmatrix} \begin{bmatrix} x^h_{ij} \\ x^v_{ij} \end{bmatrix} + \begin{bmatrix} B_1 \\ B_2 \end{bmatrix} u_{ij}$$

$i, j \in Z_+$ (the set of nonnegative integers) is defined as follows:

$$T_{ij} = \begin{cases} I_n \text{ for } i = j = 0 \\[6pt] \begin{bmatrix} A_1 & A_2 \\ 0 & 0 \end{bmatrix} \text{ for } i = 1, \ j = 0 \\[12pt] \begin{bmatrix} 0 & 0 \\ A_3 & A_4 \end{bmatrix} \text{ for } i = 0, \ j = 1 \quad \text{and} \\[12pt] T_{10}T_{i-1,j} + T_{01}T_{i,j-1} \\ \text{ for } i, j \in Z_+ \ (i + j \neq 0) \\[6pt] 0 \text{ for } i < 0 \quad \text{or/and} \quad j < 0 \end{cases}$$

where $x_{ij}^h \in R^{n_1}$ and $x_{ij}^v \in R^{n_2}$ are the horizontal and vertical state vectors, $u_{ij} \in R^m$ is the input vector, $A_1, A_2, A_3, A_4, B_1, B_2$ are real matrices.

transition rate   rate at which atoms undergo transitions from one level to another due to stimulated and spontaneous processes as well as inelastic collisions; reciprocal of transition lifetime.

transition region   the region of the I-V curve(s) of a device between the ohmic region and the current source region, in which the slope of the I-V curve(s) is rapidly changing as it transitions from the resistance region to the current source region.

translation   a geometric transformation which simply adds an offset to the pixel coordinates of an image.

translation lookaside buffer (TLB)   essentially a small fully associative address-cache used to provide fast address translation for the most used virtual addresses. The TLB is associatively searched on a virtual address, and in the event of a hit, it returns the corresponding real address. In the event of a miss, if the addressed page is in main memory, then a TLB entry is made for it; otherwise the page is first brought in after a page fault and then the TLB entry is made. In either case, the TLB eventually returns a real address. The TLB may be fully associative, set associative, or hashed.

translator   an unattended television or FM broadcast repeater that receives a distant signal and retransmits the picture and/or audio locally on another channel.

transmission   (1) the act of sending information from one location to another.

(2) transformation of an optical wave incident on a surface that passes a portion of the wave to the medium behind the surface.

(3) that class of electric power system work which is concerned with the transport of electric power from the generator to the area of consumption. The circuits of interest typically extend at the generating station and terminate at the local substation.

transmission coefficient   a number that describes the relative amplitude and phase of the transmitted wave with respect to the incident wave. The term is usually used in the context of wave transmission at a material interface or transmission line.

transmission grating   a diffraction grating that operates in transmission, i.e., the diffracted light is obtained by shining light through the grating.

transmission line   (1) an arrangement of two or more conductors used to convey electromagnetic energy from one point to another.

(2) conductive connections that guide signal power between circuit elements.

transmission line coupler   passive coupler composed of two or more transmissions spaced closely together where the proximity of the transmission lines allows signals to be coupled or transferred in part from one line to the other. The electrical length of the transmission lines is usually one quarter of a wavelength.

transmission line filter   a microwave device that is made up of sections of transmission lines so as to act as a filter in the microwave frequency range.

transmission line measurement (TLM) an experimental technique to measure the specific contact resistance of a metal ohmic contact on a semiconductor with a set of variable spaced transmission lines.

transmission line parameters parameters that describe the electrical response of a transmission line. These consists of one describing the characteristic impedance (ζ) and another describing the complex propagation constant (γ). The complex propagation constant is sometimes defined independently in terms of two separate parameters, one defining the real part (α) and a second one defining the imaginary part (β) of the complex constant.

transmission line resonator resonator formed by a transmission line.

transmission matrix for a microwave network, a matrix that gives the output quantities in terms of the input quantities, when the input/output quantities are linearly dependent. When voltages and currents are chosen, it is called voltage–current transmission matrix. If incident and reflected waves are chosen, it is called wave amplitude transmission matrix.

transmission system a transmission system transfers the motion from the actuator to the joint.

transmissivity a property that describes the transmitted energy as a function of the incident energy of an EM wave and a material body. The property may be quantified in terms of the magnitude of the transmission coefficient or the ratio of the incident to the transmitted field.

transmit/receive (T/R) switch a single-pole double-throw (SPDT) switch, connected to the antenna feed. It is used to prevent destruction of the receiver from the transmit RF power.

transmittance ratio of the complex amplitude of a transmitted wave to the complex amplitude of the corresponding incident wave at a transmitting surface.

transmitter equipment used to generate an RF carrier signal, modulate this signal and radiate it into space.

transparent code a code in which the complement of every codeword also is a codeword.

transparent mode a mode of a bistable device where an output responds to data input signal changes.

transposition (1) the practice of twisting a three-phase power line so that, for example, phase A takes the place on the tower formerly occupied by phase B, phase B takes the place of phase C, and phase C occupies the former position of phase A.

(2) a point on a three-phase electric power line where the conductors are physically transposed for purposes of improving circuit balance

transputer a class of CPU designed and manufactured by Inmos Corporation. The transputer was specifically designed to be used in arrays for parallel processing.

transverse electric (TE) referring to fields or waves in which the electric field has nonzero vector components only in the plane perpendicular (transverse) to a specified axis, usually a coordinate axis.

transverse electric mode mode having no longitudinal component of the electric field (no component in the direction of propagation). Generally referred to as TE mode.

transverse electric polarization polarization state of a transverse electric (TE) mode. Also called TE polarization.

transverse electric wave electromagnetic waves polarized so that the electric field intensity vector is perpendicular to the direction of propagation. The wave solutions have zero electric field component in the direction of propagation. Also known as TE-wave or H-mode.

transverse electromagnetic (TEM) referring to fields or waves in which both the electric and magnetic fields have nonzero vector components only in the plane perpendicular (transverse) to a specified axis, usually a coordinate axis. In a TEM wave, the electric and magnetic fields are perpendicular to each other and to the direction of propagation.

transverse electromagnetic mode electromagnetic wave propagation mode in which electric and magnetic fields are transverse to the direction of propagation (i.e., no radial field components). TEM mode propagation is a characteristic of antenna radiation in the far-field and of transmission-line propagation below the cutoff frequency of the higher order modes.

transverse electromagnetic wave the electric and magnetic field components along in the direction of propagation are zero. Abbreviated TEM wave.

transverse excitation laser pumping process in which the pump power is introduced into the amplifying medium in a direction perpendicular to the direction of propagation of the resulting laser radiation.

transverse magnetic (TM) referring to fields or waves in which the magnetic field has nonzero vector components only in the plane perpendicular (transverse) to a specified axis, usually a coordinate axis.

transverse magnetic wave the wave solutions with zero magnetic field component in the direction of propagation. Also known as TM wave and E-modes.

transverse mode term used in referring to the transverse structure or indices of the mode of a laser oscillator.

transverse mode-locking forcing the transverse modes of a laser to be equally spaced in frequency and have a fixed phase relationship; useful for obtaining a scanning output beam oscillator.

transverse resonance a technique used in order to find the modes of closed waveguides.

trap (1) in microelectronics, an imperfection in a semiconducting material that can capture a free electron or hole.

(2) in computers, a machine operation consisting of a hardware-generated interrupt or subroutine call that is invoked in the case of some error condition, for example, encountering an unimplemented instruction code in the instruction stream. *See also* exception.

TRAPATT diode acronym for trapped plasma avalanche transit time, a microwave diode that uses a high field generated electron-hole plasma and the resulting diffusion of these carriers to the contacts to create a microwave negative resistance, used as high power, high efficiency RF power sources.

trapezoidal pattern a signal produced on an oscilloscope by applying an amplitude modulated signal to the horizontal input and the modulating signal to the vertical input. By measuring the maximum and minimum height of the resulting trapezoid, the modulation index may be obtained.

trapped wave *See* bound mode.

traveling wave an electromagnetic signal that propagates energy through space or a dielectric material.

traveling wave amplifier the principle of traveling-wave amplification is a technique for increasing the gain-bandwidth product of an amplifier. The input and output capacitances of discrete transistors are combined with lumped inductors and form artificial transmission lines, coupled by the transductances of the devices. The amplifier can be designed to give a flat, lowpass response up to very high frequencies. Sometimes called a traveling wave tube amplifier (TWTA).

traveling wave excitation a method of pumping in which the excitation travels along the lasing axis at or near the velocity of light in order to pump short-lived transitions just prior to stimulation.

traveling wave tube amplifier (TWTA) *See* traveling wave amplifier.

tree (1) a connected subgraph of a given connected graph G which contains all the vertices of G but no circuits.

(2) a form of microscopic cracking which forms in flexible cable insulation. It is typically a precursor of insulation failure. *See* water tree, electrical tree.

tree code a code produced by a coder that has memory.

tree coding old name for *convolutional coding*.

tree network limited connection of subscriber nodes to a central control or distribution unit via other subscriber nodes in the network.

tree structure robots a set of rigid bodies connected by joints forming a topological tree is called a tree structure robot.

tree structured vector quantization scheme to reduce the search processes for finding the minimum distortion codevector using a tree structured codebook, where each node has m branches and there are $p = \log_m N_c$ levels of the tree. Abbreviated tree structured VQ.

tree structured VQ *See* tree structured vector quantization.

tree wire a thinly-insulated conductor used in distribution work along tree-lined streets. The insulation is suffficient to withstand a brush from a tree branch.

tree-search algorithms for searching through a tree-structured problem based on a certain cost function, or metric, increment associated with each branch of the tree.

trellis code a (channel) coding scheme in which the relation between information symbols and coded symbols is determined by a finite-state machine. The current block of information symbols and the state (in the finite-state machine) uniquely determine the block of coded output symbols as well as the next state. Thus, trellis coding can be viewed as generalized block coding for which the encoder function depends on the current as well as previous blocks of non-overlapping information symbols. In the class of trellis codes we, for example, find convolutional codes, trellis-coded modulation and continuous-phase modulation. *See also* convolutional code.

trellis coded modulation a forward error control technique in which redundancy is introduced into the source stream through an increase in number of symbol values rather than an increase in the number of symbols. Developed by G. Ungerboeck in the late 1970s, this approach has found widespread use in systems with limited bandwidth.

trellis coding *See* trellis code.

trellis diagram in convolutional codes and trellis coded modulation, a graphical depiction of all valid encoded symbol sequences, and the basis for the Viterbi decoding algorithm.

trellis search an algorithm for searching through a trellis-structured problem based on a certain cost function, or metric, increment associated with each branch of the trellis.

trellis vector quantization a method for structured vector quantization, where the input signal is classified and coded in a manner described by a mathematical structure known as a "trellis." Abbreviated trellis VQ.

trellis VQ   *See* trellis vector quantization.

trench isolation   condition in which parasitic MOSFETs are formed between transistors sharing a common substrate when polysilicon or metal layers run between the two.

tri-state circuit   (1) a logic circuit that can assume three output states, corresponding to ZERO, ONE, and OFF. Tri-state circuits are used to place signals on a bus where only one signal source is allowed to be active at a time. Using this type of device, there may be several sources that have the ability to send signals to the same receiver, at different times.

(2) a circuit branch that is in a high impedance state and effectively is aloted from the circuit.

tri-state logic   *See* tri-state circuit.

triac   a power switch that is functionally a pair of converter-grade thyristors connected in anti-parallel. Triacs are mainly used in phase control applications such as dimmer switches for lighting. Because of the integration, the triac has poor reapplied dv/dt, poor gate current sensitivity at turn-on, and longer turn-on time. They are primarily used for AC power control with resistive loads, such as in light dimmers.

triangle function   a binary operation τ on distance distribution functions that is commutative, associative, and nondecreasing in each place, i.e., satisfying the following relationships:

$$\tau(F, G) = \tau(G, F);$$

$$\tau(F, G) \leq \tau(H, K)$$

$$\Leftarrow F \leq H, G \leq K;$$

$$\tau(\tau(F, G), H) = \tau(F, \tau(G, H)); \tau(F, \mathbf{1}) = F;$$

for any distance distribution functions F, G, H, K and the unit step function $\mathbf{1}$. The triangle functions defines a triangle inequality in the given probabilistic metric space. The most important triangle functions are defined by convolutions or are induced by triangular norms.

triangular co-norm   denoted $\dot{+}$, is a two place function from $[0, 1] \times [0, 1]$ to $[0, 1]$. The most used triangular co-norm is the union $\vee$ defined as

$$\mu_{A \vee B} = \max\{\mu_A(x), \mu_B(x)\},$$

where A and B are two fuzzy sets in the universe of discourse X.

The triangular co-norm is used as disjunction in the approximate reasoning, and reduces to the classical OR, when applied to two crisp (non-fuzzy) sets.

See also approximate reasoning, fuzzy set.

triangular norm   (1) in fuzzy systems, denoted $\star$, a two place function from $[0, 1] \times [0, 1]$ to $[0, 1]$. The most used triangular norms are the intersection $\wedge$ and the algebric product $\cdot$ defined as

$$\mu_{A \wedge B} = \min\{\mu_A(x), \mu_B(x)\}, \quad \text{and}$$

$$\mu_{A \cdot B} = \mu_A(x)\mu_B(x),$$

where A and B are two fuzzy sets in the universe of discourse X.

The triangular norm is used as conjunction in the approximate reasoning, and reduces to the classical AND. when applied to two crisp (non-fuzzy) sets.

See also approximate reasoning, fuzzy set.

(2) in control, a function T from the closed unit square into the closed unit interval endowed with the following properties:

$$T(a, b) = T(b, a); T(a, b) \leq T(c, d)$$
$$\Leftarrow a \leq c, b \leq d;$$
$$T(a, 1) = a; T(T(a, b), c) = T(a, T(b, c));$$

for all $a, b, c, d \in [0, 1]$. The most important triangular norms (*t*-norms) are defined by product of two numbers, operation of taking the smaller one of two numbers and a family of T_s given by

$$T_s(a, b) = \log_s \left(1 + \frac{(s^a - 1)(s^b - 1)}{s - 1} \right)$$

t-norms are used to define a triangle inequality in the family of probabilistic metric spaces called statistical metric spaces, which may be considered as uncertain counterparts of metric spaces. Different triangular norms are used as composition operators in fuzzy systems, although the standard operation is defined by minimum.

triangular window *See* Bartlett window.

trigger in oscilloscopes and logic analyzers, the "trigger" signal is used to notify the system that a certain event has occurred, and data acquistion should commence.

trim condition *See* set point.

trinocular vision a vision model in which points in a scene are projected onto three different image planes.

triode region *See* ohmic region.

trip coil a solenoid in a circuit breaker that initiates breaker opening when energized.

triple transit echo (TTE) a multiple transit echo received at three times the main SAW signal delay time. The echo is caused due to the bidirectional nature of SAW transducers and the electrical and/or acoustic mismatch at the respective ports. This is a primary delay signal distortion which can cause filter distortion, especially in bidirectional transducers and filters.

triplen harmonics the frequency components which have a frequency of multiple of three times the frequency of the fundamental. These voltages are in phase in all three windings of a three-phase transformer and peak simultaneously. Delta connection on the other side provides a closed path for the flow of triple harmonic currents.

triplex cable a cable used for residential or commercial service drops consisting of two or three insulated conductors spiralled around a bare neutral wire which provides support for the cable.

tristimulus value one value in tristimulus color theory. Tristimulus color theory stems from the hypothesis that the human eye has three types of color receptors (cones) that have peak sensitivity in the red, green and blue visible light wavelengths respectively. The tristimulus color values are a set of three values X, Y, and Z which replace the red, green, and blue intensities with a weighted integral which calculates a spectral energy over the range of visible wavelengths of light for each value; the integrals allow for colors to be represented purely additively, while representing colors via red, green and blue intensities often requires a subtractive interaction between the "primaries."

troposphere the region of the atmosphere within about 10 km of the Earth's surface. Within this region, the wireless radio channel is modified relative to free space conditions by weather conditions, pollution, dust particles, etc. These inhomogeneities act as point scatterers, which deflect radio waves downward to reach the receiving antennas, thereby providing tropospheric scatter propagation.

trouble ticket a complaint made by a customer regarding service. Its origins trace back to a paper work request which was given to the service representative.

true complement in a system that uses binary (base 2) data negative numbers, the true complement can be represented as 1's complement of the positive number. This is also called a radix complement.

true concurrency   *See* concurrency. *Compare with* apparent concurrency.

true data dependency   the situation between two sequential instructions in a program when the first instruction produces a result that is used as an input operand by the second instruction. To obtain the desired result, the second instruction must not read the location that will hold the result until the first has written its results to the location. Also called a read-after-write dependency and a flow dependency.

true efficiency   *See* total efficiency.

truncation   for a function $x(t)$, truncation is another function, usually denoted $x_T(t)$, defined as follows:

$$x_T(t) = \begin{cases} x(t) & t \leq T \\ 0 & t > T \end{cases}$$

See also extended spaces.

truncation error   when numerically solving the differential equations associated with electrical circuits, approximation techniques are used. The errors associated with the use of these methods are termed truncation error. Controlling the local and global truncation errors is an important part of a circuit simulator's task. Limits on these errors are often given by the user of the program.

trunk   a communication line between two switching nodes.

truth model   a very detailed mathematical description of a process to be controlled. The truth model is also called the *simulation model*, since it is used in simulation studies of the process. *See also* design model.

truth table   a listing of the relationship of a circuit's output that is produced for various combinations of logic levels at the inputs.

TS fuzzy model   *See* Takagi–Sugeno–Kang fuzzy model.

TSK fuzzy model   *See* Takagi–Sugeno–Kang fuzzy model.

TTE   *See* triple transit echo.

TTL   *See* transistor–transistor logic.

tube leak   a crippling mishap in a steam power plant. High-pressure steam leaks from a cracked boiler tube with sufficient energy to cut adjacent tubes.

tunable laser   laser in which the oscillation frequency can be tuned over a wide range.

tuned-circuit oscillators   *See* LC-oscillators.

tuner   a circuit or device that may be set to select one signal from a number of signals in a frequency band.

tuning elements   generally lossless elements (probes, screw, etc.) of adjustable penetration extending through the wall of the waveguide or cavity resonator. By changing their position the reflection coefficient is adjusting. *See also* matching elements.

tunnel diode   a PN diode structure that uses band to band tunneling to produce a terminal negative differential resistance, also called an Esaki diode after its inventor L. Esaki.

tunneling   a physical phenomenon whereby an electron can move instantly through a thin dielectric.

tunneling modes   in an optical fiber, modes that are intermediate in attenuation between propagating modes and cladding modes and may be sustained between 10's to 100's of meters.

turbidity   inverse of the length over which the energy of the light transmitted in the forward direction decays to e^{-1} times its incidence value.

turbo code   the parallel concatenated convolutional coding technique introduced by Berrou, Glavieux, and Thitimajshima in 1993. These codes achieve astonishing performance through parallel encoding of the source symbol sequence and iterative serial decoding of the demodulated symbol sequence.

turbogenerator   a generator driven by a steam-turbine engine.

turn-off snubber   an auxiliary circuit or circuit element (consisting of a resistor and capacitor) used in power electronic systems to reduce the dv/dt during turn off.

turn-on snubber   an auxiliary circuit or circuit element (usually an inductor) used in power electronic systems to reduce the rate of rise or fall of the turn-on or turn off current to protect the power electronic device.

turns ratio   the ratio of the number of turns between two coupled windings, e.g., for a transformer, it is the ratio of number of turns of the primary winding to the number of turns of the secondary windings. For an induction machine, it is the ratio of the number of turns of the stator winding to the number of turns of the rotor winding.

twelve-pulse converter   the combination of two 6-pulse converters connected through a Y-Y and a delta-Y transformer in order to cancel the characteristic 5th and 7th harmonics of the 6-pulse converters. The lowest characteristic harmonics with twelve-pulse converters under balanced conditions are the 11th and 13th harmonics. The converters are connected in parallel on the AC side and in either series or parallel on the DC side, depending on the required DC output voltage.

twenty-six connected   *See* voxel adjacency.

twin-T bridge   represents a parallel connection of two T-shape two-ports. Each such two-port includes three impedances: two impedances are connected in series between input and output of the two-port, and the third impedance is connected between ground and the common point of two series impedances. The most important in applications is a twin-T bridge where one two-port is a series connection of two resistors and a capacitor is connected between their common point and ground; the other two-port is a series connection of two capacitors and a resistor is connected between their common point and ground. This circuit is used as a passive rejection filter, and as a feedback circuit of some active filters and oscillators.

twin-T bridge oscillator   an oscillator where the twin-T bridge is used as a feedback circuit of an amplifier. The twin-T bridge has a very steep phase-frequency response in the vicinity of the bridge rejection frequency, and, with proper design, this may provide high indirect frequency stability. The most problem-free design requires that the bridge transfer function has complex-conjugate zeros slightly shifted from $j\omega$-axis into the right half of s-plane. The amplifier should have a negative gain compensating the bridge losses at the oscillation frequency, which is close to the bridge rejection frequency.

twisted nematic   the alignments of the nematic planes are rotated through 90 degrees across the crystal by constraining alignments to be orthogonal at the boundaries.

twisted nematic liquid crystal   layered electro-optic material that can be switched between an electric-field-aligned state and the natural state with progressive rotation of polarization direction between layers, used in thicknesses to rotate light polarization by 90 degrees between polarizers, to produce light modulation.

two-address computer   one of a class of computers using two or fewer address instructions.

two-address instruction   a class of assembly language ALU instruction in which the two operands are located in memory by their memory addresses. One of the two addresses is also used to store the result of the ALU operation.

two-and-a-half-D sketch   a representation of the input image which is augmented at every position by information relating to 3-D structure and which is deemed to constitute a significant step on the way to human image interpretation. The name arises as the basic representation is still 2-D, whereas it is tagged with all available 3-D information: it is important in forming a bridge between early (i.e. low level) visual processes and high level vision.

two-antenna gain measurement method
technique based on Friis transmission formula in which the gain of each of two assumed to be identical antennas is calculated from a measurement of the insertion loss between the two antennas and the calculated propagation loss.

two-band filter bank   *See* two-channel filter bank.

two-beam coupling   any of several nonlinear optical processes involving two optical beams in which energy is transferred from one beam to the other.

two-channel coding   a coding scheme in which a signal is decomposed into two parts: low frequency and high frequency components. The low frequency component is undersampled and the high frequency component is coarsely quantized, thus saving data. It can be viewed as a special example of subband coding.

two-channel filter bank   a filter bank that has one high frequency band and one low frequency band in both analysis and synthesis filters.

two-degree-of-freedom system   a linear robust controller can consist of two independent parts. The feedback part's transfer function G_{fb} is typically chosen so that disturbances acting on the process are attenuated and the closed-loop system is insensitive to process variations. The feedforward part's transfer function G_{ff} is then chosen to the desired response to command signals. Such a system is called a two-degree-of-freedom system because the controller has two transfer functions that can be chosen independently.

two-dimensional acousto-optic processor   an acousto-optic signal processing system typically utilizing the two orthogonal dimensions (e.g., x and y) of Cartesian space to implement space and/or time based integration.

two-dimensional correlator   a correlator where both input signals are two-dimensional, such as images.

two-dimensional Fourier transform   an operation performed optically by a lens on an image placed at the front focal plane of the lens; the complex Fourier transform output is represented by the light at the back focal plane.

two-dimensional joint transform correlator   a type of optical correlator that employs two parallel paths, one for the input image and one for the reference image, instead of an in-line cascade. *See also* joint transform correlator.

two-dimensional memory organization   memory organization in which the arrangement on a single chip reflects the logical arrangement of memory. In the most straightforward case, each address is presented at once in its entirety and decoded by a single decoder. However, to reduce the number of pins required for addressing, the address may be split into two parts that are then sent in sequence on the same lines. Then during the row access strobe, one part is used to select a "row" of the memory, and during the column access strobe the other part is used to select a "column" of the selected "row." The "row" output may be held in a buffer and the "column" access then applied to the contents of the buffer.

In a two-and-a-half dimensional organization, the bits of each word are spread across several chips — one bit per chip in the most extreme case. Each chip is then equipped with two decoders, each of which decodes part of a split address in order to carry out a selection on the chip.

two-lamp synchronizing the process to connect two three-phase power systems in parallel using the same procedure as for three-lamp synchronizing except that lamps are placed across only two phases of the switch. *See also* three-lamp synchronizing.

two-pass assembler an assembler program that makes two passes through the source code to produce its output. The first pass determines all the referenced addresses and the second pass produces the assembled code. A two-pass assembler can produce directly loadable object code because all the label values are determined in the first pass.

two-phase clock having two separate clock signals, one high while the other is low, and vice-versa.

two-photon absorption a nonlinear optical process in which two photons are removed simultaneously from a laser beam as an atom makes a transition from its ground to its excited state. The rate at which such events occur is proportional to the square of the intensity of the laser beam.

two-port a network with four accessible terminals grouped in pairs, for example, input pair, output pair.

two-port memory a memory system that has two access paths; one path is usually used by the CPU and the other by I/O devices. This is also called dual port memory.

two-quadrant a drive that can operate as a motor as well as a generator in one direction.

two-quadrant operation operation of a motor with a controller that can provide power to run the motor and absorb power from the motor during deceleration (regenerative braking). Two quadrant operation provides improved efficiency if the motor is started and stopped frequently.

two-scale relation in general, a linear combination of the scaling functions of a wavelet. It shows clearly how the function behaves at different resolutions. A general two-scale relation could be expressed as

$$\psi\left(\frac{t}{2}\right) = \sum_{k=-\infty}^{+\infty} h_k \psi(t-k)$$

where ψ is the scaling function and the h are the coefficients that define the wavelet.

two-terminal a network with two accessible terminals.

two-wave mixing a nonlinear optical process in which two beams of coherent light interact inside nonlinear media or photorefractive crystals. When two beams of coherent electromagnetic radiation intersect inside a nonlinear medium of a photorefractive crystal, the periodic variation of the intensity due to interference will induce a volume refractive index grating. The presence of such a refractive index grating will then affect the propagation of these two beams. This may lead to energy coupling. The coupling of the two beams in nonlinear media is referred to as two-wave mixing.

two-way channel two terminal channel in which both terminals simultaneously transmit and receive using the same channel, thus disturbing each other's transmissions.

two-way interleaved in memory technology, a technique that provides faster access to memory values by interleaving memory values in two separate modules.

TWTA   traveling-wave tube amplifier. *See* traveling wave amplifier.

type-N connector   named after P. Neill of Bell Labs, this coaxial connector has both a male and female version. The outer diameter of the female connector is approximately 5/8 inch with an upper frequency range of about 18 GHz.

typical sequence   for a given probability mass function $p(x)$, a particular sequence of length n chosen i.i.d. according to $p(x)$ is typical if its empirical distribution is deemed close to the true distribution. This notion is also generalized to include the comparison of functions of the true and empirical distributions. For example, entropy–typical sequences are used in proving coding theorems in information theory.

U

UART *See* universal asynchronous receiver/transmitter.

ubitron a millimeter wave high-power quasi-quantum generator with relativistically high-speed electron beam. Millimeter waves are generated due to quantum transition between two energy states of electrons and amplified due to the velocity modulation and kinetic energy transfer principles.

UCA *See* uniform circular array.

UDT *See* unidirectional transducer.

UEESLA *See* uniformly excited equally spaced linear array.

UEP code *See* unequal error protection code.

Ufer ground a term, named for engineer Herb Ufer, used to describe concrete-encased earth electrodes (e.g., rebars in a building's foundation footers).

UHF *See* ultra-high frequency.

UHF power in television, the band of frequencies ranging from 300MHz to 3 GHz.

UL *See* Underwriters Laboratory.

ULA *See* uniform linear array.

UL classes a classification system established by Underwriters Laboratory (UL) for the purpose of defining certain operating characteristics of low voltage fuses. UL classes include G, J, L, CC, T, K, R, and H.

ULSI *See* ultra-large-scale integration.

ultimate boundedness the property of the solutions of a system equation, guaranteeing that for "small" perturbations in the equation, the solution will eventually be "small" in time.

A solution $x : [t_0, \infty] \rightarrow R^n; x(t_0) = x_0$ is said to be uniformly ultimately bounded with respect to a given set S if there exists a nonnegative constant time $T(x_0, S)$, ∞ independent of t_0 such that $x(t) \in S$ for all $t \geq t_0 + T(x_0, S)$. In other words, ultimate uniform boundedness guarantees that the state of the system enters and remains in the given neighborhood of the origin after a finite time interval.

ultra-high frequency (UHF) electromagnetic spectrum with frequencies between 300 MHz and 3000 MHz or wavelengths between 10 cm and 100 cm. Also called as decimetric waves.

ultra-large-scale integration (ULSI) an integrated circuit made of millions of transistors.

ultrasonic memory obsolete form of memory, in which data was stored as ultrasonic sound recirculating through a column of mercury. Also called mercury delay-line memory.

ultrasound an imaging modality that uses reflected high-frequency sound energy to image the interface between materials with different acoustic impedances.

ultraviolet a term referring to wavelengths shorter than 400 nm, but longer than 30 nm. The region 400–300 nm is the near ultraviolet, 300–200 is the middle ultraviolet; and 200–30 nm is the far ultraviolet or vacuum.

ultraviolet laser laser producing its output in the ultraviolet region of the spectrum.

umbra transform a morphological tranform used for visualization of operations on gray-level images Let the gray-level image F be a function

$E \rightarrow T$, where E is the set of points and T the set of gray-levels. The umbra of F is the set $U(F) = \{(h, v) \in E \times T | v \leq F(h) \text{ and } v \neq \pm\infty\}$. The behavior of a morphological operator on gray-level functions can be visualized by applying the corresponding operator for sets to the umbras of the gray-level functions. *See* morphological operator.

umbrella cell   a cell that covers the same geographical area as a number of micro- or picocells, and is aimed at supplying seamless service to subscribers with high mobility in these areas.

unary operation   an operation a computer performs that involves only one data element. The complement and increment operations are examples of such an operation while ADD is an example of an operation that requires two data elements.

unbalanced line   refers to a signal carrying line where one of the conductors is connected to ground. *Contrast with* balanced line.

unbalanced magnetic pull   a phenomenon in electric machines arising from the rotor not being symmetrical with respect to the stator or the axis of the rotor and stator not being coincident. Results in a higher pulling force on the side with the smaller airgap, resulting in additional bearing stresses.

unbalanced operation   in an n-phase system ($n > 1$), a condition in which the phase voltages (currents) are either

(1) not equal-amplitude sinusoids or

(2) have phase angles displaced by a value other than that specified for balanced operation.

The term "unbalanced" is also used to describe a machine that has unsymmetrical phase windings. *See also* balanced operation.

unbiased estimate   an estimate $\hat{x}$ of x which is not subject to any systematic bias; that is,

$$E[\hat{x} - x = 0]$$

See expectation, bias.

unbiased estimator   an estimator whose mean value is equal to the true parameter value.

unbundling   a feature of utility de-regulation in which services which were formerly bundled together are sold separately to the customer.

uncertain dynamical system model   a mathematical model of a dynamical system that includes the system's uncertainties or disturbances. A possible tool to model an uncertain system is to use a *differential inclusion*,

$$\dot{\mathbf{x}} \in \mathbf{F}(t, \mathbf{x}).$$

Another example of an uncertain dynamical system model is

$$\dot{\mathbf{x}} = \mathbf{A}\mathbf{x} + \mathbf{B}\mathbf{u} + \mathbf{h}(t, \mathbf{x}, \mathbf{u}),$$

where the vector function $\mathbf{h}$ models the system's uncertainties. *See also* matching condition and unmatched uncertainty.

unconditional branch   an instruction that causes a transfer of control to another address without regard to the state of any condition flags.

uncontrolled rectifier   a rectifier circuit employing switches that do not require control signals to operate them in their "on" or "off" states.

underexcited   a condition of operating a synchronous machine, in which the current to the DC field winding is insufficient to establish the required magnetic flux in the airgap. As a result, the machine requires reactive power from the AC system. An underexcited synchronous motor operates at a lagging power factor, as it appears as an inductive load to the AC system. An underexcited synchronous generator operates with a leading power factor, since it must deliver power to a leading (capacitive) system.

underflow a condition in a floating-point system where the result of an operation is nonzero yet too small in absolute value to be properly represented in the system.

underfrequency relay a protection device that curtails loads in an area that is deficient in generation. Lower generation compared to load demands give rise to lower frequency and a frequency threshold can be used by the relay to initiate load shedding in order to balance generation and demand.

underground distribution a class of electric power distribution work, typically used in densely-populated urban business districts, in which conductors are carried in conduits under streets between manholes and submersible distribution transformers are mounted in underground vaults.

underground residential distribution practices involved in the underground distribution of electric power to residential subdivisions through direct-buried cables and pad mound transformers.

underlay in a wireless communication system this refers to a system where a transmitter which covers a small area (small cell) transmits a signal that occupies the same spectrum as the main system. Such an underlay cell may block out coverage of the main system service in the small cell.

undervoltage a voltage that is less than nominal for a time greater than 1 minute.

undervoltage relay a protective relay that operates on low voltage or loss of voltage.

Underwriters Laboratory an insurance industry testing agency that establishes standards for and conducts testing of electrical equipment.

undetected error probability in a linear block code, the probability that a receiver will not be able to detect the presence of transmission errors in received codewords. The transmission of codewords from a linear block code, C, via a communication channel can be expressed as follows:

$$\overline{x} = \overline{y} + \overline{e}$$

with $\overline{y}$ codewords from the code C and $\overline{x}$ words received via the channel. $\overline{e}$ denotes error words generated by the channel during transmission of the codewords. The receiver will only detect the presence of errors if $\overline{x} \notin C$, i.e., if the received words are not codewords from the code C. Undetected errors occur only if $\overline{e} \in C$, in which case the linearity property of the codes causes $\overline{x} \in C$ although $\overline{x} \neq \overline{y}$. The undetected error probability is, therefore, strongly related to the nature of errors in the communication channel as well as the particular block code used. *See also* block codes.

unequal error protection (UEP) code a code in which certain digits of a codeword are protected against a greater number of errors than other digits in the codeword.

unfolding the technique of transforming a program that describes one iteration of an algorithm to another equivalent program that describes iterations of the same algorithm.

unforced system a dynamic system where all of the external sources of excitation are identically zero.

ungrounded system an electrical distribution system in which there is no intentional connection between a current-carrying conductor and ground.

uniaxial medium a medium whose permittivity and/or permeability is characterized by a 3×3 diagonal matrix where two of the three elements are the same.

Unibus bus standard used by Digital Equipment Corporation for its PDP and VAX computers.

unidirectional bus   a group of signals that carries information in one direction. Example: The address bus of the microprocessor is unidirectional; it carries address information in one direction — from the microprocessor to memory or peripheral.

unidirectional laser   ring laser in which either the clockwise or counter-clockwise circulating wave is negligible.

unidirectional resonator   ring resonator in which the electromagnetic waves circulate in either the clockwise or counter-clockwise direction but not both.

unidirectional transducer (UDT)   a transducer capable of launching energy from primarily one acoustic port over a desired bandwidth of interest.

unified cache   a cache that can hold both instructions and data. *See also* cache.

unified power flow controller   a device for both series and shunt reactive compensation, brought about by thyristor-based control.

unified transaction   a transaction, which can be a hardware instruction or a program segment, that performs a read-modify-write operation, that is allowed to complete without interruption by other processes.

uniform array   an array where the antenna elements that make up the array are uniformly spaced. Typical examples of this are the linear array and circular array.

uniform boundedness   a property of dynamical systems possessing solutions of state equations uniformly bounded.

The solutions of $\dot{\mathbf{x}} = \mathbf{f}(t, \mathbf{x})$ are uniformly bounded if for any given $d > 0$, there exists a $b = b(d) < \infty$ so that for each solution $\mathbf{x}(t)$ starting from the initial condition $\mathbf{x}(t_0)$ such that

$$\|\mathbf{x}(t_0)\| \leq d,$$

$$\|\mathbf{x}(t)\| \leq b \text{ for all } t \geq 0$$

where $\| \cdot \|$ is any Hölder norm on $\mathbb{R}^n$.

uniform chromaticity scale   a chart, which allows for the quick calculation of the X, Y, and Z tristimulus values of the CIE colorimetry system.

uniform circular array (UCA)   in array processing, an array with evenly spaced sensors placed on the perimeter of a circle.

uniform controllability   a dynamical where it is controllable in any time interval $[t_0, t_1]$.

uniform distribution   a probability distribution in which all events are equiprobable, i.e., $p(x) = k$ subject to $\int_{-\infty}^{\infty} p(s)ds = 1$.

uniform length coding   a coding scheme that assigns the same number of bits to different messages no matter what probabilities the messages assume.

uniform linear array (ULA)   in array processing, an array with evenly spaced sensors placed on a straight line.

uniform memory access   refers to a class of shared memory multiprocessor systems in which accesses to all parts of the shared memory take the same time independently of which processor makes the access.

uniform plane wave   a special class of electromagnetic problems where the E and H field are locally contained in a plane, and in each local plane, the E and H fields have a constant value over all that plane.

uniform sampling   the sampling of a continuous signal at a constant sampling frequency.

uniform scalar quantization   a structured scalar quantizer where the distance between reproduction levels is a given fixed number. Also known as uniform SQ.

uniform SQ   *See* uniform scalar quantization.

uniform stability   the constant equilibrium solution $\mathbf{x}_{eq} \in \mathbb{R}^n$ of $\dot{\mathbf{x}} = \mathbf{f}(t, \mathbf{x})$ or $\mathbf{x}(k + 1) = \mathbf{f}(k, \mathbf{x})$ is uniformly stable, in the sense of Lyapunov, if for any $\varepsilon > 0, t_0 \geq 0$, there corresponds $\delta = \delta(\varepsilon)$ independent of t_0 such that

$$\left\| \mathbf{x}(t) - \mathbf{x}_{eq} \right\| < \varepsilon$$

for $t \geq t_0$ whenever

$$\left\| \mathbf{x}(t_0) - \mathbf{x}_{eq} \right\| < \delta \,,$$

where $\| \cdot \|$ is any Hölder norm on $\mathbb{R}^n$.

uniform ultimate boundedness   the solutions of $\dot{\mathbf{x}} = \mathbf{f}(t, \mathbf{x})$, $\mathbf{x} \in \mathbb{R}^n$, are uniformly ultimately bounded, or practically stable, with respect to the ball

$$B_r = \{\mathbf{x} : \|\mathbf{x}\| \leq r\} \,,$$

if for every $d > 0$, there exists $T(d) > 0$ such that for each solution $\mathbf{x}(t)$ starting from the initial condition $\mathbf{x}(t_0)$ such that $\|\mathbf{x}(t_0)\| \leq d$,

$$\mathbf{x}(t) \in B_r \text{ for } t \geq T(d) \,,$$

where $\| \cdot \|$ is any Hölder norm on $\mathbb{R}^n$. *See also* uniform boundedness.

uniform variate   pseudo-random variate generated by computer that is equally likely to be any place within a fixed interval, usually $[0, 1]$.

uniformly asymptotically stable state   the equilibrium state of a dynamic system described by a first-order vector differential equation is said to be uniformly asymptotically stable if it is both uniformly convergent and uniformly stable. *See also* uniformly stable state, and uniformly convergent state.

uniformly convergent state   the equilibrium state of a dynamic system described by a first-order vector differential equation where there exists a δ, independent of the t_0, such that

$$\| x(t_0) - x_e \| < \delta \Rightarrow \lim_{t \to \infty} x(t) = x_e$$

See also convergent state.

uniformly excited equally spaced linear array (UEESLA)   an antenna array in which all the centers of the antennas are collinear and equally spaced and each antenna has a constant harmonic value but each antenna can have a unique phasing.

uniformly stable equilibrium   an equilibrium point of a nonautonomous system where the solutions that start "sufficiently close," stay "close" in time irrespective of the choice of initial time.

uniformly stable state   the equilibrium state of a dynamic system described by a first-order vector differential equation where if given $\epsilon > 0$ there exists a $\delta = \delta(\epsilon)$, independent of the initial t_0, such that

$$\| x(t_0) - x_e \| < \delta \Rightarrow \| x(t) - x_e \| < \epsilon \quad \forall t \geq t_0$$

See also stable state.

unilateral gain   special case of the transducer power gain of a 2-port network. The transducer power gain is the ratio of the power delivered to the load to the power available from the source. The unilateral gain is the nonreciprocal case of transducer power gain ($S_{12} = 0$).

unilateral transducer power gain   a special case of transducer power gain, G_{Tu}, where $S_{12} = 0$.

unimplemented instruction   (1) a numeric pattern in an instruction stream that does not correspond to any defined machine instruction.

(2) a type of trap operation executed by a processor when an unimplemented instruction is encountered.

uninterrupted power supply (UPS)   (1) a power supply designed to charge an energy storage medium, while providing conditioned output power, during the presence of input power and to continue providing output power for a limited time when the input to the supply is removed. These power supplies are typically used in critical applications to prevent shut-down of these systems during power failures, power surges, or brownouts.

(2) a device that provides protection for critical loads against power outages, overvoltages, undervoltages, transients, and harmonic disturbances. A typical UPS is a rectifier supplied battery bank for energy storage, and a PWM inverter-filter system to convert a DC voltage to a sinusoidal AC output. UPS systems can be on-line, as shown in the figure, where the UPS inverter powers the load continuously, or off-line where the load is connected directly to the utility under normal operation and emergency power is provided by the UPS.

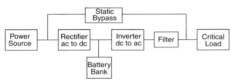

A typical UPS.

union   *See* union operator.

union of fuzzy sets   the fuzzy analogy to set the set-theoretic union.

Let A and B be two fuzzy sets in the universe of discourse X with membership functions $\mu_A(x)$ and $\mu_B(x)$, $x \in X$. The membership function of the union $A \cup B$, for all $x \in X$, is

$$\mu_{A \cup B}(x) = \max\{\mu_A(x), \mu_B(x)\}$$

See also fuzzy set, membership function.

union operator   a logical OR operator.

In a crisp (nonfuzzy) system, the union of two sets contains elements which belongs to either one or both of the sets. In fuzzy logic, the union of two fuzzy sets is the fuzzy set with a membership function which is the larger of the two. *See also* union of fuzzy sets.

unipolar neuron   neuron with output signal between 0 and +1.

uniprocessor   *See* single-instruction stream, single-data stream.

uniquely decodeable   a channel code where the correct message sequence can always be recovered uniquely from the coded sequence as observed through the channel. Of particular interest for multiple access channels. *See also* multiple access channel, zero-error capacity.

uniqueness of solution to generalized 2-D model   the generalized 2-D model with variable coefficients

$$E_{i+1,j+1} x_{i+1,j+1} = A_{ij}^0 x_{ij} + A_{i+1,j}^1 x_{i+1,j}$$

$$+ A_{i,j+1}^2 x_{i,j+1} + B_{ij}^0 u_{ij}$$

$$+ B_{i+1,j}^1 u_{i+1,j} + B_{i,j+1}^2 u_{i,j+1}$$

$i, j \in Z_+$ (the set of nonnegative integers) has the unique solution in the rectangle $[0, N_1] \times [0, N_2]$ for any sequence u_{ij} for $0 \leq i \leq N_1, 0 \leq j \leq N_2$ and any boundary conditions x_{i0} for $0 \leq i \leq N_1$ and x_{0j} for $0 \leq j \leq N_2$ if and only if the matrix F' is nonsingular, where $x_{ij} \in R^n$ is the local semistate vector, $u_{ij} \in R^m$ is the input vector $E_{ij}, A_{ij}^k, B_{ij}^k$ $(k = 0, 1, 2)$ are real matrices with E_{ij} possibly singular (rectangular), F' is defined by the equation $F' x' = Gu + Hx_0$ which follows from the model for $i = 0, 1, \ldots, N_1 - 1$;

$$j = 0, 1, \ldots, N_2 - 1,$$

$$x' := \left[x'^T_1, x'^T_2, \ldots, x'^T_{N_1} \right]^T,$$

$$x_i := \left[x^T_{i1}, x^T_{i2}, \ldots, x'^T_{iN_2} \right],$$

$$u = \left[u^T_0, u^T_1, \ldots, u^T_{N_1} \right]^T,$$

$$u_i := \left[u^T_{i0}, u^T_{i1}, \ldots, u^T_{iN_2} \right]^T,$$

$$x_0 := \left[x^T_{00}, x^T_{01}, \ldots, x^T_{0N_2}, x^T_{10}, x^T_{20}, \ldots, x^T_{N_10} \right]^T$$

uniqueness theorem of electromagnetics   a theorem stating that the solutions to Maxwell's equations are unique, given certain boundary conditions (for the differential form of the equations) or certain initial conditions (for the integral form of the equations). As an example, if the electric field and magnetic field satisfy Maxwell's equations in some volume V of lossy material and if the tangential fields satisfy a prescribed set of boundary conditions on the surface S that defines V, then the solution for the electric and magnetic fields is the only one (i.e., the unique one). *Note:* The uniqueness theorem only requires one of the following three conditions:

(1) knowledge of the tangential electric field on all of S,

(2) knowledge of the tangential magnetic field on all of S, or

(3) knowledge of the tangential electric field on part of S and the tangential magnetic field on the remaining part of S.

unit commitment problem   the task of minimizing the cost of production by deciding which of several thermal generating plants in a power system should be kept generating, on hot reserve or on cold reserve.

unit delay   in discrete time systems the delay of a signal by a single sample interval, i.e., $x(n - 1)$. Under the z-transform, $z^{-1}X(z)$.

unit impulse   a very short pulse such that its value is zero for $t \neq 0$ and the integral of the pulse is 1.

unit impulse function   In discrete time: $\delta(n) = 1$ for $n = 0$, $\delta(n) = 0$ for $n \neq 0$. In continuous time: $\delta(t) = 0$ for $t \neq 0$ and $\int_\varepsilon^\varepsilon \delta(s)ds = 1$ for $\varepsilon > 0$. *See also* Dirac delta function, Kronecker delta function

unit step   *See* unit step function.

unit step function   a mathematical function whose amplitude is zero for all values of time prior to a certain instant and unity for all values of time afterwards. The unit step signal is the integral of the unit impulse function. That is, the function $u(t)$ which is 1 for all $t \geq t_0$ and 0 for all $t < t_0$.

unitarity in scattering   law expressing the conservation of energy from the incident light to that scattered and absorbed by an inhomogeneous medium.

unitary transform   a transform whose inverse is equal to the complex conjugate of its transpose.

unity feedback   the automatic control loop configuration shown in the figure.

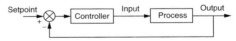

Unity feedback configuration.

universal asynchronous receiver/transmitter (UART) a standard interface often used in small computer systems, to buffer and translate between the parallel word format used by the CPU and the aynchronous serial format used by slow I/O devices.

universal coding   coding procedure that does not require knowledge of the source statistics and yet is asymptotically optimal. A typical example is Ziv–Lempel coding.

universal function approximation property uniform approximation of any real continuous nonlinear function to arbitrary level of accuracy in a compact set. It has been demonstrated that some relevant model of soft-computing (such as multilayer perceptrons, radial basis functions networks, and fuzzy systems) hold this property.

See also fuzzy systems, multilayer perceptron, radial basis functions.

universal fuzzy approximator a fuzzy system approximator in a sense that it can approximate any nonlinear function to any degree of accuracy on any compact set.

universal fuzzy controller a fuzzy controller in a sense that it can control any nonlinear plant as long as the plant can be controlled by a smooth nonlinear control law.

universal source coding refers to methods for source coding that do not rely on explicit knowledge of the source statistics. One important method for universal lossless source coding is Ziv–Lempel coding.

universal synchronous/asynchronous receiver/ transmitter (USART) a logic device that performs the data link layer functions, such as serializing, deserializing, parity generation and checking, error checking, and bit stuffing, of a serial transmission protocol for either synchronous or asynchronous transfer modes.

universe of discourse term associated with a particular variable or groups of variables, it is the total problem space encompassing the smallest to the largest allowable nonfuzzy value that each variable can take.

unloaded Q dimensionless ratio of the average over any period of time ($T = 1$/frequency) of the ratio of the maximum energy stored (Umax) to the power absorbed or dissipated in a passive component or circuit, discounting any external effects. An ideal resistor has an unloaded Q of zero, and ideal capacitors or inductors have an uloaded Q of infinity. For most applications, the higher the unloaded Q the better the part.

unmatched uncertainty a dynamical system model in which the vector function that models uncertainties in the system does not satisfy the *matching condition* as, for example, in the system model

$$\dot{\mathbf{x}} = \mathbf{A}\mathbf{x} + \mathbf{B}\mathbf{u} + \mathbf{h}(t, \mathbf{x}, \mathbf{u}),$$

where the function **h** may model system uncertainties such as modeling uncertainties, input connection uncertainties, or additive disturbances.

unmodeled dynamics in control systems design, it is often assumed that the true process is compatible with the model used in parameter estimation. However, it frequently happens that the true process is more complex than the estimated model. This is often referred to as unmodeled dynamics. The problem is complex, and a careful analysis is lengthy. If a controller is able to control processes with unmodeled dynamics and/or disturbances, we say that the controller is robust.

unpolarized if the amplitude of the wave in plane perpendicular to the direction of propagation appears to be oriented in all directions with equal probability, the wave is said to be unpolarized. Unpolarized electromagnetic waves are generated by atomic processes.

unprivileged mode one of two CPU modes, the other being privileged mode. Sometimes called user mode, this mode prohibits access to certain instructions, data, and registers.

unrelated color color perceived to belong to an area of object seen in isolation from other colors. *See* related color.

unsharp masking   an edge enhancement technique that subtracts a blurred version of an image from the input image. *See* edge enhancement.

unsigned integer   an integer numeric representation in which only positive numbers are represented. For example, a 16-bit unsigned integer has a range of 0 to 65535.

unstable   a circuit or circuit element that is likely to change state spontaneously within a short period of time.

unstable resonator   resonator in which an axial light ray is unstable with respect to perturbations, and ray trajectories become unbounded; does not imply unstable modes (unstable resonators have stable modes); high-loss resonator.

unstable state   the equilibrium state of a dynamic system described by a first-order vector differential equation is said to be unstable if it is not stable. *See also* stable state.

unstructured uncertainty   high-order variations or unknown disturbances characterized usually by a set of norm-bounded perturbations introduced into a fixed nominal plant.

For linear systems with a model in frequency domain, an uncertain system is represented by a class of plants composed of a nominal known rational matrix and an unknown but stable one whose norm is bounded by an absolute value of known rational function.

unsupervised learning   learning from unlabeled data. The learning system seeks to identify structure in the data by clustering similar input patterns.

unsupervised neural network   neural network that does require predetermined output to form the interconnection weight matrix of a network. If no input–output pairs are known and a number of inputs are available, we can only memorize them in an organized order. Similar inputs are memorized in locations close to each other and different inputs are stored in locations far from each other. The network is able to recognize an input that has already appeared before. If the input has never appeared, it will be stored in an appropriate location, which is close to similar inputs and far from different inputs. An example of unsupervised neural network is the Kohonen self-organizing map, which can be implemented using an adaptive correlator. A correlator has been long used in optical processing for measuring the similarity between two inputs.

unsymmetrical fault   a fault on a three-phase power line in which the fault current is not equal in all three phases, e.g., a single-line-to-ground, double-line-to-ground or line-to-line fault.

unsymmetrical load   a load which forces the currents in the three-phase power line which supplies it to be unequal.

up-down converter   *See* buck-boost transformer.

up-down counter   a register that is capable of operating like a counter and can be either incremented or decremented by applying the proper electronic signals.

up-down transformer   *See* buck-boost transformer.

upconversion   a nonlinear optical process in which a beam of light is shifted to higher frequency, for instance, through the process of sum-frequency generation.

upconverter   special type of microwave mixer that outputs the sum frequency signals of the input microwave RF frequency and the LO frequency. The up-conversion is useful if the microwave carrier frequency needs to be altered.

uplink   in a cellular system, the communication link from the mobile to the central base station. *See also* downlink. Also called reverse link.

upper frequency band edge   the upper cutoff frequency where the amplitude is equal to the maximum attenuation loss across the band.

upper side frequency   the sum frequency that is generated during the heterodyning process or during the amplitude-modulating process. For example, if a 500 kHz carrier signal is amplitude-modulated with a 1 kHz frequency, the upper side frequency is 501 kHz.

UPS   *See* uninterrupted power supply.

upsampling   a system that inserts $L - 1$ zeros between the samples of an input signal to form an output signal. An L-fold upsampler followed by an appropriate lowpass filter produces an output signal that is an interpolated form of the input signal, at L times the sampling rate. Upsampling also often refers to the operation of the upsampler and the lowpass filter together.

URD   *See* underground residential distribution.

USART   *See*   universal   synchronous/asynchronous receiver-transmitter.

use bit   in a paging system, a bit associated with a page entry in a lookup table which indicates that the page has been referenced since the last time the bit was reset. The bit is reset when it is read.

user mode   in a multitasking processor, the mode in which user programs are executed. In user mode, the program is prevented from executing instructions that could possibly disrupt the system and also from accessing data outside the user's specified area.

user state   a computer mode in which a user program is executing rather than a systems program.

Some computers have two modes of operation: the system state is used when parts of the computer's operating system are executing and the user state is used when the computer is executing application programs.

user-visible register   an alternative name for general purpose registers, emphasizing the fact that these registers are accessible to the instructions in user programs. The counterpart to user-visible registers are registers that are reserved for use by privileged instructions, particularly within the operating system.

utility interface   the interface of the utility with power electronic systems. Utility interface issues include maintaining power quality with the proliferation of large power electronic loads in power system networks and the utility applications of power electronics for flexible AC transmission systems (FACTS).

UV cure   a post-development process by which the resist patterns are exposed to deep-UV radiation (and often baked at the same time) in order to harden the resist patterns for subsequent pattern transfer. The UV cure is a replacement for the hard bake step.

utilization factor   the ratio of the maximum demand on the system vs the rated capacity of the system.

utilization voltage   the voltage across the power input terminals of a piece of electrical equipment.

UVPROM   *See* erasable programmable read-only memory.

V

V parameter   describes the number of modes M in an optical fiber.

$$V = 2\ pi\ \frac{a}{\lambda_o} NA$$

where a is the fiber core radius, λ is the wavelength, and NA is the fiber's numerical aperture. $V < 2.405$ for single-mode operation. Often referred to as normalized frequency.

V system   *See* vee system.

V$_{\mathrm{IN}_{p-p}}$ common notation for peak-to-peak input voltage in volts.

V$_{\mathrm{OUT}_{p-p}}$ common notation for the peak-to-peak output voltage in volts.

V$_{DS}$ common notation for the FET drain-to-source voltage.

V$_{GD}$ common notation for the FET gate-to-drain voltage.

V$_{GS}$ common notation for the FET gate-to-source voltage.

V-curves   the characteristic curves that show the variation of the armature current versus the field current in synchronous motors.

V-V transformer   *See* open-delta transformer.

vacuum capacitor   a capacitor with a vacuum between its plates.

vacuum circuit breaker   a power circuit breaker where a vacuum chamber is used as an insulating and arc clearing medium.

vacuum insulation   any insulation scheme which depends upon the dielectric capabilities of a high vacuum.

vagueness   a property indicating the lack of specifics and clarity and which is allied to imprecision and fuzziness.

valence band   the lower of the two partially filled bands in a semiconductor. *See also* conduction band.

valid bit   a bit used in caches and virtual memories that records whether the cached item or page contains valid data.

validation   (1) in electronic active and passive device modeling, the pass/fail process in which a completed, ready to use model is used in a simulation, then compared to an intended application, and is determined to suitably predict reality.
(2) a review to establish the quality of a software product for its operational purpose.

validation set   the set of data to evaluate the performance of a system which was trained on a separate set of data.

valuation   in electronic active and passive device modeling, valuation is a measure of the intrinsic value of a model in predicting a new application, condition or change in the device. The most valuable model (assigned a valuation coefficient of 1.0) would be a complete omnipotent physical based model that could be utilized to predict exact circuit response. The least valuable model (assigned a valuation coefficient of 0.0) such as a lookup table, would only be able to predict a circuit response to a specific set of conditions for a very specific device excited in a specific way.

van de Graf generator   a high-voltage device that generates high static voltages on a sphere. It is driven by a mechanical belt, which delivers the charges.

van der Pol oscillator   an oscillator or oscillating system described by the equation

$$\frac{d^2x}{dt^2} - \mu(1 - \epsilon x^2)\frac{dx}{dt} + x = 0$$

This equation is mentioned in almost any book on oscillators or on nonlinear mechanics. The reason is that this equation is relatively simple, yet it successfully lumps together two rather complex processes associated with oscillators, namely, the process of generation of periodic waveform and the process of automatically stabilizing the amplitude of this wave. The attempts to improve the solutions of the van der Pol equation and to apply to design of low-distortion oscillators can be traced to recent publications.

Vander Lugt filter   encoded optical mask for representing, in the Fourier-transform domain, the reference or library functions needed in an image correlator; encoding is performed holographically.

vanishing point   the point in the perspective projection plane in which a system of 3-D parallel lines converge.

Vapnik–Chervonenk (VC) dimension   a measure of the expressive power of a learning system with binary or bipolar outputs. For neural networks it is closely related to the number of weights in the network. For single-layer networks it is simply equal to the number of weights (including biases) but for multilayer and other networks, analytic expressions for VC dimension are not available.

vapor cooling   a cooling technique for power vacuum tubes utilizing the conversion of hot water to steam as a means of safely conducting heat from the device and to a heat sink.

VAR   *See* volt-ampere reactive.

varactor   a reverse biased PN or Schottky diode that uses the voltage variable depletion region

as a tuning element or as a nonlinear frequency multiplier.

varactor diode   a diode designed to have a repeatable and high capacitance vs. reverse voltage characteristic. A two terminal semiconductor device in which the electrical characteristic of primary interest is the voltage dependent capacitance.

varactor tuner   a tuning circuit at the input of a television receiver that uses a varactor diode. The tuning capability comes from the characteristic of a varactor, or varicap, to function as a voltage-sensitive capacitance.

variable bit rate (VBR)   describes a traffic pattern in which the rate at which bits are transmitted varies over time; such patterns are also referred to as bursty. VBR sources often result from compressing CBR sources, for example, a 64 kbps voice source in its raw form has a constant bit rate; after compression by removing the silence intervals, the source becomes VBR.

variable frequency drive   electric drive system in which the speed of the motor can be varied by varying the frequency of the input power.

variable length code   to exploit redundancy in statistical data, and to reduce average number of bits per word luminance levels having high probability are assigned short code words and those having low probability are assigned longer code words. This is called variable length coding or entropy coding. *See also* entropy coding.

variable loss   machine loss that changes with a change in the mode of machine operation such as loading, temperature and current. For example, in a transformer, the winding losses are a function of the load current, while the core losses are almost independent of the load current.

variable polarity plasma arc welding (VP-PAW)   a welding process that produces coalescence of metals by heating them with a constricted variable polarity arc between an electrode and the parts to be joined (transferred arc) or between the electrode and the constricting nozzle (transferred arc). Shielding is obtained from the hot, ionized gas issuing from the torch as well as from a normally employed auxiliary shielding gas source. Pressure is not applied and filler metal may or may not be added.

variable reflectivity mirror   mirror in which the reflection profile varies across the mirror surface; useful for discriminating against high-order transverse modes in a resonator. *See also* tapered reflectivity mirror.

variable reluctance machine   salient pole machine consisting of stators having concentrated excitation windings and a magnetic rotor devoid of any windings, commutators, or brushes. The machine operates on the principle of varying reluctance along the length of the air gap. Torque is produced by the tendency of the rotor to align itself with the stator produced flux waves in such a way that the stator produced flux-linkages are maximized. The motor operates continuously in either direction of rotation with closed loop position feedback.

variable resolution hierarchy   an approach where images corresponding to the levels of the hierarchy vary in spatial resolution. This results in a pyramid structure where the base of the pyramid represents the full resolution and the upper levels have lower resolution.

variable speed AC drive   an AC motor drive that is capable of delivering variable frequency AC power to a motor to cause it to operate at variable speeds. Induction motors and synchronous motors are limited to operation at or near synchronous speed when a particular frequency is applied. Variable speed drives rectify the incoming AC source voltage to create a DC voltage that is then inverted to the desired frequency and number of phases.

variable speed DC drive   a DC motor controller that allows the DC motor to operate over a wide speed range. A common type of variable speed DC drive uses a separately excited DC motor. Armature voltage control is used to provide operation below base speed, and field weakening is used to provide operation above base speed.

variable speed drive   *See* variable speed AC drive or variable speed DC drive.

variable structure system   a dynamical system whose structure changes in accordance with the current value of its state. A variable structure system can be viewed as a system composed of independent structures together with a switching logic to switch between each of the structures. With appropriate switching logic, a variable structure system can exploit the desirable properties of each of the structures the system is composed of. A variable structure system may even have a property, such as asymptotic stability, that is not a property of any of its structures.

variable-length instruction   the fact that the machine language instructions for a computer have different numbers of bits with the length dependent on the type of instruction.

variance   the mean-squared variability of a random variable about its mean: $\sigma^2 = \int_{-\infty}^{\infty} (x - \mu)^2 p(x)dx$ where σ^2 is the variance, μ is the mean and $p(x)$ is the probability density function. *See* covariance, correlation.

variational formula   a formula that provides the sought unknown quantity a in terms of another unknown quantity b. The advantage is in the fact that, for variational formulas, an error in b provides only a modest error in a. Hence, by approximating b we can get a fairly good estimate of a.

variational problem   a problem in which solving a differential equation is equivalent to seeking a function that minimizes an integral expression.

variational similarity between two vectors $x = (x_1, \ldots, x_n)$ and $y = (y_1, \ldots, y_m)$ is the match score in the following dynamic matching procedure:

Suppose the current point is (x_i, y_j). The next point to match should be selected from (x_{i+1}, y_j), (x_{i+1}, y_{j+1}), and (x_{i+1}, y_{j+2}). If there is one match only among them, take it as the current match point and go on in the same manner. If there is no match, select (x_{i+1}, y_{j+1}). If there is more than one match including (x_{i+1}, y_{j+1}), select (x_{i+1}, y_{j+1}), otherwise, select (x_{i+1}, y_{j+2}). This process continues until at least one of the vectors is exhausted.

During the above procedure, if there is at least one match in one step increase the match score by 1.

Varsharmov–Gilbert bound the lower bound on the minimum distance of linear (n, k) code asymptotically satisfies

$$\frac{d_{\min}}{n} \geq \alpha$$

where α is given by

$$\frac{k}{n} = 1 + \alpha \log_2 \alpha + (1 - \alpha) \log_2 \alpha \quad 0 \leq \alpha \leq \frac{1}{2}$$

vault *See* transformer vault.

VBR *See* variable bit rate.

VC dimension *See* Vapnik–Chervonenk dimension.

VCO *See* voltage-controlled oscillator.

VCO gain the ratio of the VCO output frequency to the DC control input level. This is usually expressed in units of radians per second per volt.

VCP *See* visual comfort probability.

VDS *See* drain-to-source voltage.

VDSL *See* very high-speed digital subscriber line.

vector a quantity having both magnitude and direction.

vector controlled induction motor a variable speed controller and motor in which the magnetizing and torque producing components of current are controlled separately. Some vector drives require rotor position sensors. Vector controlled induction motors can operate over a wider speed range, and may produce rated torque even at zero speed, much like a DC motor. Thus, vector controlled induction motors are often used for applications that might otherwise require a DC motor drive.

vector field when the field needed to describe some physical phenomenon has several components, it is customary to represent such a field by a vector function $\mathbf{V}(x, y, z)$ which depends on the space coordinates x, y, z.

vector image an image consisting of mathematical descriptions of the objects in the scene, e.g., equations for lines and curves. The image is independent of resolution so it can be stretched, rotated and skewed with no degradation. Vector images are often used in CAD applications. *See* bitmapped image, CAD, image.

vector network analyzer a microwave receiver designed to measure and process the magnitude and phase of transmitted and reflected waves from the linear network under test.

vector operation a hardware instruction that performs multiple similar operations on data arranged in one or more arrays.

vector processor a computer architecture with specialized function units designed to operate very efficiently on vectors represented as streams of data.

vector quantization (VQ)   quantization applied to vectors or blocks of outputs of a continuous source.

Each possible source block is represented by a reproduction vector chosen from a finite set (the "codebook"). According to rate-distortion theory, vector quantization (VQ) is able to perform arbitrarily close to the theoretical optimum if the lengths of the input blocks are permitted to grow without limit. The method was suggested by Claude Shannon in his theoretical work on source coding (during the late 1940s and the 1950s), but has found practical importance first in recent years (during the 1980s and 1990s) because of the relatively high complexity of implementation and design compared to scalar methods. Also referred to as "block source coding with a fidelity criterion."

vector quantization encoding   an encoding scheme whereby an image is decomposed into n dimensional image vectors. Each image vector is compared with a collection of representative templates or codevectors from a previously generated codebook. The best match codevector is chosen using a minimum distortion rule. Then the index of the codevector is transmitted. At the receiver this is used with a duplicate codebook to reconstruct the image. Usually called VQ encoding.

vector quantizer (VQ)   a device that performs vector quantization.

vector space   an algebraic structure comprised of a set of elements over which operations of vector addition and scalar multiplication are defined. In a linear forward error control code, code words form a vector space when addition and multiplication are defined in terms of element-wise operations from the finite field of code symbol values.

vector stride   the number of consecutive memory addresses from the beginning of one element to the next of a vector stored in memory. Also used to refer to the difference in vector index between two consecutively accessed vector elements.

vector wave   wave that cannot be adequately described in terms of a single field variable.

vector wave equation   an equation (or more specifically, a set of scalar equations) governing the various components of a vector wave, the Maxwell–Heaviside equations, for example.

vectored interrupt   an interrupt request whereby the processor is directed to a predetermined memory location, depending on the source of the interrupt, by the built-in internal hardware. In the X86 processors, the addresses are stored in an array in memory (a mathematical vector) and indexed by the interrupt number. In the 8080 and Z80, the interrupt number becomes part of a CALL instruction with an implied address that is executed on an interrupt cycle.

vectorscope   an oscilloscope-type device used to display the color parameters of a video signal. A vectorscope decodes color information into R-Y and B-Y components, which are then used to drive the X and Y axis of the scope. The total lack of color in a video signal is displayed as a dot in the center of the vectorscope. The angle, distance around the circle, magnitude, and distance away from the center indicate the phase and amplitude of the color signal.

vee system   a 3-level system in which the highest two energy states are coupled by electromagnetic fields to a common, intermediate, lower state. This system is so named because schematic representations of it often look like a capital letter V.

velocity error   the final steady difference between a ramp setpoint and the process output in a unity feedback control system. Thus, it is the asymptotic error in position that arises in a closed loop system that is commanded to move with constant velocity. *See also* acceleration error and position error.

velocity error constant   a gain K_v from which the velocity error e_v is readily determined. It is a concept that is useful in the design of unity feedback control systems since it transforms a constraint on the final error to a constraint on the gain of the open loop system. The relevant equations are

$$e_v = \frac{1}{K_v} \text{ and } K_v = s \mathop{\lim}_{\to \infty} sq(s)$$

where $q(s)$ is the transfer function model of the open loop system, including the controller and the process in cascade. *See also* acceleration error constant and position error constant.

velocity filtering   means for discriminating signals from noise or other undesired signals because of their different apparent velocities.

velocity flow field   the velocity field calculated in optical flow computation.

velocity manipulability ellipsoid   an ellipsoid that characterizes the end-effector velocities that can be generated with the given set of joint velocities, with the manipulator in a given posture.

Consider a set of generalized joint velocities $\dot{q}$ of constant, unit norm $\dot{q}^T \dot{q} = 1$. Taking into account the differential kinematics J and properties of the pseudo-inverse of the geometrical Jacobian, the unit norm can be rewritten as follows: $v^T (JJ^T)^{-1} v = 1$, which is the equation of the points on the surface of an ellipsoid in the end-effector velocity space.

velocity of light   in vacuum, a constant equal to 2.997928×10^8 meters/second. In other media, equal to the vacuum value divided by the refractive index of the medium.

velocity saturation   a physical process in a semiconductor where the carrier velocity becomes constant independent of the electric field due to high energy scattering and energy loss, compared to low electric field transport where the velocity is linearly related to the field by the mobility.

verification   the process of proving that the implementation of hardware or software meets the published system requirements.

verification kit   known impedance standards traced to NIST, other than calibration standards, used to verify the calibrated performance of a vector network analyzer system.

Verilog   a hardware description language for electronic design and gate-level simulation. *See* VHDL, VHSIC.

Versa Module Europe bus (VME bus)   a standardized processor backplane bus system originally developed by Motorola. The bus allows multiple processors to share memory and I/O devices.

vertical cavity laser   semiconductor laser in which the electromagnetic fields propagate in a direction perpendicular to the amplifying plane (the vertical direction).

vertical deflection   the direction of an entity is caused to move by some physical action; commonly describes the vertical movement of an electron beam caused by electrostatic or magnetic forces applied to produce a required scan. Magnetic deflection is frequently used with a CRT video display and requires a large deflection angle.

vertical microinstruction   a field that specifies one microcommand via its op code. In practice, microinstructions that typically contain three or four fields are called vertical.

vertical polarization   a term used to identify the position of the electric field vector of a linearly polarized antenna or propagating EM wave relative to a local reference, usually the ground or horizon. A vertically polarized EM wave is one with its electric field vector aligned perpendicular to the local horizontal.

vertical roll   in television, the apparent continuous upward or downward movement of the picture, resulting from the lack of synchronization between the transmitter and receiver.

vertical sync pulse   a signal interval of the NTSC composite video signal provided for the synchronization of the vertical deflection system; the vertical sync interval has a duration of three horizontal lines and is serrated with six pulses. The vertical sync interval starts after six equalizing pulses (3 horizontal line periods) that identify the beginning of the vertical blanking interval. The vertical serration preserves the horizontal line synchronization information during the vertical sync pulse interval with the one-half horizontal line time-signal transition from the composite video blanking signal level to the sync signal level. The serrated vertical pulse duration is at the blanking level for $7 \pm 1\%$ of the horizontal line time.

vertically integrated utility   a utility in which generation, transmission, and distribution divisions are all owned by a single entity.

very high-speed digital subscriber line (VDSL)   a digital subscriber line (DSL) that provides very high rates (13 Mbps, 26 Mbps, and 52 Mbps) through short subscriber loops (1 to 3 kft). A VDSL may support asymmetric rates between the customer premise and the central office.

very-large-scale-integration (VLSI)   (1) a technology that allows the construction and interconnection of large numbers (millions) of transistors on a single integrated circuit.

(2) an integrated circuit made of tens of thousands to hundreds of thousands of transistors.

very long instruction word (VLIW)   a computer architecture that performs no dynamic anal-

ysis on the instruction stream and executes operations precisely as ordered in the instruction stream.

very small aperture terminal (VSAT)   a small earth station suitable for installation at a customer's premises. A VSAT typically consists of an antenna less than 2.4 m, an outdoor unit to receive and transmit signals, and an indoor unit containing the satellite and terrestrial interface units.

vestigial sideband   (1) a portion of one sideband in an amplitude modulated signal, remaining after passage through a selective filter.

(2) Amplitude modulated signal in which one sideband has been partially or largely suppressed.

(3) The small amount of energy emitted in the unused sideband in a single-sided transmitter.

VGA   *See* video graphics adapter.

VHDL   acronym for very high speed integrated circuit (VHSIC) hardware description language. *See* VHSIC, Verilog.

VHF   very high frequency. See VHF power.

VHF power   in television, the band of frequencies ranging from 30MHz to 300MHz.

VHSIC   (1) acronym for very high speed integrated circuit.

(2) acronym for very high speed integrated circuit, a type of digital logic circuit. *See* VHDL.

via   a hole in the insulator between two metal layers on a multilayer integrated circuit that is etched and filled with a conducting material so that the two metal layers are electrically connected. Via resistance is typically less than 10 ohms.

via hole   hole chemically etched from the back of a MMIC wafer and filled with metal in such a way as to allow an electrical connection between the backside of a wafer and the topside of the wafer.

vibration damper any of a number of devices mounted on a power line to reduce vibrations caused by wind.

vibrational transition transition between vibrational states of a molecule.

video (1) representation of moving images for storage and processing. Often used interchangeably with television. In particular, "video signal" and "television signal" are synonyms.

(2) a particular stored sequence of moving images, e.g., on a tape or within a database.

video amplifier (1) in television, the wideband stage (or stages) that amplifies the picture signal and presents it to the picture tube.

(2) A similar wideband amplifier, such as an instrument amplifier or preamplifier having at least a 4-MHz bandwidth.

video coding compression of moving images. Coding can be done purely on an Intraframe (within-frame) basis, using a still image-coding algorithm, or by exploiting temporal correlations between frames (interframe coding). In the latter case, the encoder estimates motion between the current frame and a previously-coded reference frame, encodes a field of motion vectors that describe the motion compactly, generates a motion-compensated prediction image and codes the difference between this and the actual frame with an intraframe residue coder—typically the 8×8 discrete cosine transform. The decoder receives the motion vectors and encoded residue, constructs the prediction picture from its stored reference frame and adds back the difference information to recover the frame. *See* MPEG.

video compression *See* video coding.

video graphics adapter (VGA) a video adapter proposed by IBM in 1987 as an evolution of EGA. It is capable of emulating EGA, CGA, and MDA. In graphic mode, it allows to reach 640×480 pixels (wide per high) with 16 colors selected from a pallet of 262144, or 320×240 with 256 colors selected from a pallet of 262144.

video RFI undesired radio-frequency signals that compete with the desired video signal.

video signal the video signal in the U.S. is defined by the NTSC standard. *See* National Television System Committee.

video signal processing The area of specialization concerned with the processing of time sequences of image data, i.e., video.

video transmission the combined amplitude-modulated carrier, sync, and blanking pulses that make up a video signal.

virtual address (1) an address that refers to a location of virtual memory.

(2) the address generated by the processor in a paging (virtual memory) system. *Compare with* real address.

virtual channel a concept used to describe unidirectional transport of ATM cells associated by a common unique identifier value.

virtual circuit an abstraction that enables a fraction of a physical circuit to be allocated to a user. To a user, a virtual circuit appears as a physical circuit; multiple virtual circuits can be multiplexed onto a single physical circuit.

virtual connection a representation of the circuit between the input leads of an ideal op-amp. The voltage across and the current through a virtual connection are both zero. If one input lead of an ideal op-amp is connected to ground, the virtual connection is often termed a virtual ground.

virtual DMA DMA in which virtual addresses are translated into real addresses during the I/O operation.

virtual instrument an instrument created through computer control of a collection of instrument resources with analysis and display of the data collected.

virtual machine a process on a multitasking computer that behaves as if it were a stand-alone computer and not part of a larger system.

virtual memory main memory as seen by the processor, i.e., as defined by the processor-generated addresses, in contrast with real memory, which is the memory actually installed or that is immediately addressable.

The virtual memory corresponds to the secondary storage, and data is automatically transferred to and from real memory as needed. In paged virtual memory, secondary memory is divided into fixed-size pages that are automatically moved to and from page frames of real memory; the division is not logical and is usually invisible to the programmer. In segmentation, the divisions (known as segments) are logical and of variable-sized units that are much larger than pages: 16–24 KB versus 0.5–4 KB. Many machines combine both paging and segmentation.

Since secondary memory is much larger than main memory, virtual memory presents the programmer with the view of a main memory that appears to be larger than it actually is. Virtual memory also facilitates automatic transfer of data, protection, accommodation of growing structures, efficient management of main-memory, and long-term storage.

virtual memory interrupt interrupt that occurs when an attempt is made to access an item of virtual memory that is not loaded into main memory.

virtual page number in a paged virtual memory system, this is the part of the memory addresses that points to the page that is accessed, while the rest of the address points to a particular part of that page.

virtual path a concept used to describe the unidirectional transport of virtual channels that are associated by a common identifier value.

virtual reality three or more dimensionality of computer-generated images, which gives the user a sense of presence (i.e., a first-person experience) in the scene.

virtual register one of a bank of registers used as general purpose registers to hold the results of speculative instruction execution until instruction completion. Virtual registers are used to prevent conflicts between instructions that would normally use the same registers. *See also* speculative execution.

virtually addressed cache a cache memory in which the placement of data is determined by virtual addresses rather than physical addresses. This scheme has the advantage of decreasing memory access times by avoiding virtual address translation for most accesses. The disadvantage is that data stored in the cache may have different virtual addresses in different processes (aliasing).

visible associated with the wavelength region that can be seen by the human eye; often considered to range from about 400 to 700 nanometers.

visible light electromagnetic radiation in the visible portion of the spectrum, roughly 400 to 700 nanometers.

visual comfort probability (VCP) this rating is based in terms of the percentage of people who will be expected to find the given lighting system acceptable when they are seated in most undesirable locations.

visual display unit a common means of input/ouput to/from a computer. Consists of a CRT and a keyboard.

visual perception the perception of a scene as observed by the human visual system: it may differ considerably from the actual intensity image because of the nonlinear response of the human visual system to light stimuli.

visual space   the complete set of all possible images on a specific set of sampling and quantization parameters. Any specific image would be a member of this large space. For a 2×2 bi-level image, the space contains 16 members. Allowing all 3×3 bi-level images increases the size of the space to 512 (number of quantized levels raised to the power M, where M is the total number of pixels in the image).

Viterbi algorithm   an algorithm for finding the most probable sequence given that data can be modeled by a finite-state Markov model. For example, used in maximum likelihood decoding of trellis codes and in equalization.

VLIW   *See* very long instruction word.

VLSI   *See* very large scale integration.

voice   means for enabling a computer or data processing system to recognize spoken commands and input data and convert them into electrical signals that can be used to cause the system to carry out these commands or accept the data. Various types of algorithms and stored templates are used to achieve this recognition.

voice activity   stimuli that can be used to optimize channel capacity. The human voice activity cycle is typically 35%. The rest of the time we are either listening or pausing. In a multiple access scenario such as CDMA, all users are sharing the same radio channel. When users assigned to the channel are not talking, all other users on the same channel benefit with less interference. Thus, the voice activity cycle reduces mutual interference by 65%, tripling the true channel capacity. CDMA is not the only technology that takes advantage of this phenomenon.

voice coil   the bobbinless coil transducer element of a dynamic microphone.

voice compression   *See* speech compression.

voicing   classification of a speech segment as being voiced (i.e., produced by glottal excitation), unvoiced (i.e., produced by turbulent air flow at a constriction), or some mix of those two.

volatile   pertaining to a memory or storage device that loses its storage capability when power is removed.

volatile device   a memory or storage device that loses its storage capability when power is removed.

volatile memory   memory that loses its contents when the power supply is removed. Examples include most types of RAM.

volt (v)   the SI unit for electrical potential differents and electromotive force.

volt-ampere-reactive (VAR)   a unit of power equal to the reactive power in a circuit carrying a sinusoidal current when the product of the root-mean-square value of the voltage (expressed in volts), the root-mean-square value of the current (expressed in amperes), and the cosine of the phase angle between the voltage and the current, equals one; the unit of reactive power in the International System. Also expressed as megavars and kilovars.

voltage collapse   the rapid and uncontrollable drop of bus voltage due to a slight increase in load at the bus, generally characterized by inadequate reactive support in a high-load area.

voltage variation — long duration   a change of voltage RMS value from nominal for a time period greater than 1 minute, and can be used with the words showing a magnitude change such as overvoltage, or undervoltage.

Volta, Alessando (Corte) (1745–1827) Born: Como, Italy

Best known for the invention of a number of practical devices including the first battery (voltaic pile), a simple electrometer for measuring current and electrophorus. Volta was not a theoretical physicist, but a good researcher. He was able to follow up Benjamin Franklin's early work and that of Luigi Galvani by devising devices and experiments that allowed him to explore the physics. Volta is honored by having his name used as the unit of electromotive force, the volt.

voltage the potential to do work, voltage is the ratio of the energy available to the charge, expressed in volts.

voltage coefficient of resistance the change in resistance per unit change in voltage, expressed as a percentage of the resistance at 10% of rated voltage.

voltage and current transmission matrix a matrix representation for a two port network that provides the voltages and current at one port as a function of the voltages and current at the other port. Also known as chain matrix.

voltage change a deviation of the peak or RMS voltage between two levels that are of some fixed duration.

voltage controlled oscillator *See* voltage-controlled oscillator.

voltage dip *See* sag.

voltage distortion a change from a nominal clean sinusoidal waveform.

voltage drop the difference in potential between the two ends of the resistor measured in the direction of flow current. The voltage drop is $V = IR$, where V is voltage across the resistor, I is the current through the resistor, and R is the resistance.

voltage fed inverter *See* voltage source inverter.

voltage feedback op-amp an op-amp in which the output voltage is controlled by the differential input voltage multiplied by the open-loop gain. A voltage feedback op-amp has very high input resistance (the current in either input is ideally negligibly small), low output resistance, and large open loop gain. Ideally, the bandwidth would be infinite; in practice, the finite bandwidth leads to a gain–bandwidth tradeoff in the closed loop performance of amplifiers using voltage feedback op-amps. Thus increasing the closed-loop gain causes a proportional decrease in the closed-loop bandwidth.

voltage fluctuation refers to a series of voltage variations.

voltage gain dimensionless ratio of the peak-to-peak RF output voltage versus the peak-to-peak RF input voltage.

voltage instability proximity index an index that gives an indication of the amount of real or reactive power margin available in the system before a voltage collapse occurs.

voltage interruption the removal of the supply voltage from any phase, which is of momentary, sustained, and of temporary duration.

voltage multiplier an electronic device or circuit for multiplying the peak DC value of an input AC signal. A rectifying circuit that produces a direct voltage whose amplitude is approximately equal to a multiple of the peak amplitude of the applied alternating voltage. Voltage doublers are commonly used in consumer electronic products that are designed for use in both U.S. and European markets.

voltage protection the output voltage is limited to protect the load from an over-voltage condition. This can be accomplished by shunting the power-supply output or shutting down the drive circuit for the active switches in a switching supply if the output voltage exceeds a preset value.

voltage rating the maximum voltage that may be applied to the resistor.

voltage reference a functional block that ideally provides a constant output voltage independent of external influences such as supply voltage, loading, or temperature. Commonly used voltage references are based on the bandgap voltage of silicon (bandgap reference) or the reverse breakdown of a zener diode.

voltage regulating relay a voltage regulating relay senses RMS voltage level and issues commands to devices such as load tap changers, which then adjust the tap position to bring the voltage back to the desired level.

voltage regulation the change in delivered voltage from a generator or transformer from no-load to full-load. Voltage regulation is usually expressed as a percentage of the no-load voltage. For a DC generator, the voltage will always drop as the load increases and the voltage regulation will be a positive quantity. For AC generators and transformers, voltage regulation is the difference in the magnitude of the no-load and full-load voltages (ignoring phase angles). For capacitive (leading power factor) loads, the full-load voltage may have a higher magnitude than the no-load voltage, resulting in negative voltage regulation. Such a condition may lead to instability and is undesirable.

voltage regulator similar to a voltage reference, but provides more output current at a less precisely controlled voltage. Primarily used to "clean up" (regulate) a varying input voltage to provide circuitry with a constant power supply voltage.

voltage source inverter a power converter that takes a DC voltage from a battery or the output of a

rectifier and supplies a voltage of controllable and variable frequency and magnitude to a single or multiphase load. *See also* current source inverter.

voltage spread the difference between a power system's specified maximum and minimum voltages.

voltage stability a measure of power system stability which considers the system's capacity to support a given load.

voltage standing wave ratio (VSWR) another way of expressing impedance mismatch resulting in signal reflection. With respect to reflection coefficient G, the VSWR may be expressed mathematically as

$$VSWR = \frac{1+ \mid G \mid}{1- \mid G \mid}$$

See also reflection coefficient.

voltage transfer function any function of input to output voltage in ratio form, expressed as a dimensionless ratio. The input voltage may be the source voltage or the input voltage, which differ due to mismatch. The output voltage may be the load voltage or maximum output voltage, which also differ due to mismatch. Other voltages could also be ratioed, such as the input and output voltages of a MESFET. The voltage gain of a device is a specific case of a voltage transfer function. Regardless of the voltages used, they should be explicitly specified or confusion will result.

voltage transformer an instrument transformer specially designed and optimized for voltage measurement and power metering applications. The primary winding is rated to match the system voltage and the secondary is typically rated at a standard value to match common meters and display units. Also called a potential transformer.

voltage unbalance refers to the greatest change of the polyphase voltages from the average polyphase voltage, divided by the average polyphase voltage.

voltage unit   a protective unit (in protective relaying) whose operation depends exclusively on the magnitude of voltage.

voltage variation – short duration   a change of the voltage RMS value from nominal for a time period from 0.5 cycles to 1 minute, and can be used with the words sag, swell, and interruption for magnitude changes, and the words instantaneous, momentary, of temporary for showing duration.

voltage-behind-reactance model   a representation of a machine in which the stator voltage equations are modeled as a voltage source in series with a reactance (and typically a resistance). The voltage source represents the back emf present on the stator windings due to the coupling between the stator and rotor circuits. In synchronous machine modeling, several different voltage-behind-reactance models have historically been used, wherein approximations are used to represent the machine in various detail.

voltage-controlled bus   in power-flow analysis of an electric power system, a bus at which the real power, voltage magnitude, and limits on reactive power are specified. A bus connected to a generator will be so represented.

voltage-controlled oscillator (VCO)   (1) an oscillator where the frequency can be controlled by an external voltage. These oscillators can be divided into three categories:

1. Oscillators based on analog computer simulation of an oscillating system with quasi-static variable frequency.
2. Oscillators based on quasi-static variation of frequency using voltage-controlled capacitance of semiconductor diodes or varactors.
3. Oscillators based on control of charge–discharge currents in current-controlled multivibrators with further nonlinear waveshaping of the triangular wave.

(2) an oscillator whose frequency is designed to be controlled primarily by the amplitude of the applied voltage. The VCO is one of the building blocks making up the phase-locked-loop circuit.

voltage-source inverter (VSI)   an inverter with a DC voltage input.

Volterra integral equation   a linear integral equation wherein the limits of integration are functions of position.

Volterra series   a series expansion of a nonlinear function around a point. The Volterra series method is a generalization of the power-series method useful for analyzing harmonic and intermodulation distortion due to frequency-dependent nonlinearities in a device.

voltmeter   an instrument for measuring a potential difference between different points of an electrical circuit. Units are volts.

volts/hertz control   a method of speed control of induction machines, used below rated speed. When the volts/hertz ratio is kept constant, the current through the stator windings remains almost the same, except for very low speeds; hence, the available torque remains constant, but the speed changes due to change in frequency.

volume plasmon   a volume polariton in a plasma medium.

volume polariton   a polariton that propagates in unbounded medium, also referred to as a bulk mode, wave by 180 degrees.

volume scattering   the reflection of electromagnetic waves from a collection of particles or transitions in media properties distributed throughout a three-dimensional region. Particles may or may not be immersed or imbedded in a dielectric medium. *See also* surface scattering.

volumetric scattering   *See* volume scattering.

von Neumann architecture a stored program computer design in which data and instructions are stored in the same memory device and accessed similarly. *See also* Princeton architecture, single-instruction stream, single-data stream.

von Neumann, John (1903–1957) Born: Budapest, Hungary

Best known for his role in the development of the theory of stored program flexible computers. He is honored by the reference to von Neumann machines as a theoretical class of computers. von Neumann also invented the idea of game theory. As a mathematician, von Neumann published significant work on logic, the theory of rings, operators, and set theory. His work, *The Mathematical Foundations of Quantum Mechanics,* was significant in the mathematical justification of that field. von Neumann was a brilliant mathematician and physicist whose theoretical contributions are fundamental to modern physics and electrical engineering. He was the youngest member of the Institute of Advanced Studies at Princeton, and did important work on the Manhattan project.

voting circuit a circuit that provides fault-tolerance by comparing its inputs and taking a majority vote in case of disagreement.

vowel diagram the articulation of different vowels is strongly based on the position of the tongue, which can be high/low and front/back. The diagram defined by these two dimensions is called vowel diagram.

voxel the 3-D analogue of a pixel; abbreviation of Volumetric Picture Element. Mathematically it is a point in 3-D space having integer coordinates; concretely, it can also be interpreted as a cube of unit size centered about that point. *See* pixel.

voxel adjacency one of three types of adjacency relations defined on voxels:

1. 6-adjacency: Two voxels are 6-adjacent if they differ by 1 in one coordinate, the other two coordinates being equal; equivalently, the two unit cubes centered about these voxels have one face in common.

2. 18-adjacency: Two voxels are 18-adjacent if they differ by 1 in one or two coordinates, the remaining coordinates being equal; equivalently, the two unit cubes centered about these voxels have one face or one edge in common.

3. 26-adjacency: Two voxels are 26-adjacent if they differ by 1 in one, two, or three coordinates, the remaining coordinates being equal; equivalently, the two unit cubes centered about these voxels have one face, one edge, or one vertex in common.

In these definitions, the numbers 6, 18, and 26 refer to the number of voxels that are adjacent to a given voxel. *See* pixel adjacency.

VP-PAW *See* variable polarity plasma arc welding.

VQ *See* vector quantization or vector quantizer.

VQ encoding *See* vector quantization encoding.

VRAM *See* video random access memory.

VSAT *See* very small aperture terminal.

VSI *See* voltage-source inverter.

VSWR *See* voltage standing wave ratio.

VT *See* potential transformer.

wafer   a thin slice of semiconductor material on which semiconductor devices are made. Also called a slice or substrate.

wafer fab   the facility (building) in which semiconductor devices are fabricated. Also called a semiconductor fabrication facility.

wafer scale integration   most integrated circuits are cut from a large slice of material called a wafer. With wafer scale integration, the entire slice of material is used to create a complex circuit.

wafer sort   a preliminary electrical test of each die while still on the wafer to eliminate most of the bad die before they are assembled.

wait state   a bus cycle during which a CPU waits for a response from a memory or input-output device.

wall clock   a device providing the time of day; contrast processor clock. Elapsed wall clock time for a process does not correspond with processor time because of time used in system functions.

Walsh cover   mutually orthogonal sequences used in direct-sequence code division multiple access, obtained from the rows of a Hadamard matrix. *See also* Hadamard matrix.

Walsh transform   *See* Walsh–Hadamard transform.

Walsh-Hadamard transform (WHT)   a transform that uses a set of basis functions containing values that are either $+1$ or -1, and are determined from the rows of the Hadamard matrices.

This has a modest decorrelation capability and is simple to implement.

Waltz filtering   also termed "Boolean constraint propagation"; a method of simplifying certain tree-search problems. It was originally developed to solve the computer vision problem of labeling each edge of a line drawing in order to give a 3-D description of the represented object.

WAN   *See* wide-area network.

Ward–Leonard drive   an adjustable voltage control drive system for the speed control of DC machines, whereby variable voltage is supplied to the armature, while maintaining constant voltage across the shunt or separately excited fields. The variable voltage is obtained from a motor-generator set. The Ward–Leonard drive was frequently used in elevators.

warm start   (1) reassumption, without loss, of some processes of the system from the point of detected fault.

(2) the restart of a computer operating system without going through the power-on (cold) boot process.

watchdog processor   a processor that observes some process and signals an alert if a certain event happens or fails to happen.

watchdog timer   a simple timer circuitry that keeps track of proper system functioning on the basis of time analysis. If the timer is not reset before it expires, a fault is signaled, e.g., with an interrupt.

watercourse   a line on a surface $f(x, y)$ which represents a watershed of the inverted surface $-f(x, y)$. The line of steepest descent from a saddle point to a minimum is a watercourse. Watercourses meet watersheds at saddle points.

water resistivity   a measure of the purity of cooling liquid for a power tube, typically measured in megohms per centimeter.

watershed   a line on a surface $f(x, y)$, typically an image, which divides it into "catchment areas." Within a catchment area, lines of descent all connect to the same minimum point. The line of steepest ascent from a saddle point to a maximum is a watershed. Watersheds often correspond to ridges.

water tree   a microscopic cracking pattern which forms in the insulation of cables which are immersed in water or direct-buried in the earth. *See* tree.

Watson, Thomas J., Jr.   Watson is best known as the president of IBM who led the company into a dominant position in the computer industry. Watson took over his father's company, changed the structure, and moved the company away from the card tabulating business in which they held a dominant position. Watson Jr. oversaw the development of the IBM System/360 machines, which were to give the company a dominant position in computing.

Watson, Thomas J., Sr.   (1874–1956) Born: Cambell, New York

Best known as the president of IBM (International Business Machines). While Watson was not a technical person, his position as head of IBM put him in a position of supporting the development of a number of devices, both electronic and mechanical, leading to the development of the modern computer industry.

Watson-Watt, Robert Alexander   (1892–1973) Born: Brechin, Angus, Great Britain

Most famous for his pioneering work in the development of radar. Watson-Watt's work is based on the principles elucidated by Faraday, Maxwell, and Hertz. A German physicist, Christian Hulsmeyer had filed a patent in 1904 for an earlier device. Lack of enthusiasm from the German government and the governments in France and the United States gave the English, who supported Watson-Watt, a clear edge in this field. Watson-Watt filed his patent application in 1919.

His device proved invaluable to the Allies in World War II.

watt   unit of power in the SI system of units.

Watt, James   (1736–1819) Born: Greenock, Scotland

Best known for his work in the development of efficient steam power. Watt began his career as an instrument maker. When asked to fix a troublesome Newcomen engine, he began to make improvements. Watt eventually partnered with industrialist Matthew Boulton to form a steam engine company. Watt is credited with having devised the horsepower system. The unit of power, the watt, is named in his honor.

watt-VAR meter   meter capable of simultaneously measuring the real and reactive power delivered to an AC load.

wattmeter   an instrument for measuring electric power in watts. A wattmeter requires connections to measure both the current through and the voltage across the load being measured.

wave equation   equation governing the evolution of a wave; in electromagnetics any of several equations or equation sets starting from the most general, nonlinear multivariable differential Maxwell–Heaviside equations and ranging down to the simplest first-order rate equations.

wave impedance   the ratio of the transverse electric and magnetic fields inside a waveguide.

wave optics   formalism for optics in which the fields are represented as wave phenomena, in contrast to other ray or particle optics models.

wave plate   transparent anisotropic medium that introduces polarization-dependent phase shifts on an optical wave.

wave polarization a description of the time-varying behavior of the electric field vector as some fixed point in space. Elliptical polarization is the most general polarization and special cases include linear and circular polarizations.

wave propagation the transfer of energy by electromagnetic radiation.

wave winding an armature winding on a DC machine in which the two ends of each coil are connected to bars on opposite sides of the commutator ring. The wave winding provides two parallel paths through the armature winding, regardless of the number of poles in the machine.

waveform coding refers to the class of signal compression methods that are based on a criterion where the input waveform is to be resembled as closely as possible according to some criterion, e.g., minimum squared error, by the reproduced coded version. Waveform coding contrasts parametric coding techniques.

waveform distortion refers to a deviation from a steady-state clean sine waveform.

waveform interpolation coding parametric speech coding method where a characteristic waveform, a prototype waveform, is extracted from the speech signal at regular time instants and the intermediate signal is interpolated. Waveform interpolation coding is mostly used in low bit rate speech coding.

wavefront front of a wave; often a surface of constant phase.

waveguide a system of conductive or dielectric materials in which boundaries and related dimensions are defined such that electromagnetic waves propagate within the bounded region of the structure. Although most waveguides utilize a hollow or dielectric filled conductive metal tube, a solid dielectric rod in which the dielectric constant of the rod is very much different from the dielectric constant of the surrounding medium can also be used to guide a wave. Waveguides rapidly attenuate energy at frequencies below the waveguide lower cut-off frequency, and are limited in bandwidth at the upper end of the frequency spectrum due to wave attenuation as well as undesired mode propagation.

waveguide interconnect interconnect that uses a waveguide to connect a source to a detector. A waveguide is used for implementing a bus. The merits are large bandwidth, high speed of propagation, and compatibility with integrated optics and opto-electronics.

waveguide laser a laser in which amplification occurs within a waveguide that is confining the laser modes in the transverse direction.

wavelength a constant that describes the distance a periodic wave must travel in order to repeat itself. For example, if $v(z, t)$ is a periodic wave and if the wave travels a distance λ, then $v(z + \lambda, t) = v(z, t)$.

wavelength division multiplexing (WDM) a technique to increase capacity and throughput of systems by using a number of wavelength channels simultaneously.

wavelet a basis function that is obtained by translating and dilating a mother wavelet; it has such properties as smoothness, time-frequency localization, orthogonality, and/or symmetry.

wavelet coding coding a signal by coding the coefficients of the wavelet transform of the signal. The discrete wavelet transform is often used in image compression.

wavelet packet a family of scaling functions and wavelets by translation and dilation of a mother wavelet and a scaling function following a binary tree structure.

wavelet shrinkage a non-parametric estimation method in order to remove noise from a signal by shrinking wavelet coefficients of a signal towards zero.

wavelet transform a computational procedure which, in order to represent a given function $x(t)$ by basis function ϕ, calculates $x(a, b) = 1\sqrt{|a|}\int_{-\infty}^{\infty} x(t)\phi(t - ba)dt$, where a and b are real numbers. *See* inverse wavelet transform.

wavenumber a constant that relates the spatial rate of change of phase for a propagating wave. The wavenumber is mathematically equal to $2\pi/\lambda$, where λ is the wavelength. SI units are radians per meter. *See also* phase constant.

WDM *See* wavelength division multiplexing.

weak interconnection a connection between two power systems which has a high impedance and thus allows local disturbances at either end to threaten the synchronization of the two systems.

weak localization the name given to a process of self-interference of carriers in a mesoscopic system in which the transport is quasi-ballistic. A significant fraction of the carriers can be scattered by impurities back to their initial position in phase space, at which point they interfere with each other leading to an additional resistance. Since the scattering path can be traversed in either direction (which are time reversed paths of one another), it is said that the additional resistance is made up of continual interference between the two time-reversed paths. A small magnetic field breaks up this equivalence of the two paths and eliminates the weak localization contribution to the resistance.

weak localization of light enhanced backscattering; sometimes also called opposition effect.

weak SPR function *See* weak strictly positive real function.

weak strictly positive real function a rational function $H(s) = n(s)/d(s)$ of the complex variable $s = \sigma + j\omega$ that satisfies the following properties:

(1) $a(s)$ is a Hurwitz polynomial.
(2) $Re[H(j\omega)] > 0$ for all $\omega \geq 0$.
(3) The degrees of the numerator and denominator polynomials differ by, at most, one.
(4) If $\partial(b) > \partial(a)$, then $\lim_{\omega \to \infty}[H(j\omega)]/j\omega > 0$, where $\partial(b)$ denotes the degree of the polynomial b and similarly for a.

wearout failure failure mechanism caused by monotonic accumulation of incremental damage beyond the endurance of the product.

Weber's law an experimental result that states that the smallest luminance increment ΔL at which a region of luminance $L + \Delta L$ is just discernible from a background of luminance L is such that the ratio $\Delta L/L$ is constant. *See* brightness constancy, simultaneous contrast.

Weber, Wilhelm (1804–1891) Born: Wittenberg, Germany

Best known as the person who deduced that electricity consists of charged particles. Weber held several university appointments including professorships at Gottingen, where he had a very productive collaboration with Karl Gauss. Weber insisted on precision in his mathematical and experimental work. He developed a number of very precise measurement instruments. His efforts helped establish a sound foundation for the study of electricity and magnetism. He is honored by having his name used as the SI unit of magnetic flux density, the weber.

wedge ring detector a special photodetector structure consisting of wedge elements and annular half-ring shaped elements, each set covering a semicircle. This structure detects feature without regard to scale with the wedges, and without regard to rotational orientation with the annuli.

weight decay a technique employed in network training that aims to reduce the number of interconnections in the final, trained network. This is achieved by penalizing the weights in some way such that they have a tendency to decay to zero unless their values are reinforced.

weight initialization the choosing of initial values for the weights in a neural network prior to training. Most commonly small random values are employed so as to avoid symmetries and saturated sigmoids.

weight sharing a scheme under which two or more weights in a network are constrained to maintain the same value throughout the training process.

weighted Euclidean distance for two real valued vectors $x = (x_1, x_2, \ldots, x_n)$ and $y = (y_1, y_2, \ldots, y_n)$, defined as

$$D_\psi(x, y) = \sqrt{(x - y)^T \psi (x - y)},$$

where ψ is the inverse of the covariance matrix of x and y, and T denotes the transpose. *See also* Mahalanobis distance.

weighted mean-squared error (WMSE) a generalization of the mean squared error.

weighted residual a different form of the moment method. *See also* moment method.

weighted similarity *See* weighted Euclidean distance.

weighting filter a standardized filter used to impart predetermined characteristics to noise measurements in an audio system.

weightless network networks that are trained, not by changing weight values, but by modifying the contents of a memory device, usually a RAM.

Welch bound lower bound on the total squared cross correlation of a multi-set of sequences. For

N complex–valued sequences $s_i, i = 1, 2, \ldots, N$, each of energy $s_i^* . s_i = L$,

$$\sum_{i=1}^{N} \sum_{j=1}^{N} \left| s_i^* . s_j \right|^2 \geq K^2 L.$$

Westinghouse, George (1846–1914) Born: Central Bridge, New York

Best known as a financier and industrialist during America's age of great commercial expansion. What is less known today is that Westinghouse's fortune was based on his early inventions in the railroad industry. His braking system was eventually adopted in most rail cars. Westinghouse went on to secure over 400 patents in the rail and the gas distribution industries. Before hiring Tesla and buying his patents, Westinghouse had been a champion of alternating current for power distribution. His company provided illumination for the great Chicago Exposition of 1893. Before his death, Westinghouse was to lose control of the companies that bear his name. Undaunted, he returned to the laboratory for a number of additional years of invention.

wet etching a process that uses liquid chemical reactions with unprotected regions of a wafer to remove specific layers of the substrate.

wet withstand test a withstand test that is conducted under conditions which include simulated rain or fog.

Weyl identity an expansion of a spherical wave in terms of plane waves is known as the Weyl identity and may be written as

$$\frac{e^{-jkr}}{r} = -\frac{j}{2\pi} \int_{-\infty}^{\infty} \int_{-\infty}^{\infty} \frac{e^{-j(k_x x + k_y y) - jk_z|z|}}{k_z} \, dk_x \, dk_y$$

Wheatstone bridge a bridge circuit where all arms are resistors. The condition of balance in the circuit is used for precise measurement of resisors. In this case, one of the arms is an unknown resistor, another arm is a standard resistor (usually a variable resistor box), and two other arms (called ratio arms) are variable resistors with a well determined ratio. When the condition of balance is achieved, one can calculate the unknown resistor multiplying the standard resistor value by the ratio of ratio arms resistors. The precision of measurements is 0.05% for the range 10 ohms to 1 megohm. The Wheatstone bridge is used for resistor measurements at DC and AC (in the universal impedance bridges).

Moreover, the Wheatstone bridge is widely used in resistive transducers where one or more arms is substituted by resistors the resistance of which depends on a physical variable (temperature, pressure, force, etc.). In these applications, the deflection from balance is used for measurement of the physical variable.

whetstone   the speed of a processor as measured by the Whetstone benchmark.

Whetstone benchmark   a benchmark test program for scientific computers originally written in Fortran at Whetstone Laboratories, England.

whisker contact diode   a technique for mounting very high frequency diodes in a waveguide that involves a thin pointed wire or whisker that acts as both an antenna into the guide and as a bias contact.

whitening filter   a filter which whitens noise, i.e. one which brings noise whose power spectrum is not white into this condition, e.g. by means of a frequency dependent filter. Noise whitening is a vital precursor to matched filtering.

white noise   the noise that in its spectrum contains constant energy per unit bandwidth independent of frequency. *See also* thermal noise.

WHT   *See* Walsh–Hadamard transform.

wide band   property of a tuner, amplifier, or other device that can pass a broad range of frequencies.

wide band FM   frequency modulation scheme where the ratio of peak frequency deviation to the frequency of modulating signal is larger than 0.2.

wide sense stationary uncorrelated scattering (WSSUS) channel   a randomly time-variant channel whose first- and second-order statistics (means and correlation functions) are independent of time and frequency. The frequency independence translates into the uncorrelated scattering requirement. In a WSSUS multipath channel, the random process pertaining to a signal caused by any resolvable scatterer (reflector) is:

(1) wide sense stationary, i.e., its mean and correlation functions are independent of time, and

(2) uncorrelated with any other scatterer's contribution.

wide-area network (WAN)   a computer communication network spanning a broad geographic area, such as a state or country.

wide-sense stationary process   a stochastic process $x(t)$ for which the mean $m(t) = m =$ constant and the covariance $C(t_1, t_2)$ is a function of only $| \ t_1 - t_2 \ |$. In this case, we write $C(t_1, t_2) = C(T)$ where $T = t_1 - t_2$.

Widrow–Hoff learning rule   a gradient descent learning rule for calculating the weight vector $\vec{w}$ for a linear discriminant which minimizes the squared error objective function. The vector $\vec{w}$ is modified as $\vec{w}(n + 1) = \vec{w}(n) + \alpha(n)(d - \vec{w}^T(n)\vec{x})\vec{x}$ where $\vec{x}$ is an input vector, d is the desired output, and $\vec{w}(n)$ is the weight vector at iteration n.

Wien bridge   a bridge circuit where one arm is a series connection of a resistor and capacitor, another arm is a parallel connection of a resistor and capacitor (these two arms are called reactance arms), and two other arms (called ratio arms) are resistors. The balance detector is connected between the common point of ratio arms and the common point of reactance arms, a sinusoidal voltage source is connected to another bridge diagonal. The Wien bridge was initially designed as a frequency measuring circuit; now the main part of its application is the Wien bridge oscillator.

Wien bridge oscillator (1) an oscillator where the Wien bridge is used in the amplifier feedback. The frequently used circuit includes equal resistors and equal capacitors in reactance branches of the bridge; this arrangement provides easy continuous tuning of the oscillation frequency. Tuning of the bridge providing high indirect frequency stability of oscillations is easily combined with application of an operational amplifier as oscillator active element. The circuit of amplitude control is also easily attached. All these advantages provide wide spread of Wien bridge oscillators in high and low radio frequency ranges of applications.

(2) a form of feedback oscillator that uses a noninverting amplifier along with a feedback path that produces a phase shift of zero degrees at the operating frequency. The feedback network contains only two reactive elements of the same type.

Wiener filter filter that attempts to reduce signal noise by separating and suppressing the power spectrum of the noise from the power spectrum of the signal. The uncertainty in the estimation of the noise power spectrum will cause the signal to be smoothed. Also known as the least-mean square filter.

Wiener, Norbert (1894–1964) Born: Columbia, Misseuri

Mathematician whose contributions include: Brownian motion, stochastic processes, generalized functions, harmonic analysis, control theory, and optimal filtering. Established the field of cybernetics, author of "Cybernetics: or Control and Communication in the Animal and the Machine."

Wilkinson coupler a coupler that splits a signal into a number of equiphase and equiamplitude parts. It provides isolation between output terminals by connecting resistors between each output terminal and a common junction. A coaxial type coupler was first proposed by Dr. Wilkinson. In recent years, not only coaxial type but also MIC (microwave integrated circuit) type Wilkinson couplers are practically used for various kinds of microwave circuits.

Williams tube memory a memory device based on electric charges being stored on the screen of a cathode ray tube. Now obsolete.

Wilson central terminal reference point for forming most of the standard ECG leads. It is the average of the right arm, the left arm, and the left potentials. It is a time-varying reference.

Winchester disk a type of magnetic disk for data storage. Its characteristic property is that the disk and the read-write head are placed in a hermetically sealed box. This allows higher recording density as the read-write head can be moved closer to the disk surface. *See also* disk head.

wind–electric conversion the process by which wind (mechanical) energy is converted to electrical energy, usually by the use of wind turbine.

wind farm a plot of land on which several power-generating windmills are placed.

wind power generator a system that utilizes the energy in the wind to generate electricity. The energy in the wind drives a wind turbine which acts as the prime mover for the generator. A wind turbine operates at a variable speed, and an appropriate electric machine and controller converts the mechanical energy into electrical energy and pumps it into a utility grid.

winding a conductive path, usually wire, inductively coupled to a magnetic core or cell.

winding factor   a design parameter for electric machines that is the product of the pitch factor and the distribution factor.

window   any appropriate function that multiplies the data with the intent to minimize the distortions of the Fourier spectra.

window operation   an image processing operation in which the new value assigned to a given pixel depends on all the pixels within a window centered at that pixel location.

windowing   (1) a term used to describe various techniques for preconditioning a discrete-time waveform before processing by algorithms such as the discrete Fourier transform. Typical applications include extracting a finite duration approximation of an infinite duration waveform.

(2) the process of opening a window. In signal processing, it is common to open only a certain restricted portion of the available data for processing at any one time: such a portion is called a window or sometimes a mask or neighborhood. For instance, in *FIR* filter design, a technique known as windowing is used for truncation in order to design an FIR filter. The design of window becomes crucial in the design. In image processing, it is a common practice that a square window of (for example) 3×3 pixels is opened centered at a pixel under consideration. In this window operation, the gray level of the pixel is replaced by a function of its original gray level and the gray levels of other pixels in the window. Different functions represent different operations: in particular, they will be suitable for different filtering or shape analysis tasks. *See* median filter, thinning.

Windscale incident   a nuclear power plant accident at the Windscale plant in Great Britain.

winner-take-all network   a network in which learning is competitive in some sense; for each input a particular neuron is declared the "winner" and allowed to adjust its weights. After learning, for any given input, only one neuron turns on.

wiped joint   a fused joint used in splicing lead-sheathed cables.

wipe system   in television, a system that allows the fading in of one channel of video as a second channel of video is faded out without loss of sync.

wired OR   a circuit that performs an OR operation by the interconnection of gate outputs without using an explicit gate device. An open collector bus performs a wired OR function on active-low signals.

wireframe   a model that approximately represents a solid object by using several hundreds of triangles. It is used in applications such as facial coding, facial recognition and industrial component mensuration.

wireless local area network (WLAN)   a computer network that allows the transfer of data without wired connections.

wireless local loop   a wireless connection (using a radio link) between a subscriber terminal (for example, a telephone) and the local exchange of the public switched network.

withstand rating   the maximum voltage that electrical equipment can safely withstand, without failure, under specified test conditions.

withstand test   a test of an insulator's ability to withstand a high voltage of some specified waveform.

WLAN   *See* wireless local area network.

WMSE   weighted mean-squared error.

word parallel   processing of multiple words in the same clock cycle.

wordspotting detection or location of keywords in the context of fluent speech.

work flow management the process to monitor work progress through any number of departments.

work function amount of energy necessary to take out an electron from a material.

working set the collection of pages, $w(t, T)$, referenced by a process during the time interval $(t - T, t)$.

working-set policy a memory allocation strategy that regulates the amount of main memory per process, so that the process is guaranteed a minimum level of processing efficiency.

workstation a computer system designed for engineering design calculations, characterized by (comparatively) large main memory, high floating point computational speed, and a high resolution graphic display system. It is used primarily in engineering and scientific applications.

world modeling describes the geometric and physical properties of the object (including the robot) and represents the state of the assembly of objects in the workspace. World modeling makes it possible to implement many of the features of a task-level programming system. *See also* object-oriented programming.

WORM *See* write once read many.

worst-case design a family of control design algorithms in which parameter perturbations and/or disturbances are estimated to behave in the most unfavorable way from control objective point of view. This assumption leads usually to various min-max control algorithms based on static min-max, noncooperative game theory or H infinity design. Since the worst-case estimates are conservative, the resulting controllers, although robust,

may be in some sense too pessimistic. *See also* robust controller design.

worst-case measure of sensitivity for the multiparameter sensitivity row vector with components x_i having tolerance constants ϵ_i (these are considered to be positive numbers), i.e.,

$$x_{i0}(1 - \epsilon_i) \le x_i \le x_{i0}(1 + \epsilon_i)$$

we have

$$M_W = \int_{\omega_1}^{\omega_2} \left(\sum_{i=1}^{n} \left| \mathrm{Re} \mathbf{S}_{x_i}^{F(j\omega, \mathbf{x})} \right| \epsilon_i \right) d\omega$$

wound rotor induction motor an induction motor in which the secondary circuit consists of a polyphase winding or coils connected through a suitable circuit. When provided with slip rings, the term slip-ring induction motor is used.

wraps pre-formed wire grips or ties for mechanically joining overhead conductors to insulators.

wraparound (1) a phenomenon in signal processing that occurs in the discrete case when signals are not properly manipulated. For instance, in circular convolution, if the length of signals is not properly chosen, i.e., there are not sufficient zeros appended at the end of the signals, the so-called wraparound error will take place. That is, the contributions from different periods will overlap.

(2) a condition code or indicator that may be set, or a program segment that is executed when a register wraps around.

Wratten filter a light filter for separating colors. It is available in transparent sheets of various colors and is useful in photography and in several phases of electronics, including the operation of color meters and color matchers.

wrist for a manipulator, refers to the joints in the kinematic linkage between the arm and hand (or end-effector). Usually, wrist allows an orienting of the manipulator. Therefore, the main role of the wrist is to change the orientation of the hand (or end-effector). *See also* spherical wrist.

write allocate   part of a write policy that stipulates that if a copy of data being updated is not found in one level of the memory hierarchy, space for a copy of the updated data will be allocated in that level. Most frequently used in conjunction with a write-back policy.

write broadcast   a protocol for maintaining cache coherence in multiprocessor systems. Each time a shared block in one cache is updated, the modification is broadcast to all other caches. Also referred to as write update.

write buffer   a buffer that stores memory write requests from a CPU. The write request in the buffer are then served by the memory system as soon as possible. Reduces the number of processor wait cycles due to long latency write operations.

write instruction   a processor instruction that stores information into memory from a processor register or a higher level cache.

write invalidate   a protocol for maintaining cache coherence in multiprocessor systems. Each time a shared block in one cache is updated, a message is sent that invalidates (removes) copies of the same block in other cache memories. This is a more common alternative than write broadcast protocols.

write once read many (WORM)   used to refer for memory devices that allow data to be written once after device fabrication, and to be read any number of times. A typical example is PROM.

write policy   determines when copies of data are updated in a memory hierarchy. The two most common write policies are write through and write back (copy back).

write through   a write policy that stipulates that when a copy of data is updated at one level of a memory hierarchy, the same data are also updated in the next outer level. Write through is usually only used in low-level caches. Its advantages are that it is fast and simple to implement, and that it always guarantees that the next level of the memory hierarchy has a valid copy of all data. Its main disadvantage is that it generates much data traffic to the next level.

write update   *See* write broadcast.

write-after-read hazard   *See* antidependency.

write-after-write hazard   *See* output dependency.

write-back   *See* copy-back.

write-back cache   *See* copy-back.

write-through cache   when a location in the cache memory is changed, the corresponding location in main memory is also changed.

written-pole motor   a single-phase motor that uses a coil to write poles on the magnetic rotor. The advantage of the written-pole motor is that it draws much lower starting current, allowing much larger single-phase motors. The development of this motor has been sponsored by the Electric Power Research Institute.

Wronskian   matrix whose determinant is used to test the linear independence of solutions to differential equations (such as Maxwell's equations).

WSSUS channel   *See* wide sense stationary uncorrelated scattering channel.

wye connection   *See* Y connection.

wye-connection   a three phase source or load connected in the form of Y.

wye-delta starter   a motor starter that starts a three-phase AC motor in wye or star configuration so that the motor starts on approximately 58% of normal voltage, with a two-thirds reduction in starting current. As the motor approaches operating speed, the windings are reconfigured in delta configuration so that full voltage is applied for normal operation. The transition from star to delta is performed with the help of timer settings and contactors.

wye-delta transformer   a connection of a three-phase transformer with one primary and one secondary which can be considered as three similar single-phase transformers. The primary is connected in wye, that is one terminal from each phase is connected to neutral and one to a line voltage. The secondary is connected in delta, with each phase connected between two line voltages.

wye-wye transformer   a three-phase transformer with both the primary and secondary coils connected in wye. This connection is considered undesirable, due to the triple harmonics in the exciting current. To maintain balance load voltages under varying loads, it is necessary to solidly connect the primary and secondary neutrals to ground. This may allow some secondary current to flow on the primary neutral and may also cause interference with parallel communication lines.

wye-wye-delta transformer   a three-phase transformer in which the primary and secondary coils are connected in wye. In order to overcome the problems with the wye-wye connection, a set of tertiary coils are connected in delta to provide a path for the triple harmonic components of the exciting current to circulate.

X

X-raser   *See* X-ray laser.

X-ray   short wavelength electromagnetic radiation; often considered to range from about 0.1 to 100 Å.

X-ray image   a digital image whose pixels represent intensities of X-rays. The X-rays may come from artificial sources (medical images) or arise naturally (astronomy). Also important in modern inspection systems. *See* imaging modality, medical imaging.

X-ray laser   a laser that has emission shorter than 30 nm (soft X-rays) or shorter than ≈ 1 Å (hard X-rays); laser producing its output in the X-ray region of the spectrum. Also called an X-raser.

X-ray lithography   lithography using light of a wavelength in the range of about 0.1 to 5 nm, with about 1 nm being the most common, usually taking the form of proximity printing.

XOR   *See* exclusive OR.

XOR gate   a logic gate that performs the exclusive-OR function. Exclusive OR is defined for two inputs as one or the other being true but not both.

Y

Y connection a three-phase source or load which is connected such that the elements are connected in parallel and are thus represented in a schematic diagram in a Y or star–shaped configuration.

Y-bus a matrix which contains the admittance of each element in an electric power system.

y-parameters the input and output admittances that are used to characterize a two port device (network).

YAG *See* yttrium aluminum garnet.

Yagi–Uda array a wire antenna array consisting of three key components:

(1) a dipole antenna (roughly a half wavelength in length) that connects the antenna to a source or load,

(2) a reflector element (slightly longer than the dipole antenna), which is a wire that is placed behind, but not connected to, the dipole antenna, and

(3) director elements (slightly shorter than the dipole antenna), which are wires that are placed in front of, but not connected to, the dipole antenna. The Yagi–Uda array is commonly used in the reception of television signals.

yield percentage of acceptably good chips to the total chips considered at a certain level of a MMIC process. High yield is one of the most important parameters of a cost-efficient process. DC yield refers to the percentage of chips that behave appropriately to the application of DC biasing voltages and currents (*See also* bias voltage or current). RF yield refers the percentage of chips that properly process RF/microwave signals.

YIG *See* yttrium iron garnet.

YIG filter *See* yttrium iron garnet filter.

YIG resonator *See* yttrium iron garnet resonator.

YIQ the standard format used in U.S. color television. In this standard the image is coded as; Y (luminance), I(phase) and Q (quadrature phase) can be calculated from RGB values as follows: $Y = 0.30R + 0.59G + 0.11B$, $I = 0.28G + 0.59R - 0.32B$ and $Q = -0.53G + 0.21R + 0.31B$.

yttrium aluminum garnet (YAG) host material for rare-earth ions, such as neodymium, used for laser systems. Basis for important laser media when doped with appropriate ions, especially neodymium; output frequencies mostly in the near infrared. Also written YAlG.

yttrium iron garnet (YIG) a ferrite often used in microwave devices. This material ($Y_3Fe_5O_{12}$) has a complicated cubic crystal structure with eight formula units per cell. The five Fe ions per formula unit are distributed between antiparallel sublattices, giving the material its ferrimagnetic structure.

yttrium iron garnet filter a tunable filter employing externally biased YIG and operating near its ferrimagnetic resonance.

yttrium iron garnet resonator a tunable resonator employing externally biased YIG and operating near its ferrimagnetic resonance.

Z

Z   *See* impedance.

Z_L   common symbol for load impedence.

Z_S   common symbol for source impedance.

Z-bus   a matrix which contains the impedance of each element in an electric power system.

z-parameters   the input and output impedances that are used to characterize a two port device (network).

z-transform   a mathematical transformation that can be applied to a differential equation of a system in order to obtain the system's transfer function. The z-transform is defined as

$$Z\{f(n)\} = F(Z) = \sum_{n=0}^{\infty} f(n)z^{-n}$$

where n is the series of discrete samples, $f(n)$ is the series of sample values corresponding to the discrete samples.

The z-transform can also be written in terms of the Laplace transform variable (s) and the sampling period (T) as

$$z = e^{Ts}$$

ZCS   *See* zero current switching.

Zeeman broadening   inhomogeneous spectral broadening of a transition in a laser medium due to Zeeman shifts that vary among the laser atoms or molecules in the medium.

zener breakdown   the electrical breakdown occurring on the reverse biasing of a zener diode.

Zener breakdown occurs when the electric field in the depletion layer increases to the point where it can break covalent bonds and generate electron-hole pairs.

zener diode   a pn-junction diode that has an abrupt rise in current at a reverse-bias voltage V_z, which is usually between 3 to 6 volts. Zener diodes are deliberately fabricated to operate in the reverse breakdown region at a specified voltage and are often used in voltage reference or voltage regulator circuits.

zero   the values of a complex function which cause the value of the funtion to equal zero. The zeros are all natural frequencies of vibration, or resonances of the circuit described by the equation. They are influenced by all elements in the circuit, and will move with any circuit element change (susceptible to load pulling).

zero coprimeness of 2-D polynomial matrices
2-D polynomial matrices $A \in F^{p \times m}[z_1, z_2]$, $B \in F^{q \times m}[z_1, z_2]$ $(p + q \geq m \geq 1)$ are called zero right coprime if there exists a pair (z_1, z_2) that is a zero of all $m \times m$ minors of the matrix

$$\begin{bmatrix} A \\ B \end{bmatrix}$$

2-D polynomial matrices $A \in F^{m \times p}[z_1, z_2]$, $B \in F^{m \times q}[z_1, z_2]$ are called zero left coprime if the transposed matrices A^T, B^T are zero right coprime.

zero crossing   a point in which a function changes sign; in a digital image I, a point $\mathbf{x}$ for which $I(\mathbf{x}) > 0$ and $I(\mathbf{x} + \Delta\mathbf{x}) < 0$ for some $\Delta\mathbf{x}$, or vice versa.

zero current switching (ZCS)   the control of converter switches such that the switch is turned on or off only when the current through it is zero at the switching instant. This is typically achieved through the use of some form of LC resonance.

zero divide   *See* divide by zero.

zero flag   a bit in the condition code register that indicates whether the result of the last arithmetic or logic instruction is zero (1 for zero, 0 for not zero).

zero input response (ZIR)   the response of a system to initial conditions (i.e., to the initial energy present in the system) only. For example, the zero input response (ZIR) of the RC circuit shown in the figure is the signal $y_{zir}(t)$, $t \geq 0$ to the initial voltage across the capacitor, with zero voltage applied at the source.

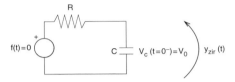

RC circuit zero input response.

zero of 2-D transfer matrix   a pair of complex numbers (z_1^0, z_2^0)

$$T(z_1, z_2) = \frac{N(z_1, z_2)}{d(z_1, z_2)},$$

$$N(z_1, z_2) \in R^{p \times m}[z_1, z_2]$$

that satisfy the condition that the rank of the matrix $N(z_1^0, z_2^0)$ drops below the normal rank of the polynomial matrix $N(z_1, z_2)$, i.e.,

$$\text{rank } N\left(z_1^0, z_2^0\right) < \min(m, p)$$

where $R^{p \times m}[z_1, z_2]$ is the set $p \times m$ polynomial matrices in z_1 and z_2 with real coefficients.

zero of generalized 2-D linear system   a pair of complex numbers (z_1^0, z_2^0)

$$Ex_{i+1,j+1} = A_0 x_{ij} + A_1 x_{i+1,j} + A_2 x_{i,j+1}$$
$$+ B_0 u_{ij} + B_1 u_{i+1,j} + B_2 u_{i,j+1}$$
$$y_{ij} = C x_{ij} + D u_{ij}$$

with the system matrix

$$S(z_1, z_2) = \begin{bmatrix} G(z_1, z_2) & -B(z_1, z_2) \\ C & D \end{bmatrix}$$

$$G(z_1, z_2) := E z_1 z_2 - A_0 - A_1 z_1 - A_2 z_2$$

$$B(z_1, z_2) := B_0 + B_1 z_1 + B_2 z_2$$

$$\text{if rank} \quad S\left(z_1^0, z_2^0\right) < n + \min(m, p)$$

where $x_{ij} \in R^n$ is the semistate vector, $u_{ij} \in R^m$ is the input vector, $y_{ij} \in R^p$ is the output vector, A_k, B_k ($k = 0, 1, 2$), C, D are real matrices with E possibly singular. A zero of 2-D transfer matrix. (*See also* zero of 2-D transfer matrix) is always zero of the system.

zero sequence   the set of in-phase components used in symmetrical component analysis. Zero sequence currents are closely associated with ground current in a grounded wye system, and do not directly flow in an ungrounded delta system.

zero state response (ZSR)   the response of a system with zero initial conditions (i.e., zero initial energy present in the system) to an applied input. For example, in the following circuit, the zero state response (ZSR) is the signal $y_{zsr}(t)$, $t \geq 0$ when the input voltage $f(t)$ is applied, and there is zero initial voltage across the capacitor.

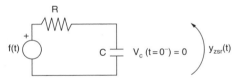

RC circuit zero state response.

zero voltage switching (ZVS)   the control of converter switches such that the switch is turned on or off only when the voltage across it is zero at the switching instant. This is typically achieved through the use of some form of LC resonance.

zero-address computer   a class of computer based on zero-address instructions. Stack-based calculators use zero-address computers and can be programmed using postfix notation.

zero-address instruction   a class of assembly language ALU instruction in which the operands are kept on a first-in-first-out stack in the CPU, and thus require no explicit addresses.

zero-coefficient sensitivity   analysis technique used for evaluation of circuit functions strongly dependent on zero locations (some bridge circuits and bridge oscillators). Zero-coefficient sensitivity is introduced in a way similar to pole-coefficient sensitivity.

zero-error capacity   for a given channel, the highest information transmission rate, such that there exists channel codes with decoding error probability identically zero. *See also* capacity region.

zero-order hold (ZOH)   a procedure that samples a signal $x(t)$ at a given sampling instant and holds that value until the succeeding sampling instant.

zero padding   technique where a discrete finite length signal is padded by adding some number of zeros to the end of the signal. The discrete Fourier transform of a zero padded signal has more frequency samples or components than that of a nonzero padded signal, although the frequency resolution is not increased. Zero padding is also sometimes used with the discrete Fourier transform to perform a convolution between two signals.

Zero-phase filter   a filter whose Fourier transform is purely real. In this way the phase response is zero. A filter considered as a signal has zero phase if it is an even signal.

zero-sequence reactance   the reactive component of the zero sequence impedance. *See also* symmetrical component.

zero-sum game   one of a wide class of noncooperative two-person games in which the sum of the cost functions of the decision makers is identically zero. In the zero-sum games, cooperation between players is impossible because the gain of one player is a loss of the other one. Thus, the game is characterized by only one cost function, which is minimized by the first player and maximized by the second one. To the zero-sum game one could also transform a constant-sum game in which the sum of the cost functions is constant. The solution in the zero-sum games has a form of saddle-point equilibrium, and roughly speaking it exists for problems in which max and min operations on the cost function commute. In zero-sum games without equilibrium in pure strategies it is possible to find saddle point in mixed strategies if the game is played many times in the same conditions. The resulting outcomes are average gains or losses of the players.

zig-zag ground   (1) a grounding arrangement which is used to supply single phase grounded circuits from an ungrounded three-phase delta connected electric power line.

(2) the winding arrangement within a grounding transformer.

zinc oxide arrester   a lightning arrester that consists of a stack of ZnO disks stacked within a vented porcelain tube. *See* gapless arrester.

zip   a file format and a set of data compression algorithms used to store one or more files in a single file. Originally devised by Phil Katz and placed in the public domain.

ZIR   *See* zero input response.

Ziv–Lempel (ZL) coding   a method for lossless source coding, due to J. Ziv and A. Lempel (1977). ZL coding is capable of achieving the bound given by the source coding theorem. Commonly used to compress computer files. *See* LZ77, LZ78, and Lempel-Ziv-Welch coding.

ZL coding   *See* Ziv–Lempel coding.

ZOH   zero-order hold.

zonal coding   a coding scheme in transform coding in which only those transform coefficients located in a specified zone in the transform domain are coded. For its counterpart, refer to threshold coding.

zonal sampling   in threshold sample selection it is dificult to transmit to the receiver which co-efficients were sent and which were not. In zonal coding, all coefficients are transmitted in order of increasing spatial frequency or some other prede-termined order and a end-of-block code-word is sent when all code words are below the threshold. *See also* threshold sampling.

zone of protection   the area of a power system for which a particular set of protective relays has primary protection responsibility. In typical cases, operation of any of these relays will open circuit breakers which will isolate this zone. Each major power system component (line, transformer, bus, generator) has a separate zone of protection.

zone plate   a Fresnel optical circular grating that produces the spatial frequencies to measure the resolution and the performance of telephoto-graphic or television systems. A moving Fresnel zone plate can measure the performance of line- or frame-based comb filters, the performance of scan converters, and the scan aperture of televi-sion images.

zone recording   a technique that allows the number of sectors per track on a magnetic disk to vary with the radius of the track. The tracks are divided into several zones, such that the number of sectors per track is determined by the maximum possible bit density on the innermost track in each zone.

ZSR   *See* zero state response.

ZVS   *See* zero voltage switching.